T0135386

Studies in Applied Philosophy, Epistemology and Rational Ethics

Volume 27

About this Series

Studies in Applied Philosophy, Epistemology and Rational Ethics (SAPERE) publishes new developments and advances in all the fields of philosophy, epistemology, and ethics, bringing them together with a cluster of scientific disciplines and technological outcomes: from computer science to life sciences, from economics, law, and education to engineering, logic, and mathematics, from medicine to physics, human sciences, and politics. It aims at covering all the challenging philosophical and ethical themes of contemporary society, making them appropriately applicable to contemporary theoretical, methodological, and practical problems, impasses, controversies, and conflicts. The series includes monographs, lecture notes, selected contributions from specialized conferences and workshops as well as selected Ph.D. theses.

Advisory Board

More information about this series at http://www.springer.com/series/10087

Lorenzo Magnani · Claudia Casadio
Editors

Model-Based Reasoning in Science and Technology

Logical, Epistemological, and Cognitive Issues

Editors
Lorenzo Magnani
Department of Humanities, Philosophy Section, and Computational Philosophy Laboratory
University of Pavia
Pavia
Italy

Claudia Casadio
Department of Philosophy, Education and Economical-Quantitative Sciences
University of Chieti and Pescara
Chieti
Italy

ISSN 2192-6255 ISSN 2192-6263 (electronic)
Studies in Applied Philosophy, Epistemology and Rational Ethics
ISBN 978-3-319-81784-2 ISBN 978-3-319-38983-7 (eBook)
DOI 10.1007/978-3-319-38983-7

Printed on acid-free paper

This Springer imprint is published by Springer Nature
The registered company is Springer International Publishing AG Switzerland

Preface

This volume is a collection of selected papers that were presented at the international conference *Model-Based Reasoning in Science and Technology. Models and Inferences: Logical, Epistemological, and Cognitive Issues* (MBR015_ITALY), held at the Centro Congressi Mediaterraneo, Sestri Levante, Italy, June 25–27, 2015, chaired by Lorenzo Magnani.

A previous volume, *Model-Based Reasoning in Scientific Discovery*, edited by L. Magnani, N.J. Nersessian, and P. Thagard (Kluwer Academic/Plenum Publishers, New York, 1999; Chinese edition, China Science and Technology Press, Beijing, 2000), was based on the papers presented at the first "model-based reasoning" international conference, held at the University of Pavia, Pavia, Italy, in December 1998. Other two volumes were based on the papers presented at the second "model-based reasoning" international conference, held at the same place in May 2001: *Model-Based Reasoning. Scientific Discovery, Technological Innovation, Values*, edited by L. Magnani and N.J. Nersessian (Kluwer Academic/Plenum Publishers, New York, 2002), and *Logical and Computational Aspects of Model-Based Reasoning*, edited by L. Magnani, N.J. Nersessian, and C. Pizzi (Kluwer Academic, Dordrecht, 2002). Another volume, *Model-Based Reasoning in Science and Engineering*, edited by L. Magnani (College Publications, London, 2006), was based on the papers presented at the third "model-based reasoning" international conference, held at the same place in December 2004. The volume *Model-Based Reasoning in Science and Medicine*, edited by L. Magnani and P. Li (Springer, Heidelberg/Berlin 2006), was based on the papers presented at the fourth "model-based reasoning" conference, held at Sun Yat-sen University, Guangzhou, P.R. China. The volume *Model-Based Reasoning in Science and Technology. Abduction, Logic, and Computational Discovery*, edited by L. Magnani, W. Carnielli and C. Pizzi (Springer, Heidelberg/Berlin 2010), was based on the papers presented at the fifth "model-based reasoning" conference, held at the University of Campinas, Campinas, Brazil, in December 2009. Finally, the volume *Model-Based Reasoning in Science and Technology. Theoretical and Cognitive Issues*, edited by L. Magnani, (Springer,

Heidelberg/Berlin 2013), was based on the papers presented at the sixth "model-based reasoning" conference, held at Fondazione Mediaterraneo, Sestri Levante, Italy, June 2012.

The presentations given at the Sestri Levante conference explored how scientific thinking uses models and explanatory reasoning to produce creative changes in theories and concepts. Some speakers addressed the problem of model-based reasoning in technology and stressed issues such as the relationship between science and technological innovation. The study of diagnostic, visual, spatial, analogical, and temporal reasoning has demonstrated that there are many ways of performing intelligent and creative reasoning that cannot be described with the help only of traditional notions of reasoning such as classical logic. Understanding the contribution of modeling practices to discovery and conceptual change in science and in other disciplines requires expanding the concept of reasoning to include complex forms of creativity that are not always successful and can lead to incorrect solutions. The study of these heuristic ways of reasoning is situated at the crossroads of philosophy, artificial intelligence, cognitive psychology, and logic: that is, at the heart of cognitive science. There are several key ingredients common to the various forms of model-based reasoning. The term "model" comprises both internal and external representations. The models are intended as interpretations of target physical systems, processes, phenomena, or situations. The models are retrieved or constructed on the basis of potentially satisfying salient constraints of the target domain. Moreover, in the modeling process, various forms of abstraction are used. Evaluation and adaptation take place in light of structural, causal, and/or functional constraints. Model simulation can be used to produce new states and enable evaluation of behaviors and other factors. The various contributions of the book are written by interdisciplinary researchers who are active in the area of modeling reasoning and creative reasoning in logic, cognitive science, science and technology: the most recent results and achievements about the topics above are illustrated in detail in the papers.

The editors express their appreciation to the members of the scientific committee for their suggestions and assistance: Atocha Aliseda, Instituto de Investigaciones Filosoficas, Universidad Nacional Autónoma de Mexico (UNAM); Tommaso Bertolotti, Department of Humanities, Philosophy Section, University of Pavia, Italy; Silvana Borutti, Department of Humanities, Philosophy Section, University of Pavia, Italy; Otávio Bueno, Department of Philosophy, University of Miami, Coral Gables, USA; Mirella Capozzi, Department of Philosophy, University of Rome Sapienza, Rome, Italy; Walter Carnielli, Department of Philosophy, Institute of Philosophy and Human Sciences, State University of Campinas, Brazil; Claudia Casadio, Department of Philosophy, Education and Economical-Quantitative Sciences, University of Chieti-Pescara, Italy; Carlo Cellucci, Department of Philosophy, University of Rome Sapienza, Rome, Italy; Sanjay Chandrasekharan, Homi Bhabha Centre for Science Education, Tata Institute of Fundamental Research, India; Roberto Feltrero, Department of Logic, History and Philosophy of Science at UNED (Spanish Open University), Madrid, Spain; Steven French, Department of Philosophy, University of Leeds, Leeds, UK; Marcello Frixione, Department of Communication Sciences,

University of Salerno, Italy; Dov Gabbay, Department of Computer Science, King's College, London, UK; Marcello Guarini, Department of Philosophy, University of Windsor, Canada; Ricardo Gudwin, Department of Computer Engineering and Industrial Automation, The School of Electrical Engineering and Computer Science, State University of Campinas, Brazil; Albrecht Heeffer, Centre for History of Science, Ghent University, Belgium; Michael Hoffmann, School of Public Policy, Georgia Institute of Technology, Atlanta, USA; Décio Krause, Departamento de Filosofia, Universidade Federal de Santa Catarina, Florianópolis, SC, Brazil; Ping Li, Department of Philosophy, Sun Yat-sen University, Guangzhou, P.R. China; Giuseppe Longo, Centre Cavaillès, République des Savoirs, CNRS, Collège de France et Ecole Normale Supérieure, Paris, France and Department of Integrative Physiology and Pathobiology, Tufts University School of Medicine, Boston, USA; Angelo Loula, Department of Exact Sciences, State University of Feira de Santana, Brazil; Shangmin Luan, Institute of Software, The Chinese Academy of Sciences, Beijing, P.R. China; Rossella Lupacchini, University of Bologna, Bologna, Italy; Lorenzo Magnani, Department of Humanities, Philosophy Section and Computational Philosophy Laboratory, University of Pavia, Italy; Joke Meheus, Vakgroep Wijsbegeerte, Universiteit Gent, Gent, Belgium; Luís Moniz Pereira, Departamento de Informática, Universidade Nova de Lisboa, Portugal; Woosuk Park, Humanities and Social Sciences, KAIST, Guseong-dong, Yuseong-gu Daejeon, South Korea; Claudio Pizzi, Department of Philosophy and Social Sciences, University of Siena, Siena, Italy; Demetris Portides, Department of Classics and Philosophy, University of Cyprus, Nicosia, Cyprus; Joao Queiroz, Institute of Arts and Design, Federal University of Juiz de Fora, Brazil; Shahid Rahman, U.F.R. de Philosophie, University of Lille 3, Villeneuve d'Ascq, France; Oliver Ray, Department of Computer Science, University of Bristol, Bristol, UK; Colin Schmidt, Le Mans University and ENSAM-ParisTech, France; Gerhard Schurz, Institute for Philosophy, Heinrich-Heine University, Frankfurt, Germany; Cameron Shelley, Department of Philosophy, University of Waterloo, Waterloo, Canada; Nik Swoboda, Departamento de Inteligencia Artificial, Universidad Politécnica de Madrid, Madrd, Spain; Paul Thagard, Department of Philosophy, University of Waterloo, Waterloo, Canada; Barbara Tversky, Department of Psychology, Stanford University and Teachers College, Columbia University, New York, USA; Ryan D. Tweney, Emeritus Professor of Psychology, Bowling Green State University, Bowling Green, USA; Riccardo Viale, Scuola Nazionale dell'Amministrazione, Presidenza del Consiglio dei Ministri, Roma, and Fondazione Rosselli, Torino, Italy; John Woods, Department of Philosophy, University of British Columbia, Canada; and also to the members of the local scientific committee: Tommaso Bertolotti (University of Pavia), Selene Arfini (University of Chieti/Pescara), Pino Capuano (University of Pavia), and Elena Gandini (Across Events, Pavia).

Special thanks to Tommaso Bertolotti and Selene Arfini for their contribution in the preparation of this volume. The conference MBR015_ITALY, and thus indirectly this book, was made possible through the generous financial support of the MIUR (Italian Ministry of the University) and of the University of Pavia. Their support is gratefully acknowledged. The preparation of the volume would not have been possible without the contribution of resources and facilities of the Computational Philosophy Laboratory and of the Department of Humanities, Philosophy Section, University of Pavia.

Several papers concerning model-based reasoning deriving from the previous conferences MBR98 and MBR01 can be found in special issues of Journals: in *Philosophica*: Abduction and Scientific Discovery, 61(1), 1998, and Analogy and Mental Modeling in Scientific Discovery, 61(2) 1998; in *Foundations of Science*: Model-Based Reasoning in Science: Learning and Discovery, 5(2) 2000, all edited by L. Magnani, N.J. Nersessian, and P. Thagard; in *Foundations of Science*: Abductive Reasoning in Science, 9, 2004, and Model-Based Reasoning: Visual, Analogical, Simulative, 10, 2005; in *Mind and Society*: Scientific Discovery: Model-Based Reasoning, 5(3), 2002, and Commonsense and Scientific Reasoning, 4(2), 2001, all edited by L. Magnani and N.J. Nersessian. Finally, other related philosophical, epistemological, and cognitive-oriented papers deriving from the presentations given at the conference MBR04 have been published in a special issue of the *Logic Journal of the IGPL*: Abduction, Practical Reasoning, and Creative Inferences in Science, 14(1) (2006), and have been published in two special issues of *Foundations of Science*: Tracking Irrational Sets: Science, Technology, Ethics, and Model-Based Reasoning in Science and Engineering, 13 (1) and 13(2) (2008), all edited by L. Magnani. Other technical logical papers presented at MBR09_BRAZIL have been published in a special issue of the *Logic Journal of the IGPL*: Formal Representations in Model-Based Reasoning and Abduction, 20(2) (2012), edited by L. Magnani, W. Carnielli, and C. Pizzi. Finally, technical logical papers presented at MBR12_ITALY have been published in a special issue of the *Logic Journal of the IGPL*: Formal Representations in Model-Based Reasoning and Abduction, 21(6) (2013), edited by L. Magnani.

Other more technical formal papers presented at (MBR015_ITALY) will be published in a special issue of the *Logic Journal of the IGPL*, edited by L. Magnani and C. Casadio.

Pavia, Italy
Chieti, Italy
February 2016

Lorenzo Magnani
Claudia Casadio

Contents

Part I
Models, Mental Models, and Representations

Visual Reasoning in Science and Mathematics

Otávio Bueno

Abstract Diagrams are hybrid entities, which incorporate both linguistic and pictorial elements, and are crucial to any account of scientific and mathematical reasoning. Hence, they offer a rich source of examples to examine the relation between model-theoretic considerations (central to a model-based approach) and linguistic features (crucial to a language-based view of scientific and mathematical reasoning). Diagrams also play different roles in different fields. In scientific practice, their role tends not to be evidential in nature, and includes: (i) highlighting relevant relations in a micrograph (by making salient certain bits of information); (ii) sketching the plan for an experiment; and (iii) expressing expected visually salient information about the outcome of an experiment. None of these traits are evidential; rather they are all pragmatic. In contrast, in mathematical practice, diagrams are used as (i) heuristic tools in proof construction (including dynamic diagrams involved in computer visualization); (ii) notational devices; and (iii) full-blown proof procedures (Giaquinto 2005; and Brown in Philosophy of mathematics. Routledge, New York, 2008). Some of these traits are evidential. After assessing these different roles, I explain why diagrams are used in the way they are in these two fields. The result leads to an account of different styles of scientific reasoning within a broadly model-based conception.

1 Introduction

The sematic view of theories emphasizes the role played by models in scientific practice, and it tends to downplay the corresponding role for linguistic considerations. As Bas van Fraassen, one of the major advocates of that view, points out:

> The syntactic picture of a theory identifies it with a body of theorems, stated in one particular language chosen for the expression of that theory. This should be contrasted with

O. Bueno (✉)
Department of Philosophy, University of Miami, Coral Gables,
FL 33124-4670, USA
e-mail: otaviobueno@mac.com

L. Magnani and C. Casadio (eds.), *Model-Based Reasoning in Science and Technology*, Studies in Applied Philosophy, Epistemology and Rational Ethics 27, DOI 10.1007/978-3-319-38983-7_1

> the alternative of presenting a theory in the first instance by identifying a class of structures as its models. In this second, semantic, approach the language used to express the theory is neither basic nor unique; the same class of structures could well be described in radically different ways, each with its own limitations. The models occupy centre stage. (van Fraassen 1980, p. 44)

This puts pressure on the semantic view to accommodate those aspects of scientific practice that rely on linguistic considerations, such as various styles of scientific reasoning.

In contrast, the received view, in the hands of Rudolf Carnap, for instance, emphasizes the importance of linguistic considerations for the proper understanding of science.

> Apart from the questions of the individual sciences, only the questions of the logical analysis of science, of its sentences, terms, concepts, theories, etc., are left as genuine scientific questions. We shall call this complex of questions the *logic of science*. […] According to this view, then, once philosophy is purified of all unscientific elements, only the logic of science remains. […] [*T*]*he logic of science takes the place of the inextricable tangle of problems which is known as philosophy* (Carnap 1934/1937, p. 279).

Since a theory is identified with a set of statements, it becomes a linguistic entity, and the crucial role played by models in scientific practice is diminished.[1] A more nuanced account, which acknowledges the significance of linguistic considerations while still preserving the proper emphasis on models, is in order.

In this paper, I offer such an account. To motivate it I focus on the different roles played by diagrams in scientific and mathematical practice. Diagrams are hybrid entities, which incorporate both linguistic and pictorial traits, and are central to any account of scientific reasoning. Thus, they provide a rich source of examples to examine the relation between model-theoretic considerations (central to the semantic approach) and linguistic features (crucial to the received view).

Diagrams play different roles in different fields. In scientific practice, their role tends not to be evidential in nature, and includes: (i) highlighting relevant relations in a micrograph (by making salient certain bits of information); (ii) sketching the plan for an experiment; and (iii) expressing expected visually salient information about the outcome of an experiment. None of these traits are evidential; rather they are all pragmatic.

In contrast, in mathematical practice, diagrams are used as (i) heuristic tools in proof construction (including dynamic diagrams involved in computer visualization); (ii) notational devices; and (iii) full-blown proof procedures (Giaquinto 2005; Brown 2008). Some of these traits are evidential.

After assessing these different roles, I explain why diagrams are used in the way they are in these two fields. The result leads to an account of different styles of scientific reasoning within a broadly semantic conception, and provides a first step toward a reconciliation of the received and the semantic views, properly re-conceptualized and formulated.

[1]For a survey of the semantic and the received views, see Suppe (1977a, b) and references therein.

The semantic and the received views can be compared under many dimensions. For the purposes of this paper, I will focus on only two of them: the role of linguistic features in the proper understanding of central aspects of scientific and mathematical practice, and the role played by models—broadly understood to include diagrams—in this practice. I will also consider strong formulations of these views: for the semantic approach (on this strong reading), linguistic considerations are largely irrelevant, and what are crucial are the relevant models, whereas for the received view (also on a strong reading), linguistic considerations—including, in particular, the requirement of formalization—are crucial, whereas models are not so central.

2 Diagrams in Scientific Practice

Diagrams, I just noted, play multiple roles in scientific practice. I will start by providing some examples to illustrate these roles. The examples, of course, are not meant to be comprehensive, but they highlight significant roles played by diagrams in this context.

(i) *A diagram may express expected visually salient information about the outcome of an experiment*. DNA nanotechnology involves the use of DNA as a biomimetic component for self-assembly (Seeman 2003, 2005; Seeman and Belcher 2002). One interesting experiment involved the construction of certain arrangements of DNA strands in predetermined shapes (Ding et al. 2004). First, a triangular arrangement was designed, followed by a hexagon formed by six DNA triangles suitably positioned. Finally, with multiple such DNA hexagons properly arranged, a DNA honeycomb is formed. A diagram is designed to express what researchers expect to detect with the output of the experiment (see Fig. 1, on the left). The outcome of the experiment, which was conducted with an atomic force microscope (AFM), seems to support the intended result, as honeycomb structures can be "seen" on the AFM image (see Fig. 2, on the right). (For additional discussion of this case, see Bueno 2011.)
It is important to note that what provides evidence for the intended conclusion, in this instance, is the AFM image rather than the diagram, which is only a representation of what the researchers expected to detect when the experiment was conducted.

(ii) *A diagram may highlight relevant information, for instance, on a micrograph*. Sometimes a diagram is drawn on a micrograph to make salient certain bits of information. This is seen in Fig. 1 (on the right). Consider the bottom right corner micrograph: in order to highlight the intended hexagon, a diagram of this shape is drawn on the AFM image. As a result, the hexagon is made salient (for further discussion, see Lynch 1991; Bueno 2011).

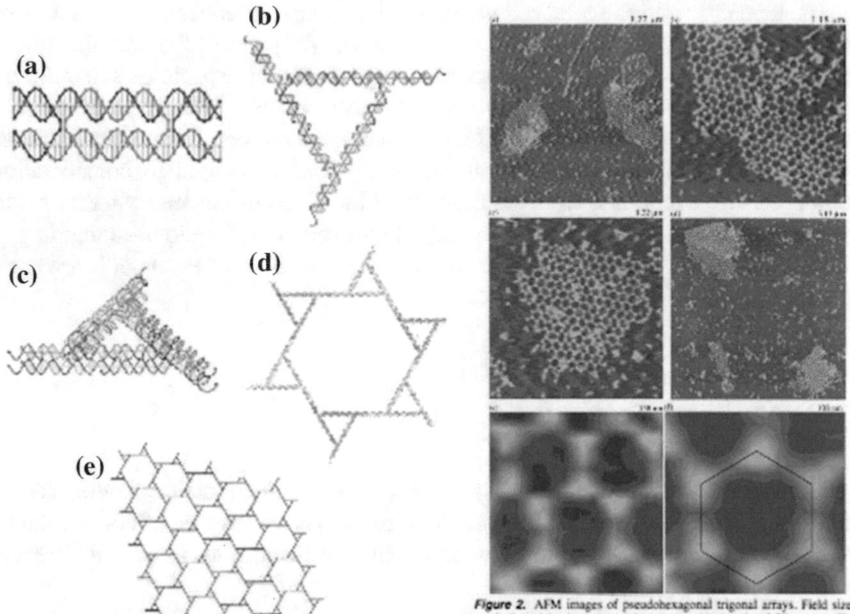

Fig. 1 The diagram of the DNA strand experiment (*left*) and, the corresponding AFM image (*right*) (Ding et al. 2004, p. 10230)

(iii) *Diagrams can also be used to sketch the plan of an experiment*. L.A. Bumm and a group of collaborators once tried to construct a single-molecular wire by establishing a conducting single-molecular current through a non-conducting medium (Bumm et al. 1996). A diagram was created to sketch the plan of the experiment (see Fig. 2, left). After dropping a conducting material through a non-conducting medium, researchers expected that single-molecular wires would be formed. The tip of the scanning-tunneling microscope (STM) would then establish a current through the non-conducting material via each 'single'-molecular wire all the way to the gold substratum. The STM micrograph exhibits the 'single'-molecular wires, viewed from the top: the little white blobs on the micrograph (Fig. 2, right) indicate 'single'-molecular wires; big white blobs indicate multiple-molecular wires.

In the end, however, it was unclear that just a single molecule was involved, rather than, say, a couple of molecules, given that a virtually indistinguishable STM micrograph would have been produced in each case. In this case, the evidence didn't support the conclusion, since it was unable to rule out possibilities that undermine the conclusion (for additional discussion, see Bueno 2011).

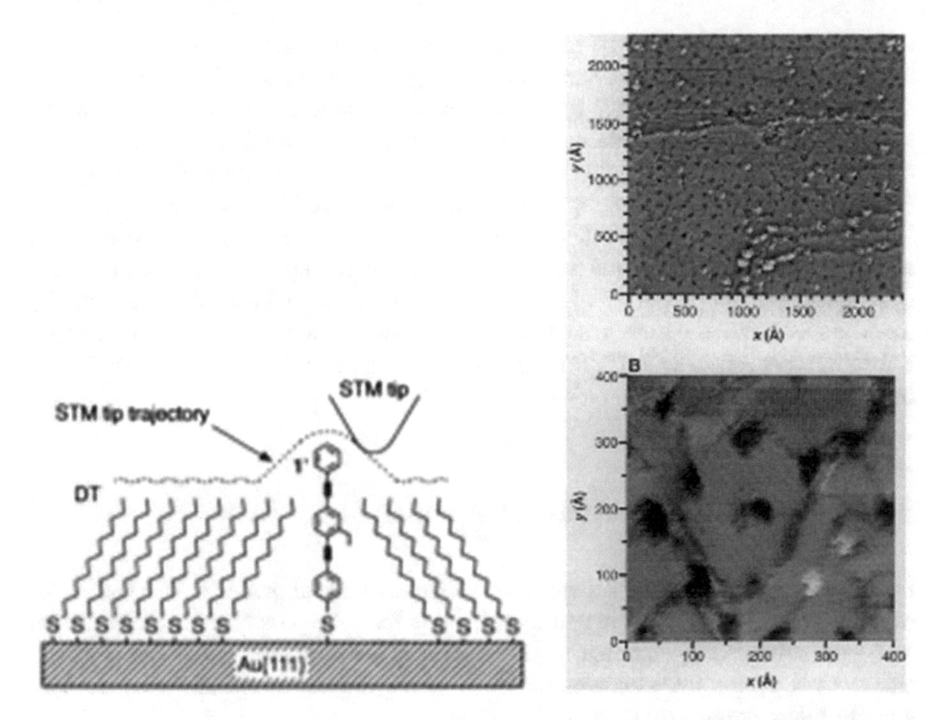

Fig. 2 'Single'-molecular wire experiment: diagrammatic sketch (*left*) and STM output (*right*) (Burnm et al. 1996)

Despite the multiple roles played by diagrams in scientific practice, it is important to note that none of them involve evidential considerations. The expression of visually salient information about the output of an experiment is just that: an expression; it is not something that provides evidence for (or against) the result of the experiment. One does not turn to the diagram in Fig. 1 (on the left) in support of the conclusion of the experiment—any more than one produces as evidence that Cyber robbed a bank a painting in which he is depicted robbing the bank. The evidence, causally produced by the interaction of the AFM and the sample under investigation (the suitable configuration of DNA), is offered by the relevant micrograph (Fig. 1, on the right).

Similarly, to highlight graphically certain bits of information in order to make certain aspects of a micrograph more salient is a pragmatic feature of diagrams. It is not an evidential trait they bear. In Fig. 1, the evidence is not on the diagram drawn on the bottom-right micrograph, but it emerges from the micrograph itself. The diagram is a useful device to highlight the hexagonal shape of the DNA strands, but it is incapable of producing the evidence: it is simply made-up by the researchers for a pragmatic purpose.

Finally, to sketch the plan of an experiment is to depict the steps involved in the attempt to obtain a given result. This is very different from actually obtaining and registering the result. A diagram provides the former, but the latter requires more:

the relevant result needs to be produced by the phenomenon in question. Hence, a micrograph, or some other mechanically generated information, needs to be invoked. In Fig. 2, the sketch of the diagram (on the left) does not show that the intended result has been obtained (in fact, as noted, it is unclear that the intended phenomenon did in fact obtain). For evidence, one needs to turn to the micrograph (on the right). It is the micrograph that exhibits that something has happened—even if what happened was not what was intended. Since double-molecular wires could not be unquestionably eliminated, given the micrograph, the single-molecular hypothesis was not properly supported. In the end, evidential considerations require micrographs (or some such mechanically generated device), given that the relevant information emerges causally from the phenomena under study rather than from a made-up diagram.

3 Diagrams in Mathematical Practice

Diagrams play significantly different roles in mathematical practice. Although some of these roles are pragmatic in nature, there is at least the possibility of an evidential role—something that, as noted, doesn't seem to be the case in the empirical sciences. Once again, the examples provided below are not meant to be comprehensive, but they emphasize some salient roles.

(i) *Mathematical diagrams can provide explanations of certain mathematical relations*. Consider Euler's formula (Giaquinto 2005, pp. 79–80):

$$e^{i\theta} = \cos\theta + i\sin\theta.$$

This formula is frequently introduced as a definition of the exponential function on complex numbers. But why should this particular definition be introduced? A geometrical diagram provides the explanation in question, since it exhibits the relevant relation between cos θ and *i* sin θ. In the diagram (see Fig. 3), the angle is θ; cosθ is represented as a vector on the horizontal axis, and sinθ is represented as a vector on the vertical axis. By reasoning with these two vectors, it becomes clear that $e^{i\theta}$ is the addition of both of them.

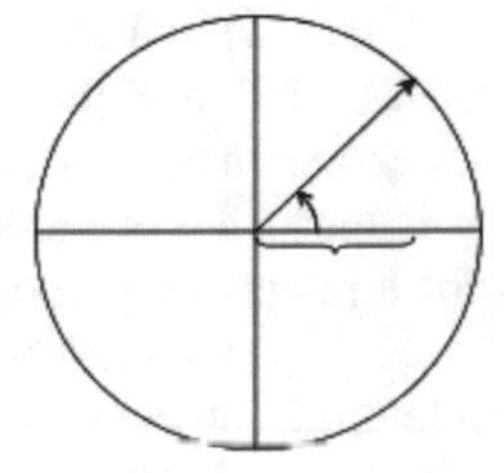

Fig. 3 Mathematical diagrams provide explanations of certain mathematical relation

(ii) *Mathematical diagrams can also provide heuristic tools in proof construction.* Consider, again, Euler's formula. It is clear that if we expand or contract the coordinates of the $e^{i\theta}$ vector by a real magnitude r (that is, 'r' stands for a real number), so that we have $r\cos\theta$ and $r\sin\theta$, the corresponding vector will expand or contract by a factor r. We thus have the geometrical significance of the following simple result (Giaquinto 2005, pp. 79–80):

$$re^{i\theta} = r\cos\theta + ri\sin\theta.$$

This result is clearly expressed in the diagram (see Fig. 4). By reasoning with the rule of vector addition, the resulting vector $re^{i\theta}$ is obtained by adding the expanded vectors $r\cos\theta$ (represented in the horizontal axis) and $r\sin\theta$ (represented in the vertical axis).

(iii) *A mathematical diagram can arguably provide a full-blown proof—at least in some cases*. Consider, for instance, the following theorem from number theory (Brown 2008, p. 36):

Theorem

$$1 + 2 + 3 + \cdots + n = n^2/2 + n/2.$$

This result can be straightforwardly established by mathematical induction. After showing that the result holds for the base case (n = 1), it is easy to show that if it holds for an arbitrary natural number i, it also holds for i + 1. By mathematical induction, the result then holds for every natural number.

But it has been argued that a diagram (see Fig. 5*) also establishes the result* (Brown 2008, p. 36).

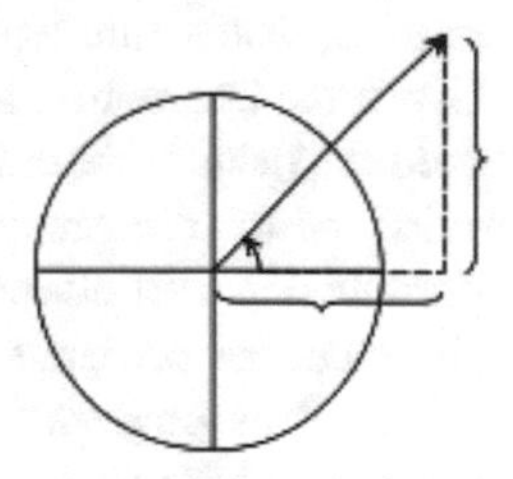

Fig. 4 Mathematical diagrams can also provide heuristic tools in proof construction

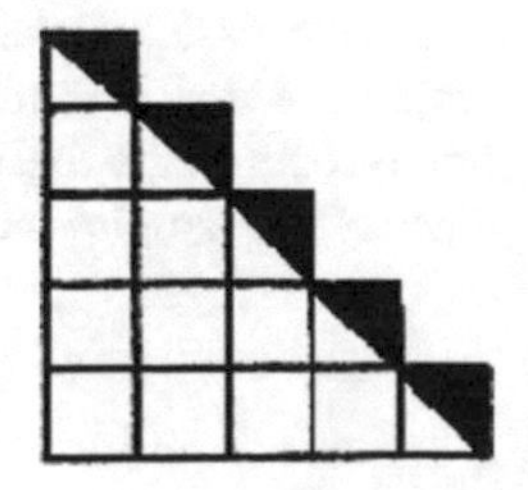

Fig. 5 A mathematical diagrams can arguably provide a full-blown proof

Proof Admittedly, the diagram only exhibits the result for $n = 5$. It represents each natural number as a square, and addition is represented as concatenating and pilling up such squares. There are 15 squares in the diagram, which corresponds to the addition of the first 5 natural numbers ($n = 5$). With a little bit of reasoning about the diagram, one can see that another way of getting the same configuration of 15 squares would be by squaring $n = 5$, which would yield a diagram with 25 squares, and excluding half of them ($n^2/2$), thus yielding a diagram with 12.5 squares; finally, to complete 15 squares, one adds $n/2$ (= 2.5) squares back. These operations, which correspond to the terms in the right-hand side of the identity sign in the theorem's statement (namely, $n^2/2 + n/2$), are perfectly general and can be performed for any natural number. Thus, although the diagram itself only exhibits the result for a particular instance ($n = 5$), it can be generalized without loss.

It is important to note that the diagrammatic proof incorporates both visual (pictorial) and linguistic (reasoning-based) traits. While the diagram itself emphasizes visual elements, some reasoning about the diagram is needed to establish that the intended result holds and can be generalized beyond the particular case the diagram depicts. Diagrams are, thus, hybrid objects that include the visual and the linguistic, and in the context of mathematical practice, as opposed to the sciences, they may be used as sources of evidence.

4 Diagrams and Styles of Reasoning

Why do diagrams have such different roles in scientific and mathematical practice? Diagrams cannot play an evidential role in the sciences given that this requires a causal relation between the objects in the sample under study and the corresponding image—such as the one provided by a microscope. It is in virtue of this causal relation that a micrograph provides evidence for what takes place in the sample. Given the interaction of the microscope with the sample, the objects in the sample produce (cause) the microscope image, which, as a result, can be taken as offering evidence for the presence of the relevant objects. If the image has been properly produced, it will allow researches to rule out (likely) possibilities that, should they obtained, the presence of the phenomena displayed in the image would be undermined. This process of elimination of undermining possibilities, such as artifacts and confounding factors, is sometimes achieved by combining the results generated by the microscope with those of additional instruments.

In the case of the AFM micrograph in Fig. 1 (on the right), researchers emphasize that it is the presence in the sample of DNA strands configured in a honeycomb shape that produces the resulting microscope images. Similarly, in the case of the 'single'-molecular wire experiment (Fig. 2, on the right), researchers also insist that a current from the surface of the sample to the gold substratum produces the small blobs on the STM micrographs.[2]

No such causal relations are required in mathematics. This opens up the possibility of having diagrams as full-fledged evidential devices, conveying the content of the theorem's statement without presupposing any causal relation between the diagram and the configuration among mathematical objects that the theorem describes. As opposed to what happens with a micrograph, it is not the case that relations among mathematical objects produce the corresponding diagram. There is simply no causal connection among these objects, and none is expected, given that mathematical objects are causally inert.

Moreover, one need not be a platonist about mathematics to recognize the potential evidential role of diagrams in mathematical practice. Even if mathematical objects did not exist at all (as nominalists insist they don't), diagrams could still be used as sources of evidence. What is important is that a diagram conveys properly the *relevant* conceptual relations described in the statement of the theorem—whether the objects involved exist or not.[3] The non-relevant relations need not be properly represented at all: diagrams often misrepresent many features of the objects under consideration. In the case of the number-theoretic diagram just discussed (Fig. 5), clearly natural numbers are not squares and to add such numbers is not strictly to pile squares up. Several relations in the diagram cannot, thus, be taken literally. The diagram is a representational device that does not convey faithfully every aspect of the relations among the relevant mathematical objects. What is crucial is that the central, relevant relations—those explicitly stated in the theorem—are properly displayed in the diagram.

This point also highlights the hybrid nature of diagrams: they are non-linguistic entities that can have informational content. If the content is right, it may convey all the information required to establish the truth of the theorem under consideration, suitably augmented by proper bits of reasoning. By displaying the intended result (the theorem's content) in a particular case, and by indicating, by a suitable reasoning, the possibility of extending that instance to any relevant case, the diagram provides the relevant content. This is illustrated in the diagram of Fig. 5 for the case in which $n = 5$, since the ability to take the diagram as a source of evidence requires the additional reasoning that generalizes the result beyond that case.

[2]Whether a realist reading of these images is justified or not is not a topic for this paper (see Bueno 2011 for some critical discussion). The point about the evidence requirement stands independently of this issue.

[3]Someone may complain that if mathematical objects don't exist, then there is no distinction between proper and improper ways of describing them. Given their nonexistence, it doesn't matter how they are described. But this is not right. Sherlock Holmes doesn't exist, but it is proper to describe him as a detective rather than a milkman; it is improper to describe him as (literally) a mouse rather than a person. The same point applies to numbers and other mathematical objects: even if numbers don't exist, it is proper to describe the sum of two natural numbers as a natural number; it is improper to describe the cumulative hierarchy of sets as (literally) a blacksmith.

Moreover, those diagrams that increase one's understanding of the theorem in question also rely on a suitable reasoning (with the diagram). As noted in the case of Euler's formula, the relevant diagrams allow one to see why the theorem in question is true (see Figs. 3 and 4). The geometrical interpretation displayed in the diagram, in which the theorem is formulated in terms of suitable vectors, relies on reasoning with such vectors so that the result expressed in the theorem's statement can be obtained. Augmented by such reasoning, the geometrical interpretation exhibited in the diagram provides a significant explanatory device. The diagram's explanatory role then becomes one of its significant traits: one can see why the relevant result holds—as long as one can also reason properly with the vectors displayed in the diagram.

Underlying the different roles of diagrams in scientific and mathematical practice, we find two different styles of reasoning: one is a heuristic style; the other is a warranting style. The *heuristic style* invokes diagrams as merely theoretical aids, as props that help to pave the way to an evidential claim, but which never become evidence themselves. The use of diagrams in scientific practice is of this kind. Of course, some diagrams can be used in this way in mathematical practice too. The diagrams used in the case of Euler's formula are employed as heuristic devices to help one's understanding of the result rather than as full-blown proof procedures.

In contrast, the *warranting style* takes diagrams as items that can convey the required information to establish the relevant results. Diagrams are themselves sources of evidence. Some uses of diagrams in mathematical practice are of this kind, but interestingly no use of diagrams in scientific practice invokes a warranting style. Given the way diagrams are produced—they are simply made up by researchers rather than causally generated by the relevant phenomena—they are not the kind of entity that can be a source of evidence.

It may be argued that this is not so. When Robert Hooke presented his microscope images in *Micrographia* (1665), the images were averages of a multitude of observations so that the nuances and details of the objects under study could be properly depicted. Thus the images were not causally produced by the instrument, but were only based on what the instrument allowed Hooke to see. He tells us:

> What each of the delineated Subjects are, the following descriptions annext to each will inform. Of which I shall here, only once for all add. That in divers of them the Gravers have pretty well followed my directions and draughts; and that in making of them I have endeavored (as far as I was able) first to discover the true appearance, and next to make a plain representation of it. This I mention the rather, because of these kinds of Objects there is much more difficulty to discover the true shape, then of those visible to the naked eye, the same Object seeming quite differing, in one position of Light, from what it really is and may be discover'd in another. And therefore, I never began to make any draught before by many examinations in several lights, and in several positions to those lights, I had discover'd the true form. For it is exceeding difficult in some Objects, to distinguish between a prominency and a depression, between a shadow and a black stain, or a reflection and a whiteness in the colour. Besides, the transparency of most Objects renders them yet much more difficult then if they were opacous (Hooke 1665).[4]

[4]For further discussion of this issue, see Wilson (1995) and Pitt (2004), who quote the passage from Hooke.

The process of learning how to use a microscope requires some care and training (just as the learning of how to use our eyes does, even if we no longer remember how that happened). It is by comparing and contrasting a number of views of the same specimen that one learns how to visualize the object under study.

One issue that Hooke considered was how to present the content of what he experienced through the microscope. The technology that would allow one to take a photograph of the visually salient features of the specimen would not have been created for almost couple of centuries. Hooke had to devise a way of conveying that information to the readers of his book. He would draw an image of each specimen he saw after varying the conditions of observation in order to appreciate the contrasts and to obtain salient traits of the sample. He would then try to reproduce these features in a typical image: one that incorporates, into a single image, the variety of traits of the specimen in order to capture the relevant look of the specimen. The series of interactions with the sample paves the way to the drawing of a typical case. The drawing itself is then the evidence that the sample in fact had those features that the image represents it as having: features that were seen with the microscope at a certain stage. Thus, in such conditions, so the argument goes, a drawing can be a source of evidence.

The idea that a mechanically generated image is a source of evidence emerged from the understanding of objectivity according to which in order to guarantee that information about the specimen is properly captured and recorded one needs to implement the process mechanically. This process is expected to ensure that no illicit interpretation of the data is inadvertently introduced. If, however, there are mechanisms to guarantee that no unintended interpretation is advanced and that the data are properly conveyed, then mechanical image generation may not be required. In this case, a drawing of a typical case can be the source of evidence. However, these drawings are *not* diagrams, but are supposed to be faithful representations that capture the features of the phenomena. They are produced according to specified rules that ensure that the drawings are sensitive to the visually salient traits of the phenomenon under study. It is this counterfactual dependence between the specimen studied with the instrument and the corresponding drawing that guarantees that the drawing is a source of evidence. Later, when mechanical image generation became possible and was established as the norm in scientific research, it is arguably the counterfactual dependence that supports the resulting images as evidence (see Bueno 2011).

It may be argued that drawings are used to represent generic types, whereas machine-made images represent particulars.[5] For drawings have a significant trait: selectivity. Since they are intentionally created, drawings can convey selected features of their target, and highlight traits that could otherwise be missed. In contrast, machine-made images simply reproduce those features of the objects that are present at the scene before the machine and to which the machine is sensitive. The result is then the representation of the particulars: those before the machine.

---

[5]For a critical discussion of this distinction, see Lopes (2009).

However, this divide between the representational traits of drawings and machine-made images is questionable. For machine-made images can represent types: the properties of the objects to which the relevant instrument is sensitive. Rather than particulars, instruments can represent selective properties: those that the objects the instrument interacts with are detected by it as having. Scientific instruments are built in order to be sensitive to a range of properties and are entirely insensitive to a number of others (see Humphreys 2004). For instance, scanning tunneling microscopes are sensitive to the surface of conducting materials at the nanoscale, whereas transmission electron microscopes interact with the inner structure of properly prepared samples. None are sensitive to sound properties or, when used to study objects at the nanoscale, to color properties, which are not present at that scale in any case. The properties a scanning tunneling microscope detects are generic since the images it produces are the result of multiple interactions between the sample under study and the instrument. The image itself is an aggregate of multiple interactions. Choices are also present in the process of construction of the relevant images. (This echoes Hooke's old practice with his drawings, although now the generic construction of the microscope image is machine made.) With a transmission electron microscope, one can focus on certain parts of the sample, change the contrast to highlight certain features of the specimen, or tilt the sample to obtain contrastive information about it. Thus, selectivity is clearly involved from the design to the use of these instruments. Since they are produced to detect certain properties and not others, not surprisingly, when instruments are used, these are the properties they detect.

Moreover, drawings can also represent particulars. Lucian Freud's portrait of Queen Elizabeth II represents the queen, not any generic person. It may be said that the portrait doesn't represent the queen exactly as she looked in any particular moment, but rather it is the outcome of an aggregation of many particular looks over time. In this way, nuances about her can be captured that no single photograph could apprehend. Portraits, at least good ones, do have this capacity. But it is still the queen, this particular person, who is thereby represented. Moreover, the aggregative feature of some machine-made images has already been noted, and this is something that drawings (portraits, in particular) share with some images that are mechanically generated. Thus, one cannot maintain that there is a distinction between drawings and instrument-made images along the lines of the representation of types and particulars, respectively. This is too simple a distinction to capture the complexities involved in these images (see also Lopes 1996, 2009).

Properly understood, mathematical diagrams also challenge this simple account since they manage to represent both particulars and types. As we saw, such diagrams are significant in that they represent both a particular case and, given the generalizability of that case based on a particular reasoning, a generic instance. We literally see that the theorem holds in a particular case (such as, in Fig. 5, for $n = 5$), but also realize that particular case is perfectly arbitrary: any other particular case could have been represented just as well with perfectly obvious adjustments. This dual role of diagrams is significant, since, as the simple, traditional view would have it, due to their pictorial features, diagrams would tend to convey content about particulars rather than generic traits.

Some illustrations in archeology, namely, lithic illustrations, also challenge this divide between drawings and machine-generated images, since these illustrations are drawings that convey information about particulars (Lopes 2009). Lithic illustrations are produced in accordance with very strict rules to guarantee that the relevant features of the stones that need to be studied are properly displayed and the right traits are highlighted. These rules guarantee the counterfactual dependence between the traits on the surface of a stone and the corresponding drawing. As a result, lithic illustrations are taken as sources of evidence in archeology.

Note, however, that lithic illustrations are *not* diagrams, given that diagrams are schematic renderings of certain aspects of their targets and need not preserve visually salient features of the objects they represent. If lithic illustrations were diagrams, they wouldn't be sources of evidence, since they need not be produced counterfactually from the target. In contrast, lithic illustrations are faithful representations of their targets, and are sensitive to and properly capture the traits of the relevant stones.

By considering the relations between the heuristic and the warranting styles important traits of diagrams can also be highlighted. The warranting style acknowledges that in order for one to determine that a certain diagram establishes the intended result, some bit of reasoning is typically required. One needs to reason with the diagram, in the way illustrated in the discussion of Fig. 5 above, in order to determine that the relevant theorem does hold.

The warranting style is significant for providing a way of using diagrams in which they convey evidential information about the objects under consideration. The heuristic style is important for accommodating the way in which diagrams provide resources to motivate and understand certain results.

There is a significant connection between these two styles. On the one hand, the warranting style presupposes the heuristic style in that, in order for diagrams to provide justification for a given result, they need first to be the kind of thing that can yield some insight about the subject matter. On the other hand, although the heuristic style takes diagrams in such a way that they need not provide justification for the result under consideration, they can still help in the interpretation of certain results as well as in suggesting and motivating them.

5 The Semantic and the Received Views Reunited

As we saw, central to the divide between the semantic and the received views is the role of linguistic features in scientific practice. But ultimately both views require linguistic considerations. Theories, even if thought of as nonlinguistic devices (e.g. as a family of models), have informational content, which, in turn, involves some linguistic structure.

In this context, diagrams provide a suitable setting to bring the semantic and the received views together. Diagrams are non-linguistic devices, but they do have informational content: (a) they *represent a certain situation*, (b) and the representation is implemented *in a given way* (according to a certain *form*). In Fig. 5, the diagram *represents* the process of adding *n* natural numbers *as* one of pilling up squares: each square represents a particular number and addition is represented by the concatenation of piles of squares. In Fig. 2 (on the left), the diagram *represents* the sketch of the experiment, with the 'single'-molecular wire, represented *as* a string of bounded atoms, connecting the tip of the STM with the gold basis. In Fig. 1 (on the left), the expected outcome of the DNA strand experiment is sketched, in which different DNA configurations shaped as triangles, hexagons and honeycombs are displayed. The particular geometrical objects in the diagram *represent* the configuration of the expected DNA arrangements *in space*.

In all of these cases, the representation is achieved, in part, by a structural similarity (such as partial isomorphism, partial homomorphism etc.[6]) between the diagrams and their targets. It is due to the appropriate structural similarity that the former can be *used* to represent the latter in each scientific context. In the case of the number-theoretic theorem (Fig. 5), the relevant structural similarity is found between the concatenation of piles of squares and the addition of natural numbers. In the 'single'-molecular wire experiment (Fig. 2), there is structural similarity in the position of the 'single'-molecular wire relative to the STM tip, the enveloping substratum, and the gold basis as represented by the configurations on the diagram and as arranged by the appropriate molecules in the sample. In the DNA strand experiment (Fig. 1), the structural similarity is established between the geometrical configurations of lines on the diagrams and DNA strands in the sample.

As noted, diagrams are hybrid entities, very similar—in this respect—to scientific theories. Even if, following the semantic view, such theories are thought of as families of models (which are nonlinguistic objects) they are models that satisfy certain conditions, namely, those specified in the formulation of the theories. Newtonian theory may be presented as a family of models, but these models satisfy Newtonian laws of motion, which are specified linguistically. Without such linguistic specification it would be unclear what these models are models of. These conditions need to be presented linguistically, although the particular language in which they are formulated is, for the most part, immaterial. Similarly, linguistic and nonlinguistic considerations are brought to bear in thinking through the status of diagrams. Nonlinguistic considerations enter due to the diagrams' pictorial nature: the information they convey is presented in such a way that the shape and relative position of each line has significance. And it is through their pictorial features that diagrams have informational content. As we saw above, a sequence of piled-up squares can be used to represent the addition of numbers, with the number of squares in the concatenated geometrical object representing the resulting number.

---

[6]For further discussion of these concepts in the context of scientific representation, see Bueno and French (2011).

Moreover, the informational content of a diagram has linguistic structure. But, as opposed to what the proponents of the semantic approach suggest, the form that such linguistic structure takes does matter. Some presentations of the relevant information may be unusable to those who want to employ them: the relevant information, although there, may not be salient enough. Consider, for instance, a purely set-theoretic formulation of the diagram used to prove the theorem in number theory discussed above (Fig. 5). In this formulation, the diagram would not be of much use, since its pictorial content (what is specifically diagrammatic about it) would be entirely lost. Information presented *in the right form* is needed. The right form, in this instance, is the way in which the content is presented. That is a linguistic choice: the choice of a certain language, even if it is one that involves, as in the case of diagrams, nonlinguistic components. This means that linguistic features cannot be entirely disregarded, as they seem to be in some instances by those who favor the semantic approach. These are the kinds of consideration that are needed to bring together the semantic and the received views.

Proponents of the received view, in turn, emphasize the role of language considerations in the reconstruction of scientific practice. The capacity of formulating theories in a suitable formal language was often taken as a criterion of adequacy for the proper understanding of the relevant theory. It is this requirement of rational, formal reconstruction of the sciences that was so central to the received view. It was also this requirement that ultimately forced the received view to provide so much emphasis on linguistic considerations: the particular features of the formal languages in which the reconstruction of scientific theories and other aspects of scientific practice were supposed to display.

In contrast, the semantic view tends to take scientific practice literally, as something that is in no need for reconstruction. One of the goals of a philosophical account of the sciences is to understand features of the science in its own terms rather than to reformulate it in a formal language. This aspect of the semantic view is significant and was perceived by many as a significant step in the right direction, freeing the philosophy of science from a particular formalized setting.

To reconcile both views requires resisting extreme versions of either of them. The complete rejection of linguistic considerations suggested by the strong form of the semantic view is untenable. To properly formulate a scientific theory linguistic considerations are needed. The same goes for the proper understanding of diagrams, which, as noted, require linguistic considerations in the styles of reasoning that are associated with them, both of which include the capacity of reasoning with them.

But from the acknowledgement of the role played in linguistic considerations, one should not conclude that the requirement of formalization in a given language is justified. The strong form of the received view, which insists on the need for formalization in a particular language (first-order or higher-order; with modal operators or without), cannot be maintained either. For diagrams to play their role *as* diagrams, it is crucial that their nonlinguistic, pictorial content be preserved. Otherwise, one cannot use or make sense of them.

The proposed reconciliation sketched here tries to bring together the strengths of both the received and the semantic views. With the semantic approach, the proposal acknowledges that theories should be thought of as a family of models. With the received view, it notes that in order for such models to be properly presented and used, they need to be formulated in some language, and it is here that the specification of certain conditions for each model is introduced. Throughout this process, diagrams provide an insightful illustration, since they indicate a salient aspect of scientific practice in which both linguistic and nonlinguistic considerations are so clearly present.

6 Conclusion

We have here an account of diagrams and different styles of scientific reasoning within a broadly semantic conception. But the view also acknowledges, as it should, the role played by linguistic considerations, such as the informational content in diagrams and theories, and the form that such content takes. Thus, the resulting view provides a first step toward a reconciliation of the received and the semantic views, by highlighting the need for both linguistic and nonlinguistic components, and by noting that these components can be brought together without tension.

References

Baird, D., Nordmann, A., & Schummer, J. (Eds.). (2004). *Discovering the nanoscale*. Amsterdam: IOS Press.

Brown, J. R. (2008). *Philosophy of mathematics* (2nd ed.). New York: Routledge.

Bueno, O. (2011). When physics and biology meet: The nanoscale case. *Studies in History and Philosophy of Biological and Biomedical Sciences, 42*, 180–189.

Bueno, O., & French, S. (2011). How theories represent. *British Journal for the Philosophy of Science, 62*, 857–894.

Bumm, L.A., Arnold, J.J., Cygan, M.T., Dunbar, T.D., Burgin, T.P., Jones II, L., et al. (1996). Are single molecular wires conducting? *Science 271*, 1705–1707.

Carnap, R. (1934/1937). *The logical syntax of language* (S. Amethe, Trans) London: Kegan Paul.

Ding, B., Sha, R., & Seeman, N. (2004). Pseudohexagonal 2D DNA crystals from double crossover cohesion. *Journal of the American Chemical Society, 126*, 10230–10231.

Giaquinto, M. (2005). Mathematical activity. In P. Mancosu, K. Jørgensen & Pedersen S. (Eds.), pp. 75–87.

Humphreys, P. (2004). *Extending ourselves: Computational science, empiricism, and scientific method*. New York: Oxford University Press.

Hooke, R (1665). *Micrographia*. London: Royal Society.

Lopes, D. (1996). *Understanding pictures*. Oxford: Oxford University Press.

Lopes, D. (2009). Drawing in a social science: Lithic illustration. *Perspectives on Science, 17*, 5–25.

Lynch, M. (1991). Science in the age of mechanical reproduction: Moral and epistemic relations between diagrams and photographs. *Biology and Philosophy, 6*, 205–226.

Mancosu, P., Jørgensen, K., & Pedersen, S. (Eds.). (2005). *Visualization, explanation and reasoning styles in mathematics*. Dordrecht: Springer.

Pitt, J. (2004). *The epistemology of the very small.* In D. Baird, A. Nordmann & J. Schummer (Eds.), pp. 157–163.
Seeman, N. (2003). DNA in a material world. *Nature, 421*, 427–431.
Seeman, N. (2005). From genes to machines: DNA nanomechanical devices. *Trends in Biochemical Sciences, 30*, 119–125.
Seeman, N., & Belcher, A. (2002). Emulating biology: Building nanostructures from the bottom up. In *Proceedings of the National Academy of Science* Vol. 99, pp. 6451–6455.
Suppe, F. (1977a). *The search for philosophic understanding of scientific theories*. In F. Suppe (Ed.), pp. 1–232.
Suppe, F. (Ed.). (1977b). *The structure of scientific theories* (2nd ed.). Urbana: University of Illinois Press.
van Fraassen, B. C. (1980). *The scientific image*. Oxford: Clarendon Press.
Wilson, C. (1995). *The invisible world*. Princeton: Princeton University Press.

Bas van Fraassen on Success and Adequacy in Representing and Modelling

Michel Ghins

Abstract In his *Scientific Representation. Paradoxes of Perspective* (2008), Bas van Fraassen offers a pragmatic account of scientific representation and representation *tout court*. In this paper I examine the three conditions for a user to succeed in representing a target in some context: identification of the target of the representational action, representing the target as such and correctly representing it in some respects. I argue that success on these three counts relies on the supposed truth of some predicative assertions, and thus that truth is more fundamental than representation. I do this in the framework of a version of the so-called "structural" account of representation according to which the establishment of a homomorphism by the user between a structure abstracted from the intended target and some relevant structure of the representing artefact is a *necessary* (although certainly not sufficient) condition of success for representing the target in some respects. Finally, on the basis of a correspondence *view* (not *theory*) of truth, I show that it is possible to address what van Fraassen calls "the loss of reality objection".

In his book *Scientific Representation. Paradoxes of Perspective* (2008) Bas van Fraassen offers a philosophical analysis of representation which is both empiricist and pragmatic. To represent is to perform some kind of action, and actions are evaluated with respect to their success or failure in attaining specific goals. Moreover, in science success or failure in representing a target must be assessed on the basis of observable phenomena.

As is well known, according to van Fraassen, science aims at empirical adequacy, that is, at saving the phenomena. Thus, a scientific theory is successful if we

M. Ghins (✉)
Centre Philosophie des Sciences et Sociétés (CEFISES),
Institut Supérieur de Philosophie, Place du cardinal Mercier, 14, B-1348,
Louvain-la-Neuve, Belgium
e-mail: michel.ghins@uclouvain.be

L. Magnani and C. Casadio (eds.), *Model-Based Reasoning in Science and Technology*, Studies in Applied Philosophy, Epistemology and Rational Ethics 27, DOI 10.1007/978-3-319-38983-7_2

have good reasons to believe that it saves the phenomena within its domain. Since for van Fraassen scientific theories are models in the first place, a theory is successful if its empirical parts, called "empirical substructures", adequately represent the phenomena within its domain.

A satisfactory account of representation is thus central to van Fraassen's philosophy of science. The success of a representational action can be evaluated on three counts. First, the user of a representing artefact must succeed in identifying its target or referent. Second, the target is always represented from a certain perspective *as* having such and such properties. Third, we may ask if our representing activity conveys some reliable information about its target or, in other words, if our representing artefact (a scientific model for example) is adequate or accurate to its target in some respects.

I will argue that the three criteria of success in representing always rely on the truth of some predicative assertions or statements and thus that truth is more fundamental than success. I will also defend that successful representation necessarily involves the institution by the user of a homomorphism between what is represented and its representing artefact. Given this, it is possible to show that, contrary to what van Fraassen defends, what he calls the "loss of reality objection" is not dissolved but solved.

1 Success in Representing a Target

van Fraassen's ambition is not to delineate a set of necessary and sufficient conditions which would allow a user to declare that his *representational action* (called "representation" in what follows) is successful or not. His main objective is *not* to provide a definition of representation but to identify the circumstances in which a representation succeeds. Thus, his main query is not "*What* is a representation?" but "*When* is a representation successful?" (van Fraassen 2008, p. 21).

Since representation is an action, it presupposes someone who acts, the "user" who employs a representing artefact, which I will call the "representor", in order to represent a specific target, at least partially. When the target is observable and perceptually present, the user can identify the target by means of some observable properties presumably belonging to it. Such referring action would be *external* to the representor. Generally, the user intends to represent some target. Intentionality is essential to any kind of representation.[1] Indeed, it is crucial to distinguish denotation (or reference) from representation. Denotation of the target is a preliminary, necessary condition for representation, but it is far for being sufficient. In

[1]See Chakravartty (2010, p. 206). Intentionality is essential to the success of all kinds of representation, not only scientific representation.

order to have a relation of representation by a user U in some context C between a representor R and a represented target T, some additional conditions must obtain. One of my main contentions is that some kind of mapping—specifically a homomorphism (see below)—between structures abstracted from the target and the representor is a *necessary* condition—but certainly *not sufficient*—for representation to occur. On this point at least, I agree with the so-called "structural" or "informational" accounts of representation defended by Suppes (1967, 2002), Da Costa and French (2003), Bartels (2005, 2006) and Chakravartty (2010), among others.

Now, one of the most important claims made by van Fraassen is that there are no properties of a thing which make it ipso facto a representor of a specific target. On this, he approvingly quotes Nelson Goodman:

> The most naïve view of representation might perhaps be put something like this: "A represents B if and only if A appreciably resembles B". Vestiges of this view, with assorted refinements, persist in most writing on representation. Yet more error could hardly be compressed into so short a formula. (Goodman 1976: 3–4) (van Fraassen 2008, p. 11)

Certainly, both Goodman and van Fraassen are right on this. We could use a photograph of the Atomium in Brussels to represent the Eiffel tower in Paris provided we make explicit some conventions which would obviously depart from the conventions implicitly agreed upon in our culture when we look at a postcard. Success in representing a target depends on specifying a *code*, be it implicit or explicit. But what is a code? A code is a *mapping* which institutes a correspondence between some characteristics of a representor and some characteristics of its target. So, "czmfdq" written on a piece of paper can represent the word "danger" given a certain code (which I leave for the reader to find as an exercise…). True, a different code could have been used so that "czmfdq" represents "change" (another exercise…).

Thus, van Fraassen states what he calls the *Hauptsatz* of his approach to representation:

> There is no representation except in the sense that some things are used, made, or taken, to represent some things as thus or so. (van Fraassen 2008, p. 23)

Success in representing a target then presupposes a mapping between *selected relevant* ingredients of whatever thing you like to use as a representing artefact and selected ingredients of an intended targeted thing. Once the code has been instituted, some things acquire the status of representors and other things the status of represented targets. Of course, some artefacts and codes are more manageable and practical than others for representing some targets. But this is not the point. The point is that the code is *external* to an artefact: it is brought from outside to bestow on a thing the role of a representing artefact. And this is why the representor deserves to be called an "artefact" even if it is a natural object such as a shell or a pebble collected on a beach.

Since the code is *external* to the thing used as a representor, the success of a representation does not have to trade on some resemblance between the representor R and its target T. Resemblance or likeness can certainly play a representational role in some cases such as pictures and portraits, but this happens because it has been freely decided and agreed upon that some colour or shape in the representor is relevant for representing a given target. More generally there are no inner properties of a thing or relations between its parts that make it ipso facto the representor of a specific target. Consequently, *anything can be used to represent anything* (van Fraassen 2008, p. 23).

Technically the code is stipulated by a specific mapping between structures extracted by abstraction from the target and the representing artefact. Thus, except in pure mathematics, the target and its representor are not structures; they are *concrete* things (whether imagined or real). For representation to occur, the user must select some elements and relations among them in order to construct relational structures.

Now, following Dunn and Hardegree (2001) (thereafter D&H) let us give some definitions. A relational structure[2] **A** is a couple of two ingredients: a domain A of individual elements and a family $\langle R_i \rangle$ of relations on A. For some natural number n, a n-place relation R_i or a relation R_i of degree n is a set of n-tuples of elements of A (D&H, p. 10).[3]

Take two (relational) structures $\mathbf{A} = \langle A, R_i \rangle$ and $\mathbf{B} = \langle B, S_i \rangle$. A *homomorphism* from **A** to **B** is any function h from A into B satisfying the following condition for each.

$$\text{(ST) If } \langle a_1, \ldots, a_n \rangle \in R_i, \text{ then } \langle h(a_1) \ldots, h(a_n) \rangle \in S_i$$

In this case, the homomorphic function h achieves a *structural transfer* (ST) from **A** to **B**.

(ST) does not require that h is surjective. We say that **B** is a *homomorphic image* of **A** if there exists a *homomorphism* from **A** to **B** that is *onto* B [in symbols $\mathbf{B} = h^*(\mathbf{A})$].[4] A function h maps **A** *onto* **B** if for every $b \in B$ there exists an a in A such that $h(a) = b$. (Ibidem, p. 15) In other words, h is *surjective*. If h is also injective and thus bijective, then we have an *isomorphism*.

Suppes (2002, p. 56) uses a stronger definition of homomorphism since he replaces (ST) above by:

$$\text{(PS)} \langle a_1, \ldots, a_n \rangle \in R_i \text{ if and only if } \langle h(a_1) \ldots, h(a_n) \rangle \in S_i$$

[2]We use bold font to refer to structures, e.g. **A,** and italic to denote the domains, e.g. *A*.

[3]If some of the elements belonging to the domains do not stand in any relation, we have what Da Costa and French call a "partial structure" (2003, p. 19).

[4]Dunn and Hardegree give the definition for *similar* structures, namely structures of the same *type*, that is, whose families of degrees of their respective relations are the same (p. 10). Our philosophical discussion will implicitly be restricted to representations which involve structures of the same type. For example, two structures which contain only one-place relations (properties) and two-place relations are similar.

Suppes remarks that a weaker notion of homomorphism is generally used in algebra (this weaker definition is the one provided with (ST) by Dunn and Hardegree). "However (…) in the philosophy of science, the definition here used is more satisfactory" (Suppes 2002, p. 58, Footnote 5).

The further condition adduced by Suppes is what D&H call *absolute faithfulness*.[5]

A homomorphism h from **A** to **B** is *absolutely faithful* if for each i

$$\text{(AF) If } \langle h(a_1)\ldots, h(a_n)\rangle \in S_i, \quad \text{then} \langle a_1, \ldots, a_n\rangle \in R_i$$

Take the simple example of two structures **A** and **B** with their respective domains

$A = \{a_1, a_2\}$ and $B = \{b_1, b_2\}$ and two 2-place relations R and S on A and B respectively.

$\mathbf{A} = \langle A, R\rangle$ and $\mathbf{B} = \langle B, S\rangle$ are *homomorphic* according to (PS) if and only if there is a function h such that:

1. The domain of h is A and the range of h is B. $h: A \rightarrow B$
2. h is surjective
3. If x_1 and x_2 are in A then x_1Rx_2 if and only if $h(x_1)Sh(x_2)$ (Suppes 2002, p. 56).

In a discussion of representation it is useful to introduce a weaker notion of fidelity, which D&H call *minimal fidelity*.

A homomorphism h from **A** to **B** is *minimally faithful* if for each i

(MF) If $b_1,\ldots,b_n$ are in the range of h, then if $\langle b_1,\ldots,b_n\rangle \in S_i$, then there are elements $a_1, \ldots, a_n \in A$ such that $h(a_1) = b_1 \ldots, h(a_n) = b_n$, and $< a_1, \ldots, a_n > \in R_i$. (Dunn and Hardegree 2001, p. 16).

Thus, minimal fidelity requires that the image has no "gratuitous" structure but contains only the amount of structure necessary for the structure of the source to be transferred to the image. "Beyond the structure required by the structural transfer condition, the image has no further structure." (Dunn and Hardegree 2001, p. 16).

In mathematics all is well and clear. But in representational acts, we use concrete representors to represent concrete targets. It is thus important to distinguish, as we said above, concrete things on the one hand, and the structures we abstract from them to perform representations on the other hand. In accordance with what is called the "structural account of representation" I will defend that in order to use a thing, a representor, to represent another thing, its target, we must establish a *minimally faithful homomorphism* between structures which we selectively abstract from the target and its representor. Before I do that, some further formal observations are in order.

The notion of isomorphism is very weak. It only requires that the two domains A and B have the same number of elements (cf. Newman's theorem). The relations

[5] A structure can be a *faithful* homomorphic image of another structure, without being *accurate* or *exact*. We come back to this important point below.

on the two domains can be very different. A relation such as "x is higher than y" can correspond to the relation "x is brighter than y" or 'x is stronger than y' and so on. Remember that anything can be used to represent anything… Yet, the relation of isomorphism between two structures has some formal properties, namely symmetry, transitivity and reflexivity, which are independent of the specific relations in the respective structures. Certainly, as it has often been observed, representation cannot be *reduced* to isomorphism, since representation is not symmetric, not reflexive and not transitive. Asymmetry for example, must come from outside by means of a *referential action*: the agent uses the representor with the intention of representing a given target and not the other way around.[6]

The definition of homomorphism is a weakening of the definition of isomorphism. Whereas isomorphism has to be bijective (one-one), a homomorphism is a surjective (many-one) and not an injective mapping. The standard example is a two-dimensional photograph which represents a three-dimensional subject. A set of elements of the three-dimensional object is sent by the function h to only one element of the two-dimensional picture. Moreover, not all characteristics of the target are sent to elements of the picture. For example, colour is not taken into account in the representor in the case of a black and white picture. "Going to the subject to its image involves (so to speak) compressing the three-dimensional subject into two dimensions" (Dunn and Hardegree 2001, p. 15). Homomorphism is not symmetric and not reflexive. However, representation cannot be reduced to the establishment of a homomorphic function between a targeted structure and its representor, if only because representational success necessitates a prior referential intentional act.

Given this, I maintain that what is called "mistargeting" (Suárez 2003; Pero and Suárez 2015) amounts to *incorrect representation* of an intended target. Intuitively, mistargeting is to take the representor to represent target T when in fact the representor represents target U, which is distinct of T. How can this happen? First of all, we must never lose sight of the fact that representing is an action performed by a user or agent. The user must identify what she intends to represent. And what she intends to represent is in the first place a thing, a concrete object which can be *identified in a referential act*. Typically, the user identifies some specific properties allegedly belonging to the intended target, thereby presupposing the truth of some predicative statements attributing these properties to it. Ostension is not enough in most contexts to convey to others what the user is referring to. At this stage, we do *not* have representation yet but denotation only. This is not the place to discuss the various philosophical theories of reference, and I will assume that reference is by and large unproblematic for common observable things.

Thus suppose I intend to represent my desk, that is, an object which is used to write, say. Certainly, in order to construct a representation of my desk under some

[6]On this I disagree with Bartels (2006, p. 12) who claims that "causal relations" between two things, such as between an object and a photograph of it, can play a role in determining the direction, and thus the asymmetry, of a representation. The asymmetry is determined by the intention of the user only.

perspective or point of view, I must point out some of its properties which, for some reason, I consider to be of interest or relevant. I have several options. Suppose I wish to construct a geometrical representation. Then, I assume that the desktop has, say, a circular shape. In other words, I suppose that the proposition "The top surface of the desk has a circular shape" is true. I might be wrong about that. Perhaps, my desktop doesn't have a circular shape. And I assume the truth of other propositions of this kind. Then I draw a geometrical picture (with a circle in it) of my desk and claim: this picture represents my desk. Then I show the picture to my friend Lucy who, without having seen my desk, says: this is not a picture of a desk, but the picture of a table. Desks are not circular, but tables can be, so she argues.

Does this remark show that I have misrepresented my desk in the sense of mistargeting it? No, not in my regimented use of the word "representation". Identification of a target rests on an intentional referring action, which is not a representational action. Such referring act is *independent* of the kind of homomorphism I will establish to construct a representation of the target I referred to, and on this I agree with Suárez. However, in order to be able to speak of a *representation of* a specific target I necessarily must attribute (rightly or wrongly) properties to the target, which will be put in correspondence with properties of the representor by some homomorphism. If not, we have denotation only and not representation yet. If Lucy claims that the picture represents a table and not a desk, she says so because she relies on conventions which are a matter of course in some cultural milieu (ours…). Equivalently, she presupposes the truth of some other propositions than the ones I have been using in constructing my representation of the desk. If she is right, then I have *misrepresented* the desk in the sense that I have *incorrectly* represented it, but it is still a (wrong) representation *of* my desk and not of something else.

Goodman (1976) and van Fraassen (2008) mention the example of "the painting *of* the Duke of Wellington which everybody agreed resembled the Duke's brother much better" (van Fraassen 2008, p. 19). If a user bestows to the painting the role of representing the Duke of Wellington by a referential intentional action, the painting does indeed represent the Duke of Wellington (and not his brother). Further, success in representing relies on an established a minimally faithful homomorphism between some structure abstracted from the target and some structure abstracted from the painting. It might be true that the painting *misrepresents* the Duke of Wellington. But such a judgement relies on the supposed truth of some predicative statements or assertions. For example, the Duke's brother has some specific facial features and these features correspond to shapes and colours in the portrait which resemble the Duke's brother features according to our implicitly accepted codes better than the Duke's himself.

Thus, mistargetting cannot be, as Pero & Suárez claims, using a model "as a representation of a system or object that is not intended for" (2015), since the target is determined by an intentional referring action. What can happen, of course, is that the user employs a model which incorrectly represents its intended target. But in order to make such a claim representation must occur in the first place and this necessarily involves the establishment of some homomorphism by the user as will be further explained below.

So far, we have reached three important conclusions. The successful identification of a target of a representation is achieved firstly by a referential action which isolates what the user intends to represent by picking out some properties allegedly belonging to the concrete target. At this stage, we do not have representation yet, but only reference. Next, the user must establish some homomorphism between a selected structure abstracted from the concrete target and a structure abstracted from the representor, thereby specifying a code. Only then can we speak of *representation*, and not mere denotation. However, and this is the third conclusion, although there are no intrinsic properties of a thing which impose its use as a representor of a specific target, representors and their targets do possess characteristics independently of their possible selection to play a representational role in some context. Some predicative statements assert that representors and their targets (whether the latter are real or imagined) possess specific characteristics, truly or wrongly.

2 Success in Representing a Target as Such and Such

Users aim at representing a target *as* having some properties. In science, when a model (representor) is proposed by a scientist, it is put forward as adequate, at least possibly so. Even if scientific models are structures (specifically, they are relational structures of *properties*: see below) they implicitly convey some claims about the properties possessed by their targets. However, and I wish to insist on this, models and statements belong to different categories. It would be a category mistake to say that a model is true or false; only statements can be true or false. Thus, only statements permit to ground inference, that is, a reasoning which goes step by step from one statement to another according to a rule. If models can perform an inferential function (as contended by Suárez 2004) it is because the user realizes that the success of the model relies on the supposed truth of some statements.

Let us look more closely at this through one of van Fraassen's examples, namely the caricature of Margaret Thatcher *as* draconian (van Fraassen 2008, pp. 13–15). For the caricature to function as a representor *of* Margaret Thatcher its target must first be identified. In this case, the identification of the target is not achieved by means of an ostensive act, but the referential intention is sustained by statements, which are supposed to be true by the user and which assert that Mrs.T. has specific physical traits.[7] Some of her facial features are mapped by means of a homomorphism into some corresponding parts of the caricature (representor) which preserve some spatial relations, for example the relation that her mouth is below her nose. The spatial arrangement of selected elements of Mrs. Thatcher's visage resembles

[7]Again, a user could use the so-called caricature of Mrs. Thatcher to represent Bismarck say, by means of another referential action.

the spatial relations of the corresponding selected elements in a structure extracted by the agent from the caricature. In a certain context C, the user U of the caricature takes it as a representator R of the target T, namely Mrs. T. So, a representation is a four place relation. It is not a two place relation between a representor and a target. There is no inner structure of the caricature that necessarily makes it a caricature of Mrs. T. The thing which we take as a caricature by relying on implicit conventions and codes could have been used to represent anything else.

Yet, it certainly is true that the caricature-thing does possess some properties. These facts can be expressed by predicative statements such as "This part of the picture resembles a nose". On the other hand, it is also true that Mrs. Thatcher does have a nose of a certain form and facts like these can also be described by true predicative statements. To achieve the identification of a *represented*—and not only denoted—target, the user must institute some kind of mapping between a referent or target and its purported representor. Such mapping is called the "representational function" and is a homomorphism (many-one correspondence) because it preserves the structure or the form which we—the users—consider relevant. Of course, an infinite number of mappings between a given thing and another thing could be generated but each mapping must be based on a selection of properties and structures which do belong to the things involved in a representational correspondence.

A representation is an action which is performed with a certain *aim* in sight. In the example of the caricature of Mrs. Thatcher, the aim of the caricaturist is this: whoever appropriates the representor and knows the code will understand that the target is Mrs. T. and will use it as representing her as draconian. Moreover, the user is supposed to laugh. The caricaturist certainly surmises that Mrs. T. *is* draconian, otherwise this other intended aim would not be reached. But success in representing Mrs. T. *as* draconian does not depend on whether she was in fact draconian or not.

Suppose now that I have never heard of Mrs. Thatcher and that I happen to have a neighbour whose name is Grace and resembles her very much. Then, when looking at the drawing, and given the context and the codes implicitly in place, I will appropriate the drawing and make it a representor of Grace *as* draconian, even if she is a very gentle and amiable person. My referential act will attribute a referent distinct from Mrs. T., but I will represent my target in the same way, namely as draconian, given the mapping and the code implicitly in operation.

van Fraassen insists that representational success of Mrs. T. as draconian is achieved because of the *distortion* of some of Mrs. Thatcher's features into some properties of a mythical animal—a dragon—which western legends portray as mean and mischievous. In a caricature some exaggeration must be present… The distortion, which is necessary to represent the target *as* having some property, entails that representation here is a *misrepresentation*, as van Fraassen contends. Obviously, we all know that Mrs. T. did not physically resemble a dragon. Thus, the qualification of the caricature as a misrepresentation rests on the improbable situation in which a user would believe in the truth of some statements attributing to Mrs. T. facial properties resembling the ones attributed to dragons. Yet, the caricaturist manages to represent Mrs. T. *as* draconian by means of a mapping between some features of imaginary dragons and some parts of the caricature. Again, establishing a homomorphism is necessary in order to attain the caricaturist's goal. The distortion of some of Mrs. T.'s physical traits into a dragon's features allows an user who knows the code to understand that the caricature is meant to attribute (rightly or wrongly) to Mrs. Thatcher's specific psychological characteristics which in our culture are associated with dragons.

Also in this example, the success of the representation rests on the supposed truth of statements attributing properties to Mrs. T. and dragons, and a homomorphic mapping between properties which bestows to the artefact its representational function. Specifically, the spatial relations between some colours and shapes of an imagined dragon are put into correspondence with some colours and shapes in the caricature. On the basis of statements supposed to be true in some context, a user might take as true a statement attributing to the caricaturist a specific intention. The user may also attribute a psychological property to the target, which is the relation of possession between a property and its bearer, specifically between the property D (draconian) and the target T (Mrs. T.). More generally, I submit that the institution of some homomorphism between the purported target and the representor is a necessary condition for the success of a representation, even if resemblance is not. *I call this necessary, but not sufficient, condition of success "structural similarity"*. (Structural similarity is not to be confused with resemblance or likeness.)

This is not to say that a representation is a form of predicative act (Ghins 2010). The caricature of Mrs. Thatcher is not a statement which could be true or false. True, as Goodman and van Fraassen contend, we could interpret the caricature as predicating a psychological trait to the target. But this would be an *external* ingredient added to the representation. Since representing is an action, the categories of true or false are *not* applicable; only success and failure are. To succeed in representing a target *as* such and such is independent of the physical existence of the target or its actually having such and such properties. If we add a cross or a green flag in a representor to indicate that it is meant to convey correct information about the target, this move is tantamount to constructing *another* artefact which would require a new act of appropriation by a user.

On the other hand, success in representing a target as such and such presupposes the assumed truth of some predicative statements about the target and other entities, as we saw. But predicating is not representing. Granted, predicating is also an action. It consists in attributing a property to an identified subject, which can be called the "target" of the predicating action if you wish. The result of such an action is the production of a statement of the form "S is P". But a statement is *not* a representation (Ghins 2010, 2011). First, because it states that some situation or state of affairs obtains. It is an assertion. It then possesses an illocutionary force, namely an assertive force, which is a characteristic representations lack. If I intend to assert that a portrait is beautiful by drawing a cross on it, I only have manufactured another artefact which is deprived of any illocutionary force. The second reason why to predicate is not to represent is that a statement needn't exhibit some structural similarity with a possible state of affairs which would make it true to perform its function, which is to assert that some state of affairs obtains.

Granted, it is always possible to use a sentence sign, written or spoken, as a representation, since anything could be used as a representor by attributing a representative role to it. In doing so, we represent but do not assert. Famously, Wittgenstein's picture theory of meaning developed in his *Tractatus* mainly foundered on its inability to account for any kind of illocutionary force because it attempted to reduce meaning to structural similarity (see Ghins 2011). In the *Tractatus* a proposition is meaningful if and only if there is a isomorphism between its components (called "names") and the components (called "objects") of a possible state of affairs the existence of which makes the proposition true. Such isomorphism was supposed to preserve the logical form common to the proposition and the possible state of affairs it represents.

If some instituted structural similarity between a representing artefact and its target is taken to be a necessary condition for the success of a representation, then statements do not represent. Thus, language does not represent the world, according to this restricted, regimented sense of representation. Although I maintain that there is some kind of correspondence between statements and facts when they are true (see below) such correspondence cannot be construed as structural similarity and, as a consequence, is not representational.

3 Success in Adequately Representing the Identified Target

Scientists aim at constructing models which adequately represent at least observable phenomena. But scientists are not the only ones who strive to construct correct representations. A map user relies on the information that he manages to extract from a particular artefact in order to find his way. This is what Micronesian navigators did when they used artefacts such as this one[8]:

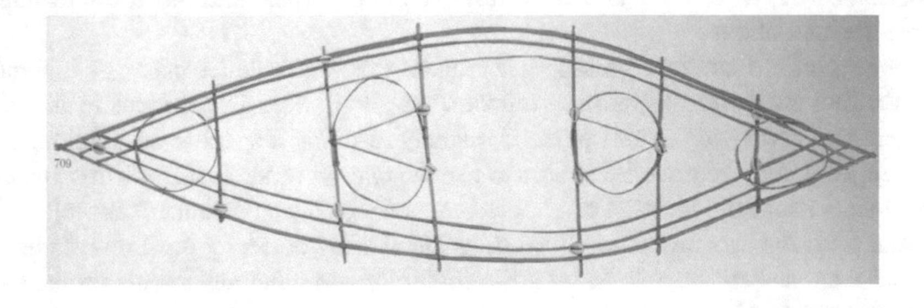

[8]Meyer (1995, p. 616, Fig. 709). The map is part of the collection of the Linden-Museum in Stuttgart. I wish to thank Anthony Meyer and Dr. Ingrid Heermann, curator of the Oceanic art section of the Linden-Museum for their kind authorization to reproduce this photograph. [I here revisit an example discussed in Ghins (2011)].

To gather correct, and therefore useful, information from this artefact I must know the code, which is external to it. But this Micronesian object has internal properties: it is made of wooden sticks bound together by knots and shells. The environment specifies which properties are pertinent to navigation, such as dominant winds, sea currents, stars, locations of the islands etc. But to make good purpose of this object as a map, I must know the code, i.e. the kind of homomorphism between the properties of the target which the map manufacturer intended to convey to a potential sailor and the properties of the artefact. Equivalently, I must know which relevant statements the craftsman took to be true, such as: With respect to island I current C flows in the direction of the polar star.

In order to make efficient use of this navigational map, a sailor must also locate himself with respect to the map. A map in itself is "impersonal" in the sense that it can be read by different people at several locations. Localization involves both position and orientation. If I locate myself erroneously on the map, I will be unable to utter correct statements such as: sailing in direction D will get me to island I. Adequate use of the map presupposes the truth of an indexical statement which says: I am here on the map.

As far as correctness is concerned, we reach the same conclusions as above when we discussed success in identifying a target and in representing the target as such and such. Correct information about the target can be gathered by the user only when he brings in information which does not belong to the artefact itself and which typically is expressed by assertions about—monadic or structural—properties of its target and characteristics of the context. It is not sufficient for the map to be a "faithful homomorphic image" of the target, since the Micronesian artefact does not have an inner structure that makes it ipso facto a maritime map, let alone a correct one. It could be used to represent the lamp on my desk, and still be, given some instituted homomorphism, a faithful image of it, and even an incorrect or inexact one. If I take the spatial relation of the bulb and the plug of my lamp to correspond to the spatial coincidence of a specific shell with the intersection of wooden sticks, then my representation is incorrect, *albeit* faithful in the technical sense defined above.

Successful utilization of a map necessitates that the traveller manages to handle it as a tool for collecting correct information about what is pertinent in its environment relatively to the aim pursued, namely reaching a specific destination. This presupposes that the traveller is able to use the map in order to formulate some true statements about the location of selected elements of his environment as well as the spatial relations among them. Given this, and also by correctly positioning himself on the map, he will be able to infer the truth of some other statements on the right way to safely arrive at his chosen destination.

At each of the three stages of the representational activity (identification of the target, representing it as such and such and representing it correctly in some respects), the user adopts some standpoint or perspective on what she aims at representing. Selection of a concrete target and its relevant properties is crucial at each stage. Selecting implies neglecting. Traditionally, such a way of proceeding has been called "abstraction". In what follows, I will briefly examine the way in which abstraction operates in scientific representation.

4 Scientific Representation

Examining representing practices in various domains such as caricature and travel, as van Fraassen does, is supposed to shed light on the way representation works in science. Scientific activity starts with what I call an "original", "inaugural" or "primary" abstraction (Ghins 2010, p. 530), which consists in looking at phenomena as *systems*, that is, sets of properties standing in some relations. The poet looks at the night sky as a magnificent whole and expresses the awe it inspires in beautifully sounding words. The religious person sees the celestial vault as the work of God. Both attitudes are *holistic*. By embracing them, the poet and the religious see the sky as a unified totality with which they attempt to personally and closely connect in a particular way. The scientist, on the contrary, adopts an *objective* attitude and sets himself at a distance from the phenomena by seeing them as systems. Such systems are not only posited as external to the scientist but are estranged from his human nature as a person.[9]

In performing an original abstraction, an astronomer intentionally isolates in the sky luminous spots which move relatively to other apparently stable bright points. He calls the former "planets" and the latter "stars". Planets are identified by their properties of brightness and motion. Then, in a next abstracting move, the astronomer decides to take orbital periods (the durations of the complete revolutions of planets) and apparent distances to the Sun as the relevant properties of interest. This second abstracting move, which actually occurs simultaneously with the first, I call the "secondary abstraction". The observed properties of orbital periods and distances to the Sun can be organized in a system by means of an ordering relation. The orbital period of a planet is proportional to its distance to the Sun. The astronomer has then constructed a system of properties organized by relations which I call a "phenomenal structure".

In science, the properties of interest usually are susceptible of being quantified. This is the case of course for orbital periods and distances to the Sun. Whereas the phenomenal structure is constructed on the basis of crude observations, a data model is a structure of carefully measured properties. Since the data model and the phenomenal structure are both systems of properties organized by relations, a homomorphism can be instituted between them. Such a homomorphism is a representational function which captures the intended structural similarity between the target—the phenomenal structure and its representor—the data model.

[9]For a presentation of the distinction between the holistic and the objective attitudes, see Ghins (2009). The scientific objective attitude is extensively discussed by van Fraassen (2002).

"Target" is thus an ambiguous word. On the one hand, it refers to concrete objects, e.g. the planets and their successive positions, which are observable phenomena. On the other hand, it denotes the phenomenal structure which has been abstracted from phenomena. Thus, it would be useful to use different words, namely "concrete target" and "targeted structure" to refer to the former and the latter, respectively. As van Fraassen rightly stresses, observable phenomena are not abstract. Instead, they are concrete entities. He is very clear about this:

"*Phenomena* will be observable entities (objects, events, processes). Thus 'observable phenomenon' is redundant (…) *Appearances* will be the contents of observation or measurement outcomes." (van Fraassen 2008, p. 20).

Thus, the concrete phenomenal targets are the planets but these are represented *as* having some abstracted properties organized into a structure of appearances, namely the phenomenal structure, of which the data model is a homomorphic image. A data model also is a structure of appearances according to van Fraassen's terminology. In my regimented use of the word "representation", success is achieved only when some homomorphism has been established by a user between what is represented and the representor. Therefore, only systems or structures can be represented by other structures or systems. The concrete system is represented in a *derivative* sense only. Strictly speaking the concrete target is *not* represented but denoted or referred to by the representor, i.e. the data model. What is represented by the data model, namely the phenomenal targeted structure, also denotes or designates the concrete target.

Notice that the scientist can succeed in representing a planetary system as having specific properties even it does not actually possess these properties, such as being inhabited by intelligent beings like Martians. Of course, scientists aim at constructing models of properties which are actually possessed by concrete phenomena. Individual planets are observationally identified by means of visual properties. Having done this, an astronomer attempts to represent their arrangement with respect to some relevant properties which are abstracted from them, such as an orbital period and a distance to the Sun. Of course, a scientist might err in attributing to planets characteristics they do not possess, such as producing musical notes as the Pythagoreans believed.

Constructing a data model the domain of which only contains properties which belong to concrete phenomena is not sufficient for the data model to be accurate. The institution of a representational function between the data model and the phenomenal structure is not sufficient either. Just imagine that a systematic error has occurred when measuring some property. Then, the representational function will be in place, but the data model wouldn't be adequate. Faithfulness does not imply correctness. We fall back to the same point: the correctness or adequacy of a model rests on the truth of predicative statements. If the properties organized in the representor, the data model, do not belong to the concrete target, the representation is inaccurate.

Scientists do not stop at the level of data models in their representational activity. As van Fraassen says, they manage to embed the data in theoretical structures which provide a unifying view of the domain, deliver explanations and satisfy some useful aims, such as permitting the calculation of future data in a deterministic way (van Fraassen 2008, pp. 36–37). Embedding phenomena means constructing theoretical models which contain, in a minimal set-theoretical sense, empirical substructures homomorphic to data models, or surface models [which are smoothed out data models (van Fraassen 2008, p. 143)]. Since empirical substructures are substructures of theoretical models, they are theoretical as well. Phenomena can be said to be embedded in a theory to the extent that an accurate data model is homomorphic to an empirical substructure of the theory. The overall situation can be summarized in the following table:

Real phenomena
@ *Inaugural abstraction*: system of properties
Secondary abstraction: selection of properties
Phenomenal structure: appearances (crude naked eye observations)
↓ *Homomorphism*: representational function
Data model—Surface model: appearances (measurement results)
↓ *Homomorphism*
Empirical substructure
∩ *Set theoretical inclusion*
Theoretical models: embedding the phenomena.

5 The "Loss of Reality" Objection

Since phenomenal structures and data models are abstract structures, how can we use them to represent concrete targets? *Stricto sensu*, as we saw above, representation is successful only if a homomorphism has been established between a targeted structure and a representor, which is also seen as a relational structure. Structural similarity is an essential condition for the success of a representation. If this is so, a wide gap opens between the representing artefacts and the concrete targets we aim to represent.

> How can an abstract entity, such as a mathematical structure, represent something that is not abstract, something in nature? (van Fraassen 2008, p. 240)

Such question echoes the problem faced by the founding fathers of modern science, such as Descartes, who were at pains to prove that our mental geometrical representations or ideas adequately represent external realities. Surely, some distancing from the things immediately given in perception was the price to pay to achieve the mathematizing of the world. The objective attitude essentially consists in seeing a thing as a system, i.e. as a domain of properties standing in mathematical relations. Then, a mathematical representation becomes possible because targets are systems which are structures just as mathematical representations are. Initially, at the birth of modern science, things were not only seen as mechanisms but identified with mechanisms, namely systems of geometrical parts in relative spatial motions which could accurately be represented by geometrical ideas.

For the philosophers of modern times, to know is to represent. Once this epistemological posture had been embraced, a wedge was driven between our ideas and the real things in the world. While mathematical ideas are structures, concrete things—phenomena—are not. The latter certainly cannot be reduced to mechanisms as the initiators of modern science believed. Moreover, they cannot even be identified with any kind of single system because the same thing can be seen as a *different* system, depending on the perspective adopted. If a thing could be reduced to a unique system, there wouldn't be any difficulty to represent it, because both the target and its representor would belong to the same category: the category of systems.

What I call the "idealistic predicament" consists precisely in the quandary of bridging the gap between our abstract mathematical structures and concrete things. This is not the place to look at the diverse sophisticated ways scientists and philosophers since Galileo and Descartes grappled with this issue, yet without reaching any satisfactory solution. I just want to submit that the loss of reality objection is a revival of the idealistic predicament clad in a new garment. This objection brings back an ancient difficulty which takes its roots in what Michel Foucault appropriately named the *épistèmè de la représentation*.

Surely, van Fraassen is right to insist that scientific models are not mental ideas.

> I will have no truck with mental representation, in any sense. [This] view (…) has nothing to contribute to our understanding of scientific representation—not to mention that it threw some of the discussion then back into the Cartesian problem of the external world, to no good purpose. (2008, pp. 16–17)

However, the model-theoretic approach to theories emphasizes that theories are foremost classes of models. If this is so, the cognitive role is mainly carried out by models and their representational function. If models "take centre stage" as van Fraassen puts it (1980, p. 44) statements are relegated behind the scene and carry less cognitive weight.

So, how does van Fraassen address the "loss of reality objection"? As a genuine empiricist, it is natural for him to resort to pragmatics. His answer is simple but quite ingenious.

> *For us* the claims:
> (A) that the theory is adequate to the phenomenon
> (B) that it is adequate to the phenomenon as represented, i.e. *as represented by us* are the same! (2008, p. 259)

The claims (A) and (B) are both assertions made by the user who aims at representing a targeted concrete phenomenon by means of a representing artefact, a theory in this case. Certainly, I cannot assert (A) without also asserting (B) since claiming that a theory—a model—is adequate to a given phenomenon, is tantamount to saying that it contains an empirical substructure that is homomorphic to a data model containing measurement results, i.e. numerical properties supposedly carefully gathered from the phenomenon. Representation always is indexical. It is impossible for me to climb on some kind of overarching platform from which I could contemplate phenomena on the one hand and my model on the other hand in order to compare them and check whether they correctly match. A godlike point of view or a view from anywhere, which would bracket my own perspective, lies beyond our reach.

van Fraassen's contention can be reformulated at a more basic level in the following way:

> *For us* the claims:
> (A′) that the *phenomenal structure* is adequate to the phenomenon
> (B′) that it is adequate to the phenomenon as represented, i.e. *as represented by us*, are the same!

Quite remarkably, van Fraassen offers his pragmatic move not as a *solution* but as a *dissolution* of the loss of reality objection. Given the unavoidability of the indexical ingredient in any representational activity, it makes no sense for him to ask if a proposed model hits on something external. Such external reality would be a metaphysical posit, devoid of empirical meaning. If this kind of *ding an sich* exists, a possibility which is not excluded after all, it is definitely beyond our ken.

6 The Loss of Reality Objection Solved

I agree with van Fraassen that denying (B′) while asserting (A′) would be a pragmatic inconsistency. It would be tantamount for me to assert "p is true" and at the same time say "I don't believe p" (van Fraassen 2008, p. 212). But the main question is the following: what reasons do I have to believe that a model correctly represents a concrete target? Pragmatically, if I subscribe to the representational way of knowing, there is no way to deny that I represent the concrete target when I claim that my representation is adequate to it.

At this point, two questions can be raised. First, what does it *mean* for a model to represent a concrete entity? Second, what *reasons* do we have to believe that the concrete target is adequately represented?

First, as I emphasized, adequacy relies on the truth of some predicative statements which assert that planets, for example, possess some specific quantitative properties such as an orbital period of a certain value. Although van Fraassen doesn't give pre-eminence to statements, he acknowledges that adequacy rests on the truth of some claims.

> To offer something X as a representation of Y as F involves making claims about Y, and the adequacy of the representation hinges on the truth of those claims, but that point does not put us in the clutches of a metaphysics of 'truth makers'. (2010a, b, pp. 513–514)

Unlike van Fraassen, I maintain that true statements do have truth-makers, namely facts which make them true, and that there is some sort of correspondence between facts and true statements. For example, it is a fact that planets are in motion and this fact can be ascertained by simple observation. The strongest argument in favour of a correspondence *view* (not a *theory*) of truth, is our experience of error, when we are forced to change some of our beliefs when confronted to new evidence. In such occasions, we realize that there are facts external to us, which we don't control and exist independently of our wishes, language and models. There is no need to resort to a metaphysics of things in themselves to account for this quite common experience. The occurrence of some facts can be ascertained on the basis of immediate perceptual experience, while not eliminating any risk of error.

Admittedly, I am unable to explicate in what consists the correspondence between statements and their truth-makers. Such an explication could perhaps be provided by a full-fledged correspondence *theory* of truth which would detail the characteristics of such correspondence. As we saw above, a famous example of a (failed) correspondence *theory* of truth is given in Wittgenstein's *Tractatus*. To my knowledge, no satisfactory correspondence *theory* of truth has been devised so far. But this situation, doesn't prevent us to defend a minimal correspondence *view* of truth, which makes the quite limited claim that some kind of correspondence obtains between the facts and the statements they make true, while remaining silent on the exact nature of such correspondence.

Since van Fraassen subscribes to a deflationary theory of truth, he cannot rely on the truth of statements to warrant that our adequate representations do represent our intended targets. For a deflationist to say that "snow is white" is true is simply to assert that snow is white. That is all there is to it, and there is no need to gloss on what truth is and the specific relation, should there be one, between an assertion and a fact. Searching the nature of truth is a will-o'-the-wisp. But if we do gloss (just a little bit) on the relation between statements and facts, simply by claiming that there is a truth-maker, a fact, which is external and independent of what we may assert about it, then we are in a position to identify the concrete target of our representations. The concrete target is just what we talk about, namely the things to which we attribute some properties in an act of predication resulting in a statement. Again, to admit the existence of truth-makers doesn't commit us to a lofty metaphysics remote from perceptual experience, but to facts to which we have epistemic access in perception, independently of our wishes, language and modelling activity.

A correspondence view of truth is part and parcel of a realist position in epistemology, *already at the empirical level*. (In this paper, I leave aside the issue of the existence of unobservable entities posited by some scientific theories.) On the contrary, a deflationary theory of truth implies the following:

Asserting Snow is white means the same thing as asserting that "Snow is white" is true.

Certainly, pragmatically if we assert that snow is white, we must also assert that "snow is white" is true, as Tarski instructed us a while ago. But the two statements do not have the same *meaning*, contrary to what the deflationary theory of truth claims. If we accept this, we are invited to tell what the word 'true' means. According to the correspondence view it means that there is some relation between a statement and something distinct from it, namely its truthmaker. If this is correct, a true predicative statement identifies what it talks about—its target—without ambiguity by mentioning some of its properties. Then, some of its other properties can be employed to construct representors and models as I explained above in the example of the planetary system.

The loss of reality objection is solved because true predicative statements provide firm ground of contact with the concrete targets from which our representations are constructed. Predication is not representation however. When I attribute a property to a thing, I do not represent the thing as having a property. I simply attribute a property (rightly or wrongly) to the concrete targeted thing. Representation proceeds next in organizing properties in systems, structures and models. If a representation succeeds in representing a concrete entity it is only derivatively so, since its success is parasitic on the truth of predicative statements which hit on targets in their concreteness. Abstract representors are organized sets of properties supposedly pertaining to concrete targets (whether fictional or real).

Now, to briefly address the second question raised above, let us simply point out that the adequacy of a representation depends on the possession by the target of the properties involved in the representational activity. In the example of the planetary motions, an astronomical model is adequate if we have reasons to believe that planets have the properties used to construct the model and that these properties are arranged in the planetary system in a way which is correctly and structurally similar to the way the corresponding elements in the model are. This is all.

Is such a solution of the loss of reality objection committed to the view that there is some unique fundamental structure in the world which somehow "carves nature at its joints" just as the mechanistic conception of nature of modern times assumed? No, not at all.

Although the scientist certainly carves a targeted phenomenon into properties that belong to it (at least he so believes), he always operates from a certain point of view and accepts that there are other ways to look at the phenomenon. If he is correct in doing so, predicative statements attributing properties and relations among them are true. In this limited sense, there is some structure intrinsic to the phenomenon that is capable of being represented by a user. However, there is no inner structure in a phenomenon that makes it ipso facto representable by a specific representor, such as a photograph, as we saw. The fact that a phenomenon possesses a certain structure (among other ones) does not determine the nature of the representing artefact which could be employed to represent it. Conversely, there is no intrinsic structure of a thing that makes it a potential representor of a specific target or class of targets,[10] since anything can represent anything. Any entity could be used to represent some characteristics of planetary motions, provided some homomorphism is conventionally established between the representing artefact and its targeted system. Some properties and relations in the representing thing are chosen by the user as relevant, but this doesn't prevent the representor from intrinsically having those properties and relations. On the contrary, it is because the representor really possesses some identifiable properties that a representational

[10]On this I disagree with Bartels (2006, p. 14).

function can be bestowed on it by the user. Of course, the same thing can be endowed with a large variety of representational roles when used as a representor.

Given that a concrete thing can be looked at as having some structure S from a certain point of view, and as having another structure S' from some other perspective, a realist must demand that the properties and relations that are believed to actually belong to the target be logically compatible. A concrete thing cannot have contradictory properties at the same time. However, in science, some representations of the same target appear at first sight to be incompatible. This situation especially occurs when various models are offered to represent unobservable things, such as atoms. According to some models, molecules and atoms contain no parts, and according to other models they are composed of protons, neutrons and electrons. For quantum mechanics, particles can be entangled, whereas in classical contexts they don't.

Such a situation surely raises a problem for the realist. But here we can't examine this issue in depth. Let me just make three brief observations. First, a model always neglects some properties of the target. In these cases, the realist should refrain to attribute to them properties which they would always possess, beyond a specific context of investigation, such as indivisibility in the case of atoms. Instead, the realist should only claim that molecules and atoms do not break, and behave as if they were indivisible, in some particular context such as the emulsions studied by Perrin. Second, some relevant properties may be approximately exemplified by the target as in the liquid drop model of the atomic nucleus (Da Costa and French 2003, pp. 50–51). Third, some properties of the representor may play a representative role without being put in correspondence with actual properties of the target. Mrs. Thatcher doesn't have dragon's wings, but the caricature aims at representing her *as* having a specific trait of character. Such a procedure is typical of graphs widely used in many scientific disciplines in which the abscissa and ordinate axis do not have correlates in the target.

7 Conclusion

Success in representing crucially rests on predicative statements which are true in accordance with a correspondence *view* of truth. These statements play a decisive role in the three aspects involved in the success of a representation. The user of a representing artefact intentionally identifies her target by relying on supposed properties of the target, be it real or only fictional, e.g. being a bright spot moving to apparently immobile bright spots in the sky. She successfully represents the target *as* such and such by instituting a homomorphism between some relevant supposed properties of the target and properties of the representor, in some context. Finally, her representation is correct or adequate if the target actually possesses the relevant properties attributed to it. Truth, or at least supposed truth in the two first stages, is therefore more fundamental than success, since the latter is achieved on the basis of the former.

Such a conception of representation does *not* involve the heavy metaphysical commitments to "things in themselves" or a unique "carving of nature at its joints". Phenomenal things at least are directly accessible to human sensory perception. We are then in a position to ascertain (or not) the occurrence of facts which are the truth makers of (true) predicative statements. Moreover, several perspectives can be adopted when attempting to represent things. In doing so, the user selects in the phenomenon some properties which are organized in a certain manner. The targeted phenomenon is then seen as a system which can be represented by another thing, an artefact, which is also seen as a system. Despite the various possible perspectives and the leeway allowed in choosing the relevant conventions, both the target and the representor can be said to actually possess some intrinsic properties. When various perspectives are taken on the same target, the realist must certainly avoid to attribute contradictory properties to it.

Acknowledgement I would like to thank Otávio Bueno, Marco Giunti, Dimitris Kilakos, Diego Marconi, Stathis Psillos and Alberto Voltolini for their useful comments at oral presentations of some parts of this paper, and Peter Verdee for his suggestions on the more technical parts.

References

Bartels, A. (2005). *Strukturale Repräsentation*. Paderborn: Mentis.

Bartels, A. (2006). Defending the structural concept of representation. *Theoria, 55*, 7–19.

Chakravartty, A. (2010). Informational versus functional theories of representation. *Synthese, 172*, 197–213.

Da Costa, N., & French, S. (2003). *Science and partial truth. A unitary approach to models and scientific reasoning*. Oxford: Oxford University Press.

Dunn, J. M., & Hardegree, G. M. (2001). *Algebraic method in philosophical logic*. Oxford: Clarendon Press.

Ghins, M. (2009). "Realism", entry of the online. *Interdisciplinary Encyclopaedia of Religion and Science*. http://www.inters.org

Ghins, M. (2010). Bas van Fraassen on Scientific Representation. *Book Symposium on van Fraassen's Scientific Representation. Analysis, 70*, 524–536. http://analysis.oxfordjournals.org/cgi/reprint/anq043?ijkey=8y8OZR6C4lhmJCy&keytype=ref

Ghins, M. (2011). Models, truth and realism: Assessing Bas Fraassen's views on scientific representation. In *Science, Truth and Consistency: A Festschrift for Newton da Costa-Proceedings of the CLE/AIPS Event 2009*. Ed. by Evandro Agazzi, Itala D'Ottaviano and Daniele Mundici. *Manuscrito* Vol. 34, pp. 207–232. http://www.cle.unicamp.br/manuscrito/

Goodman, N. (1976). *Languages of art: An approach to a theory of symbols*. Hackett Publishing Company.

Meyer, A. J. P. (1995). *Oceanic Art - Ozeanische Kunst - Art océanien*. Cologne: Könemann.

Pero, F., & Suárez, M. (2015). Varieties of misrepresentation and homomorphism. *European Journal for Philosophy of Science*. http://link.springer.com/article/10.1007/s13194-015-0125-x

Suárez, M. (2003). Scientific representation: Against similarity and isomorphism. *International Studies in the Philosophy of Science, 17*, 226–244.

Suárez, M. (2004). An inferential conception of scientific representation. *Philosophy of Science, 71*, 767–779.

Suppes, P. (1967), What is a scientific theory? In S. Morgenbesser (Ed.), *Philosophy of Science Today* (pp. 55–67). New York: Basic Books.

Suppes, P. (2002). *Representation and invariance of scientific structures*. Stanford: CLSI.
van Fraassen, B. (1980). *The scientific image*. Oxford: Oxford University Press.
van Fraassen, B. (2002). *The empirical stance*. New Haven: Yale UniversityPress.
van Fraassen, B. (2008). *Scientific representation. Paradoxes of perspective*. Oxford: Oxford University Press.
van Fraassen, B. (2010a). "Summary". *Book Symposium on van Fraassen's Scientific Representation. Analysis, 70*, 511–514. http://analysis.oxfordjournals.org/cgi/reprint/anq043?ijkey=8y8OZR6C4lhmJCy&keytype=ref
van Fraassen, B. (2010b). Reply to Contessa, Ghins and Healey. *Book Symposium on van Fraassen's Scientific Representation. Analysis, 70*, 547–556. http://analysis.oxfordjournals.org/cgi/reprint/anq043?ijkey=8y8OZR6C4lhmJCy&keytype=ref

Ideology in Bio-inspired Design

Cameron Shelley

Abstract Bio-inspired design refers to the use of the natural world as a source of models for the design of artifacts. For example, Velcro is a fastening system made from nylon that deliberately imitates the structure of burrs, which are adapted to stick to animal fur. In general, in selecting and adapting models to design problems, designers can also seek to satisfy their ideological goals. That is, designers seek not only to solve technical problems but also to respect and promote professional and societal values that they find important. This observation applies to bio-inspired design. This paper examines two ideologies that are present in bio-inspired design. The first ideology examined is *biomimicry*, on which the natural world is characterized as the result of natural selection, a competition for survival that produces rugged and individualistic organisms. The second ideology is *biosynergism*, on which the natural world is characterized as interdependent systems integrated into a larger whole that operates in a sustainable manner. Both of these ideologies are explicated and their effects on design work examined.

1 Introduction

Designers have long drawn inspiration for their work from the organic world. Perhaps the best-known example of such bio-inspired design, as it is sometimes known, is Velcro (Vogel 1998, pp. 268–270). Swiss engineer George de Mestral investigated how seed pods of the burdock plant stuck so tenaciously to his dog's fur. The exterior of the pods are covered with spines ending in small hooks. The hooks become entangled in loops in the fur, causing the pods to come loose from the plant and adhere to it. Using these observations, de Mestral devised an artificial version made from nylon that has become very popular as a fastener under the trade name of Velcro.

C. Shelley (✉)
Centre for Society, Technology and Values, University of Waterloo,
Waterloo, ON, Canada
e-mail: cam_shelley@yahoo.ca

L. Magnani and C. Casadio (eds.), *Model-Based Reasoning in Science and Technology*, Studies in Applied Philosophy, Epistemology and Rational Ethics 27, DOI 10.1007/978-3-319-38983-7_3

One important aspect of bio-inspired design is its ideological component. The selection and adaptation of biological models for artifactual design problems is influenced not only by the technical demands of the problem but also by the broader views and goals of the designers involved. Shelley (2013) points out that ideology is evident in the handling of models in design within the design movements of Gothic Revivalism, Modernism, and 20th century industrial design. Designers within each of these movements selected and applied models not only to deal with technical challenges in their design problems but also to acknowledge and promote ideological values that were important to them. A.W.N. Pugin, for example, built churches closely based on medieval models in order to promote a return of sorts to medieval Christianity among his fellow Britons.

The same pattern may be observed among bio-inspired designers. Besides embodying technical design principles, biological models can reflect broader views and goals that designers seek to promote. An interesting, additional complexity that arises with bio-inspired design is that it can be used to promote different ideologies. That is, designers of different ideological views exploit the biological world for models in the pursuit of their work. For example, some designers emphasize a view of nature in which the survival of the fittest is the most salient fact about the natural world. They look to the fruits of natural selection to provide inspiration for artifacts that embody a kind of rugged, individualistic view of good design. Other designers emphasize a view of nature in which the biological world consists of a set of interdependent networks of organisms and communities of organisms. They look to nature to inspire the design of artifacts that operate in relationships with other artifacts or people as integral parts of greater wholes. Views about the natural world and about the social realm tend to inform one another and influence how designers select and adapt models from nature in their work.

The purpose of this article is to explore different ideologies evident in bio-inspired design. The notion of ideology as it figures in design is clarified first. Then, three different design ideologies are examined that have appeared in bio-inspired design. The first ideology is *biomimicry*, on which the biological world is characterized by natural selection and thus lends itself to the design of rugged and independent designs. The second ideology is *biosynergism*, on which the biological world is characterized by interdependent natural communities and thus lends itself to design for interactive and sustainable systems. The term bio-inspired is used here to cover both these ideologies, as well as any others in which natural models are crucial.

Each of these design ideologies is characterized and then compared and contrasted. This exploration shows that multiple design ideologies cannot necessarily monopolize a given source domain. Of more central importance to each ideology is the treatment of the domain, consisting in the ideological values that are applied in the selection and adaptation of source materials.

2 Ideology in Design

A design ideology is a set of principles and practices that constrain what counts as a good design. More particularly, ideological constraints reflect the social values that a design is meant to reflect and to promote. Forty (1986) provides an analysis and several examples. One example concerns the presence of genderism in Victorian watch design. Genderism is the view that there exist two genders, the *masculine* and the *feminine*, and that this distinction is an important cultural value. Forty points out that Victorian watch designs were divided into two groups, that is, men's and women's watches, with consistent design features to distinguish them (p. 65):

> In wristwatches, the disparity in size between those for gentlemen and those for ladies exceeded that between male and female wrists, and a lady's watch usually had more delicate features and face. Being smaller, ladies' watches have generally been more expensive, but when they can be compared to men's watches of a similar price, the ladies' models are still more ornamented. In the 1907 Army and Navy Stores catalogue, the men's watches were all calibrated with Roman numerals, while the ladies' watches all had Arabic numerals, whose form—curvilinear rather than angular—may be judged more delicate.

The differences in watch types have less to do with any physical differences between men and women and more to do with differences in their stereotypical social roles. Furthermore, these design differences are not merely a reflection of the social importance of gender but also a reinforcement of it. Having gender displayed in the design of watches and other goods tends to make it appear to be a sort of brute fact about the world, like the difference between the sea and the sky. Such an appearance then reinforces the naturalness and perhaps irrevocability of masculine and feminine roles in society.

3 Models and Design Ideology

The example of genderism in watches illustrates how a broad social agenda can be taken up and applied by designers as a criterion of good design. This illustrates what is meant by an ideology in design. In the case of watches, Victorian designers selected among possible watch elements that were consistent with the identity of the watch as masculine or feminine. However, ideology can also influence the use of *models* in design.

Models tend to play an important role in design work. Finding a solution to a design problem may be made more straightforward with access to model solutions to analogous problems. A key step in this process is the selection of a model to emulate. Like the selection process involved in design in general, as described above, the selection of models to emulate is open to ideological influence.

In bio-inspired design, elements of the organic world serve as a pool of models for potential use in solving problems. Given the richness and complexity of that world, there is an embarrassment of choice for the designer. This embarrassment, as it were, is partly overcome by the application of ideological criteria, helping designers to focus in on only a few possible models. In the following sections, two different design ideologies found in bio-inspired design are discussed, namely biomimicry, and biosynergism.

4 Biomimicry

The first paradigm of bio-inspired design examined here is *biomimicry*. For present purposes, biomimicry is defined as the use of organisms as models of engineering design. Engineers have long found organic forms useful in the solution of engineering problems, especially as they embody the engineering values of optimality and efficiency (Nachtigall 1974).

The example of Velcro, noted earlier, provides a good illustration. Swiss engineer Georges de Mestral noted the tenacity with which burdock seed pods stuck to the fur of his pet dog. He examined the pods and found that their exterior was covered with a field of small hooked structures, which allowed the pods to grip tangles in his dog's hairs. He realized that he could reproduce this grip between two surfaces of nylon fibers, one containing small hooks and the other small loops.

This illustration displays the values of biomimicry. It is intended to solve a functional problem, namely designing a physical fastener, in an efficient way through imitation of a natural structure evolved by plants for reproduction through exploitation of the physical properties of fur.

In this section, I will examine the ideology of biomimicry as it is described by French (1988) because his exposition is particularly clear on this point. To begin, French characterizes good engineering by emphasizing its *analytical and dispassionate* method, for which evolution provides an instructive example (xii):

> Living organisms are examples of design strictly for function, the product of blind evolutionary forces rather than conscious thought, yet far excelling the products of engineering. When the engineer looks at nature he sees familiar principles of design being followed, often in surprising and elegant ways.

The forces of evolution are blind in the sense that they respond only to the problems of the present and are not guided by sentiment or concern for the future. This emphasis on the dispassionate character of evolution accords, in his view, with the cool and strictly analytic approach to design appropriate for engineering.

This *functionalist* stance of biomimicry focuses the attention of the designer on the products of evolution through natural selection. In many treatments, natural selection is the only kind of evolutionary scheme discussed, e.g., (Vogel 1998) while other possibilities such as sexual selection are set aside. Darwin (1871) argued that certain parts of organisms could be explained as responses to sexual preferences. The extravagant tail of the peacock, for example, could not be explained by a struggle for existence as it hardly improves the peacock's ability to fight, fly or feed. However, it could be explained by the need of male birds to impress choosy females. The focus on aesthetics involved in sexual selection is not so compatible with the functionalist outlook of biomimicry.

Another functionalist aspect of biomimicry is its emphasis on *efficiency*. French notes how organisms display adherence to a principle of economic efficiency, namely the division of labor. In classical economics, an economy functions most productively when each of its members simply does what they do best, and nothing else. A similar principle holds in the organic world (French 1988, p. 3):

> The division of labour is but a special case of a more general principle of functional design, the *separation of functions*. Thus, simple single-celled organisms have to provide all their functions in one cell, whereas higher animals and plants have many different kinds of cell for special purposes, carrying sap, extracting water and minerals from the soil, transmitting signals, secreting digestive juices, etc. The early steam-engines, following Newcomen's design of 1712, had a cylinder in which the steam did work and in which it was also condensed. Watt's engine, fifty years or so later, separated the functions of working cylinder and steam condenser, so greatly increasing the efficiency.

Evolution through natural selection also results in organisms that display this economy of organization. Thus, evolution provides engineers with many examples that reinforce this approach to engineering design.

Note that this characterization of the natural world diverts attention from organic structures that serve multiple functions. The feathers on a bird's wings, for example, may help that structure to create lift but also play a role in maneuvering, in addition to supplying a platform for decorative features important in the competition for mates.

Often, biomimicry involves suspicion of ornament and shows aesthetic concern only through attention to *elegance*. A design, like an organism, can achieve elegance simply by being thoroughly functional (French 1988, p. 14):

> One characteristic of functional design is elegance. Most people find a buttercup beautiful, and many would say that the locomotive was at least pleasant to look at. However, the buttercup has an essential elegance, much more fundamental than its mere appearance. It is an elegant solution to a difficult problem in functional design.

This approach to aesthetics is *reductionist* in the sense that it holds that a kind of beauty, namely elegance, can be achieved by adhering to non-aesthetic values such as dispassionate analysis and economy of organization. This kind of beauty is exemplified by organisms like buttercups but also, to a lesser degree perhaps, by artifacts like locomotives. This reductionist stance of biomimicry can be viewed as a resolution to the cognitive dissonance of advocating a dispassionate approach to design while also admitting admiration for the beauty of organic structures such as the flower of the buttercup.

Another ideal of biomimicry as described by French may be termed *masculinity*. Hofstede (1984) defines masculinity in a culture as a gender role that emphases self-orientation, assertiveness, and ambition over feminine values such as relationships, communication and caring. The preference for masculine features of evolution in biomimicry is suggested by the emphasis placed on natural selection as a competition for survival amongst individuals, a kind of war of all against all.

Attention is paid mostly to features that animals have for obtaining food, fleeing predators or fighting rivals. Features that animals have for more feminine tasks such as forming groups, cooperating with others and raising offspring are less readily considered. Note the following description of the severe demands of maturation to which natural designs respond (French 1988, pp. 265–266):

> The difficulties of a hostile environment are added to by those of growth; many organisms must fend for themselves from an early stage; for example, fish may be all on their own when only a centimeter or so long, though they may eventually reach a meter in length. The caterpillar-pupa-butterfly and tadpole-frog metamorphoses are familiar. If the caterpillar is eaten there will be no butterfly; each stage must be viable. Insects and other arthropoda, such as crabs and shrimps, have a hard outer skin which cannot grow with them, and must be moulted periodically as it becomes too tight; until the soft new armour hardens, they are relatively defenceless.

Adaptations that animals possess for feminine functions are mentioned briefly and as an afterthought (French 1988, p. 266):

> One of the devices used to cope with the extreme severity of the design problem of living creatures is a very high production rate, so that out of millions of embryos a handful may survive. But some less wasteful approaches have appeared in the course of evolution, principally, the protection of the young by adults among the higher animals and termites, ants and bees.

This last sentence is the only one in the book devoted to more feminine adaptations for survival. Thus, it is clear that biomimcry takes a masculine view of biological adaptation as the default for purposes of addressing design problems.

As French has characterized it, biomimicry is a design ideology that emphasizes the fruits of evolution by natural selection. As such, it values organic structures that are functional, efficient, elegant, and masculine. This approach commits the designer to an analytic and reductionist method and also to opposition to methods with competing values, as exemplified by architecture.

5 Biosynergism

The second paradigm of bio-inspired design examined here may be called *biosynergism*. As with biomimicry, biosynergism involves the use of biological systems as models for solving design problems. However, biosynergism differs in the how those systems are characterized and, thus, how they inform design solutions. In brief, biosynergism centers on the values of organicism, environmentalism, sustainability and femininity.

A recent and instructive example of biosynergism comes from the design of the Visitor Centre at Vancouver's VanDusen Botanical Garden. The design of the Centre has sought to integrate it profoundly with its environment. For example, waste from the toilets is harvested and mixed with composted food waste and then applied as fertilizer in the gardens. Wastewater is separated and purified for use in irrigation. The buildings were constructed partly from rammed-earth components that were formed from materials dredged from ponds on the site. Steps and ramps provide easy access to the green roof from ground level to encourage wildlife to inhabit that space. Coyotes have been spotted on it (Flint 2015).

Designed by architect Peter Busby of Perkins + Will, the Centre is an example of "Regenerative Design," to which we will return. It also reflects the fact that biosynergism has its historical roots not so much in engineering but in urban planning, architecture, and landscape design.

The first point to observe about the Visitor Centre's design is that it illustrates the value of *organicism*. To begin with, organicism is a view that the biological world is an ecological system, in which each individual organism occupies a place in a "web of life" characterized by complex interdependencies and cooperation in addition to competition. For a simple example, plants are eaten by small animals that are then eaten by predators, and the feces of all form fertilizer that promotes the growth of the plants. As a result of the complex relationships among organic populations, the biological world is not merely an aggregate of competing individuals but a community that is an integrated whole. The Visitor Centre reflects this organicist principle in the way that each of its internal systems provides support for the others. Waste from the toilets is used to support the growth of the plants in the Gardens, rather than being isolated from them as would usually be the case in conventional building design.

Scottish biologist and pioneering urban designer Patrick Geddes described the tendency of ecosystems to be more than merely the sum of their parts as *synergy* (Casillo 1992). There is a tendency in nature, he felt, for individuals and species to gravitate towards such relationships. He argued that good cities also displayed design based on this principle (Welter 2002):

> Like a flower and butterfly, city and citizen are bound in an abiding partnership of mutual aid.

The reference to "mutual aid" also reflects the influence of the political philosophy of Kropotkin (1902), a Russian biologist who argued that cooperation and mutual help were important factors in evolution and should also be cultivated in a civilized society.

In addition to the ecological concept, organicism also makes analogies to medicine. In particular, Geddes noted the cooperation of organs within the bodies of individuals as a model for the design of cities. Just as different bodily organs cooperate to support the metabolism of the whole, components of the civic body should cooperate to support the metabolism of the entire city. Geddes developed tables to track and compare the input and output of cities in terms of the resources that cities consume and the products they manufacture. In the 1960s, this idea was developed as the concept of *urban metabolism* by Abel Wolman and has now become a common method of analysis in urban design and civil engineering (Wolman 1965).

Besides displaying integration of internal systems, the Visitor Centre is also integrated with its external environment. In doing so, it illustrates the value of *environmentalism*, on which artificial designs are good to the extent that they preserve the surrounding environment as possible. The Visitor Centre observes this value in various ways. It was built as far as possible using materials actually derived from the site itself, such as the earth for the rammed-earth structural elements. Also, pre-existing ponds were adapted to the needs of the Centre's irrigation system, rather than being replaced by tanks and pumps. The physical profile of the building was designed to fit in with the surroundings to the extent that local wildlife may continue to utilize it. Also, the Centre was designed to reprocess its own waste and thus minimize the waste stream that results from its operation.

The environmentalist emphasis on external integration derives from the ecological values of biosynergism. Not only should a good design consist of mutually supporting parts but it should itself be a supportive part of a greater whole to which it belongs. External integration has been measured using another ecological concept, namely the *ecological footprint*. Also known as the carrying capacity, the ecological footprint of a city is the size of land area that it would take to sustain its existence (Wackernagel and Rees 1996). Consider all the agricultural acreage, forest, mines, lakes, etc. it would take to support a given city through resource extraction and waste recycling. That measurement provides a notion of the environmental impact that a given city has.

Also, the Visitor Centre design exhibits the value of *sustainability*. Systems within the Centre are designed in a circular fashion, minimizing the need for external inputs and the impact of external outputs. In this way, the Centre is designed to be highly efficient and self-sufficient. Through these measures, it becomes easier to continue operating the Centre for a longer time, given a fixed budget of inputs and output handling capacity.

Biosynergism enjoys great currency today. As noted above, the design of the Visitor Centre is an example of *regenerative design*, a term coined by landscape architect John Tillman Lyle (1996). Lyle defines the concept in this way (p. 10):

> Regenerative design means replacing the present linear system of throughput flows with cyclical flows at sources, consumption centers, and sinks [endpoints].
>
> …
>
> A regenerative system provides for continuous replacement, through its own functional processes, of the energy and materials used in its operation.

A regenerative design is thus "environmentally friendly" and self-sustaining at the same time.

Biosynergism is also reflected in the *cradle-to-cradle* design paradigm espoused by McDonough and Braungart (2002). Influenced by Lyle, McDonough and Braungart argue that sustainable design is best achieved by cyclical flows of resources within a system, minimal reliance on external inputs, and a synergistic relationship with the external world. They apply this approach not only to the design of buildings but to commercial goods in general.

Finally, biosynergism is notably *feminine* in its cultural orientation. Unlike a masculine culture, a feminine one is characterized by cooperation, caring, and mutual support. Biosynergism clearly reflects these values. It emphasizes interdependence and relationships of mutual support among the components of designed systems.

The importance of caring is also apparent in the work of Patrick Geddes. For example, Geddes argued for an approach to civic improvement that he called "conservative surgery", a term taken from medicine. He argued that improvements to a city should be made in a way that is minimal and that utilizes its existing social and cultural heritage. This approach stands in contrast to the wholesale razing and redevelopment employed by Baron Haussmann in Paris, for example (Goist 1974). Conservative surgery requires the urban designer to learn about a civic area through first-hand experience of its site and people. Only with a strong sense of place and the values of the inhabitants can an urban designer come up with plans to improve an urban area appropriately. Thus, biosynergism professes caring through taking to heart the interests of people most directly affected by changes in design.

6 Discussion

6.1 Analogical Versus Schematic Transfer

Biomimicry and biosynergism display different methodological tendencies. Biomimicry is typically case-based. That is, practitioners proceed by identifying a particular organic adaptation and then adapting it for use in a design through analogical transfer. By contrast, biosynergism is typically schema-based.

In the case of Velcro, de Mestral studied the method of adhesion of burrs to animal fur. Upon discovering the use of multiple small hooks that entangle with loops in dog fur, de Mestral then set out to find a way to produce a similar structure in an artificial material. De Mestral used a single instance of adhesion in the organic world, analogized it with artificial systems of interest to him, and then adapted the result for manufacturing.

This process of selecting individual adaptations and then emulating them is a hallmark of biomimicry. Consider a more recent example. Researchers at the ETH in Zürich designed a system to defend ATM machines from attack by emulating the defense mechanism of the bombardier beetle (Vonarburg 2014). When attacked, the beetle emits a caustic chemical from a chamber in its abdomen. The spray is produced when two inert chemicals, stored in separate body cavities, are mixed in the special nozzle-like body cavity. Researchers at the ETH found a way to make a similar system that could be installed, for instance, in ATM machines to spray attackers with an unpleasant mixture of hydrogen peroxide and manganese dioxide.

By contrast, biosynergism tends not to emphasize case-based reasoning. Instead, practitioners elaborate upon and appeal to the concept of synergism. Noting how, in general, populations in nature tend to support each other through a set of synergistic interdependencies, advocates show how similar ideas can be embodied in design methodology or, at least, in the design of individual artifacts.

The Visitor Centre displays this tendency. The Centre was not designed using any particular organism as a model. Instead, the general schema of synergistic relationships was applied.

For another example, consider the work of Henk Jonkers, a microbiologist at Delft University of Technology (Jonkers 2007). He and his colleagues have developed a way of embedding capsules of limestone-producing bacteria into concrete. As concrete ages, intrusion of moisture and air causes cracking, which degrades the integrity of concrete structures. However, upon exposure to air and moisture, bacteria in the embedded capsules within Jonkers' concrete produce limestone, which seals cracks and thus enhances the integrity of structures. Jonkers calls this design "self-healing" concrete to denote the inspiration that the process of bone healing played in its development (European Patent Office 2015). In this case, the general concept of healing was the schema applied to the design of "bio-concrete" without reference to any particular, organic healing mechanism.

This difference in methodology seems to arise from different, ideological orientations towards sustainability. Biomimicry does not include a commitment to any overarching social goals. As such, practitioners are free to make comparisons on a case-by-case basis without the cases necessarily sharing anything in common. Biosynergism includes a commitment to sustainability, which constrains the kinds of cases that are relevant to practitioners. They interpret sustainability to imply organicism and environmentalism. Any specific, biological cases that they consider will therefore display essentially the same conceptual structure. In that event, there is no need for detailed analogical transfer.

6.2 Tension Between Ideologies

Thagard (2014) notes that, in the political case, ideologies are characterized not only by commitment to their own values but also by their opposition to other ideologies committed to different values. Evidence of tension between design ideologies is exhibited in the above discussion.

For example, French makes explicit mention of how the values of biomimicry contrast with those of architects, who otherwise would appear to be engaged in a very similar pursuit (pp. 5-6):

> However, in much architecture the functional aspects are very secondary to aesthetic ones, and, moreover, rather readily met (or indeed, neglected altogether, as in some badly-designed buildings which have nonetheless won awards). Another defect of architecture as a training ground for functional design was that often the economic constraint, so powerful throughout nature and engineering, was virtually absent, the patrons caring more for glory than the public good (which was the public's loss then, but is sometimes our gain now).

French admits that architectural works are sometimes worthy of praise but usually on aesthetic rather than structural grounds. In his view, engineering through biomimicry is a very different pursuit than architecture, although it may not be apparent to the casual observer of their works.

There is a clear tension between the ideologies of biomimicry and biosynergism. As noted above, biomimicry may be described as *masculine* in orientation. That is, it emphasizes features of biology that suit pursuit of self-interest, competition among individuals, and survival of the fittest. Biosynergism notably emphasizes features of biology that are *feminine* in orientation. That is, biosynergism emphasizes the pursuit of common interests, cooperation and communication among individuals, and caring and mutual support. These values are grounded in biological ecology and also in medicine.

In practice, advocates of biosynergism do not pay much attention to biomimicry as such. Instead, they see their approach as in opposition to what Patrick Geddes called the *paleotechnic* paradigm. Geddes held the view that civilization was progressing through three technological phases, which he called *paleotechnic*, *neotechnic*, and *biotechnic*. The paleotechnic phase is characterized by mechanical technology, running on dirty fuels like coal, based on an economy of exploitation of natural resources for private gain. The neotechnic phase, which Geddes thought was emerging in his own time, is characterized by clean electricity as a power source and is based on an economy of conservation of natural resources for the common benefit (Geddes 1915). In the later, biotechnic phase, technology would operate in a manner that is biosynergistic in its organization and would allow civilization to exist in harmony with nature (Geddes 1929).

However, the notion of a biotechnic civilization contains a curious ambiguity. It was understood by some of Geddes' followers as the use of biological organisms for the benefit of humankind (Bud 1991). This understanding became associated with the term *biotechnology*, which, today, refers to something akin to the assimilation of nature to the demands of what Geddes would consider to be a paleotechnic civilization. Certainly, this view seems to be shared by many practitioners of biosynergism, who see their work as being in the service of the conservation of nature and not its further industrialization; cf. (Lyle 1996, p. 13).

Although biomimicry is not the same as biotechnology, it shares with the latter a niche in an economy based on the exploitation of natural resources without any special commitment to organicism, environmentalism or sustainability. Thus, biomimicry and biosynergism are two ideologies in tension, in spite of their superficial similarities.

6.3 The Function of Design Ideologies

In general, ideologies have many functions. They help to define a group of practitioners into a coherent whole, capable of organizing resources, prioritizing projects, judging their value, and promoting their worth to others. Many of these functions are visible in the discussion above.

Bud (1991, p. 418) makes an interesting observation about biosynergism as developed by Patrick Geddes. At the outset of the 20th Century, many biologists wanted to promote their science both within academia and to the public at large. They sought to emulate the development of chemistry, which, in the 19th Century, had become very prominent in part through its application to industrial processes.

In particular, German chemists in the dye industry had developed many new chemicals that became important in society, in areas such as textiles, medicine, and munitions. To promote the importance of their discipline, biologists like Geddes pointed out how it was relevant to commercial technology.

This effort has ultimately been successful. Biosynergism is widely acknowledged to be relevant to modern life, both in academia and in general, due to increasing concerns in society about sustainability. Biosynergism promotes the usefulness of biological knowledge to achieve sustainable, technological development.

Biomimicry has been successful as well, although in a different way. Biomimicry clearly aims to promote the importance of biological knowledge within the field of engineering. Modern engineering has come to emphasize the importance of physics and, to a lesser extent, chemistry. Training in this subject is central to education in this profession. Promoters of biomimicry make a case for their field by pointing out that evolutionary biology is full of examples of how organisms have evolved to deal with physical and chemical challenges in sophisticated ways. Their emphasis on functionalism, elegance and even masculinity in the paradigm helps advocates of biomimicry to promote their value within the engineering profession specifically and thence to society in general.

6.4 Insight and Blindness

Besides their social and organizational functions, the ideologies of biomimicry and biosynergism also provide practitioners with a fund of insights for the advancement of their work. There is no doubt that the biological world provides designers with a fertile source of useful models. However, ideologies are also associated with a kind of willful blindness. When we say that a decision is "ideological", we usually mean that it is based on convictions that are fanatical and impervious to the blandishments of reason. If design paradigms are ideological, then it seems to follow that they promote a similar fanaticism and blindness among their practitioners.

Thagard (2014) explains that ideologies influence people's thinking under the pressure of emotions. A positive feeling about a belief can motivate people to accept it uncritically. Something similar can be said about designs (Shelley 2011). Likewise, negative emotions such as fear or rage can lead people to uncritical acceptance or rejection of ideas. There are many examples of regrettable and fanatically held ideologies.

It is possible that bio-inspired design experiences some regrettable, ideological issues. Given the currency of biosynergism in sustainable design, for example, it may be hard for designers who do not subscribe to biosynergism to get a fair hearing for their work on sustainability. (This is a hypothetical example; I am not aware of any research showing this situation to be the case.)

However, ideologies become blinkered only when the emotions involved are extreme and the ideologies become embedded in powerful and aggressive institutions. Thagard (2014) notes that Nazi ideology became especially virulent when it became the official ideology of the German state, backed up by a violent propaganda machine. There is no evidence that either biomimicry or biosynergism have become emotionally charged or institutionalized in such a way.

7 Conclusions

Bio-inspired design encompasses a variety of design practices and ideologies. Two of those ideologies may be captured under the terms *biomimicry* and *biosynergism*. Biomimicry denotes a design ideology emphasizing the values of functionalism, efficiency, elegance, and masculinity. As such, practitioners tend to focus on biological adaptations produced by natural selection for fitness and success in the struggle for survival. By contrast, biosynergism emphasizes values of organicism, environmentalism, sustainability, and femininity. As such, practitioners tend to focus on structures of interdependence and cooperation in biological systems or bodily organs.

As would be expected, these ideological differences are manifested in various ways. Methodologically, biomimicry is characterized by case-based reasoning between specific source and target analogs. Biosynergism is characterized by the application of general schemata arising in ecology and medicine to artifacts such as cities, buildings, and other products.

Some tension between these ideologies is also displayed in the presence or absence of broad, social commitments. For the most part, biomimicry concentrates on achieving success as defined in the industrial world, that is, through greater ease and efficiency in converting resources into work. With its emphasis on sustainability, biosynergism concentrates on achieving useful work without degradation of the surrounding environment. This difference in ideologies is captured by Geddes' distinction between paleotechnic and biotechnic civilizations.

Both ideologies have been successful in achieving their goals of forming practitioners into coherent groups capable of organizing resources, prioritizing projects, judging their value, and promoting their worth to others. Biomimicry has become established as an acceptable, engineering practice while biosynergism has become central to the broad, modern movement focused on sustainable design.

Finally, neither biomimicry nor biosynergism appear to have become dogmatic. Despite acknowledged differences in these forms of bio-inspired design, neither group has become obviously fixed or inflexible in its thinking, nor hostile or aggressive towards others. The fact that the distinction between these groups is normally overlooked may be evidence that this is the case.

As such, biomimicry and biosynergism form an instructive study in the role of ideology in model-based design.

References

Bud, R. (1991). Biotechnology in the twentieth century. *Social Studies of Science, 21*(3), 415–457.

Casillo, R. (1992). Lewis Mumford and the organicist concept in social thought. *Journal of the History of Ideas, 53*(1), 91–116.

Darwin, C. (1871). *The descent of man and selection in relation to sex*. London: John Murray.

European Patent Office. (2015, May 13). *Hendrik Marius Jonkers (The Netherlands): Finalist for the European inventor award 2015*. Retrieved July 1, 2015, from, European inventor award: http://www.epo.org/learning-events/european-inventor/finalists/2015/jonkers.html

Flint, A. (2015, May 19). *Can regenerative design save the planet?* Retrieved June 4, 2015, from, The Atlantic: Citylab: http://www.citylab.com/design/2015/05/can-regenerative-design-save-the-planet/393626/

Forty, A. (1986). *Objects of desire: Design and society from Wedgwood to IBM*. New York: Pantheon Books.

French, M. J. (1988). *Invention and evolution: Design in nature and engineering*. Cambridge: University of Cambridge Press.

Geddes, P. (1915). *Cities in evolution*. London: Williams & Norgate.

Geddes, P. (1929). Social evolution: How advance it? *The Sociological Review, 21*(4), 334–341.

Goist, P. D. (1974). Patrick Geddes and the city. *Journal of the American Institute of Planners, 40* (1), 31–37.

Hofstede, G. (1984). *Culture's consequences: International differences in work-related values* (2nd ed.). Beverly Hills, CA: SAGE Publications.

Jonkers, H. (2007). Self-healing concrete: A biological approach. In S. van der Zwaag (Ed.), *Self healing materials. An alternative approach to 20 centuries 195 of materials science* (pp. 195–204). Berlin: Springer.

Kropotkin, P. (1902). *Mutual aid: A factor in evolution*. London: William Heinemann.

Lyle, J. T. (1996). *Regenerative design for sustainable development*. New York: Wiley.

McDonough, W., & Braungart, M. (2002). *Cradle to cradle: Remaking the way we make things*. New York: North Point Press.

Nachtigall, W. (1974). *Biological mechanisms of attachment: The comparative morphology and bioengineering of organs for linkage, suction, and adhesion*. (M. A. Beiderman-Thompson, Trans.) Berlin: Springer.

Shelley, C. (2011). Motivation-biased design. In *Proceedings, 33rd Annual Conference of the Cognitive Science Society* (pp. 2956–2961).

Shelley, C. (2013). Models and ideology in design. In L. Magnani (Ed.), *Model-based reasoning in science and technology: Theoretical and cognitive issues* (pp. 609–623). Berlin: Springer.

Thagard, P. (2014). The cognitive-affective structure of political ideologies. In B. Martinovski (Ed.), *Emotion in group decision and negotiation*. Berlin: Springer.

Vogel, S. (1998). *Cats' paws and catapults: Mechanical worlds of nature and people*. New York: W. W. Norton & Co.

Vonarburg, B. (2014, April 11). *The ATM strikes back*. Retrieved 5 1, 2015, from ETHzürich: https://www.ethz.ch/en/news-and-events/eth-news/news/2014/04/der-bancomat-schlaegt-zurueck.html

Wackernagel, M., & Rees, W. (1996). *Our ecological footprint: Reducing human impact on the Earth*. Gabriola Island, BC: New Society Publishers.

Welter, V. M. (2002). *Biopolis: Patrick Geddes and the city of life*. Cambridge, MA: The MIT Press.

Wolman, A. (1965). The metabolism of cities. *Scientific American, 213*(3), 179–190.

Thought Experiments and Computer Simulations

Marco Buzzoni

Abstract The main purpose of this paper is to investigate some important aspects of the relationship between thought experiment (hereafter TE) and computer simulation (hereafter CS), from the point of view of real experiment (RE). In the first part of this paper, I shall pass in critical review four important approaches concerning the relationship between TE and CS. None of these approaches, though containing some important insights, has succeeded in distinguishing between CS and TE, on the one hand, and REs, on the other. Neither have they succeeded in distinguishing TEs and REs (Sect. 1–4). In Sect. 5, the paper briefly outlines an account of CSs as compared with TEs that takes REs as a central reference point. From the perspective of the analysis of the empirico-experimental intensions of the concepts of TE, CS, and RE—considering their empirical content and actual performance within a discipline—the attempts to find a distinction in logical kind between TEs, CSs and REs breaks down: for every particular characteristic of one of these notions there is a corresponding characteristic in the two others. From an epistemological-transcendental point of view, the only difference in kind between TEs and CSs consists in the fact that any simulation, even a computer one, involves a kind of *real* execution, one that is not merely psychological or conceptual. In TEs the subject operates concretely by using mental concepts in the first person; in contrast, real experiments and simulations involve an 'external' realisation. As shown in Sect. 6, this manifests itself in the higher degree of complexity often found in CSs as compared with TEs.

1 Introduction

The main purpose of this paper is to investigate some aspects of the relationship between thought experiment (hereafter TE) and computer simulation (hereafter CS), from the point of view of real experiment (RE). Four approaches have been

M. Buzzoni (✉)
Section of Philosophy and Human Sciences, Department of Humanistic Studies, University of Macerata, Via Garibaldi 20, I-62100 Macerata, Italy
e-mail: marco.buzzoni@unimc.it

L. Magnani and C. Casadio (eds.), *Model-Based Reasoning in Science and Technology*, Studies in Applied Philosophy, Epistemology and Rational Ethics 27, DOI 10.1007/978-3-319-38983-7_4

prominent in the literature: (1) TE is only a particular form of the more general concept of 'simulative model-based reasoning'; (2) Following Norton's doctrine that TEs can be reconstructed as arguments with tacit and explicit assumptions, some authors have maintained that CSs also are arguments; (3) The main difference between TEs and CSs lies in the much greater opacity and/or complexity of the latter as compared with the former: CSs are 'opaque' TEs; (4) CSs share so many important aspects with REs that they could be considered as falling under the more general concept of experiment. The first four Sections are devoted respectively to these four approaches.

These approaches, though containing some important merits, have not succeeded in distinguishing between CS and TE, on the one hand, and REs, on the other. Neither have they succeeded in distinguishing TEs and REs (Sect. 1–4).

Section 5 and 6 briefly outline an account of CS as compared with TE that takes RE as a central reference point that can provide an answer to the difficulties raised in the preceding sections. In Sect. 5, I shall maintain that—from the perspective of the analysis of the empirico-methodological intensions of the respective concepts—it is a hopeless task to find a particular difference in logical kind which applies *exclusively* to TE, CS, or RE: for every particular characteristic of one of these notions there is a corresponding characteristic in the two others. However, from another point of view, there is between TE and CS on the one hand, and RE on the other, an epistemological (or transcendental) difference which we must not overlook. In a different sense, this is also true of CS and TE. The only difference in kind between TE and CS consists in the fact that any simulation, even a computer one, involves a kind of *real* execution, one that is not merely psychological or conceptual, even though it must be distinguished from that of a RE. As in real experiments, this execution depends on us for its realisation only in the initial moment when we set off its 'mechanism'. In TEs the subject operates concretely by using mental concepts in the first person; in contrast, real experiments and simulations involve an 'external' realisation, which is initially practical-experimental and tied to the experimenter's body, and thereafter develops independently of the subject and 'impersonally'.

Two preliminary remarks are in order before we plunge *in medias res*:

1. Here I am interested above all in empirical TEs and CSs, whose results are liable to correction by new experimental findings. TEs and CSs of *pure* logic and mathematics deserve a separate treatment.[1] Also philosophical TEs will not be considered here.

[1]This distinction also applies to particular methods of CS, such as, for example, the Monte Carlo Method, which may be of both kinds. This method is in some cases purely mathematical (as for example in the case of the calculus of π or of a definite integral), but in others it is of an empirical kind (as for example in the case of its application to high-energy physics). There is not space in the present paper to examine this point at the length that it deserves. For two opposite accounts of Monte Carlo simulations, cf. Dietrich (1996) (Monte Carlo simulations are "ordinary experiments") and Beisbart and Norton (2012) (Monte Carlo simulations are "are merely elaborate arguments"). Concerning TEs in mathematics as distinct from empirical TEs, cf. Buzzoni (2011).

2. I shall presuppose the broad sense of CS defined by Winsberg in his authoritative entry for the Stanford Encyclopedia of Philosophy: according to which a CS "includes choosing a model; finding a way of implementing that model in a form that can be run on a computer; calculating the output of the algorithm; and visualizing and studying the resultant data. The method includes this entire process—used to make inferences about the target system that one tries to model—as well as the procedures used to sanction those inferences." (Winsberg 2013. There are other definitions of "CS" that are compatible with the main theses of this paper: cf. for example Humphreys 1991, p. 501, and Hartmann 1996, p. 5 (in the wider sense of his definition, which also includes *static* models or systems), but I will focus on Winsberg's).

2 TE and CS as Simulative Model-Based Reasoning: Promissory Eliminativism About TE

According to the first view to be examined, the cognitive role played by TEs is a form of simulative model-based reasoning carried out with mental models. This has led some to maintain that there is no essential difference between TE and CS (Nersessian 1992, 1993; Miščević 1992, 2007; Palmieri 2003; Gendler 2004; Misselhorn 2005; Cooper 2005; Morrison 2009; Chandrasekharan et al. 2013; cf. also Galison 1987, p. 779, 1996, whose position has inspired very different accounts, ranging from that which we are discussing to that which emphasizes the similarity in kind between CS and RE).

In a nutshell, the central claim is that both in TEs and CSs we gain knowledge to the extent that we fruitfully manipulate mental models. It is worth noting that this account was already latent in Mach, who wrote:

> (E)ven instinctive knowledge of so great logical force as the principle of symmetry employed by Archimedes, may lead us astray [...] Everything which we observe in nature imprints itself *uncomprehended* and *unanalysed* in our percepts and ideas, which, then, in their turn, *mimic* [nachahmen] *the processes of nature in their most general and most striking features*. (Mach 1883: 26–27; last italics added, Engl. transl. pp. 27–28)

As Häggqvist puts it with reference to TEs (but, *mutatis mutandis*, this also holds true of CS), the main difficulty with this account consists in the fact that it seems to assume mental mediators in order to explain how words relate semantically to the world. In both cases, "something must be said about how they acquire their semantical properties" (Häggqvist 1996, p. 81). In other words, this account clearly and crucially depends on what reasons there may be for judging that mental mediators are anchored to reality.

The upholders of this view have attempted to answer this question with the introduction of some "engineering" (Nersessian 2006, p. 699) or manipulative (cf. Miščević 2007, p. 194) constraints in their theorizing. As Miščević writes:

> the use of mental models demands rules for manipulation. Some constraints on manipulation come directly from the geometry of the model. It is essential that manipulation of elements mobilizes the spatial skills of the subject, his "knowledge how" which is generally not verbalisable. (Miščević 2007, p. 194)

But this is only a metaphorical and loose way of speaking: we cannot manipulate mental models in the same way in which we manipulate objects and processes of everyday life. Someone who, before moving to a new house, tries to figure out beforehand how the furniture might fit in, can manipulate a scale models of the furniture, but she/he cannot manipulate his/her own mental models in the same sense. In both cases, moreover, the final and decisive test can only consist in the technical and experimental realization of the plan in everyday life. Only if, in the end, any piece of furniture will occupy the right place in the new home, are our mental models adequate to the task.

Anyway, the insistence that "simulation devices" are "systems [...] possessing engineering constraints" Nersessian (2006, p. 704), though not capable of solving the aforementioned problem, has led some authors to treat TEs as a sort of provisional stopgap that someday will be abandoned in favour of CSs:

> computational modeling is largely replacing thought experimenting, and the latter will play only a limited role in future practices of science, especially in the sciences of complex nonlinear, dynamical phenomena. (Chandrasekharan et al. 2013, p. 239)

The reason presented for this is the complexity of the natural systems that scientists and engineers are modelling. This complexity is such that the relationship between the different elements of natural systems cannot be captured through TE, but only by the new computational visualization tools that are being developed in computer science.

It must be admitted that CSs are *usually* more complex than TEs, but this is not enough to support the claim in question. Apart from minor grounds on which this claim may be criticized, there is one main objection against any attempt of this kind. This claim is a prediction about human knowledge, and therefore, in analogy with Popper's "promissory materialism", it may be considered as a kind of "promissory eliminativism" concerning TEs. As such, it is undermined by Popper's argument according to which, "if there is such a thing as growing human knowledge, then we cannot anticipate today what we shall know only tomorrow" (Popper 1957 [1961], p. xii).

3 A Second Position About TE and CS: The Argument View

Following Norton's doctrine that TEs can be reconstructed as arguments based on both tacit and explicit assumptions, some authors have drawn a detailed comparison between TE and CS (Stäudner 1998; Stöckler 2000; Velasco 2002; Beisbart 2012, Beisbart and Norton 2012).

As Beisbart (2012) has aptly pointed out, the fundamental question of TEs has its counterpart in the debate about the nature of the CS: How do scientists gain new knowledge about a target system by simulating that very system using a computer? According to Beisbart "[t]he argument view answers this question by saying that computer simulations are also arguments" (Beisbart 2012, p. 429). Notwithstanding some differences, the view taken by El Skaf and Imbert 2013 may be classified into this group to the extent that they admit that "arguments are also a way to extract information from premises" (El Skaf and Imbert 2013, p. 3466).

In fact, equating TEs and arguments, as Norton does, has led some authors to a detailed comparison between TEs and CSs. As for example Stäudner has made:

> The initial equations that we are striving to solve, together with the relevant boundary values, form a set of 'premises'. The numerical procedure by means of which we calculate the solutions we are looking for corresponds to a 'logical type of inference', that is to a determinate form of argument. The result of the calculation is the 'conclusion'. As in valid arguments true conclusions follow from true premises, we may consider the result of the calculation of a simulation as an adequate description of nature if the 'premises' contain adequate descriptions of nature, in the sense that they are empirically confirmed and therefore belong to the well-established 'theoretical patrimony' of the natural sciences. (Stäudner 1998, p. 157).

However, no matter how many similarities between TE and CS may be found, this does not remove the fundamental difficulty of Norton's view about TEs, which consists in reducing *empirical* TEs to *logical* arguments, and which we must briefly examine.[2]

According to Norton, TEs "are arguments which (i) posit hypothetical or counterfactual states of affairs, and (ii) invoke particulars irrelevant to the generality of the conclusion" (Norton 1991, p. 129, see also 2004a, b). For Norton, TEs can always be reconstructed as deductive or inductive arguments ("reconstruction thesis") and, more importantly, they must always be evaluated as such:

> The outcome is reliable only insofar as our assumptions are true and the inference valid [...] [W]hen we evaluate thought experiments as epistemological devices, the point is that we should evaluate them as arguments. A good thought experiment is a good argument; a bad thought experiment is a bad argument. (Norton 1996, p. 336)

Norton's view has been widely criticised for a number of mostly unconvincing reasons. Some criticisms simply miss the target, since they do not take into account Norton's explicit insistence that the argument may be inductive.[3]

However, among the many objections that have been raised against Norton, one is very instructive for our purposes. According to this objection, the translation of a TE into formal terms (i.e., the elimination of pictures and diagrams concomitant

---

[2]The following objection is taken from Buzzoni (2008), pp. 65–71, to which I must refer the reader for more details.

[3]See, e.g., Norton (1991, p. 129, 1996, p. 335, 2004a, pp. 52–53, 2004b, p. 1144) ("thought experiments are arguments that exploit the familiar deductive and inductive logics."). Among the authors that make this mistake are Miščević (1992, p. 217), and Gooding (1994, p. 1035).

with their translation into propositional contents) causes a partial loss of the original meaning or content, to the point that the experiment becomes unintelligible (cf. especially Brown 1997 and Arthur 1999, p. 219).

Taken literally, the objection is inconclusive since it sets forth no reason in *principle* why this must happen. Therefore, Norton is right in countering this objection by de facto reconstructing many paradigmatic TEs as arguments: It is plausible to assume that the same may be done even for experiments that have not yet undergone such a reconstruction (Norton 2004a, p. 50).

However, it is not difficult to find a sense in which the objection is well taken. My main objection is as follows. Even though in a TE this or that particular empirical element may be "irrelevant to the generality of the conclusion" (for example, in Einstein's lift experiment it is irrelevant whether the observer is or is not a physicist), it is not irrelevant that TEs are generally performed by constructing particular cases, which need concrete elements that are in principle reproducible in specific spatio-temporally individuated situations. TEs, stripped of any necessary reference to concrete experimental situations, are confined to a domain of purely theoretical statements and demonstrative connections. As a result, empirical TEs are reduced to logico-mathematical arguments.

According to Mike Stuart, when Norton says that TEs "invoke particulars irrelevant to the generality of the conclusion", he does not mean (at least in the case of inductive TEs, where some 'relevant' particular are necessary) that no particulars are necessary, or that the particulars play no role in the argument at all. According to him, if thought experiments are arguments, some of the content of their premises does matter for their identification and individuation. [...] In fact Norton argues elsewhere for an account according to which "inductive inferences will be seen as deriving their license from facts. These facts are the material of the inductions" (Stuart 2015, p. 41).

However, this defense fails, since Norton by "inductive argument" means an argument that is valid within a formal calculus. He writes, "If we are to recognize the logic as delimiting the successful thought experiments, there must be something in the logic that evidently confers the power of a thought experiment to justify its conclusion. For example, deductive logics are characterized by their preservation of truth and inductive logics by the preservation of its likelihood, so that a thought experiment using these logics will have a justified outcome if it proceeds from true premises" (Norton 2004a, p. 54). This restriction of the meaning of the term "inductive", far from solving the difficulty and enabling us to distinguish between empirical and logical or mathematical thought experiments, just confirms that difficulty: in this sense, an inductive argument, like a deductive one, may be tested as to its logical validity (suitably understood: for example, as to its capacity of preserving probability), but not as to the experimental 'correspondence' of its conclusions with reality.

Norton's equation of empirical with logico-mathematical argumentation is not only untenable, but—what is more important for our present purposes—prevents us from seeing any distinction between TE and CS. Beisbart himself is aware of this difficulty: he meets it essentially in two ways, either by replying that "running a computer simulation may be thought of as the execution of an argument" (Beisbart 2012, p. 423), or by insisting on the greater complexity and opacity of CSs as compared with TEs. As we shall see later, the first alternative, which regards the running of a CS as the execution of an argument, is in one sense true, but Beisbart interprets it in the light of the extended mind hypothesis (cf. Beisbart 2012, p. 395; for this hypothesis in its more general interpretation, cf. Clark and Chalmers 1998; as far as CS is concerned, cf. Charbonneau 2010), and this produces an internal contradiction in Beisbart's account. According to the extended mind thesis, cognitive systems may extend beyond a human being. But if the construction of an 'experimental machine' which extends the original operativity of our organic body is treated as a part of the mind, there can be no difference in principle between TE and CS on the one hand and RE on the other. This is in contradiction with Beisbart's explicit rejection of the idea that CSs produce new knowledge because they are real world experiments (Beisbart 2012, p. 425), a thesis which indeed would have undermined his whole *argument* view.

According to the second alternative, in order to avoid the complete reduction of CSs to logico-mathematical arguments (and hence to TEs interpreted in accordance with Norton's argument view), Beisbart adheres to the idea that CSs are more complex and opaque than TEs. This, however, is an autonomous line of interpretation of the relationship between TE and CS, which we have now to examine separately.

4 A Third Position About TE and CS: CSs as Opaque Thought Experiments

The third kind of comparison between TE and CS that I shall now discuss is to be found in many incidental remarks of a wide range of different authors, but it was elaborated by Di Paolo et al. (2000) in a particularly influential way. The main idea is that CSs, unlike TEs, "are opaque and must be explored" (Di Paolo et al. 2000, p. 503). As Di Paolo et al. (2000) write:

> A thought experiment has a conclusion that follows logically and clearly, so that the experiment constitutes in itself an *explanation* of its own conclusion and its implications. [...] In contrast, a simulation can be much more powerful and versatile, but at a price. This price is one of *explanatory opacity:* the behaviour of a simulation is not understandable by simple inspection. (Di Paolo et al. 2000, p. 502)

A similar view was formulated by Mark Bedau: CSs are a kind of TE, which he calls "emergent" because they allow us to draw conclusions about (emergent) properties of complex systems—such as biological ones—that could not be intuited and/or justified by armchair TEs (Bedau 1999. Similar statements can be found in the work of many other writers: cf. for example Buschlinger 1993, p. 75; Lenhard and Winsberg 2010; Beisbart 2012).

This position is open to many objections, and all support the conclusion that one has to relativize opacity to a background context and cannot treat it as an absolute concept.

To begin with, 'opacity' is no hallmark of CSs (or REs) in contrast with TEs. Indeed, in this regard there is only a difference in degree between CSs and TEs, that is, a difference that may be turned upside down in particular cases: a very simple CS may be less opaque than many TEs (such as Einstein's black body radiation TE: cf. Norton 1991). Moreover, this thesis presupposes that TEs have a kind of almost Cartesian clearness, which, at least apparently, like that of Descartes' *cogito*, would be static and without a history. This presupposition has probably been inspired by Hacking's claim that, while REs "have a life of their own", TEs "are rather fixed, largely immutable" (Hacking 1993). However, this thesis is untenable. To see this, it is sufficient to recall the history of the interpretations of the most important TEs (such as Maxwell's Demon or Galileo's freely falling bodies).[4]

Second, both TEs and CS are 'opaque' in the sense that both are liable to error and uncertainty. In both cases, REs are the antidote in the sense that, if we have some doubt, we must turn to REs, which remain the ultimate criterion for all empirical TEs and CSs. In this sense too, the similarity in kind between TEs and CSs is easily proved more fundamental than the differences in the degree of complexity.[5]

Lenhard's version of the theory that we have been considering escapes both of these objections (cf. Lenhard *forthcoming*). He distinguishes two types of iteration, namely the "convergence" and the "atlas" type. The first "is involved in cases like exploring a new pathway that eventually becomes your routine way". Repeating exercises in music or sports, for example, becomes a routine that one may inattentively perform. This kind of iteration is distinctive of TEs, and "repeated execution eliminates initial intransparency or ambiguity." (Lenhard *forthcoming*). The second kind of iteration "works rather by exhausting the possibilities" (it is in the sense that produces a "compendium", or "atlas"). This kind of iteration is distinctive of CSs: "Repeated, and slightly varied, model runs do not eliminate opacity, but rather explore the space of possible model behaviour." Based on this distinction, he sums up his view as follows:

> Thought experiments and simulation experiments are similar in that both make use of iterations. However, they differ fundamentally in the types of iterations they use, and in the functions those iterations fulfill. In thought experiments, the iterations of the convergence-type eventually produce a cognitive tool that is sufficiently transparent to run in human intuition. Simulation experiments, on the other hand, do not remove, but rather circumvent or compensate opacity with the help of atlas-type iterations. Iterative algorithms utilize computational power and can work where thought experiments cannot. In particular, if iterations are incompressible, there is hardly a chance to render the results epistemically transparent. (Lenhard, *forthcoming*).

---

[4]For the claim that TEs have a history and must be interpreted, see Buzzoni (2008, pp. 107–108), and *passim*.

[5]For more details on this general point, cf. Buzzoni (2008, for example p. 96), and (2013).

To the extent that Lenhard's distinction is relevant to our discussion, it does not mark any distinction in principle between TEs and CSs. What takes place in a TE is not only an iteration in the sense of a psychological or sociological phenomenon—which leads to the formation of a habit and in this sense is not relevant in our present discussion,[6] but also an iteration in the second sense, which according to Lenhard should be the distinguishing characteristic of CS. As Lenhard concedes, iteration may efface the opacity of a TE only because it is based on an intersubjectively testable and reproducible process (at least for a group of educated people). However, this intersubjective reproducibility is a demand that cannot be confined to any one particular field of study because it is an essential ingredient of any scientific procedure (and, more generally, not only of any piece of theoretical knowledge, but also of any practical conclusion).

The last consideration leads to a third and more important objection that applies to all the versions of the theory that I have been discussing in this section. In order to understand and then test the truth of a scientific statement, we always must retrace in the first person, even operational-experimentally and technically, the steps that led to that statement being asserted. That is, by means of a personal act we must re-appropriate the reasons of that statement's truth. In the end, it is only an individual person who, at any given time, can challenge the value of any claim by retracing the steps that led to it. Whenever the truth of a conclusion is called into question—regardless of the question whether the conclusion was obtained by TE or CS—it is necessary to assess the data upon which the theory is founded, by reconsidering and retracing thc procedural steps that led (and still lead) to that theory.

This is the exact meaning of the methodological principle of the reproducibility of experiments, according to which any person with sufficient basic knowledge can recognize the validity of a scientific statement by repeating certain fixed operations. In science, this is tantamount to the fundamental principle of objectivity: no sentence will be accepted as truly scientific if it is not accompanied by the means that allow it to be tested intersubjectively.

Accordingly, we must reject Popper's belief that genetic questions are irrelevant for ascertaining whether a statement is true or false. Because we have no direct revelation of the truth of a statement, we are forced to find and retrace the 'paths' that led to its being accepted or rejected. Certainly, we can use Pythagoras's Theorem in a practical way without recalling the procedural steps that led to its discovery. Nevertheless, if someone challenged the validity of this theorem, we ought to reconsider and retrace in the first person the procedural steps that led (and still lead) to that theorem being asserted.

---

[6]In this sense, Lenhard's distinction is very important in another context, that is, when trying to distinguish the human from the natural sciences. According to Buzzoni (2010), the typical law-like object of the human sciences is the unconscious, prima facie understood as the repository of the routines and quasi-automatisms, which largely constitute our lives. In this sense, however, Lenhard's distinction is irrelevant to our present discussion.

Returning now to the question with which the discussion of the point began, it is clear that, in view of what has been said, CSs ought to consist of concrete methodical procedures, which we may, at least in principle, reconstruct, re-appropriate and evaluate in the first person. A particular truth-claim resulting from a CS may be considered as scientific only under the condition that it is in principle intersubjectively testable. No matter how complicated the 'mechanization' of cognitive performances may be, if we accept the results of a CS, we presuppose that any change concerning the hardware/software may be in principle reconstructed and re-appropriated in the first person (as is well known, this is also true of a random number generation).

In this general sense, the conclusions reached in a TE are not different in kind from that of a CS, and this applies to Lenhard's distinction between two kinds of iteration. In the last case, it might be objected that this misses the point. According to Lenhard's distinction, TEs are different from CSs because in the first case there is only one 'iteration' (in the general sense just explained), followed by many (psychological) iterations, without any intention to limit and exhaust a complex range of possibilities, while in the second case a complex chain of 'iterations' (again in the general sense just explained) is present.

However, this is not always true. Some TEs contain iterations in Lenhard's second sense. Stevin's famous TE, to mention only one example, is demonstratively powerful because we must exclude certain theoretical possibilities. When we exclude that the chain draped over two differently inclined frictionless planes will move in one direction or the other (otherwise we would have a perpetual motion machine), what remains is that the setup will be in static equilibrium.

As a result, no difference in kind between TE and CS follows from the notion of "iteration", taken in the sense that is relevant to our discussion. What distinguishes CS from TE is not so much a particular type of iteration, which would take place in the CS, but not in TE, as its application to much more complex situations, which require much more powerful instruments (in the main, in computation), such as digital computers. However, given that CSs must be translated into intersubjectively reconstructible form before they can be accepted epistemologically, their opacity must be faced and left behind. This removes their opacity as a distinguishing factor between CSs and TEs, at least in terms of their epistemologies.

5 A Fourth Position About TE and CS: CSs as Real Worlds Experiments

Recently, the relationship between CS and traditional experimentation has attracted more and more attention. Already Galison pointed out that simulations have become an essential part of many REs, as in high energy physics where certain simulation techniques have become indispensable tools for all experimenters (cf. Galison 1996, p. 120, 1987, p. 266). But many other authors tried to secure the specificity of CS by comparing it with real-world experiments (cf. for example

Keller 2003; Parker 2009, 2010; Morrison 2009; Chandrasekharan et al. 2013; Guala 2002, 2005; Morgan 2002; Norton and Suppe 2001; Winsberg 2003, 2013; Küppers and Lenhard 2005a, b; Lenhard 2007).

The thesis most frequently recurring is that a CS is, as Winsberg (2003, p. 220) has called it, a "hybrid of experiment and theory". We can, of course, undertake no minute discussion of the many varieties of this approach. I shall confine myself to criticising the authors who have emphasized the experimental aspects of CSs to such an extent that they are considered as falling under the more general concept of experiment. According to this view, CS models are measuring instruments which, as happens with traditional measurements, involve causal contact with the real-world systems about which information is sought. For example Norton and Suppe 2001 write:

> Simulations often are alleged to be only heuristic or ersatz substitutes for real experimentation and observation. This will be shown false. Properly deployed simulation models are scientific instruments that can be used to probe real-world systems. Thus, simulation models are just another source of empirical data. (Norton and Suppe 2001, p. 87; for a similar view see Morrison 2009)

It must be conceded that CS and RE would coincide completely *if* CSs could be considered as particular REs, whose experimental set-up is the computer taken in its material aspect or structure, that is, as an object of sensible perception with which we can interact as agents.

But the analogy, though correct within certain limits (as we shall see later), fails in a point essential to the argument. As Hughes said, CSs are not performed to learn something about computers (cf. Hughes 1999[2010], p. 203), and "we will know whether or not the theory of cosmic defects is adequate, not via computer experiments, but through the use of satellite-based instruments" (Hughes 1999[2010], p. 209. A similar objection has been made, among others, by Winsberg 2003, p. 115; Muldoon 2007, p. 882; Frigg and Reiss 2009; Beisbart 2012, p. 245).

In my opinion, the fundamental reason for this is that the kind of contact with the world that is made possible by CS is not the same relation in which our organism stands to the surrounding world. A computer simulation may give us information about the actual world, only because we have independent and empirical–experimental evidence of the model's *meaning*. More precisely, in a RE, the construction of an 'experimental machine' which extends the original operativity of our organic body is connected not only to the 'method of variation', but also to the principle of causality. On the contrary, in the case of CS, our contact with reality is always mediated by models, to which real objects may or may not correspond. Differently stated, CSs would be indistinguishable from REs only if we might change our judgments in the same way in which we are able to change outside reality, that is, by interacting with it through our organic body. In fact this is not so, or, at least, not always so.

6 CSs and TEs Versus Real World Experiments

This section will briefly outline an account of CSs as compared with TEs that takes RE as a central reference point. The same considerations which lead me to reject the equation of CS and stated in Sect. 4 seem to contain a positive implication of great importance for the comparison between CS and TE.

Any two things, which are from one point of view similar, may be dissimilar from another point of view. Therefore, to avoid comparisons that are insignificant or of little importance for the philosophy of science, it is necessary to adopt a fruitful point of view, that is a point of view which opens up new prospects for research and does not divert thought into side issues.

In our context, both historical and systematic reasons lead us to expect that it is fruitful to take RE as a central reference point for contrasting not only CS, but also TEs with REs.

The history of TE and CS provides a natural and effective guide to finding relations of similarity and difference between TE and CS with one another and both with real world experiments. On the one hand, according to Mach, the principle of economy is not only the source of science as such (and hence of REs), but also of thought experimentation:

> We experiment with thought, so to say, at lower prices because our own ideas are more easily and readily at our disposal than physical facts (Mach 1905a, p. 187, English transl., p. 136).

Moreover, both REs and TEs are based on the "method of variation" (*Methode der Variation*), whereby some variables are systematically modified to establish which relation of dependence, if any, holds between them. While in REs this variation works upon natural circumstances, in TEs it is representations that are made to vary in order to see the consequences of those variations (cf. Mach 1883, 1905a, b, c).

On the other hand, CSs also share this similarity between TEs and REs that Mach stressed. First, historically speaking, CSs were also motivated by 'economical' reasons in Mach's sense. As is easily seen without any further comment, the principle of economy has been at work in all three more-or-less distinct stages which Keller identified in the history of CS:

> (1) the use of the computer to extract solutions from pre-specified but mathematically intractable sets of equations by means of either conventional or novel methods of numerical analysis; (2) the use of the computer to follow the dynamics of systems of idealized particles ('computer experiments') in order to identify the salient features required for physically realistic approximations (or models); (3) the construction of models (theoretical and/or 'practical') of phenomena for which no general theory exists and for which only rudimentary indications of the underlying dynamics of interaction are available. (Keller 2003, p. 202)

Second, it is difficult to deny that CSs are also based on the "method of variation". For instance, Galison has pointed out that, "as in an experiment, the Monte Carlo writer can vary the inputs to see that the corresponding output is well behaved"

(Galison 1987, p. 266). And the intense debate concerning robustness—from Levins till the present time—has shown the usefulness of the notion of variation as applied to models (cf. Levins 1966; Wimsatt 1981; Weisberg 2006; Muldon 2007).

TE, CS and RE all obey the principle of economy, and all apply Mach's method of variation. However, it is very easy to find many other similarities. For instance:

(a) TEs, CSs and REs are constituted by a theory and a particular, well-specified experimental situation;
(b) All of them ask questions about nature and its laws;
(c) All of them can do this only in a theory-laden way, so that the meaning of all of them must always be interpreted;
(d) All of them involve idealizations;
(e) In all cases visualisation, perspicuity, intuitive appeal, and clarity are important because TEs, CSs and REs apply hypotheses to particular cases that are relevant for testing their truth or falsity. (For the importance of visualisation in CSs, cf. for example Winsberg 2003, p. 111; Beisbart 2012, p. 426. As far as TE and RE are concerned, I have discussed all this in detail in Buzzoni 2008, *passim*).

For this reason, there is a prima facie ground for maintaining a much more radical thesis. From the perspective of the analysis of the empirico-experimental intensions of the respective concepts, REs, TEs, and CSs show only differences in degree, not in kind.

In order to understand appropriately the meaning of this thesis, it will be best to discuss a further point of contact between TE and CS. As Mach pointed out, when faced with the slightest doubt about the conclusions of a TE, we have to resort to REs:

> The result of a thought experiment [...] can be so definite and decisive that any further test by means of a physical experiment, whether rightly or wrongly, may seem unnecessary to the author. [...] The more uncertain and more indefinite the result is, however, the more the thought experiment necessitates the *physical experiment* as its natural continuation which must now intervene to complete and determine it. (Mach 1905a, pp. 188–189, Engl. transl., pp. 137–138; translation slightly modified.)

On this point there is essentially agreement between Mach and Kant:

> In experimental philosophy the delay caused by doubt may indeed be useful; no misunderstanding is, however, possible which cannot easily be removed; and the final means of deciding the dispute, whether found early or late, must in the end be supplied by experience. (KrV B 452–453, AA III 292, lines 27–31)

According to this, the particular content of any empirical TE or CS must be, at least *in principle*, ultimately reducible to *empirico-experimental interventions on reality,* that is, *to experimentations*. Whatever resists this reduction thereby shows itself to be an arbitrarily introduced factor, which is legitimate only if this factor disappears in the final result. In this sense, a TE or a CS would be devoid of *empirical meaning* (that is, it would not be a TE or a CS proper to empirical science) if, in formulating and evaluating them, we did not in principle assume an at least implicit reference to a RE, which in its most general sense of the word is both the starting point and the ultimate testing criterion for the truth of the conclusions both of (empirical) TEs and CSs.

It is true that TEs and CSs have a certain autonomy as regards experience in the sense that both anticipate an answer to theoretical problems without resorting directly to REs. Empirical TEs and CSs anticipate, at the linguistic-theoretical or representational level, a hypothetical experimental situation so that, on the basis of previous knowledge, we are confident that certain interventions on some variables will modify some other variables, with such a degree of probability that the actual execution of a corresponding RE may be superfluous (for this definition, as applied to TEs, see Buzzoni 2013, p. 97).

The autonomy of empirical TEs and CSs is not different from that of applied mathematics: if someone puts two coins, and then two more coins into an empty money box, I know that there are now four coins in that money box, and I will persist in that knowledge even if, say, the money box immediately afterwards falls into a deep lake so that I—for whatever reason—will never be able to count how many coins it contains.

But this knowledge can never outstrip our initial knowledge as to its certainty or degree of justification: for example, if the person that put the coins into the money box was a conjurer, this might cause doubts about the box's content that could be dispelled only by resorting to experience. Similarly, if in the simulation of a hurricane there appeared objects that my background knowledge told me should not appear, I would be faced by a difficulty that only a real test, in the last analysis, could solve in the most reliable way.

TEs and CSs (in so far as the latter can be traced back to the former) are therefore only *relatively* autonomous with regard to real world experiments; their autonomy consists in the ability to follow paths that for long stretches cannot be directly compared with experimental results, even though in the end they must reconnect with them.

This just-mentioned difference between TEs and CSs on the one hand and REs on the other, is a very important exception to the rule that, from the perspective of the analysis of the empirico-methodological intensions of TEs, REs and CSs—considered in their determinate way of being within a discipline, as particular mental contents occurring in the mind of concrete persons—, REs, TEs, and CSs do not essentially differ. Nevertheless, strictly speaking this difference is not an exception to the empirico-experimental identity of TEs, REs and CSs after all, because it is only an *epistemological or transcendental* (as Kant would have called it) difference separating TE and CS on the one hand, from REs on the other.

More precisely, this epistemological-transcendental difference has two distinct, but related, aspects, one subjective and one objective. The subjective side consists in the capacity of the mind to anticipate a hypothetical or counterfactual experimental situation. From this point of view, w*hat TEs and CSs have over and above real ones is only the fact that they exist in a purely hypothetical sphere.* This capacity of the mind to anticipate a hypothetical experimental situation underpins the difference in principle—a properly transcendental difference—between TEs and CS on the one hand and REs on the other. This difference cannot be suppressed since it is the same distinction between the hypothetical-reflexive domain of the mind (which is in principle able to enter into contradiction with itself) and reality (which is able to develop in only one way).

However, this transcendental difference has also an objective-ontological counterpart: *what REs have over and above TEs and CSs is only the fact that they* are causal-experimental interactions between our body (or our instruments, understood as logical-practical extensions of our body) and the surrounding reality. For this reason, simply to imagine that the experimental apparatus, counterfactually anticipated in a TE or in a CS, has been realized is sufficient to erase any empirico-experimental or methodical difference between TE, CS and RE.

In this connection, Kant's example of a hundred dollars is very instructive. On the one hand, "the real contains no more than the merely possible. A hundred real thalers do not contain the least coin more than a hundred possible thalers." On the other hand, "My financial position is, however, affected very differently by a hundred real thalers than it is by the mere concept of them (that is, of their possibility). For the object, as it actually exists, is not analytically contained in my concept, but is added to my concept (which is a determination of my state) synthetically" (KrV B 627, AA III 401. From my point of view, of course, "synthetically" is to be understood in the sense of "by means of real interactions between our body and the surrounding world").

It is interesting to note that the epistemological-transcendental difference between TE and CS on the one hand and RE on the other is the true reason for the fact that the intensions of the concepts of TEs, CSs, and REs coincide, as do the hundred real dollars and the hundred merely thought ones. Every (empirical) TE or CS corresponds to a real one that satisfies the same conceptual characteristics, and vice versa. *All REs may also be thought of as realisations of TEs or CSs; conversely, all empirical TEs and CSs must be conceivable as preparing and anticipating a RE: They must, that is, anticipate a connection between objects that, when thought of as realised, make the TE and CS coincide completely with the corresponding RE.* Here we find the fundamental reason why, for every particular characteristic of one of the three notions, there is a corresponding characteristic in the two others.

And note another point that is brought out by Kant's example of the hundred dollars: Thought dollars, like TEs and CSs, exist only in the sphere of the possible, while real dollars, like REs, occupy a specific place among the interactions between our bodies and the surrounding reality; but *neither thought nor real entities are what they are for us outside their mutual relationship*. Paraphrasing Kant, (empirical) TEs and CSs without REs are empty; REs without TEs and CSs are blind. Thought dollars, like TEs and CSs, exist only in the sphere of the possible, while real dollars, like REs, occupy a specific place among the interactions between our bodies and the surrounding reality; but *neither thought nor real entities could exist outside their mutual relationship.*

As we shall see in the last part of my paper, an aspect of the difference between CS and TE on the one hand and RE on the other reverberates in the relationship between TE and CS.

7 The Peculiarity of CSs

As we shall see in this section, there is an element of truth both in the attempts to distinguish between CS and TE and the accounts that regard CS as a form of RE. Briefly stated, this element of truth is as follows: In TEs the subject uses in the first person concepts, inferences, etc.; in contrast, CSs, as also REs do, involve an 'external' realisation, so that strictly speaking we can reconstruct them only *ex post*. Not that they cannot be in principle reconstructed and intersubjectively controlled *ex ante* (which we have seen to be untenable in discussing the attempt of Di Paolo et al. 2000 to see a difference in principle between CS and TE in the inevitable opacity of the first). But if they are reconstructed *ex ante*, they coincide, in the last analysis, with TEs again!

I think this is the only possible way to distinguish *in principle* CS from TE. This distinction, when stated in this quite general manner, is extremely subtle, but it will perhaps become a bit clearer in the course of its application. The central point is that any simulation, even a computer one, involves a kind of *real* execution, one that is not merely psychological or conceptual. In a CS, the striking of certain keys is followed by a sequence of actual physical steps, i.e. the operations carried out by the hardware and the software, with the appearance of certain signs on the screen or in the print-out.

As in REs, this execution depends on us for its realisation only in the initial moment when we set off its 'mechanism'. The initial action is followed by a real process that occurs independently of a perceiving mind and ends, for example, with a pointer moving on a dial.

As I have already mentioned, in RE knowledge is obtained by the construction of an 'experimental machine' which extends the original operativity of our organic body and directly interacts with the surrounding reality. As we have seen, this does not hold true of CS because computers, as machines for processing symbols under the guide of a programme, are not tools that give us operational-experimental access to the world. Therefore, access to reality must be presupposed with respect to a computer's starting point, and then tested with respect to its point of arrival.

However, unlike TEs, CSs cannot dispense with any technical realisation. In this case, our ability to objectify our thought contents in real objects does not materialize in instruments that extend the operativity of our body and give information about reality by causal interaction with the surrounding sensible world, but in instruments that are embodied, or 'physically' realized, ways of representing, analysing, exploring and understanding this world.

According to what we have been saying, CSs have two distinct aspects: on the one hand, as TEs do, they anticipate an answer to some theoretical problems without resorting directly to experience. On the other hand, the similarities between the two should not obscure the distinction between the *hypothetical-counterfactual* context where the test of a hypothesis is planned, and the *real* context where this plan is actually carried out. A CS shares the first aspect with TEs, and the second with real ones. A plan for testing the relevant hypothesis must have been devised before CSs get under way (this holds also for "experimental simulations", such as that of a car prototype in a wind tunnel). But a CS involves an application of logics and mathematics to reality which is, in the last analysis, a technical-practical execution.

Finally, in the light of the results reached in our discussion, we may recognize certain elements of truth both in Frigg and Reiss's challenging article about CSs (Frigg and Reiss 2009) and in the claim that CS provides a new and different methodology for the physical sciences (Rohrlich 1991, p. 507. Cf. also the literature cited by Frigg and Reiss 2009, pp. 594–595). According to Roman Frigg and Julian Reiss, even though CSs constitute interesting and powerful new science, "[t]he philosophical problems that do come up in connection with simulations are not specific to simulations and most of them are variants of problems that have been discussed in other contexts before" (Frigg and Reiss 2009, p. 593).

This claim is untenable in a strict sense since the philosophical problems that usually arise are practically *always* "variants of problems that have been discussed in other contexts before." For example, the philosophical problems raised by the most recent atomic quantum theory are also variants of problems that have been discussed in much older historical contexts.

This not fully satisfactory formulation notwithstanding, there is some truth in Frigg and Reiss's claim, if it is understood in a very general sense, which we may call epistemological-transcendental: in this sense, for example, I have said above that there is no difference in principle between CS and TE because also a particular truth-claim resulting from a CS may be considered as scientific only under the general condition that it is in principle intersubjectively testable, that is, if we may, at least in principle, reconstruct, re-appropriate and evaluate it in the first person.

But over and above this general sense of the word, there are many particular, empirico-methodical senses in which we may speak of a greater methodological complexity of CSs in comparison with TEs. The mathematical-technical execution of a CS makes a very peculiar difference in principle, which carries with it many other differences in degree, and these may often play a very important part in particular scientific contexts.

It becomes clear that CSs, though qualitatively similar to TEs in the sense that they do not involve fundamentally different epistemic processes and share a common representational-instrumental character (which makes them belong to an important genus as far as the prediction and control of nature is concerned), provide a *relatively* new methodology for the empirical sciences. Thus, the execution or realisation involved in a CS, although distinct in meaning from the causal interactions occurring in REs, is the reason for the de facto greater methodological complexity of CSs in comparison with TEs. Accuracy, error analysis, calibration, and in general the management of uncertainty, though not peculiar to CSs, are de facto concepts that we encounter more frequently in discussing CSs than TEs.

There is a certain similarity between this claim and Winsberg's reply to Frigg and Reiss (2009) based on Galison's idea that CS is a hybrid of experiment and theory (cf. Winsberg 2009, p. 2013). However, even if it is true that the greater de facto complexity of CS in comparison to TE is connected with the fact that CSs involve an 'external' realisation, it is also true that this realisation is different in kind from that of a real world experiment: on the one hand, it is a real subsistent external 'thing', on the other it is something that only on our reading relates to something else and has a truth-value.

8 Conclusion

The main conclusions, at which we have arrived so far, may be briefly summed up as follows.

In the first part of this paper (Sects. 1–4), I have passed in critical review some important ways of comparing CSs and TEs: (1) The 'promissory eliminativism' concerning TE based on 'simulative model-based reasoning'; (2) Norton's argument view of TE and its extension to CS; (3) The view according to which CSs are opaque TEs; (4) The claim that CSs fall under the more general concept of real world experiment. No one of these accounts, though containing some important merits, has succeeded in distinguishing CS from TE and RE.

In the second part of my paper, I briefly outlined an account of CSs as compared with TEs that at least removes the difficulties of the four approaches which I have been discussing. Section 5 has tried to show some important aspects of the connection, made up of both unity and distinction, between TE and CS. From the perspective of the analysis of the empirico-methodological intensions of the concepts of TE, CS and RE, it is a hopeless task to find a particular methodological mark or difference that applies to only one of the three notions. For every particular characteristic of one of these notions, a corresponding characteristic in the two others is easily found. However, there is a difference in kind (an epistemological-transcendental difference) between TEs and CSs on the one hand and REs on the other (which, on reflection, is necessary for explaining their empirico-methodical similarities). From a pre-operational, epistemological or transcendental point of view, *what TEs and CSs have over and above real ones is only the fact that they exist in a purely hypothetical sphere. And* vice versa: *what REs have over and above TEs and CSs is only the fact that they are causal and empirical-experimental interactions between our bodies and the surrounding reality.*

As shown in Sect. 6, an aspect of this last difference reverberates in the relationship between TE and CS. The similarities between CS and TE should not obscure the distinction between the hypothetical context where the test of a hypothesis is planned, and the real context where this plan is actually carried out: to run a simulation is not only a representational, but also a real process. In TEs the subject operates concretely by using mental concepts in the first person; in contrast, REs and simulations involve an 'external' realisation, which is initially practical-experimental and tied to the experimenter's body, and thereafter develops independently of the subject. CSs share the first aspect with TEs, and the second with real ones. The fact that CSs involve an 'external' realisation explains the important difference in degree between TEs and CSs consisting in the usually greater methodological complexity of the latter.

Acknowledgments Thanks to Mike Stuart for constructive criticism and helpful comments that improved the paper, and the participants at the international conference "Model-Based Reasoning in Science and Technology: Models and Inferences: Logical, Epistemological, and Cognitive Issues" at Sestri Levante (Italy) in June 25–27, 2015, where an earlier version has been presented.

References

Arthur, R. (1999). On thought experiments as a priori science. *International Studies in the Philosophy of Science, 13*, 215–229.

Bedau, M. (1999). Can unrealistic computer models illuminate theoretical biology? In A. S. Wu (Ed.), *Proceedings of the 1999 Genetic and Evolutionary Computation Conference Workshop Program* (pp. 20–23). San Francisco: Morgan Kaufmann.

Beisbart, C. (2012). How can computer simulations produce new knowledge? *European journal for philosophy of science, 2*, 395–434.

Beisbart, C., & Norton, J. D. (2012). Why Monte Carlo simulations are inferences and not experiments. *International Studies in the Philosophy of Sciences, 26*, 403–422.

Brown, J. R. (1997). Proofs and pictures. *The British Journal for the Philosophy of Science, 48*, 161–180.

Buschlinger, W. (1993). *Denk-Kapriolen? Gedankenexperimente in Naturwissenschaften, Ethik und Philosophy of Mind*. Würzburg: Königshausen & Neumann.

Buzzoni, M. (2008). *Thought experiment in the natural sciences*. Würzburg: Königshausen & Neumann.

Buzzoni, M. (2010). The unconscious as the object of the human sciences. In E. Agazzi & G. Di Bernardo (Eds.), *Relations between natural sciences and human sciences* (pp. 227–246). Special Issue of Epistemologia. An Italian Journal for the Philosophy of Sciences.

Buzzoni, M. (2011). On mathematical thought experiments. *Epistemologia. An Italian Journal for Philosophy of Science, 34*, 5–32.

Buzzoni, M. (2013). Thought experiments from a Kantian point of view. In J. R. Brown, M. Frappier, & L. Meynell (Eds.), *Thought experiments in philosophy, science, and the arts* (pp. 90–106). Routledge: London.

Chandrasekharan, S., Nersessian, N. J., & Subramanian, V. V. (2013). Computational modeling. Is this the end of thought experiments in science? In M. Frappier, L. Meynell, & J. R. Brown (Eds.), *Thought experiments in philosophy, science, and the arts* (pp. 239–260). London: Routledge.

Charbonneau, M. (2010). Extended thing knowledge. *Spontaneous Generations, 4*, 116–128.

Clark, A., & Chalmers, D. J. (1998). The extended mind. *Analysis, 541*, 7–19.

Cooper, R. (2005). Thought experiments. *Metaphilosophy, 3*, 328–347.

Dietrich, M. R. (1996). Monte Carlo experiments and the defense of diffusion models in molecular population genetics. *Biology and Philosophy, 11*, 339—356.

Di Paolo, E. A., Noble, J., & Bullock, S. (2000). Simulation models as opaque thought experiments. In M. Bedau (Ed.), *Artificial life VII. Proceedings of the Seventh International Conference on Artificial Life* (pp. 497–506). Cambridge MA, MIT Press.

El Skaf, R., & Imbert, C. (2013). Unfolding in the empirical sciences: Experiments, thought experiments and computer simulations. *Synthese, 190*, 3451–3474.

Frigg, R., & Reiss, J. (2009). The philosophy of simulation: Hot new issues or same old stew? *Synthese, 169*, 593–613.

Galison, P. (1987). *How experiments end*. Chicago: University of Chicago Press.

Galison, P. (1996). Computer simulations and the trading zone. In P. Galison & D. Stump (Eds.), *The disunity of science: Boundaries, contexts, and power*. Stanford: Stanford University Press.

Gendler, T. (2004). Thought experiments rethought—and reperceived. *Philosophy of Science, 71*, 1152–1164.

Gooding, D. (1994). Imaginary science. *The British Journal for the Philosophy of Science, 45*, 1029–1045.

Guala, F. (2002). Models, simulations, and experiments. In L. Magnani & N. Nersessian (Eds.), *Model-based reasoning: Science, technology, values* (pp. 59–74). Kluwer: New York.

Guala, F. (2005). *The Methodology of experimental economics*. Cambridge: Cambridge University Press.

Hacking, I. (1993). Do thought experiments have a life of their own? *Proceedings of the Philosophy of Science Association, 2*, 302–308.

Häggqvist, S. (1996). *Thought experiments in philosophy*. Stockholm: Almqvist & Wiksell International.

Hartmann, S. (1996). The world as a process: Simulations in the natural and social sciences. In R. Hegselmann, U. Müller, & K. Troitzsch (Eds.), *Modelling and simulation in the social sciences from the philosophy of science point of view* (pp. 77–100). Kluwer: Dordrecht.

Hughes, R. I. G. (1999)[2012]. The Ising model, computer simulation, and universal physics. In M. S. Morgan & M. Morrison (Eds.), *Models as mediators* (pp. 97–146). Cambridge: Cambridge University Press. Quotations are from R. I. G. Hughes, *The theoretical practices of physics: philosophical essays* (pp. 164–209). Oxford: Oxford University Press, 2010.

Humphreys, P. (1991). Computer simulations. In A. Fine, M. Forbes, & L. Wessels (Eds.), *PSA 1990* (Vol. 2, pp. 497–506), East Lansing, MI: Philosophy of Science Association.

Keller, E. F. (2003). Models, simulation, and computer experiments. In H. Radder (Ed.), *The philosophy of scientific experimentation* (pp. 198–215). Pittsburgh: University of Pittsburgh Press.

Küppers, G., & Lenhard, J. (2005a). Computersimulationen: Modellierungen 2. Ordnung. *Journal for General Philosophy of Science, 36*, 305–329.

Küppers, G., & Lenhard, J. (2005b). Validation of simulation: Patterns in the social and natural sciences. *Journal of Artificial Societies and Social Simulation, 8*(4). http://jasss.soc.surrey.ac.uk/8/4/3.html

Lenhard, J. (2007). Computer simulation: The cooperation between experimenting and modeling. *Philosophy of Science, 74*, 176–194.

Lenhard, J. (forthcoming). Thought experiments and simulation experiments: Exploring hypothetical worlds. In J. R. Brown, Y. Fehige, & M. Stuart (Eds.), *The routledge companion to thought experiments*. London: Routledge.

Lenhard, J., & Winsberg, E. (2010). Holism, entrenchment, and the future of climate model pluralism. *Studies in History and Philosophy of Science Part B, 41*, 253–262.

Levins, R. (1966). The strategy of model building in population biology. *American Scientist, 54*, 421–431.

Mach, E. (1883)[1933]. *Die Mechanik in ihrer Entwickelung. Historisch-kritisch dargestellt*, Leipzig, Brockhaus (9th ed. Leipzig 1933); *The science of mechanics. A critical and historical account of its development*, transl. J. J. McCormack, Chicago and London: Kegan Paul and Open Court, 1919 (4th ed.).

Mach, E. (1905a). Über Gedankenexperimente, in *Erkenntnis und Irrtum*, Leipzig, Barth, 1905 (5th ed. 1926, pp. 183–200); On thought experiments. In *Knowledge and error* (pp. 134–147), transl. T. J. McCormack, Dordrecht (Holland) and Boston (U.S.A.): Reidel, 1976.

Mach, E. (1905b). Das physische Experiment und dessen Leitmotive, in *Erkenntnis und Irrtum*, Leipzig, Barth (5th ed. 1926, pp. 201–219); Physical experiment and its leading features. In E- Mach, *Knowledge and error* (pp. 148–161), transl. T. J. McCormack, Dordrecht (Holland) and Boston (U.S.A.): Reidel, 1976.

Mach, E. (1905c). Die Hypothese, in *Erkenntnis und Irrtum*, Barth, Leipzig, 1905, (5th ed. 1926, pp. 232–250); Hypothesis. In *Knowledge and Error* (pp. 171–184), transl. T. J. McCormack, Dordrecht (Holland) and Boston (U.S.A.): Reidel, 1976.

Miščević, N. (1992). Mental models and thought experiments. *International Studies in the Philosophy of Science, 6*, 215–226.

Miščević, N. (2007). Modelling intuitions and thought experiments. *Croatian Journal of Philosophy, 7*, 181–214.

Misselhorn, C. (2005). *Wirkliche Möglichkeiten—Mögliche Wirklichkeiten,* Paderborn: Mentis.

Morgan, M. (2002). Model experiments and models in experiments. In L. Magnani & N. Nersessian (Eds.), *Model-based reasoning: Science, technology, values* (pp. 41–58). New York: Kluwer.

Morrison, M. (2009). Models, measurement and computer simulation: The changing face of experimentation. *Philosophical Studies, 143*, 33–57.

Muldon, R. (2007). Robust simulations. *Philosophy of Science, 74*, 873–883.

Nersessian, N. (1992). How do scientists think? Capturing the dynamics of conceptual change in science. In R. Giere (Ed.), *Cognitive models of science* (pp. 3–44). Minneapolis: University of Minnesota Press.

Nersessian, N. (1993). In the theoretician's laboratory: Thought experimenting as mental modeling. In D. Hull, M. Forbes, & K. Okruhlik (Eds.), *PSA 1992* (Vol. 2, pp. 291–301). East Lansing, MI: Philosophy of Science Association.

Nersessian, N. (2006). Model-based reasoning in distributed cognitive systems. *Philosophy of Science, 73*, 699–709.

Norton, J. D. (1991). Thought experiments in Einstein's work. In T. Horowitz & G. J. Massey (Eds.), *Thought experiments in science and philosophy* (pp. 129–144). Savage, MD: Rowman and Littlefield.

Norton, J. D. (1996). Are thought experiments just what you thought? *Canadian Journal of Philosophy, 26*, 333–366.

Norton, J. (2004a). Why thought experiments do not transcend empiricism. In C. Hitchcock (Ed.), *Contemporary debates in the philosophy of science* (pp. 44–66). Oxford: Blackwell.

Norton J. (2004b). On thought experiments: Is there more to the argument? *Philosophy of Science, 71*, 1139–1151 (In *Proceedings of the Biennial Meeting of the Philosophy of Science Association, Milwaukee, Wisconsin*).

Norton, S., & Suppe, F. (2001). Why atmospheric modeling is good science. In C.A. Miller & P. Edwards (Eds.), *Changing the atmosphere: Expert knowledge and environmental governance* (pp. 67–106). Cambridge, MA: MIT Press.

Palmieri, P. (2003). Mental models in Galileo's early mathematization of nature. *Studies in History and Philosophy of Science, 34*, 229–264.

Parker, W. S. (2009). Does matter really matter? Computer simulations, experiments, and materiality. *Synthese, 169*, 483–496.

Parker, W. S. (2010). An instrument for what? Digital computers, simulation and scientific practice. *Spontaneous Generations: A Journal for the History and Philosophy of Science, 4*, 39–44.

Popper, K. (1957)[1961]. *The poverty of Historicism*. London: Routledge (quotations are from the 3th ed., 1961).

Rohrlich, F. (1991). Computer simulation in the physical sciences. In *PSA 1990* (Vol. 2, pp. 507–518). East Lansing, MI: The Philosophy of Science Association.

Stäudner, F. (1998). *Virtuelle Erfahrung. Eine Untersuchung über den Erkenntniswert von Gedankenexperimenten und Computersimulationen in den Naturwissenschaften*. Erlangen: Diss. Friedrich-Schiller-Universität.

Stöckler, M. (2000). On modeling and simulations as instruments for the study of complex systems. In: M. Carrier, G. J. Massey, & L. Ruetsche (Eds.), *Science at the century's end: Philosophical questions on the progress and limits of science* (pp. 355–373). Pittsburgh, PA: University of Pittsburgh Press.

Stuart, M. (2015). Thought experiments in science. PhD Thesis, Institute for the History and Philosophy of Science and Technology, University of Toronto.

Velasco, M. (2002). The use of computational simulations in experimentation. *Theoria (new series), 17*, 317–331.

Weisberg, M. (2006). Forty years of 'The Strategy': Levins on model building and idealization. *Biology and Philosophy, 21*, 623–645.

Wimsatt, W. C. (1981). Robustness, reliability, and overdetermination. In M. B. Brewer & B. E. Collins (Eds.), *Scientific inquiry and the social sciences* (pp. 124–163). San Francisco: Jossey-Bass.

Winsberg, E. (2003)[2010]. Simulated experiments: Methodology for a virtual world. *Philosophy of Science*, *70*, 105–125, rist. In E. Winsberg, *Science in the age of computer simulations*. Chicago: University of Chicago Press, 2010.

Winsberg, E. (2009). Computer simulation and the philosophy of science. *Philosophy Compass, 4*(5), 835–845.

Winsberg, E. (2013). Computer simulations in science. In E. N. Zalta (Ed.), *Stanford encyclopedia of philosophy*. http://plato.stanford.edu/archives/sum2015/entries/simulations-science/

Approaches to Scientific Modeling, and the (Non)Issue of Representation: A Case Study in Multi-model Research on Thigmotaxis and Group Thermoregulation

Guilherme Sanches de Oliveira

Abstract Recent contributions to the philosophical literature on scientific modeling have tended to follow one of two approaches, on the one hand addressing conceptual, metaphysical and epistemological questions about models, or, on the other hand, emphasizing the cognitive aspects of modeling and accordingly focusing on model-based reasoning. In this paper I explore the relationship between these two approaches through a case study of model-based research on the behavior of infant rats, particularly thigmotaxis (movement based on tactile sensation) and temperature regulation in groups. A common assumption in the philosophical literature is that models represent the target phenomena they simulate. In the modeling project under investigation, however, this assumption was not part of the model-based reasoning process, arising only in a theoretical article as, I suggest, a post hoc rhetorical device. I argue that the otherwise nonexistent concern with the model-target relationship as being representational results from a kind of objectification often at play in philosophical analysis, one that can be avoided if an alternative form of objectification is adopted instead.

Keywords Scientific modeling · Model-based reasoning · Representation · Thigmotaxis · Thermoregulation · Robotics · Agent-based modeling

1 Introduction: Two Approaches to Scientific Modeling

Scientists routinely use models such as computer simulations, concrete replicas and mathematical equations to study target systems and phenomena as complex and varied as inflation and unemployment rates, DNA structures, deforestation and reforestation processes, and predator-prey relations. Models do not only supplement

G.S. de Oliveira (✉)
University of Cincinnati, 2700 Campus Way 206 McMicken Hall,
Cincinnati, OH 45221, USA
e-mail: sanchege@mail.uc.edu

L. Magnani and C. Casadio (eds.), *Model-Based Reasoning in Science and Technology*, Studies in Applied Philosophy, Epistemology and Rational Ethics 27, DOI 10.1007/978-3-319-38983-7_5

traditional forms of experimentation, but sometimes even replace them altogether when the target phenomena are too poorly understood, or when manipulating and intervening upon the real-world objects of interest is impractical and potentially dangerous or unethical. In these and other cases philosophers have suggested that models provide alternative grounds for scientific research, enabling a kind of "surrogate reasoning" (Suarez 2004) and "indirect theoretical investigation of a real-world phenomenon" (Weisberg 2007a, p. 209).

In the burgeoning literature on scientific modeling, two main themes or modes of investigation have dominated the discussion in the past couple of decades. In the first approach, authors tend to focus on traditional philosophical issues arising from scientific modeling. Questions about the nature of models, about the relationship between models and target phenomena, and about how we can learn through modeling, among others, occupy center stage in this approach, and are addressed more or less formally, depending on the context and objectives of the account. In the second approach, authors emphasize the cognitive aspects of modeling, and accordingly focus on model-based *reasoning* as their object of study. Drawing from the history of science, from ethnographic observation, and from empirical research on human problem-solving more generally, work within this second approach prioritizes considerations about how scientists build and utilize models to generate novel hypotheses, explanations, and predictions. These two approaches have been compared as emphasizing different aspects of the phenomenon, with the second, more "psychologistic" approach falling *within* the first, broader philosophical one (Godfrey-Smith 2006a). In this first section I will review some of the central features of each of the two approaches and highlight some of their particularities. Section 2 will be devoted to a brief case study of a model-based research project on movement and group temperature regulation in infant rats. This case study will then serve as illustration for my interpretation, in Sect. 3, of the tension between the two philosophical approaches to scientific modeling. As I will suggest, some of the issues that arise in this literature—particularly questions about *representation*—are not actual concerns for the modelers and do not correspond to identifiable aspects of their modeling practice. I argue that these concerns arise from a kind of *objectification* or *reification* at play in philosophical analysis, and can be circumvented if an alternative form of objectification is adopted instead. But let us first consider the two approaches in a little more detail.

In an influential paper that is representative of the first approach, Godfrey-Smith (2006a) talks about the "strategy of model-based science" as one among various other strategies scientists have at their disposal. Using models in science, he says, is "a distinctive style of theoretical work, which yields particular kinds of representations, explanations, and patterns of change" and which can be used to complement or substitute other more traditional strategies, such that "a single scientific problem can be approached both with and without models functioning as the currency of theorizing and explanation" (Godfrey-Smith 2006a, p. 739). Accounts of model-based science in line with this "tradition" have tended to focus on the semantics, ontology and metaphysics of models (Frigg and Hartmann 2012). As such, authors have dealt with, among other issues, questions about: the nature of models in contrast with alternative research strategies and scientific products (Godfrey-Smith 2006a, b; Weisberg 2007a, 2013); the different types of idealizations and abstractions and their role in model-based science (Godfrey-Smith 2009a; Weisberg 2006, 2007b, 2013); the nature of model-based representation and of the relationship between model and target as representational (Frigg 2006; Suarez 2003, 2004, 2010; Weisberg 2013); and the parallel between scientific modeling and representations in various non-scientific practices, e.g., in the visual arts and in fiction (Godfrey-Smith 2006b, 2009b; Frigg 2010a, b, c; Magnani 2012; Woods 2014). Peculiar to this approach to conceptual issues arising from scientific modeling is the attempt to articulate *formally* aspects of model-based science. Weisberg's (2013) "weighted feature-matching" mathematical account of model-target relations is one such example, and can itself be considered a *mathematical model* of model-based science, containing its own assumptions, scope, idealizations, abstractions, and other characteristic features of scientific models.

A distinct approach to modeling can be found in a line of investigation that Godfrey-Smith has characterized, perhaps almost pejoratively, as "psychologistic" (2006a). In this second approach, the focus of investigation is not on semantics, ontology and epistemology as much as on the cognitive processes involved in constructing and using models in problem-solving, in theorizing and in the generation of explanations, predictions, novel hypotheses, and so on. Work within this approach is varied, utilizing, for example, historical records of the work conducted by scientists such as Michael Faraday and James Clerk Maxwell (Nersessian 1992, 1999, 2002, 2008; Tweeney 1985, 1989, 2014), as well as ethnographic observation of current model-based research projects and laboratories (Nersessian and Patton 2009; Nersessian 2009; MacLeod and Nersessian 2013; Osbeck et al. 2013). While in the first approach "model-based science" is taken to be one among various strategies or styles of work *within science*, what is distinctive about this second approach is the explicit connection it draws *between* scientific reasoning and similar kinds of reasoning *outside* of science. It is with this in mind that Nersessian articulates the "continuum hypothesis"—that is, the fundamental assumption that cognitive practices in science are only "very sophisticated and refined outgrowths of ordinary reasoning and representational processes" (Nersessian 1992, p. 5), such that there is continuity between ordinary and scientific cognitive abilities and strategies: "the cognitive practices of scientists are extensions of the kinds of practices humans employ in coping with their physical and social environments and in problem-solving of a more ordinary kind" (Nersessian 2002, p. 135). This perspective on the continuity between ordinary and scientific reasoning has implications that connect the philosophy of science with various other disciplines: on the one hand, authors within the second approach tend to draw from contemporary experimental research in cognitive science for their philosophical investigations (see, e.g., Carruthers et al. 2002; Thagard 2012); and, in the opposite direction, the "model-based reasoning" approach also tends to have a more practical orientation, offering insights that are applicable to fields such as STEM education (see, e.g., Raghavan and Glaser 1994; Stephens et al. 1999; Gilbert 2005). The interdisciplinary flavor of the "model-based reasoning" approach is also made evident in the resemblance, and in some cases actual thematic overlap, with empirical and theoretical work within the "psychology of science" and the "cognitive science of science" (see, e.g., Gholson et al. 1989; Feist 2006; Proctor and Capaldi 2012; Feist and Gorman 2013).

As indicated earlier, these two approaches have been compared and differentiated in terms of having distinct emphases, with cognitive considerations corresponding only to a "component" of the broader philosophical issues (Godfrey-Smith 2006a). But I think Godfrey-Smith misses the real tension between the two approaches. It is certainly the case that the two are not mutually exclusive. That is, despite the many differences between the first, more "conceptual" approach to *modeling,* and the second, more "cognitive" approach to *model-based reasoning*, a given author can quite naturally be interested in and contribute to research on both "traditions." This is the case with Ronald Giere. Having published influential work articulating a decidedly cognitive approach to science (Giere 1988, 1992), Giere has also provided applications of this approach to broader issues in the philosophy of science (see Giere 1999, 2002, 2004, 2005), bringing his framework closer to the first approach while still exerting influence on both, as Godfrey-Smith (2006a) explicitly recognizes. Still, I think it is a mistake to frame the cognitive approach as inferior to or narrower than the "more philosophical" one. After examining in Sect. 2 the foundations, methods and results of a specific modeling research project, I will, in Sect. 3 offer an alternative interpretation of how the two approaches relate to one another and to the phenomenon of scientific modeling.

2 Case Study: Model-Based Investigation of Thigmotaxis and Group Thermoregulation in Rat Pups

The present case study focuses on a model-based research project on the behavior of infant rats ("rat pups"), particularly thigmotaxis—i.e., motion based on tactile perception—and group thermoregulation—i.e., control and maintenance of body temperature through contact and coordination with other individuals. The project started in the late 1990s and continues today. It is based at the University of California, Davis, and is led by Jeffrey Schank from the Department of Psychology, in collaboration with colleagues from the same department as well as from the Department of Mechanical and Aeronautical Engineering. Besides being characterized by its *interdisciplinarity*, the project can also be described as *multi-model*, *multimodal*, and *incremental*. That is because the project has involved, respectively, the use of multiple "incompatible" models, built with distinct tools and exploiting different modeling modalities, and which are further developed and evolved over multiple iterations. In order to highlight these aspects of the project, the goal of this section is to outline the research context, methodology and the results obtained. Most of the observations come directly from technical papers published in scientific journals, except for the most recent article, published in an edited collection aimed at an interdisciplinary audience of a more conceptual and philosophical bent. As I will suggest in Sect. 3, the orientation of the latter publication was significant for the content of the article itself, in a way that is particularly relevant for our purposes.

The driving question of the research project is centered on the huddling behavior of rat pups. As had already been observed before the beginning of the project, rat pups tend to group together, nestling closely so as to maintain physical contact with other individuals in the litter. This behavior is understood to be correlated with temperature regulation and energy conservation: individual rat pups can adjust their temperature by changing their location in the huddle, moving toward the center of the huddle if feeling cold or outward if hot; as a result, they increase their energy efficiency by enabling the allocation of more resources to developmental processes rather than to endogenous thermoregulation. An early paper in the project (Schank and Alberts 1997), published when Schank was a post-doctoral researcher at Indiana University, reviews this background knowledge and poses the question that would guide the project for years: can the emergence of a group behavior such as huddling in rat pup litters be explained simply in terms of the behavior of individual pups? Schank and Alberts point out that the perceptual modalities of rat pups are seriously restricted: they are born blind, and with very limited capacity to process olfactory and auditory cues, relying almost entirely on tactile sensation to guide their behavior. This made the researchers' job both easier and more challenging. On the one hand, with fewer possibilities of sensory input and fewer variables to manipulate, it was easier to simulate individuals and to test the hypothesis that individual behavior suffices to generate the observed group phenomenon of huddling. On the other hand, however, the serious challenge was that of explaining how a complex social behavior—one that continues throughout the rats' lifespan, the authors point out—could come about solely from simple individual-level factors, namely the preferences and very restricted abilities of individual pups. The researchers summarized their problem and hypothesis thus: "simple sensorimotor rules for activity and preferences for objects in an infant rat's environment can largely explain the complex patterns of aggregation, cohesion, and dynamics of contact observed among pups," and moreover, "autonomous individuals interacting locally in accordance with simple rules can generate complex and group regulatory behaviors without assuming group-level regulatory mechanisms" (Schank and Alberts 1997, p. 12). The seemingly *reductionist* rationale that this formulation could suggest was soon corrected by Schank in a theoretical essay (Schank 2001). Reductionist approaches, Schank says, attempt to "explain individual behavior by using one or a few levels of analysis (e.g., genetic)"; but the project, he clarifies, focuses instead on "individuals as complex systems—consisting of and embedded in other complex systems," where the *interactions* between components at various levels and time scales are integral to the behavioral outcome (Schank 2001, p. 33). In the remainder of this section I will review two of the modeling strategies the researchers took in the project and summarize the results they reported, leaving the discussion for Sect. 3. It is important to note, however, that both types of models were preceded by observations of the behavior of actual rat pups. At various stages of the research project, and particularly before and during the development of each of the modeling frameworks, Schank and colleagues conducted experiments by placing real rats in an arena with controlled temperature and incline, and by having their individual behavior, social behavior, and individual–environment interactions video-recorded, coded and analyzed; the data was then used both as input for constructing the models and as contrast case for evaluating the models' performance (Schank and Alberts 1997; Schank et al. 2004).

2.1 Computer Simulations

The simulation component of this project followed the agent-based paradigm presented by Schank and Alberts (1997). Their hypothesis was that the behavior of individual rat pups corresponded to the interaction of two parameters, namely "the probability of a pup moving or reorienting itself in the arena" and "preferences for objects in a pup's local environment" (Schank and Alberts 1997, p. 17). Accordingly, they built a simulated rectangular arena with discrete cells to be occupied by individual virtual agents, who could move to an adjacent cell according to bodily orientation, movement preferences and activity state. Bodily orientation determined which of the adjacent cells were accessible: at any given time unit, a simulated rat pup had eight immediate cells surrounding the one it was currently in (namely the central cell of a 3×3 grid); still, movement at each time unit was only possible toward the three cells "in front" of the individual, namely the cell immediately on top of the simulated agent, and the two cells to the left and right of the top one, which could be accessed diagonally. Varying values were assigned to movement preferences according to factors such as whether the adjacent cell was next to a wall, and whether it was occupied by another pup or next to a cell occupied by another pup. And movement was dependent on activity state, that is, if a pup is active or inactive. Activity state was calculated with conditional probabilities using a binomial model; this assumed that activity state at given time t depends only on the individual's prior activity states, regardless of the activity of other agents. Following these principles (whose mathematical properties are described in detail in Schank and Alberts 1997), the investigators ran Monte Carlo simulations for the duration of 120 time units, at each time unit generating a probability matrix for each of the 8 simulated agents in the virtual arena. Depending on activity state and according to preference values, at each time unit the agents would attempt to move to one of the three accessible adjacent cells; in case of failed movement (because, e.g., cells were occupied or walled), the simulated rat pups would remain in the same cell but reorient clockwise so as to change the possible movement-space to a different set of three adjacent cells (see Fig. 1).

The first paper in the project reported overall positive results. Given the data obtained through observation of real rat pups, simulated agents within a specific range of probability and preference values exhibited behavior that was significantly similar to that of real rats, resulting in "a very good match between simulation and

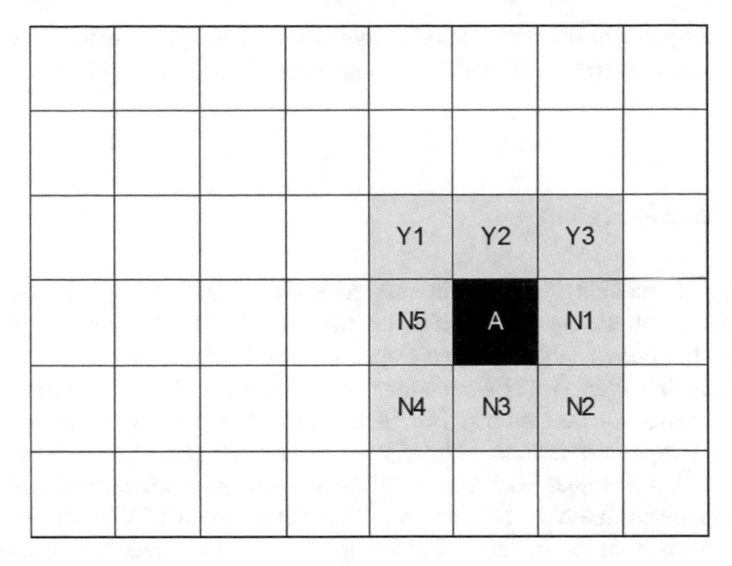

Fig. 1 A depiction of the agent-based simulation array, with the *black-background* cell indicating where the agent (*A*) is located at a given time, and the *gray-background* cells indicating the cells adjacent to the agent at that time. The agent can choose to move to one of the accessible cells (Y1, Y2, Y3) or choose instead to reorient itself and thereby shift the set of accessible cells (e.g. by turning clockwise, N1 and N2 would become accessible, while Y1 and Y2 would no longer be accessible). In some of the simulations, the presence of another agent in one of the cells adjacent to A could delimit A's movement possibilities (i.e., if the other agent was located in an otherwise accessible cell, namely Y1, Y2, or Y3). Based on Fig. 5 in Schank and Alberts (1997)

data" (Schank and Alberts 1997, p. 23). Particularly on the frequency of contact with the walls of the arena, the authors claim that the values were "almost identical to the best fit found" and "well within the range of error for these simulations" (ibid.). The comparison between the behavior of simulated agents and that of infant rats, particularly of 7-day-old rats, supported the hypothesis that the group behavior of huddling could emerge from individual behavior based on sensorimotor rules, particularly through the individuals' use of spatial-tactile stimuli. And yet, the researchers indicated that the fit did not apply to the comparison with 12-day-old rats. In a matter of a few days, real infant rats become less independent and more sensitive to the activity levels of other individuals, modulating their behavior in accordance with what others are doing. This was one of various other limitations of the first model, and as the project progressed over the years, several changes were introduced. For example, in this first agent-based model, simulated pups would only move to empty cells, and they displayed preference for proximity to other pups but not for proximity to walls. As Schank et al. (2014) discuss, only the second assumption, of proximity to others, corresponded to behavior observed in real pups; in reality, they point out, rat pups do not only move to empty spaces (they actually do climb on one another) and they are not indifferent to walls either, more likely opting for resting around corners than in open space. Subsequent developments in the model paradigm addressed some of these limitations by, for example, changing preference parameters and introducing patterns of random movement (Schank 2008).

2.2 Robotic Agents

The second modeling strategy in this research project consists in the use of autonomous robotic agents to simulate the behavior of rat pups. This line of modeling has unavoidably involved elements from the previous one as well, in that researchers first built virtual simulations of the robots and only then proceeded to construct actual concrete robots. The mathematical foundations as well as issues in design and implementation that I briefly review here are described in detail by Bish et al. (2007). The researchers used MATLAB™ for the mathematical calculations and the graphical interface of Simulink™ (an extension of MATLAB) to evaluate the dynamics from the mathematical input.[1] In later experiments they used other simulation softwares, such as the open-source platform Breve (May et al. 2011). As in the previous cases, this modeling project began with observations of actual rat pups. Individual rats were placed in walled rectangular arenas with controlled temperature, light and incline, and had their movements and body position video-recorded and coded. The coordinates of nose tip and tail were recorded and later used for comparison with model outputs.

The design specifications of both simulated robots and concrete robots mirrored some characteristics of rat pups 7–10 days of age, including size ratio, and motor and perceptual limitations. Simulated and concrete robots followed the body proportions of actual rat pups, namely 3:1 length to width ratio. They also displayed similarly limited mobility: real rat pups propel their body forward by pushing with their hind legs, which was modeled in the robots' use of two propelling back wheels and a passive front wheel for support. Finally, the rat pups' aforementioned perceptual constraints—namely their reliance solely on tactile stimuli—were simulated by endowing robots with touch sensors spread around their perimeter. As the researchers emphasize, the robots' head was doubly biased, first because they contained more sensors, and second because the input from those sensors received more priority weight in processing, thereby mirroring the relative importance of sensory stimulation around a rat pup's snout (Bish et al. 2007, pp. 984, 997). Besides these specifications, the design and implementation of the simulated robotic pups involved the use of three complementary subsystems. An *environment subsystem* modeled the robot's contact with arena walls, calculating the sensory and impact inputs to be sent to the other subsystems. The wall impact input from the environment subsystem was fed into the *dynamics subsystem*, which also used wheel force values to calculate the position, velocity and orientation of the robot at

[1]MATLAB and Simulink are trademarks of the Mathworks Company, Needham, MA.

any given time unit. And sensor inputs from the environment subsystem were sent to the *behavior subsystem*, which implemented behavioral algorithms and converted their outputs into wheel movement force, whose output was in turn calculated by the dynamics subsystem (Bish et al. 2007, p. 991–993). These general design and architectural specifications were utilized both in simulated robots and in actual concrete robots, with the further methodological assumption that differences in coding had no significant impact in the simulated and concrete results, i.e., that there were "no significant numerical differences in the calculations performed by the Java code executed by the robot and the MATLAB code executed by a PC" (Bish et al. 2007, p. 993).

The results from experimentation with this first generation of robotic models were mixed. The overall strategy involved comparing the observed behavior of rat pups, as captured by nose-tail coordinates, with the corresponding outputs for the trajectory of simulated and concrete robotic agents. Concerning the fit between model and the data obtained from real rat pups, researchers noted that individual robots "behaved qualitatively in some respects like a single rat pup moving in an arena" (Schank et al. 2014, p. 157), but they added that important limitations would have to be addressed in subsequent generations. The four robots used in the first experiments had been designed and built to be physically identical to one another (other than having distinct wiring color to enable telling them apart) and they were given the same algorithms for processing; still, the robots displayed incompatible behavior, tracing very distinct trajectories from one another. This divergence, the researchers hypothesized, was due to unintended differences between the robots: "slight differences in motors, small body shape construction differences, gradual deterioration of robot parts (e.g. wheels, supports, sensor mounts), and subsequent replacement of parts" as well as, possibly, "imperfections in the floor, coupled with dust or debris on the robot wheels" which could have affected performance. Whatever the reasons, the divergent trajectories made it so that the *simulated* robots accurately corresponded to the behavior of only one of the four real robots used.

But beyond the unintended differences between robots, further developments in subsequent generations concerned changes in design which were associated with interesting insights. For example, although rat pups have flexible joints, the first robots had rigid bodies; as a result, when encountering a corner, while real rat pups are able simply to flex to turn to one side or another, robots had to back up in order to change direction. This mismatch, necessitated by the design constraints of first-generation robots, was indicated early on as a limitation that could potentially be overcome (Schank et al. 2004; May et al. 2006; Bish et al. 2007). A natural way to address this limitation would be to make the robotic models more complex by introducing body flexibility. Before this was pursued, however, the researchers explored an alternative route, namely by experimenting with randomized robot behavior. Instead of following the standard procedure described above, the robots had their sensors turned off and would at each time unit perform one random body movement out of a list of ten possibilities. This strategy yielded surprising results, with the researchers noting: "It is striking how well the trajectory plots of random robots visually matched the plots for both 7- and 10-day-old pups" (May et al.

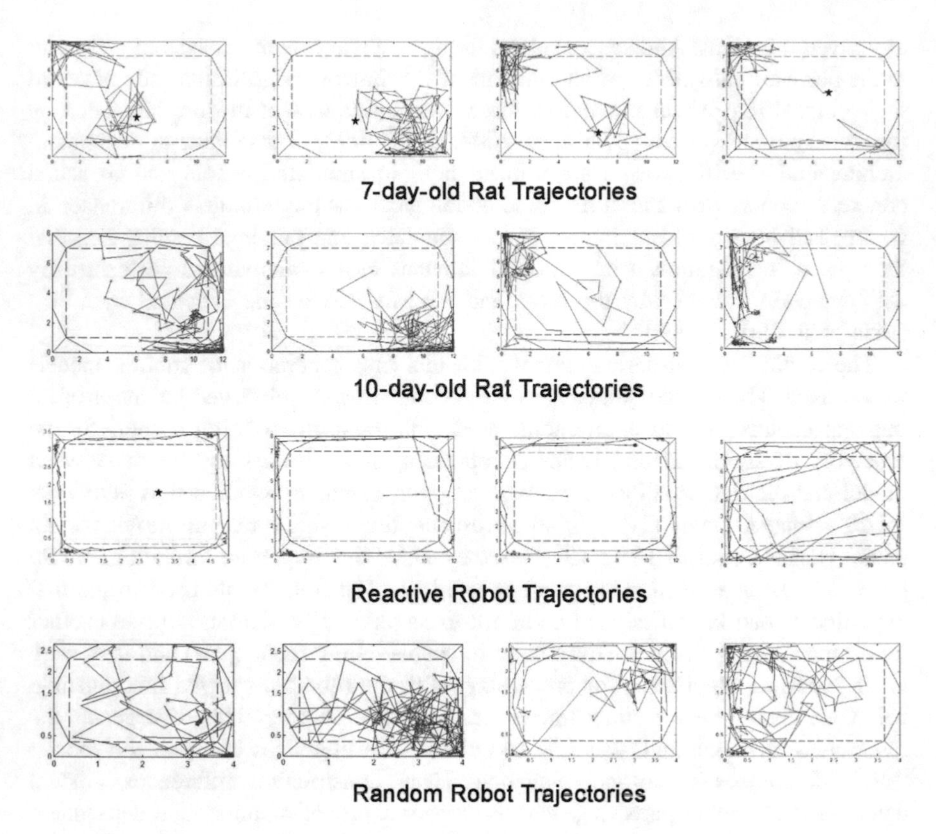

Fig. 2 Trajectory plots based on head coordinates obtained from trials with 7-day-old rat pups, 10-day-old rat pups, robots with reactive architecture, and robots with randomized behavior. Reproduced from May et al. (2006) with permission from John Wiley and Sons

2006, p. 59; see Fig. 2). This applied not only to the individualized trials already reviewed, but was the case even for novel "social" experiments in which experimenters placed eight robots in the arena in order to observe aggregation patterns in comparison with litters of eight rat pups (see Fig. 3). May et al. (2006) point out, for example, that in the group experiments robots aggregated significantly more than 7-day-old pups but less than 10-day-old pups; this placed the robots somewhere between the two age stages, with the latter corresponding to the stage when developmental changes not only improve the rat pups' motor and perceptual abilities, but also make them responsive to the behavior of surrounding individuals. After successfully experimenting with randomized robotic behavior, the researchers built subsequent generations of robots with increasing degrees of flexibility, first with two-segmented bodies, and then with three-segmented bodies. Although flexible body morphology in robots resulted in more similar behavior to that of real rat pups, interestingly enough the increase was not gradual: as expected, the most

Fig. 3 Aggregation patterns observed in trials with real rat pups (*a*, *c*, *e*) and robotic agents (*b*, *d*, *f*). Reproduced from Schank et al. (2014) with permission from the MIT Press

realistic robots with three-segmented bodies presented the best fit; yet, two-segmented robots performed worse even than rigid-bodied robots (May and Schank 2009; May et al. 2011; Schank et al. 2014).

3 Representation and Objectification

This paper started with a review, in Sect. 1, of two popular trends in the recent philosophy of science literature on models: the first deals with semantic, metaphysical and epistemological questions, while the second focuses instead on cognitive issues in scientific modeling. Godfrey-Smith (2006a) has proposed that the difference between the two approaches is only a matter of distinct emphasis, with the "narrower" cognitive approach consisting of a "component" of the broader philosophical approach. This suggests that the choice of one over the other will be dictated by how fine-grained the analysis is, i.e., if it is general and philosophical or if specific to the select cognitive aspects of the phenomenon. It is my contention that this comparison between the two approaches is mistaken, and in this section I use Schank and colleagues' model-based research to motivate a different interpretation. Although both are, in broad terms, approaches to the same general phenomenon of scientific modeling, I propose that the two approaches *objectify* the phenomenon differently, and that this difference is to a great extent responsible for contemporary philosophical perplexities about *representation*. Lastly, I conclude with a cautionary note on how philosophers of science who are serious about understanding scientific practice are to interpret what scientists say.

3.1 Much Ado About Representation

Godfrey-Smith has described model-based science as "fundamentally a strategy of indirect *representation* of the world" (Godfrey-Smith 2006a, p. 730, emphasis added) and an "indirect approach to *representing* complex or unknown processes in the real world" (Godfrey-Smith 2006b, p. 7, emphasis added). If these statements seem unproblematic, that is because the underlying assumption is indeed very familiar, perhaps even one of the orthodoxies in contemporary philosophy of science. The idea that a model is a *representation* of some target phenomenon has been the unquestioned starting point of much of the recent philosophical work on modeling. Much attention has been devoted, for example, to the question whether there is a "special problem about scientific representation" (Callender and Cohen 2006) and whether theories of representation from other areas of human activity can be applied to science as well (Suarez 2010; Toon 2010; Chakravartty 2010a, b). Philosophers have also discussed several competing accounts of the nature of representation—that is, what makes something a representation of something else. One tendency has been to move away from accounts that treat representation as a mind-independent feature of model and target, and toward the inclusion of an agential component, namely the scientists' role in determining the "representational mapping" between model and target. This movement includes the update of traditional accounts based on *isomorphism* (van Fraassen 2008) and *similarity* (Giere 2004, 2006, 2010), as well as accounts that have been dubbed *inferential* (Suarez 2003, 2004), *interpretational* (Contessa 2007), and *semiotic* (Knuuttila 2010), among many others. Some, like Suarez (2015) and Morrison (2015), opt for deflationary approaches, treating models as representational but refraining from offering a full-blown account of representation. Whether deflationary or not, philosophers holding these triadic views of representation will accept that what makes X a model of some target phenomenon Y is that X represents Y, further adding that X represents Y "as thus and such" or "for researchers R in context Z," and so on.

As a quick examination of the literature makes clear, disagreements abound concerning what representation is. Still, most take for granted that models *depict* or *describe* more or less accurately what some target is like. Indeed, there is wide acceptance of the idea that models are never *entirely* accurate: all models contain abstractions, simplifications, approximations and idealizations, such that all models—even the ones we deem explanatory or otherwise epistemically worthy—are, strictly speaking, *false* (Elgin 2004; Kennedy 2012). But saying that a model or parts of a model are true/false implies that there is something other than itself that the model or its parts are true/false *about*, that is, some object that the model describes or depicts more or less successfully. This basic *intentional* assumption that models are *representations* of some target remains largely unquestioned in the philosophical literature.[2] The sheer number of representational accounts of modeling can give the impression that representation must be a central aspect of scientific modeling: even if there is no consensus about the *nature* of representation, it is hard to doubt that the "phenomenon exists" and think that philosophers were concerned with a pseudo-problem all along. Or is it? While my goal here is not to make this case in any conclusive way, I hope to motivate this possibility by considering the case reviewed in Sect. 2. For the particular model-based project at hand, it seems safe to conclude that considerations about whether the models stand in a representational relationship to the target of investigation did not play the role one would expect given the inordinate amount of attention philosophers have devoted to representation in their accounts of modeling.[3]

Schank and colleagues were surely concerned with similarities and differences between model and target. In the technical reports they make very explicit their interest in "simulating" or "emulating" rat pup behavior, generating models that "corresponded" in some way or were "analogous" to what they observed in direct experimentation. They claim, for example, that their goal was to build "four identical robots that are morphologically *analogous* to rat pups" (Schank et al. 2004, p. 164, emphasis added) with the "primary aim of *emulating* some of the relevant physical and sensorimotor characteristics of rat pups" (p. 165, emphasis added). In another paper they give a little more detail, saying: "Robots were designed to have a morphology *similar* to a pup, with a long body tapering to a rounded snout, and [they] moved using two rear-driven wheels and a front stabilizing ball" (May and Schank 2009, p. 310, emphasis added). Still, this idea of analogy or similarity should not be interpreted as corresponding to the philosophical conception of the model-target relationship as representational—that is, as *intentional*, with models describing their target and somehow being *about* them. Beyond comparisons between model and target, Schank and colleagues used the

[2]Knuuttila (2011) calls this assumption the "representational view of models" and she explores an alternative approach, although it is unclear that her account succeeds in providing a real alternative since it maintains some role for representations in modeling.

[3]I think this insight can be applied more broadly to motivate an anti-representationalist view of scientific modeling, but this is a topic I pursue in more detail elsewhere.

very same language to talk about comparisons between one model and another: "Results for three-segmented agents contained both *similarities* with and *differences* from the results for rigid-bodied agents" (May and Schank 2009, p. 323, emphasis added). However, such similarities and differences do not imply that three-segmented robots "represented" rigid-bodied robots in the intentional sense of "representation" at play in the philosophical literature, nor does it imply that the *results* from one kind of model "represented" the results from the other in that sense. Measurements of coordinates or trajectories and experimental results can be "similar" in the sense of being numerically equivalent or proximate in a statistically significant way. Likewise, the morphology of robots and pups can resemble one another without it being the case that the researchers' intention was to *describe* what rat pup morphology is like through the morphology of the robot.

The researchers also talk about "realism" and "accuracy," but again we are far from the flavor of intentionality and aboutness those terms seem to have when used by philosophers in the context of representational views of modeling. Consider this discussion of their results:

> Our metrics do not allow us to determine why two-segmented agents were a poorer fit. But these results do suggest that "more" is not necessarily better, as the morphologically more realistic two-segmented agents did not fit 7-day-old pups as well as the morphologically less realistic one-segmented agents. Three-segmented agents matched 7-day-old pups on one more metric (wall frequency) than did one-segmented agents. They may be better models of 7-day-old rats than one-segmented agents, but further research is needed to tease apart the behavioral implications of one- versus three-segmented morphologies. (May et al. 2011, p. 288)

Here the researchers concluded that a closer match in morphology was not correlated with a closer match in behavioral results. In the gradient from rigid-bodied robots up to two- and finally three-segmented agents, the increase in morphological similarity did not result in a corresponding increase in how the behavior of robots matched the behavior of actual rat pups. But the "morphological realism" of one or another type of robot need not be understood *representationally* as meaning "more accurately describing the morphology of rat pups," but rather it can be framed in terms of biological viability and resemblance with target. In the first, broader sense, morphological realism concerns the biological feasibility of certain body shapes and structures, the "naturalness" of certain organizations and forms. Humanoid robots, for example, are robots built to display features similar to those of humans, such as bipedal motion, facial expressions, etc. Although they are built to be realistic and human-like, it is hard to substantiate the claim that they are built to "represent" what humans are like: the goal is not to *describe* humans in general or any individual human being, but rather to create something that is similar in ways of interest. In some cases the intended similarity will be merely behavioral, and in other cases morphological as well; either way, humanoid robots are not descriptions of *what humans are like* even if they are *like* humans in some ways (and *unlike* humans in others)—which suggests that similarity does not require the intentionality implied by talk of representation. This brings us to the second sense of realism as pertaining to the similarity between the morphology of robots and of rat pups in particular. As is well known, any two things are similar in various respects, regardless of whether one is used to represent the other or not. In the case at hand, the different degrees of similarity between types of robots and rat pups were obviously not accidental. But, in manipulating the particular ways in which models were similar and different from the target, the researchers were able to isolate possible difference-making factors. They say:

> Dynamic simulation models allow systematic analysis of potentially relevant parameters, but not all physical parameters are known or can be accurately modeled. Robots allow for the explicit instantiation of physical parameters and allow experimental studies that vary several parameters, but whether robotic parameters represent the animal or are artifacts is an empirical question (Schank et al. 2004, p. 163).

This is one of the rare instances of the researchers' use of the word "represent," and even here there is good reason to take them to be thinking in terms more common to the statistical analysis of experimental data than in the intentional sense found in the philosophy of science literature. In experimentation and statistical analysis, scientists are concerned with whether the sample is "representative" of the population of interest—that is, if the phenomenon the data captures is the same as the phenomenon that is meant to be explained or understood. For the present case, in the investigation of, for example, the relationship between thigmotaxis, on the one hand, and motion trajectory and frequency of wall contacts, on the other, body color obviously does not play any role, and that is why the robots did not have to match rat pups in that respect. But if trajectory and wall contact results covary with distinct body morphologies, then morphology needs to be accounted for in order for the interaction of interest both in robots and in rat pups to be considered, to an extent, part of the same population. In this sense both rat pups and robots are "representative" of the same phenomenon, but this does not imply that one is "representing" the other (say, that robots represent rats), and much less so that both robots and rat pups are "representing" some third thing.

Unlike what is implied by philosophical discussions of modeling, the epistemic value of Schank and colleagues' model-based scientific work did not stem from any assumption by the researchers that their models were representations of real rat pups. Simulations and robots can be adjusted and tweaked in ways that are impossible, impractical or unethical to do with living organisms, and they allow scientists to investigate otherwise unstudied theoretical and practical possibilities: "Surrogate experimental analyses allow us to generate new theoretical and testable hypotheses, which facilitate the refinement and construction of multilevel models" (Schank 2001, p. 38). Further, the divergence between robot and pup behavior in some of the experiments exemplifies another way in which modeling was epistemically useful for the researchers: observing the behavioral mismatch led to "the discovery of a previously unrecognized pattern of behavior in rat pups," which they took to illustrate "the value of models in leading to discovery of new patterns of behavior in the system modeled" (Schank et al. 2004, p. 161). In this case, the behavior observed in the robots directed the researchers' attention to a particular aspect of rat behavior that had not been noticed in previous research. And the mismatch between the two supported the idea that, while thigmotaxic architectures were intimately related to the behaviors observed in rat pups, other contributing factors needed to be identified and further studied.

3.2 Objects, Discourse, and a Conclusion for the Philosophy of Science

Schank and colleagues' model-based research, from what we have seen, appears to have been successful independently of any assumption that their models stood in a representational relationship to the target rat pups. This is in direct conflict with the emphasis philosophers have given to the supposedly representational basis of scientific modeling. In light of the amount of attention devoted in the literature to issues surrounding the notion of representation, one would expect those issues to capture some central aspect of scientific model-based practice. Despite its popularity, however, the representational view of models does not appear to hold up under a closer examination of how models are actually built and used. Although in this paper I cannot make the stronger case for how this applies to all scientific modeling, I want to propose that this possibility needs to be taken seriously. My suggestion is that the philosophical issue of representation is a non-issue for scientists and that it does not correspond to a readily identifiable aspect of scientific practice, but it is only a pseudo-problem and an artifact of philosophical analysis. As we will see, this poses challenges to the philosophy of science from within and from without.

The mismatch between the importance philosophers claim representation to have in scientific modeling and the real importance it has for scientific practice is due to characteristics of philosophical analysis that bring us back to the tension between the two approaches discussed earlier. It is widely accepted nowadays that philosophy of science is not concerned with a rational reconstruction of science, nor with the logical form of scientific theories, nor (exclusively) with normative ideals; rather, it is often assumed that a comprehensive philosophical understanding of science needs to account for science as we know it, science as it is practiced. This general orientation is illustrated in the modeling literature by the increasingly frequent consideration of real historical and contemporary cases of modeling. Still, discussions about the model-target relationship as representational seem to persist without the proper empirical anchoring. By focusing on semantic, metaphysical and epistemological questions around scientific modeling, the "mainstream" philosophical approach has tended to isolate the phenomenon of scientific modeling in order to analyze it. In contrast, the more "cognitive" or "psychologistic" approach to model-based reasoning is explicitly aimed at investigating how scientists *reason* through the use of models, and accordingly it attempts to examine models in context. This corresponds to a difference in how the two approaches "objectify" models, that is, two different ways in which they treat models as *objects*.

Philosophers in the mainstream approach "reify" models or turn them into objects of study by treating them as entities in and of themselves, whose semantics, metaphysics and epistemology can be philosophized about in isolation from practice and from what scientists take their practice to involve. In more cognitive approaches, however, models are treated as objects in the sense of being seen as epistemic *tools* (Knuuttila 2011), *instruments* of investigation (Morrison and Morgan 1999), that is, *scientific objects* that are built and used by certain people in specific situations for particular purposes. In this second way of treating models as (scientific) objects it is just pointless to remove a model from the context it was built in, and to ignore who built it, what for, and how it was used—that would be like trying to understand the differences between basketball and soccer by focusing on the physical properties of basketballs and soccer balls without relating those properties to how the games are played.

The first kind of objectification, treating models as isolated objects of philosophical scrutiny, is in great part responsible for the otherwise nonexistent concerns with the model-target relationship as being representational: it assumes that models describe *what the world is like*, while missing the fact that scientists simply treat models as being *like the world*. In order to understand the case described in Sect. 2 and discussed in Sect. 3.1, there is no need to talk about a representational model-target relationship because most if not all of the model-based reasoning processes depended simply on analogy and similarity. In the cases we have seen, models were tools for exploring hypotheses (such as that group behavior can be accounted for by individual preferences, cf. Schank and Alberts 1997), directing attention (discovery of behavioral pattern in rats, cf. Schank et al. 2004), and so on. The mathematical modeling discussed by Bish et al. (2007) was not a mathematical description of actual rats, but a blueprint for the simulated robot, with the results then feeding back into the research and informing the construction of actual robots. The overall goal, that is, was not to describe what the animal is like, but rather to build something that is like (and different from) the animal in relevant respects. If this insight generalizes to other model-based projects as I believe it does, it results in scientific modeling being more like "tool-building" than "representing" or "representation-making."

The example of model-based research I have considered in this paper is interesting for an additional twist it brings to the conclusion I have been putting forward. In a recent book chapter, Schank and colleagues explicitly endorsed the representational view I have been criticizing here. They affirmed that they treat "models as *representations* of physical animate systems (e.g., animals) that *support* our understanding of those systems" (Schank et al. 2014, p. 147, italics original). They further added that "No model is a perfect representation of another physical system" but that models *are* representations: "Many representational properties of a model do not correspond directly to properties of the system modeled because of our uncertainty about whether a system has the properties in question," and still "A model is always false as a representation of all or many properties of a system (Schank et al. 2014, p. 147). I see that these quotes could be taken to contradict my interpretation of the project and to cast doubt on my skepticism about representationalism in models. Yet, I do not think that conclusion is warranted because, I propose, Schank and colleagues' portrayal of their models as representations was only a post hoc rationalization and rhetorical device. The quotes are from a chapter published in an edited volume targeted at a philosophically-inclined audience. In that chapter the authors cite many of the philosophers I mentioned earlier as advocates of the representational view of models. As a result, it seems safe to assume that the authors' goal with that chapter was to participate in the philosophical conversation, and not to give an impartial description of their research (that was the goal of the technical papers published in scientific journals). Their adoption of the philosophical representationalist discourse, then, seems to be motivated by their desire to contribute to the philosophical debate more than from a strictly rigorous and independent assessment of their own work. For, as we have seen, in their previous research, representation did not pose an important conceptual question nor was it a practical consideration in model-building and model-using. There were, as already discussed, questions about similarity, realism and accuracy—but those concerned whether the simulations and robots were similar to one another and to the target (real rat pups) in important ways, not whether they stood in intentional representational relationships in any deep philosophical sense. Their view wasn't, in other words, that in building simulations and robots they were describing what the target is like; rather, their goal seemed to be to construe simpler tools that, among other results, afford the generation of hypotheses about the target phenomenon.

From what I suggested above, the traditional approach philosophers adopt to investigate phenomena of interest poses a challenge from *within*: the way we objectify a phenomenon such as modeling by isolating a model from its context constrains our ability to see how the nature of the phenomenon is shaped by what brings it about (the individual scientists, the research context, disciplinary traditions, and technological possibilities in addition to properties of the target); this, in turn, limits our understanding of the role the phenomenon plays in some practice (namely, the fact that model-based investigations are not simply *of* some target but also *for* specified purposes). The challenge from within is that our own philosophical mode of investigation makes it so that we get caught up on ethereal metaphysical concerns that have nothing to do with the phenomenon in the real world of scientific practice. But this challenge from within, intrinsic to the conceptual or philosophical approach, also generates a challenge from *without*. As certain assumptions become widespread within the philosophical community and a corresponding discourse is established, there is always the danger that others outside the community will try to adopt that same language if they wish to participate in the philosophical conversation. This is what I take to have happened in the case of Schank and colleagues' recent pronouncement in the philosophically-oriented book publication. Motivated by their awareness of developments in the philosophy of science, they reinterpreted their practice using the "right" vocabulary for the intended philosophical audience. Cross-pollination between science and philosophy is definitely something to be celebrated and furthered, but the case at hand illustrates an additional difficulty for the philosophy of science. If philosophers are to try to avoid the internal constraints of philosophical assumptions by accounting for scientific *practice*, an added challenge is that we have to be careful with the weight we give to what scientists *say* about their practice. The influence, illustrated by this case, of philosophical assumptions on how scientists interpret their own work creates a potentially dangerous loop in which philosophers can take the scientists' interpretation of their research to corroborate the philosophical assumptions without noticing that the philosophical assumptions were what generated the scientists' interpretation in the first place.

Contrary to Godfrey-Smith's (2006a) view, I propose that the cognitive approach is not a subset of the broader philosophical approach, but rather it is a complementary alternative that provides tools that can help philosophical analysis avoid the threats from within and from without. The cognitive approach is still subject to internal biases—as can be seen, for example, in the surprising persistence of cognitive accounts that ignore embodied and extended theories of cognition (see Thagard 2012 for an example where embodiment seems to play no significant role). Cognitive approaches are also subject to having those internal biases feed back onto the investigation from without; still, the way the approach objectifies the phenomenon seems to put it in a better position than the more conceptual approach. In the cognitive approach what is studied is model-based reasoning, that is, how scientists design, build and use models to aid in their *reasoning* processes and their investigations *of* target phenomena *for* some purpose. This anchoring in the real world of practice precludes considerations of models as separated from the complex context of scientific modeling. Moreover, the use of theoretical as well as empirical tools to study how people solve problems by building and utilizing models gives the cognitive approach more content and evidential basis, beyond having to rely on mere philosophical speculation or the report of scientists, both of which, as we have seen, can be dangerously biased. Rather than relegating the cognitive approach to a lowly position as an optional, more fine-grained component of the broader philosophical approach, I suggest that the cognitive approach is central to a cautious philosophy of science that aims to understand real science as it is practiced. As the present case study illustrated, considering how scientists *build* and *use* models to think about targets of interest (but not necessarily considering how scientists *talk* about that process) challenges the representational view. Assuming that models depict or describe their target as being one way or another was unnecessary for the results obtained and was perhaps only a post hoc rationalization and rhetorical device. Rather than being a representational endeavor, model-based reasoning seems to be fundamentally dependent upon comparison and analogy between independent entities, with no further assumption of intentionality being required. The philosophical issue of representation, then, seems to be a non-issue, an artifact of a biased approach to the phenomenon, something that could be avoided if philosophical investigation was informed by the cognitive aspects of the object of study.

References

Bish, R., Joshi, S., Schank, J., & Wexler, J. (2007). Mathematical modeling and computer simulation of a robotic rat pup. *Mathematical and Computer Modelling, 45*(2007), 981–1000.
Callender, C., & Cohen, J. (2006). There is no special problem about scientific representation. *Theoria, 21*(1), 67–85.
Carruthers, P., Stich, S., & Siegal, M. (2002). *The cognitive basis of science*. Cambridge University Press.
Chakravartty, A. (2010a). Informational versus functional theories of scientific representation. *Synthese, 172*, 197–213. doi:10.1007/s11229-009-9502-3
Chakravartty, A. (2010b). Truth and representation in science: Two inspirations from art. In R. Frigg & M.C. Hunter (Eds.), *Beyond mimesis and convention* (Vol. 262). doi:10.1007/978-90-481-3851-7_3 (Boston Studies in the Philosophy of Science)
Contessa, G. (2007). Scientific representation, interpretation, and surrogative reasoning. *Philosophy of Science, 74*, 48–68.
Elgin, C. (2004). True enough. *Philosophical Issues, 14*(1), 113–131.
Feist, G. J. (2006). *The psychology of science and the origins of the scientific mind*. New Haven: Yale University Press.
Feist, G. J., & Gorman, M. E. (2013). *Handbook of the psychology of science*. Berlin: Springer.
Frigg, R. (2006). Scientific representation and the semantic view of theories. *Theoria, 55*, 49–65.
Frigg, R. (2010a). Fiction in science. In J. Woods (Ed.), *Fictions and models: New essays* (pp. 247–287). Munich: Philosophia Verlag.
Frigg, R. (2010b). Models and fiction. *Synthese, 172*(2), 251–268.
Frigg, R. (2010c). Fiction and scientific representation. In R. Frigg & M.C. Hunter (Eds.), *Beyond mimesis and convention* (Vol. 262). doi:10.1007/978-90-481-3851-7_3 (Boston Studies in the Philosophy of Science)
Frigg, R., & Hartmann, S. (2012). Models in science. In E. N. Zalta (Ed.), *The stanford encyclopedia of philosophy* (Fall 2012 Edition). http://plato.stanford.edu/archives/fall2012/entries/models-science/
Gholson, B., Shadish, Jr., W. R., Neimeyer, R. A., & Houts, A. C. (1989). *Psychology of Science: Contributions to Metascience*. Cambridge: Cambridge University Press.
Giere, R. (1988). *Explaining science: A cognitive approach*. Chicago: The University of Chicago Press.
Giere, R. (1992). *Cognitive models of science*. Minneapolis: University of Minnesota Press.
Giere, R. (1999). Using models to represent reality. In L. Magnani, N. J. Nersessian, & P. Thagard (Eds.), *Model-based reasoning in scientific discovery* (pp. 41–57). New York: Kluwer/Plenum.
Giere, R. (2002). Models as parts of distributed cognitive systems. In L. Magnani & N. Nersessian (Eds.), *Model based reasoning: Science, technology, values* (pp. 227–41). New York: Kluwer.
Giere, R. (2004). How models are used to represent reality. *Philosophy of Science, 71*(5), 742–752.
Giere, R. (2005). Scientific realism: Old and new problems. *Erkenntnis, 63*(2), 149–165.
Giere, R. (2006). *Scientific perspectivism*. Chicago: University of Chicago Press.
Giere, R. (2010). An agent-based conception of models and scientific representation. *Synthese, 172*, 269–281.
Gilbert, J. K. (2005). *Visualization in science education*. Berlin: Springer.
Godfrey-Smith, P. (2006a). The strategy of model-based science. *Biol Philos, 21*, 725–740. doi:10.1007/s1053900690546
Godfrey-Smith, P. (2006b). Theories and models in metaphysics. *Harvard Review of Philosophy, 14*(2006), 4–19.
Godfrey-Smith, P. (2009a). Abstractions, idealizations, and evolutionary biology. In A. Barberousse, M. Morange, & T. Pradeu (Eds.), *Mapping the future of biology: Evolving concepts and theories* (pp. 47–56). Berlin: Springer (Boston Studies in the Philosophy of Science).
Godfrey-Smith, P. (2009b). Models and fictions in science. *Philosophical Studies, 143*, 101–116.

Kennedy, A. G. (2012). A non representationalist view of model explanation. *Studies in History and Philosophy of Science, 43*, 326–332.

Knuuttila, T. (2010). Not just underlying structures: Towards a semiotic approach to scientific representation and modeling. In: Bergman, M., et al. (Eds.), *Ideas in Action: Proceedings of the Applying Peirce Conference* (pp. 163–172).

Knuuttila, T. (2011). Modelling and representing: An artefactual approach to model-based representation. *Studies in History and Philosophy of Science, 42*, 262–271.

MacLeod, M., & Nersessian, N. J. (2013). Building simulations from the ground-up: Modeling and theory in systems biology. *Philosophy of Science, 80*, 533–556.

Magnani, L. (2012). Scientific models are not fictions: Model-based science as epistemic warfare. In: L. Magnani & P. Li (Eds.), *Philosophy and Cognitive Science* (pp. 1–38). Berlin, Heidelberg: Springer (SAPERE 2).

May, C. J., & Schank, J. C. (2009). The interaction of body morphology, directional kinematics, and environmental structure in the generation of neonatal rat (*Rattus norvegicus*) locomotor behavior. *Ecological Psychology, 21*(4), 308–333. doi:10.1080/10407410903320975

May, C. J., Schank, J. C., & Joshi, S. (2011). Modeling the influence of morphology on the movement ecology of groups of infant rats (*Rattus norvegicus*). *Adaptive Behavior, 19*(4), 280–291. doi:10.1177/1059712311413476

May, C. J., Schank, J. C., Joshi, S., Tran, J., Taylor, R. J., & Scott, I. (2006). Rat pups and random robots generate similar self-organized and intentional behavior. *Complexity, 12*(1). doi:10.1002/cplx.21049

Morrison, M. (2015). *Reconstructing reality: Models, mathematics, and simulations*. Oxford: Oxford University Press.

Morrison, M., & Morgan, M. (1999). *Models as mediators*. Cambridge: Cambridge University Press.

Nersessian, N. J. (1992). How do scientists think? Capturing the dynamics of conceptual change in science. In R. N. Giere (Ed.), *Cognitive models of science* (pp. 3–45). Minneapolis: University of Minnesota Press.

Nersessian, N. J. (1999). Model-based reasoning in conceptual change. In L. Magnani, N. J. Nersessian, & P. Thagard (Eds.), *Model-based reasoning in scientific discovery* (pp. 5–22). New York: Kluwer Academic/Plenum Publishers.

Nersessian, N. J. (2002). Maxwell and "the method of physical analogy": Model-based reasoning, generic abstraction, and conceptual change. In D. Malament (Ed.), *Essays in the history and philosophy of science and mathematics* (pp. 129–166). Lasalle, Il: Open Court.

Nersessian, N. J. (2008). *Creating scientific concepts*. Cambridge: The MIT Press.

Nersessian, N. J. (2009). How do engineering scientists think? Model-based simulation in biomedical engineering laboratories. *Topics in Cognitive Science, 1*, 730–757.

Nersessian, N. J. & Patton, C. (2009). Model-based reasoning in interdisciplinary engineering. In A. W. M. Meijers (Ed.), *The handbook of the philosophy of technology & engineering sciences* (pp. 678–718). Berlin: Springer.

Osbeck, L. M., Nersessian, N. J., Malone, K. R., & Newstetter, W. C. (2013). *Science as psychology: Sense-making and identity in science practice*. Cambridge: Cambridge University Press.

Proctor, R. W., & Capaldi, E. J. (2012). *Psychology of science: Implicit and explicit processes*. Oxford: Oxford University Press.

Raghavan, K., & Glaser, R. (1994). Studying and teaching model-based reasoning in science. In S. Vosniadou, E. Corte, & H. Mandl (Eds.), *Technology-based learning environments: Psychological and educational foundations* (pp. 104–111). Berlin, Heidelberg: Springer.

Schank, J. C. (2001). Beyond reductionism: Refocusing on the individual with individual-based modeling. *Complexity*, *6*(3), 33–40.

Schank, J. C. (2008). The development of locomotor kinematics in neonatal rats: An agent-based modeling analysis in group and individual contexts. *Journal of Theoretical Biology, 254* (2008), 826–842. doi:10.1016/j.jtbi.2008.07.024

Schank, J. C., & Alberts, J. R. (1997). Self-organized huddles of rat pups modeled by simple rules of individual behavior. *Journal of Theoretical Biology, 189*(1), 11–25. doi:10.1006/jtbi.1997.0488

Schank, J. C., May, C. J., & Joshi, S. S. (2014). Models as scaffolds for understanding. In L. R. Caporael, J. R. Griesemer, & W. C. Wimsatt (Eds.), *Developing scaffolds in evolution, culture, and cognition*. Cambridge: MIT Press.

Schank, J. C., May, C. J., Tran, J. T., & Joshi, S. S. (2004). A biorobotic investigation of norway rat pups (*Rattus norvegicus*) in an arena. *Adaptive Behavior, 12*(3–4), 161–173. (1059-7123 (200409/12) 12:3–4; 161–173; 048910).

Stephens, S. A., McRobbie, C. J., & Lucas, K. B. (1999). Model-based reasoning in a year 10 classroom. In *Research in science education* (Vol. 29, Issue 2, pp 189–208). New York: Kluwer Academic Publishers. doi:10.1007/BF02461768

Suarez, M. (2003). Scientific representation: Against similarity and isomorphism. *International Studies in the Philosophy of Science, 17*(3), 225–244.

Suarez, M. (2004). An inferential conception of scientific representation. *Philosophy of Science, 71*(5), 767–779.

Suarez, M. (2010). Scientific representation. *Philosophy Compass*, 5(1), 91–101.

Suarez, M. (2015). Deflationary representation, inference, and practice. *Studies in History and Philosophy of Science Part A, 49*, 36–47.

Thagard, P. (2012). *The cognitive science of science: Explanation, discovery, and conceptual change*. Cambridge: MIT Press.

Toon, A. (2010). Models as make-believe. In R. Frigg & M. C. Hunter (Eds.), *Beyond mimesis and convention* (Vol. 262). doi:10.1007/978-90-481-3851-7_3 (Boston Studies in the Philosophy of Science)

Tweney, R. D. (1985). Faraday's discovery of induction: A Cognitive Approach. In D. Gooding & F.A.J.L. James (Eds.), *Faraday rediscovered: Essays on the life and work of Michael Faraday, 1791–1867*. New York: Stockton Press/London: Macmillan, pp. 189-210.

Tweney, R. D. (1989). Fields of enterprise: On Michael Faraday's thought. In D. Wallace, & H. Gruber (Eds.), *Creative people at work: Twelve cognitive case studies* (pp. 91–106). Oxford University Press.

Tweney, R. D. (2014). Metaphor and model-based reasoning in maxwell's mathematical physics. In L. Magnani (Ed.), *Model-based reasoning in science and technology: Studies in applied philosophy, epistemology and rational ethics*. Berlin, Heidelberg: Springer, pp. 395–414. doi:10.1007/978-3-642-37428-9_21

van Fraassen, B. (2008). *Scientific representation*. Oxford: Oxford University Press.

Weisberg, M. (2006). Robustness analysis. *Philosophy of Science, 73*, 730–742. doi:10.1086/518628

Weisberg, M. (2007a). Who is a modeler? *British Journal for Philosophy of Science, 58*, 207–233.

Weisberg, M. (2007b). Three kinds of idealization. *Journal of Philosophy, 104*(12), 639–659.

Weisberg, M. (2013). *Simulation and similarity: Using models to understand the world*. Oxford University Press.

Woods, J. (2014). Against fictionalism. In: L. Magnani (Ed.), *Model-based reasoning in science and technology* (Vol. 8). Berlin, Heidelberg: Springer. doi:10.1007/978-3-642-37428-9_2 (Studies in Applied Philosophy, Epistemology and Rational Ethics).

Philosophy Made Visual: An Experimental Study

Nevia Dolcini

Abstract The advent of experimental philosophy has recently expanded the domain of philosophical debates so as to include discussions about survey-based methodology and the validity of its employment in philosophical inquiry. One of the main criticisms of this approach questions the alleged response-intuition equation, by claiming that 'pragmatic cues' might prevent the subjects from reporting their genuine intuitions about the survey scenarios and questions. The pragmatic cues discussed by the literature include aspects of a quite different nature, ranging from thinking-styles to semantic ambiguities. In order to distinguish between language-related pragmatic cues, and other features not strictly dependent on language, the distinction between the 'response problem' and the 'interpretation problem'—potentially triggered by language-related pragmatic cues—is introduced. By employing an *illustrated* survey, this study aims at revealing the extent to which the use of 'non-linguistic' vignettes might constitute a valid aid to the traditional 'linguistic' vignette. A positive response would encourage the usage of illustrations and other non-linguistic or minimally-linguistic models in survey-based studies, which considerably restricts the liability of surveys for the interpretation problem.

1 Introduction

This work addresses some of the methodological issues raised by the recent adoption of non-traditional practices in philosophical inquiry, that is, the recourse to survey-based studies for gathering data about folk-intuitions. While survey studies are well-established research tools in many disciplinary areas, especially in the social sciences, they certainly represent an unexpected novelty for philosophers. The use of this empirical research method is common in 'experimental philosophy', a controversial approach loosely understood as a study of people's intuitions about

N. Dolcini (✉)
Faculty of Arts and Humanities, University of Macau,
Avenida da Universidade, Macau, Taipa, China
e-mail: ndolcini@umac.mo

L. Magnani and C. Casadio (eds.), *Model-Based Reasoning in Science and Technology*, Studies in Applied Philosophy, Epistemology and Rational Ethics 27, DOI 10.1007/978-3-319-38983-7_6

philosophically relevant scenarios (Sytsma 2014). While testing ordinary intuitions, also called '*folk* intuitions', experimental philosophers attempt to describe people's ordinary attributions of different kinds ('what we would say', or 'how things seem to us to be') by applying the methods of the social and cognitive sciences to the study of philosophical cognition (Alexander 2012), often combined with the use of statistical tools and analysis techniques such as mediation analysis or structural equation modeling.

The advent of experimental philosophy, around fifteen years ago, revitalized the discussion about a major meta-philosophical issue: what are the proper methods, aims and ambitions of philosophy? In this debate, the traditional intuition-based a priori method, involving massive recourse to thought experiments and reflective equilibrium, is confronted with a sort of methodological naturalism,[1] which explores empirically oriented practices of addressing philosophical problems through a posteriori methods. Some authors suggest that, as long as philosophers' intuitions are inconsistent with folk intuitions, the traditional methodology and the central epistemic role of intuitions, as a support for hypothesis testing and judgment making, should be questioned. Thus, studies in experimental philosophy[2] aim—at the very least—to provide evidence regarding whether the cognitive process of formation of people's intuitions and judgments about particular scenarios is sensitive to various factors, such as culture, socio-economic status, level of expertize in philosophy, affective context, and order of presentation. These data would suggest that intuitions are not, after all, as *stable* as philosophers have traditionally tended to think. If people's intuitions and judgments about philosophical puzzles and concepts vary across cultures, as well as through other parameters, then philosophers should reassess the role of intuitions in philosophical practices, and of course, if this is the case, then the appeal to intuitions might be methodologically mistaken.

Although controversial, the challenge posed by experimental philosophy to 'methodological rationalists' cannot be easily dismissed. Many voices have joined the debate in order to defend traditional methods from the incursion of the 'methodological naturalists', whose criticism ranges from disagreement on the interpretation of the research data (Sosa 2009), to questioning the accuracy of the survey methodology (Kauppinen 2007; Cullen 2010). In addition to methodological issues, some consider experimental philosophy as a mere attack on a straw man (Williamson 2007; Deutsch 2010, 2015; Cappelen 2012), since—they argue—traditional philosophers don't in fact rely on intuitions, and neither philosophy nor philosophical methodology has anything to fear from intuitions' cultural variability. Despite numerous opponents, the effects of this novel direction in philosophical

[1]I here shape the debate in terms of 'rationalists' *versus* 'naturalists' after Fisher and Collins (2015). Here 'naturalism' is to be regarded as a methodological stance totally independent from the metaphysical position.

[2]Since the survey study conducted by Weinberg et al. (2001), the literature in experimental philosophy has sharply increased.

inquiry extend (to various degrees) well beyond its original terrain,[3] and experimental philosophy is attracting the attention of an increasingly wider audience.

In this paper, I address these issues in a twofold way. Firstly, I examine a particular *methodological issue* related to survey studies, namely the criticism that pragmatic cues might prevent subjects from reporting genuine intuitions, and propose that there are in fact two separate problems that need to be distinguished: the 'response problem' and the 'interpretation problem'. Pragmatic cues, I will argue, have a non-homogeneous nature: while some of them are the source of the interpretation problem, others are responsible for the response problem. In particular, I will claim that the interpretation problem, but not the response problem, is strictly dependent on the linguistic medium typically employed in survey studies, so that its effects can be sensibly limited by the recourse to non-linguistic models, which function as an aid to the traditional linguistic vignette. I then report an experiment designed to investigate whether changing the format of the vignettes typically used in experimental philosophy from purely verbal to primarily pictorial, might allow us to avoid the force of this criticism in respect to the interpretation problem. In the last two sections, I offer an overview of the survey study design, its content and procedures, and results.

2 The 'Interpretation Problem' in Philosophical Scenarios

Vignettes are traditionally employed as a way of collecting data across a variety of disciplinary fields, such as, among others, social psychology, business science, and health science. Not only are vignette-based studies far less expensive than other lab experimental methodologies, but they also have the advantage of allowing for the collection of people's responses to scenarios too difficult, if not impossible, to reproduce in a laboratory. For example, the typical philosophical 'story' about a counterfactual scenario or a thought experiment (brain-in-a-vat, twin Earth, and the like), seems to be no good candidate for being presented, represented, and described in a survey study other than in a 'vignette'. While reading a vignette, subjects are invited to imagine and simulate a certain hypothetical scenario, and on that basis they are asked to produce a 'cognitive' output as a response, or judgment.

Since vignette-based studies have made their appearance on the philosophical scene, some have questioned their methodological validity in the specific field of philosophy. In general, the validity of vignettes as a tool for research studies has long been discussed in the context of social sciences (Faia 1980; Rossi and Alves 1980; Collett and Childs 2011), but philosophers have recently raised more specific concerns in relation to the effect of *pragmatic cues* on intuition generation.

Some authors (Kauppinen 2007; Cullen 2010) have challenged the equation between survey responses and intuitions by considering that various factors

[3]For example, see Michael Devitt's recent appeal to linguists' intuitions about linguistic usage, in addition to philosophers' intuitions (Devitt 2012, 2014).

belonging to the *context of the experiment* might influence people's responses to a survey's vignettes, while not simultaneously influencing the subjects' actual intuitions about them. For example, subjects might be pressed by experiment context-related 'task demands', which would activate background beliefs as an effect of what they think the experimenters are interested in, or are looking for. If that were the case, it is argued, the presumed relation between subjects' concepts and their responses, or linguistic behaviors, might be undermined. Most researchers, however, consider such a problem to be a non-fatal illness of survey-based studies, and vignettes are still widely regarded as a useful and reliable tool for testing the way in which people would behave or make judgments in hypothetical circumstances, which are otherwise very hard to present (e.g., Hughes 1998).

Other authors have concerns about the way in which subjects may properly understand the vignette, or fail to do so to various degrees. In case of poor interpretation or utter misinterpretation of the vignette, subjects' responses in survey studies should not be taken as mirroring their genuine intuitions with respect to the presented scenario or concept. For example, culture-related conversational or response styles (Johnson et al. 2005; Hofstede 2001), are also regarded as problem triggers, given the role played by such pragmatic aspects in the subjects' process of interpretation of a given vignette. Subjects' implicit beliefs about the experimenter's expectations, conversational norms, and culture-dependent survey response styles are all elements of a pragmatic nature to be taken into consideration when evaluating survey data, especially in the case of cross-cultural studies.

In general, all these problems have so far been addressed as instances of either the *response problem*, or the *interpretation problem*. Roughly, the former questions the very nature of people's responses to survey data (*what* the responses exactly are, and whether or not they represent people's intuitions), whereas the latter regards people's interpretation of philosophical vignettes (*how* people read the vignettes). Issues related to people's interpretation of the vignette might be regarded as intertwined—or overlapping—with the response problem, insofar as for one to interpret a vignette implies tailoring one's understanding of a described scenario to what one assumes the experimenter's goal or expectation to be.[4] Culture-related conversational or response styles (Johnson et al. 2005; Hofstede 2001), are also thought to cause the response problem, given the role played by these pragmatic aspects in the subjects' process of interpreting a given vignette. Subjects' implicit beliefs about the experimenter's expectations, conversational norms, and culture-dependent survey response style certainly are all elements of a pragmatic nature to be taken into consideration when evaluating survey data, especially in the case of cross-cultural studies. However, the distinction between the 'response problem' and the 'interpretation problem' is too vague, and a sharper analysis of the pragmatic cues

[4]This worry can be seen in Cullen (2010), where he argues that people's judgments are a kind of behavior generated by inputs of various sorts, as for example the subjects' "beliefs about what the researchers are interested in" (pp. 277–278).

triggering the two problems would be of great benefit for the debate concerning the methodology adopted in many studies in experimental philosophy.

The domain accorded to pragmatic cues in the literature is especially broad, and it ranges from semantic ambiguities to culture-dependent response styles. The category of 'pragmatic cues' includes a variety of aspects belonging to the concrete circumstances in which the survey is administered, as well as assumptions or background beliefs held by subjects before and/or at the time of the survey administration, culture-related thinking or response styles, subjects' education level, emotions triggered by the proposed scenarios, culture-related attitudes with respect to thinking of possible, past or future events (Guo et al. 2012), and much more. In addition to that, pragmatic cues are also instantiated in features of the particular language employed by the questionnaire, as in the case of syntactic, lexical and pragmatic ambiguities, implicit elements, implicatures, with also presuppositions typically playing a role along the interpretation process of a linguistic text.[5]

Within the broad category of pragmatic cues, some of the features belong strictly to the 'linguistic' context, whereas others depend more (or exclusively) on the 'extra-linguistic' context. I therefore propose a reading of the 'response' *vs.* 'interpretation' distinction as deriving from two different sorts of pragmatic cues: the 'response problem' depends on extra-linguistic pragmatic cues, whereas the 'interpretation problem' depends on linguistic cues, that is, it originates from the characteristics of the language employed in the survey. Not only do the pragmatic cues potentially responsible for the interpretation problem belong to language, but some of them also specifically depend on *the* particular language employed (e.g., English, Chinese, Japanese, Italian, etc.). In a similar vein, language-based pragmatic cues are also responsible for the majority of the difficulties related to survey translations.

If and when something goes wrong during the subjects' process of interpretation of a given vignette *because* of the pragmatic cues related to the linguistic context, then the 'interpretation problem' arises. The vignette's misinterpretation, or poor interpretation, is not sufficient for justifying the claim that the subjects' responses are not the subjects' intuitions on the matter. Indeed, such responses *are* the subject's genuine intuitions about what she has read, and indeed the response is based on her interpretation of the vignette: as such, this does *not* qualify as an occurrence of the 'response problem', where the subject's response does not correspond to her genuine intuition. Responses affected by the interpretation problem would change in the case that the subject is given the opportunity to revise her interpretation.

In the case of vignettes depicting complex philosophical scenarios, the significant cognitive effort required by the readers might amplify the risk that their

[5]In the experimental philosophy literature, a case of 'perspectival ambiguity' has been recently discussed by Sytsma and Livengood (2011).

interpretation falls short, especially if the administered questionnaire is not written in the reader's mother language (a common occurrence in cross-cultural studies). Is there any way in which the 'interpretation problem' can be contained? This study addresses the question of whether *illustrated vignettes* can be employed in survey studies. A positive response would indicate a way to contain the potential danger of the pragmatic cues—the source of the interpretation problem—often present in the narrative texts. For this purpose, I have designed a survey for testing the hypothesis that the employment of 'models'—in this specific case, black-and-white hand-drawings—in surveys does not trigger systematic (non accidental) variations in subjects' responses as compared to the subjects' responses to *purely* linguistic surveys.

3 Illustrated Survey and Study Design

Typically, survey studies in experimental philosophy exclusively employ the linguistic medium: vignettes narratively describe philosophically relevant scenarios, and are followed by statements or questions to be answered by the subjects. As previously discussed, some pragmatic features of the text might trigger the interpretation problem, especially in the case of cross-cultural studies in which subjects are confronted with a survey in their second language, or when the survey is translated from one language to another. One possible way to contain the challenges intrinsic to language with respect to users from various cultural and linguistic backgrounds might be the recourse to 'models' of the scenario depicted in the vignette. These models, in the form of illustration, might be *purely non-linguistic* if no linguistic element is employed (this is the case of illustrations or diagrams which can exhaustively and non-ambiguously offer a representation of the scenario). However, illustrations would be more typically *minimally linguistic*; in fact, in order for the 'model' to represent the scenario well, sometimes a few simple linguistic elements (names, dates, etc.) are required.

The present study has been designed as a vignette-based survey, which comes in two versions: the *Verbal survey* is a purely linguistic one; the *Illustrated survey* features black-and-white hand-drawings in addition to the narrative text. The illustrations from the Illustrated survey are sometimes enriched with 'labels' so as to facilitate the reader's understanding of the story.[6] The subjects participating in the study were either presented with the Verbal survey or with the Illustrated survey (none of the subjects were presented with both versions). The semiotic nature of the survey is analyzed as the main factor and potential variation-trigger. The expected

[6]The minimal recourse to linguistic elements in the illustrations—let's call them 'minimally-linguistic illustrations'—does not hamper the very goal of the illustrated survey, and dates or names stand in the drawing as a sort of 'Peircean index'.

result is that the introduction of illustrations does not trigger any sensitive variation to subjects' responses.

The study hypothesis, if confirmed, would suggest that the recourse to models in survey studies does not jeopardize the subjects understanding of the vignettes. Yet, the visual inputs help the subject to effectively resolve the potential issues triggered by pragmatic cues of linguistic origin and nature, which are deemed responsible for the interpretation problem. The main advantage of employing visual inputs derives from the consideration that illustrations, given their non-linguistic nature, are immune to the interpretation problem as defined above. In order to test the study hypothesis, linguistic vignettes have been written in the form of mainly descriptive statements so as to avoid possible interpretation-problem triggers.

In line with the typical experimental philosophy survey study, the vignettes tackle a philosophical problem, namely the problem of object identity and personal identity through time and change: the scenarios are based on the classical Theseus ship puzzle, and the brain transplant thought experiment.[7]

The survey includes five vignettes, followed by either two or three statements/questions. Two out of the five vignettes are about the problem of objects' identity through time and change, and the scenarios present the readers with the Theseus ship paradox; two other vignettes remodel the Theseus ship paradox into a puzzle about personal identity; one vignette depicts a 'brain transplant' scenario. The present study includes two survey versions and numerous questions; since excessive space would be required for a full and comprehensive report, I will present two out of the five vignettes: the 'Brain Transplant' scenario (story B) and the related full set of questions; the 'Theseus Ship' scenario (story C) and the related questions.

Here is the brain transplant scenario (story B):

B: Mike and John undergo intertwined brain transplantation together. During the surgery, Mike's brain is transplanted into John's skull, while John's brain is transplanted into Mike's skull.

In order to make sure that the participants correctly understand the content of the stories, each vignette is followed by a check-question. Note that the subjects' responses to the check-question are extremely important for the purpose of the present study; in fact, check-questions uniquely monitor the quality of the readers' comprehension of the vignette.

Here is the check-question (B1) for the brain transplant scenario:

B1. Mike has his brain removed and replaced with John's brain; whereas, John has his brain removed and replaced by Mike's brain.
A. Yes, I agree B. No, I disagree

[7]Given the focus on the methodological problem, the data analysis will here be limited to the *medium* as a possible variation trigger. While the content of the vignettes is widely irrelevant with respect to the methodology issue, the collected data offer an occasion to check folk intuitions about the problem of object/person identity through time and change. The discussion of these data will be presented in a forthcoming paper.

After the check question, the subjects are asked to answer the question (B2):

B2. Select the statement that you believe to be true.

A. After the intertwined brain transplant, John is the individual who possesses John's brain and Mike's body; and Mike is the individual who possesses Mike's brain and John's body.
B. After the intertwined brain transplant, John is the individual who possesses John's body and Mike's brain, while Mike is the individual who possesses Mike's body and John's brain.
C. After the intertwined brain transplant, neither John nor Mike exists any longer. In their place there are two entirely new individuals. It is like, thanks to the intertwined brain transplantation, two entirely new persons are brought to life.
D. After the intertwined brain transplant, the two resulting individuals are hybrid persons. Namely, both of the individuals are partially Mike, and partially John.

In the Illustrated survey, story B is illustrated as follows:

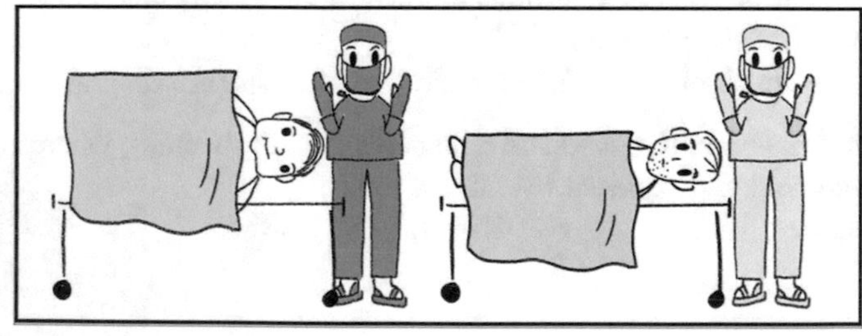

Mike and John undergo intertwined brain transplantation together

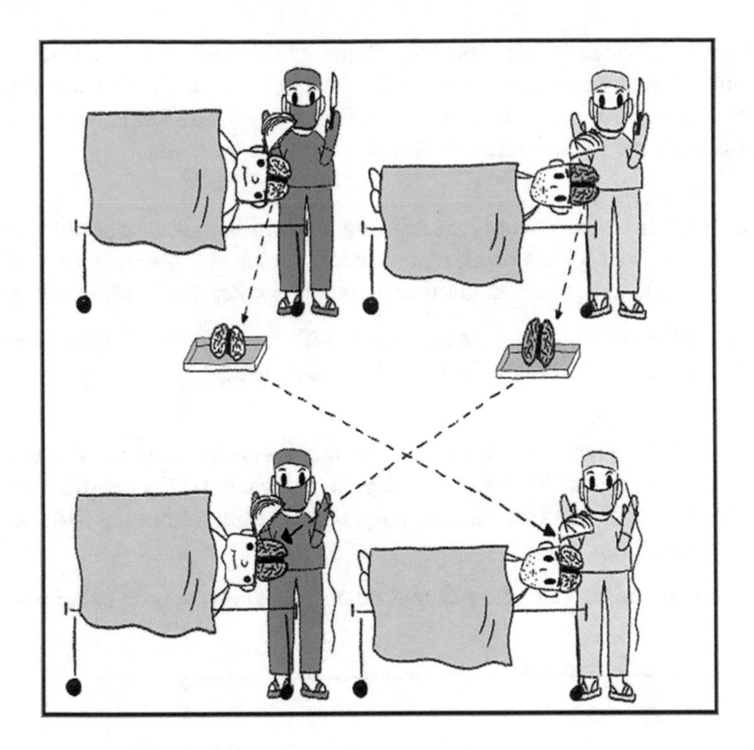

During the surgery, Mike's brain is transplanted into John's skull, while John's brain is transplanted into Mike's skull

The illustrations for questions (B1) and (B2) simply repeat the vignette illustrations.[8] Participants in the survey were asked to judge the statement/answer to the questions, and to choose *only one* answer, which was mostly consistent with their intuitive judgments.

Here is the Theseus Ship scenario, or story C, from the Verbal survey, followed by the check-question (C1), and question (C2):

C: The ship Santa Maria was built 1000 years ago, and made of 1000 wooden planks. Since the very first year of life, the ship components start decaying due to natural consumption. Therefore, its wooden planks have been taken out, and replaced with brand new wooden planks. In the first year, the Santa Maria had its first plank replaced. In its second year, it had its second plank replaced. And so on at the rate of one plank at a year. Eventually, after 1000 years, the ship Santa Maria had its last original plank replaced, too.

[8]For reasons of space, vignettes and their related questions in the illustrated survey are placed on different sheets, which is why the repetition of the illustrations is needed on the question page.

C1. After 1000 years of life, the ship Santa Maria had all its original wooden planks taken away and replaced with new ones. Thus, after 1000 replacement operations, there is no plank in the 1000-year-old Santa Maria, which was also a plank of the Santa Maria at its inaugural day.

A. Yes, I agree.
B. No, I disagree. There is at least one plank of the 1000-year-old ship Santa Maria (after its last plank replacement), which was also a plank of the ship Santa Maria at its inaugural day (before its first plank replacement).

C2. The 1000-year-old ship Santa Maria is still the same ship Santa Maria at its inaugural day, and before its first plank replacement.

A. Yes, I agree.
B. No, I disagree. The 1000-year-old ship Santa Maria is not the same ship Santa Maria at its inaugural day. In fact, the 1000-year-old ship Santa Maria is now an entirely new ship, totally distinct from the Santa Maria at its inaugural day.

In the Illustrated survey, story C and related questions appear as follows:

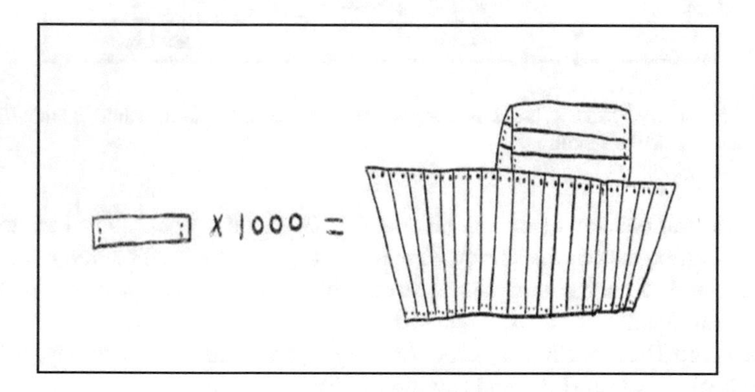

The ship Santa Maria was built 1000 years ago, and made of 1000 wooden planks. Since the very first year of life, the ship components start decaying due to natural consumption. Therefore, its wooden planks have been taken out, and replaced with brand new wooden planks

In the first year, the Santa Maria had its first plank replaced. In its second year, it had its second plank replaced. And so on at the rate of one plank at a year

Eventually, after 1000 years, the ship Santa Maria had its last original plank replaced, too

C1. After 1000 years of life, the ship Santa Maria had all its original wooden planks taken away and replaced with new ones. Thus, after 1000 replacement operations, there is no plank in the 1000-year-old Santa Maria, which is also a plank of the Santa Maria at its inaugural day.

A. Yes, I agree.
B. No, I disagree. There is at least one plank of the 1000-year-old ship Santa Maria (after its last plank replacement), which is also a plank of the ship Santa Maria at its inaugural day (before its first plank replacement).

C2. The 1000-year-old ship Santa Maria is still the same ship Santa Maria at its inaugural day, and before its first plank replacement.

A. Yes, I agree.

B. No, I disagree. The 1000-year-old ship Santa Maria is not the same ship Santa Maria at its inaugural day. In fact, the 1000-year-old ship Santa Maria is now an entirely new ship, totally distinct from the Santa Maria at its inaugural day.

4 Procedures and Experimental Evidence

The study compares the data gathered by the responses of two homogeneous groups of Asian subjects. A first group was presented with the typical linguistic survey ('Verbal'), whereas a second group was presented with the illustrated survey ('Illustrated'). Overall, 602 undergraduate and graduate students (250 males and 352 females) from the University of Macau participated in the study. The sample presents a mild variation with respect to the familiarity with philosophical knowledge from both the Western and the Eastern tradition: 238 participants have never taken any philosophy class (39.5 %), 290 participants have taken one or two philosophy classes (48.2 %), and 10 participants are trained in philosophy at either undergraduate or graduate level (1.6 %). The age range of the sample is from 18 to 33 ($M = 21$; $SD = 1.09$); subjects in between 18 and 25 years cover the 97.7 % of the whole sample.

In classroom settings, students were randomly given either the Illustrated or the Verbal survey; participants joined the study on a voluntary basis, and they were informed that they were allowed to quit responding to the survey at anytime, and at their will.

The survey used in this study is originally designed and developed by the researcher and is in English, the participants' second language. Surveys were administered in a paper-and-pencil form, and time-allowance was of about twenty to thirty minutes. Participants were clearly instructed to choose only one answer, the one that suited their intuitions the most, and responses with more than one selected choice have been excluded from the data analysis. In order to limit or deactivate the possible influence of the previous story on the subject's judgment of the following stories, both Verbal and Illustrated surveys were arranged in six different vignette-sequence types.

For each set of questions related to the five vignettes, a chi-squared test consistently yielded no significant variation between the responses to the Illustrated survey and the responses to the Verbal survey. As predicted, subjects did not give significantly different answers to the survey questions across conditions; given the wide sample tested, such results strongly confirm the study hypothesis. Figures 1 and 2 report the results to questions B1 and B2, whereas Figs. 3 and 4 report the

Crosstab

			B1		Total
			A agree	B disagree	
Version	Illustrated	Count	238	55	293
		Expected Count	234.8	58.2	293.0
		% within Version	81.2%	18.8%	100.0%
		% within B1	49.6%	46.2%	48.9%
		% of Total	39.7%	9.2%	48.9%
	Verbal	Count	242	64	306
		Expected Count	245.2	60.8	306.0
		% within Version	79.1%	20.9%	100.0%
		% within B1	50.4%	53.8%	51.1%
		% of Total	40.4%	10.7%	51.1%
Total		Count	480	119	599
		Expected Count	480.0	119.0	599.0
		% within Version	80.1%	19.9%	100.0%
		% within B1	100.0%	100.0%	100.0%
		% of Total	80.1%	19.9%	100.0%

Fig. 1 $X^2 = .432^a$, $p = .511$

results to question C1 and C2. In both cases, the homogeneity of the responses from the Illustrated and Verbal surveys is striking.

As can be seen in the figures above, there is no significant variation across the two conditions for any of the questions in line with the hypothesis under consideration: that the subjects' responses to the survey would not vary if the survey was enriched with accompanying illustrations, which function as a *model* with respect to the traditional linguistic vignette representing a philosophical scenario.

The results clearly show that, despite the information added by drawings to the linguistic narration (e.g., characters' appearance, shapes of described objects, etc.), such information does not negatively interfere with the subjects' process of intuition formation.

Crosstab

			B2				Total
			A brain define identity	B body define identity	C entirely new individuals	D hybrid persons	
Version	Illustrated	Count	68	91	74	61	294
		Expected Count	74.6	84.4	71.2	63.8	294.0
		% within Version	23.1%	31.0%	25.2%	20.7%	100.0%
		% within B2	44.7%	52.9%	51.0%	46.9%	49.1%
		% of Total	11.4%	15.2%	12.4%	10.2%	49.1%
	Verbal	Count	84	81	71	69	305
		Expected Count	77.4	87.6	73.8	66.2	305.0
		% within Version	27.5%	26.6%	23.3%	22.6%	100.0%
		% within B2	55.3%	47.1%	49.0%	53.1%	50.9%
		% of Total	14.0%	13.5%	11.9%	11.5%	50.9%
Total		Count	152	172	145	130	599
		Expected Count	152.0	172.0	145.0	130.0	599.0
		% within Version	25.4%	28.7%	24.2%	21.7%	100.0%
		% within B2	100.0%	100.0%	100.0%	100.0%	100.0%
		% of Total	25.4%	28.7%	24.2%	21.7%	100.0%

Fig. 2 $X^2 = 1.918^a$, $p = .590$

Crosstab

			C1		Total
			A agree	B disagree	
Version	Illustrated	Count	186	109	295
		Expected Count	184.7	110.3	295.0
		% within Version	63.1%	36.9%	100.0%
		% within C1	49.3%	48.4%	49.0%
		% of Total	30.9%	18.1%	49.0%
	Verbal	Count	191	116	307
		Expected Count	192.3	114.7	307.0
		% within Version	62.2%	37.8%	100.0%
		% within C1	50.7%	51.6%	51.0%
		% of Total	31.7%	19.3%	51.0%
Total		Count	377	225	602
		Expected Count	377.0	225.0	602.0
		% within Version	62.6%	37.4%	100.0%
		% within C1	100.0%	100.0%	100.0%
		% of Total	62.6%	37.4%	100.0%

Fig. 3 $X^2 = 1.041^a$, $p = .308$

Crosstab

			C2		
			A agree	B disagree	Total
Version	Illustrated	Count	140	151	291
		Expected Count	146.2	144.8	291.0
		% within Version	48.1%	51.9%	100.0%
		% within C2	46.7%	50.8%	48.7%
		% of Total	23.5%	25.3%	48.7%
	Verbal	Count	160	146	306
		Expected Count	153.8	152.2	306.0
		% within Version	52.3%	47.7%	100.0%
		% within C2	53.3%	49.2%	51.3%
		% of Total	26.8%	24.5%	51.3%
Total		Count	300	297	597
		Expected Count	300.0	297.0	597.0
		% within Version	50.3%	49.7%	100.0%
		% within C2	100.0%	100.0%	100.0%
		% of Total	50.3%	49.7%	100.0%

Fig. 4 $X^2 = 2.540^a$, $p = .111$

5 Conclusion

I have argued that experimental philosophers are faced with two distinct problems that arise from the non-homogeneous nature of pragmatic cues: the 'response problem' and the 'interpretation problem'. In order to address the interpretation problem it is suggested that models, in the form of minimally linguistic diagrams, be used to accompany vignette text. A possible risk of adding content to the vignettes is that this might interfere with the process of intuition formation. The study presented here suggests that this is not the case.

The data analysis compared the results from both the Verbal and the Illustrated versions of the survey. The data here presented and discussed only focus on the medium parameter, and no systematic variations is shown in the subjects' responses when answering the Illustrated survey as compared to the Verbal survey. This output suggests the validity of making recourse to illustrations or, more generally, to models in survey-based methodology; the main advantage of models' employment pertains to the significant restriction of the domain of errors related to the interpretation problem, as defined in the present work. Moreover, models of this sort constitute an effective aid for the readers, as they help improve the overall quality of the subjects' comprehension of the vignette. Given the non-linguistic nature of illustrations, they are immune to the interpretation problem, thus illustrated surveys should be regarded as less exposed than linguistic surveys to the challenges posed by language-dependent pragmatic cues.[9]

Acknowledgments The author wishes to thank Angel Qiantong Wu for her assistance with the data collection and analysis, Sara Huang Yang for the drawings for the illustrated survey, as well as Mog Stapleton, Alessandra Fermani, and two anonymous referees who provided constructive comments for improving the manuscript.

References

Alexander, J. (2012). *Experimental philosophy. An Introduction*. Cambridge: Polity Press.

Cappelen, H. (2012). *Philosophy without intuitions*. Oxford: Oxford University Press.

Collett, J. L., & Childs, E. (2011). Minding the gap: Meaning, affect, and the potential shortcomings of vignettes. *Social Science Research, 40*, 513–522.

Cullen, S. (2010). Survey-driven romanticism. *Review of Philosophy and Psychology, 1*, 275–296.

Deutsch, M. (2010). Intuitions, counter-examples, and experimental philosophy. *Review of Philosophical Psychology, 1*, 447–460.

Deutsch, M. (2015). *The myth of the intuitive: Experimental philosophy and philosophical method*. Cambridge Massachusetts: MIT Press.

Devitt, M. (2012). Whither experimental semantics? *Theoria, 27*, 5–36.

Devitt, M. (2014). Linguistic intuitions and cognitive penetrability. In *The baltic international yearbook of cognition, logic, and communication, Vol. 9. Perception and Concepts* (pp. 1–14).

Faia, M. A. (1980). The vagaries of the vignette world: A comment on Alves and Rossi. *American Journal of Sociology, 85*, 951–954.

Fischer, E., & Collins, J. (2015). Rationalism and naturalism in the age of experimental philosophy. In E. Fischer & J. Collins (Eds.), *Experimental philosophy, rationalism, and naturalism* (pp. 3–33). Routledge, London and New York: Rethinking Philosophical Method.

Guo, T., Ji, L. J., Spina, R., & Zhang, Z. (2012). Culture, temporal focus, and values of the past and the future. *Personality and Social Psychology Bulletin, 38*, 1030–1040.

Hofstede, G. H. (2001). *Culture's consequences: Comparing values, behaviors, institutions, and organizations across nations*. London: Sage.

Hughes, R. (1998). Considering the vignette technique and its application to a study of drug-injecting and HIV risk and safer behavior. *Sociology of Health & Illness, 20*, 381–400.

Johnson, T., Kulesa, P., Young, I. C., & Shavitt, S. (2005). The relation between culture and response styles. *Journal of Cross-Cultural Psychology, 36*, 264–277.

Kauppinen, A. (2007). The rise and fall of experimental philosophy. *Philosophical Explorations, 10* (2), 95–118.

Rossi, P. H., & Alves, W. M. (1980). Rejoinder to Faia. *American Journal of Sociology, 85*, 954–955.

[9]This paper offers the presentation and the results of the first phase of a cross-cultural research, and the same survey will be conducted with Western subject as to verify that the responses' insensitivity to the non-linguistic medium is a cross-cultural one.

Sosa, E. (2009). A defense of the use of intuitions in philosophy. In M. Bishop & D. Murphy (Eds.), *Stich and his critics* (pp. 101–112). Wiley-Blackwell: Malden, MA.
Sytsma, J. (Ed.). (2014). *Advances in experimental philosophy*. London: Bloomsbury.
Sytsma, J., & Livengood, J. (2011). A new perspective concerning experiments on semantic intuitions. *Australasian Journal of Philosophy, 89*, 315–332.
Weinberg, J. M., Nichols, S., & Stich, S. (2001). Normativity and epistemic intuitions. *Philosophical Topics, 29* (1&2), 429–460.
Williamson, T. (2007). *The philosophy of philosophy*. Malden, MA and Oxford: Blackwell.

Mathematical Models of Time as a Heuristic Tool

Emiliano Ippoliti

Abstract This paper sets out to show how mathematical modelling can serve as a way of ampliating knowledge. To this end, I discuss the mathematical modelling of time in theoretical physics. In particular I examine the construction of the formal treatment of time in classical physics, based on Barrow's analogy between time and the real number line, and the modelling of time resulting from the Wheeler-DeWitt equation. I will show how mathematics shapes physical concepts, like time, acting as a heuristic means—a discovery tool—, which enables us to construct hypotheses on certain problems that would be hard, and in some cases impossible, to understand otherwise.

Keywords Formalism · Heuristics · Time · Mathematics · Modelling

1 Introduction

Understanding time in physics is a typical problem at the frontier of knowledge, where the inquiry is so hard that we need to rely on very tentative and risky heuristics. So this understanding can contribute both to the on-going tradition of studies on heuristics[1] and to mathematical physics on time. In this paper I argue that mathematics can serve as a heuristic tool in tackling this problem. To this end, I examine two case studies: the first is from elementary mathematics (the numbers line), i.e. the construction of orthodox mathematical modelling of time in classical

[1] See Polya (1954), Hanson (1958), Lakatos (1976), Laudan (1977), Simon (1977), Nickles (1980a, b), Simon et al. (1987), Gillies (1995), Grosholz and Breger (2000), Grosholz (2007), Abbott (2004), Darden (2006), Weisberg (2006), Magnani (2001, 2013), Magnani and Li (2007), Nickles and Meheus (2009), Magnani et al. (2010), Gertner (2012), Cellucci (2013), Ippoliti (2011, 2014).

E. Ippoliti (✉)
Department of Philosophy, Sapienza University of Rome, Via Carlo Fea, 2, 00161 Rome, Italy
e-mail: emi.ippoliti@gmail.com

L. Magnani and C. Casadio (eds.), *Model-Based Reasoning in Science and Technology*, Studies in Applied Philosophy, Epistemology and Rational Ethics 27, DOI 10.1007/978-3-319-38983-7_7

physics, and the second is from advanced mathematics, i.e. the modelling of time in the Wheeler-DeWitt equation.

The mathematical modelling of time exhibits exemplary features of the construction of hypotheses and theories. In particular it shows at a fine grain how conceptualisations are built, how we can go from a conceptualisation to a formalisation, and what is the heuristic role of formalisms[2] and mathematical modelling. In effect, these two examples show well how mathematical modelling shapes physical concepts and theories by actively adding content to them, by attributing properties to the entities contained in the data in order to generate plausible interpretations and explicative frameworks.

More specifically, I will examine how:

(1) mathematics supplies physics with extra ways of generating and testing hypothesis by means of known results, which are especially useful when experiments cannot be performed using current technology.
(2) mathematics offers solutions-awaiting-problems to be used in physics when a mathematical framework for the problem at issue can be found. Much of the mathematics produced does not have a direct physical application, but it can be thought as an engine that continually generates frameworks to be used in problem-solving.

2 The Orthodox Mathematical Modelling of Time in Classical Physics

The heuristic use of a seemingly simple piece of mathematics, the real number line, shapes the way classical physics conceptualizes and expresses time. More specifically the construction of orthodox mathematical modelling of time in classical physics draws on Barrow's analogy between time and the real number line, and shows how mathematics heuristically models the conceptualization of time. Such conceptualization of time can be summarized in the following, analogical, way: "time as the independent variable. Time, as an aggregate of instants, compared with a line, as the aggregate of points" (Barrow 1660, 35).

Of course there are different kinds of lines, that is aggregations of points, that can be associated with the notion of time conceived as an aggregation of instants: straight, circular, etc. In order to produce a conceptualisation and then a mathematical theory of time, Barrow argues for the use of the real line as cogent way of modelling time. He provides reasons and a rational basis for this by means of a list of similarities between time and lines. In this way he constructs an interpretation, which associates features we can identify for 'time' with items in a mathematical

[2]That is the strings of symbols representing objects and operations and assembled according to certain syntactic rules.

model. More specifically he states that time shows "many analogies with a line, either straight or circular, and therefore may be conveniently represented by it; for time has length alone, is similar in all its parts, and can be looked upon as constituted from a simple addition of successive instants or as from a continuous flow of one instant; either a straight or a circular line has length alone, is similar in all its parts, and can be looked upon as being made up of an infinite number of points or as the trace of a moving point" (Barrow 1660, 37).

Then Barrow refines his conceptualization and choses a specific model (straight line over the circular one) on the basis of further analogical arguments and evaluations:

> for as Time consists of parts altogether familiar, it is reasonable to consider it as a quantity endowed with one dimension only; for we imagine it to be made up, as it were, either of the simple addition of rising Moments, or of the continual flux of one Moment, and for that reason ascribe only length to it, and determine its quantity by the length of the line passed over: As a line, I say, is looked upon to be the trace of a point moving forward, being in some sort divisible by a point, and may be divided by Motion one way, [...]; so Time may be conceived as the trace of a Moment continually flowing, having some kind of divisibility from an Instant, and from a successive flux, [...] as it can be divided some how or other. And like as the quantity of a line consists of but one length following the Motion, so the quantity of Time pursues but one succession, stretched out, as it were, in length, which the length of the space moved over shows and determines. We therefore shall always express Time by a right line (Barrow 1660, 14)

This mathematical idea of time and its refinement, in turn, shapes Newtonian mechanics and in general the orthodox mathematical treatment of time in classical physics—see e.g. Windred (1933a, b, 1935) for a nice account of it. As a matter of fact, in classical physics this conceptualization produces a passage from analogy to identity: the mathematical model of time *is* the real number line—time is a line of real numbers, *literally*. The physical puzzle is absorbed by the mathematical entity and the hypothetical and analogical scaffolds employed for its construction are dismantled, so that all the choices made to build it are put aside and not questioned anymore.

Now, modelling time in this way implies that we are transferring to it a set of properties that are only in part deductively or empirically contained in the data, while other properties are extended to time by the mathematical model. Moreover, these properties are ascribed to time all together even if *separable* in principle. So under this specific model, time turns out to be absolute, non-ramified, one-dimensional, oriented, not closed, infinite (in both directions), continuous (see also Fabri 2005 on this point).

These properties are hard to test experimentally and the mathematical model (the real number line) goes strictly beyond the data by ampliating the information, i.e. introducing properties that are not in the data at the beginning of the inquiry. These properties are contained in, and required by, the mathematical model. Therefore the choice of the mathematical representation is crucial for the interpretation of the data, the development of hypotheses about time and the resulting growth of knowledge. In order to discuss this point, I will focus on one of the most controversial, and least experimentally verifiable properties (along with linearity[3]) of time as conceived in classical physics, that is continuity.

2.1 Continuity

Continuity is a nice example of a property extended to the data in classical physics because a specific mathematical model requires it in order to deal with certain problems. In this respect it is strictly heuristic, in the sense that it is adopted because it enables us to solve a class of problems. As matter of fact, real numbers and rational numbers are both dense sets and the former rather than the latter are employed since they completely solve a range of problems via the Intermediate Value Theorem. But in principle there is no need for the use of real numbers in order to account for time: also the set of rationals, as dense, can do that. In effect, also by means of rational numbers we can build intervals arbitrarily small. So continuity is a requirement strictly coming from the mathematical model. This is clear from a simple example: the law of falling bodies, which shows that the rational numbers are not enough to solve the problem—i.e. to perform the calculations.

Let us take the well-known equation for calculating the distance d travelled by an object o falling for a period of time t,

$$d = \frac{1}{2}gt^2$$

If we want to find for any d a solution t, we need a set of numbers where it is always possible to extract the square root—precisely a continuous set. The rational numbers are not enough to perform this calculation. As a matter of fact, under the hypothesis that the space covered is a function of a *continuous* time, via the intermediate value theorem (Bolzano 1817) and the other results on continuous functions, we are guaranteed that for any given interval of time we can always find at least an instant t where a body has a specific position between the starting and the end point of d. But there is no evidential base for this. As a matter of fact, it is the heuristic fruitfulness of the theory developed on this basis, that is its capability of solving problems, that speaks for the fact that such a basis is plausible and cogent to some extent, and not vice versa.

Since the mathematical model is partly suggested by data and experience, and in part goes beyond them, the orthodox mathematization of time embeds a set of properties, some of which are supported by data and experience, while others deliberately amplify them. And we accept them, like continuity, since they give us a way of solving problems, drawing inferences and advancing more and more plausible hypotheses about time and its structure. Moreover, the properties

[3]See e.g. Devito (1997) for a non-linear account of time.

introduced by the mathematical model enable us to test the theory. In the case of continuity, even if we do not have strong evidence that time is continuous, once continuity is extended to time in virtue *of* the mathematical model, we can use other pieces of mathematics to advance conjectures about the characteristics of time.

In particular, we can use known mathematical results about discreteness and continuity to build hypotheses about the characteristics of time and in order to specify what would be a challenge to that hypothesis. For instance, if time were *not* continuous, i.e. discrete,[4] we should expect space to be discrete too (if we accept that space and time are interconnected). On this basis, we can use known mathematics to deepen this analogy with the help of the available formal results. More specifically, space is many-dimensional, and would have to be arranged as a kind of lattice (as made up of discrete units). Accordingly, since a discretization of space alone would be essentially non-relativistic, we can use lattice theory in order to advance conjectures about space-time and its properties by means of its mathematic models. In other worlds, we can develop a scenario under the hypothesis of a "lattice world" (see Heisenberg 1930; Carazza and Kragh 1995), or better a lattice structure into three-dimensional space, which commits us to accept the notion of a 'fundamental length'.[5] For instance, we can have a lattice that is homogeneous, meaning that space is the same at every location (point), and in that case it makes no difference along which straight line it is expressed. On the other side, we can have a lattice that is isotropic—roughly, meaning that from the origin the space looks the same in every direction[6]—and then it has spherical symmetry and can be rotated about its origin. But in that case the origin has a different status from the other points, and there is a difference according to which straight line it is expressed. Conclusion: in that case we cannot have a space-time lattice that is both homogeneous and isotropic. This example shows how we can draw conclusions about homogeneity and isotropy of space-time, which tell us how it should look like under the hypothesis of discreteness. We have built a way to better understand and test our physical theories, a way provided by mathematical modelling under certain physical constraints.

Of course, there are caveats in this way of proceeding and we have to be aware of them. Not only are all these properties of time hard to test, and not only is the evidence speaking for them really scarce (at least in some cases), but we need to keep in mind that this notion of time has been developed and works well for a very limited portion of reality and on the basis of very specific experimental settings and data. Newtonian mechanics and parts of classical physics are based on this set of properties for time, which accounts well for such a (very) limited portion of reality. If we move up or down along orders of magnitude (at a cosmological scale or at

[4]Note that this would also imply the replacement of the differential equations with difference equations, so the time derivative would be replaced by a finite difference.

[5]See Hagar (2014), pp. 64–75, for a careful analysis of the issue.

[6]Formally this means that the space is invariant under rotations in space, since rotations can map any direction onto any other direction.

quantum scale), beyond the limits of this carefully cut portion of reality, all those properties of time would not to be tenable anymore. We could require a quite different notion of time, and this notion could even be not compatible with other notions of time developed for different field of phenomena and experimental settings. A unifying framework for the notion of time could even be not possible. This problem arises even more strongerly in our next example—the Wheeler-DeWitt equation.

3 The Modelling of Time in the Wheeler-DeWitt Equation

The Wheeler-DeWitt equation is the outcome of an attempt to reconcile in a consistent way Quantum Mechanics (QM) and General Relativity (GR). The construction of the Wheeler-DeWitt equation, that is

$$\left((\hbar G)^2 \left(q_{ab}q_{cd} - \frac{1}{2}q_{ac}q_{bd}\right)\frac{\delta}{\delta q_{ac}}\frac{\delta}{\delta q_{bd}} - \det\, qR[q]\right)\Psi(q) = 0 \tag{1}$$

sheds light on how mathematics helps us to discover and understand supposedly unintelligible physical ideas. In particular I will argue that the main characteristic of (1), the seemingly puzzling absence of the variable t (a temporal dependence and the consequent ‘problem of time’), is a consequence of the mathematics employed in its construction and the partial interpretation that it encapsulates. There are several attempts to reconcile QM and GR: three influential ways are the covariant line of research, the canonical line of research, and the sum over histories line of research (see e.g. Rovelli 2001).

I will focus on the canonical line of research that has been put forward by using a specific mathematical procedure—the so-called canonical quantization. The starting point of the inquiry here is that “the application of the conventional formulation of QM to the space-time geometry itself is especially problematic in the case of a closed universe, since there is no place to stand outside the system to observe it” (Everett 1957, p. 455). The main hypothesis to handle this is “to regard pure wave mechanics as a complete theory. It postulates that a wave function that obeys a linear wave equation everywhere and at all times supplies a complete mathematical model for every isolated physical system without exception […]. The wave function is taken as the basic physical entity with no a priori interpretation. Interpretation only comes after an investigation of the logical structure of the theory. Here as always the theory itself sets the framework for its interpretation” (*ibid.*). This conceptualization, in a nutshell, aims at rewriting Einstein’s equations for gravity in the same form as the equations for electromagnetism. Since GR and electromagnetism are both classical field theories, and since canonical quantization has worked for the latter, it seemed reasonable to use this technique, which ended up with a quantum theory of electromagnetism, also for relativity.

But the approach to gravity and relativity in that way, as we will see, ends up with a strange equation. More specifically, DeWitt (1967) produced a controversial equation by modelling a quantum wave mechanics (i.e. the Schrödinger equation) so to deal with a domain (like GR) that is invariant under diffeomorphism.[7] He starts from the idea, explicitly stated in the 1966 preprint of his paper, that as long as we can conceive the wave function as having physical reality independently of any 'classical' observers, it is plausible to ascribe a quantum state (and hence a wave function) to the universe as a whole. I will offer just a sketch of the formal construction put forward by DeWitt, which should be enough for the purposes of this paper.

First of all the use of the wave mechanics, and the wave function Ψ, of course implies the use of the Schrödinger equation:[8]

$$\frac{\partial^2\Psi}{\partial x^2} + \frac{8\pi^2 m}{h^2}(E - V)\Psi = 0$$

The first step to combine GR and QM in this way is to impose constraints: Einstein's empty-space field equations may be obtained by taking the Poisson bracket of the various dynamical variables with the Hamiltonian and then imposing external constraints, that is restrictions on coordinates and on dynamical freedom of the field, in particular quantum constraints for diffeomorphism in order to express covariance. Once a Hamiltonian constraint has been imposed, you can obtain a wave function like this

$$\left(G_{ijkl}\frac{\delta}{\delta\gamma_{ij}}\frac{\delta}{\delta\gamma_{kl}} + \gamma^{1/2\,(3)}R\right)\Psi[^{(3)}\mathcal{G}] = 0, \tag{2}$$

where

$$G_{ijkl} \equiv \frac{1}{2}\gamma^{-1/2}(\gamma_{ik}\gamma_{jl} + \gamma_{il}\gamma_{jk} - \gamma_{ij}\gamma_{kl}). \tag{3}$$

After a suitably interpretation and manipulation of the Hamiltonian constraint condition, we get

$$\left(\frac{1}{48\pi^2}R^{-1/4}\frac{\partial}{\partial R}R^{-1/2}\frac{\partial}{\partial R}R^{-1/4} - 12\pi^2 R + Nm\right)\Psi = 0, \tag{4}$$

[7]Diffeomorphism invariance is a way of mathematically expressing general covariance—the invariance of the form of physical laws under arbitrary differentiable coordinate transformations and then the background independence of a theory.

[8]Where ∂^2 is the second derivative w.r.t. x; x the position; Ψ the Shroindger wave function; E the energy; V the potiantial energy.

from this it follows

$$-(3/64\pi^2)\partial^2\Phi/\partial X^2 + 12\pi^2 X^{2/3}\Phi = Nm\Phi. \tag{5}$$

which is a particular Schrödinger equation. From here we can write down the following two general Schrödinger equations:

$$i\partial\mathbf{\Phi}/\partial\tau = \Im C(q, -i\partial/\partial q)\mathbf{\Phi}, \tag{6}$$

$$i\partial\Phi/\partial\tau = \Im C(R, -i\partial/\partial R)\Phi. \tag{7}$$

Now we can rewrite them in an extended way, obtaining the following functional differential equation

$$\left(\left(\hbar G\right)^2\left(q_{ab}q_{cd} - \frac{1}{2}q_{ac}q_{bd}\right)\frac{\delta}{\delta q_{ac}}\frac{\delta}{\delta q_{bd}} - \det qR[q]\right)\Psi(q) = 0$$

that is the Wheeler-Dewitt equation (1), which can be read as saying that the functional over all the possible universe's phase states, expressed as a wave function, has a Hamiltonian Constraint equal to zero, in short

$$\widehat{H}(x)|\psi\rangle = 0$$

Bottom line: this construction shows that the Schrödinger equation, when extended to a diffeomorphism invariant context—like gravity—generates (1). It is worth noting here that the mathematical modelling involves a meager structure, namely a Hilbert space, a Hamiltonian, and a Schrodinger equation for vectors in the Hilbert space. Tellingly, Eq. (1) is not a well-defined mathematical object: it not a rigorous as, in general, the physical variables do not have a continuum limit. But the most debated[9] feature of this equation is the absence of the variable t (a temporal dependence). In effect, while the wave packets in QM spread in 'time', the packet ψ in the (1) does not and:

> it is for this reason that we may say that 'time' is only a phenomenological concept, useful under certain circumstances. It is worth remarking that it is not necessary to drag in the whole universe to argue for the phenomenological character of time. If the principle of general covariance is truly valid then the quantum mechanics of every-day usage, with its dependence on Schrodinger equations of the form (6) or (7), is only a phenomenological theory. For the only 'time' which a covariant theory can admit is an intrinsic time defined by the contents of the universe itself (Dewitt 1967, 160)

More precisely, "any intrinsically defined time is necessarily non-Hermitian, which is equivalent to saying that there exists no clock, whether geometrical or material, which can yield a measure of time which is operationally valid under all circumstances, and hence there exists no operational method for determining the

[9] See Parentani (1997), Peres (1999).

Schrödinger state function with arbitrarily high precision" (Dewitt 1967, *ibid.*). This raises the well-known 'problem of time', which is sometimes interpreted as saying that since in (1) there is nothing about time evolution, the universe is frozen in time—at its fundamental level may be static. More specifically the equation says that in a quantum diffeomorphism-invariant universe time is meaningless. In turn, this scenario opens up another problem, namely how the 'time' that we experience emerges from a static universe.[10] These last points show how and to what extent mathematical modelling is a heuristic tool by solving problems and posing new ones. (1) offers a hypothesis to manage the reconciliation of GR and QM, and in doing this it also opens up new problems—namely the problem of time and the problem of emergence of 'time' in a 'timeless' universe, which in turn require solutions, interpretations and new hypotheses.

Even if there are several possible ways of approaching this problem (e.g. imposing constraints, etc.), a preliminary and intended interpretation of the physical problem shapes the mathematical model and the formalism. In the construction of (1) the intended interpretation is the one that uses a wave function in order to account for gravity. But also the converse holds—that is the formalism drives interpretations. The mathematical model generates and shapes also unintended interpretation and knowledge. And reasoning about and manipulating the formal expressions do reveal properties and relations about the issue at stake.

In this sense it is not so surprising the fact that the Schrödinger equation, when extended over a diffeomorphism invariant context like gravity, ends up with an equation which simply says nothing about time evolution: it follow from the formalism employed in its construction. The formalism shapes part of the content of the relations and the objects involved in the problem. In effect, any procedure for quantizing field theories that are invariant under diffeomorphism will result in the removal of any temporal dependence, in other words the variable t. In this sense the absence of t derives from the very construction of the mathematical model and the choices made to build it. More in detail, the metric field here is a dynamic variable and not a parameter and, since it is this that describes how the fields evolve, it is not possible to specify a unique external frame of reference to define a temporal evolution.

So, on one side the mathematical models and formalism are the end-point of a conceptualization, and on the other side, once they have been built, they can serve as starting-points for new conceptualizations and interpretations of the initial problem. The mathematical model and its formalism add pieces of information to the data, but do not exhaust them, leaving room for several possible interpretations, and putting constraints on further conceptualizations.

[10]See for instance the answer provided by the termal time hypothesis in Connes and Rovelli (1994).

4 Remarks on the Employed Heuristics and Ways of Advancing Knowledge

The notion of time shows how the use of mathematical modelling is a tool for ampliating knowledge in at least two ways:

(1) mathematics supplies physics with extra ways of generating and testing hypothesis by means of known results, which are especially useful when experiments cannot be performed using current technology. As observation and experimentation become more and more difficult to perform (as beyond the limits of current technology), mathematics becomes an increasingly necessary tool to enlarge human understanding of the problem at issue: it plays a heuristic role and becomes a 'laboratory' where inquiry takes place. More in detail, mathematics provides constraints, and shapes the form and content of certain relations, which cannot be determined simply from the phenomena themselves, or better from the data we have collected about a given phenomenon.
(2) mathematics offers solutions-awaiting-problems to be used (in physics in this case) when a suitable mathematical framework for the problem at issue can be found. Much of the mathematics produced does not have a direct physical application, but it can be thought of as an engine that continually generates frameworks and tools to be used in problem-solving.

More in detail, the two examples above display that once a conceptualization has been developed and mathematically set up—that is an interpretation that associates terms in the formalism to entities in a domain has been built—, we have that:

- mathematical modelling provides constraints, shapes the form and the content[11] of certain entities and relations, which cannot be determined simply from the data that we have about a certain phenomenon;
- reasoning about or manipulating the formalisms can reveal deep properties of the original situation and can qualify what would count as an empirical test for the theory.
- the simple rewriting by means of different formalisms of a problem can create new viewpoints and interpretations.

In the construction of a mathematical hypothesis to solve a problem, several concepts are selected, combined and pushed to the extreme, so to build scenarios with their own logic and constraints—scenarios that might become 'real' or that will remain physical possibilities. In this way mathematics is a means of cultural evolution that offers a fast and effective way of solving problems, providing us with an increasing set of tools to apply to our conceptualization of physics puzzles. It is a mode of anticipating solutions. It is worth noting here that it is not a mere, passive

[11]See Tzanakis (2002a, b) on the way mathematical modelling can determine the content of physical entities and theories.

correspondence, or an alleged isomorphism, between the available mathematics and the conceptualized physics puzzle that enables their application, but a process of approximation and accommodation. In effect, this application requires an interpretation and a translation into formalism, which is an active process whereby something is added to and something is lost from the phenomenon that we are trying to model. What is added is a set of new determinations, objects, properties, and relations, which the chosen piece of mathematics brings in, and which are not all contained in the data at he beginning of the process. What can be lost is a set of characteristics of the problem that are not treatable using known mathematics and conceptualisation and that are thus to be ignored or abandoned.

Obviously, the application of these mathematical tools is possible only when they can be adjusted so to provide a reasonable approximation to our conceptualization of available data about a given phenomenon. And "when no suitable mathematical tools are available, we try to develop them by inventing new mathematics. If we are unable to do so, we cannot express a law of nature in mathematical terms and must content ourselves with expressing it in ordinary language" (Cellucci 2015, 36).

In order to conceptualize a physics problem, that is a limited portion of the world, you have to make a selection of the relevant aspects of the phenomenon under investigation and cut off other aspects. The adequacy of this selection, whenever can be figured out, varies according to the scope and the purpose of it. For example, treating planets as points can be acceptable in the design of an interplanetary trip, but not in the case of a landing on one of them; Newtonian mechanics is good for calculations and predictions over a comparatively short period of time, but not over a time scale of millions of years. More in detail, a selection of variables accounting for a given phenomenon extrapolates and isolates not simply those features that are relevant by themselves, if anything like this it is possible,[12] but more often it extrapolates and isolates those ones that can be approached by means of already available theoretical frameworks and mathematical tools. In a sense they are selected just in view of, or because of the employment of these frameworks and tools. This means that are selected those aspects of objects and relations that can be treated in terms of the properties, constraints, or relations of these existing tools and frameworks. That can appear as a problematic and conservative approach, but that is just the way human conceptualization tends to proceed, and, in a specific sense, has to.

In effect, the only way human understanding has to approach what is new (unknown) and unfamiliar is by means of what is known and familiar[13]—for instance by using straight lines in order to make sense of some aspects of time, as we have seen, or waves and particles in order to understand phenomena at the quantum scale. The simple act identifying an (new) entity as well as an attempt at its

[12]See e.g. Ippoliti (2006) for a critical examination of the notion of relevance.

[13]See e.g. Brown (2015), Bunge (1981), Turner (1988, 2005), Johnson (1988) on this point.

interpretation require an assimilation of new phenomena to existing conceptualizations. It is by means of a continual confrontation between what is new to what is known that we proceed in building knowledge—for a part of knowing something is to know what it is not. The basic idea here is that this process of assimilation and accommodation in the long run can generate new conceptualizations (see Piaget and Cook 1952) that might eventually lead to an altered versions of the existing conceptualizations so as to produce new ones that can be used to understand new phenomena, that is phenomena that cannot be understood in terms of existing conceptualisation.[14] Of course this process is risky and problematic, but it offer the basis on which we can construct formal expressions and conceptual systems.

The relation between conceptualizing and formalising can be critical[15] and, tellingly, it is not unidirectional. That is, we can also start from formalism in order to produce a new conceptualization of a problem, e.g. simply rewriting it. Of course the preliminary conceptualization, and the consequent interpretation of the available data, drives and shapes the mathematical modelling and its formalism. For instance in (1), the choice of the wave function ψ and the wave packet, the Hamiltonian, the several constraints etc., are the result of a preliminary conceptualization aiming at expressing the hypothesis that a quantum state (and hence a wave function) can be applied to the universe as a whole and set up a specific formalism. Mathematical models and formalism are tools to express the meaning that comes out of a conceptualization. They are languages[16] and the manipulations that can be performed on their formal expressions are means to deepen our understanding of the phenomena under inquiry. A mathematical model and its formalism embed a partial interpretation of the phenomenon under inquiry, intended or not. More precisely, they encode information about a set of features of the phenomenon under investigation selected after the preliminary conceptualisation. As I have stressed, most of the time this selection of properties is made via an accommodation to existing models and formalisms: that is, we choose to focus on certain properties because they can be approached and expressed by existing mathematical models and formalisms—and not necessarily because (we think) they are the crucial ones for its understanding.

But mathematical tools and formalisms are underdetermined by the physical phenomenon under inquiry, or better by our conceptualisations of it. In effect, these tools can be put in use when they offer an acceptable approximation to our conceptualization of a portion of reality, and since they are just approximations, "other mathematical tools can always be found that provide a better approximation to that conceptualization" (Cellucci 2015, 37). In fact there are several formalisms that can be used to treat a given problem, and they can produce quite different outcomes. For instance in the case of the Wheeler-DeWitt equation, by using a different formalism,

[14]For instance Feynman integrals can be regarded as an altered version of the notion of integral—see Ippoliti (2013).

[15]See Kvasz (2008) for a deep account of this point.

[16]See Butterworth (1999), Devlin (2000), Lakoff and Nunez (2001) on this point.

namely the Hamilton-Jacobi one, it is possible to construct a dynamical time (that is an internal 'clock') and then employ it for a consistent canonical quantization of relativity, with the correct classical limit, and therefore avoiding the main problems in (1)—see Peres (1999).

Thus, on one side a formal-mathematical treatment conveys a partial interpretation of a physics problem, which is necessary to set up the formalism. This interpretation shapes the form taken by part of relations, properties, constraints, etc. occurring in the formalism. But on the other side, the formalism is not exhausted by the preliminary interpretation of the data, as it also embeds *other* properties that can be used to infer or examine further issue of the problem. A formal expression deductively implies other properties and formal expressions, or it can be derived by other formal expression, so that it can produce new determinations and solutions about the problem at issue, which can be ascribed to the phenomenon under inquiry. Since the construction of a particular formalism conveys a partial interpretation and embeds other properties that can be extended to the phenomenon under investigation, the construction and the choice of a specific formalism is never a 'neutral' act. It is a mode of conceptualising. This crucial and controversial step is an engine of ampliation of knowledge—capable of generating new interpretations and hypotheses. But it does not come cheap.

4.1 Epistemic Risks, Benefits and Forced Choices

This heuristic use of mathematical modelling and the resulting formalisms of course have epistemic risks, gains and forced choices. Let us start with the epistemic risks.

First of all, the fact that the construction of a conceptualisation moves from the attempt to assimilate the 'new' to existing conceptualizations might lead to inadequate outcomes. In this sense, quantum theory might be exemplary: here the attempt to approach quantum phenomena in terms of conceptualization developed for classical physics (i.e. waves and particles) leads to very strange results—'entities' acting both like waves and particles, waves or particles interfering with themselves, or particles following infinite numbers of paths at the same time, just to mention a few. As noted by Feynman:

> [...] the rules for the motions of particles were incorrect. The mechanical rules of "inertia" and "forces" are wrong—Newton's laws are wrong—in the world of atoms. Instead, it was discovered that things on a small scale behave nothing like things on a large scale. That is what makes physics difficult—and very interesting. It is hard because the way things behave on a small scale is so "unnatural"; we have no direct experience with it. Here things behave like nothing we know of, so that it is impossible to describe this behavior in any other than analytic ways. It is difficult, and takes a lot of imagination" (Feynman 1963, p. 33).

In this sense, it is not surprising that the quantum world shows such 'oddities'. The theoretical frameworks and the formalism employed in the current approaches to quantum world have been developed to account for specific entities, i.e. waves and particles, and so they could not work for 'entities' and proprieties at atomic scale. This does not mean that we cannot calculate things with these tools, but that these conceptualizations and formalisms could not provide a good "machinery",[17] quoting Feynman (1977), of the quantum world. The endpoint of this conceptualisation and formalisation is good enough for certain calculations and predictions, but not for a deep understanding and explanation (see Morrison 2000 on this point). The mathematical frameworks could not even offer understanding of the specific details of a phenomenon, leaving open the question of how to interpret a formal model, and whether or not part of the concepts in that interpretation have a counter-part in reality. As a matter of fact, the mathematical model could even hinder the understanding of a problem, since the problem could require genuinely new conceptualizations, mathematical models and formalisms.

A further complication comes from the very nature of formalisms. As I have pointed out, not simply the passage from conceptualizing to formalising, but also simply playing with formalisms, such as rewriting with a different formalism the original situation, can be problematic, since the construction and choice of formalism is essentially a conceptual act, intended or not. Now, this step is controversial in many ways, as you have to take on several risks. First, as we move forward in the mathematical conceptualization of a problem by expressing it formally and manipulating such formal expression, we run the risk of a progressive disconnection between the formal expression and the conceptualisation of the phenomenon that it aims at expressing.

Second, just as in the case of (1), you can end up with a non-rigorous, not well-defined mathematical object, a 'bad' piece of mathematics that requires foundation and explanation, as we do not know if it has contradictions, caveats and oddities to be fixed.

Third, using a formalism to understand the machinery of the phenomenon might lead to a dead end. Whilst whole parts of a phenomenon might be obscure, a formalism is a glass-box, for we can determine what is going on inside it. So a common heuristic and inferential move is to use the machinery of the formalism as a guide to understand the machinery of the phenomenon. This move has plenty of risks. A clear lesson from quantum mechanics, and also the Eq. (1), is that formalisms are multivalent objects. More precisely:

(a) given formalism can be conceptualised and interpreted in multiple, alternative ways. This means that the empirical success of formalism might say nothing about the machinery of the phenomenon expressed by it.
(b) we can build alternative formalisms for the same phenomenon, each generating different interpretations for it.
(c) a formalism conveys properties derived from conceptualisation of other phenomena. When we employ such a formalism in the attempt to understand a new phenomenon, we are implicitly assimilating at least portions of the new

[17]The "machinery" is the term used by Feynman to express the mechanism that explains why, but especially how, certain processes take place—see also Morrison (2000, p. 3) on this point.

phenomenon to portions of the known ones, as part of these properties, entities, or relations are built-in the formalism. But in that case we could argue that the formalism is inadequate to account for the phenomenon if it is really new, that is radically different from what we known.

Bottom line: “we must be careful not to mistake such tools for the world itself. Although scientific theories concern reality, there is no isomorphism between them and reality” (Cellucci 2015, 37). Certain concepts or relations occurring in our formal expressions might not have counter-parts in reality at all and, additionally, they are subjected to alternative interpretations. Thus, the multivalent nature of formalisms seems to pose a big threat to the heuristics of approaching a problem by means of existing tools and formalisms. Not only does there seem to be no isomorphism, but also no convergence between formalism and reality. So, the idea that by assimilating and accommodating a new phenomenon to existing conceptualisations we build an altered version of it that in the long run could generate genuinely new conceptualization capable of accounting for the new phenomena is very weak and risky. Since we can have both alternative interpretations of a given formalism and alternative formalisms for the same conceptualization, if we begin the process “from a particular formalism, we may well end up with multiple, competing interpretations. And we should expect different outcomes if we begin from different formalisms. There is little reason for thinking that the Piagetian process will converge to a single account of the underlying reality” (Brown 2015, 6).

At this point, the status of Eq. (1) becomes very vague. What is (1) saying? Maybe the problem of time is not ‘real’, for it is simply a scenario resulting from the kind of mathematics employed in the modelling of the problem—i.e. canonical quantization over diffeormophic invariant domains. Moreover, different formalisms for canonical quantization of relativity, as we have seen, produce two opposite outcomes—one timeless and one with a ‘clock’ via the Hamilton-Jacobi formalism. Or maybe time in QM is a different notion than time in GR, and they cannot be consistently reconciled in a mathematical framework.

An answer to all this could be that “models and concepts are to be understood as constructive images that either represent physical possibilities or are used for purely heuristic purposes. When there is independent experimental evidence available for these hypotheses or quantities, the debate about their realistic status can begin” (Morrison 2000, 107).

In the end, all these critical remarks are reasonable. These risks are all there. But we have to take them on if we aim at ampliating knowledge, as we do not seem to have a choice. They are forced choices, but there are rational ways to handle them, and a few ways are better than others. We need to employ something that we already know in the attempt of understanding something new or unknown, by comparing it with the ‘old’ and the ‘known’. ‘New’, ‘problem’, ‘data’ are all positional notions: they are such only at a given time in relation to existing knowledge and conceptualisations. Moreover, there is nothing to protect us from the fact that in approaching a problem we could find out that almost nothing of what we already know would help us in its understanding—that is to face a pure negative knowledge scenario. In this case, that something will remain unknown to us as member of the human race. A price must paid: epistemic gains require risks. There is no free lunch in heuristic logic.

Acknowledgments I would like to thank Sergio Caprara, Angleo Vulpiani and the other friends at the Dept. of Physics—Sapienza University of Rome for the fruitful dialogues about technical as well as theoretical issues. I would also like to thank the two anonymous referees for their comments and suggestions that helped me to improve the paper.

References

Abbott, A. (2004). *Method of discovery*. New York: W.W. Norton & Company Inc.

Barrow, I. (1660). Letiones geometricae (J. M. Child, Trans., 1916). *The geometrical lectures of isaac barrow*. London: The Open Court Publishing Company.

Bolzano, B. (1817). Rein analytischer Beweis des Lehrsatzes dass zwischen je zwey Werthen, die ein entgegengesetztes Resultat gewaehren, wenigstens eine reele Wurzel der Gleichung liege. Prague. (S. B. Russ "A Translation of Bolzano's Paper on the Intermediate Value Theorem", Trans.) *History of Mathematics, 7*, 156–185, 1980.

Brown, H. (2015). *Against interpretation in mathematical physics*. Draft.

Bunge, M. (1981). Analogy in quantum theory: From insight to nonsense. *The British Journal for the Philosophy of Science, 18*(4), 265–286.

Butterworth, B. (1999). *The mathematical brain*. London: Macmillan.

Carazza, B., & Kragh, H. (1995). Heisenberg's lattice world: The 1930 theory sketch. *American Journal of Physics, 63*, 595.

Cellucci, C. (2013). *Rethinking logic*. New York: Springer.

Cellucci, C. (2015). Naturalizing the applicability of mathematics. *Paradigmi Rivista di critica filosofica, 2*, 23–42.

Connes, A., & Rovelli, C. (1994). Von neumann algebra automorphisms and time-thermodynamics relation in general covariant quantum theories. *Classical and Quantum Gravity, 11*, 2899–2918.

Darden, L. (Ed.). (2006). *Reasoning in biological discoveries: Essays on mechanisms, inter-field relations, and anomaly resolution*. New York: Cambridge University Press.

Devito, C. L. (1997). A non-linear model for time. In W. G. Tifft & Cocke, W. J. (Eds.), *Modern mathematical models of time and their applications to physics and cosmology* (pp. 357–370). Springer.

Devlin, K. (2000). *The math gene: How mathematical thinking evolved and why numbers are like gossip*. New Tork: Basic Books.

DeWitt, B. S. (1967). Quantum theory of gravity. I. The canonical theory. *Physical Review, 160*, 1113.

Everett, H. (1957). Relative state formulation of quantum mechanics. *Reviews of Modern Physics, 29*, 454–462.

Fabri, E. (2005). *Insegnare relatività nel XXI secolo*. Bollettino trimestrale dell'Associazione Italiana per l'insegnamento della fisica. XXXVIII (2). Supplemento.

Feynman, R. (1963). *Six easy pieces*. New York: Basic Books.

Feynman, R. (1977). *Lectures on physics*. Addison Wesley.

Gertner, J. (2012). *The idea factory*. New York: Penguin Press.

Gillies, D. (1995). *Revolutions in mathematics*. Oxford: Oxford University Press.

Grosholz, E. (2007). *Representation and productive ambiguity in mathematics and the sciences*. Oxford: Oxford University Press.

Grosholz, E., & Breger, H. (Eds.). (2000). *The growth of mathematical knowledge*. Dordercht: Springer.

Hagar, A. (2014). *Discrete or continuous? The quest for fundamental length in modern physics*. New York: Cambridge University Press.

Hanson, N. (1958). *Patterns of discovery: An inquiry into the conceptual foundations of science*. Cambridge: Cambridge University Press.

Heisenberg, W. (1930). The self-energy of the electron. In A. Miller (Ed.), *Early quantum electrodynamics* (pp. 121–128). Cambridge: Cambridge University Press 1994.

Ippoliti, E. (2006). Demonstrative and non-demonstrative reasoning by analogy. In C. Cellucci & P. Pecere (Eds.), *Demonstrative and non-demonstrative reasoning in mathematics and natural science* (pp. 309–338). Cassino: Edizioni dell'Università di Cassino.

Ippoliti, E. (2011). Between data and hypotheses. In C. Cellucci, E. Grosholz, & E. Ippoliti (Eds.), *Logic and knowledge*. Newcastle Upon Tyne: Cambridge Scholars Publishing.

Ippoliti, E. (2013). Generation of hypotheses by ampliation of data. In L. Magnani (Ed.), *Model-based reasoning in science and technology* (pp. 247–262). Berlin: Springer.

Ippoliti, E. (Ed.). (2014). *Heuristic reasoning*. London: Springer.

Johnson, M. (1988). Some constraints on embodied analogical understanding. In D. H. Helman (Ed.), *Analogical reasoning: Perspectives of artificial intelligence, cognitive science, and philosophy*. Dordrecht: Kluwer.

Kvasz, L. (2008). *Patterns of change. Linguistic innovations in the development of classical mathematics*. Basel: Birkhäuser.

Lakatos, I. (1976). *Proofs and refutations: The logic of mathematical discovery*. Cambridge: Cambridge University Press.

Lakoff, G., & Nuñez, R. (2001). *Where mathematics come from: How the embodied mind brings mathematics*. New York: Basics Books.

Laudan, L. (1977). *Progress and its problems*. Berkeley and LA: University of California press, California.

Magnani, L. (2001). *Abduction, reason, and science. Processes of discovery and explanation*. New York: Kluwer Academic.

Magnani, L. (2013). *Model-based reasoning in science and technology*. Berlin: Springer.

Magnani, L., & Li, P. (Eds.). (2007). *Model-based reasoning in science, technology, and medicine*. London: Springer.

Magnani, L., Carnielli, W., & Pizzi, C. (Eds.). (2010). *Model-based reasoning in science and technology: Abduction, logic, and computational discovery*. Heidelber: Springer.

Morrison, M. (2000). *Unifying scientific theories*. New York: Cambridge University Press.

Nickles, T. (Ed.). (1980a). *Scientific discovery: Logic and rationality*. Boston: Springer.

Nickles, T. (Ed.). (1980b). *Scientific discovery: Case studies*. Boston: Springer.

Nickles, T., & Meheus, J. (Eds.). (2009). *Methods of discovery and creativity*. New York: Springer.

Parentani, R. (1997). Interpretation of the solutions of the Wheeler-DeWitt equation. *Physical Review D, 56*(8), 4618–4624.

Peres, A. (1999). Critique of the Wheeler-DeWitt equation. In A. Harvey (Ed.), *On Einstein's path* (pp. 367–379). New York: Springer-Verlag.

Piaget, J., & Cook, M. T. (1952). *The origins of intelligence in children*. New York: International University Press.

Polya, G. (1954). *Mathematics and plausible reasoning*. Princeton: Princeton University Pres.

Rovelli, C. (2001). *Notes for a brief history of quantum gravity*. arXiv:gr-qc/0006061v3.

Simon, H. (1977). *Models of discovery*. Dordrecht: Reidel.

Simon, H., Langley, P., Bradshaw, G., & Zytkow, J. (Eds.). (1987). *Scientific discovery: Computational explorations of the creative processes*. Boston: MIT Press.

Turner, M. (1988). Categories and analogies. In D. H. Helman (Ed.), *Analogical reasoning: Perspectives of artificial intelligence, cognitive science, and philosophy*. Dordrecht: Kluwer.

Turner, M. (2005). The literal versus figurative dichotomy. In S. Coulson & B. Lewandowska-Tomaszczyk (Eds.), *The literal and nonliteral in language and thought* (pp. 25–52). Frankfurt: Peter Lang.

Tzanakis, C. (2002a). On the relation between mathematics and physics in undergraduate teaching. In I. Vakalis, et al. (Eds.), *2nd international conference on the teaching of mathematics (at the undergraduate level)* (p. 387). New York, NY: Wiley.

Tzanakis, C. (2002b). Unfolding interrelations between mathematics and physics, in a presentation motivated by history: Two examples. *International Journal of Mathematical Education in Science and Technology, 30*(1), 103–118.

Weisberg, R. (2006). *Creativity: Understanding innovation in problem solving, science, invention, and the arts*. Hoboken (NJ): John Wiley & Sons.

Windred, G. (1933a). The history of mathematical time: I. *Isis, 19*(1), 121–53.

Windred, G. (1933b). The history of mathematical time: II. *Isis, 20*(1), 192–219.

Windred, G. (1935). The interpretation of imaginary mathematical time. *The Mathematical Gazette, 19*(235), 280–290.

Modelling Scientific Un/Certainty. Why Argumentation Strategies Trump Linguistic Markers Use

Luigi Pastore and Sara Dellantonio

Abstract In recent years, there has been increasing interest in investigating science communication. Some studies that address this issue attempt to develop a model to determine the level of confidence that an author or a scientific community has at a given time towards a theory or a group of theories. A well-established approach suggests that, in order to determine the level of certainty authors have with regard to the statements they make, one can identify specific lexical and morphosyntactical markers which indicate their epistemic attitudes. This method is considered particularly appealing because it permits the development of an *algorithmic model* based on the quantitative analysis of the occurrence of these markers to assess (almost) *automatically* and *objectively* the opinion of an author or the predominant opinion of a scientific community on a topic at a given time. In this contribution we show that this line of research presents many kinds of problems especially when it is applied to research articles (rather than to popular science texts and basic research reports). To this aim, we propose two main lines of reasoning. The first one relies generally on the argumentative structure of scientific articles and shows that certainty/uncertainty markers are used differently in different argument forms and that therefore their number/frequency of use does not offer reliable indications for determining whether the topic at issue is considered by the authors to be more or less factual/speculative. The second one is based on the analysis of a sample of psychiatric research articles on homosexuality written over a long time span and taken from *The British Journal of Psychiatry*. Since the psychiatric perspective on homosexuality changed radically during the decades in which these articles were published, they offer an inventory of various kinds of argumentative strategies directed both at defending and confuting dominant as well as marginal positions. We focus especially on uncertainty markers and show that frequently the positions stated using expressions indicating uncertainty are actually not considered as

L. Pastore (✉)
Università degli Studi di Bari Aldo Moro, Bari, Italy
e-mail: luigi.pastore@uniba.it

S. Dellantonio
Università degli Studi di Trento, Trento, Italy
e-mail: sara.dellantonio@unitn.it

L. Magnani and C. Casadio (eds.), *Model-Based Reasoning in Science and Technology*, Studies in Applied Philosophy, Epistemology and Rational Ethics 27, DOI 10.1007/978-3-319-38983-7_8

conjectural or speculative by their authors, but that the use of uncertainty markers is motivated by a number of different and often incongruent rhetorical strategies.

Keywords Hedging · Boosting · Argumentation · Certainty/uncertainty in science · Scientific writings

1 Introduction

In recent years, there has been increasing interest in examining and modeling the pragmatic aspects of information exchange in real communication dynamics (Kamio 1997; Ifantidou 2005) as well as in investigating how these models might be applied to the analysis of oral and written science communication (Hyland 1995, 1996a, 1996b, 2002; Aguilar 2008; Heritage 2011). This work pursues a number of different aims: some research investigates scientific communication per se (see e.g. Crismore and Farnsworth 1990; Hyland 1997, 2011; Heritage 2011); while other research addresses the possible differences in communication across scientific disciplines (see e.g. Skelton 1997; Agarwal and Yu 2010; Chao and Hu 2014; Afsar et al. 2014); still other research relies on the analysis of communication to gain new insights into scientific and/or nonscientific reasoning (see e.g. Origgi and Sperber 2002; Wilson 2000, 2005; Durik et al. 2008; Winter et al. 2015), etc. However, some studies addressing scientific communication seek a different and possibly more ambitious objective, i.e. to find a (possibly automatic) method to distinguish between certain (i.e. factual) and uncertain (i.e. speculative) information in texts. The contention is that this method assures that information can be correctly detected and extracted from the source. The goal of this line of investigation is to develop a model to determine the level of confidence that an author or a scientific community (that expresses itself through corpora of scientific texts or oral communications) has at a given time towards a theory or a complex of theories.

One of the methods considered to be the most *objective* for assessing the degree of certainty/uncertainty regarding the various hypotheses embedded in a scientific text consists in analyzing the occurrences of specific lexical or morphosyntactical certainty or uncertainty markers used by the authors (which are also called respectively *boosters* and *hedges*; see e.g. Hyland 1996b, 2004). Indeed, this approach apparently offers both a quantitative and a qualitative measure of the degree of certainty/uncertainty conveyed by a scientific text, since it quantifies the number/frequency of marker use to determine what information is certain and what is uncertain. This method is considered particularly promising because it could lead to the development of an *algorithmic model* based on the analysis of the occurrence of specific linguistic markers to assess (almost) *automatically* and *objectively* the opinion of an author or the predominant opinion of a community on a topic or at least the general attitude of one or more authors towards an issue at a given time. Thus, this model could also be used to take practical decisions based on scientists' confidence with respect to specific views. Recently, this project has gained the attention of the research field known as *Natural Language Processing* (NLP) whose aim is to make Turing's dream come true, i.e. to develop computational systems capable of processing (summarizing, translating, interpreting, answering questions, etc.) natural languages. (see e.g. Auger and Roy 2008; Vincze et al. 2008; Morante and Daelemans 2009; Kim et al. 2009; Farkas et al. 2010; Choi et al. 2012).

In this contribution we aim to show that this line of research on scientific texts is not as promising as it seems and presents a number of relevant problems. Since research articles are the most important source of 'fresh', 'first hand' scientific knowledge and therefore are also the most interesting locus in which to carry out this analysis, we focus specifically on research articles and we show that this kind of linguistic analysis is not adequate to assess whether authors consider the hypotheses they address in their papers to be more or less reliable or uncertain. To support this position, we discuss the relationship between two types of analysis that can be carried out on scientific papers. Specifically, we compare those analyses that aim to assess the degree of certainty/uncertainty of specific positions expressed in scientific papers by quantifying hedges and boosters with those analyses that use the classical techniques of argument evaluation (adopted e.g. by informal logic and critical thinking). On the basis of this comparison, we propose two challenges to the two different lines of argumentation presented by researchers who support the use of boosters and hedges as a measure of certainty/uncertainty.

(i) The first argument is theoretical and takes into account how hedges and boosters are used in various argument forms. Its aim is to show that in the case of texts with an argumentative structure the global number/frequency of occurrence of hedges and boosters does not indicate whether the author considers the matter s/he deals with—i.e. the conclusions s/he draws—to be factual/certain or to be speculative/uncertain (*Thesis1*).
(ii) The second argument relies on the examination of a sample of scientific articles on psychiatry taken from *The British Journal of Psychiatry* (BJP). In particular, we focus on papers addressing a very controversial psychiatric issue that has undergone many conceptual revolutions from the beginning to the end of the 20th century—i.e. *homosexuality*. We examine these articles using both linguistic markers assessment and manual reconstructions of their argumentative structure. Then, we compare the outcome of these two kinds of analysis in order to show that both the number of markers and their kind (hedges vs. boosters) are irrelevant with respect to the issue of assessing the degree of confidence or of commitment authors have with respect to the hypotheses they discuss. On the basis of this analysis we conclude that it would be a mistake to trust linguistic markers of certainty and uncertainty to distinguish factual from speculative knowledge (*Thesis2*).

2 Linguistic Means to Express Un/Certainty and the Dream of an Automatic Method for Distinguishing Factual from Uncertain Knowledge

Studies that aim at distinguishing between speculative (i.e. uncertain) knowledge and factual (certain) information on the basis of specific linguistic cues originate both historically and theoretically from two main lines of research. The first line arises from the anthropological and cross-linguistic investigations initiated by Boas (1938: 133) on the grammatical expressions used in various languages to specify the source of conveyed information and is centered in particular on the issues of *evidentiality* and *epistemicity* or *epistemic position(ing)*. The notion of *evidentiality* is used to describe the problem of determining the means used in different languages to specify what kind of evidence supports specific statements or even to specify whether there is any evidence for making a specific claim—i.e. whether someone saw something in the first person, or whether s/he heard this information from someone else, etc. (for an overview on these studies see e.g. Aikhenvald 2003a, 2003b). The related notion of *epistemic position(ing)* refers to the ways in which languages express a speaker's attitude regarding the reliability of sentence content (whether s/he believes it, knows it, doubts it, etc.); in other words: whether s/he considers it more or less likely or reliable.[1] A second line of investigation was initiated by Lakoff (1973: 471) and concerns *hedges*, i.e.—according to Lakoff's definition—"words whose meaning implicitly involves fuzziness—words whose job is to make things fuzzier or less fuzzy." In spite of the fact—acknowledged later by Lakoff himself (see Lakoff 1987: 122)—that thc hcdging dcvices identified in his 1973 paper are inadequate, this work gave rise to extensive research on hedging strategies, i.e. on the linguistic markers used in various languages to express caution (uncertainty) regarding specific claims. The opposite linguistic strategy, which aims to confer a high degree of reliability to statements or to communicate that they are based on well-grounded knowledge, is called *boosting* while the expressions used for boosting are considered *boosters* (for some early literature on these two notions see e.g. Holmes 1983, 1984, 1990).

Despite their differing theoretical roots, these two lines of study can be equated in terms of their common research objective to investigate (a) *how* certainty and doubt are conveyed in scientific communication (i.e. what are the expressions or markers adopted to achieve this aim) and (b) *why* authors use these stylistic devices to 'hedge' or to 'boost' their claims (i.e. what is the exact function of these expressions in scientific discourse).

---

[1]For an overview on epistemicity and for a discussion of its affinity to evidentiality see e.g. Chafe and Nichols 1986, Bednarek 2006, Cornillie 2009. Even though most authors agree that these two notions are positively related to each other (that they overlap, or that epistemicity is included in/derived from evidentiality), the debate is not unanimous on this aspect: for a brief overview of the different ways in which the relationships between these two notions have been interpreted see e.g. Dendale and Tasmowski (2001).

(a) The communication of a writer's confidence regarding the contents s/he presents is achieved using specific linguistic means, which serve as caveats for the reader and indicate whether a claim must be taken with caution or whether it should be considered to be reasonably reliable. Literature on scientific writing tried therefore to identify—i.e. to classify and list—these linguistic markers in order to study their function in scientific papers. The various studies rely on more or less detailed lists of markers, but they all include approximately the same (kinds of) items (markers). Hyland (1996b, 2004) is one of the authors that has addressed the issue of classifying hedges and boosters most systematically. He extracted from the available literature a set of linguistic devices used to express an authors' commitment to the truth of the claims s/he makes and organized them into specific groups according to their kind. To produce a list of these markers that is as complete as possible is necessary particularly for studies that aim at investigating large corpora of articles using automatic search (for a short list of the most important lexical markers see Hyland 2004: 188–189). Manual search allows one to identify hedging devices that were not included in the initial catalogue.

According to Hyland (1996b), in spite of the fact that the literature has focused mainly on *modal verbs* (e.g. 'could', 'may', 'might', 'would', 'should'), there are also a large number of other devices used in scientific articles to express certainty and uncertainty. Moreover, caution towards a claim is often expressed using *epistemic lexical verbs* like 'indicate', 'suggest', 'appear', 'propose'. These verbs can be further differentiated into three main groups: *speculative* (e.g. 'suggest', 'propose', 'believe', 'speculate', etc.); *inferential*[2] ('infer', 'indicate', 'imply', etc.); and *evidential*. So-called evidential verbs include verbs expressing *quotative evidence*, i.e. they specify the source of an information and the reliability of this source or the commitment of the quoted authors with respect to the truth of a claim (e.g. author X *suggested* that …; or the authors Y and Z *speculate* that … etc.); and verbs expressing *sensory evidence* ('sensory' is used here in a metaphorical sense to indicate verbs describing sensory experience, especially visual experience, like 'appear', 'seem', 'observe', 'attempt to gain insight' etc.). Other means to express un/certainty are *epistemic adjectives* ('un/likely', 'possible', 'probable' etc.) or *epistemic adverbs* ('apparently', 'probably', 'essentially', 'relatively', 'presumably', etc.). Also so-called *'downtoners', adverbs* which lower the effect of a verb (like 'quite', 'almost', 'usually', 'partially', 'rarely', 'virtually'. 'probably', 'generally', 'approximately', 'somehow' etc.) function as hedges. Even though Hyland does not mention it, we can hypothesize that there are corresponding '*uptoners', adverbs* (like 'completely', 'always', 'exactly', etc.) that function as boosters (for a concise list of the devices that fall into the classes mentioned above see Hyland

[2]Actually Hyland (1996b) calls this verb class 'deductive', but we prefer not to use this term since we consider it to be incorrect for the kind of verbs that he has in mind. In fact, he specifies that the verbs included in this class are related to *inferential* reasoning, while in another article he admits that science is mainly inductive: "In fact, my data indicates that few knowledge claims are presented in unmitigated form: induction and inference rather than deduction and causality characterize most arguments in scientific discourse" (Hyland 1996a: 435).

2004: 188–189). Finally, linguistic means to hedge or boost specific claims include so-called *strategic hedges* which consist of more complex expressions like 'we do not know whether…', 'one cannot exclude …', 'we are aware of the concerns expressed by' etc. Markers that are related to the structure of a sentence including interrogative or hypothetical forms also belong in this class.

We will refer to this general class of linguistic means to express doubt and certainty indiscriminately as hedges/boosters, or as hedging/boosting devices or as certainty/uncertainty markers. Furthermore, we will speak of lexical and morphosyntactical markers to designate respectively the words and the structural elements in a sentence (verbal tenses, moods, interrogative or hypothetical forms, etc.) that serve as linguistic means to express certainty and uncertainty. From Hyland's discussion it emerges that hedging in scientific writing is more important than boosting. This is not surprising. In fact, in science we often do not need to emphasize or to stress our certainty with respect to specific claims. To express factual statements that we consider to be reliable we can simply use the indicative present (see e.g. Hyland 1996a: 435). However, other authors point out that the use of boosting strategies in scientific literature must also be considered relevant since they allow us to identify information which is explicitly presented as certain (see e.g. Rubin et al. 2006).

(b) The study of the specific linguistic strategies people use to express certainty and uncertainty has important applications in the field of academic writing for a number of different reasons. On the one hand, they might help in understanding the rhetoric of scientific writings and the motivations that drive the authors to weaken or enforce their commitment to the truth of the hypotheses they discuss. On the other hand, they are also interesting to understand how science actually works (see e.g. Hyland 1996b: 252). As the literature of the rhetoric of science makes clear, hedging strategies in particular are used for a number of different reasons (for an overview see e.g. Hyland 2012). Sometimes they are used to *persuade* colleagues of the strength of their position and at the same time they aim to express their positions in a *cautious* and *modest* way, in the awareness that the search for truth in science is never definitive or conclusively confirmed. In this sense, hedging strategies can also be a means to show *accuracy* and *precision* (Hyland 1996b; Skelton 1997). Sometimes authors use hedging strategies to obtain the *consensus* of the community, i.e. to explicitly comply with certain views (Hyland 2012: 195–210). In other cases, hedges accomplish other stylistic functions like being *polite*; showing deference towards the scientific community; they are also used as a means of *captatio benevolentiae* (see e.g. Brown and Levinston [1978](1987); Hyland 2005) or to avoid personal commitment (Coates 1983; Chafe and Nichols 1986; Liddicoat 2005).

To these observations on the general functions accomplished by hedging (and to some extent boosting) strategies, it must be added that their use varies depending on many factors; among these at least the subjective style of the writer and the disciplinary field of the paper must be considered particularly relevant (see e.g. Crismore et al. 1993; Hyland 1999, 2009). As for this second aspect, a number of studies carried out on corpora belonging to various fields state that there are consistent differences in the use of hedges and boosters made by different disciplines: in general, texts in the humanities tend to use these linguistic markers more liberally than those on more technical subject matters (see e.g. Coffin et al. 2003; Hyland 2009). Other authors instead call attention to the composition of research articles and specifically to their organization into sections (Introduction, Methods, Results, Discussion, Conclusion) arguing that the use of linguistic markers and more generally the style adopted in the different sections also varies widely, because it aims to accomplish different rhetorical functions (Lemke 1998; Hyland 2006).

However, while the studies cited above limit the scope of their research to the study of the rhetorical functions accomplished by linguistic markers for certainty and uncertainty in scientific writings, there are also different kinds of analysis addressing the usages of hedges and boosters that do not focus on the style or on the rhetoric of scientific writings per se, but rather pursue a different and much more ambitious project. This line of study starts from the assumption that certainty and (to a greater extent) uncertainty are conveyed using specific markers. Thus, hedges and boosters can help us identify which scientific hypotheses/views an author considers to be less reliable or more uncertain. As is suggested e.g. by Auger and Roy "From a linguistic point of view, the identification and automatic tagging of expressions of certainty/uncertainty in textual data is a *sine qua non* condition to enable the empirical study and modeling of how humans assess certainty through their use of language" (Auger and Roy 2008: 1860).

Furthermore, if we could develop particularly sophisticated methods of search and algorithms (possibly machine-learning algorithms) which are able to identify all and only the markers that actually indicate uncertainty, ruling out ambiguities and false positive results,[3] the distinction between certain and uncertain claims could be carried out automatically also on very large corpora of literature. This kind of natural language processing is for example the goal Agarwal and Yu set for their work: "Hedging is frequently used in both the biological literature and clinical notes to denote uncertainty or speculation. It is important for text-mining applications to detect hedge cues and their scope; otherwise, uncertain events are incorrectly identified as factual events. However, due to the complexity of language, identifying hedge cues and their scope in a sentence is not a trivial task. Our objective was to develop an algorithm that would automatically detect hedge cues and their scope in biomedical literature" (Agarwal and Yu 2010: 953). A very similar project is sketched by a number of authors who pursue the NLP program: "Identifying hedged information in biomedical literature is an important subtask in information extraction because it would be misleading to extract speculative information as

[3]As examples of possible false results obtained identifying uncertain statements merely on the basis of uncertainty markers Agarwal and Yu (2010: 954) cite sentences like: "We can now study regulatory regions and functional domains of the protein in the context of a true erythroid environment, experiments that have not been possible heretofore." In addition, they point out that—in the case of complex sentences in which only one part/aspect is qualified as uncertain through the association with a marker like "Right middle and (probable right lower) lobe pneumonia"—we need to be able to distinguish between the certain and the uncertain information (indicated in the example using the square brackets).

factual information. In this paper we present a machine learning system that finds the scope of hedge cues in biomedical texts" (Morante and Daelemans 2009: 28). And further: "The CoNLL-2010 Shared Task was dedicated to the detection of uncertainty cues and their linguistic scope in natural language texts. The motivation behind this task was that distinguishing factual and uncertain information in texts is of essential importance in information extraction" (Farkas et al. 2010: 1).

This generic reference to 'literature' or 'texts' as the locus where this kind of research can be carried out offers a clue to understanding what the final ambition or the ultimate goal of this kind of research might be: the conversion of the subjective certainty/uncertainty expressed by single authors into a general form of certainty/uncertainty. In fact, if science is interpreted, as Kuhn's sociological account (1970) suggests, as a collective and democratic enterprise, then the truth of a hypothesis is decided collectively depending on whether most scientists agree with it. If so, since scientific literature is the means researchers use to express their views and their results, hedges and boosters could help us determining the degree of reliability attributed by a scientific community to a specific hypothesis at a certain time. We could indeed check large corpora of scientific literature in a specific field and 'calculate' how many (i.e. what percentage of) researchers consider a certain hypothesis to be un/certain. If this were possible, the project of natural language processing would have achieved a great success, providing society with a method for determining what scientific results are trusted by the most scientists in a given time and thus what scientific results should be trusted by most by everyone at that time.

3 Scientific Articles as Arguments

A research program that aims at identifying an automatic method based on lexical and morphosyntactical markers to determine whether a position expressed in a single constituent of text (one sentence, or one part of a sentence or a small cluster of sentences) is considered to be (more or less) reliable (factual) or uncertain (speculative) is not unproblematic. In fact, positions like the one we mentioned in the previous section according to which hedges and (to a lesser extent) boosters are highly variable rhetorical means that accomplish many functions and that "can only be understood in terms of a detailed characterization of the institutional, professional and linguistic contexts in which they are employed" (Hyland 1996a: 434) already represent a criticism of these views and a possibly insurmountable obstacle to their fulfilment as they argue that hedging and boosting strategies are *not* used primarily to express certainty or uncertainty (we will go back to this below).

However, in order to consider the very possibility of pursuing this kind of project, let us assume, for the sake of argument, that hedges and boosters are indeed used at least mostly to express un/certainty. Even in this case—this is the point we will argue for in this section (*Thesis1*)—the project of using the kind and number of linguistic markers included in a text to determine whether the author was more or less certain of the positions expressed cannot be successful. The reason for this must be traced back to the argumentative structure of scientific writings and to their argumentative character as opposed to the descriptive nature of more simple kinds of writings like e.g. brief research reports or popular science which merely communicate observations carried out by the researchers.

Scientific articles do not present mere expositions of facts, but they are written in order to support or to challenge certain hypotheses. More precisely, they are long and complex arguments whose aim is to assess what views should be sustained or dismissed. Arguments consist in a number of statements logically related to each, consisting respectively in premises and conclusions. Their purpose is to give reasons for a conclusion. Sometimes they are used to justify or to refute a point of view on the basis of reasons, at other times they are used to offer explanations for something in terms of causes.[4] We have deductive explanations in which the conclusion (the *explanandum*) is derived from some statements (the *explanans*) describing general principles or laws and the specific condition concerning the phenomenon to explain[5]; we have inductive explanations or inferences to the best explanation, in which we infer possible causes from a number of situations.[6] Reasons and causes might also be combined in the same argument when we need both to individuate causes that make something happen and reasons to justify why certain hypotheses involving the causes are stronger than others.

The complex arguments scientific articles consist of usually start from observational data and/or hypotheses accepted in the literature which they use as premises to infer the conclusions. Some intermediate conclusions are then used in connection with some supplementary hypotheses as additional premises to draw further conclusions. Confirmation is the simplest form of argumentation: an author has some evidence that s/he considers more or less certain and s/he uses this evidence to support a certain conclusion. In the case of confirmation, the certainty of the conclusion depends on the certainty of the premises. For this reason, we might expect that a high number of hedges in the premises will indicate that the conclusion is uncertain and that the topic considered is basically speculative. Conversely, it can also be assumed that a high number of boosters (or at least the absence of hedges) will show that the conclusion is certain and that the topic is generally considered factual.

However, scientific research is rarely only about confirmation. Scientific inquiries frequently aim to refute some well-established view or to argue for a position through the refutation of alternative possibilities. Most often, scientific writings consist of complex arguments including different augmentative forms (refutation; confirmation through liked or convergent premises, a chain of inferences, etc.). All

[4]Even though reasons and causes are clearly not one and the same thing, from the point of view of the logical structure of the argument we can consider them as equal (see Sinott-Armstrong and Fogelin 2009: 3–16).

[5]According to the *Deductive Nomologic Model* developed by Carl Gustav Hempel and Paul Hoppenheim this is the only appropriate way to develop scientific explanations. See Hempel and Hoppenheim 1948, Hempel 1965, Ladyman 2002.

[6]See Harman 1965, Boyd 1991, Psillos 1999, Ladyman 2002.

of these argumentative forms differ in the way the premises provide support for the conclusion and one cannot take for granted that the conclusion is asserted with a high degree of certainty only when the premises are considered to be certain as well or, conversely, that the conclusion is asserted with a low degree of certainty only when the premises are considered to be uncertain. This means that the number and the quality of hedging or of boosting devices used in an argument to characterize the statements that are used to describe the premises as more or less reliable are not necessarily predictive of the strength (the degree of certainty/uncertainty) with which the conclusion is asserted starting from these premises.

Let us consider several cases more closely.

(I) We could have e.g. an argument in which the premises are doubtful, and therefore the conclusion must be rejected. In this situation, the author might use a high number of hedges in the premises to indicate their implausibility. However, the argument can be strong: the author can be very sure that—because of the weakness of the premises—the conclusion must be rejected. Thus, globally we will have a high number/frequency of hedges, even though the argument and its conclusion are considered as strong.

(II) A similar but opposite situation can occur when the link between premises and conclusion is weak, i.e. when the premises offer only indirect evidence for the conclusion. Here the premises might be presented as certain (using corresponding markers), but the conclusion can still be very uncertain because of its loose link to the premises. In this case, even though the global number/frequency of hedges is extremely low and/or the number/frequency of boosters is high, the author might be uncertain of his/her conclusion.

(III) Analogously, in the case of linked or convergent arguments (arguments in which more premises are combined in order to support a conclusion or in which more premises provide independent reasons for supporting the conclusion), it is possible that several uncertain premises—presented as such using corresponding markers—give rise to a fairly strong argument because taken together they offer strong support for the conclusion. Also in this case, the global number of hedging/boosting devices is not predictive of the strength of the argument or indicative of the certainty of the author on the matter.

Because in all these cases there is no correspondence between the kind/number of linguistic markers and the actual strength of the argument, these argumentative strategies represent a serious challenge for the project of identifying an automatic method based on linguistic markers to assess whether specific hypotheses discussed in scientific writings are considered by the authors to be more or less reliable.

(*Thesis1)* More specifically, the use of complex arguments in scientific research articles argues against the possibility of using the *quantity* of certainty/uncertainty markers—i.e. statistical analyses on the frequency of their use—to draw conclusions on whether specific articles are more or less speculative or whether they adopt a more or less assertive/cautious style or whether the style of scientific articles has changed over time becoming more or less assertive/cautious (see e.g. Bongelli et al. 2012, 2014; Rosenthal et al. 2010; Kiyavitskaya et al. 2005; Vincze et al. 2008; Szarvas et al. 2008). Indeed, this last conclusion is coherent with the findings of e.g. da Graça Pinto et al. 2014 who do not notice any significant variation in the frequency of the use of certainty/uncertainty markers in medical papers of different periods, and conclude that—even though there are differences among disciplines in their use of hedges/boosters—in general the adoption of a more cautious or confident style by the authors depends on their individual preferences.

These considerations on the discrepancy between the certainty/uncertainty of the various claims made in an argument and its strength offer a *theoretical* argument against the possibility of using the frequency of automatically detected certainty/uncertainty markers to draw conclusions on the text, and on whether its style is cautious rather than assertive. However, it is still possible to argue for a different hypothesis: one could consider individual parts (claims) of a text—be they premises or conclusions of the arguments that compose it—and maintain that, if they are hedged, they must be considered speculative, while if they are boosted (or at least do not include hedges), they must be considered factual.

At first glance, the idea that an author hedges a claim s/he makes when s/he wants to express his/her uncertainty about it and that, on the contrary, an author boosters (or at least does not hedge) a claim s/he makes when s/he wants to express his/her certainty about it seems to be obvious, but in fact, the matter is not so straightforward. As we discussed in the previous section (point b), literature on hedging and boosting devices points out that authors use them for a range of rhetorical reasons other than the specification of their degree of commitment to the truth of the positions they discuss. Argument analysis can help to clarify this point because it offers an independent means to assess whether an author is more or less certain about the various claims s/he makes and thus it can help in determining whether in general hedges and boosters are used consistently and reliably in arguments to indicate whether a premise or a conclusion is considered to be certain or uncertain.

We tried to address this issue empirically analyzing a sample of research articles and comparing the arguments devised by the authors from the point of view of their strength on the one hand and the hedging and boosting strategies they adopt on the other. To have a variegated pattern of argumentation strategies, we consider articles in the field of psychiatry, which is a particularly interesting discipline for this kind of discussion for a number of reasons. On the one hand, psychiatry is linked to medicine, belongs to the biomedical sciences and relies, at least in part, on the same empirical methods and techniques; however, on the other hand, it is also strongly linked to psychology and to philosophy and it makes extensive use of argumentation to assert its theses. Furthermore, its theories and hypotheses have often been subject to strong disagreements among scholars and to radical revisions over time. Because of these characteristics, psychiatric research articles—especially when they consider controversial topics—are particularly interesting with respect to the issue of certainty/uncertainty in scientific communication.

To make sure that the articles are homogeneous and thus synchronically comparable with respect to their definition of the psychiatric approach, and that the differences among them can be traced back solely to factors related to personal divergences on the suggested hypotheses, we chose only papers published in a single Journal: *The British Journal of Psychiatry* (BJP). BJP was originally founded in 1853 and has continued to publish without interruption up until the present.[7] This is considered as one of the world's leading psychiatric journals and it represents a link between European and North-American research. Since we are especially interested in complex arguments which also contain refutations and confirmation through refutation, we decided to consider a subclass of psychiatry articles focusing on a particularly controversial topic which underwent many radical revisions and changes over time, and where any revision or change presupposed the refutation of previous points of view (for a general philosophical perspective on how scientific research proceeds during scientific revolutions see Thagard 1992).

The topic we consider most suitable for this aim is homosexuality. Homosexuality was pathologized in the 19th century and thus included in psychiatric theory and in the psychiatric lexicon as a form of mental 'illness' (Beyer 1987; Rosario 2002). Its official classification by the psychiatric community has changed radically over time until 1987 when it definitively ceased to be considered as a psychiatric pathology (Drescher and Merlino 2007; Porcher 2014: 27–28). Its psychiatric description along with its changes across time were codified in the five *Diagnostic and Statistical Manuals of Mental Disorders* (DSMs) published by the American Psychiatric Association from 1952 till 2013. These define the standard criteria for the classification of the mental disorders accepted at the time of publication of the respective manual.[8] In the first DSM edition (DSM-I) published in 1952 homosexuality was described as a sociopathic personality disturbance. In the second DSM edition (DSM-II) whose first edition appeared in 1968 sexual 'deviations' were no longer considered as a sociopathic personality disturbance, but were categorized as sexual 'deviations' due to a personality disorder. Only in the seventh printing of DSM-II in 1974 homosexuality ceased to be interpreted as an 'illness' and started to be considered as a 'sexual orientation disturbance': in virtue of this definition homosexuality was considered as a sexual disorder only in those cases where people experienced a conflict with their sexual orientation. In the third DSM edition which appeared in 1980 (DSM-III) this classification was adjusted again and the label 'sexual orientation disturbance' was changed to 'ego-dystonic homosexuality' to indicate that sexual orientation might become a disorder in cases where it is not compliant with one's ideal self-image. Since the issue of one's ideal self-image appeared to be due to cultural factors while the reason why one might

[7]From 1855 till 1858 the periodical was known as Asylum Journal and from 1858 to 1863 it took the name of Journal of Mental Science.

[8]DSM-I was published in 1952; DSM-II appeared in 1968, but it underwent some important changes in the seventh printing of 1974; DSM-III was published in 1980 but in 1987 a revisited version (DSM-III-R) came out; DSM-IV appeared in 1994; a revisited version was then published in 2000. The fifth and last version of DSM was published in 2013.

experience his/her homosexuality as ego-dystonic could be traced back to the widespread cultural homophobia, this classification was canceled in the revision of DSM-III-R published in 1987. DSM-III-R only mentions "sexual disorders not otherwise specified", without considering homosexuality as a particular category. As specified by DSM-IV (1994) the diagnosis of "not otherwise specified disorders" had to rely on symptoms that "cause(s) clinically significant distress or impairment in social, occupational, or other important areas of functioning" (DSM-IV 1994: 7). Since DSM-IV homosexuality has definitively been depathologized and in the revision of 2000 "reparative", "conversion", and/or "aversion" therapies were condemned.

115 articles considering (more or less specifically) homosexuality (till the beginning of the 20th century also called "inversion") have been published in JBP since its foundation in 1853 up until today. 26 of them are research articles; the others are brief articles (mainly, medical notes or literature reviews). We focused on the research articles which were published in 1921, 1957, 1962, 1964 (one article for each year), 1965 (two articles), 1968 (one article), 1969 (3 articles), 1970, 1971, 1972 (one article for each year), 1973 (3 articles), 1974 (3 articles), 1980 (2 articles), 1981, 1983, 1986, 1999, 2001 (one article for each year).[9] We carried out (manually) two separate analyses on this data. (a) For the first analysis, we

[9]Here we report the complete list of the titles and authors making up the sample. 1921 (*Homosexuality*, by C.S. Read); 1957 (*Psychometric Aspects of Homosexuality*, by T.G. Grygier); 1962 (*Homosexuality and Genetic Sex* by M. Pritchard); 1964 (*Homosexuality in Twins: A Report on Three Discordant Pairs* by N. Parker); 1965 (2 articles of the same author, *On the Genesis of Female Homosexuality* and *On the Genesis of Male Homosexuality: An Attempt at Clarifying the Role of the Parents* by E. Bene); 1968 (*Studies in Female Homosexuality IV. Social and Psychiatric Aspects* by F.E. Kenyon); 1969 (*Parental Age of Homosexuals* by K. Abe and P. A.P. Moran); 1969 (*Aversion Therapy of Homosexuality. A pilot study of 10 cases* by J. Bancroft); 1969 (*Homosexuality, Exhibitionism and Fetishism-Transvestism. Some Experiences in the Use of Aversion Therapy in Male* by B.H. Fookes); 1970 (*Subjective and Penile Plethysmograph Responses to Aversion Therapy for Homosexuality: A Follow-up Study* by N. McConaghy); 1971 (*A Male Monozygotic Twinship Discordant for Homosexuality. A Repertory Grid Study* by K. Davidson, H. Brierley and C. Smith); 1972 (*Parent-Child Relationships and Homosexuality* by G. Robertson); 1973 (*Heterosexual Aversion in Homosexual Males*) by K. Freund, R. Langevin, S. Cibiri and Y. Zajac; 1973 (*Classical, Avoidance and Backward Conditioning Treatments of Homosexuality* by N. McConaghy and R.F. Barr); 1973 (*Doctors' Attitudes to Homosexuality* by P.A. Morris); 1974 (*Sex Chromosome Abnormalities, Homosexuality and Psychological Treatment* by A. Orwin, S.R.N. James and R.K. Turner); 1974 (*Parental Background of Homosexual and Heterosexual Women* by M. Siegelman); 1974 (*Personality Characteristics of Male Homosexuals Referred for Aversion Therapy: A Comparative Study* by R.K. Turner, H. Pielmaier, S. James and A. Orwin; 1980 (*Social and Psychological Functioning of the Ageing Male Homosexual* by K.C. Bennett and N.L. Thompson); 1980 (*Homosexuality and Parental Guilt* by B. Zuger); 1981 (*Neuroendocrine Mechanisms and the Aetiology of Male and Female Homosexuality* by M.J. MacCulloch and J.L. Waddington; 1983 (*Homosexuality and Lesbianism* by D.J.West); 1986 (*Homosexuality in Monozygotic Twins Reared Apart* by E.D. Eckert, T. J. Bouchard, J. Bohlen and L. Heston); 1999 (*British psychiatry and homosexuality* by M. King and A. Bartlett); 2001 (*Straight talking: an investigation of the attitudes and practice of psychoanalysts and psychotherapists in relation to gays and lesbians* by A. Bartlett, M. King and P. Phillips).

identified the linguistic lexical or morphosyntactical markers that are commonly used to assess the certainty/uncertainty of texts and speeches (Hyland 1996b, 2004: 188–189). (b) For the second analysis, we used basic tools from informal logic to reconstruct the structure of the argument in each article. Specifically, we identify the basic and the non-basic premises as well as the intermediate and the final conclusions, rewriting the extended argument presented in each paper in standard form and diagramming it in order to make its structure clearly visible. These two methods of analysis were then compared in order to show that they produce diverging results.

(*Thesis2*) The result we obtained from this comparison was fairly homogeneous for all the articles we considered and it univocally suggests that the use of hedging and boosting devices is not consistent with respect to the degree of certainty that emerges from the overall argument presented by the papers and that hedges and boosters are not used only or primarily to indicate certainty/uncertainty but fulfill different rhetorical aims. Thus, it would be a mistake to trust the linguistic markers of certainty and uncertainty to distinguish factual from speculative knowledge.

For reasons of space it would be impossible to illustrate this comparison for all the articles we considered. However, it is important to give a precise idea of how hedging and boosting strategies are used in research articles as well as to specify how argument analysis can be concretely applied with the aim of determining how certain/uncertain (i.e. strong/weak) arguments presented by authors in scientific articles may be. Thus, in the following section of this work we will first focus on three of the articles we selected which were published in critical years for psychiatric research on homosexuality (1957, 1962, 1974).

We should point out that this selection of articles is not meant to be an actual 'corpus' since the set of articles we analyzed (26 research papers) is much too small for this aim. This selection is intended to give some significant or prototypical examples of research writings in which the degree of certainty/uncertainty signaled by lexical markers do not correspond to the degree of certainty with which an author asserted a position as assessed on the basis of an independent evaluation of the strength of the arguments and sub-arguments developed in the article. In this sense, the selected papers should be considered more as a list of meaningful counterexamples which challenge (falsify) the claim that linguistic markers can be used as an automatic means to determine the degree of certainty attributed by authors to a certain thesis in corpora of *scientific research papers*.

4 Certainty Identification—Arguments Versus Un/Certainty Markers: Some Psychiatry Research Papers

The first article we analyzed, written by T.G. Grygier, is dated 1957, and is on the *Psychometric Aspects of Homosexuality*. Its aim is "to discuss some attempts at measuring the psychological characteristics of the homosexual, and the direction and intensity of his impulses." (Grygier 1957: 514). The article addresses several psychological tests that "have been used in the diagnosis of homosexuality and in research concerned with this problem" (Grygier 1957: 514) and must be considered as extremely revolutionary for its time because the underlying position that motivates it is that homosexuality is not an illness for which we can find an etiology, but it is a personality trait and it has to do with personal history and personal relationships (Grygier 1957: 522). This is the reason why psychometric tests alone are considered an inadequate means for its assessment.

Here we address in particular two parts of this article. (A) First of all, we analyze the first argument presented by the author which concerns one of the most frequently used tests in clinical practice, i.e. the Minnesota Multiphasic Personality Inventory. (B) Secondly, we consider the conclusions of the paper, in which the author summarizes his criticisms of the various tests he considers and the conclusions that can be drawn from them with respect to psychometric measures of homosexuality.

(A) The first argument reads as follows (hedges are in bold, there are no boosters):

> Among the inventories now used in clinical practice the most popular is **most probably** the Minnesota Multiphasic Personality Inventory […]. There is a vast literature on its uses and abuses, and conflicting results about its factorial composition and validity have been reported […]. The masculinity-femininity scale of the M.M.P.I. gives a particular picture of **supposedly** masculine characteristics.
> A 'typical' male comes out as a cynical opportunist, prejudiced and superstitious, spiteful and revengeful, tough and cruel to animals, a slow thinker but quick to start a fight, enjoying primitive pleasures and having no manners.
> By contrast, an effeminate man (or a woman) is sentimental, has aesthetic, artistic and cultural interests, is concerned with feelings and has insight into them, but also suffers from phobias, anxieties and worries.
> In spite of the empirical basis of the test one **cannot escape a suspicion** that the above pictures represent stereotypes rather than reality. They are, moreover, Mid-Western American stereotypes which follow the familiar pattern of cultural superiority of the female sex. It **seems** that the masculinity-femininity scale has been so loaded with cultural factors that people with high intelligence and good education are bound to appear rather effeminate in the test profiles (Grygier 1957: 515–516).[10]

This argument is a mode of refutation in which the author shows that the scale he addresses has absurd implications (because it depicts all people with high intelligence and good education as effeminate) and is therefore not reliable for clinical practice. The premises provide very strong evidence for the conclusion and the conclusion comes out as very certain. However, from the point of view of the style and especially the hedges used this argument it is nearly impossible to appreciate the strength and indisputability of its conclusion. Uncertainty markers only suggest that the masculine characteristics implied by the test *might be disputable*; that they *might be* due to prejudices; and that *probabl*y the masculinity-femininity scale is

[10]The square brackets do not indicate omitted text, we left out only supporting literature; bold indicates the hedges.

loaded with cultural factors. All the main points addressed are associated with uncertainty markers (we have seven claims, four uncertainty marker, no boosters). However, the analysis of the argument makes it clear that the author is very certain of all the points he makes and of his conclusion which is asserted without reserve or doubts. The reason why the author might have hedged claims that he considered to be perfectly certain can be traced back to *irony*. The absurd implications of the scale are ironically stressed using uncertainty markers.

(B) The analysis of this individual argument also provides us with a basis for addressing the overall argument presented in the paper which is summarized by the author in the concluding part and is quoted in full below (hedges are in bold, boosters are underlined; the numbers indicate the parts of the arguments: they are put in standard form and are diagrammed below).

(i) It **appears** that no exact measures of the direction of sexual attraction are available at present [1]. Some indications of homosexual attraction or heterosexual revulsion **may** be obtained by means of projective tests, especially the Draw-a-Person Test, the Rorschach and the thematic tests [2], but the validity of the indicators is **very uncertain** [3].

(ii) With regard to the peculiarities of the homosexual's attitudes, behavior and personality traits more precise data are available [4]. <u>It must be remembered</u>, however, that the homosexual is not simply a man behaving like a woman [5], and for this reason tests of masculinity and femininity are **not always** very enlightening in the understanding of homosexual behavior [6]. As Terman and Miles (1936b) say, a 'most emphatic warning is necessary against the assumption that an extremely feminine score for males or an extremely masculine score for female can serve as an adequate basis for the diagnosis of homosexuality, either overt or latent'. This is particularly true of high masculinity scores in women [7].

(iii) The concepts of masculinity and femininity are **not always clearly** understood [8]. Psychological masculinity is a more complex concept than the difference between the male and the female sexes [9]. The usual technique in measuring masculinity and femininity is a questionnaire, but more comprehensive tests, such as the Terman-Miles (Terman and Miles 1936a), or those measuring personality in developmental terms **may be** more promising [10].

(iv) All the projective and developmental tests **lay claim** to their ability to detect the dynamic aspects of homosexuality [11] and therefore to help in the understanding of the origin of sexual inversion in individual cases and **possibly** in more general terms [12]. **In our present state of knowledge**, it is difficult to validate these claims [13], as understanding extends beyond measurement [14], and therefore beyond the usual sphere of experimentation and validation [15]. Precision of measurement is an important asset in this field, but the clinical psychologist <u>must be more</u> than a mere psychometrician [16]. (Grygier 1957: 523–524).[11]

[11]In the original article the author uses Arabic numerals (1)–(4) instead of Roman ones (i)–(iv). We changed this in order to distinguish more clearly between the points made by the authors and the numbers used to diagram the argument.

This argument can be diagrammed as follows:

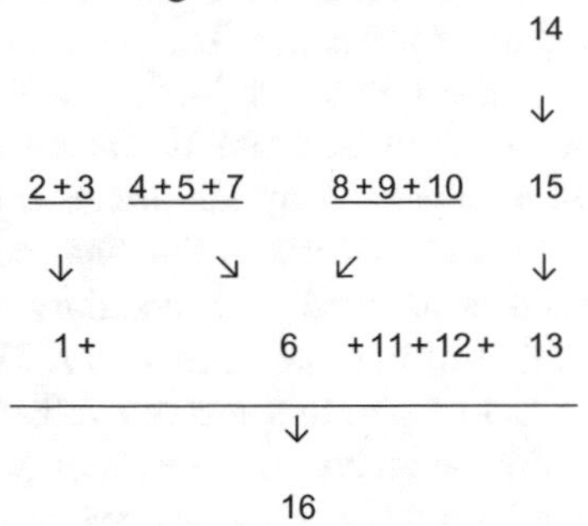

Point (i) presents a linked argument in which the conclusion [1] is derived by the combination of two premises [2] and [3]. Both premises and the conclusion are hedged using corresponding markers, even though the conclusion [1] (i.e. no exact measures of the direction of sexual attraction is available at present) is quite strongly supported by the arguments given in the previous part of the paper. Points (ii) and (iii) offer a converging argument which, starting from the linked premises [4]–[5]–[7] and [8]–[9]–[10]—lead to the conclusion [6] (i.e. tests of masculinity and femininity are not enlightening in the understanding of homosexual behavior). This conclusion is hedged even though it is derived from a number of premises that are mostly presented as certain (six premises/three hedges/one booster—if we consider the arguments presented previously in the paper, the hedged conclusion appears to be considered as very certain by the author). Point (iv) is more complicated and it offers both a chain argument (leading from [14] to [15] and then to [13]) and two additional premises [11]–[12] which—in combination with the conclusion of the other lines of reasoning—allow us to infer the final conclusion [16]. The intermediate conclusion [13] is hedged as it is part of the premise [12]. Only the final conclusion is boosted, even though the marker is not referred directly to a theoretical claim, but concerns the role clinical psychologists should take considering all the conclusions reached in the previous part of the text.

This final part of the paper could be discussed starting from several points of view. We will examine the most important ones. In general, the claims made in the premises and in the intermediate conclusions of the argument are expressed in a cautious manner. However, this does not mean that the author considers his position to be speculative rather than factual. On the contrary, his argument is extremely strong and his conclusion is asserted with absolute certainty. Since the final conclusion is reached through converging and linked arguments, in line with the thesis we presented in Sect. 3 we could say that the conclusion is asserted with a high degree of certainty in spite of the fact that the individual premises are hedged because the conclusion is supported by all of them jointly. However, in this case this explanation seems not to be appropriate and the hedging devices used in the premises seem mainly to express the style of the author. As a matter of fact, he puts forward very strong arguments (analogous to the one illustrated before in A) to argue for the limitations of the various psychometric tests; thus, the sub-conclusion of the first arguments that are then used as a premise for this final conclusion are considered by the author to be very certain. In spite of this, he makes use of hedges to summarize them in the final part of the paper. The point here is that the use of hedging and boosting devices in the text seems to be incoherent and does not correspond with the degree of certainty that can be inferred from the author's arguments. The factual knowledge expressed by the premises of this final conclusion is hedged. If these premises were uncertain, the final conclusion should be uncertain too. But the conclusion is asserted with certainty, confirming that the hedges in the premises de facto do not indicate uncertainty. The point that can be drawn from these observations is that markers are not reliable indicators of the degree of un/certainty attributed by the author to the various parts of his argument.

A possible reason why the author uses hedging devices in this unusual way can be inferred from the overall discussion in the article. Indeed, the author does not oppose the idea that psychometric tests can offer "clinically useful quantitative data for individual use" (p. 522) He rather insists that it would require more than this quantitative data for clinical psychologists to arrive at a meaningful qualitative interpretation. Thus, the uncertainty markers used in this final part of the article to describe the potential of psychometric tests do not indicate that the author is actually uncertain about whether they can or cannot give useful results. On the contrary, the author is quite certain that they can give useful results when adopted in combination with a qualitative analysis, while they give absurd outcomes when they are used alone, without the support of qualitative diagnostic means.

The second article we would like to discuss, written by M. Pritchard, is dated 1962 and is on *Homosexuality and Genetic Sex*. The aim of the paper is to challenge the theory of the genetic determination of homosexuality advanced by Theo Lang in 1940 which was still widely accepted in the early sixties. According to Lang, some male homosexuals have male morphological sex characteristics but a female chromosomal pattern. The evidence for Lang's view was taken from observations made in the 30s on certain species of insects and on statistical studies on the abnormal sex ratio in the siblings of male homosexuals. In those years, researchers could not rely on tests to directly determine the genetic sex of an individual. However, Pritchard refers to a 1961 study that validates a method to establish genetic sex in a direct way. Pritchard's arguments appeal to data produced by this new test as well as to other evidence deriving from studies on people born with an anomalous sexual anatomy (neither female nor male, the word he uses is 'intersexes') and on syndromes describing intersexes chromosomal conditions (like Klinefelter's and Turner's syndromes) to argue that "Lang's hypothesis can no longer be regarded as tenable." (Pritchard 1962: 623) Sexual conditions of chromosomal origin are discussed in the article because homosexuality was considered a genetic condition and Pritchard's goal is to show that—since chromosomal alterations do not necessarily lead to homosexuality—homosexuality must have a different origin.

Pritchard's article assumes the form of a classical convergent argument in which all sections are meant to assert intermediate conclusions that are than used as independent reasons in support of the final conclusion—i.e. the refutation of Lang's view. For reasons of space, it is not possible to quote the full article; we will analyze in detail only the 'Summary and conclusions' (hedges are in bold, boosters are underlined; the numbers indicate the parts of the arguments: they are put in standard form and are diagrammed below):

> The evidence for and against Lang's hypothesis that some male homosexuals are genetically female has been reviewed. Studies of the sibling sex ratio of homosexuals have **generally** supported the hypothesis but provide only indirect evidence [1]. Investigation of the intersexes, Klinefelter's and Turner's syndromes, has been shown to be irrelevant to this problem, but evidence from studies of other developmental sex anomalies does not support the hypothesis [2]. The nuclear sex[12] of homosexuals has always proved to be consistent with the phenotypic sex and this is convincing evidence against Lang's hypothesis [3]. An investigation of the somatic chromosomes in a small group of six male homosexuals is described. The results confirmed the sex chromosome constitution as male in each case, providing further direct evidence in line with the findings of nuclear sex [4]. It is therefore concluded that Lang's hypothesis can no longer be regarded as tenable. [5] (Pritchard 1962: 622–623)

[1]–[4] are the preliminary conclusions reached in the initial parts of the paper. This final part presents a convergent argument in which these are used as independent premises to argue for a further conclusion [5].

1 2 3 4
↘ ↘ ↙ ↙
5

If we consider the number of hedges and boosters used in the parts of the article that we could not report in full—i.e. the parts in which the preliminary conclusions have been reached—we notice that globally the balance is slightly in favor of hedges *H* over boosters *B* (2*H* vs. 7*B* for [1]; 7*H* vs. 3*B* for [2]; 8*H* vs. 2*B* for [3]; 9*H* vs. 7*B* for [4]) even though all the arguments that lead to these preliminary conclusions are quite strong. In fact, hedges are particularly high in the argument that leads to the conclusions [3] and [4] in which empirical data resulting from the application of the test are reported. This is very surprising because these data should univocally express factual knowledge and therefore be presented using certainty markers. On the contrary, they are presented in a hedged form as tendencies, while only the very final sentence which summarizes their meaning is expressed with a high degree of certainty: "these results do confirm the findings of nuclear sexing and afford further direct evidence against Lang's hypothesis." (Pritchard 1962: 622).

As for the final part of the article reported in full above, the premises of the argument—which are the sub-conclusions reached in the previous sections—constitute a *crescendo* of boosting strategies. The first premise is mildly hedged, but refers to the evidence which supports Lang's theory indicating that it is weak. Premise [2] includes one booster in the first part of the claim, but the positive

[12]By "nuclear sex" the author means the sex as determined by using the presence or absence of sex chromatin in somatic cells. Its presence usually indicates the female genotype XX; while its absence indicates the male genotype XY.

statement made in the second part using the indicative mood is also presented as factual. Premise [3] is boosted with two markers while premise [3] includes one booster only, even though it is presented univocally as factual through the specification that there is direct evidence for it. In sum, the premises give very strong support for the conclusion, which is asserted with a high degree of certainty.

The problem with the use of these hedging strategies is that they are used incoherently in the paper in at least two different ways. Factual knowledge (empirical data) is presented as uncertain. Moreover, the very same claims are presented as uncertain in the body text where they constitute the intermediate conclusion reached in the various sections, while they are presented as certain when they are used as premises to draw the final conclusion in the 'Summary and conclusions' section. A possible explanation of this inconsistent use of hedging strategies might be that the author wanted to be careful and cautious in criticizing Lang's view which in 1962 was still a well-established and widely accepted position. Since the empirical data were the strongest and most factual element against this view, he is most careful when he is presenting them. Whatever the reasons which motivated the author, what we have in this article is an incoherent use of hedging strategies which would be completely misleading if we tried to rely on it to assess what knowledge the author considers to be factual or instead speculative.

The third article we would like to directly address here was written by Orwin et al. in 1974 and is on *Sex Chromosome Abnormalities, Homosexuality and Psychological Treatment*. The article discusses one single case of sexual reorientation of a male homosexual with Klinefelter's syndrome using electric aversion therapy.[13] Klinefelter's syndrome is a chromosomal condition that leads to hypogonadism and other analogous dysfunctions. Most patients have less sexual interest; when they do, they are usually heterosexuals, confirming that homosexuality is not due to a chromosomal condition. The reason why the authors consider the case of their homosexual patients with Klinefelter's syndrome to be particularly interesting is that the psychological treatment (i.e. aversion therapy) was successful in spite of the patient's condition confirming that "abnormal sex chromosomes are relatively unimportant as determinants of psychosexual disorders." (Orwin et al. 1974: 295).

The article cannot be reported in full. For this reason, we will focus on the final part of the *Discussion* in which the authors summarize the core of their argument (hedges are in bold, boosters are underlined; the numbers indicate the parts of the arguments: they are put in standard form and are diagrammed below):

[13]While this was considered a pathology, a wide variety of techniques had been used in the treatment of homosexuality. Among them there were several types of "aversion therapies" which made use of various kinds of aversive stimuli to change the preference of so-called inverts towards heterosexuality. In this paper, the authors used an 'anticipatory avoidance therapy' analogous to that described by MacCulloch and Feldman (1967: 594). Basically, this is a technique similar to classical conditioning: the patient receives an electric shock when he is watching at pictures of attractive males.

> The picture from these descriptive papers is of considerable variation but of no **definite** correlation between Klinefelter's syndrome and specific psychosexual disorders [1]. […][14]
>
> In this case of Klinefelter's syndrome there was in fact ample evidence to **suggest** that the homosexuality was psychogenic in origin (e.g. early absence of father figure; maternal dominance causing repression of masculine assertiveness; seduction by stepfather; and small testes leading to feelings of sexual inadequacy with consequent avoidance of women) [2]. It **could** therefore be hoped that the sexual orientation would be altered by psychological treatment despite the presence of the sex chromosome abnormality, and there was a successful outcome [3]. This **tends** to confirm the relative unimportance of this abnormality, at least in the development of sexual preferences [4], and **may** indicate more generally that such abnormalities present no bar to the psychological treatment of psychosexual disorders [5]. (Orwin et al. 1974: 294)

The authors propose essentially a chain argument of the form:

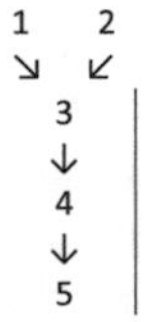

```
1   2
 ↘ ↙
  3  |
  ↓  |
  4  |
  ↓  |
  5  |
```

Each step of the argument consists of views which are well-established in the literature and depend more or less directly on the refutation of Lang's argument which was already discussed on the basis of the 1962 article. As Orwin and colleagues acknowledge already in the very first sentence of their article: "The theory that some male homosexuals might have female chromosomal constitution (Lang 1940) became untenable when techniques for studying sex chromosomes were developed." If this position is untenable, then it must be certain that sex chromosome abnormality and homosexuality are not related and that homosexuality has a psychogenic origin. In spite of the fact that at the time in which the authors write this appears to have been factual knowledge, their argument is studded with hedges. All steps of the argument present one uncertainty marker, including the last conclusion. This is very surprising since, in the light of the other positions discussed in the paper and of the references quoted in it, all the sub-conclusions of this argument as well as the final conclusion are considered as factual knowledge and should be asserted with a high degree of confidence.

The reason why they are hedged can probably be traced back to the authors' intent to make their hypothesis more appealing. Indeed, the point the authors want to make in their paper is quite predictable and unsurprising considering the knowledge available at that time; however, hedging strategies are used to artificially create some 'suspense'. Thus, the authors seem to use hedges as a rhetorical strategy to make normal science and its activities of puzzle-solving (in Kuhn's sense) appear more extraordinary. Also in this case, it would be a mistake to trust the linguistic markers of certainty and uncertainty to distinguish factual from speculative knowledge (*Thesis2*).

---

[14]In the text we left out the authors merely present further literature which confirms the point.

In general, our analysis on the whole sample shows that, when research challenges a well-established position, hedges are used primarily with the aim of being cautious, to avoid personal commitment or analogous (in our sample this seems to be the case for Bancroft 1969; Fookes 1969; McConaghy 1970; McConaghy and Barr 1973; Freund et al. 1973). On the contrary, when a paper discusses a view that can be considered as mainstream, hedges are mainly used in order to increase the interest of the reader and to make the results more appealing (in our sample this is the case for Davidson et al. 1971; Orwin et al. 1974).

Furthermore, there is another element that stands out in the analysis we carried out on our sample. Indeed, this element shows that the so-called paratext—e.g. tables, diagrams, figures etc.—is essential for understanding the authors' hedging and boosting strategies (Lemke 1998; da Graça Pinto et al., 2014). Our analysis reveals that the use of non-linguistic devices in papers became increasingly frequent in more recent times confirming that paratextual (non-linguistic) structures gained importance in psychiatric articles when data and empirical proof started to be considered as central also for this discipline, to the detriment of its connection with the human sciences. Initially used as a support for the text, with time they become part of the text and of its argumentative structure. In the sample we considered (which includes articles largely written during a period of time from the end of the 60 s to the beginning of the 80 s) the function of paratextual elements (above all tables) is to present the available data as evidence that speaks for itself. Unexpectedly, the increase in the use of these means is accompanied by an increase in the use of hedging strategies and especially of lexical uncertainty markers. In this way, the style becomes more impersonal, the authors do not take a binding position and their participation in what they report is neutralized. Thus, the hedging strategies became a 'distinctive mark' of scientific writings, in line with a Weberian view of science and of scientific communication, according to which scientific description should be as objective as possible and therefore should not depend on some preconception or theoretical conviction of the researchers (Crismore and Farnsworth 1990: 135).

In the articles we analyzed, the increased use of both hedging devices and tables is not aimed at indicating that the authors do not consider their results to be factual or certain (Bene 1965a; 1965b; Kenyon 1968; Freund et al. 1973; McConaghy and Barr 1973; Siegelman 1974). It is rather meant to minimize their responsibility for what they report, making the scientific results appear to be independent of the views of the persons who report them. This use of hedging devices in research which relies on empirical results presented through the paratext shows once more that hedging (and boosting) devices cannot be considered as reliable indicators of the degree of certainty/uncertainty attributed by the authors to the hypotheses they consider. Indeed, this confirms that they perform a primarily rhetorical function. In this sense, an actual evaluation of the argument is the only means we have to determine whether the knowledge conveyed in a scientific paper is meant by the authors to be factual or speculative.

5 Concluding Remarks

The issue we address in this work is whether the analysis of linguistic markers is a suitable method to assess the degree of certainty/uncertainty attributed by authors to the scientific hypotheses they discuss. We focus on scientific research articles only, which we consider to be both more difficult and more interesting to work with than other kinds of texts such as popular science texts or short research reports. The reason for this is that they do not merely illustrate specific scientific conclusions, but rather present complex arguments in which the final position is supported on the basis of many kinds of elements: positive evidence as well as refutations of conflicting, alternative views.

The hypothesis we started with in this paper is that the number (the frequency of use) and the type of the markers occurring in a paper (hedges vs. boosters) does not offer any reliable index to assess whether the arguments put forward by authors in their papers are more or less strong (i.e. whether the authors are more or less certain of the hypotheses they discuss in their scientific articles). The analysis we carried out on the sample of 26 articles on homosexuality taken from the *British Journal of Psychiatry* and exemplified by the three papers discussed in the previous section fully confirm this hypothesis. In fact, we showed that there is no relationship between the linguistic markers used in the various steps of an argument and its strength or the actual certainty of the premises and conclusions presented in it.

Hedges are considered to be more important than boosters in scientific literature because often to express factual knowledge in science we simply make claims in the indicative mood (see above Sect. 2). However, it is exactly these hedging kinds of markers which appear to be most problematic. In fact, our analysis confirms that an extensive use of hedges does not necessarily or even usually indicate uncertainty towards a position or a cautious commitment to its truth. Hedges are used for a number of other rhetorical reasons. Beyond the reasons that are most often pointed out in the literature (see above Sect. 2), the examples we discussed show that hedging strategies might also be used to make an argument more interesting (less routine and/or more extraordinary) or to be ironical and to show e.g. how absurd the implications of a certain viewpoint may be (on this aspect of scientific communication see also Knorr-Cetina 1981; Gilbert and Mulkay 1984; Latour and Woolgar 1979; Myers 1989; Liddicoat 2005; Thompson 2001) or to make scientific results appear to be neutral, objective and independent of the opinions/preconceptions of the authors.

As for the project of developing an automatic method based on the search for hedges/boosters which could assess the commitment of authors to the truth of the theses they discuss, our analysis indicates that this is doomed to fail at least if the idea is to use such a method with scientific research articles, which are arguably the most interesting target for such an application. Since scientific research involves complex arguments devised to support specific conclusions, the automatic analysis of lexical and morphosyntactical markers is of no help in assessing the degree of certainty with which specific positions are affirmed, because there is no correspondence between the number and kinds of markers used and the strength of the argument. This has various implications for various relevant issues. First of all, linguistic markers alone are not an adequate way to distinguish factual and speculative information in a text. Secondly, an approach based on these markers is inadequate to track down the thinking processes underlying scientific reasoning. Finally, assessing the degree of certainty/uncertainty with which a hypothesis is put forward in a scientific paper can only be done on the basis of the (manual) reconstruction of the arguments it presents. Moreover, using this method it is also possible to establish the basis for this certainty/uncertainty and this is an essential piece of information in the dialectic and cooperative dynamic of science.

Acknowledgements We would like to thank the anonymous reviewers for their valuable comments. This research was supported by the ORBA10H82A grant to LP and by the 40201691 grant to SD.

References

Afsar, H. S., Moradi, M., & Hamzavi, R. (2014). Frequency and type of hedging devices used in the research articles of humanities, basic sciences and agriculture procedia. *Social and Behavioral Sciences, 136*, 70–74. doi:10.1016/j.sbspro.2014.05.290.

Agarwal, S., & Yu, H. (2010). Detecting hedge cues and their scope in biomedical text with conditional random fields. *Journal of Biomedical Informatics, 43*, 953–961. doi:10.1016/j.jbi.2010.08.003.

Aguilar, M. (2008). *Metadiscourse in academic speech: A relevance-theoretic approach*. Berlin: Peter Lang.

Aikhenvald, A. Y. (2003a). Evidentiality in typological perspective. In A. Y. Aikhenvald (Ed.), *Studies in evidentiality* (pp. 1–31). Amsterdam/Philadelphia: John Benjamins Publishing.

Aikhenvald, A. Y. (2003b). Preliminaries and key concepts. In A. Y. Aikhenvald (Ed.), *Evidentiality* (pp. 1–22). New York: Oxford University Press.

Auger, A., & Roy, J. (2008). Expression of uncertainty in linguistic data. In *Information Fusion, IEEE, 11th Conference on Information Fusion* (pp 1860–1866).

Bancroft, J. (1969). Aversion therapy of homosexuality: A pilot study of 10 cases. *British Journal of Psychiatry, 115*, 1417–14321.

Bednarek, M. (2006). Epistemological positioning and evidentiality in English news discourse: A text-driven approach. *Text & Talk, 26*(6), 635–660.

Bene, E. (1965a). On the genesis of male homosexuality: An attempt at clarifying the role of parents. *British Journal of Psychiatry, 111*, 803–813.

Bene, E. (1965b). On the genesis of female homosexuality. *British Journal of Psychiatry, 111*, 815–821.

Beyer, R. (1987). *Homosexuality and American psychiatry: The politics of diagnosis*. Princeton (NJ): Princeton University Press.

Boas, F. (1938). Language. In F. Boas (Ed.), *General anthropology* (pp. 124–145). Boston/New York: D. C. Heath and Company.

Bongelli, R., et al. (2012). A corpus of scientific biomedical texts spanning over 168 years annotated for uncertainty. In N. Calzolari, K. Choukri, T. Declerck, M. Uğur Doğan, B. Maegaard, J. Mariani, A. Moreno, J. Odijk & S. Piperidis (Eds.), Proceedings of the Eight International Conference on Language Resources and Evaluation (LREC) vol. 12 (pp. 2009–2014). Istanbul: European Language Resources Association—ELRA.

Bongelli, R., Riccioni, I., Canestrari, C., Pietrobon, R., & Zuczkowski, A. (2014). Biouncertainty: A historical corpus evaluating uncertainty language over a 167-year span of biomedical scientific articles. In A. Zuczkowski, R. Bongelli, I. Riccioni, & C. Canestrari (Eds.), *Communicating certainty and uncertainty in medical, supportive and scientific contexts* (pp. 309–339). Amsterdam: John Benjamins Publishing.

Boyd, R. (1991). Explanation, explanatory power and semolicity. In R. Boyd, R. Gasper, & D. Trout (Eds.), *The philosophy of science* (pp. 349–377). Cambridge (MA): The MIT Press.

Brown, P., & Levinson, S. C. [1978](1987). Politeness: Some universals in language usage. Cambridge: Cambridge University Press.

Chafe, W., & Nichols, J. (1986). Introduction. In W. Chafe & J. Nichols (Eds.) (vii–xi), *Evidentiality: The linguistic coding of epistemology*. Praeger (NJ): Ablex.

Chao, F., & Hu, G. (2014). Interactive metadiscourse in research articles: A comparative study of paradigmatic and disciplinary influences. *Journal of Pragmatics, 66*, 15–31.

Choi, E., Tan, C., Lee, L., Danescu-Miculescu-Mizil, C., & Spindel, J. (2012). Hedge detection as a lens on framing in the GMO debates: A position paper. In *ACL-2012 Workshop on Extra-propositional Aspects of Meaning in Computational Linguistics* (pp. 70–79).

Coates, J. (1983). *The semantics of the modal auxiliaries*. London-Camberra: Croom Helm.

Coffin, C. J., Curry, M. J., Goodman, S., Hewings, A., Lillis, T., & Swann, J. (2003). *Teaching academic writing: A toolkit for higher education*. London: Routledge.

Cornillie, B. (2009). Evidentiality and epistemic modality. On the close relationship between two different Categories. *Functions of Language, 16*(1), 44–62. doi:10.1075/fol.16.1.04cor.

Crismore, A., & Farnsworth, R. (1990). Metadiscourse in popular and professional science discourse. In W. Nash (Ed.), *The writing scholar: Studies in academic discourse* (pp. 119–136). Newbury Park (CA): Stage.

Crismore, A., Markkanen, R., & Steffensen, M. S. (1993). Metadiscourse in persuasive writing. *Written Communication, 10*(1), 39–71.

da Graça Pinto, M., Osorio, P., & Martins, F. (2014). A theoretical contribution to tackling certainty and uncertainty in scientific writing. In A. Zuczkowski, R. Bongelli, I. Riccioni, & C. Canestrari (Eds.), *Communicating certainty and uncertainty in medical, supportive and scientific contexts* (pp. 291–305). Amsterdam: John Benjamin (Four research articles from the journal *Brain* in focus).

Davidson, K., Brierley, H. & Smith, C. (1971). A male monozygotic twinship discordant for homosexuality: A repertory grid study. *British Journal of Psychiatry, 118*, 675–682.

Dendale, P., & Tasmowski, L. (2001). Introduction: Evidentiality and related notions. *Journal of Pragmatics, 33*, 339–348. doi:10.1016/S0378-2166(00)00005-9.

Drescher, J., & Merlino, J. P. (2007). Introduction. In J. Drescher & J. P. Merlino (Eds.), *American psychiatry and homosexuality: An oral history* (pp. 1–11). New York/London: Harrington Park Press.

Durik, A. M., Britt, M. A., Reynolds, R., & Storey, J. (2008). The effects of hedges in persuasive arguments. A nuanced analysis of language. *Journal of Language and Social Psychology, 27* (3), 213–234.

Farkas, R., Vincze, V., Móra, G., Csirik, J., & Szarvas, G. (2010). The CoNLL-2010 shared task: Learning to detect hedges and their scope in natural language text. In *Proceedings of the 14th Conference on Computational Natural Language Learning: Shared task* (pp. 1–12).

Freund, K., Langevin, R., Cibiri, S. & Zayac, Y. (1973). Heterosexual aversion in homosexual males. *British Journal of Psychiatry, 122*, 163–169.

Fookes, B.H. (1969). Some experiences in the use of aversion therapy in male homosexuality, exibitionism and fetishism-transvestitism. *British Journal of Psychiatry, 115*, 339–341.

Gilbert, G. N., & Mulkay, M. (1984). *Opening pandora's box: A sociological analysis of scientific discourse*. Cambridge: Cambridge University Press.

Grygier, T. G. (1957). Psychometric aspects of homosexuality. *The British Journal of Psychiatry, 103*, 514–526. doi:10.1192/bjp.103.432.514.

Harman, G. H. (1965). Inference to the best explanation. *Philosophical Review., 74*(1), 88–95.

Hempel, C. G. (1965). *Aspects of scientific explanation*. New York: Free Press.

Hempel, C. G., & Hoppenheim, P. (1948). Studies in the logic of explanation. *Philosophy of Science, 15*(2), 135–175.

Heritage, J. (2011). Territories of knowledge, territories of experience: Emphatic moments in interaction. In T. Stivers, L. Mondada, & J. Steensig (Eds.), *The morality of knowledge in conversation* (pp. 159–183). Cambridge: Cambridge University Press.
Holmes, J. (1983). Speaking English with the appropriate degree of conviction. In C. Brumfit (Ed.), *Learning and teaching languages for communication: Applied linguistics perspectives* (pp. 100–121). London: BAAL.
Holmes, J. (1984). Modifying illocutionary force. *Journal of Pragmatics, 8*, 345–365.
Holmes, J. (1990). Hedges and boosters in women's and men's speech. *Language & Communication, 10*(3), 185–205.
Hyland, K. (1995). Author in the text. Hedging scientific strategies. *Hong Kong Papers in Linguistic and Language Teaching, 18*, 33–42.
Hyland, K. (1996a). Writing without conviction? Hedging in scientific research articles. *Applied Linguistics, 17*(4), 433–454.
Hyland, K. (1996b). Talking to the academy: Forms of hedging in science research articles. *Written Communication, 13*(2), 251–281.
Hyland, K. (1997). scientific claims and community values: Articulating an academic culture. *Language & Communication, 17*(1), 19–32.
Hyland, K. (1999). Academic attribution: Citation and the construction of disciplinary knowledge. *Applied Linguistics, 20*(3), 341–367.
Hyland, K. (2002). Authority and invisibility: Authorial identity in academic writing. *Journal of Pragmatics, 34*, 1091–1112.
Hyland, K. (2004). *Disciplinary discourses: Social interactions in academic writing*. USA: University of Michigan Press.
Hyland, K. (2005). *Metadiscourse*. London: Continuum.
Hyland, K. (2006). Disciplinary differences: Language variations in academic discourses. In K. Hyland & M. Bondi (Eds.) *Academic discourses across disciplines* (pp. 17–45). Bern Frankfurt a.M. London New York: Peter Lang.
Hyland, K. (2009). Writing in the disciplines: Research evidence for specificity. *Taiwan International ESP Journal, 1*(1), 5–22.
Hyland, K. (2011). Disciplines and discourses: Social interactions in the constructions of knowledge. In D. Starke-Meyerring, A. Paré, N. Artemeva, M. Horne, & L. Yousoubova (Eds.), *Writing in the knowledge society* (pp. 193–214). West Lafayette (IN): Parlor Press.
Hyland, K. (2012). Disciplinary identities: Individuality and community in academic discourse. Cambridge, (UK): Cambridge University Press.
Ifantidou, E. (2005). The semantics and pragmatics of metadiscourse. *Journal of Pragmatics, 37*, 1325–1353.
Kamio, A. (1997). *Territory of information*. Amsterdam: John Benjamin.
Kenyon, F.E. (1968). Studies in female homosexuality: IV. Social and psychiatric aspects. *British Journal of Psychiatry, 114*, 1337–1350.
Kim, J.-D., Ohta, T., Pyysalo, S., Kano, Y., & Tsujii, J. (2009). Overview of BioNLP'09 shared task on event extraction. In *Proceedings of the BioNLP 2009 Workshop* (pp. 1–9). Boulder, Colorado.
Kiyavitskaya, N., Zeni, N., Cordy, J. R., Mich, L., & Mylopoulos, J. (2005). Semi-automatic semantic annotations for web documents. *Proceedings of SWAP, 2005*, 14–15.
Knorr-Cetina, K. (1981). *The manufacture of knowledge: An essay on the constructivist and contextual nature of science*. Oxford: Pergamon.
Kuhn, T. (1970). *The structure of scientific revolutions*. Chicago: University of Chicago Press.
Ladyman, J. (2002). *Understanding philosophy of science*. London: Routledge.
Lakoff, G. (1973). Hedges: A study in meaning criteria and the logic of fuzzy concepts. *Journal of Philosophical Logic, 2*, 458–508.
Lakoff, G. (1987). *Women, fire, and dangerous things. What categories reveal about the mind*. Chicago: Chicago University Press.
Lang, T. (1940). Studies on the genetic determination of homosexuality. *Journal of Nervous and Mental Desease, 92*, 55–64.

Latour, B., & Woolgar, S. (1979). *Laboratory life: The social construction of scientific facts*. Beverly Hills (CA): Sage.
Lemke, J. (1998). Multiplying meaning. Visual and verbal semiotics in scientific text. In J. R. Martin & R. Veel (Eds.), *Reading science. Critical and functional perspectives on discourse of science* (pp. 87–113). London: Routledge.
Liddicoat, A. J. (2005). Writing about knowing in science: Aspects of hedging in french scientific writing. *LPS & Professional Communication, 5*(2), 8–27.
MacCulloch, M. J., & Feldman, M. P. (1967). Aversion therapy in management of 43 homosexuals. *British Medical Journal, 2*, 594–597.
McConaghy, N. (1970). Subjective and penile plethysmograph responses to aversion therapy for homosexuality: A follow-up study. *British Journal of Psychiatry, 117*, 555–560.
McConaghy, N. & Barr, R.F. (1973). Classical, avoidance and backward conditioning treatments of homosexuality. *British Journal of Psychiatry, 122*, 151–162.
Morante, R., & Daelemans, W. (2009). Learning the scope of hedge cues in biomedical texts. In *Proceedings of the Workshop on BioNLP* (pp. 28–36).
Myers, G. (1989). The pragmatics of politeness in scientific articles. *Applied Linguistics, 10*, 1–35.
Origgi, G., & Sperber, D. (2002). Evolution, communication, and the proper function of language. In P. Carruthers & A. Chamberlain (Eds.), *Evolution and the human mind: Language, modularity, and social cognition* (pp. 140–169). Cambridge (MA): Cambridge University Press.
Orwin, A., James, S. R. N., & Turner, K. (1974). Sex chromosome abnormalities, homosexuality and psychological treatment. *The British Journal of Psychiatry, 124*, 293–295. doi:10.1192/bjp.124.3.293.
Porcher, J. E. (2014). A note on the dynamics of psychiatric classification. *Minerva, 18*, 27–47.
Pritchard, M. (1962). Homosexuality and genetic sex. *The British Journal of Psychiatry, 108*, 616–623. doi:10.1192/bjp.108.456.616.
Psillos, S. (1999). *Scientific realism*. London New York: Routledge.
Rosario, V. A. (2002). Homosexuality and science. A guide to debate. Santa Barbara (CA): ABC-CLIO.
Rosenthal, S., Lipovsky, W. J., McKeown, K., Thadani, K., & Andreas, J. (2010). Towards semi-automated annotation for prepositional phrase attachment. In *Proceedings of the 7th Conference on International Language Resources and Evaluation LREC 2010*. http://www.lrec.conf.org.
Rubin, V. L., Liddy, E. D., & Kando, N. (2006). Certainty Identification in texts: Categorization model and manual tagging results. In J. G. Shanahan, Y. Qu, & J. Wiebe (Eds.), *Computing attitude and affect in text: Theory and applications* (pp. 61–76). Dordrecht (NL): Springer.
Siegelman, M. (1974). Parental background of homosexual and heterosexual women. *British Journal of Psychiatry, 124*, 14–21.
Sinott-Armstrong, W., & Fogelin, R. J. (2009). *Understanding arguments: An introduction to informal logic*. Cengage: Wadsworth.
Skelton, J. (1997). The representation of truth in academic medical writing. *Applied Linguistics, 18* (2), 121–140.
Szarvas, G., Vincze, V., Farkas, R., & Csirik, J. (2008). The Bioscope corpus: Annotation for negation, uncertainty and their scope in biomedical texts. *Association for Computational Linguistics, BioNLP: Current Trends in Biomedical Natural Language Processing*, 38–45.
Terman, L.M. & Miles, C.C. (1936a). *Attitude-interest analysis test*. New York: McGraw-Hill.
Terman, L.M. & Miles, C.C. (1936b). *Sex and personality*. New York: McGraw-Hill.
Thagard, P. (1992). *Conceptual revolutions*. Princeton: Princeton University Press.
Thompson, G. (2001). Interaction in academic writing: Learning to argue with the reader. *Applied Linguistics, 22*, 58–78.
Vincze, V., Szarvas, G., Farkas, R., Móra, G., & Csirik, J. (2008). The bio-scope corpus: Biomedical texts annotated for uncertainty. *Negation and their Scopes. BMC Bioinformatics, 9* (11), S9. doi:10.1186/1471-2105-9-S11-S9.

Wilson, D. (2000). Metarepresentation in linguistic communication. In D. Sperber (Ed.), *Metarepresentations: A multidisciplinary perspective* (pp. 411–448). Oxford: Oxford University Press.

Wilson, D. (2005). New directions for research on pragmatics and modularity. *Lingua, 115*, 1129–1146.

Winter, S., Krämer, N. C., Rösner, L., & Neubaum, G. (2015). Don't keep it (too) simple—How textual representations of scientific uncertainty affet layperson' attitude. *Journal of Language and Social Psychology, 34*(3), 251–272.

Extending Cognition Through Superstition: A Niche-Construction Theory Approach

Tommaso Bertolotti

Abstract Superstitious practices have been considered since the ancient times as signs of deviating cognitive forms (such as the elders'), concerned with irrelevant causal relationships, and/or reducible to religious beliefs (and hence explained away). Recent theories such as the extended mind and cognitive niche construction, though, can shed new light on superstition and its apparently unreasonable success. The trigger is to observe how most superstitions are not mere "beliefs" (such as religious beliefs could be) hosted in a naked mind, but rather involve a strong coupling between the mind and some external props allowing its extensions away from the skull: from bodily gestures, to artifacts and other agents (human and animal). The mind's capability to extend into the environment supports the related theory of cognitive niche construction, suggesting that human agents achieved better and better performances by creating external structures (cognitive niches) able to provide better and persistent scaffoldings for their cognitive performances. When it is not possible to detect and exploit the presence of a cognitive niche in the environment, superstitious practices can be identified as the possibility to deploy an emergency-cognitive niche projected by the superstitious agent into the world by means of a superstitious prop (item, ritual, gesture). It is poorer and less reliable but preferable to utter blank (and the consequent inaction), and most important it is still coupled with the external world (be it the body or its ecology in forms of artifacts and other agents), thus maintaining the fundamental characteristic of cognitive niches, that is distribution.

1 Introduction

Since the ancient times, a clear demarcation between *religion* and *superstition* has been enforced. Christian philosopher Isidore of Seville composed in the 7th century the *Etymologies* (or *Origins*), an encyclopedic treatise summarizing hundreds of

T. Bertolotti (✉)
Department of Humanities, Philosophy Section and Computational Philosophy Laboratory, University of Pavia, Pavia, Italy
e-mail: bertolotti@unipv.it

L. Magnani and C. Casadio (eds.), *Model-Based Reasoning in Science and Technology*, Studies in Applied Philosophy, Epistemology and Rational Ethics 27, DOI 10.1007/978-3-319-38983-7_9

sources that maintained a referential status until the publication of Diderot and D'Alembert's *Encyclopédie*. As the title suggests, Isidore professed that the knowledge concerning a concept could be drawn from and coincided with the etymology of the word standing for a concept.[1] The entry concerning superstition is revealing:

> Superstition (*superstitio*) is so called because it is a superfluous or superimposed (*superinstituere*) observance. Others say it is from the aged, because those who have lived (*superstites*) for many years are senile with age and go astray in some superstition through not being aware of which ancient practices they are observing or which they are adding in through ignorance of the old ones. And Lucretius says superstition concerns things 'standing above' (*superstare*), that is, the heavens and divinities that stand over us, but he is speaking wrongly (Barney et al. 2006, VIII.iii.6–iii.7, p. 174).

Three things are to be retained from Isidore's (etymological) understanding, inasmuch as they still resonate in the contemporary theoretical understanding of superstition:

1. Superstition is superfluous. It is a lesser accident of human cognitive performances and it is superimposed, in the sense that it is not a true manifestation of human cognition.
2. Superstition is a form of deviated cognition, such as the elders' according to Isidore.
3. Superstition should *absolutely* not be confused with religion.

These biases still affect the will to achieve a cognitive understanding of superstition, but recent theories such as the extended mind and cognitive niche construction can shed new light on superstition and its apparently unreasonable success. The trigger is to observe how most superstitions are not mere "beliefs" (such as religious beliefs could be) hosted in a naked mind, but rather involve a strong coupling between the mind and some external props allowing its extensions away from the skull: from bodily gestures, to artifacts and other agents (human and animal).

The mind's capability to extend into the environment supports the related theory of cognitive niche construction, suggesting that human agents achieved better and better performances by creating external structures (cognitive niches) able to provide ever-improving and persistent scaffoldings for their cognitive performances. My contention will be that when an environment does not offer the possibility to individuate a proper, reliable cognitive niche, individuals may still extend their cognitive systems in unwarranted ways via superstitious props, creating individually-projected cognitive niches that reduce anguish from cluelessness and the related impaired performances or impossibility to initiate a course of action.

[1]Many entries are quite accurate, and their etymology is indeed illuminating as for the knowledge of the concept, while some other entries are outright hilarious.

2 Superstition as Extended Cognition

Cognitive science has been dealing with superstition as if it was a minor sister of religion, or a mere subpart of the latter. Most milestones in cognitive studies of religion (Boyer 2001; Atran 2002; Dennett 2006) either mention superstition only incidentally or they reduce it to religion itself. As pointed out by Willem B. Drees, "academic definitions of superstition as distinct from genuine religion are hard to come by; the point of the label 'superstition' is dismissive" (Drees 2010, p. 70). This is not to say that cognitive science has totally overlooked superstition, but it attracted far less interest than religion, notwithstanding the fact that in our life we are most likely to engage in much more superstitious behavior than actually religious one.

The origin of superstition has been usually drawn back to associative learning, namely from the observation of coincidences that provoked a wrong idea about external reality (Beck and Forstmeier 2007). Although religion and superstition are surely related inasmuch as they testify the universal appeal of magical thought on the human mind, they can be told one from the other quite significantly. As meaningful as this may appear in the extended cognitive framework (Menary 2010) I will adopt in this paper, religion concerns primarily (celestial) beliefs while superstition is mostly about (local) practices. In the second edition of his monograph about superstition, Vyse (2014) elaborates on Judd Marmor's definition of superstition as "beliefs or practices groundless in themselves and inconsistent with the degree of enlightenment reached by the community to which onc bclongs" (Marmor 1956, p. 119). Vyse suggests that "beliefs or practices that are inconsistent with our common scientific understanding would apply to both believing in ghosts and wearing a lucky bracelet to a job interview, but most people would sat that only the lucky bracelet is a superstition:" this leads to his defining superstition as "the subset of paranormal beliefs that are pragmatic: used to bring about good luck or avoid bad" (Vyse 2014, p. 24).

The pragmatic relevance of superstition is sometimes shared by religion: we often interact with religious artifacts and rituals in ways that do not concern eternal salvation but rather mundane issues. At the same time, one needs not be religious at all to entertain (and enact) superstitious beliefs.[2] The epitome of superstitious belief is exemplified by Skinner's famous "superstitious pigeon", an animal with no existential anxiety nor will for salvation (that we know of), supposing that a behavior it enacts is somehow connected to the good outcome of a free meal (Skinner 1948). Interestingly, Donald Norman refers to this conception of superstition when dealing with the coping between one's mental models and the target of those models:

[2]Religious authorities and scholars have always fought a theological and intellectual battle against superstition, as interestingly shown by the dedicated entry in the *Routledge Dictionary of Religious and Spiritual Quotations* (Parrinder 2000, p. 24).

> When people attribute their actions to *superstition* they appear to be making direct statements about limitations in their own mental models. The statement implies uncertainty as to mechanism, but experience with the actions and outcomes. Thus, in this context, superstitious behavior indicates that the person has encountered difficulties and believes that a particular sequence of actions will reduce or eliminate the difficulty (Norman 1983, pp. 10–11).

While religion coopts natural routines for magical thinking to cope with existential issues, superstition routes them for coping with *local difficulties* and *uncertainties*. Malinowski's anthropological findings were clear with this respect: superstition is a way of managing uncertainty and the connected anxiety. Seeing the relationship between sailors and the sea as the archetype of uncertainty, he argued as follows:

> But even with all their *systematic knowledge*, methodically applied, they are still at the mercy of *powerful and incalculable tides*, sudden gales during the monsoon season and unknown reefs. And here comes in their *magic*, *performed* over the canoe during its construction, carried out at the beginning and in the course of expeditions and resorted to in moments of real danger. If the modern seaman, entrenched in science and reason, provided with all sorts of safety appliances, sailing on steel-built steamers, if even he has a *singular tendency to superstition* –which does not rob him of his knowledge or reason, nor make him altogether pre-logical– can we wonder that his savage colleague, under much *more precarious conditions*, holds fast to the safety and comfort of magic?
>
> An interesting and crucial test is provided by fishing in the Trobriand Islands and its magic. While in the villages on the inner lagoon fishing is done in an easy and absolutely reliable manner by the method of poisoning, yielding abundant results without danger and uncertainty, there are on the shores of the open sea dangerous modes of fishing and also certain types in which the yield greatly varies according to whether shoals of fish appear beforehand or not. It is most significant that in the Lagoon fishing, where man can rely completely upon his knowledge and skill, magic does not exist, while in the open-sea fishing, full of danger and uncertainty, there is extensive magical ritual to secure safety and good results. (Malinowski 1948, pp. 13–14, emphasis added)

Malinowski further corroborates his contention by stating how, in the Trobriand Islands, lagoon fishing (yielding abundant results without danger and uncertainty) did not involve superstitious practices, while uncertain and risky open-sea fishing did. Seafaring has already been explored as paradigmatic for distributed cognition (Hutchins 1995), as the extreme uncertainty requires the distribution of complex cognitive tasks between several cognizants and their artifacts. Open-sea fishing exponentially increases the uncertainty, adding to that of sea-faring the search for catch and its abundance.

As stated by Malinowski, the superstitious magic is not only *performed* during the fishing expedition, but also constructed into the fishing artifacts that are at the same time an extension of the fishers embodied cognitive systems.

Most forms of everyday superstition, Vyse's extensive review shows, are to some extent distributed into the environment by some kind of props. Sportsmen have lucky items and complex rituals they enact before every game, just as gamblers; furthermore, students facing exams display a bewildering array of lucky garments (sweatshirts, scarves, etc.), objects (namely pens), and practices (finding a coin before the exam) that either indicate or foster a positive outcome (Vyse 2014,

Chap. 2). Also, if we consider well-known forms of superstition such as tossing spilled salt beyond one's back, reacting to black cats and broken mirrors, jinxing, crossing fingers or other gestures, spitting, carrying lucky amulets such as a rabbit paw and so on, we can see that all of those instances share a *distributed* nature.

Let's recall Norman's vision of superstition as the acknowledgement of a mismatch between one's mental model and its target: this mismatch is uncertainty. You don't need to be *lucky* to find your way from work to home, you have experience with action, outcomes and mechanisms, so your mental model perfectly matches its target. Conversely, if you are for instance a gambler, a fisherman, or a student undergoing an examination you have some experience in the domain, you know the actions and the desired outcome, but your mental model cannot fully account for the mechanism, which is dominated by uncertainty. Here is where superstition comes in: it compensates a mismatch between the cognizant and its target by a "magical" distribution of cognition into the environment.

This distribution has two main effects: first, "a causal effect of an activated good-luck-associated superstition on subsequent performance […] mediated by an increase in perceived level of self-efficacy" (Damisch et al. 2010, p. 1018) and second, in cases of utter uncertainty it averts the risk of paralysis and fosters the possibility to discover unseen chances for action (Bardone 2011). Last, but not least, the distribution of a practice (as an object, or a gesture, or a phrase) into the external world allows that it is *socially learnt* and *repeated*. In order to better frame the action of superstition as a distributed cognitive practice to cope with uncertainty, I will introduce a broader theory concerning the relationship between cognition and the environment.

3 Cognitive Niche Construction: A Wider Frame for Extended Cognition

We can wonder: what is, pragmatically speaking, the ultimate goal of cognition? a stimulating way to answer might be the following: *cognition aims at predicting what is predictable, and making predictable what is uncertain*. The relationship between an organism and its environment is indeed one of coping with the uncertainty of its system, and that is what sparked the origin of the multifaceted phenomenon known as cognition. An interestingly essential definition of cognition as the one offered by the *Stanford Philosophy Encyclopedia* is a good starting point: "Cognition is constituted by the processes used to generate adaptive or flexible behavior."

The adaptive behavior implied by cognitive capabilities is a response to the uncertainty of the environment. The uncertainty one has to cope with can be biologically translated as the environment's *selective pressure*. According to a cluster of theories describing *ecological* and *cognitive niche construction*, organisms have always attempted to reduce the selective pressure (and so the uncertainty) by modifying their environments into better-suited *niches* (Tooby and DeVore 1987; Odling-Smee et al. 2003; Day et al. 2003; Pinker 2003; Clark 2005; Magnani 2007, 2009).

Niche theories consist of a theoretical framework that is proving extremely profitable in bridging evolutionary biology, philosophy, cognitive science, and anthropology by offering an inter-disciplinary ground causing novel approaches and debates to crucial issues in all of the aforementioned fields (Clark 2006; Iriki and Taoka 2012; Sinha 2015; Bertolotti and Magnani 2015).

As far as human beings are concerned, niche construction goes beyond modifications aimed at reducing environmental pressure (i.e. shelters, dams, nests etc.) even though they are of course still present. Human niche construction is characterized by a pervasive presence of models, representations, and other various mediating structures between one's cognition and the environment, all of which are massively distributed into the environment in a cognitive niche construction activity (Clark 2005; Iriki and Taoka 2012; Bertolotti and Magnani 2015). This can be seen as a comprehensive frame to individual instances of extended cognition (Clark and Chalmers 1998). Consider writing: writing is an example of how we extend our mind beyond our brains in order to achieve cognitive tasks that would not be possible otherwise, or much more difficult. At the same time, more or less ample sets of written material consist in cognitive niches with the aforementioned goal of deflecting environmental uncertainty. For instance, a code of laws perfectly reflects this dual nature: in being written, it can couple with individual cognitions offering an archival support unmatched by "naked minds," but in its being accessible it offers a series of guidelines and prescriptions that can be used as heuristics to navigate our social settings and dramatically reduce the uncertainty about what to expect from others.

The point this reflection starts from is the acknowledgment that one of the key factors that produced this is human beings' ability to continuously delegate and distribute cognitive functions to the environment to lessen their bio-cognitive limits. One of the explanations is that human beings create models, representations and other various mediating structures, that are thought to be of aid for thinking: such intense activity of distributing cognition into the environment is described as *cognitive niche construction* (Magnani 2009; Clark 2005, 2006; Iriki and Taoka 2012; Magnani and Bertolotti 2013; Bertolotti and Magnani 2015).

In order to further clarify what "cognitive niche construction" means, it will be necessary to briefly review and critically combine the various notions contributing to its definition. *Cognitive niche construction* is an extremely rich theory advocated by distributed cognition supporters such as Clark (2005, 2006) and Magnani (2009, Chap. 6). Prima facie, cognitive niche construction lays at the intersection of two larger theories: (ecological) niche construction (Odling-Smee et al. 2003) and cognitive niche theories (Tooby and DeVore 1987; Pinker 2003). Albeit both Magnani and Clark refer to Tooby and DeVore and to Pinker as initiators of cognitive niche theories, it should be observed that the meaning of cognitive niche in the latter and in the former views is quite different: whereas Clark and Magnani emphasize a local (or even better ecological) view of cognitive niches—true to the tenets of distributed cognition theories—, Tooby and De Vore and Pinker present a rather temporal-anthropological view of the cognitive niche, to the point that their description is more similar to an *era* of cognition. Indeed, they describe the cognitive niche as something that has been "entered" by human beings at some point of their evolutionary history. Tooby and DeVore (1987) and Pinker (2003) suggest that this jump in human cognitive history can be said, in other words, to coincide with human beings becoming "informavore."

> Gathering and exchanging information is, in turn, integral to the larger niche that modern Homo sapiens has filled, which Tooby and DeVore (1987) have called 'the cognitive niche' (it may also be called the 'informavore' niche, following a coinage by George Miller). Tooby and DeVore developed a unified explanation of the many human traits that are unusual in the rest of the living world. They include our extensive manufacture of and dependence on complex tools, our wide range of habitats and diets, our extended childhoods and long lives, our hypersociality, our complex patterns of mating and sexuality, and our division into groups or cultures with distinctive patterns of behaviour. Tooby and DeVore proposed that the human lifestyle is a consequence of a specialization for overcoming the evolutionary fixed defences of plants and animals (poisons, coverings, stealth, speed, and so on) by cause-and-effect reasoning. Such reasoning enables humans to invent and use new technologies (such as weapons, traps, coordinated driving of game, and ways of detoxifying plants) that exploit other living things before they can develop defensive countermeasures m evolutionary time. This cause-and-effect reasoning depends on intuitive theories about various domains of the world, such as objects, forces, paths, places, manners, states, substances, hidden biochemical essences, and other people's beliefs and desires (Pinker 2003, p. 27).

Pinker, elaborating on Tooby and DeVore's analysis, suggests that humans *filled* the cognitive niche as a "consequence of a specialization for overcoming the evolutionary fixed defences of plants and animals by cause-and-effect reasoning," (p. 27) which enabled the invention of new technologies and a whole series of enhancements that were quicker than evolutionarily-evolved countermeasures.

The element of *construction* I highlighted earlier is what, notwithstanding obvious commonalities, discriminates between "cognitive niche" theories and "cognitive niche construction" theories. The latter, in fact, stress how any particular cognitive niche is *constructed* by an active modification of the environment by a cognizant:

> […] "cognitive niche construction" [is] hereby defined as the process by which animals build physical structures that transform problem spaces in ways that aid (or sometimes impede) thinking and reasoning about some target domain or domains. These physical structures combine with appropriate culturally transmitted practices to enhance problem-solving, and (in the most dramatic cases) to make possible whole new forms of thought and reason (Clark 2005, pp. 256–257).

This definition, offered by Clark, can still work accepting Tooby and DeVore's seminal contention that entering the cognitive niche is ultimately about uncovering and exploiting, in a persistent way, cause-effect models of the external world:

> At the core of this lies a causal or instrumental intelligence: the ability to *create and maintain cause-effect models* of the world as guides for prejudging which courses of action will lead to which results. […] Our cognitive system is knowledge or information driven, and its models filter potential responses so that newly generated behavioral sequences are appropriate to bring about the desired end. (Tooby and DeVore 1987, p. 210, added emphasis).

I agree with Laland and his colleagues in their contention that "while the capacity for [ecological] niche construction is universal to living creatures, human niche construction is extraordinarily powerful" (Laland and Hoppitt 2003, p. 158): the reason for this success must rely in some peculiarity of niche construction activities, and not in the definition of cognition that we could use to determine the cognitive niche construction in favor of human animals. Referring to Tooby and DeVore's description of the cognitive niche (as a stage, entered according to them by human beings alone), the following triple identity can be proposed: if *cognition* amounts to the *processes used to generate adaptive or flexible behavior*, then it also supports the *flexible capability to detect and exploit, at least in an occasional and embodied/tacit way, cause-effect relationships in one's environment*.

Summing this up, it seems that cognitive niche construction is the process by which organisms modify their environment to affect their evolutionary fitness[3] by introducing structures that facilitate (or sometimes impede) the persistent individuation, the modeling, and the creation of cause-effect relationships within some target domain or domains. These structures may combine with appropriate culturally transmitted practices to enhance problem-solving, and (in the most dramatic cases) they afford potential whole new forms of thought and reason.

4 Superstition as Special Niche Construction

A reflection has to be made explicit before tackling the relationship between cognitive niche construction, uncertainty and superstition: both in our reworked definition of cognition, and in Tooby and De Vore's considerations about the cognitive niche, the ability to model cause-effect relationships must not be assumed as the ability to model *true* (in a correspondence sense) cause-effect relationships. It suffices that the cognizant candidate is able to apprehend such relationships and consequently modify the behavioral response: in this sense also a superstitious association is a rightful instance of one's cognitive capacities. As far as the biological and pre-linguistic levels are concerned, it can be argued that the detected cause-effect relationships do not matter for their *truth-reliability* but rather for their *fitness-reliability* (Sage 2004)—understanding "fitness" as comprehending the less rigorous notion of "welfare." While our human language-dominated world accounts for the fact that we consider the notion of truth, naïvely, as correspondence, from a biological perspective (often enacted by human beings as well) the *favored* detected relationship is the *most successful*, the one leading to survival.

It is the same kind of argument underlying evolutionarily positive appreciations of religion. At the lowest level, our cognitive systems are biased in favor of false

[3]In agreement with (Odling-Smee et al. 2003, pp. 256–257), the concept of fitness here has to be intended as *loosely Darwinian* because of the following reasons: extragenetically informed behavior patterns are broadly adaptive and maladaptive; variants occurring during genetic evolution are random, whereas those of extra-genetic information are not.

positives rather than false negatives: it is better to see a snake when it is just a branch than taking a branch for a snake (Barrett 2009). In the bigger picture, wrong or unwarranted (correspondence-wise) beliefs concerning the existence of supernatural beings may have a positive impact on the welfare of individuals and societies (Wilson 2002; Johnson and Bering 2006; Bulbulia 2009). In this sense, religion can be easily seen as a complex, multi-leveled cognitive niche distributing beliefs about supernatural agents in words, rituals and artifacts (Mithen 1999; Bertolotti and Magnani 2010) that foster positive outcomes (namely cooperation) and so reduce environmental uncertainty. What about superstition?

We can start by remembering the difference between religion and superstition, a distinction that religion has always enforced—one-sidedly if necessary. Superstition occurs when there is a mismatch between our mental models and their targets: paradoxically, religions are not affected by such mismatch. As they encode strict moral guidelines, religious practices preach a perfect correspondence between what happens in our mental models and Norman's aforementioned triad of *outcome*, *experience* and *mechanism* (Magnani 2011). Even if the relevant mechanism (e.g. God's will) is unknowable, many religions are understood as a list of prescriptions that, if correctly followed, will grant bliss, success, eternal life and so on.[4] No one goes to heaven because of a lucky charm. But as I said since the beginning, the proper domain of superstition are *terrestrial* issues, and not *celestial* ones.

A mismatch between how well our mental models represent the desired outcome and the mechanism to obtain it signifies a failure or the impossibility to individuate persistent cause-effect relationships to be exploited in the external world. Connecting this to the definition of cognitive niche I worked out in the previous section, superstition is individuated in those "environments" (broadly conceived) over which it is not possible to structure a cognitive niche, or at least not a satisfactory one. Consider fishing, or gambling, or anything that significantly depends on hazard such as lotteries or sports. These are clearly cognitive niches. Games and lotteries have rules, just as fishing has strategies and equipments: in all of these cases there are distributions of knowledge that exploit the affordances of what is at stake, or create new ones (Gibson 1977). Still, at the chore of what determines the outcome, utter uncertainty rules: I can have the best boat, the best fishing lines and the strongest fishermen, but none of this will grant that there will be fish where I drop the anchor. The same goes for gamblers: players can be as good as it gets, but this will not affect the shuffling of the cards.

Optimism is not enough to make up for the fact that one is in uncharted (or *un-niched*) waters. *Optimism has to be distributed* into the word just as knowledge is distributed in the cognitive niche in order to be picked up and create a bond between the mind and its possibility to extend into the environment.

[4]I explicitly refer to how religion is received: this is not the lieu for an analysis of theological debates such as those relating to God's grace, the relationship between faith and works, and obviously predestination. It is worth noticing, though, that also in theologies preaching predestination the fact that salvation is *pre*-established automatically excludes it from the kind of luck that can be affected by superstitious practices.

Even if superstition can consist of shared practices, either identical in culturally transmitted form (think of rabbit paws, not chicken feet) or in nature (having a *lucky something*), they are enacted individually. A cognitive niche is exploitable to the maximum because it is persistent and ecologically *available*: it is a constant repair against uncertainty. Superstition only "works" when it is deployed by the superstitious agent: it can be seen as *fleeting* and *agent-projected*. By activating a superstitious belief through a practice that connects to the world, the agent *projects* a kind of cognitive niche onto the immediate surroundings. Cognitive niches can be seen and learnt as repositories of environmental chances and guidelines on how to use them. The individually-projected cognitive niche that emanates from the coupling between the mind and the superstitious props is mostly about guidelines for actions. The environmental chances are created by superstition as a one-way projection: sometimes they are really there, sometimes they are not, but that is preferred to the inaction caused by total cluelessness.

5 Conclusion

Superstition is a framework of magical thinking that cognitive science explored far less than religion. Psychological literature is used to the observation that superstition actually increases performances: I suggest that the reason is indeed the reduced anxiety and the increased perception of self-efficacy. Superstition is a form of self-delusion into believing one is within the safe behavioral pattern constituted by an established and maintained cognitive niche (which indeed reduces anxiety), when one is actually relying on a self-projected cognitive niche instantiated just by the presence of the superstitious prop.

Framing the analysis of superstition within cognitive niche construction theory affords a unified toolbox explaining its connection with the (apparent) reduction of uncertainty allowing agents to enact a pragmatical course of action as if they were in the presence of an actual cognitive niche. Further developments of this analysis should concern the relationship between distribution of superstitious practices as cognitive niche projection and their cultural sharing. Finally, from this analysis it seems to follow that an individually-projected superstitious cognitive niche and a well-established, shared one may coexists in those cases where the agent is unable to detect the latter, or is unwilling to accept its guidelines and its resources: in these cases, refusing the existing cognitive niche opens up a space of uncertainty where the individual niche can be projected. An example could be a terminal patient who, unhappy with the results provided by the medical cognitive niche, denies them and vows herself to superstition practices hoping to repel her sickness.

This final observation opens up an interesting space for further reflection about the possibility of adopting alternative behaviors with respect to those belonging to the orthodoxy of the niche. In this sense, the superstitious agent and the cheater interact with the niche in different ways, though both embedding an element of *rejection* and *messing-with*. As I suggested, the superstitious agent may suspend an established cognitive niche in order to project her own (through an ideal superstitious prop), in order to reduce the anxiety caused by her impossibility (or unwillingness) to cope with the established niche.[5] Conversely, the cheater performs an aware deactivation of the orthodoxy of the cognitive niche, but only as far as she is concerned. In order to perform her goals, indeed, the cheater must rest assured that her partners (or opponents, as in a card game) are sticking to the orthodoxy of the niche:[6] this is in line with the everyday observation that superstitious people usually encourage the diffusion of their practices, whereas cheaters do not aim at making adepts in their attack against the cognitive niche.

References

Atran, S. (2002). *In gods we trust: The evolutionary landscape of religion*. Cambridge: Oxford University Press.

Bardone, E. (2011). *Seeking chances: From biased rationality to distributed cognition*. Berlin/Heidelberg: Springer.

Barney, S. A., Lewis, W. J., Beach, J. A., & Berghof, O. (Eds.). (2006). *The etymologies of Isidore of Seville*. Cambridge: Cambridge University Press.

Barrett, J. (2009). Cognitive science, religion and theology. In J. Schloss & M. J. Murray (Eds.), *The believing primate* (pp. 76–99). Oxford: Oxford University Press.

Beck, J., & Forstmeier, W. (2007). Superstition and belief as inevitable by-products of an adaptive learning strategy. *Human Nature, 18*(1), 35–46.

Bertolotti, T., & Magnani, L. (2010). The role of agency detection in the invention of supernatural beings: An abductive approach. In L. Magnani, W. Carnielli, & C. Pizzi (Eds.), *Model-based reasoning in science and technology. Abduction, logic, and computational discovery* (pp. 195–213). Heidelberg/Berlin: Springer.

Bertolotti, T., & Magnani, L. (2015). Contemporary finance as a critical cognitive niche. *Mind and Society, 14*(2), 273–293.

Boyer, P. (2001). *Religion explained*. London: Vintage U.K. Random House.

Bulbulia, J. (2009). Religiosity as mental time-travel. In J. Schloss & M. J. Murray (Eds.), *The believing primate* (pp. 44–75). Oxford: Oxford University Press.

Clark, A. (2005). Word, niche and super-niche: How language makes minds matter more. *Theoria, 54*, 255–268.

Clark, A. (2006). Language, embodiment, and the cognitive niche. *Trends in Cognitive Science, 10* (8), 370–374.

Clark, A., & Chalmers, D. J. (1998). The extended mind. *Analysis, 58*, 10–23.

[5]This can be linked to confabulatory behavior (Hirstein 2009) towards the exploitation of cognitive niches, inasmuch as an agent is unaware of her own projection—especially when superstition needs to discard a preexistent niche. Still, I uphold this consideration only to the extent that it does not involve a pathologization of superstition, which would utterly go against the spirit of this article.

[6]This is akin to the difference between *bullshitting* (sic.) and lying in their relationship with truth (cf. Magnani 2011, Chap. 4.6; Frankfurt 2005). However noxious, the bullshitter's lack of commitment towards the truth is usually mirrored by an equal lack of commitment towards *the others' commitment*. A liar, instead, has to ensure that the others adhere to the commitment to truth she is denying.

Damisch, L., Stoberock, B., & Mussweiler, T. (2010). Keep your fingers crossed! How superstition improves performance. *Psychological Science, 21*(7), 1014–1020.
Day, R. L., Laland, K., & Odling-Smee, F. J. (2003). Rethinking adaptation. The niche-construction perspective. *Perspectives in Biology and Medicine, 46*(1), 80–95.
Dennett, D. (2006). *Breaking the spell*. New York: Viking.
Drees, W. B. (2010). *Religion and science in context*. New York, NY: Routledge.
Frankfurt, H. (2005). *On bullshit*. Princeton, NJ: Princeton University Press.
Gibson, J. J. (1977). The theory of affordances. In R. E. Shaw & J. Bransford (Eds.), *Perceiving, acting and knowing*. Hillsdale, JN: Lawrence Erlbaum Associates.
Hirstein, W. (2009). Introduction. What is confabulation? In W. Hirstein (Ed.), *Confabulation: Views from neuroscience, psychiatry, psychology and philosophy* (pp. 1–12). Oxford: Oxford University Press.
Hutchins, E. (1995). *Cognition in the wild*. Cambridge, MA: The MIT Press.
Iriki, A., & Taoka, M. (2012). Triadic (ecological, neural, cognitive) niche construction: A scenario of human brain evolution extrapolating tool use and language from the control of reaching actions. *Phil Trans R Soc B, 367*, 10–23.
Johnson, D., & Bering, J. (2006). Hand of god, mind of man: Punishment and cognition in the evolution of cooperation. *Evolutionary Psychology, 4*, 219–233.
Laland, K. N., & Hoppitt, W. (2003). Do animals have culture? *Evolutionary Anthropology, 12*, 150–159.
Magnani, L. (2007). Creating chances through niche construction. The role of affordances. In B. Apolloni (Ed.), *Knowledge-based intelligent information and engineering systems: 11th international conference, KES 2007, Vietri sul Mare, Italy, September 12–14, 2007, Proceedings, Part II*. Lecture Notes in Computer Science. Berlin/Heidelberg: Springer.
Magnani, L. (2009). *Abductive cognition: The epistemological and eco-cognitive dimensions of hypothetical reasoning*. Berlin/Heidelberg: Springer.
Magnani, L. (2011). *Understanding violence. Morality, religion, and violence intertwined: A philosophical stance*. Berlin/Heidelberg: Springer.
Magnani, L., & Bertolotti, T. (2013). Selecting chance curation strategies: Is chance curation related to the richness of a cognitive niche? *International Journal of Knowledge and System Science, 4*(1), 50–61.
Malinowski, B. (1948). In R. Redfield (Ed.), *Magic, science and religion and other essays*. Glencoe, IL: Free Press.
Marmor, J. (1956). Some observations on superstitions in contemporary life. *American Journal of Orthopsychiatry, 26*, 119–130.
Menary, R. (Ed.). (2010). *The extended mind*. Cambridge, MA: The MIT Press.
Mithen, S. (1999). Handaxes and ice age carvings: Hard evidence for the evolution of consciousness. In A. R. Hameroff, A. W. Kaszniak, & D. J. Chalmers (Eds.), *Toward a science of consciousness III. The third Tucson discussions and debates* (pp. 281–296). Cambridge: MIT Press.
Norman, D. A. (1983). Some observations on mental models. In D. Gentner & Stevens A. L. (Eds.), *Mental models* (pp. 7–14). New York and London: Psychology Press.
Odling-Smee, F. J., Laland, K., & Feldman, M. W. (2003). *Niche construction. A neglected process in evolution*. New York, NJ: Princeton University Press.
Parrinder, G. (2000). *The Routledge dictionary of religious and spiritual quotations*. New York, NY: Routledge.
Pinker, S. (2003). Language as an adaptation to the cognitive niche. In M. H. Christiansen & S. Kirby (Eds.), *Language evolution* (pp. 16–37). Oxford: Oxford University Press.
Sage, J. (2004). Truth-reliability and the evolution of human cognitive faculties. *Philosophical Studies, 117*, 95–106.
Sinha, C. (2015). Ontogenesis, semiosis and the epigenetic dynamics of biocultural niche construction. *Cognitive Development*. Online first, doi:10.1016/j.cogdev.2015.09.006
Skinner, B. F. (1948). Superstition in the pigeon. *Journal of Experimental Psychology, 38*, 168–172.

Tooby, J., & DeVore, I. (1987). The reconstruction of hominid behavioral evolution through strategic modeling. In W. G. Kinzey (Ed.), *Primate models of hominid behavior* (pp. 183–237). Albany: Suny Press.

Vyse, S. (2014). *Believing in magic: The psychology of superstition* (2nd ed.). Oxford: Oxford University Press.

Wilson, D. S. (2002). *Darwin's cathedral*. Chicago and London: Chicago University Press.

Neurophysiological States and Perceptual Representations: The Case of Action Properties Detected by the Ventro-Dorsal Visual Stream

Gabriele Ferretti

Abstract Philosophers and neuroscientists often suggest that we perceptually represent objects and their properties. However, they start from very different background assumptions when they use the term "perceptual representation". On the one hand, sometimes philosophers do not need to properly take into consideration the empirical evidence concerning the neural states subserving the representational perceptual processes they are talking about. On the other hand, neuroscientists do not rely on a meticulous definition of "perceptual representation" when they talk about this empirical evidence that is supposed to show that we perceptually represent such and such properties. It seems that, on both sides, something is missed. My aim is to show that, in the light of empirical evidence from neuroscience, the case of action properties is a good candidate in order to properly talk of perceptually represented properties. My claim is that the neurophysiological states encoding action properties are perceptual processes and that these perceptual processes are representational processes. That is, in the case of those neurophysiological states involved in the detection of action properties, it is correct to speak of perceptual representational states, and hence, *ipso facto*, of perceptually represented properties. First, I describe a reasonable and widely agreed upon conception of perceptual representation in the philosophical literature. Then, I report evidence from vision and motor neuroscience concerning the perception of action properties, which is subserved by the ventro-dorsal stream, a portion of the dorsal visual system. Finally, I show that a strong connection can be found between the philosophical idea of perceptual representation I have reported before and the neuroscientific evidence concerning the activity of the ventro-dorsal stream, whose job is, as said, to detect action properties.

G. Ferretti (✉)
Department of Pure and Applied Science, Università di Urbino Carlo Bo,
Via Timoteo Viti, 10, Urbino (PU) 61029, Italy
e-mail: gabriele.ferretti88@gmail.com

G. Ferretti
Centre for Philosophical Psychology, University of Antwerp,
Lange Sint Annastraat 7, Antwerp 2000, Belgium

L. Magnani and C. Casadio (eds.), *Model-Based Reasoning in Science and Technology*, Studies in Applied Philosophy, Epistemology and Rational Ethics 27, DOI 10.1007/978-3-319-38983-7_10

1 Introduction

It is widely agreed in the philosophical literature, that we perceptually represent objects as having a lot of properties (Siegel 2006; Nanay 2010, 2013b: 3.2; Jacob and Jeannerod 2003; Peacocke 1992; Siewert 1998; Clark 2000; Crane 1992). For instance, we can perceive objects as being pencils and cups (Siegel 2006), as being present (Matthen 2005) and familiar (Dokic 2010), or absent (Farennikova 2012), as sounds (O'Callaghan 2014), or colors (Campbell 1993), as having spatial properties (Tye 2005; Peacocke 1992, 2001), or dispositional properties (Nanay 2011b), or as being tropes (Nanay 2012). Moreover, someone has claimed that we perceive objects as affording us the possibility of performing specific actions (Gibson 1979; Jacob and Jeannerod 2003; Zipoli Caiani 2013; Pacherie 2000, 2002, 2011; Jeannerod 1997, 2006; Siegel 2014; Jacob 2005; Prosser 2011), i.e. as being edible, climbable, or Q-able in general (the expression "Q-able" is by Nanay 2011a), etc. (for a review see Nanay 2013b).

The issue at stake here is twofold. On the one hand, sometimes philosophical accounts about perceptually represented properties do not need to properly take into consideration the technical empirical evidence concerning the neural states subserving the representational perceptual processes they are talking about. On the other hand, those who offer a more empirically-informed account do not rely on a meticulous definition of "perceptual representation" when they talk about this empirical evidence that is supposed to show that we perceptually represent such and such properties. It seems that, on both sides, something is missed.

My aim is to show that, in the light of empirical evidence from neuroscience, the case of action properties is a good candidate in order to properly talk of perceptually represented properties. My claim is that the neurophysiological states[1] encoding action properties are perceptual processes and that these perceptual processes are representational processes. That is, in the case of those neurophysiological states involved in the detection of action properties, it is correct to speak of perceptual representational states, and hence, *ipso facto*, of perceptually represented properties. In what follows, I develop this idea.

Important specification: although considering perceptual states as representations has been considered the most reliable way of describing our perceptual system, this framework has recently been criticized (Hutto and Myin 2013; Chemero 2009; Noë 2004). I will not take part to the contemporary divide in the literature between those who believe that perception is the construction of internal representations and those who reject this idea (for a review see Nanay 2012, 2013b, 2014b; Jacob and Jeannerod 2003). Apparently, the balance is effectively in favor of representationalism (Pautz 2010; Nanay 2013a, 2014b) but I will not survey the literature reporting these advantages here. However, my paper implicitly suggests a good example of a

[1]I am using the word "mental" and I will use the expression "cortical" or "mental" or neurophysiological state/representation as synonyms, leaving aside the issues concerning the mind-body problem.

rigorous representational interpretation of empirical evidence with respect to the case of the perception of action properties.

Furthermore, note that I cannot develop the proposal in full detail here and that my account only gestures towards such an argument. A fully fledged argument will have to wait for another occasion.

2 Representing Action Properties Between Philosophy and Neuroscience

Action properties are those object properties the representation of which is necessary for the performance of the subject's motor act[2] (this definition is by Nanay 2013b: 39; see also Ferretti 2016). Everyday objects exhibit geometrical properties such as size, shape, and spatial location. These geometrical properties are, from the motor point of view of the subject, action/motor properties, in that they afford to the subject a precise action possibility satisfiable with a precise motor act. For example, the geometrical features of a mug can be seen as action properties which open an action possibility (grasping) and which can be satisfied by a proper motor act: a power grip. The concept of action possibility has been in the spotlight of both neuroscience (see Borghi and Riggio 2015) and philosophy (see Jacob and Jeannerod 2003; Nanay 2013b) in the last twenty-five years.

On the one hand, as said at the beginning, some philosophers have claimed that we perceptually represent action possibilities (for a review see Siegel 2014; Jacob 2005; Pacherie 2002; Nanay 2013b; Jacob and Jeannerod 2003). For example, according to Nanay (2011a), it is not the case that we first represent some object property, say size and shape, perceptually, and then we infer, from the deliverances of our perception, that the object is Q-able. In other words, we perceptually represent Q-ability. On the other hand, a massive amount of evidence from visual and motor neuroscience has shown which processes are performed by our visual system in order to perceive the action possibilities offered by the objects in the environment (a mug as graspable, an apple as touchable, etc.). Neuroscientists usually talk about this detection performed by perception using the word "representation" (Borghi and Riggio 2015).

However, philosophers and neuroscientists start from very different background assumptions when they use the term "representation". On the one hand, neuroscientists seem to use the verb "representing" as a near synonym of "encoding".[3] So,

[2]Representations of these properties are usually called in literature "pragmatic/motor representations" (see Nanay 2013b, 2014a; Jacob and Jeannerod 2003; Jeannerod 2006; Pacherie 2000; Butterfill and Sinigaglia 2014). So, one might think that another way of formulating my claim would be to say that motor representations really are perceptual representations. However, in order to avoid problems with the different interpretations of the expression "motor representation" within the general literature, I am not committed to this reformulation here, see (Sect. 5).

[3]The famous "Principles of Neural Science" by Kandel et al. (2013) mentions the different expressions "to represent" and "representation" more than 200 times in the book. Concerning the

they talk about portions of the brain, especially of the cortical mantle, as representing something in the external world. The problem with this usage is that it seems to ignore the importance of the philosophical implications of the word "representation". Too often, they talk about a simple detection based on a causal covariation/co-occurrence between the neural firing of a brain area and an object being presented in the external environment at the same time. In the words of Pacherie (2002: footnote 12, pp. 66–67), "Neuroscientists often make a more liberal use of the term 'representation' than philosophers are wont to do, thus making the philosophers wary that what the neuroscientists call "representations" really qualify as "mental representations". (…) One may wonder whether these so-called representations are really representations, where for something to qualify as a representation it must have correctness conditions, be capable of misrepresenting, and exhibit some degree of intentionality".

On the other hand, even those philosophers who are more careful when they talk about perceptual representations when describing some mental process are usually guided by philosophical arguments that do not need to take into account the specific empirical brain-based counterpart of the mental processes they are talking about. At best, they need to take into account the bare minimum empirical evidence in order to argue that a minimal confirmation for their claim is held from neuroscience. This is reasonable insofar as some philosophical claims in the literature—I have mentioned in (Sect. 1)—do not need a precise empirical model, but just some empirical sketch, in order to be defended. However, the result is that it is hard to find a rigorous philosophical representational interpretation of the technical empirical data from neuroscience, concerning such and such mental phenomenon. Yet, it would be desirable, for several mental phenomena, to have an account that is, at the same time, empirically well-founded and philosophically rigorous. The aim of the present paper is exactly to provide this account, in particular with respect to the claim that action properties are perceptually represented.

It is important to underline that motor and vision neuroscientists regularly follow Gibson (1979) when, speaking of action properties/possibilities, they use the term "affordance" as a near synonym. However, the latter term is philosophically demanding. Indeed, it implies a precise conception of how the visual system works. I cannot survey the debate here, but several important arguments have suggested that we should not talk about affordances both in vision and motor neuroscience (Jacob and Jeannerod 2003) and in philosophy (Nanay 2013b)—see footnote 12. This is a further symptom of how neuroscientists usually do not care about the epistemological implications of their interpretation of the empirical results and the debate on action properties is the badge (see footnote 12). Hence I prefer to avoid the word "affordance" and simply use the term action property, or action possibility.

(Footnote 3 continued)

presence of representational terminology in neuroscience, see also (Brooks and Akins 2005; Bennett and Hacker 2003, 2012; Bennett et al. 2007; Bickle 2013; Jacob 2005; Mandik 2005).

So, the paper proceeds as follows. I describe a reasonable and widely agreed upon conception of perceptual representation in the philosophical literature (Sect. 3). I report evidence from vision and motor neuroscience concerning the perception of action properties, which is subserved by the ventro-dorsal stream (henceforth V-D), a portion of the dorsal visual system (Sect. 4). Then, I show that a strong connection can be found between the philosophical idea of perceptual representation I have reported before and the neuroscientific evidence concerning the activity of V-D, whose job is, as said, to detect action properties (Sect. 5).

The argument is the following: first, V-D is involved in the detection of action properties; second, V-D turns out to operate through perceptual representations; therefore, action properties are not simply detected, but perceptually represented. That is, V-D is a neurophysiological state that is both a perceptual state and a representational state; it is a perceptual representational state.

3 Perception as Representation in the Philosophical Domain: Perceptually Represented Properties

In this section I state the constraints that a mental state has to satisfy in order to be considered a perceptual representation.

Usually, perceptual states are considered representations because they have content. When I see a pumpkin, my perceptual state is about this pumpkin, it refers to this pumpkin. As suggested by Nanay (2012: 2), in general, a perceptual state "represents this particular thing as having a (limited) number of properties and the content of the perceptual state is the sum total of these properties" (I slightly reworded Nanay 2012: 2).[4]

When it comes to neuroscience, we can talk about neurophysiological states with a content (Jacob and Jeannerod 2003) following the common idea that representations are (bio)physical structures with the function of carrying information about something in the environment (Jacob and Jeannerod 2003; Dretske 1988, 1995). Often, these neurophysiological representational states pertain to cortical structures linked to perceptual systems; this is the reason why we can talk about them as performing perceptual representations. Moreover, representations are useful in order to explain behaviors that otherwise would be difficult to explain (Orlandi 2011: Sect. 1): "we tend to ascribe representational states to a system when the system's behavior and capacities cannot be easily (or at all) understood by making reference solely to environmental or organic conditions" (Orlandi 2011: 3; see also Brooks 1991; Fodor and Pylyshyn 1988, quoted by Orlandi). Finally, in order to be considered as such, representations have to have the possibility to fail, leading to misrepresentation (Jacob and Jeannerod 2003).

---

[4]Of course here we are talking of a perceptual state in general, as the whole outcome of our perceptual systems: the representation of colors, shapes, smells by different representational states etc. When talking of a particular representational state, it is not problematic to argue that the content of the representational state is the property it represents.

Following other analyses of the case, we can take stock of the philosophical constraints that a mental state has to satisfy in order to be a perceptual representation (I follow, in particular, the interesting review by Orlandi 2011: Sect. 1; but see also Dretske 2006; Wilson 2010; Burge 2010; Jacob and Jeannerod 2003). A subject is perceptually representing something if and only if he is in a state that:

(a) is derived from an accredited perceptual system;
(b) carries information about an object or state of affairs being a certain way (the state has content/accuracy conditions).
(c) given (b), can misrepresent by occurring in the absence of what it carries information about.
(d) gives us a description of the performance, e.g. planning to act.

In particular, philosophers have taken (b) and (c) to be crucial for the definition of a perceptual representational state: perceptual representations have to have content and have to be able to misrepresent (Burge 2010; Fodor 1987: 131, 2; Millikan 1984, 1993; Jacob and Jeannerod 2003). In what follows, I examine these four points.

3.1 (a) Derivation from a Perceptual System

Following (a) the fact that a subject represents what he/she represents perceptually, rather than in other ways, e.g. conceptually, means that the state is derived from an accredited receptor system (for a review see Orlandi 2011: 2).

As said, that means that it is not the case that we do represent some object property, say, size and shape perceptually, and then we infer, from the deliverances of our perception, that the object offers such and such action possibilities. We perceptually represent action possibilities (Nanay 2011a).

Moreover, as Nanay specifies, while describing a perceptual state is to describe the properties it attributes, not all properties that we represent objects as having are perceptually attributed/represented. I perceive a cup of coffee as black, as spatially located and as big. I can represent this as having the property of being the same cup of coffee I have used yesterday to drink water (for a complete review see Nanay 2011b; see also Ferretti 2016). Then, (a) does not deal with these cases of non-perceptual attribution/representation of properties.

3.2 (b) Carried Information, Content and Accuracy Conditions

The second constraint imposes to have a perceptual state with a content. But things are not so easy here. "If a state (*S*) carries information about property *F* and *F* is correlated with property *G*, then *S* carries information about *G*. The informational relation is transitive, information is ubiquitous and, unlike semantic content, informational content is indeterminate (e.g. metal bar carries information about the temperature, variations in temperature correlated with variations in atmospheric pressure, metal bar carries information about atmospheric pressure: representing the temperature, however, is not representing atmospheric pressure)" (Jacob and Jeannerod 2003: 5). As Jacob and Jeannerod note, "the length of the metal bar carries information about both the temperature and atmospheric pressure, so it cannot *represent* the temperature at all because it cannot misrepresent" (2003: 5). We have a similar neuroscientific case when we talk about several "stages on the way from the retina through the optic nerve to the higher levels of information-processing in the visual cortex, which carries information about the distal stimulus and about everything the distal stimulus stands in some non-accidental correlation with. Here, neither the retina nor the optic nerve represent everything they carry information about" (2003: 5).

An important specification is needed here. One thing is to ask:

(1) Which properties are represented in (visual) perception?

Another thing is to ask:

(2) What properties does our visual perceptual system respond to/covary with respect to the perception of objects?

(1) is about what properties are attributed by the perceptual system (or what it represents entities as having) and not about what properties are out there (presumably in a causal relation with our perceptual system), as in the case of (2). The properties our perceptual system responds to or tracks may not be the same as the ones it represents objects as having (see Nanay 2011b; see also Ferretti 2016). The answer to (2) is fundamental to get the answer to (1).

Note that, informational semantics, i.e. semantics based on non-coincidental correlations/co-occurrence, is an important tool of neuroscience. Neuroscientists map the electrophysiological activity of specific brain areas looking for cortical activations and neural discharges carrying the information about such and such properties in the environment. However, correlational relation is not shellproof. Reliability does not mean infallibility: misfiring may occur at some stage in the system (Jacob and Jeannerod 2003: 8). But crucially, in order to obtain the constraint (b), we need that constraint (c) is satisfied.

3.3 (c) Misrepresentation

Following (c) it is not possible to have a representation without the possibility of having a *mis*representation. Unless a signal could *mis*represent what it indicates, it cannot represent it. Mental and non-mental representations must be able to go wrong (Dretske 1988; Jacob and Jeannerod 2003: 5, 6; Burge 2010). But we have two different kinds of misrepresentation in literature.

The first one is very simple. There could be cases in which a device is damaged, so it cannot represent what it is supposed to represent normally, on the basis of its proper function.[5]

The second case is more complicated. As Jacob and Jeannerod suggest, usually, in an ecological situation, property *G* matters to the survival of the animal. However, the animal's sensory mechanism, responds to instantiations of property *F*, not property *G*. Often enough in the animal's ecology, instantiations of *F* coincide with instantiations of *G*. So detecting an *F* is a good cue if what enhances the animal's fitness is to produce a behavioral response in the presence of a *G*. But the animal does not represent *G* as such. This device has the proper function of representing instantiations of property *F*, not property *G* as such, to the extent that instantiations of *F* coincide with instantiations of *G*: detecting an *F* is a good cue in order to perform a behavioral response in the presence of a *G*. The misrepresentation here occurs when the device is representing *F* (often coinciding with instantiations of *G*), but in this precise case the instantiation of *F* does not coincide with the instantiation of *G* (I slightly rephrased the idea proposed by Jacob and Jeannerod 2003: 5–8, but see all the section 1; see also Fodor 1987; Schulte 2012).

In other words, in relation to (b), sometimes our perceptual system covaries with some property in the environment, but it might happen that the instantiation of those properties does not always coincide with the instantiations of the properties we can usually represent on the basis of this covariation.

3.4 (d) Usefulness in Performance Description-Explanation

Finally, "we tend to ascribe representational states to a system when the system's behavior and capacities cannot be easily (or at all) understood by making reference solely to environmental or organic conditions" (Orlandi 2011: 3; see also Brooks 1991; Fodor and Pylyshyn 1988). In other words, we say that "a system represents something when the system's behavior and capacities result at least in part, from the manipulation of an internal model of the situation, rather than from the situation itself" (see Orlandi 2011: 3, see also p. 14). "If, by contrast, we could explain everything that the system does without appeal to contentful representational states we might avoid invoking the presence of perceptual representations for reasons of parsimony" (Orlandi 2011: 3; see also Brooks 1991; Burge 2010; Segal 1989; Van Gelder 1995, quoted by Orlandi). That is, "we should think of perceptual states as representations, only if such states are available to figure in some otherwise unexplainable cognitive activity", as for example beliefs, learning, reasoning, planning to act etc. (Orlandi 2011: 3).

Now we have a philosophical definition of perceptual representation. The next step is a look to empirical evidence about the detection of action properties.

[5]"Arguably, unless a device has a function to carry information about some property, it makes no sense to say that it is misfunctioning" (Jacob and Jeannerod 2003: 6; see also Millikan 1984).

4 The Detection of Action Properties Subserved by the Ventro-Dorsal Stream. A Look at Motor Neuroscience

Following the two visual systems model, the visual system of humans and other mammals can be divided in two main visual pathways, grounded on distinct anatomical structures (Milner and Goodale 1995): the ventral stream for visual object recognition, and the dorsal stream for visually guided action. These pathways can be dissociated due to cortical lesions. Lesions in the dorsal stream (the occipito-parietal network from the primary visual cortex to the posterior parietal cortex) impair one's ability to use what one sees to guide action (optic ataxia), but not object recognition; lesions in the ventral stream (the occipito-temporal network from the primary visual cortex to the inferotemporal cortex) impair one's ability to recognize things in the visual world (visual agnosia), but not action guiding vision (see Jacob and Jeannerod 2003). Accordingly, the dissociation is also suggested from the results of behavioral studies, in normal subjects, involving visual illusions that can deceive the ventral stream but not the dorsal one (Bruno and Battaglini 2008).

While this division has been deeply challenged (see Briscoe 2009; Bruno and Battaglini 2008; McIntosh and Schenk 2009; Clark 2009), mentioning the distinction is important to start my discussion. Indeed, it has been famously shown that the dorsal stream is further divided into two: the dorso-dorsal stream, (D-D, following the division of the intraparietal sulcus, which subdivides the posterior parietal lobe, it is related to the superior parietal lobule (SPL)—also known as the dorso-medial circuit, which projects to the dorsal premotor cortex) and the ventro-dorsal stream (V-D, related to the inferior parietal lobule, (IPL)—also known as the dorso-lateral circuit, which projects to the ventral premotor cortex) (Gallese 2007; Turella and Lignau 2014; Rizzolatti and Matelli 2003). Though—both for human and non-human primates—the dorsal visual stream is crucial for the visual guidance of actions, the transformation of intrinsic object properties into action properties, and then in motor acts relies on a well defined cortical network lying between the parietal and the ventral premotor cortex (Gallese et al. 1994; Fogassi et al. 2001; Rizzolatti and Luppino 2001), that is, V-D, whose main components for this task are the anterior intraparietal (AIP) area and F5 (in the most rostral part of the ventral premotor cortex): that is, the parieto-premotor network AIP-F5 (Gallese 2007).

Empirical data (Baumann et al. 2009; Srivastava et al. 2009; Theys et al. 2012a, b, 2013; for a review see Romero et al. 2014; Chinellato and del Pobil 2016; Graziano 2009; Castiello 2005; Castiello and Begliomini 2008; Kandel 2013: Chap. 19; concerning the lesion studies, see Andersen et al. 2014) suggest that, since many AIP neurons respond selectively to objects during both passive fixation and grasping,[6] it is AIP that extracts visual object information concerning action

[6]A large part of neurons in this area discharges during object fixation and is selective for object properties, such as shape, size, and orientation (Verhoef et al. 2010).

properties—that is, it is AIP that translates/reads geometrical properties in/as action properties for grasping purposes and sends this information to neurons in the area F5, with which it is directly connected (Borra et al. 2008)—which then activate the primary motor cortex.

Accordingly, visuomotor[7] canonical neurons in F5 encode the information about—object geometrical attributes read as—action properties received by AIP and use this information in order to compute the appropriate motor commands for motor interaction with the object. Also, canonical neurons respond during object fixation, regardless of the actual execution of an action (Murata et al. 1997, 2000; Sakata et al. 1995; Raos et al. 2006). In experiments with monkeys, just as the subject looks at the object its neurons fire, activating the motor program that *would be* involved were the observer actively interacting with the object. The evoked (visuo)motor response is just a potential act (for a review Castiello 2005; Turella and Lignau 2014).

Moreover, this evidence seems to show that seeing an object is getting at the same time its visuomotor priming (i.e. the visuomotor representation of its "action property"), and the internal *simulation*[8] of one of the actions we could perform upon it (i.e. the most suitable motor program required to interact with it in the light of the action possibility offered in a given context), regardless actual action execution (Gallese 2000; Jeannerod 2006; Jacob and Jeannerod 2003).

In other words, when I am looking at the cup of coffee on my desk, the V-D circuit is crucial for my purpose of catching the handle of the cup[9]: V-D extracts, from layout objects' properties, those action properties necessary to grasp an object, responding to those 3D geometrical properties of objects that serve such visuomotor tasks as grasping them (Shikata et al. 2003). In the next section I will show that the activity of V-D with respect to this detection of action properties can be modeled in the representational terms expressed in (Sect. 3).

Important specification: for the sake of coherence toward neuroscience, I should specify that, of course, whereas my analysis is accurate for the point I am defending here, this is not a neuroscientific review of the neural underpinnings of action-guiding vision (Chinellato and del Pobil 2016). Accordingly, though both the V-D and the D-D and the ventral pathways are strictly interconnected and are involved in different stages of action computation concerning object properties

[7]Those "visuomotor" neurons showed a specific selectivity, discharging more strongly during the fixation of certain solids as opposed to others, the difference between them depending on the kind of grip afforded by those objects (e.g. precision grip, finger prehension, etc.).

[8]There are different notions of simulation: here I mean an automatic mechanism with perceptual function to facilitate the motor preparation (Gallese 2000, 2009; Gallese and Sinigaglia 2011; Jeannerod 2006; Borghi and Cimatti 2010; Decety and Grèzes 2006; Borghi et al. 2010; Ferretti 2016). The fact that simulation involves several bodily changes (Jeannerod 2006) is another reason for talking not about neural states but neurophysiological states.

[9]First, AIP detects the geometrical features of the handle that exhibit precise motor characteristics with respect to my motor repertoire. This means that shape, texture, size are encoded as action properties. Thus, this information is sent to F5, which computes the most suitable motor act (say, a power grip) in order to catch the handle of the cup. Though, things may be fuzzier then this, see footnotes 15–19.

(Borghi and Riggio 2015: 351; Turella and Lignau 2014), data—that I cannot review here—suggest the leading role of the AIP-F5 circuit in the V-D stream for what concerns the visuomotor transformation (for a critical review see Turella and Lignau 2014; Kandel et al. 2013: 871).

Also, the experimental evidence I report is based on studies concerning both human and non-human primates (for a review see Orban and Caruana 2014; Borghi and Riggio 2015; Shikata et al. 2003; Martin 2007; Chinellato and del Pobil 2016).

5 V-D Perceptually Represents Action Properties

I have listed the constraints that a mental state has to satisfy in order to be a perceptual representation (Sect. 3) and reported evidence concerning the detection of action properties subserved by V-D (Sect. 4). Here I will show that the four representational constraints listed in (Sect. 3) are satisfied by the activity of V-D reported in (Sect. 4). Following (Sect. 3), a V-D state is a perceptual representational state if:

(a) it is derived from an accredited perceptual system;
(b) it carries information about an object or state of affairs being a certain way (the state has content/accuracy conditions).
(c) given (b), the state can misrepresent by occurring in the absence of what it carries information about.
(d) it gives us a description of the performance, e.g. planning to act.

Important clarification: there is a huge debate on whether dorsal vision—and its V-D portion—is normally taken as being unconscious. I cannot focus on this debate (see Brogaard 2011; Briscoe 2009; Kravitz et al. 2011, 2013; Bruno and Battaglini 2008; McIntosh and Schenk 2009). Accordingly, someone has argued that action guiding-vision is not necessarily unconscious (Nanay 2013a, b; Briscoe 2009) and that the representation of action properties may underlie visual awareness. This is because, as a matter of fact, no crucial evidence that dorsal processing cannot underlie visual awareness (Wallhagen 2007; Jacob and de Vignemont 2010) and that V-D is completely unconscious (Rizzolatti and Matelli 2003; Gallese 2007) has been addressed. Moreover, the contribution of ventral vision to dorsal vision might give awareness to action-guiding vision (Briscoe 2009) and thus to our motor representations (Nanay 2013a, b). I have no problem here insofar: (a) I am considering the V-D stream—and not motor representations in general (footnote 2)—as the neurophysiological state which can be defined as a perceptual representational state[10]; (b) my argument is about the subpersonal level of perceptual representations concerning the visuomotor transformation; (c) the debate on the possibility for a

[10]Of course, it is difficult in a neural geography to isolate portions of our cortical systems. This is a crucial practice in the neuroscientific analysis, though.

conscious action-guiding vision is still open. However, there are good arguments to show that the same holds for the personal level of the perception of action properties (see my Sect. 5.3: V-D and (c); see also Siegel 2014; Prosser 2011).

5.1 V-D and (a): Derivation from a Perceptual System

In order to show that V-D's activity respects (a), it would be sufficient to argue that V-D is deeply involved in the activity of the perceptual visual system insofar it manipulates visual information which starts from the retinotopic map and, through different cortical stages, arrives at the motor cortex.

But we have other reasons for arguing that this is a perceptual state. Data from motor neuroscience are compatible with the idea that we perceptually represent action properties. First, it has been recently suggested by Zipoli Caiani (2013) that the V-D activity does not require inferential processes because the detection of patterns for motor-related possibilities of action in the perceptual stimulus can be effective even if brain lesions limit the use of other abilities in categorizing and locating objects (for example Humphreys and Riddoch 2001a, b; see Zipoli Caiani 2013 for a discussion of this empirical evidence). This gives empirical support to the idea that the perception of action possibilities is a process that makes no use of inferences based on action-independent forms of categorization. Moreover, it has been suggested, and it is widely agreed, that the visuomotor phenomena described in (Sect. 4) are an interesting case of motor perception[11] (Fadiga et al. 2000).[12] The information is transformed at cortical level, and does not depend on higher

[11]Perceptual features of objects are read as contents of a (sensori) motor nature and visual stimuli are 'motorically' encapsulated (Jacob and Jeannerod 2003: 177). This is an internal state in which perception and action are not precisely delimited (Jeannerod 2006): for example, the discharge of visuomotor neurons is neither purely visual nor purely motor, codifying a potential motor action (Fadiga et al. 2000: 176).

[12]Zipoli Caiani (2013) has defended the idea that V-D detects affordances—that, in gibsonian terms, means that V-D does not use representations in the detection of action possibilities—insofar this perceptual process does not involve any detached representation of the target, being involved in the direct detection of sensorimotor patterns in the stimulus. This is because it automatically maps the information contained in the perceptual stimulus on a specific motor plan for action, insofar the perceptual stimulation conveys enough information to somatotopically activate the sensorimotor system. However, it should be noted that what Zipoli Caiani has in mind are inferential representations *a là* computationalism and following his interpretation, Gibsonian anti-representationalism only rejects this kind of representations. The rejection of this particular kind of representations is not difficult to agree with (Nanay 2013a: 1056, 2013b: 3.1; Jacob and Jeannerod 2003: Chap. 6) and the same holds for me. The problem is that usually the interpretation of gibsonian anti-representationalism is not this and the term affordance is used in order to avoid every kind of representation (Orlandi 2011: 20; Pacherie 2002: 69). I do not care about this point here. However, several arguments suggest that, in describing the complexity of our motor interactions, we should not talk about affordances, insofar we do not visually perceive affordances in the gibsonian sense, if they are used to avoid every kind of representation (Jacob and Jeannerod 2003: Chap. 6).

cognitive processes. Once again, the visuomotor transformation is a subpersonal, automatic phenomenon. Accordingly, it's not by chance that my account focuses on the subpersonal level of analysis.

5.2 V-D and (b): Carried Information, Content and Accuracy Conditions

I said above that, concerning (b), it is important to distinguish

(1) What kinds of properties does our visual system attribute to objects?

from a very different question, namely:

(2) What properties does our perceptual system respond to/covary with during object perception?

(1) is about what properties are attributed by the perceptual system (or what it represents entities as having) and not about what properties are out there (presumably in a causal relation with our perceptual system), as in the case of (2). The properties our perceptual system responds to or tracks may not be the same as the ones it represents objects as having (Nanay 2011b; Ferretti 2016; Burge 2010). I also said that the answer to (2) is fundamental to get an answer to (1). The neural phenomena described in (Sect. 4) translate object attributes into motor commands: they represent S (geometrical objects features) as being F (action properties) and F (action properties) as Q (motor acts) resonating with the simulation of the performable motor act. But I have said that the properties our perceptual system responds to or tracks may not be the same as the ones it represents objects as having (Nanay 2011b; Ferretti 2016). Thanks to the visuomotor transformation process, our perceptual system responds to/covaries with particular "object geometrical features". Accordingly, thanks to the visuomotor transformation, our visual system attributes "action properties" to the object. Thus, the properties our perceptual system responds to or tracks (geometrical properties) are not the same as the ones it represents objects as having (action properties). Yet, the detection of particular geometrical features permits to transform (or say, read) these geometrical features in action properties, or, in other words, to attribute action properties to the object.

As suggested in (Sect. 3), a good way of arguing that (b) is the case is to show that (c) is the case.

5.3 V-D and (c): Misrepresentation

Following (c) it is not possible to have representation without *mis*representation. Unless a signal could *mis*represent what it indicates, it cannot represent it. Mental and non-mental representations must be able to go wrong (Dretske 1988; Jacob and

Jeannerod: 2003: 5, 6). V-D can undergo both cases of misrepresentations reported in (Sect. 3.3).

Recall the first: cases in which a device is damaged, so it cannot represent what it is supposed to represent normally. Neural phenomena in (Sect. 4) can be victim of misrepresentations in this sense, due to cortical diseases such as optica ataxia, ideomotor apraxia and visual agnosia (see Jacob and Jeannerod 2003; Milner and Goodale 1995).

Now, let's recall the second case carefully: usually, in an ecological situation, property *G* matters to the survival of the animal. However, the animal's sensory mechanism responds to instantiations of property *F*, not property *G*. Often enough, in the animal's ecology, instantiations of *F* coincide with instantiations of *G*. So detecting an *F* is a good cue if what enhances the animal's fitness is to produce a behavioral response in the presence of a *G*. But the animal does not represent *G* as such. This device has the proper function of representing instantiations of property *F*, not property *G* as such, to the extent that instantiations of *F* coincide with instantiations of *G*: detecting an *F* is a good cue in order to perform a behavioral response in the presence of a *G*. The misrepresentation here occurs when the device is representing *F* (often coinciding with instantiations of *G*), but in this precise case the instantiation of *F* does not coincide with the instantiation of *G* (for examples see Jacob and Jeannerod 2003; Fodor 1987; Schulte 2012).

In other words, in relation to (b), sometimes our perceptual system covaries with some property in the environment, but the instantiation of those properties does not coincide with the specific property we can usually represent on the basis of this covariation. I would like to suggest that this second case of misrepresentation is possible concerning the evidence in (4). Consider that, concerning this second case of misrepresentation, a good observation is that a geometrical property can be perceptually represented as an action property, while it is not the case that the object is offering an action possibility, insofar the geometrical arrangement pertains to a 2-D figure and not a real 3-D solid object. Let's go more slowly on this.

We have evidence that the V-D functions in a somewhat similar manner when we are perceiving depicted objects (for a complete review see Ferretti 2016). Both its components F5 (Chao and Martin 2000; see also Buccino et al. 2009; Costantini et al. 2010; Tucker and Ellis 2004; Proverbio et al. 2011; Grèzes and Decety 2002; Zipoli Caiani 2013) and AIP (Romero et al. 2012, 2014)[13] are activated during fixation of

[13]One might argue that this evidence shows only that the dorsal stream responds to pictures because it is involved in the perception of the surface of the pictures. However, evidence shows that motor related effects registered are deep related with the kind of motor act (e.g. power grip) one can perform on the depicted object (e.g. the handle of a mug), which is, in these cases, different from the act one can perform on the picture surface (precision grip). In many of the experimental settings, pictures are presented on a monitor, which, of course, cannot afford the same action afforded by the depicted object. However, looking at an image of an object triggers the activation of a suitable motor pattern for the execution of a motor act and the motor activation is highly specific to the action that is represented (see Jeannerod 2006 about this specificity). For example, in the case of Buccino et al. (2009) subjects observe virtual images of objects, in this case of handles. Here the motor-evoked potentials (MEPs) are from the right opponent pollicis and from

depicted objects. This visuomotor resonance seems to be a general feature of the dorsal stream processing. For example, neurons in dorsal intraparietal sulcus selectively respond to depicted objects exhibiting different texture gradient and linear perspective due to particular 3D shapes and precise orientations (Nelissen et al. 2009; Taira et al. 2001; James et al. 2002; Tsutsui et al. 2002, 2005,[14] Sakata et al. 2003). And this is in line with the evidence that F5 decodes grip type, while AIP encodes object orientation (Baumann et al. 2009; Gallivan and Wood 2009; Fluet et al. 2010).

So, V-D is selective for shared sensorimotor patterns based on geometrical properties, pertaining both to real and depicted targets.[15] There is no distinction concerning the distal source. So, concerning (1)/(2), we have the same answer in relation to V-D's processing for both real and depicted objects. The same covariation with depicted objects' geometrical properties is activated. The same visuomotor transformation is performed, due to the apparent possibility of motor interaction. So, we can perceptually attribute action properties also to depicted objects. Indeed, the same motor response can be fostered either by a real object (the handle of a real cup), or by the picture of an object of the same kind (the depicted handle), due to the activation of the same visuomotor processing involved in the encoding of action properties (for a complete review see Ferretti 2016).

This is a clear example of misrepresentation in the second sense: the definition given above fits this case. Indeed, sometimes, instantiations of S (geometrical pattern) correspond to instantiations of F (action property) and thus instantiations of F (action property) correspond to possibility of Q (motor act). But with picture

(Footnote 13 continued)

the first dorsal interosseous muscle. These anatomical components are crucial in grasping, and the presence of this kind of motor response shows us that the motor act encoded pertains to the handle and not, of course, to the surface of the image, since in this case the image is not a normal picture, but an image on a monitor, which cannot require grasping (for a very interesting review see Zipoli Caiani 2013). Accordingly, in the case of Chao and Martin (2000) motor response is dependent on particular pragmatic features of depicted objects (the depicted handle) (see also Grazes and Decety 2002). That is, motor responses are deep related with the kind of motor act (e.g. power grip) one can perform on the depicted object (e.g. the handle of a mug), which is, in these cases, different from the act one can perform on the picture surface (precision grip)—for a philosophical review of this empirical evidence with respect to this specific point see (Ferretti 2016).

[14]Tsutsui et al. (2002) explored the sensitivity of caudal intraparietal (CIP) neurons in the dorsal stream to texture-defined 3D surface orientation. CIP neurons are involved in high-level disparity processing (the reconstruction of 3D surface orientation through the computation of disparity gradients). Some of CIP neurons are sensitive to texture gradients, which is one of the major monocular cues. Some of them are sensitive to disparity gradients, suggesting their involvement in the computation of 3D surface orientation. Moreover, those sensitive to multiple depth cues were widely distributed together with those sensitive to a specific depth cue, suggesting CIP's involvement in the integration of depth information from different sources. The convergence of multiple depth cues in CIP seems plays a critical role in 3D vision by constructing a generalized representation of the 3D surface geometry of objects (Tsutsui et al. 2005).

[15]Arguably, this is possible because AIP, which is the stage from the visuomotor transformation starts, responds to small 2-D fragments. Since it is AIP to send the information for encoding the motor acts to F5 it is possible that thanks to this encoding, the action property, and then the potential motor act, are computed despite the distal cause of stimulus (see Romero et al. 2014).

perception this is not the case. Often, in our everyday life, the 2D structures perceived pertain not to a real object, but to a picture of that. This is the case in which two instantiations fail to co-occur. In this case, however, the visuomotor system seems to respond, misrepresenting.

An important clarification is that the activation of the motor system is not the same during simulation (in its various forms) as during execution. However, simulating is not doing, and substantial differences are observed between simulation and execution. The activation of most of the areas of the motor system during action representation is consistently weaker than during execution. This partly due to the fact that it is coupled with an additional mechanism for suppressing motor output (motor inhibition), a prerequisite for the off-line functioning of the representation. As a consequence, the muscles do not contract and the limbs do not move—in that the sensory reafferences normally produced by a movement are lacking; thus, simulating is not doing (see Jeannerod 2006: 2.3.3), insofar neural commands for muscular contractions are effectively present, but simultaneously blocked by inhibitory mechanisms.

Yet the important thing to note here is that, although during simulation we are not acting [think about the case of the frog that tries to nab, with its tongue, the small lead pellet in the example by made by Fodor 1987: (131–29)], the misrepresentation remains. Indeed, remaining at the subpersonal level of analysis, the job of our visuomotor system is not to make us act. It is rather to transform visual attributes—which reliably pertain to objects we can act upon—in motor acts for suitable action performance; that is, to translate the geometrical patterns in action properties, and then, on the basis of these action properties, in motor acts. Even in the case of depicted objects, the visuomotor transformation succeeds. Our brain translates object geometrical (layout) attributes into action properties, and then into motor commands, even when they pertain to 2D objects and even there is no overt motor response, but only covert motor simulation. It is as if the brain was encoding that a performable motor act is "in the quiver", even in those cases in which action is not possible. Our visuomotor system is calling our motor system saying: "you're ready for a suitable motor act", even in those situations in which it's not the case (for a complete review see Ferretti 2016).

One might object that this misrepresentation is systematic, while misrepresentations should occur sporadically. However, this is not a solid argument. Indeed, a lot of representations can often be turned in misrepresentations in particular environmental situations. Once again, think about the case of the frog quoted above.

The explanation of this relies on the fact that the dorsal stream distinguishes between images of graspable and non-graspable objects (Rice et al. 2007: 36)[16], but does not discriminate in a precise manner between objects seen face-to-face and

[16]The practice of using pictures in neuroscience to study how seeing tools automatically activates motor information is very widespread. I cannot offer a survey here (see Craighero et al. 2002; Ranzini et al. 2011; Borghi et al. 2012; Borghi and Riggio 2015; Ferretti 2016).

depicted objects, because this capacity is subserved by the ventral stream (Westwood et al. 2002).[17]

Moreover, coming back to the general point expressed above at the beginning of (Sect. 5), even though here my point concerns the subpersonal level of description, a similar misrepresentation might occur at the personal level as well. This is possible mentioning the case of *trompe l'oeil* paintings, which seems to deceive also the conscious dimension of our vision for action (Nanay 2015), leading our visual system to the illusion of being faced with something we are able to act upon.[18]

5.4 V-D and (d): Usefulness in Performance Description-Explanation

The last constraint we face imposes that the state has to give us a description of the examined performance.

In general, a state is a representation only if it carries out some cognitive function that is otherwise unexplainable. But we saw that carrying information about features of the environment, in the sense of, say, co-varying with some environmental quantity is not enough, by itself, for a state to be a representation: neurons that fire in the presence of a certain physical property may, in this sense, carry information about it, but they do not, *ipso facto*, represent the property. However, the neural phenomena described above not only covary with a (geometrical) property in the environment, but represent, through the visuomotor transformation, the property it covaries with in an another property, an action property.

We saw that a cortical process can be said to perceptually represent something, only if this representation is available for the production of a given behavior. (Sect 4) has shown that this cortical state perceptually represents the possibility of action. AIP translates layout properties in action properties and F5 uses this information to

[17]Also, prior to discriminating depicted objects as such, infants seem to perceive depicted objects as real objects affording action and they even grasp at the pictures as if trying to pick up the depicted objects (Deloache 2004: 68; see also Pierroutsakos and DeLoache 2003; Deloache et al. 1998). Accordingly, Westwood et al. (2002) asked a neurological patient with visual-form agnosia—a ventral impairment leaving the subject with visual dorsal encoding only—(patient D.F.) to grasp 3D objects and 2D images of the same objects and to estimate their sizes manually. D.F.'s grip aperture was scaled to the sizes of the 2D and 3D target stimuli, but her manual estimates were poorly correlated with object size. The interpretation of this evidence suggests—Westwood et al. conclude —that dorsal perception cannot discriminate between 2D and 3D objects, responding in a similar way to a 3D object and a 2D image of the same object (see Ferretti 2016).

[18]There is also a definition of misrepresentation according to which the perceptual representational state can occur even in the absence of the property which the state carries information about. The case of encoding of an action possibility when we are facing with a picture is an example of this kind of misrepresentation.

generate potential motor acts.[19] Given that this process permits to represent action properties and the "potential" motor acts to perform upon them, the following points provide an argument for the constraint (d).

First, in order to understand the attribution of those relational properties such action properties we need to posit perceptual representations. Action properties are not in the environment as trees and tables (see the (1)/(2) distinction): they depend on the representation of an action possibility given by relationship between the motor skills we have with respect to our body and in relation to what an object affords to us when coupling with it.

Second, in order to encode a suitable potential motor act which is maintained "in our motor quiver", the brain has not only to represent something which is not actually in the environment (action properties) as trees and tables, but it also has to represent (and simulate) something which might be useful: the potential motor act. This is possible because the visuomotor system not only *executes* actions but also internally *represents* them in terms of 'motor ideas' (Fadiga et al. 2000: 165; see also Gallese 2000, 2009; Gallese and Metzinger 2003). Note also that, during the representation of potential action given by the interplay between AIP and F5, different informations about action properties are encoded and different motor acts are inhibited at the expense of others, in rclation to the action possibilities we are facing with. This selection suggests the potential dimension of our visuomotor perception (for technical neurophysiological details see Rizzolatti and Sinigaglia 2008; Cisek 2007; Borghi and Riggio 2015; Sakreida et al. 2013). Accordingly, following the idea of motor simulation expressed in (Sect. 3), "the (overt) execution of an action is necessarily preceded by its (covert) representation (which is its frame is a simulation), while a (covert) representation is not necessarily followed by an (overt) execution of that action, insofar representation can be detached from execution, existing on its own" (Jeannerod 2006: p. 2; but see also Chap. 2). All this is in line with the idea that those representations are the immediate mental antecedent of actions (Nanay 2013b, 2014a).

[19]It has been argued that AIP might need the help of F5—which encodes motor acts—for the encoding action properties: it is difficult to properly discern how the representation of action properties is detached from the representation of the related motor acts and whether those two encodings are properly divided (Romero et al. 2014; Nowak and Hermsdörfer 2009; Theys et al. 2015; Janssen and Scherberger 2015; Chinellato and del Pobil 2016). Recent evidence has been offered about the complex interplay between AIP and F5, confirming the AIP/F5 union in forming a fronto-parietal network for transforming visual signals into grasping instructions (Brochier and Umilta 2007; Brochier et al. 2004). For my point, this is not relevant.

6 Conclusion

The arguments I have reported suggest that V-D works through perceptual representations. Yet V-D is the main cortical circuit involved in the detection of action properties. It follows that V-D is a clear example of a neurophysiological representational perceptual state. Accordingly, action properties are perceptually represented properties.

So my account has a twofold utility. On the one hand, whereas philosophers and neuroscientists start from very different background assumptions when they use the term "representation", my account shows that, at least concerning action properties, we can talk about perceptually represented properties. On the other hand, while I do not push the line against anti-representationalism, those who look for arguments against this view can find in my account a reason for talking about perceptual representations, at least about action properties.

Once again, my account only gestures towards such an argument, hunting forward the possible way in which we can defend this thesis. A fully fledged argument will have to wait for another occasion.[20]

References

Andersen, R. A., Andersen, K. N., Hwang, E. J., & Hauschild, M. (2014). Optic ataxia: From Balint's syndrome to the parietal reach region. *Neuron, 81*, 967–983. doi:10.1016/j.neuron.2014.02.025.

Baumann, M. A., Fluet, M.-C., & Scherberger, H. (2009). Context-specific grasp movement representation in the macaque anterior intraparietal area. *Journal of Neuroscience, 29*, 6436–6448.

[20]Different ideas concerning the topics mentioned in this paper were presented at the Italian Conference for Analytic Philosophy, University of Cagliari, at the Salzburg Conference for Analytic Philosophy, University of Salzburg, at the Italian Conference for Logic and Philosophy of Science, University of Urbino, at the International Conference for Logic and Philosophy of Science, University of Rome 3, at the European Conference for Analytic Philosophy, University of Bucharest, at the International Conference for Analytic Philosophy, University of L'Aquila, at the Consciousness and Experiential Psychology Annual Conference, Sidney Sussex College, University of Cambridge, at the European Society for Philosophy and Psychology, University of Messina in Noto, at the International Conference for Cognitive Sciences, University of Rome 3, at the Conference on Model-Based Reasoning in Science and Technology—Models and Inferences: Logical, Epistemological, and Cognitive Issues—Computational Philosophy, in Sestri Levante, and in different departmental colloquia in the Department of Pure and Applied Science in Urbino, with the research group in Science of Complexity. I would like to thank these various audiences for their comments. I also warmly thank these scholars who discussed with me, with enthusiasm, several topics mentioned in this paper and provided numerous insightful comments. Special thanks go to Bence Nanay, Mario Alai, Adriano Angelucci, Silvano Zipoli Caiani, Corrado Sinigaglia, Pierre Jacob, Riccardo Cuppini, Alfredo Paternoster, Michele Di Francesco, Pierluigi Graziani, Vincenzo Fano, Claudio Calosi, Andrea Borghini, Chiara Brozzo, Dan-Cavendon Taylor, Laura Gow, Neil Van Leeuwen, Joseph Brenner, Angelica Kaufmann and Achille Varzi.

Bennett, M., Dennett, D., Hacker, P. M. S., & Searle, J. (2007). *Neuroscience and philosophy. Brain, mind, and language*. New York: Columbia University Press.

Bennett, M., & Hacker, P. M. S. (2003). *Philosophical foundations of neuroscience*. Oxford: Wiley-Blackwell.

Bennett, M., & Hacker, P. M. S. (2012). *History of cognitive neuroscience*. Oxford: Wiley-Blackwell.

Bickle, J. (2013). *The Oxford handbook of philosophy and neuroscience*. Oxford: Oxford University Press.

Borghi, A. M., & Cimatti, F. (2010). Embodied cognition and beyond: Acting and sensing the body. *Neuropsychologia, 48*, 763–773.

Borghi, A. M., Flumini, A., Natraj, N., & Wheaton, L. (2012). One hand, two objects: Emergence of affordance in contexts. *Brain and Cognition, 80*, 64–73.

Borghi, A. M., Gianelli, C., & Scorolli, C. (2010, June 14). Sentence comprehension: Effectors and goals, self and others. An overview of experiments and implications for robotics. *Frontiers in Neurorobotics 4*(3). http://dx.doi.org/10.3389/fnbot.2010.00003.

Borghi, A. M., & Riggio, L. (2015). Stable and variable affordances are both automatic and flexible. *Frontiers in Human Neuroscience, 19*(9), 351. doi:10.3389/fnhum.2015.00351. (eCollection 2015).

Borra, E., Belmalih, A., Calzavara, R., et al. (2008). Cortical connections of the macaque anterior intraparietal (AIP) area. *Cerebral Cortex, 18*, 1094–1111.

Briscoe, R. (2009). Egocentric spatial representation in action and perception. *Philosophy and Phenomenological Research, 79*, 423–460.

Brochier, T., Spinks, R. L., Umilta, M. A., & Lemon, R. N. (2004). Patterns of muscle activity underlying object-specific grasp by the macaque monkey. *Journal of Neurophysiology, 92*, 1770–1782.

Brochier, T., & Umilta, M. A. (2007). Cortical control of grasp in non-human primates. *Current Opinions in Neurobiology, 17*, 637–643.

Brogaard, B. (2011). Conscious vision for action versus unconscious vision for action? *Cognitive Science, 35*, 1076–1104. doi:10.1111/j.1551-6709.2011.01171.x.

Brooks, R. A. (1991). Intelligence without representation. *Artificial Intelligence, 47*, 139–159.

Brooks, A., & Akins, K. (2005). *Cognition and the brain. The philosophy and neuroscience movement* (pp. 252–283). Cambridge: Cambridge University Press.

Bruno, N., & Battaglini, P. P. (2008). Integrating perception and action through cognitive neuropsychology (broadly conceived). *Cognitive Neuropsychology, 25*(7–8), 879–890.

Buccino, G. S., Sato, M., Cattaneo, L., Rodà, F., & Riggio, L. (2009). Broken affordances, broken objects: A TMS study. *Neuropsychologia, 47*, 3074–3078.

Burge, T. (2010). *Origins of objectivity*. Oxford: Oxford University Press.

Butterfill, S., & Sinigaglia, C. (2014). Intention and motor representation in purposive action. *Philosophy and Phenomenological Research, 88*(1), 119–145.

Campbell, J. (1993). A simple view of color. In J. Haldane & C. Wright (Eds.), *Reality representation and projection* (pp. 257–268). Oxford: Oxford University Press.

Castiello, U. (2005). The neuroscience of grasping. *Nature Reviews, 6*(9), 726–736. doi:10.1038/nrn1744.

Castiello, U., & Begliomini, C. (2008). The cortical control of visually guided grasping. *The Neuroscientist, 14*(2), 157–170. doi:10.1177/1073858407312080. (Epub 2008 Jan 24).

Chao, L. L., & Martin, A. (2000). Representation of manipulable man-made objects in the dorsal stream. *Neuroimage, 12*, 478–484.

Chemero, A. (2009). *Radical embodied cognitive science*. Cambridge: MIT Press.

Chinellato, E., & del Pobil, A. P. (2016). *The visual neuroscience of robotic grasping. Achieving sensorimotor skills through dorsal-ventral stream integration*. Switzerland: Springer.

Cisek, P. (2007). Cortical mechanisms of action selection: The affordance competition hypothesis. *Philosophical Transactions of the Royal Society of London B: Biological Sciences, 362*, 1585–1599. doi:10.1098/rstb.2007.2054.

Clark, A. (2000). *A theory of sentience*. Oxford: Oxford University Press.
Clark, A. (2009). Perception, action, and experience: Unraveling the golden braid. *Neuropsychologia,* doi:10.1016/j.neuropsychologia.2008.10.020.
Costantini, M., Ambrosini, E., Tieri, G., Sinigaglia, C., & Committeri, G. (2010). Where does an object trigger an action? An investigation about affordance in space. *Experimental Brain Research, 207*, 95–103.
Craighero, L., Bello, A., Fadiga, L., & Rizzolatti, G. (2002). Hand action preparation influences the responses to hand pictures. *Neuropsychologia, 40*, 492–502.
Crane, T. (Ed.). (1992). *The contents of experience*. Cambridge: Cambridge University Press.
Decety, J., & Grèzes, J. (2006). The power of simulation: Imagining one's own and other's behavior. *Brain Research, 1079*, 4–14.
Deloache, J. (2004). Becoming symbol-minded. *Trends in Cognitive Sciences, 8*, 66–70, doi:10.1016/j.tics.2003.12.004.
DeLoache, J. S., Pierroutsakos, S. L., Uttal, D. H., Rosengren, K. S., & Gottlieb, A. (1998). Grasping the nature of pictures. *Psychological Science, 9*, 205–210.
Dokic, J. (2010). Perceptual recognition and the feeling of presence. In B. Nanay (Ed.), *Perceiving the world* (pp. 33–53). New York: Oxford University Press.
Dretske, F. (1988). *Explaining behavior*. Cambridge, MA: MIT Press.
Dretske, F. (1995). *Naturalizing the mind*. Cambridge, MA: MIT Press.
Dretske, F. (2006). Perception without awareness. In T. S. Gendler & J. Hawthorne (Eds.), *Perceptual experience* (pp. 147–180). Oxford University Press.
Fadiga, L., Fogassi, L., Gallese, V., & Rizzolatti, G. (2000). Visuomotor neurons: Ambiguity of the discharge or 'motor' perception? *International Journal of Psychophysiology, 35*, 165–177.
Farennikova, A. (2012). Seeing absence. *Philosophical Studies, 166*, 1–26. doi:10.1007/s11098-012-0045-y.
Ferretti, G. (2016). Pictures, action properties and motor related effects. *Synthese*, doi:10.1007/s11229-016-1097-x.
Fluet, M.-C., Baumann, M. A., & Scherberger, H. (2010). Context-specific grasp movement representation in macaque ventral premotor cortex. *The Journal of Neuroscience, 30*, 15175–15184.
Fodor, J. A. (1987). *Psychosemantics*. Cambridge, MA: MIT Press.
Fodor, J. A., & Pylyshyn, Z. W. (1988). Connectionism and cognitive architecture: A critical analysis. *Cognition, 28*, 3–71.
Fogassi, L., Gallese, V., Buccino, G., Craighero, L., Fadiga, L., & Rizzolatti, G. (2001). Cortical mechanism for the visual guidance of hand grasping movements in the monkey: A reversible inactivation study. *Brain, 124*, 571–586.
Gallese, V. (2000). The inner sense of action. Agency and motor representations. *Journal of Consciousness Studies, 7*(10), 23–40.
Gallese, V. (2007). The "conscious" dorsal stream: Embodied simulation and its role in space and action conscious awareness. *Psyche, 13*(1), 1–20.
Gallese, V. (2009). Motor abstraction: A neuroscientific account of how action goals and intentions are mapped and understood. *Psychological Research PRPF, 73*, 486–498.
Gallese, V., & Metzinger, T. (2003). Motor ontology. The representational reality of goals, actions and selves. *Philosophical Psychology, 16*(3), 365–388.
Gallese, V., Murata, A., Kaseda, M., Niki, N., & Sakata, H. (1994). Deficit of hand preshaping after muscimol injection in monkey parietal cortex. *Neuroreport, 5*, 1525–1529.
Gallese, V., & Sinigaglia, C. (2011). What is so special with embodied simulation. *Trends in Cognitive Science, 15*(11), 512–519.
Gallivan, J. P., & Wood, D. K. (2009). Simultaneous encoding of potential grasping movements in macaque anterior intraparietal area (review of Bauman et al.). *The Journal of Neuroscience, 29* (39), 12031–12032.
Gibson, J. J. (1979). *An ecological approach to visual perception*. Boston: Houghton Mifflin.
Graziano, M. (2009). *The intelligent movement machine: An ethological perspective on the primate motor system*. Oxford: Oxford University Press.

Grèzes, J., & Decety, J. (2002). Does visual perception of object afford action? Evidence from a neuroimaging study. *Neuropsychologia, 40*, 212–222.

Humphreys, G. W., & Riddoch, M. J. (2001a). Detection by action: Neurobiological evidence for action defined template in search. *Nature Neuroscience, 4*, 84–89.

Humphreys, G., & Riddoch, M. (2001b). Knowing what you need but not what you want: Affordances and action-defined templates in neglect. *Behavioural Neurology, 13*, 75–87.

Hutto, D. D., & Myin, E. (2013). *Radicalizing enactivism: Basic minds without content*. Cambridge: MIT Press.

Jacob, P. (2005). Grasping and perceiving objects. In A. Brooks & K. Akins (Eds.), *Cognition and the brain. The philosophy and neuroscience movement* (pp. 252–283). Cambridge: Cambridge University Press.

Jacob, P., & de Vignemont, F. (2010). Spatial coordinates and phenomenology in the two-visual systems model. In N. Gangopadhyay, M. Madary, & F. Spicer (Eds.), *Perception, action and consciousness* (pp. 125–144). Oxford: Oxford University Press.

Jacob, P., & Jeannerod, M. (2003). *Ways of seeing. The scope and limits of visual cognition*. Oxford: Oxford University Press.

James, T., Humphrey, G., Gati, J., Menon, R., & Goodale, M. (2002). Differential effects of viewpoint on object-driven activation in dorsal and ventral stream. *Neuron, 35*, 793–801.

Janssen, P., & Scherberger, H. (2015). Visual guidance in control of grasping. *Annual Review of Neuroscience, 8*(38), 69–86. doi:10.1146/annurev-neuro-071714-034028.

Jeannerod, M. (1997). *The cognitive neuroscience of action*. Oxford: Blackwell.

Jeannerod, M. (2006). *Motor cognition: What actions tell the self*. Oxford: Oxford University Press.

Kandel, E. R., Schwartz, J. H., Jessell, T. M., Siegelbaum, S. A., & Hudspeth, A. J. (2013). *Principles of neural science*. New York: McGraw-Hill.

Kravitz, D. J., Saleem, K. I., Baker, C. I., & Mishkin, M. (2011). A new neural framework for visuospatial processing. *Nature Reviews Neuroscience, 12*, 217–230.

Kravitz, D. J., Saleem, K. S., Baker, C. I., Ungerleider, L. G., & Mishkin, M. (2013). The ventral visual pathway: An expanded neural framework for the processing of object quality. *Trends in Cognitive Sciences, 17*(1), 26–49.

Mandik, P. (2005). Action-oriented representation. In A. Brooks & K. Akins (Eds.), *Cognition and the brain. The philosophy and neuroscience movement* (pp. 295–305). Cambridge: Cambridge University Press.

Martin, A. (2007). The representation of object concepts in the brain. *Annual Review of Psychology, 58*, 25–45. doi:10.1146/annurev.psych.57.102904.190143.

Matthen, M. (2005). *Seeing, doing, and knowing: A philosophical theory of sense-perception*. Oxford: Oxford University Press.

McIntosh, R. D., & Schenk, T. (2009). Two visual streams for perception and action: Current trends. *Neuropsychologia, 47*, 1391–1396, doi:10.1016/j.neuropsychologia.2009.02.009

Millikan, R. G. (1984). *Language, thought and other biological categories*. Cambridge: MIT Press.

Millikan, R. G. (1993). *White queen psychology and other essays for Alice*. Cambridge: MIT Press.

Milner, A. D., & Goodale, M. A. (1995). *The visual brain in action*. Oxford: Oxford University Press.

Murata, A., Fadiga, L., Fogassi, L., Gallese V., Raos, V., & Rizzolatti, G. (1997). Object representation in the ventral premotor cortex (area F5) of the monkey. *Journal of Neurophysiology, 78*, 2226–2230.

Murata, A., Gallese, V., Luppino, G., Kaseda, M., & Sakata, H. (2000). Selectivity for the shape, size and orientation of objects for grasping in neurons of monkey parietal area AIP. *Journal of Neurophisiology, 79*, 2580–2601.

Nanay, B. (Ed.). (2010). *Perceiving the world*. New York: Oxford University Press.

Nanay, B. (2011a). Action oriented-perception. *European Journal of Philosophy, 20*(3), 430–446. 1–17. USA: Blackwell Publishing. doi:10.1111/j.1468-0378.2010.00407.x. ISSN 0966-8373.

Nanay, B. (2011b). Do we sense modalities with our sense modalities? *Ratio, 24*:299–310.
Nanay, B. (2012). Perceiving tropes. *Erkenntnis, 77*, 1–14.
Nanay, B. (Ed.). (2013a). Is action-guiding vision cognitively impenetrable? In *Proceedings of the 35th Annual Conference of the Cognitive Science Society (CogSci 2013)* (pp. 1055–1060). Hillsdale, NJ: Lawrence Erlbaum
Nanay, B. (Ed.). (2013b). *Between perception and action*. Oxford: Oxford University Press.
Nanay, B. (2014a). Every act an animal act: Naturalizing action theory. In M. Sprevak & J. Kallestrup (Eds.), *New Waves in the philosophy of mind* (pp. 226–241). Basingstoke: Palgrave Macmillan.
Nanay, B. (Ed.). (2014b). Empirical problems with anti-representationalism. In B. Brogaard (Ed.), *Does Perception have Content?* New York: Oxford University Press.
Nanay, B. (2015). Trompe l'oeil and the dorsal/ventral account of picture perception rev. *Review of Philosophy and Psychology, 6*, 181–197. doi:10.1007/s13164-014-0219-y.
Nelissen, K., Joly, O., Durand, J. B., Todd, J. T., Vanduffel, W., & Orban, G. A. (2009). The extraction of depth structure from shading and texture in the macaque brain. *PLoS ONE, 4*(12), e8306.
Noë, A. (2004). *Action in perception*. Cambridge, MA: The MIT Press.
Nowak, D. A., & Hermsdörfer, J. (2009). *Sensorimotor control of grasping: Physiology and pathophysiology*. Cambridge: Cambridge University Press.
O'Callaghan, C. (2014). Auditory perception. *The Stanford Encyclopedia of Philosophy* (Summer 2014 Edition). In N. Zalta, Edward (Ed.), URL:http://plato.stanford.edu/archives/sum2014/entries/perception-auditory/
Orban, G. A., & Caruana, F. (2014). The neural basis of human tool use. *Frontiers in Psychology, 5*, 310. doi:10.3389/fpsyg.2014.00310
Orlandi, N. (2011). Embedded seeing: Vision in the natural world. *Nous, 47*, 727–747.
Pacherie, E. (2000). The content of intentions. *Mind and Language, 15*(4), 400–432.
Pacherie, E. (2002). The role of emotions in the explanation of action. *European Review of Philosophy, 5*, 55–90.
Pacherie, E. (2011). Non-conceptual representations for action and the limits of intentional control. *Social Psychology, 42*(1), 67–73.
Pautz, A. (2010). An argument for the intentional view of visual experience. In B. Nanay (Ed.), *Perceiving the world*. New York: Oxford University Press.
Peacocke, C. (1992). *A study of concepts*. Cambridge: MIT Press.
Peacocke, C. (2001). Phenomenology and nonconceptual content. *Philosophy and Phenomenological Research, 62*, 609–615.
Pierroutsakos, S. L., & DeLoache, J. S. (2003). Infants' manual exploration of pictured objects varying in realism. *Infancy, 4*, 141–156.
Prosser, S. (2011). Affordances and the phenomenal character in spatial perception. *Philosophical Review, 120*, 475–513.
Proverbio, M. A., Adorni, R., & D'Aniello, G. E. (2011). 250 ms to code for action affordance during observation of manipulable objects. *Neuropsychologia, 49*, 2711–2719.
Ranzini, M., Borghi, A. M., & Nicoletti, R. (2011). With hands I do not centre! Action- and object-related effects of hand-cueing in the line bisection. *Neuropsychologia, 49*, 2918–2928.
Raos, V., Umilta, M. A., Murata, A., Fogassi, L., & Gallese, V. (2006). Functional properties of grasping-related neurons in the ventral premotor area F5 of the macaque monkey. *Journal of Neurophysiology, 95*, 709–729.
Rice, N. J., Valyear, K. F., Goodale, M. A., Milner, A. D., & Culham, J. C. (2007). Orientation sensitivity to graspable objects: An fMRI adaptation study. *Neuroimage, 36*, T87–T93.
Rizzolatti, G., & Luppino, G. (2001). The cortical motor system. *Neuron, 31*, 889–901.
Rizzolatti, G., & Matelli, M. (2003). Two different streams form the dorsal visual system: Anatomy and functions. *Experimental Brain Research, 153*, 146–157.
Rizzolatti, G., & Sinigaglia, C. (2008). *Mirrors in the brain. How our minds share actions and emotions*. Oxford: Oxford University Press.

Romero, M. C., Pani, P., & Janssen, P. (2014). Coding of shape features in the macaque anterior intraparietal area systems/circuits 4006. *The Journal of Neuroscience, 34*(11), 4006–4021.
Romero, M. C., Van Dromme, I., & Janssen, P. (2012). Responses to two-dimensional shapes in the macaque anterior intraparietal area. *European Journal of Neuroscience, 36*, 2324–2334.
Sakata, H., Taira, M., Murata, A., & Mine, S. (1995). Neural mechanisms of visual guidance of hand action in the parietal cortex of the monkey. *Cerebral Cortex, 5*, 429–438.
Sakata, H., Tsutsui, K., & Taira, M. (2003). Representation of the 3D world in art and in the brain. *International Congress Series, 1250*, 5–35.
Sakreida, K., Menz, M. M., Thill, S., Rottschy, C., Eickhoff, B., Borghi, A. M., Ziemke, T., & Binkofski, F. (2013). Neural pathways of stable and variable affordances: A coordinate-based meta-analysis. *F1000Posters, 4*, 663, Poster Number 3762.
Segal, G. (1989). Seeing what is not there. *The Philosophical Review, 98*, 189–214.
Schulte, P. (2012). How frogs see the world: Putting Millikan's teleosemantics to the test. *Philosophia. 40* 483–496, doi:10.1007/s11406-011-9358-x.
Shikata, E., Hamzei, F., Glauche, Koch M., Weiller, C., Binkofski, F., & Büchel, C. (2003). Functional properties and interaction of the anterior and posterior intraparietal areas in humans. *European Journal of Neuoroscience, 17*, 1105–1110.
Siegel, S. (2006). Which properties are represented in perception? In T. S. Gendler & J. Hawthorne (Eds.), *Perceptual experience* (pp. 481–503). Oxford: Oxford University Press.
Siegel, S. (2014). Affordances and the contents of perception. In B. Brogaard (Ed.), *Does perception have content?* (pp. 51–75). New York: Oxford University Press.
Siewert, C. (1998). *The significance of consciousness*. Princeton: Princeton University Press.
Srivastava, S., Orban, G. A., De Mazière, P. A., & Janssen, P. (2009). A distinct representation of three-dimensional shape in macaque anterior intraparietal area: Fast, metric, and coarse. *The Journal of Neuroscience, 29*, 10613–10626, doi:10.1523/JNEUROSCI.6016-08.2009.
Taira, M., Nose, I., Inoue, K., & Tsutsui, K. (2001). Cortical areas related to attention to 3D surface structures based on shading: an fMRI study. *Neuroimage, 14*, 956–959.
Theys, T., Pani, P., van Loon, J., Goffin, J., & Janssen, P. (2012a). Selectivity for three-dimensional shape and grasping-related activity in the macaque ventral premotor cortex. *The Journal of Neuroscience, 32*, 12038–12050.
Theys, T., Srivastava, S., van Loon, J., Goffin, J., & Janssen, P. (2012b). Selectivity for three-dimensional contours and surfaces in the anterior intraparietal area. *Journal of Neurophysiology, 107*, 995–1008.
Theys, T., Pani, P., van Loon, J., Goffin, J., & Janssen, P. (2013). Three-dimensional shape coding in grasping circuits: A comparison between the anterior intraparietal area and ventral premotor area F5a. *Journal of Cognitive Neuroscience, 25*, 352–364.
Theys, T., Romero, M. C., van Loon, J., & Janssen, P. (2015). Shape representations in the primate dorsal visual stream. *Frontiers in Computational Neuroscience, 9*, 43. doi:10.3389/fncom.2015.00043.
Tsutsui, K., Sakata, H., Naganuma, T., & Taira, M. (2002). Neural correlates for perception of 3D surface orientation from texture gradient. *Science, 298*, 409–412.
Tsutsui, K., Taira, M., & Sakata, H. (2005). Neural mechanisms of three-dimensional vision. *Neuroscience Research, 51*, 221–229.
Tucker, M., & Ellis, R. (2004), Action priming by briefly presented objects. *Acta Psychologica, 116*(2), 185–203, doi:10.1016/j.actpsy.2004.01.004.
Turella, L., & Lignau, A. (2014). Neural correlates of grasping. *Frontiers in Human Neuroscience, 8*, 686. doi:10.3389/fnhum.2014.00686.
Tye, M. (2005). Non-conceptual content, richness, and fineness of grain. In T. Gendler & J. Hawthorne (Eds.), *Perceptual experience* (pp. 504–526). Oxford: Oxford University Press.
Van Gelder, T. (1995). What might cognition be if not computation? *The Journal of Philosophy, 92*(7), 345–381.
Verhoef, B. E., Vogels, R., & Janssen, P. (2010). Contribution of inferior temporal and posterior parietal activity to three-dimensional shape perception. *Current Biology, 20*, 909–913.

Wallhagen, M. (2007). Consciousness and action: Does cognitive science support (mild) epiphenomenalism? *The British Journal for the Philosophy of Science, 58*(3), 539–561.
Westwood, D., Danckert, J., Servos, P., & Goodale, M. (2002). Grasping two-dimensional images and three-dimensional objects in visual-form agnosia. *Experimental Brain Research, 144*, 262–267.
Wilson, R. (2010). Extended vision. In N. Gangopadhyay, M. Madary, & F. Spicer (Eds.), *Perception, action and consciousness*. Oxford: Oxford University Press.
Zipoli Caiani, S. (2013). Extending the notion of affordance. *Phenomenology and the Cognitive Sciences, 13,* 275–293. doi:10.1007/s11097-013-9295-1.

Ideality, Symbolic Mediation and Scientific Cognition: The Tool-Like Function of Scientific Representations

Dimitris Kilakos

Abstract In this paper, I attempt to sketch a dialectical approach on scientific representations and their role in scientific cognition. In my understanding, scientific representations can be construed as 'tools' mediating scientific cognition. These 'tools' are products of our cognitive activity, by which we signify which features of certain objects or states of affairs should be embodied in abstractive representations of them. In such a context, I explore the merits of bringing some ideas of thinkers whose work is underestimated in the relevant discussion nowadays (such as K. Marx, E.V. Ilyenkov, L.S. Vygotsky, M. Wartofsky) in dialogue with currently discussed approaches.

Keywords Scientific representation · Models · Ideality · Symbolic mediation

1 Introduction

I maintain that any comprehensive approach to scientific representations should address at least the following two questions:

(a) *which is the role of scientific representations in scientific cognition?*
(b) *what is the nature of scientific representations?*

En route to such a comprehensive approach, I attempt to sketch an artifactual account on a representational basis, in which the epistemic function of scientific representations is attributed to their role as mediators in scientific cognition.

I would like to thank Paul Teller, Theodore Arabatzis, Stathis Psillos and two anonymous referees for their useful comments on an earlier version of this paper.

The original version of this chapter was revised: For detailed information please see Erratum. The erratum to this chapter is available at https://doi.org/10.1007/978-3-319-38983-7_38

D. Kilakos (✉)
Department of Philosophy and History of Science, University of Athens,
Panepistimioupoli, 157 71 Zografou, Greece
e-mail: dimkilakos@hotmail.com

L. Magnani and C. Casadio (eds.), *Model-Based Reasoning in Science and Technology*, Studies in Applied Philosophy, Epistemology and Rational Ethics 27, DOI 10.1007/978-3-319-38983-7_11

The first part of this paper (Sects. 2 and 3), deals with the first of the aforementioned questions. Specifically, in Sect. 2, I discuss the possible merits of an artifactual approach to scientific representations and look for insights in M. Wartofsky's work on models and representation. In Sect. 3, I discuss the models-as-mediators view and I invoke the Vygotskian understanding of mediation, in an attempt to adjust my understanding of scientific representations as artifacts (which is discussed in Sect. 2) with their function as epistemic mediators. The second part of the paper (Sects. 4 and 5) deals with the second of the aforementioned questions. Specifically, in Sect. 4 I roughly review the current discussion about the ontology of abstract models. In Sect. 5 I discuss E.V. Ilyenkov's approach to ideality and symbolic mediation which I maintain that could contribute in the attempted venture. In the concluding section, I summarize how the foregoing discussion informs my answer to the aforementioned questions.

2 Scientific Representations as Artifacts

One of the disputes among philosophers who are engaged in the relevant discussion is about the representational status of scientific models. In the last two decades, there is a growing number of philosophers who argue that models do not (or need not) represent and/or contest the view that representation presupposes (or is reduced to) a structural relation between the model and its target. Some of them find representation not only dubious or ambiguous, but also even unnecessary for the actual role that models play in scientific practice (i.e. Knuuttila 2005; Boon and Knuuttila 2009).

For example, Knuuttila argues that the emphasis on representation does not do justice to the various roles of models in science. She claims that treating models as predominantly representative entities ignores their material and interactive side, from which their heuristic, mediating and many other epistemic capabilities arise (Knuuttila 2005, p. 23). Knuuttila (2011) thinks of models as 'epistemic tools', which are built by specific representational means. Models are designed and employed in order to facilitate the scientific inquiries, and scientists learn from them by means of construction and manipulation. Knuuttila mainly contests the view that scientific models give us knowledge because they succeed in representing something external to themselves.

Although I share with Knuuttila the inclination to overcome the semantic tradition and despite the fact that I am fairly sympathetic to the prospects of an artifactual view, I disagree with her at this point. I argue that it is exactly their representational status, properly construed, that allows us to account for the role of models in scientific practice. In such an artifactual approach on a representational basis, it could be the case that the representational status of any given model which is employed in a scientific inquiry could be judged in the context of the specific inquiry.

I deem that a reassessment of M. Wartofsky's nowadays nearly neglected nowadays work on representation and models could be advantageous for my purposes.

Wartofsky contends that cognitive acquisition is necessarily mediated by representation (Wartofsky 1979, p. xix). Representations are the artifacts that accrue from our ability to represent an action by symbolic means. When producing an artifact, we are at the same time producing a representation, since these artifacts do not only have a use, but also represent the mode of action in which they are used or have been produced (Wartofsky 1979, p. xiii). Hence, representations are approached as symbolic externalizations of objectifications of possible modes of action, according to some convention (Wartofsky 1973/1979, p. 200) and models and theories turn to be representational cognitive artifacts. The detachment of the representational sign from an artifact exceeds the original interpretation of the tool as itself the symbol of the mode of action by which it is made, or in which it is used (Wartofsky 1979, p. xvii).

Therefore, our cognitive confrontation with reality is not subservient to sense-data but a tendency to interact, to actively participate in its phenomena. In such a context, cognition is construed as an active process. To know means to manipulate the object of knowledge, to transform it into a tool of action; it is active interference in objective reality by means of our cognitive artifacts, sign-systems and conceptual frameworks. Ergo, a scientific representation is an active representation of objective reality, functioning as a tool for scientific cognition (Azeri 2013).

Wartofsky persists (and I agree) that the emphasis should be shifted from what representation is to the activity of representing. A model is chosen to represent abstractly only the features of its target that we consider to be significant or valuable. Therefore, models are telic (Wartofsky 1968/1979, p. 142). Wartofsky's approach can offer an alternative to the views of Van Fraassen (2008), Suárez (2003, 2004, 2009) and others, who propose that our attention should be shifted from the semantics to the pragmatics of scientific representation. He is distinct from them in that he considers agency as a crucial aspect of the truth-hunting activity of modeling in science, which aims to conform our modes of inquiry to the real state of affairs to be actively represented via them. Thus, intentionality in modeling turns to be a guide to the objectivity of representation.

Wartofsky underlines that the discussion about models and representation imputes an epistemology in which the knowing subject confronts a surrogate object of knowledge—the model—as a representation of the external world, which is the target of our inquiries. This point is consonant with the views of several philosophers (e.g. Swoyer 1991; Suárez 2004; Contessa 2007) who appeal to surrogative reasoning in order to argue about how multiple functions of models stem from their representational function.

Let me sum up some conclusions drawn from the discussion in this section. I maintain that a model represents its target system, because and as long as it successfully conveys and/or explains (some of) its features, regardless of any previously established or acknowledged by an agent structural relation between them. Even if such a relation occurs, it should not be considered as the reason why a scientist is allowed to draw eligible inferences for the modeled system via the use of the model.

For example, when working on introducing the double-helix model, Watson and Crick were striving to make their model comply with experimental data, to embody previously acquainted knowledge in it, to increase its explanatory capacity. In this sense, they were building their model as an investigative instrument, as a workable representation of the actual structure of DNA molecules. It was not the need to establish a structural relation between the model and its target that guided them in their efforts. The representational status of the model was judged and determined in the context of the specific inquiry, since it was proved that it successfully conveys and explains (some of) the features of the target system.

3 The Mediating Role of Scientific Representations

Let me now discuss the *models-as-mediators* view, which is an influential approach on the role of models in scientific practice. My goal in this section is to complete my discussion on the epistemic function of scientific representations, on the basis of my understanding of their artifactual character, which I presented in Sect. 2.

In their (1999), Morrison and Morgan emphasize the epistemic importance of models' function as mediators, which stems from their partial independence from both theory and data. Due to their independence, models mediate in several ways and they function as tools for knowledge exactly because of their autonomous nature. Morrison and Morgan attribute much of models' epistemic significance to the processes of their construction and manipulation instead of focusing only on their purported representational status. It should be noted that Morrison and Morgan do not go all the way down in rejecting representation, since they claim that learning from models actually depends on representation: we can learn from models because they represent. They conceive of representation as a kind of rendering, in the sense of a partial representation that either abstracts from a system, or translates it into another form, which is capable of embodying only a portion of the system (Morrison and Morgan 1999, p. 27).

Nancy Cartwright argues for a similar view. According to her, models bridge the gap between the general theoretical principles and the complexity of data which phenomenological laws strive to capture: "The route from the theory to reality is from theory to model, and then from model to the phenomenological law. The phenomenological laws are indeed true of the objects of reality—or might be; but the fundamental laws are true only of objects in the model" (Cartwright 1983, p. 4). A model includes some genuine properties of the target object(s). However, it also contains properties of convenience and fiction (Cartwright 1983, p. 15), which are necessary to bring the target object(s) into the confines of the theory. Thus, model-building is a pragmatic activity, since "adjustments are made where literal correctness does not matter very much in order to get correct effects where we want to get them; and very often … one distortion is put right by another" (Cartwright 1983, p. 140).

In Morgan and Morrison's account, the logical link between theory and models is broken down by taking into consideration external elements which depend on various and diverse (mathematical, technological, economical etc.) constraints, which render models partially independent from both theory and data. In this context, the notion of "theory" is understood primarily in its opposition to observations, and hence encompasses also laws, theoretical notions and concepts. In this sense, the mediating function of models enables us to use them as instruments of inquiry about both theory and the world. This is the reason why philosophers who appeal to an artifactual approach (i.e. Knuuttila) take the models-as-mediators view into serious consideration.

Since I also attempt to articulate an artifactual account, in the rest of this section I discuss some aspects of my understanding of the mediating role of scientific representations in comparison with some features of the models-as-mediators view.

Cartwright and Morrison and Morgan argue—and I agree—that since the theory is incapable of providing everything that is required for both the construction of the model and the choice of the appropriate correcting interventions (which are necessary in order a model to accomplish the role that is reserved for it within the context of the specific scientific inquiry), models are not fully suggested by the theory. However, an important difference of my understanding of the mediating role of scientific representation from theirs stems from the place they reserve for models due to this argumentation. According to them, models populate the middle ground between theory and the world (or data). On the contrary, as I have discussed in the previous section, I agree with Wartofsky on that both models and theories are representational artifacts, which are jointly employed in our cognitive efforts to grasp features of the real world.

Previously acquired knowledge, experience, objectives, technical resources, etc. are inherent parts of the construction of a specific model and its employment in the course of scientific inquiry. These features of the modeling process are important for various aspects of the scientific endeavor, such as hypothesis testing, theory development, etc. According to the models-as-mediators view, these features highlight models' partial independency from both theory and data, due to which scientific models function as instruments. In my understanding, the artifactual character of scientific representations accounts for these features, which are inherent parts of the developing representational activity by which epistemically useful outcomes are being produced.

According to the models-as-mediators view, the mediation between theory and the world is rendered possible through the instrumental use of models, in course of which we learn about both these domains through the manipulation and/or application of models. In this sense, models convey the role of mediators for enabling the communication between these two domains, by which the inquiry advances. Given my understanding of representations in science as active representations,

I maintain that scientific representations mediate scientific cognition in a way similar to that in which tools mediate the work of technicians.[1]

My understanding of the mediating role of models as representing artifacts may be elucidated by invoking some ideas of L.S. Vygotsky. Mediation is a central concept in Vygotsky's view of cognitive development. It roughly means that human beings interpose tools between them and their environment, in order to modify it for the sake of obtaining certain benefits. It is via mediation that we learn to ascribe meaning and to internalize areas of life that are not instantly relevant to our immediate existence (for further discussion, see Vygotsky 1987; Wertsch 2007; Karpov and Haywood 1998).

According to Vygotsky, all higher mental functions are products of mediated activity. By issuing activity mediators, humans are able to modify the environment and this is exactly what is characteristic about humans' way of interacting with nature. There are various kinds of psychological tools, which enable us to perform higher mental functions. Among them, one could find physical tools, counting-systems, algebraic symbol systems, artwork, writing, schemes, diagrams, all sorts of conventional signs, etc. The common feature of these forms of mediation is that they are acquired through culture, aggregation of prior generations' acquired knowledge; they are themselves products of the socio historical context (Azcri 2013).

Vygotskian mediation radically differs from any empiricist or positivist understanding of mental representation, since it does not contradict the idea that thought can embrace an independent world: mediators are not placed in a metaphysically peculiar layer between reality and us. This is one more reason why I favor this alternative to Morrison and Morgan's or Cartwright's conception of the mediating role of models (Azeri 2013).

Following Vygotsky's understanding of mediation, I maintain that scientific representations mediate scientific cognition in a tool-like fashion (like Vygotsky's signs). They do not intervene between the human agent and the object of cognition which they are meant to represent, as if the human agent does not access or act upon the object of cognition itself; rather, they pilot and enact this cognitive activity (Azeri 2013).

Therefore, I contend that scientific representations are mediating cognitive tools, which compass scientific cognition, operating as a grapnel in our inquiries for the essence of reality. In this sense, scientific representations are introduced in scientific

[1]Nersessian (1999) suggests that the term "model-based reasoning" (MBR) indicates the construction and manipulation of various kinds of representations. Magnani (2004) argues that "many model-based ways of reasoning are performed in a manipulative way by using external tools and mediators". An interesting aspect of the discussion about manipulative abduction regards "the relationship between unexpressed knowledge and external representations". He further states that, in any case, much of MBR and abduction (both theoretical and manipulative) is about "their ability to extract and render explicit a certain amount of important information, unexpressed at the level of available data". Bringing my understanding of scientific representations as artifacts mediating scientific cognition in discussion with Magnani's understanding of epistemic mediators is an interesting topic for further research.

inquiries when our existing cognitive tools are unable to provide answers to questions which are raised in the course of our scientific inquiries.

Furthermore, I contend that scientific representations yield novel ways of thinking and acting, which were unavailable to us prior to their introduction. Hence, scientific representations not only facilitate our engagement in certain, already emerged problems, but they also contribute formulating new questions that may guide new forms of practical activity or enable us to unveil new phenomena as objects of cognition.

Let me attempt to elucidate the epistemic function of models in these terms, using once again the example of the double-helix model of DNA structure. Scientists use it to draw inferences about the real system, i.e. about the mechanism of replication. In their inquiries, scientists' reasoning is directed upon the replicating DNA molecules via the model; in that sense, the model serves as a surrogate for the real system. Furthermore, the model obviously mediates between scientists' theoretical work on the replication and the experimental data at their disposal. For example, without using the model as vehicle for inferences, scientists could not explain the accuracy of replication. In this sense, scientists draw inferences about the process of replication due to the mediation of the surrogate.

4 The Ontology of Models: Fictionalism, Realism and Counter Arguments

In the first part of this paper, I have discussed the epistemic function of scientific representations. In the following sections, I will deal with issues regarding ontological aspects and their intertwining with epistemological concerns.

Scientific knowledge is arguably a domain of abstractive inquiries. It is an essential aspect of scientists' theoretical work to abstract the problem they are dealing with by leaving out certain features of the real situation, or approximating it by consciously forging the values of several variables for practical purposes and/or using approximating mathematical techniques. Thus, in the course of our scientific inquiries, we reconstruct reality in an ideal form by means of exemplification.

Insofar as abstraction (and/or idealization—I will not discuss whether or how these processes differ in this paper) is multifariously involved in model building, there is no surprise that there is a vivid debate about the ontology of models. On the one hand, there are philosophers who argue that abstract models are fictions and on the other hand there are other philosophers who argue for a realistic understanding of models as abstract objects. My intention here is not to give a detailed account of the several views which have been proposed on the issue; a pretty rough sketch suffices for my purposes.

The attitude to regard models to be fictional entities could be traced back to the German neo-Kantian Vaihinger (1911), who emphasized the importance of fictions for scientific reasoning (Fine 1993). In the recent literature, there are accounts that deal with models as works of fiction or intellectual constructions (i.e. Cartwright 1983), views that consider scientific models as being analogous to fictional entities in literature (i.e. Frigg 2010) and approaches, according to which models are hypothetical entities that do not exist spatiotemporally, but are not pure mathematical or structural entities in the sense that they would be natural objects if they were real (i.e. Godfrey-Smith 2006), among many others. Fictionalist accounts of scientific models, in contradistinction with realistic ones, typically explain the value of engaging them in scientific inquiry without assuming the literal truth of the posited entities (in the broadest sense of the term).

Although fictionalism about models may look an attractive choice, important counter arguments have been elaborated by several philosophers. For example, Giere argues that if abstract models are considered to be fictions, then the distinction between science and science fiction is destroyed (Giere 2009, p. 251). Furthermore, Magnani claims that fictionalism about models seems to enforce a kind of "epistemic concealment", which may obliterate the actual gnoseological finalities of science (Magnani 2012, p. 7). Moreover, it is not clear what the representational function of a fictional object is and how it is sustained, given that, in my understanding, scientific models represent aspects and features of the world insofar as they are successfully employed in scientific inquiries. I find these arguments effective. After all, if models were fictions, then it could not be explained how our truth-hunting inquiries upon reality, in which models play a crucial role, could be successful if not by chance.

Invoking fictionalism is not the only option available for anyone who wants to discuss about the ontological status of scientific representations. Several philosophers argue for a realistic understanding realism, by claiming that they postulate abstract entities. The bottom line of these approaches is that they urge us to take our theoretical scientific descriptions at face-value. An insightful approach on these grounds is offered by Psillos (2011).

One of the traditional arguments against postulating abstract objects is based on Ockham's razor: if it can be proved that certain concrete objects can perform the theoretical roles usually associated with abstract objects, one should refrain from postulating abstract objects. Another classical counterargument is that we cannot have knowledge (or reliable belief) about abstract objects, because they are causally inert.

In my opinion, what is primarily at stake in this discussion is not if one is able to think of an abstract object, but how could s/he be aware of such an object. According to the view I endorse, scientific representations are employed in our attempts to grasp aspects of the real world in a manner that allows us to interfere and interact with it. If this is the case, in order to join realists, I would like to be able to explain how one can become aware of their existence in order to actively employ them in scientific inquiries.

This question about awareness points to the fact that any ontological commitment has important epistemological consequences (this could equally be reversed). Thus, ontological and epistemological concerns merge in one and the same magnifying lens which is employed in our inquiries. I maintain that E.V. Ilyenkov's approach on the problem of the ideal may offer interesting insights to address this issue.

5 Ilyenkov's Approach to Ideality and Symbolic Mediation

Ilyenkov discusses the problem of the ideal as an inquiry about the relation of the internal world of thoughts and experiences and the external world of objects. In brief, for Ilyenkov the ideal is the reflection of things emerging in objective, reality-transforming activity, existing in 'patterns and images' of object-oriented activity of man as the active agent of social production (Ilyenkov 1977a, p. 261), supervening above individuals as accumulated signs, remnants and reflections of their past practice.

According to Ilyenkov, Marx's deployment of value form is a typical and characteristic case of ideality in general (Ilyenkov 1977b, pp. 90–91), the most typical case of the idealization of actuality (Ilyenkov 1977a, p. 267). Marx distinguishes between the natural form of the commodity, which is expressed in its use-value, and its value form, which is expressed in its exchange-value (Marx 1976, p. 138). Thus, use-value and exchange-value are two distinct aspects of a single motion existing simultaneously and expressing themselves sequentially in social interactions. Exchange-value is an ideal value form and, as such, it can only exist symbolically (Marx 1973, p. 154).

An interesting implication for my purposes appears in the case of the money form of value. The commodity (i.e. gold), which plays the role of the general equivalent, and thus becomes money, also becomes the symbol of the commodity as such; of its exchange value itself. It can only be a social symbol, as it expresses nothing more than a social relation. This material sign of exchange value is a product of exchange itself, and by no means the execution of an idea which has been conceived a priori (Marx 1973, p. 144). In this case, as Ilyenkov notes, it becomes apparent that the ideal is not (and it should not be treated as) a symbol or sign of immaterial relations or connections without a material substratum; it is (and should be treated as) a symbol of the social relations between people (Ilyenkov 1977a, p. 270).

The pattern of commodity-money circulation is expressed by the formula C-M-C. In commodity-money circulation, the commodity (C) appears in it as both the beginning and the end of the cycle, and money (M) as its mediating link, as the "metamorphosis of the commodity". However, at a certain point in the cyclical movement C–M–C–M–C–M… and so on, money is no longer a simple intermediary; it "self-expands" by deploying a seemingly autonomous function within the process. This phenomenon is expressed in the formula M–C–M (Ilyenkov 1977a, pp. 244–245).

Such a symbol is the objective form of existence of an ideal form. It is ideal because, within the dynamic of the system, it acts not as the form of itself but of something else which it represents (reflects, surrogates). It represents not the sensuously perceived image of the commodity, but its very existence within the system which creates the situation being analyzed. The key to understand this symbolic process lies in the contradictory nature of the commodity as a doublet of use-value and exchange-value.

Let me now turn to some conclusions, which are relevant to my purpose.

1. In Ilyenkov's materialistic understanding of the conception, ideality is not the relation between a symbol or sign and an idea, but a specific correlation between two or more material systems (objects or processes) within which one of them, while remaining itself, becomes a representative of the other. This symbolic function does not derive from the material properties of the artifact used as a symbol, but is a social product.

Therefore, ideality is a feature of reified, objectified images of historically formed modes of human social life, which confront the conscious agent as a special kind of objective reality, as special objects comparable with material reality and situated on one and the same spatial line (Ilyenkov 1977a, p. 262). Thus, the physical (natural, material) properties of the forms of symbolic mediation should not make us defy their ideal function. Moreover, the material body of the thing is brought into conformity with its function (Ilyenkov 1977a, p. 273). The symbol is, as Marx says of money, a means to accomplish this abstraction (Marx 1973, p. 142).

In an analogous manner, as a scientific inquiry advances, the representation of the features of the real systems fulfills its role in the symbolically mediated scientific activity (which is part of social activity). In any given moment of the development of both the represented system and the representation itself (as the scientific inquiry advances), this function is being approximated. This representation is an objectified image of the current stage of our understanding of the development of the represented system. Therefore, scientists are justified to correlate, compare etc. abstract models to the specific systems (objects, processes) which are represented by them, because there are certain degrees of reality in them. This is possible only insofar as models prove to actually be representations of their targets, which is ultimately judged in terms of social practice in the forms it is expressed with regard to scientific cognition.

2. The symbolic function emerges when a thing becomes the material for the expression of features (which are highlighted via abstraction) of some other thing. A thing is converted into a symbol due to its specific emergent role as mediator within an already formed system of relations between people mediated by things (Ilyenkov 1977a, p. 272). Like in the case of money, while the symbol does not create what is represented by it, its generation and incorporation into the process which it mediates produces a new form of appearance to the objects involved in this process.

As Maidansky (2014, p. 130) notes, the logical structure of ideal representation according to Ilyenkov is this: the essence of *A* is mirrored in the form of *B*, in a way that the nature of *B* does not mingle with the nature of *A* which it represents. By themselves, these objects (systems, processes) are material, since in the world there is nothing but matter in motion. The ideal is only the special form of their motion in the course of social human practice. In other words, the ideal is the mode of the active representation of the nature of things by means of other things in the course of social human practice. Outside the activity process, the ideal form of the object is not manifested as such; it has been materialized in various objective forms. There is nothing ideal beyond human activity.

In this sense, Ilyenkov's ideal is the subjective being of an object, which emerges when its essence receives such a peculiar *other-being* within the field of activity. Thus, Ilyenkov's ideal turns to be a *subjective image of objective reality*. The mechanism of the mutual conversion of object to subject, which expresses the dialectical transition of the form of activity into the form of thing, is human labour.

These two conclusions taken together seem to point to an alternative answer to the problem of the metaphysical status of abstract models. According to it, they could be construed neither as fictions nor as abstract entities. Rather, if we read Ilyenkov's conception of ideality in terms of scientific cognition we get an account that pays equal justice to both epistemological and ontological concerns. How this is done, may be elucidated by the third conclusion.

3. Ilyenkov maintains that ideal phenomena exist objectively as aspects of the mind-independent world. Although it is sensuously imperceptible, ideal is objective, part of objective reality, since it is being objectified in human activity.[2]

Perhaps the best way to clarify the notion of 'objectification' is in the case of artifacts, which is of primary interest for me, since artifactuality is an important feature of my approach on scientific representations. An artifact (i.e. a table) is distinguished from other physical objects (i.e. wooden raw materials) in that the artifact bears a certain significance which is acquired not by virtue of its physical nature, but because it has been produced within certain social relations, for a certain use and incorporated into a system of human ends and purposes. Hence, the artifact confronts us as an embodiment of meaning, placed and sustained in it by aim-oriented human activity; it functions representatively and expresses its essence in the symbolic mediation of activity. By analogy to Ilyenkov's elaboration on the function of money, the artifacts are the 'money' of Nature (Maidansky 2014, p. 132).

[2]An anonymous referee suggested that one could counter-argue that if something is sensuously imperceptible then nothing can falsify the thesis for its existence. But if nothing can falsify the existence of ideal phenomena then it turns out that the thesis that ideal phenomena exist objectively is no different from religious theses. I am thankful for this comment. However, falsifiability is not a criterion of existence. In any case, theories in which abstractions are embedded are empirically falsifiable, hence they are not like religious theories. I would also like to take advantage of the opportunity to countersign Ghins' assertion that success in scientific representation relies on the supposed truth of some predicative assertions (cf. Ghins' contribution to this volume). I argue that Ghins' point could also inform an answer to the worry about falsifiability. If (a) Ilyenkov's approach to ideality is applicable to scientific representations, (b) the success of the inferences drawn by the scientists who employ scientific representations in their inquiries relies on the supposed truth of the relevant predicative assertions and (c) this assertions are falsifiable, then the worry about falsifiability is answered in an indirect way. At a more fundamental level, in Ilyenkov's (and mine) line of reasoning, there is nothing beyond the mind-independent world, the objective reality of which is the domain of our subjective cognitive efforts. On this basis, ideality is being objectified in human activity (through which our forms of thought are embedded into the structure of the environment) and our theoretical understanding of the outcomes of this activity is expressed in forms which are falsifiable.

Through the objectification of activity, our forms of thought are embedded into the structure of the environment.

Ilyenkov argues that all the things involved in the social process acquire a new "form of existence" that is not included in their physical nature and differs from it completely—their ideal form (Ilyenkov 1977b, p. 86). The 'things' Ilyenkov refers to have a representative function and are expressing their essence in the symbolic mediation of activity. The phenomena which have such a symbolic or ideal function are objectified in verbal expressions, sculptural, graphic and plastic forms, in the form of the 'routine-ritual' ways of dealing with things and people etc.; their objectification is expressed in language, in drawings, models and such symbolic objects as coats of arms, banners, or as money, including gold coins and paper money, credit notes etc. (Ilyenkov 1977b, p. 79). Thus, ideality is the form of social human activity represented as a thing (Ilyenkov 1977b, p. 86); it is not the whole of culture but one of its aspects, one of its dimensions, determining factors, properties (Ilyenkov 1977b, p. 96). The point, then, is that artifacts embody human aims (expressed in human activity that produces them), which are nothing but the material process and outcome of activity in ideal form.

Marx mentions that: "[e]xchange value as such can of course only exist symbolically, although in order for it to be employed as a thing and not merely as a formal notion, this symbol must possess an objective existence; it is not merely an ideal notion, but is actually presented to the mind in an objective mode" (Marx 1973, p. 145). From this point of view, Ilyenkov makes a remark that is of great importance for my discussion on the ontology of abstract models: "the ideal as a form of human activity exists only in that activity and not in its results… When an object has been created, society's need for it is satisfied; the activity has petered out in its product, and the ideal itself has died" (Ilyenkov 1977a, pp. 275–276). Thus, it is not the instruments of labour which are 'ideal'; such an understanding would amount to an inversion of the relationship of material to ideal.

6 Envoi

Let me summarize how the foregoing discussion informs my answers to the questions posed in the introductory section. With regard to the first question, I have attempted to deploy an artifactual approach on a representational basis, according to which using scientific representations to draw inferences about real systems is equivalent to them serving as surrogates for the real systems in scientific inquiries, which, in turn, amounts to them functioning as mediators in scientific cognition. With regard to the second question, my argumentation was meant to show that we can account for neither the explanatory nor the predictive success of scientific representations if we do not indicate why and how the outcomes of abstraction which are embodied in them can be correlated with worldly things that are real and concrete.

I maintain that a scientific representation embodies objective aspects and features which are highlighted by the abstractive process through which it is constructed. It serves as a symbolically mediating artifact. It turns to be a symbol through its function within a specific system of relations mediated by artifacts. The representing artifact functions as a surrogate in activity performed throughout the scientific inquiry. It is each specific scientific representation's function as such that determines the dynamics of the inquiry.

References

Azeri S. (2013). Conceptual cognitive organs: Toward an historical materialist theory of scientific knowledge. *Philosophia, 41*(4), 1095–1123.

Boon, M., & Knuuttila, T. (2009). Models as epistemic tools in engineering sciences: A pragmatic approach. In: *Philosophy of technology and engineering sciences. Handbook of the philosophy of science* (Vol. 9, pp. 687–720). Amsterdam: Elsevier/North-Holland.

Cartwright, N. (1983). *How the laws of physics lie*. Oxford: Clarendon Press.

Contessa, G. (2007). Scientific representation, interpretation and surrogative reasoning. *Philosophy of Science, 74*(1), 48–68.

Fine, A. (1993). Fictionalism. *Midwest Studies in Philosophy, XVIII*, 1–18.

Frigg, R. (2010). Models and fiction. *Synthese, 172*(2), 251–268.

Giere, R. N. (2009). Why scientific models should not be regarded as works of fiction. In M. Suárez (Ed.), *Fictions in science. Philosophical essays on modelling and idealization* (pp. 248–258). London: Routledge.

Godfrey-Smith, P. (2006). The strategy of model-based science. *Biology and Philosophy, 21*, 725–740.

Ilyenkov, E. V. (1977a). *Dialectical logic, essays on its history and theory*. Moscow: Progress Publishers.

Ilyenkov, E. V. (1977b). The concept of the ideal. In: *Philosophy in the USSR: Problems of dialectical materialism* (pp. 71–99). Moscow: Progress.

Karpov, Y. V., & Haywood, H. C. (1998). Two ways to elaborate Vygotsky's Concept of Mediation. *American Psychologist, 53*(I), 27–36.

Knuuttila, T. (2005). *Models as epistemic artefacts: Towards a non-representationalist account of scientific representations*. Philosophical Studies from the University of Helsinki 8. Helsinki: Published by the Department of Philosophy and the Department of Social and Moral Philosophy.

Knuuttila, T. (2011). Modelling and representing: An artefactual approach to model-based representation. *Studies in History and Philosophy of Science Part A, 42*(2), 262–271.

Magnani, L. (2004). Reasoning through doing. Epistemic mediators in scientific discovery. *Journal of Applied Logic, 2*(2004), 439–450.

Magnani, L. (2012). Scientific models are not fictions: Model-based science as epistemic warfar. In L. Magnani & P. Li (Eds.), *Philosophy and cognitive science: Western and eastern studies* (pp. 1–38). Heidelberg/Berlin: Springer.

Maidansky, A. (2014). The reality of the Ideal. In A. Levant & V. Oittinen (Eds.), *Dialectics of the ideal. Evald Ilyenkov and creative soviet marxism* (pp. 125–143). Leiden-Boston: Brill.

Marx, K. (1973). *Grundrisse: Foundations of the critique of political economy*. London: Penguin Books.

Marx, K. (1976). *The capital* (Vol. 1). London: Penguin Books.

Morrison, M., & Morgan, M. (1999). Models as mediating instruments. In M. Morgan & M. Morrison (Eds.), *Models as mediators. Perspectives on natural and social science* (pp. 10–37). Cambridge: Cambridge University Press.

Nersessian, N. J. (1999). Model-based reasoning in conceptual change. In N. J. Nersessian, L. Magnani, & P. Thagard (Eds.), *Model-based reasoning in scientific discovery* (pp. 5–22). New York: Kluwer Academic/Plenum.

Psillos, S. (2011). Living with the abstract: Realism and models. *Synthese, 180*, 3–17.

Suárez, M. (2003). Scientific representation: Similarity and isomorphism. *International Studies in the Philosophy of Science, 17*, 225–244.

Suárez, M. (2004). An inferential conception of scientific representation. *Philosophy of Science, 71*, 767–779.

Suárez, M. (2009). Scientific representation. *Blackwell's Philosophy Compass, 5*(1), 91–101.

Swoyer, C. (1991). Structural representation and surrogative reasoning. *Synthese, 87*(3), 449–508.

Vaihinger, H. (1911). *The philosophy of 'As If'* (German original. English translation, 1924). London: Kegan Paul.

Van Fraassen, B. (2008). *Scientific representation: Paradoxes of perspective*. Oxford: Clarendon Press.

Vygotsky, L. S. (1987). Thinking and speech. In R. Rieber & A. Carton (Eds.), *The collected works of L. S. Vygotsky* (Vol. 1, pp. 37–285). New York & London: Plenum Press.

Wartofsky, M. (1968/1979). Telos and technique: Models as modes of action. In M. Wartofsky (Ed.), *Models: Representations and the scientific understanding* (pp. 140–153). Boston & London: D. Reidel Publishing Company.

Wartofsky, M. (1973/1979). Perception, representation, and the forms of action: Towards an historical epistemology. In M. Wartofsky (Ed.), *Models: Representations and the scientific understanding* (pp. 188–210). Boston & London: D. Reidel Publishing Company.

Wartofsky, M. (1979). *Models: Representations and the scientific understanding*. Boston & London: D. Reidel Publishing Company.

Wertsch, J. (2007). Mediation. In H. Daniels, M. Cole, & J. Wertsch (Eds.), *The Cambridge companion to Vygotsky*. Cambridge: Cambridge University Press.

Embodying Rationality

Antonio Mastrogiorgio and Enrico Petracca

The world and reason are not problematical
M. Merleau-Ponty

Abstract The current notions of bounded rationality in economics share distinctive features with Simon's original notion of bounded rationality, which still influences the theoretical and experimental research in the fields of choice, judgment, decision making, problem solving, and social cognition. All these notions of bounded rationality are in fact equally rooted in the information-processing approach to human cognition, expressing the view that reasoning is disembodied and that it can be reduced to the processing of abstract symbolic representations of the environment. This is in contrast with the last three-decade advancements in cognitive psychology, where a new view on human cognition has emerged under the general label of 'embodied cognition', demonstrating that cognition and reasoning are grounded in the morphological traits of the human body and the sensory-motor system. In this paper we argue that embodied cognition might reform the current notions of bounded rationality and we propose a number of arguments devoted to outline a novel program of research under the label of 'embodied rationality': (1) reasoning is situated as it arises from the ongoing interaction between the subject and the environment; (2) reasoning, not being exclusively a mental phenomenon, constitutively relies on the physical resources provided by the environment; (3) the sensory-motor system provides the building blocks for abstract reasoning, (4) automatic thinking is rooted in the evolutionary coupling between the morphological traits of the human body and the environment.

A. Mastrogiorgio (✉)
Department of Neurosciences, Imaging and Clinical Sciences,
"G. d'Annunzio" University of Chieti-Pescara, Via Luigi Polacchi, 11,
66100 Chieti, Italy
e-mail: mastrogiorgio.antonio@gmail.com

E. Petracca
Department of Economics, University of Bologna, Piazza A. Scaravilli, 2,
40126 Bologna, Italy
e-mail: enrico.petracca2@unibo.it

L. Magnani and C. Casadio (eds.), *Model-Based Reasoning in Science and Technology*, Studies in Applied Philosophy, Epistemology and Rational Ethics 27, DOI 10.1007/978-3-319-38983-7_12

1 Introduction[1]

In 1976, the still-to-become Nobel laureate in economics and cognitive psychologist Herbert A. Simon wrote these important words: "a person unfamiliar with the histories and contemporary research preoccupations of these two disciplines [economics and cognitive psychology] might imagine that there were close relations between them—a constant flow of theoretical and empirical findings from the one to the other and back. In actual fact communication has been quite infrequent [entailing a] state of mutual ignorance" (Simon 1976, p. 65). It is with a certain embarrassment that after 40 years, notwithstanding a remarkable lip service to the necessity of integrating psychology into economics, and some not negligible effort in that direction, these words have still a quantum of truth.

Scientific exchange between economics and psychology—or, better, between economic psychology[2] and cognitive psychology—still occurs today in a fragmentary, instrumental and fundamentally time-lagged way, mostly inattentive to the current foundational debates on how human cognition actually works (see, e.g., Rabin 1998). Among economic psychologists, there seems to be no real acknowledgment that in the last 30 years cognitive psychology has been undergoing a true paradigm shift, hinging upon the hypothesis of the constitutive dependence of human cognition from the morphological traits of the human body and its sensory-motor system. A huge amount of theoretical and empirical research has supported this new hypothesis (without any pretension to be exhaustive, among the major works are Varela et al. 1991; Clark 1997; Clancey 1997; Clark and Chalmers 1998; Rowlands 1999; Lakoff and Johnson 1999; Wilson 2002; Shapiro 2004; Noë 2004; Gallagher 2005; Pfeifer and Bongard 2006; Barrett 2011).[3] This rich *corpus* of research is today identified through the label *embodied cognition*, which encompasses slightly different approaches such as *embodied* (strictly speaking) *cognition*, *distributed cognition*, *situated cognition*, *embedded cognition* and *enacted cognition.*[4]

In this essay we argue that the theoretical and empirical relationship between economic psychology and cognitive psychology deserves to move beyond such instrumental and time-lagged approach. If this relationship has to be established, as

[1]This contribution is a modified version of an essay published in Italian in the journal *Sistemi Intelligenti* (Mastrogiorgio and Petracca 2015). We would like to thank Professor Lorenzo Magnani for the invaluable encouragement and support. A particularly grateful thought goes to the memory of Werner Callebaut, who first supported us on the way to embodied rationality.

[2]With the term 'economic psychology' we mean that domain of inquiry oriented to study phenomena such as *choice*, *judgment*, *decision making*, *problem solving* and *social cognition*. In this broad definition we include also the so-called *behavioral economics*. However, we remark that there is a significant disciplinary divide between 'psychological' and 'economic' approaches to the topics above, characterized for instance by different experimental practices (Hertwig and Ortmann 2001).

[3]Research in AI also supported this point of view on cognition (see, e.g., Brooks 1990).

[4]For a panorama on these different labels and their theoretical interconnections see Goldman and de Vignemont (2009), Kiverstein and Clark (2009), Fischer (2012).

we argue, at the foundational level, it has to unavoidably take into consideration the implications of embodied cognition for economic psychology. The notion that has historically represented the privileged interface for exchanges between economics and psychology is the notion of economic *rationality*. Thus, rationality shall be our focus domain in order to inquire into a new foundational debate in economic psychology. In particular, within rationality studies in economics, our focus will be on the notion of *bounded rationality*. This choice is founded on specific and fundamental reasons. First, Herbert A. Simon, who first introduced the notion of bounded rationality in economics, was both an economist and a cognitive psychologist, founding father of the approach to cognitive psychology called *cognitivism*, object of serious critiques from embodied cognition. Furthermore, far from being a piece of archaeology, bounded rationality still constitutes the bulk of the most important developments in economic rationality. The recent notions of economic rationality stem from, as we shall see, Simon's original notion of bounded rationality, which still influences, both directly and indirectly, knowingly and more often than not unknowingly, the theoretical and experimental research in the fields of choice, judgment, decision making, problem solving, and social cognition in economics.

To the objective of setting the ground for a new foundational dialogue between economic psychology and cognitive psychology this essay is structured as follows. Section 2 offers a panorama of the main contemporary threads of inquiry into economic rationality, particularly emphasizing their Simonian roots. Section 3 is devoted to review the main (and various) conceptual pillars of embodied cognition, assessing their potential influence on economic rationality. In this essay, we use the label *embodied rationality* to convey the view of economic rationality as reformed in the light of embodied cognition.

2 State of the Art in Economic Rationality

The notion of rationality in economics has gone through a quite stylized historical development, originating from the concept of 'classical' rationality seen as the individual's ability to make optimizing choices (i.e., choices that maximize a function of interest, typically what is called 'utility', given the satisfaction of consistency criteria in individual preferences) subject to exogenous constraints (see Blume and Easley 2008). Against this background, Simon introduced the notion of *bounded rationality* as a radical conceptual shift, meant to provide an altogether new framework for economic rationality (Callebaut 2007). By using the well-known metaphor of 'scissors' is probably the most suitable way to convey the nucleus of novelty of bounded rationality. As Allen Newell and Simon himself claimed (Newell and Simon 1972, p. 55): "[j]ust as a scissors cannot cut a paper without two blades a theory of thinking and problem solving [i.e., a theory of rationality] cannot predict behavior unless it encompasses both an analysis of the structure of task and an analysis of the limits of rational adaptation to task environment". According to this definition, the psychology of the individual on the one hand, and the environment in which the individual is embedded on the other hand, represent two necessary theoretical requirements—two blades of scissors—to develop a theory of rationality. In this way, continuing with the metaphor, the scissors "have cutting power [...] only when both blades operate" (Bendor as quoted in Callebaut 2007, p. 78), that is to say, the two cores of rationality must be studied in conjunction. Alternatively, one can say that individuals and environments represent a single analytical unit.

This notion of bounded rationality has typically suffered from distorted, mostly diminutive and instrumental interpretations, forcing Simon himself, from time to time over the decades, to reaffirm the true revolutionary intentions behind the notion. In particular, the greatest misunderstanding over bounded rationality is related, as Gigerenzer and Goldstein (1996) point out, to a partial view of the scissors argument; indeed, many economists have equated, mistakenly, bounded rationality with the view of humans as 'limited' information processors. In this view, bounded rationality is reduced to just one blade of Simon's scissors. The lack of environment as key variable in the rationality framework has entirely expunged the adaptive dimension of rationality, which has thus been reduced to a static notion. Not by chance, it has been economists' reductionism to bring the revolutionary nucleus of bounded rationality into line with the framework of classical rationality, seen as optimization under constraints (see, e.g., Conslik 1996; Rubinstein 1998). But the environment is a necessary requirement for bounded rationality because, as Gigerenzer and Gaissmaier (2011, p. 457) point out, if one looks only at cognition, one is not able to understand when and why reasoning works or, alternatively, fails. In this regard, Callebaut (2007, p. 81) emphasizes that "[bounded rationality's] significance turns not on absolute cognitive levels, but on the difference between cognitive resources and task demands", that is, it turns on in terms of difference between cognitive abilities and environmental issues. This 'difference' engenders the adaptation process that is mostly visible when individuals use 'satisficing',[5] rather than 'optimizing', criteria for making decisions. This focus on heuristic rules for judgment is probably the most significant and revolutionary aspect of Simon's new paradigm. In accordance with this revolutionary nucleus, the emphasis on heuristics has been the hallmark of the most recent notions of rationality, as we are going to discuss in what follows.

[5]*Satisficing* is a neologism coined by Simon (1956), standing for the synthesis of the words *satisfying* and *sufficing*.

2.1 Heuristics and Biases Approach

The research program in *heuristics and biases*, founded by Daniel Kahneman and Amos Tversky in the 1970s, has today gained such wide consensus in economics,[6] so as to earn Kahneman the Nobel Prize in 2002. This research program builds on the possibility of experimentally identifying a positive model of human behavior that violates some normative requirements of rationality; in particular, these normative requirements are based on the assumption that individuals are able to properly use formal logic and probability calculus (in particular Bayesian probability, see Oaksford and Chater 2007) in order to formulate correct judgments. In the *heuristics and biases* program, heuristics play a central role, in so far as people's reliance on them for judgment is at the root of systematic and factual violations of rationality canons. Heuristics are therefore the main source of systematic errors (*biases*) in judgment formulation (Kahneman et al. 1982; Gilovich et al. 2002). Over the decades, a large number of heuristics and cognitive biases has been experimentally identified (Kahneman 2003). In order to explain the cognitive dynamics behind the use of heuristics, Kahneman identifies two types of cognitive process (*dual-system hypothesis*): on the one hand, the so-called System 1 is fast, automatic, emotional and involves unconscious aspects; on the other hand, System 2 is slower, deliberate, analytical and relies on conscious evaluations. System 1's imprinting over heuristics would explain in the end why humans make systematic errors in judgment (Evans and Frankish 2009; Kahneman 2011). The ability to use intentional reasoning (based on System 2) in order to formulate correct judgments would emerge only after the automatic responses (System 1-based) are suppressed. The Cognitive Reflection Test (CRT) (Frederick 2005) has been accordingly developed as a measure of humans' ability to suppress automatic responses in favor of intentional and conscious reasoning. In this perspective, the very nucleus of intelligence (conducive to rationality) would consist in the ability to control the dual-system dynamics.

2.2 Ecological Rationality

An alternative research program in economic rationality, definitely critical of Kahneman and Tversky's *heuristics and biases*, is *ecological rationality*, developed by Gerd Gigerenzer and his research group. Gigerenzer explicitly claims that Kahneman's view stems from an incautious exegesis of Simon's work. In

[6]This is mainly because the heuristic and biases approach is the theoretical foundation to *behavioral economics* (see Heukelom 2014).

particular, Gigerenzer criticizes Kahneman for having embraced the wrong prevailing interpretation of bounded rationality as cognitive limitations in information processing (just one blade of the scissors), thus not paying any attention to the role of the environment. The *ecological rationality* research program is thus intended to restore Simon's original idea that cognitive resources and environmental demands form an analytical unit.

Heuristics are, once again, the core of the analysis. Challenging the *heuristics and biases* approach, according to which heuristics are the source of judgment errors, *ecological rationality* looks at heuristics as *fast and frugal* rules able to reliably provide choice criteria in different situations. The main hypothesis of *ecological rationality* is in fact that each heuristic stems from an adaptive process to a specific environment (Gigerenzer and Gaissmaier 2011, p 456), and thus, accordingly, that heuristics allow comparatively better judgments (even with respect to the 'rational' rules of logic and probability calculus) when they are used in their own specific original environment (Gigerenzer and Brighton 2009). Heuristics under this point of view compose an *adaptive toolbox* (Gigerenzer and Selten 2002), which is the resultant of evolutionary processes (Barkow et al. 1992) and provides humans with adequate tools for accomplishing specific tasks.

Ecological rationality rejects the assumption that humans are 'Bayesian statisticians'—an assumption that would justify the use of normative standard of logic and probability for judgment assessment (as in Kahneman's view)—and posits that judgment's rationality can only be evaluated on the basis of which specific heuristics are chosen to accomplish specific tasks. Notice that *ecological rationality* does not reject the existence of errors in judgment: it simply posits that the source of errors does not lie in the cognitive limits of the subjects, but stems from the mismatch between heuristics and the environment in which they are used.

The panorama described in the two sections above is that of 'rationality wars' between Kahneman and Gigerenzer (Samuels et al. 2004), whose intensity shows no sign of abating. The research program in *embodied rationality* that we introduce in what follows, is sympathetic with ecological rationality regarding the necessity of following Simon's authentic view but, as we are going to discuss, it shall point out that *ecological rationality* is still decisively tied to a limit (if considered from today's standpoint) of Simon's framework: the 'cognitivist' view of cognition. Thus, *embodied cognition* is meant to amend what unsatisfactory is still in the cognitive psychology foundations of *ecological rationality*.[7]

[7]A further and recent thread of research in rationality is that of *grounded rationality* (Elqayam and Evans 2011). Grounded rationality conceives rationality as a set of rules embedded in specific epistemic communities. In this perspective, rationality is at first a relative and descriptive notion that, once institutionalized in a community, acquires a normative status.

3 Toward a Theory of Embodied Rationality

3.1 Cognitivism and Rationality

Technically, the 'cognitivist heaven' in which Simon placed bounded rationality is founded on the *physical symbol system hypothesis*, namely the idea that human cognition works through internal (i.e., mental) symbolic representations of the external environment processed by a centralized (i.e. in-line[8]) analytical system (Newell and Simon 1976). Residua of these cognitivist foundations are usually taken for granted—when acknowledged—by the economic rationality threads of research; in this regard, we maintain that a significant shift in economic rationality can originate only from the explicit acknowledgment, questioning and overcoming of the residual cognitivist stances; in other words, this means following in rationality studies the same path followed by cognitive psychology in its progressive detachment from cognitivism (Haugeland 1978; Johnson 1997). Erkki Patokorpi (2008) has inaugurated this path, by acknowledging the so-called 'Simon's paradox', that is, the fact that the 'bounds' of human reason are represented through an 'unbounded' tool (i.e. the digital computer), and by identifying this heritage into contemporary rationality theories.

The residual cognitivist foundations of current theories of rationality are evident in both the *heuristics and biases* and the *ecological rationality* research programs. As already mentioned, in the *heuristics and biases* approach heuristics are just mental phenomena, lacking any coupling with the environment; using the computer metaphor, somewhat foundational to the cognitivist paradigm, heuristics would be like bugged computer programs.[9] On the other hand, despite *ecological rationality* puts a decisive emphasis on the cognition-environment coupling, heuristics, conceived as formal rules for information processing, are implemented through 'computer programs', explicitly in the footsteps of Simon's tradition (Gigerenzer et al. 1999).

In the following sections, we shall discuss the ways in which the new point of view of embodied cognition can provide alternative foundations to the notion of bounded rationality. Two caveats are however in order here. Firstly, as we mentioned in the Introduction, *embodied cognition* is far from being a stable corpus of theories (see, e.g., Wilson 2002 that identifies 'six views' of embodied cognition). Due to this theoretical plurality, this essay does not focus on or embrace a specific view of embodied cognition, but rather brings out a number of new foundational possibilities. The second *caveat* is related to the fact that many topics that we discuss here may sound not so new to scholars accustomed to cognitive

[8]This assumption has later been relaxed, for instance by models of parallel processing.

[9]Fiori (2011) states that the 'dual-system' foundation of *heuristics and biases* (see Sect. 12.2.1) represents a break with respect to Simon's cognitivism. This interpretation is—according to us—not conclusive because Simon himself saw cognitivism as perfectly compatible with dual-system theories (see, e.g., Vera and Simon 1993).

psychology. However, our assumption here is that there are good reasons to think that these topics might sound interesting to many scholars involved in economic psychology, making them worthy of being discussed here.

3.2 An Enriched Environment

In the next two subsections, we discuss two approaches within embodied cognition, respectively *situated cognition* and *distributed cognition*, considered here because they provide a new point of view on the role of environment in cognitive psychology. Our specific contribution here shall be to identify how these two points of view can help to rethink the notion of 'decision-making environment', so central in economic psychology.

3.2.1 The Limits of Syntactic Representationalism and the First/Third-Person Distinction

Distinctive evidence of the cognitivist heritage in contemporary investigations on rationality—as shown in particular by ecological rationality—lies in the notion of decision-making environment as expressed through the notion of *structure of the task environment* (see Simon 1956; Bullock and Todd 1999). In particular, ecological rationality categorizes environments according to syntactic 'structural' traits such as information redundancy, rarity, etc. (see Rieskamp and Dieckmann 2012; McKenzie and Chase 2012). This characterization tends to underestimate the role of the environments' semantics: a playground or a battlefield might result in fact in the same syntactic representation. In this framework then, the adequacy of heuristics to a particular environment is measured by the *ecological correlations* of syntactic structures between heuristics and environments (e.g. McKenzie and Chase 2012).[10] A perspective that goes beyond the syntactic view of environments replaces, at a first level of approximation, the structure of the task environment—in Simon's or Gigerenzer's understanding—with that of *context*, according to which the environment presents a markedly semantic dimension. This shift has already been accomplished by economists, especially behavioral economists, who claim the

[10]We have to remark an important incongruence between the theoretical assumptions *of ecological rationality* and the actual framework through which these assumptions are implemented. A fundamental assumption of ecological rationality is that heuristics and environments are 'content-specific' and, as such, semantically characterized. But, this semantic dimension is practically lost when heuristics and environments are respectively characterized as rules and stylized structures.

necessity of running experiments in semantic environments, very similar to real-life contexts (Loewenstein 1999).

However, recent cognitive psychology claims that this 'contextual turn' is not enough. *Situated cognition* denies that the notion of 'context' is adequate to account for the phenomena of human cognition, and claims that it should be replaced by the more radical notion of 'situation' (see e.g. Rohlfing et al. 2003). In particular, while contexts are founded on the conflation between 'first-person' and 'third-person' representations, that is, between subjective and objective descriptions of environments, situated cognition refuses this unwarranted conflation (Clancey 1993).[11] The notion of situation builds on the centrality of action: action, and in particular inter-action between subjects and environments, makes first-person representations irreducible (basically because the outcome of a process of interaction is not pre-specifiable) (Greenberg 2001). The emphasis on the notion of interaction leads some researchers even to reject the ontological distinction between first and third person, and to propose a completely new ontological view in which the notion of interaction is autonomous[12] (Agre 1993).

The distinction between first and third person has major implications for the conceptualization of the decision-making environment: there is no way to a priori determine which specific trait of the environment will be 'salient' for decisions. The attempt of amending the syntactic view of decision-making environments just by adding more semantic content is therefore inadequate because it would lack the essential interactionist perspective. Real-world interactions cannot even be substituted by surrogate interactions, as conceived for instance in game theory and experimental economics, where interactions lack any material and ostensive dimension. Consider, for instance, how the notion of 'learning' is framed in game theory (see, e.g., Fudenberg 1998).

A radical alternative to the poorly-semanticized and non-interactive rationality frameworks has been proposed by the research program in *naturalistic decision making* (Klein 2008). Naturalistic decision making rejects the idea that decision making can be 'simulated' at all: real-life decisions are therefore considered the only legitimate place to investigate into human rationality.[13]

---

[11]This point was at the center of a debate in 1993 between Simon (with his colleague Alonso Vera) and *situated cognition* scholars. Vera and Simon argued that situated cognition's arguments were not sufficient to legitimately claim for a re-foundation of cognitive psychology (see Petracca 2015).

[12]A perspective in which the ontology of relations outranks the one of subjects/environments can be found, for instance, in the 'dynamic systems' approach to cognition.

[13]*Ecological rationality* has tried to integrate *naturalistic decision making* within its own theoretical framework. In fact, Todd and Gigerenzer (2001, p. 382) state that their objective is that of providing a 'content-dependent' framework to *naturalistic decision making*. In spite of their attempt, it seems that they have not fully acknowledged the first- and third- person distinction, implicit in *naturalistic decision making*.

3.2.2 A 'Distributed' and 'Extended' Reasoning

The cognitivist imprint on current economic rationality is also visible in the persistent interpretation of the task environment as a set of constraints. In the ecological rationality interpretation, for instance, these constraints exert a selective pressure on heuristics, which in response adapt to them (Bullock and Todd 1999). The cognitive psychology approach called *distributed cognition* rejects the idea of the environment as a mere set of constraints and proposes a radically different perspective, summarized by the words of Suchman that the "world's independence of [our] control is not an obstacle to be overcome but a resource to be made use of" (Suchman 1986, p. 13). In order to re-define the role of the environment in cognition, cognitive psychology has had to contend, even linguistically, with the theoretical limits imposed by the dualism cognition/environment. An attempt to amend such limits lies in the notion of 'extended mind' (Clark and Chalmers 1998), which represents an attempt to blur the boundaries between what is cognition and what is environment. Both 'distributed cognition' and 'extended mind', challenging the traditional understanding of environments, are meant to cast new light also on the reasoning process.

The central idea is that environments in which humans act systematically provide the resources that can be employed in reasoning processes. Consider the so-called *cognitive artifacts*, that is, physical objects such as a calendar, a shopping list, a computer or even fingers, which are used to support and improve reasoning (Hutchins 1999, p. 126). Cognitive artifacts are mainly used to *off-load* the cognitive load on the environment, thereby making cognitive resources available for other purposes. But the environment is also used for cognitive purposes less trivial than mere off-loading, which involve the very act of 'reasoning'. In this spirit, for instance, Kirsh and Maglio (1994) distinguish between *pragmatic action* and *epistemic action*: the former is devoted to a pragmatic purpose, i.e. to change the environment according to definite objectives, while the latter uses the environment for reasoning. Consider the game of Tetris, specifically studied by Kirsh and Maglio: to rotate figures is an action that players perform in order to facilitate the decision process of where to place figures. The rotation action—conceptually unnecessary for the purpose of the game—is an epistemic action that exploits the environment in order to make the decision process faster and more effective. Another way in which the environment can support reasoning is when material interactions provide otherwise unattainable insights. The value of this kind of environmental interaction was suggested, for instance, by the mathematician Pólya (1957), who recommended the heuristic use of pen and paper to facilitate mathematical reasoning (see also Zhang 1997). With explicit reference to inferential processes, Lorenzo Magnani introduces the notion of *manipulative abduction* (Magnani 2001). A fundamental role in abductive reasoning (i.e. that reasoning process oriented to formulate explanatory hypotheses of observed facts) is played by the so-called *epistemic mediators* (such as diagrams on sheets of paper, etc.) used for hypotheses discovering. Magnani distinguishes between 'thinking about doing' and 'thinking through doing': in particular, the latter characterizes reasoning processes in which environmental interaction serves the purpose of providing information that would not be otherwise accessible to subjects.

These arguments can arguably have a direct impact on economic psychology. They do not only imply conceptualizing the task environment as 'resource' rather than as 'constraint', as it would be natural (and right) to do. Distributed cognition would also provide an altogether new point of view on the economic agent, which could be characterized more as a 'chance seeker' than as the usual 'information processor' (Bardone 2011). More concrete implications of distributed cognition and extended mind for economic psychology can be found at the methodological level. One, for instance, concerns the current practices of experimental economics. Excessive rigidity and standardization of experiments—claimed to be distinctive traits of experimental economics' investigations (see Hertwig and Ortmann 2001)—triggers a sort of 'illusion of control' with respect to actual human behavior.[14] Further, the pervasive mediation of computer screens in experimental economics laboratories, in particular when used to study human interactions, leads us to another consideration. As stressed by Oullier and Basso (2010), an essential component of the interaction among humans relies on the materiality of the interaction. Information conveyed through the body (i.e., through the so-called 'body language') is invaluable to the extent it could not emerge saliently otherwise.

Acknowledging that human-to-human is a specific form of human-to-environment interaction—in fact, as McDermott said, "we are environment to each other" (quoted in Suchman 1987, p. 47)—implies a wider understanding of 'distributedness' (Hutchins 2006). The pioneer of this wider 'distributed' perspective in economics is Hayek (1948). Clark and Chalmers (1998, Chap. 9) introduced the concept of 'scaffold' to express how distributed and interactive mechanisms lead to establish supra-individual structures, such as routines or formal and informal rules, able to steer individuals' social action (see also Denzau and North 1994).

3.3 Rationality and Body Correlates

3.3.1 Reasoning as Simulation and the Role of Embodied Metaphors

The notion of 'procedural rationality'—one of Simon's main intellectual achievements[15]—is fundamentally linked to a 'pragmatic' interpretation of rationality

[14]It is interesting to recall, on this point, the anecdote reported by Daniel Dennett concerning a child who, not allowed to use fingers for calculations, used tongue and teeth as substitutes (Dennett 1995).

[15]Simon (1976) distinguished between 'substantive' rationality, where rationality concerns the outcome of choice, and 'procedural' rationality, where rationality concerns the process of choice. Procedural rationality, in the case for instance of consumers' choice, focuses on how consumers choose and not on what they choose.

(Harman 1993). In his early work on administrative behavior, Simon (1947) in fact understood rationality as means-ends chains, where the core of rationality consisted in evaluating means' adequacy to reach pre-specified ends.[16] The human cognitive faculties of imagining, planning, predicting things or events that are alien to the current situation (i.e. outside the strict phenomenological dimension of 'here and now') are thus necessary cognitive requirements for the notion of procedural/pragmatic rationality. Beyond emphasizing the importance of contingent cognition (*on-line cognition*), the foundational perspective of embodied cognition is able to shed a new light also on non-contingent cognition (*off-line cognition*) (Wilson 2008).[17] In what follows, we shall consider what embodied cognition is able to say on off-line reasoning in economic rationality.

The cognitive phenomenon of *simulation*, which concerns the exploitation of the sensory-motor system for understanding and reasoning (Jeannerod 2001; Hesslow 2002; Gallese and Lakoff 2005), has been identified as the main mechanism at the root of off-line reasoning (Barsalou 2008; Goldman and de Vignemont 2009). Simulation assumes a central role in contemporary embodied theories because it constitutes the fundamental mechanism through which 'mental representations' and their 'manipulations' (indeed very controversial notions in cognitive psychology) work.[18] Among the different types of emphasis on the role of *modal*[19] encoding in simulation (see Meteyard and Vigliocco 2008), the most radical embodied cognition approach—also known as 'strong embodiment'—claims that anything necessary to create and manipulate representations is embodied in the sensory-motor system, and thus it can be identified in terms of *body correlates*. An example of simulative approach to reasoning is Lakoff and Johnson (1999)'s attempt to explain inferential processes through the mechanism of *inference-preserving-cross-domains mappings*, according to which one projects the inferential structure of an original domain to a target domain, usually more abstract. For example, if we say "she is a cold person", the concept of 'cold' (source domain) will be mapped into 'lack of affection' (target domain) (Núñez 2008, p. 337). Therefore the notion of 'cold', which evokes a body dimension, provides the foundations for creating the notion of 'lack of affection', instantiated through a metaphorical process.[20] In this perspective, the morphological traits of the human body play the role of non-arbitrary constraints to the human capability of making abstractions and inferences.

---

[16]Russell and Norvig (1994) import this definition of rationality in the AI framework.

[17]While opponents to embodied cognition typically reduce it to a theory of on-line cognition, Wilson claims that offline cognition is embodied cognition's true testbed (Wilson 2008, p. 330).

[18]Whether the supporters of *situated cognition* underestimate the role of mental representations (in fact representations are almost unessential in their framework), the supporters of the 'simulation' view try to explain the very nature of those representations. This distinction is revealing of the theoretical plurality underlying embodied cognition.

[19]Modal is a representation encoded through the sensory-motor system. Conversely, a-modality pertains to representations' independence from the sensory-motor system.

[20]Notice that the metaphor of 'scissors' itself, used to define *bounded rationality*, is based on this logic.

The importance of the framework of embodied cognition to economics can in particular be appreciated when there is an overlapping of topics between disciplines, as is the case with the topic of 'ownership'. Beyond the many 'economic' explanations, ownership has been explained in the embodied cognition framework by emphasizing the determinants of contact and proximity to the owned object (see, for example, Tummolini et al. 2013). By juxtaposing the studies on embodied ownership on the one hand, and ownership-related economic phenomena—such as the *endowment effect* (Kahneman et al. 1991)—on the other hand, answers to many puzzles (Plott and Zeiler 2005) might easily be found. Further on ownership, embodied metaphors play a crucial role. In a recent study, Florack et al. (2014) shows that the physical act of hand washing—one of the most popular embodied metaphors—decreases the endowment effect.

3.3.2 Body Correlates and Dual-System Dynamics

The human body as the ultimate new foundation of cognitive processes and, in turn, of reasoning processes, can also be central in understanding dual-system dynamics in economic judgment (see, e.g., Kahneman 2011). In a recent article, Mastrogiorgio and Petracca (2014)[21] investigate the body determinants of automatic/deliberative reasoning in numerical tasks. They argue that the activation of automatic responses (System 1) is closely dependent on the use of specific numerals (and not 'numbers' as magnitudes). The idea is that, within a given numeral system, such as the common 10-based Arabic system, some numerals (e.g., 1, 2, 5, 10, 100,…) are handled in a faster and more automatic way if compared with other numerals of the same system.[22] The literature in mathematical cognition (see Cohen Kadosh and Walsh 2009) shows that the automatic use of specific numerals in numeral systems is dependent by the underlying counting systems. For instance, the 10-based Arabic system relies on the computation method based on the fingers of both hands (see Gibbs 2006). This example suggests that the identification of body correlates underlying automatic behavior in different decision domains can be of fundamental importance to a theory of economic reasoning.

3.4 *The Body as Pivot of the Scissors*

If the goal that inspires modern rationality studies in economics is the identification of "invariants of human behavior" (as suggested by Simon 1990), then one should

[21]See also Mastrogiorgio (2015) for further remarks.

[22]Wulf Albers (2001), within the *ecological rationality* framework, models heuristic calculation by means of the so-called 'prominent numbers' (i.e., numerals 1, 2, 5, 10 …) in the decimal system. Albers does not however explain why some numerals are processed faster and easier than others.

suddenly acknowledge that a true invariant in human behavior is the human body. However, Simon's scissors metaphor constitutively rules out the human body, considered as a sort of cumbersome presence. It is somehow ironic that the human body is however right there, both in Kahneman's and Gigerenzer's theories, just because they place so great emphasis on heuristics: a synonym for 'heuristic' is in fact—and of course not by chance—'rule of *thumb*'. The metaphor of the scissors is, once more somehow ironically, perfectly suited for considering the role of body as theoretical *locus* of connection between cognition and environment: as we discussed above, the two blades have cutting power only in combination, that is to say, the two blades can cut only if there is a pivot that holds them together. This pivot is the human body, which constitutes the material interface between cognition and environment.

It is important to emphasize that in this new perspective the body is a necessary theoretical *locus* for a theory of rationality, and so that it needs to be more than simply 'taken into account'. Many experimental studies already take into account body variables (e.g. temperature, blood pressure, etc.) that affect choices, decisions and judgments. What those experiments lack is, however, the attribution of a deeper theoretical status to the human body for a theory of human rationality. This is, however, a situation that is going to be amended (Spellman and Schnall 2009; Reimann et al. 2012; Mastrogiorgio and Petracca 2015).

3.4.1 Policy Implications

Some brief consideration, as economists who want to speak mainly to other economists, has to be proposed concerning the implications of embodied cognition for policy making. In recent years, research has focused primarily on *nudging*, that is, the idea of designing environments so as to drive individuals' choice toward socially desired outcomes, without modifying the structure of economic incentives and maintaining the freedom of choice (Thaler and Sunstein 2008). The notion of *architecture of choice* refers in particular to the way in which options are presented, in order to encourage socially-desired choices. Experiments on nudging are typically run in real-world contexts, and place a decisive emphasis on the embodied substratum of the architecture of choice. An example of typical nudging advice would be to place healthy foods at the 'eye level' in a self-service restaurant, so as to make them comparatively more chosen (Thorndike et al. 2012). In this regard, in a domain that is already implicitly 'embodied', explicitly considering the point of view of embodied cognition can be important in at least two directions: i) theoretically, so as to identify the body as a conceptual *locus* for nudging; ii) operatively, shaping environments in order to enhance their *ergonomics*. Ergonomics is the keyword here: all the environments of choice are ostensive at the human-body scale. Thus, an important objective of embodied rationality is to foster a programmatic link between the disciplines of ergonomics and economics (see also Hendricks 1996).

4 Conclusions

This essay has focused, on the one hand, on the identification of those traits, distinctly cognitivist, characterizing Simon's legacy in current studies on economic rationality; on the other hand, it has proposed new hypotheses to reform those traits in the light of the recent advancements in cognitive psychology. The starting point of our analysis has been the crucial notion of bounded rationality. As one of the most influential cognitive psychologists, Andy Clark, put it: "we should however distinguish the conception of reason as embodied and embedded from the important but still insufficiently radical notion of 'bounded rationality" (Clark 1997, p. 243, n. 4). This essay has accordingly attempted the 'radicalization' of bounded rationality in the light of the multiple directions suggested by the fertile and varied field of study of embodied cognition. There is much work to do but, for the moment, it has just been important to remark the existence of a plurality of roads.

It is useful to provide a brief summary of the specific ways embodied cognition might reform the notion of bounded rationality, as they have been discussed in this paper. They can be condensed in a few stylized programmatic points:

- reasoning is situated, that is, interaction is the fundamental way in which reasoning takes place: thus, it is necessary to distinguish between first-person and third-person representations of decision environments.
- reasoning is not exclusively a mental phenomenon as humans constitutively use the resources provided by the environment in order to reason; this also means that the environment should be conceptualized as a resource rather than as a constraint;
- off-line reasoning works through simulations that exploit the resources provided by the sensory-motor system, so that sensory-motor experience provides the building blocks of abstraction;
- automatic thinking stems from the (evolutionary) coupling between some morphological traits of the human body and the environment they originally fitted to.

We hope these programmatic points to be at the center of future research under the label of *embodied rationality* (see Spellman and Schnall 2009; Mastrogiorgio and Petracca 2015).

References

Agre, P. E. (1993). The symbolic worldview: Reply to vera and simon. *Cognitive Science, 17*(1), 61–69.

Albers, W. (2001). Prominence theory as a tool to model boundedly rational decisions. In G. Gigerenzer & R. Selten (Eds.), *Bounded rationality: The adaptive toolbox* (pp. 297–317). Cambridege: MIT Press.

Bardone, E. (2011). *Seeking chances: From biased rationality to distributed cognition*. Berlin and Heidelberg: Springer.

Barkow, J., Cosmides, L., & Tooby, J. (Eds.). (1992). *The adapted mind: Evolutionary psychology and the generation of culture*. New York: Oxford University Press.
Barrett, L. (2011). *Beyond the brain: How body and environment shape animal and human minds*. Princeton: Princeton University Press.
Barsalou, L. W. (2008). Grounded cognition. *Annual Review of Psychology, 59*(1), 617–645.
Blume, L. E., & Easley, D. (2008). Rationality. In V. S. N. Durlauf & L. E. Blume (Eds.), *The new Palgrave dictionary of economics*. Basingstoke: Palgrave Macmillan.
Brooks, R. A. (1990). Elephants don't play chess. *Robotics and Autonomous Systems, 6*(1–2), 3–15.
Bullock, S., & Todd, P. M. (1999). Made to measure: Ecological rationality in structured environments. *Mind and Machines, 9*(4), 497–541.
Callebaut, W. (2007). Simon's silent revolution. *Biological Theory, 2*(1), 76–86.
Clancey, W. J. (1993). Situated action: A neuropsychological interpretation (response to vera and simon). *Cognitive Science, 17*(1), 117–133.
Clancey, W. J. (1997). *Situated cognition: On human knowledge and computer representations*. Cambridge, MA: Cambridge University Press.
Clark, A. (1997). *Being there: Putting brain, body, and world together again*. Cambridge, MA: MIT Press.
Clark, A., & Chalmers, D. (1998). The extended mind. *Analysis, 58*(1), 10–23.
Cohen Kadosh, R., & Walsh, V. (2009). Numerical representation in the parietal lobes: Abstract or not abstract? *Behavioral and Brain Sciences, 32*(3–4), 313–373.
Conlisk, J. (1996). Why bounded rationality. *Journal of Economic Literature, 348*(2), 669–700.
Dennett, D. C. (1995). *Darwin's dangerous idea: Evolution and the meanings of life*. New York: Simon and Schuster.
Denzau, A. T., & North, D. C. (1994). Shared mental models: Ideologies and institutions. *Kyklos, 47*(1), 3–31.
Elqayam, S., & Evans, St.B.T. (2011). Subtracting 'ought' from 'is': Descriptivism versus normativism in the study of the human thinking. *Behavioral and Brain Sciences, 34*(5), 233–248.
Evans, J., & Frankish, K. (Eds.). (2009). *In two minds: Dual processes and beyond*. Oxford: Oxford University Press.
Fiori, S. (2011). Forms of bounded rationality: The reception and redefinition of Herbert A. *Simon's Perspective. Review of Political Economy, 23*(4), 587–612.
Fischer, M. H. (2012). A hierarchical view of grounded, embodied, and situated numerical cognition. *Cognitive Processing, 13*(1), 161–164.
Florack, A., Kleber, J., Busch, R., & Stöhr, D. (2014). Detaching the ties of ownership: The effects of hand washing on the exchange of endowed products. *Journal of Consumer Psychology, 24* (2), 284–289.
Frederick, S. (2005). Cognitive reflection and decision making. *Journal of Economic Perspectives, 19*(4), 25–42.
Fudenberg, D. (1998). *The theory of learning in games* (Vol. 2). Cambridge, MA: MIT Press.
Gallagher, S. (2005). *How the body shapes the mind*. New York: Oxford University Press.
Gallese, V., & Lakoff, G. (2005). The brain's concepts: The role of the sensorimotor system in reason and language. *Cognitive Neuropsychology, 22*(3), 455–479.
Gibbs, R. W. (2006). *Embodiment and cognitive science*. New York: Cambridge University Press.
Gigerenzer, G., & Brighton, H. (2009). Homo heuristicus: Why biased minds make better inferences. *Topics in Cognitive Science, 1*(1), 107–143.
Gigerenzer, G., Todd, P. M., & The ABC Research Group. (1999). *Simple heuristics that make us smart*. New York: Oxford University Press.
Gigerenzer, G., & Gaissmaier, W. (2011). Heuristic decision making. *Annual Review of Psychology, 62*(1), 451–482.
Gigerenzer, G., & Goldstein, D. G. (1996). Reasoning the fast and frugal way: Models of bounded rationality. *Psychological Review, 103*(4), 650–669.

Gigerenzer, G., & Selten, R. (Eds.). (2002). *Bounded rationality: The adaptive toolbox*. Cambridge, MA: MIT Press.

Gilovich, T., Griffin, D., & Kahneman, D. (2002). *Heuristics and biases: The psychology of intuitive judgment*. Cambridge, UK: Cambridge University Press.

Goldman, A., & de Vignemont, F. (2009). Is social cognition embodied? *Trends in Cognitive Sciences, 13*(4), 154–159.

Greenberg, S. (2001). Context as a dynamic construct. *Human-Computer Interaction, 16*(2), 256–268.

Harman, G. (1993). Rationality. In E. E. Smith & D. N. Osherson (Eds.), *Thinking: An invitation to cognitive science* (Vol. 3). Cambridge, MA: MIT Press.

Haugeland, J. (1978). The nature and plausibility of cognitivism. *Behavioral and Brain Sciences, 1* (2), 215–226.

Hayek, F. A. (1948). *Individualism and economic order*. Chicago: University of Chicago Press.

Hendricks, H. W. (1996). The ergonomics of economics is the economics of ergonomics. *Proceedings of the Human Factors and Ergonomics Society Annual Meeting, 40*(1), 1–10.

Hertwig, R., & Ortmann, A. (2001). Experimental practices in economics: A methodological challenge for psychologists? *Behavioral and Brain Sciences, 24*(3), 383–403.

Hesslow, G. (2002). Conscious thought as simulation of behaviour and perception. *Trends in Cognitive Sciences, 6*(6), 242–247.

Heukelom, F. (2014). *Behavioral economics: A history*. New York: Cambridge University Press.

Hutchins, E. (1999). Cognitive artifacts. In R. A. Wilson & F. C. Keil (Eds.), *MIT encyclopedia of cognitive science* (pp. 126–128). Cambridge, MA: MIT Press.

Hutchins, E. (2006). The distributed cognition perspective on human interaction. In N. Enfield & S. Levinson (Eds.), *Roots of human sociality* (pp. 375–398). New York: Berg.

Jeannerod, M. (2001). Neural simulation of action: A unifying mechanism for motor cognition. *NeuroImage, 14*(1), 103–109.

Johnson, D. M. (1997). Good old-fashioned cognitive science: Does it have a future? In D. M. Johnson, D. M & Erneling, C. E. (Eds.), *The future of the cognitive revolution* (pp. 13–31). New York: Oxford University Press.

Kahneman, D. (2003). Maps of bounded rationality: Psychology for behavioral economics. *American Economic Review, 93*, 1449–1475.

Kahneman, D. (2011). *Thinking fast and slow*. New York: Farrar, Straus and Giroux.

Kahneman, D., Knetsch, J. L., & Thaler, R. H. (1991). Anomalies: The endowment effect, loss aversion, and status quo bias. *Journal of Economic Perspectives, 5*(1), 193–206.

Kahneman, D., Slovic, P., & Tversky, A. (1982). *Judgment under uncertainty: Heuristics & biases*. Cambridge, UK: Cambridge University Press.

Kirsh, D., & Maglio, P. (1994). On distinguishing epistemic from pragmatic action. *Cognitive Science, 18*(4), 513–549.

Kiverstein, J., & Clark, A. (2009). Introduction: Mind embodied, embedded, enacted: One church or many? *Topoi, 28*(1), 1–7.

Klein, G. (2008). Naturalistic decision making. *Human Factors: The Journal of the Human Factors and Ergonomics Society, 50*(3), 456–460.

Lakoff, G., & Johnson, M. (1999). *Philosophy in the flesh: The embodied mind and its challenge to Western thought*. New York: Basic Books.

Loewenstein, G. (1999). Experimental economics from the vantage-point of behavioural economics. *The Economic Journal, 109*(453), 25–34.

Magnani, L. (2001). *Abduction, reason, and science: Processes of discovery and explanation*. New York: Kluwer Academic/Plenum Publishers.

Mastrogiorgio, A. (2015). Commentary: Cognitive reflection versus calculation in decision making. *Frontiers in Psychology, 6*, 936. doi:10.3389/fpsyg.2015.00936.

Mastrogiorgio, A., & Petracca, E. (2014). Numerals as triggers of system 1 and system 2 in the 'bat and ball' problem. *Mind & Society, 13*(1), 135–148.

Mastrogiorgio, A., & Petracca, E. (2015). Razionalità incarnata, *Sistemi Intelligenti, 27*(3), 481–504.

McKenzie, C. R. M., & Chase, V. M. (2012). Why rare things are precious: How rarity benefits inference. In P. Todd & G. Gigerenzer, the ABC Research Group (Eds.), *Ecological rationality: Intelligence in the real world* (pp. 309–334). New York: Oxford University Press.

Meteyard, L., & Vigliocco, G. (2008). The role of sensory and motor information in semantic representation: A review. In P. Calvo & A. Gomila (Eds.), *Handbook of cognitive science: An embodied approach* (pp. 293–312). San Diego, US: Elsevier Publishers Limited.

Newell, A., & Simon, H. A. (1972). *Human problem solving*. Englewood Cliffs, NJ: Prentice Hall.

Newell, A., & Simon, H. A. (1976). Computer science as empirical inquiry: Symbols and search. *Communications of the ACM, 19*(3), 113–126.

Noë, A. (2004). *Action in perception*. Cambridge, MA: MIT Press.

Núñez, R. (2008). Mathematics, the ultimate challenge to embodiment: Truth and the grounding of axiomatic systems. In P. Calvo & A. Gomila (Eds.), *Handbook of cognitive science: An embodied approach* (pp. 333–353). San Diego, US: Elsevier Publishers Limited.

Oaksford, M., & Chater, N. (2007). *Bayesian rationality: The probabilistic approach to human reasoning*. Oxford: Oxford University Press.

Oullier, O., & Basso, F. (2010). Embodied economics: How bodily information shapes the social coordination dynamics of decision-making. *Philosophical Transactions of the Royal Society B: Biological Sciences, 365*(1538), 291–301.

Patokorpi, E. (2008). Simon's paradox: Bounded rationality and the computer metaphor of the mind. *Human systems management, 27*(4), 285–294.

Petracca, E. (2015). *A tale of paradigm clash: Simon, situated cognition and the interpretation of bounded rationality*. MPRA Paper 64517, University Library of Munich. https://mpra.ub.uni-muenchen.de/64517/.

Pfeifer, R., & Bongard, J. (2006). *How the body shapes the way we think: A new view of intelligence*. Cambridge, MA: MIT Press.

Plott, C. R., & Zeiler, K. (2005). The willingness to pay-willingness to accept gap, the 'endowment effect', subject misconceptions, and experimental procedures for eliciting valuations. *American Economic Review, 95*(3), 530–545.

Pólya, G. (1957). *How to solve it: A new aspect of mathematical method*. London: Penguin.

Rabin, M. (1998). Psychology and economics. *Journal of Economic Literature, 36*(1), 11–46.

Reimann, M., Feye, W., Malter, A. J., Ackerman, J. M., Castaño, R., Garg, N., et al. (2012). Embodiment in judgment and choice. *Journal of Neuroscience, Psychology, and Economics, 5* (2), 104–123.

Rieskamp, J., & Dieckmann, A. (2012). Redundancy: Environment structure that simple heuristics can exploit. In P. Todd & G. Gigerenzer, the ABC Research Group (Eds.), *Ecological rationality: Intelligence in the real world* (pp. 187–215). New York: Oxford University Press.

Rohlfing, K. J., Rehm, M., & Goecke, K. U. (2003). Situatedness: The interplay between context (s) and situation. *Journal of Cognition and Culture, 3*(2), 132–157.

Rowlands, M. (1999). *The body in mind: Understanding cognitive processes*. Cambridge, UK: Cambridge University Press.

Rubinstein, A. (1998). *Modeling bounded rationality*. Cambridge, MA: MIT Press.

Russell, S. J., & Norvig, P. (1994). *Artificial intelligence: A modern approach*. Englewood Cliffs, NJ: Prentice-Hall.

Samuels, R., Stich, S., & Bishop, M. (2004). Ending the rationality wars: How to make disputes about human rationality disappear. In E. Renee (Ed.), *Common sense, reasoning, and rationality* (pp. 236–268). New York: Oxford University Press.

Shapiro, L. (2004). *The mind incarnate*. Cambridge, MA: MIT Press.

Simon, H. A. (1947). *Administrative behaviour: A study of decision-making processes in administrative organization*. New York: MacMillan.

Simon, H. A. (1956). Rational choice and the structure of environments. *Psychological Review, 63* (2), 129–138.

Simon, H.A. (1976). From substantive to procedural rationality. In T. J. Kastelein, S. K. Kuipers, W. A. Nijenhuis & G. R. Wagenaar (Eds.), *25 Years of economic theory: Retrospect and prospect* (pp. 65–86). Boston: Springer.

Simon, H. A. (1990). Invariants of human behavior. *Annual Review of Psychology, 41*(1), 1–19.
Spellman, B. A., & Schnall, S. (2009). Embodied rationality. *Queen's Law Journal, 35*(1), 117–164.
Suchman, L. (1986). What is a plan? *ISL Technical Note*, Xerox Palo Alto Research Center.
Suchman, L. A. (1987). *Plans and situated actions: The problem of human-machine communication*. New York: Cambridge University Press.
Thaler, R., & Sunstein, C. (2008). *Nudge: Improving decisions about health, wealth, and happiness*. New Haven, CT: Yale University Press.
Thorndike, A. N., Sonnenberg, L., Riis, J., Barraclough, S., & Levy, D. E. (2012). A 2-phase labeling and choice architecture intervention to improve healthy food and beverage choices. *American Journal of Public Health, 102*(3), 527–533.
Todd, P. M., & Gigerenzer, G. (2001). Putting naturalistic decision making into the adaptive toolbox. *Journal of Behavioral Decision Making, 14*(5), 381–383.
Tummolini, L., Scorolli, C., & Borghi, A. M. (2013). Disentangling the sense of ownership from the sense of fairness. *Behavioral and Brain Sciences, 36*(1), 101–102.
Varela, F., Thompson, E., & Rosch, E. (1991). *The embodied mind: Cognitive science and human experience*. Cambridge, MA: MIT Press.
Vera, A. H., & Simon, H. A. (1993). Situated action: A symbolic interpretation. *Cognitive Science, 17*(1), 7–48.
Wilson, M. (2002). Six views of embodied cognition. *Psychonomic Bulletin & Review, 9*(4), 625–636.
Wilson, M. (2008). How did we get from there to here? An evolutionary perspective on embodied cognition. In P. Calvo & T. Gomila (Eds.), *Handbook of cognitive science: An embodied approach* (pp. 375–395). San Diego, US: Elsevier Publishers Limited.
Zhang, J. (1997). The nature of external representations in problem-solving. *Cognitive Science, 21*, 179–217.

Analogy as Categorization: A Support for Model-Based Reasoning

Francesco Bianchini

Abstract Generally speaking, model-based reasoning refers to every reasoning that involves model of reality or physical world, and it is especially involved in scientific discovery. Analogy is a cognitive process involved in scientific discovery as well as in everyday thinking. I suggest to consider analogy as a type of model-based reasoning and in relation with models. Analogy requires models in order to connect a source situation and a target situation. A model in an analogy is required to establish salient properties and, mostly, relations that allow transfer of knowledge from the source domain to the target domain. In another sense, analogy is the model itself, or better, analogy provides the elements of model of reality that enable the processes of scientific discovery or knowledge increase. My suggestion is that some insight on how an analogy is a model and is connected to model-based reasoning is provided by recently proposed theories about analogy as a categorization phenomenon. Seeing analogy as a categorization phenomenon is a fruitful attempt to solve the problem of feature relevance in analogies, especially in the case of conceptual innovation and knowledge increase in scientific domain.

Keywords Analogy · Model-based reasoning · Concepts · Categorization · Models · Scientific discovery

1 Introduction

Analogy is a kind of model-based reasoning, at least in some senses. Model-based reasoning involves models that are used in inferences, from a formal and logical point of view (for example, the traditional expert systems of artificial intelligence[1]).

[1]See Russell and Norvig (2010).

F. Bianchini (✉)
Department of Philosophy and Communication Studies,
University of Bologna, Via Zamboni 38, 40126 Bologna, Italy
e-mail: francesco.bianchini5@unibo.it

L. Magnani and C. Casadio (eds.), *Model-Based Reasoning in Science and Technology*, Studies in Applied Philosophy, Epistemology and Rational Ethics 27, DOI 10.1007/978-3-319-38983-7_13

But, in a wider sense, model-based reasoning refers to every reasoning that involves models of reality or the physical world, and it is especially involved in scientific discovery (Magnani et al. 1999). Analogy is a cognitive process involved in scientific discovery as well as in everyday thinking. Analogy requires models in order to connect a source situation and a target situation. Roughly speaking, a model is required in an analogy to establish salient properties and relations that allow the transfer of knowledge from the source domain to the target domain. In another sense, analogy is the model itself, or better, analogy provides the elements of the model of reality that allow the processes of scientific discovery or increase in knowledge to take place. My suggestion is that some insight into how an analogy is a model and is connected to model-based reasoning is provided by recently proposed theories on analogy as a categorization phenomenon, which are consistent with a more general cognitive thesis according to which analogy-making is categorization *and* categorization is analogy-making (Hofstadter and Sander 2013).

To support my claim, in the next sections I deal with the relationship between analogy and models in general (Sect. 2); I then comment on the connection between categorization and concept creation (Sect. 3) and I illustrate the relationship between analogy-making and categorization, explaining how we can speak about analogy-making as categorization (Sect. 4). Lastly, I combinc analogy-making as categorization and models to clearly describe how to consider the role played by analogy in model-based reasoning (Sect. 5). I draw general conclusions of these connections in the last section (Sect. 6).

2 Analogy and Models

Two preliminary questions seem to be relevant and strictly related to our discussion: how is analogy connected to models? How is analogy connected to concepts? They imply, however, a more general question: how to define analogy?

There are many answers to such a question, according to different ways to see analogy. We may say that analogy is: a cognitive process, a transfer of (semantic) information, an inference—actually an inductive inference—from a formal point of view, and a type of argument—argument by analogy—from an informal viewpoint. But, from a conceptual standpoint, analogy can also be regarded as a similarity relation, or a structure alignment, or an abstraction process. Finally, we can also say that analogy is a particular kind of model-based reasoning, if we consider the known situation as the model for the analogical unknown situation; or even, an analogy can be regarded as a model itself, insofar as it involves a model structure rooted in the traditional, ancient meaning of analogy, that is, the meaning of proportion, stemming from a mathematical view of analogy as the relationship between ratios. All these elements concur to structure the concept of analogy, which makes it very hard to find a single definition—we could call it the hard problem of analogy.

A main distinction is between analogy as a fact and analogy as an act. In the former case, we may consider every analogy as an individual *fact*—an inference, an argument, an established similarity relation—that we can analyze *ex post* in order to evaluate its accuracy or goodness, or likewise to justify it, especially in the scientific context. In the latter case, we may see every analogy as an individual *act* (of thought, of reasoning), whose rising and happening we have to explain, so as to understand the *ex ante* aspects leading to the specific analogy. Usually, an analogy as a fact is mostly a matter of logic and it is studied as the outcome of analogical reasoning, in an attempt to provide a justification of it or to formalize it.[2] On the other hand, analogy as an act has been studied as a cognitive process, to model it or to explain the creative process of analogy building.[3] What's more, these two ways of dealing with analogy are usually rather separate, because they have distinct methods and aims, and they seem to be two different things and not one and the same thing. The result is that it appears unavoidable to sacrifice the understand of analogy as a fact if we want to explain the act of analogy, and vice versa—we could call it the entanglement of analogy.

How can we combine them? Maybe, by connecting analogy and models through concept creation so as to encompass the process and the outcome of analogy within a single perspective. We will see in the following sections in which sense analogy may be a support for model-based reasoning, but we first need to consider the relation between analogy and models.

For our purposes, a model is a simplified reproduction of a real phenomenon, in which relevant elements are structured to study the phenomenon itself by manipulating its elements and parameters. The key point is the *relevance* of features constituting the model. Analogy involves relevance as well. In fact, one or more relevant features of a known situation or domain are connected to corresponding features of an unknown or less known situation or domain to establish new features in it. So, the known situation or domain is modeled—i.e. becomes the model—for the unknown one, producing new knowledge through the transfer of the old knowledge. This is true both in logic and in the computational/cognitive approach to analogy. So, the informal argument by analogy (Baronett 2012) is:

A is similar to *B* in certain respects
A also has the feature P

$\Longrightarrow$ *B* has the feature P

(plausibly or with some degree of support)

where the "certain (relevant) respects" plus P of A are the model for B. In a more formal fashion, the structure of analogy as inductive inference is:

[2]A recent study on such a topic is Bartha (2010).

[3]See for example Holyoak and Thagard (1995), Gentner et al. (2001).

$$P_1(x) \wedge P_2(x) \wedge P_3(x) \wedge \ldots \wedge P_n(x)$$
$$P_1(y) \wedge P_2(y) \wedge P_3(y) \wedge \ldots \wedge P_n(y)$$
$$Q(x)$$
$$\Rightarrow Q(y)$$

where x and y are objects and P and Q are properties. For example, this kind of inference is used to ascribe properties to a species, or a medical substance, starting from the known properties. Generally speaking and with reference to a logical standpoint, we may characterize the *analogy as the use of a model to extend the application domain of a specific property*.

From another perspective, analogy itself is a model, whose structure involves four elements. In the classical meaning of analogy as a proportion, two ratios are compared. The typical scheme of analogy of ancient and medieval thought is: A : B = C : D.[4] In modern terms, analogical problems of this type require that D be found. So, for instance, problems like "Book : Reader = Instrument : ?" or "Sun : Summer = Snow : ?" most likely have solutions such as "Player" and "Winter" respectively.[5] The solution is just likely and not certain, because it depends on the context in which the analogy problem is solved, assuming that the context of four elements, or three plus a fourth to be found, can be indeterminately complex. The fact that the solution is not certain is the reason why analogy has become a kind of inductive inference in modern logic, whereas the four elements pattern underlies contemporary computational and cognitive approaches to analogy, especially for issues raised by the difficulty in establishing the relevance of features involved in an analogical cognitive process.

A model of analogy based on the four elements structure is provided by Hesse (1966) in her tabular representation. In this model, there is a distinction between a source domain (S) and a target domain (T), which subsequently became typical of cognitive approaches as well. S and T are in a horizontal relation, whereas the list of traits of each domain is the vertical part of the pattern. A new correspondence can be inferred between a feature of S and a feature of T from the known similarities between traits. This model can be described as being tabular in shape, or as a four-element pattern whose vertical and horizontal relations are arranged in a square form. The corresponding elements of the two domains are objects, properties, relations, functions, roles, etc.

Tabular representation can be regarded as a model of analogical inference, but also of analogy as a cognitive process. For example, in French (1995) this model is meant to capture the process of analogy building from a dynamic standpoint. So, from S an abstract scheme is drawn and then its variables are replaced in the process of conceptual slipping leading to a new abstract scheme, whose variables are re-constrained to attain the situation in T. In the model, the process of abstraction

[4]See Prior Analytics by Aristotle 69a1 (Aristotle 1984).

[5]For one of the first computational approach to this kind of analogy problems see Evans (1968).

and its opposite are vertical, the process of conceptual slipping is horizontal and corresponds to semantic information transportation, producing new knowledge. The process of abstraction and the relation between abstract and concrete are typical in cognitive approaches to analogy. One of their purposes is to explain how an analogy is created or built and how the contextual pressures cause the emerging of analogy relevant elements.[6] And all these processes involving abstraction involve *concepts* as well.

Models also involve concepts, but analogy seems to involve other features shared with models: representations, relational structures and their systematicity, relations of different kinds within the S and T domains (logical, causal, explanatory, functional, and/or mathematical relations), the salience of features upon which both of them—a model and an analogy—are built. For example, the SME traditional cognitive model of Gentner finds analogy in different domains (such as the atom and solar system) by exploiting the abstract relations of two domains ("greater", "cause", "attracts", etc.), which are in this way modeled and aligned by the program (Falkenhainer et al. 1989). The SME model has been criticized because it uses pre-built (by a programmer) representations. Subsequent connectionist and hybrid models have been designed to capture the dynamicity of analogical cognitive process, in particular by building their representations of elements of two domains (Chalmers et al. 1992), or by exploiting the dynamical activation of the nodes in a semantic network, as occurs in the multiconstraints satisfaction theory of analogy of Holyoak and Thagard (1995).[7]

These two main computational approaches to analogy stress different traits of the analogy phenomenon. On the one hand, computational models based on prebuilt representation want to capture the structured, deep-relation-rooted nature of analogy, its connection with inferential process, the justification of the analogy as the outcome of the process itself, and the definition of relevant and essential features of situations involved in an analogy. On the other hand, computational models based on dynamical and autonomous—by the program—representation building try to capture the spontaneity of analogy, the pressures of context, the abstraction processes and the analogy creation, without considering involved domain structures alone, but the duality of superficial and deep features involved in the analogy building. If we consider models as being involved in analogies, we may say that they share the same condition and that both are subjected to the same dichotomies. So, the opposition, in different approaches, between a given analogy and analogy building corresponds to the one between a given model and model building. In the former case, we may have an *ex post* explanation of an analogy phenomenon as well as models involved in the analogy; in the latter case, we may have an *ex ante* explanation of an analogy phenomenon as well as of the building of the models

[6]For a list of different cognitive approaches to analogy, see Kokinov and French (2003).

[7]There are many computational models of analogy. A close examination is beyond the scope of this paper. For an introduction and a discussion of the cognitive processes involved see Hall (1989) and Kokinov and French (2003).

involved. The latter case is more relevant to our purposes because it is connected with concepts and concept formation, even because finding a model in an analogy overlaps using a model to make an analogy. Below, we will see an interesting proposal that illustrates this point.

3 Analogy, Models and Concept Formation

A noteworthy theory on analogy, model-based reasoning and concept formation (in scientific domain) is the one formulated by Nersessian (2008). The main idea is that the traditional S-T scheme of analogy as a cognitive process has to be supplemented as follows: the transfer of information is not direct from source (S) to target (T), but there is a model mediation between S and T. This model mediation is due to a *hybrid model* in which connection crossing of the S and T elements takes place (Nersessian 2008; Nersessian and Chandrasekharan 2009). In this way, the hybrid model, comprised of constraints from the S and T domains, can be exploited for further refining of the model itself, on the basis of the S and T constraints *and* the constraints provided by the hybrid model. The process is dynamical because it is grounded on an increase in the number of constraints that leads to a clarification of the S-T relationship and to the solutions of the analogy problem. From this point of view, analogy is seen and explained as a mechanism of conceptual innovation.

Nersessian and Chandrasekharan (2009) describe the process that led to conceptual innovation during a research aimed at understanding undesired spontaneous "bursts" in an *in vitro* model of cortical neural network activity, which are phenomena that do not occur *in vivo* properly functioning animal brains. The *in vitro* model, named "the dish", was stimulated using different electric signals and was connected to robot devices and visualized animats moving around in simulated computational worlds. The aim of the experiment is to control this embodied cortical neural network. Since the *in vitro* model had constraints that were relevant features of real neurons and the neuron network, the dish was already an analogy, but was not enough to attain the goal. Thus, an *in silico* model, a computational neural network, was built by using constraints from the neuroscience domain and from a generic dish: it was a hybrid model. Different and more refined versions of the computational model were built until a visualized activity of a version allowed the novel notion of "center of activity trajectory" to be developed. At that point, this last version of the computational model was able to replicate the activity of the dish. The greater controllability of the computational model allowed many experiments to be conducted with the dish, thanks to the potential transfer from the *in silico* model to the *in vitro* model. So, what was first developed for the computational model was then developed, by analogy, for the *in vitro* model.

One of the most important outcomes of the whole process was the emergence of a new concept, the "center of activity trajectory". It was a consequence of the mapping between the two models (an analogy of analogies, if we consider the two models according to their analogical nature) and the visualization process.

In particular, the hybrid models, the dynamical and incremental process of (re)-building the computational model that integrates constraints from sources, targets and models themselves, are what allows the emergence of new structures, behaviors and, eventually, new concepts.

The hybrid model theory is very interesting for several reasons. First of all, it does not reject the traditional scheme of analogy and its standard cognitive explanation (Gentner 1983). Second, even though Gentner's structure-mapping model is based on the formal connection of structures of relations, which are considered unavoidable for analogy, hybrid analogy theory also assumes that semantic and pragmatic aspects are equally important, just as they are in the multiconstraints satisfaction theory of analogy of Holyoak and Thagard (1995).[8] In the hybrid analogy theory, the understanding, interpretation and goals of an analogy problem drive the selection of relevant constraints in the incremental process of hybrid model development, thereby contributing, together with the syntactical aspect, to the emergence of solutions as well as of new concepts. Third, the stress on representation building (Nersessian 2008) implies that the way in which the model representations are built is a fundamental issue that needs to be dealt with, as the constraints of T determine which constraints of S are (potentially) relevant to analogical comparison. This is the big issue of creative analogies: pinpointing relevance. The incremental process of model representation building is an attempt to deal with such a fundamental problem. As we said above, although the problem of representation building was not central in the literature on analogy and analogical reasoning in past decades, we should bear in mind that there are some hints of it in computational modeling of analogy. Hofstadter and The Fluid Analogies Research Group (1995) provide some interesting reflections on this topic in their criticism of ready-made representations of the SME cognitive model (Falkenhainer et al. 1989) based on the structured mapping theory by Gentner, which is nevertheless valid for the systematicity principle of relation interconnection in analogy explanation. Representation-building and transfer are two further processes that Kokinov and French (2003) added to the four standard approaches to the computational modelling of analogy: recognition, elaboration, evaluation, consolidation (Hall 1989). Chalmers et al. (1992) addressed the problem from the artificial intelligence methodology point of view, arguing that the only way to understand cognitive process is to consider representation as the outcome of a continuous, dynamical process of high-level perception, concerning both modal and amodal aspects in relation with concept representation. This is especially true for analogy explanation.

To summarize the important features of the hybrid models theory, which in Nersessian and Chandrasekharan give rise to hybrid analogies, we may say that hybrid models (a) allow creative analogies—i.e. analogies between analogies, which are the models used in experimental laboratory processes; (b) are used for

[8]See the description of the computational model based on the simultaneous satisfaction of a set of semantic, structural and pragmatic constraints, and the description of the ARCS program, in Thagard et al. (1990).

reasoning purposes; (c) are only models and not real world entities; (d) allow visual, imagistic, simulative and manipulative processes; (e) and lead to conceptual innovation and to new concepts. For the purposes of our investigation, a final remark by the authors is very noteworthy: "although our case might be considered extraordinarily creative, our intuition is that if analogy use 'in the wild' were to be studied systematically, the construction of such intermediary hybrid representations, making use of visualization and mental simulation, would be seen to be significant dimensions of *mundane* usage as well" (Nersessian and Chandrasekharan 2009: 187 [emphasis added]). We will now consider another proposal that connects analogy, concepts and categorization.

4 Analogy as a Kind of Categorization

When considering the relation between analogy and categorization, two elements from the previous sections are needed to establish to what extent hybrid models are connected to the categorization process: (i) the idea that a dynamical and purpose-oriented representation building process is fundamental in the model-based reasoning involving analogy, and (ii) the fact that concept innovation is strictly linked to concept formation, and thus to categorization as a sort of concept formation. The idea of analogy as categorization is not new. For example, Glucksberg and Keysar (1990) see metaphors as class inclusion statements and argue that understanding metaphors means understanding such statements, which are categorizations or, rather, category attributions. Research on teaching science subjects and the role of analogy shows that analogy creation is different from analogy interpretation and that analogies and analogy creation can be seen as categorization phenomena (Atkins 2004).

Starting from the psychological evidence that mechanisms underlying analogy and conceptual processes, especially categorization, are very similar,[9] Dietrich (2010) tries to unify analogy-making and categorization by showing that analogy-making is based on construing, which is a kind of categorization. He claims that, besides the usual incremental representation building process, which is typical of the reasoning phase, two further representational processes are present in the analogy-making phase. The two steps of the analogy-making process are rapid abstraction and construing, the latter being a type of categorization. In relevant and insightful cases of analogy construing involves very different semantic analogs.

An example of the rapid abstraction process is modeled by STRANG, a computational model that makes analogies in the letter string domain (Dietrich et al. 2003).[10] The STRANG program uses a grammar to pack strings of letters so as to

---

[9]See, among others, Ramscar and Yarlett (2003).

[10]For a discussion on letter-string domain and COPYCAT, one of most important computational models of analogy in this domain, see Mitchell (1993).

find an abstract structure that is equivalent in two strings: Target (T) and Base (B). Target is the input string and Base is the string in the long-term memory. For example, if T is ababccc and B is mnopqrhijhijhij, the program produces the following outcome: (((ab)(ab))(ccc)) and (((mno)(pqr))((hij)(hij)(hij))), putting together the two strings according to the abstract description "two same-length sequences followed by a 3-item repeating string" (Dietrich 2010). The process modeled by STRANG is an example of rapid abstraction, which is particularly interesting insofar as the program associated with this process can violate the grammar rules to create packages of letters that are not the results of a direct application of a rule, but rather the continuous application of rules until the program finds a general representational abstraction connecting T and B. Significantly, B is a *model* in the long-term memory.

What is not modeled in this version of STRANG is the second step, based on semantic distance. According to Liberman and Trope's theory on the relation between psychological and semantic distance and the abstraction process,[11] psychological distance induces abstraction.[12] So, if we start from a point of origin, a situation we are dealing with here and now, and we are provided with a *relevant psychological distance*, we can make an analogy or, in some circumstances, analogies may arise spontaneously. The relevant psychological distance enables the second rapid representational abstraction, according to a dynamical process involving a passage from concreteness to abstraction: "*concrete* representations are less structured, more contextualized, and contain more information in the form of incidental features. Higher level *abstract* representations are schematic, decontextualized, and tend to represent the *gist* of an object or event by focusing on core features and omitting incidental information" (Dietrich 2010: 338). For example, if I am thinking of my home, an analogy may arise with my country because of the psychological distance between them, which is nonetheless based on relevant abstract shared features, such as organization or place-where-I-live. The analogy stems from a representational change from concrete to abstract. In the domain of scientific discovery, this kind of explanation should hold also as regards, for example, an analogy between the Rutherford-atom model and solar-system model because of their semantic and psychological distance.

There are some problems in this view. For instance, how can we explain semantic distance without being too vague? And what makes what is relevant in an analogy relevant? If we consider the point of origin alone, we do not obtain an explanation for why some specific concepts are retrieved and other semantically distant concepts are not. To solve the semantic problem, Dietrich inverts the perspective and considers analogy-making as categorization, stating that the construing process is a process leading to meaningful categorization, which in two steps produces an analogy; and this is true for every analogy. A first construal is attained

[11]See Liberman and Trope (2008) and Liberman and Förster (2009).

[12]Even though this aspect of theory is problematic and presents some weaknesses. For a discussion see Dietrich (2010).

through a process from an initial visual stimulus to integration, by means of a mapping process, between perceptual elements and a retrieved category. The construal turns out to be the meaningful categorization of a perceived object. This is a transition from a semanticless stimulus to a semantic meaningful mental representation (in the preceding example, my home). The first construal is the base—the point of origin—for another (meta-)construal, attained by retrieving a semantic distant category (in the example, my country). Only at this point is the analogy complete and is it possible a work on its details, which, according to Dietrich, can be properly called "analogical reasoning".

Even if this attempt to unify analogy-making and categorization has some problems, it is interesting because it connects analogy-making and new knowledge production through concept exploitation. The construal is a form of categorization. It is clear in the first step of the construing process. However, if semantic connection between distant categories is also a construal, we have to conclude that this is another case of categorization, and consequently that analogy in the proper meaning of connection of abstract features between two distant domains—whose gift is what is relevant—is a sort of categorization. So, categorization turns out to be the combination of shared features within a new conceptual structure at an abstract level. In other terms, it is a dynamical building of a representation, which is, in fact, the analogy and eventually leads to the detailed analogical reasoning work.

5 Analogy as Categorization, and Its Consequences

Dietrich's view is noteworthy because it is an attempt to hold together cognitive and, at least in part, logical aspects of analogy. Another theory that goes in depth in dealing with analogy and categorization is based on the idea that "the spotting of analogies pervades every moment of our thought, thus constituting thought's core" (Hofstadter and Sander 2013: 18). In Hofstadter's perspective, it stems from seeing analogy as deeply intertwined with the process of high-level perception and representation building. It is also part of a general theory according to which analogy is the core and the essence of cognition (Hofstadter 2001). In other terms, analogy as analogy-making is what allows the general dynamics of cognition, by being an integral part of perceptual and representational processes, reasoning, learning, memory and language. It also underlies what is usually and standardly considered as analogy, the correspondence between an S and a T domain, and it is very present in the creativity process, scientific discovery, decision making, concept formation and categorization—in fact, it itself is categorization. This main thesis can be divided in two sub-theses: (1) analogy (making) is categorization; (2) categorization is analogy (making). Both of them are implied in the general idea that concepts are (formed by) analogies. In the rest of this section, I will try to explain this claim, proposing a theory of conceptual extension that is consistent with Hofstadter and Sander's view.

Analogy-making implies memory retrieval as a fundamental part. Every cognitive process is, at its core, due to a central cognitive loop that works this way:

"a long-term memory node is accessed, transferred to short-term memory and there unpacked to some degree, which yields new structures to be perceived, and the high-level perceptual act activates yet further nodes, which are then in turn accessed, transferred, unpacked, etc., etc." (Hofstadter 2001: 517). Emphasis on memory and memory retrieval is found in subcognitive models,[13] and it is in line with other general cognitive architectures. So, one root of this idea clearly lies in cognitive modelling, especially in the traditional symbolic approach, such as the total cognitive system scheme used by Allen Newell to explore his attempt to find his own unified theory of cognition (Newell 1990). The main features of the total cognitive system are a long-term memory, with different sub-processes, connected to a working memory interacting with the external environment by means of perceptual systems (i.e. the input of the systems) and motor systems (i.e. the whole system behavioral output), which can also be part of the input.

More interestingly, a second root of Hofstadter's theory lies in concept theories. For example, Barsalou asserts that "Rather than being retrieved as static units from memory to represent categories, concepts originate in a highly flexible process that retrieves generic information and episodic information from *long term memory* to construct temporary concepts in *working memory* […] This concept construction process is highly constrained by goals…[and]…context…" (Barsalou 1987: 101). There is a huge body of literature on the central role of context, similarity and dynamical concept development dating from the 1980s and 1990s.[14]

Starting from subthesis 1, which asserts that analogy is categorization, and according to Hofstadter and Sander's arguments, I will try to show that analogies involve conceptual extension, and two kinds of conceptual extensions in particular: vertical extension (VE) and horizontal extension (HE). Hofstadter and Sander refer to them as vertical category leap and horizontal categorical broadening, respectively, but this distinction is not so clear-cut in their theory because in many cases "we see that there is no sharp line of demarcation between vertical category leaps and horizontal category extensions" (Hofstadter and Sander 2013: 468). It seems to depend on the interpretation of analogy and on the context pressures considered in the explanation of an analogy. So, as contextual pressures are connected to the problem of relevance, I will propose a dynamical pattern through which analogy is produced in human mundane and scientific thought. My argument from sub-thesis 1 will proceed in this way:

1. Categorization is concept formation
2. Concept formation is concept extension (or broadening)
3. Concept extension is on a horizontal or vertical level
4. Analogy is categorization

[13]Such as Copycat, Metacat, Tabletop, Letter Spirit; see Hofstadter (1995), Mitchell (1993), French (1995), Marshall (2006).

[14]See, among others, Barsalou and Medin (1986) and Goldstone (1994).

A first conclusion is that:

5. Analogy is *both* on a horizontal *and* vertical level

from which we can draw a second conclusion:

6. Analogy involving *abstraction* (VE move) is always analogy between analogies (HE move).

Let's start with an example to show how concept extension, and thus concept formation, works. The development of concepts from childhood to adulthood is an enrichment process that usually leads a concept from a single-member to a cloud of concepts (Hofstadter and Sander 2013: 37).[15] Consider a very common concept: father. First a child get to know her/his Daddy—with the capital letter, as there is just *one* daddy as far as that child knows. Upon finding out that other children have their own daddy, the child develops the concept of daddy—with the small letter, as the concept refers to many different people, i.e. daddies. Then the child can learn that there is a more objective sense of the concept, and finds out that her/his daddy belongs to the category "father". Afterwards, the concept may be extended to embrace other forms or kinds of fatherhood, which are more distant from the initial core of the concept, such as adoptive father, father-in-law, father-to-be, father of mathematics, Fathers of the Nation, Fathers of the Church, etc., in a hierarchical level from the center to the periphery of the cloud.[16] During life our concept of father extends without ever reaching a final boundary, as it can always be extended further in an increasingly metaphorical way. Indeed, we place within the same category an increasing number of different categories and instances by analogy, that is by exploiting similarities at levels of varying abstraction. This is why analogy produces categorization and, in the end, analogy is categorization. Such an endless process is the vertical extension, which enlarges a concept and gives rise to new concepts by adding parts of its meaning.

Now, let's consider another concept, which is related to the concept of father: mother. We may imagine the same conceptual development and, consequently, a similar (vertical) extension: from Mommy to mommy (because there are mommies), then to mother, and afterwards to adoptive mother, nursing mother, surrogate mother, mother-in-law, mother-to-be, mother earth, mother country, Mother Nature, Mother Church, mother tongue, etc. If we compare these two concepts, we make another kind of conceptual extension, which leads to conceptual innovation or formation. For example, we can connect mother and father and discover the concept of parent (it is very likely something that happened many years ago in our life, when we were six or seven). And it is highly likely that we have, at a certain point later in our life, made another horizontal extension, connecting adoptive father and adoptive mother, discovering the concept of adoptive parent, under the pressure of

[15]On conceptual development, see Rakison and Oakes (2008), Carey (2009).

[16]Semantic and local neural networks can be used to model hierarchical and heterarchical structures of concept clouds in their dynamical activation pattern; see, among others, Mitchell (1993).

context of our familiar and social environment, as in the first case. The extending process has no an end point, and we may imagine that, for instance, the concepts of Fathers of the Nation and mother country, which are a long way from the core terms "mother" and "father" in our hierarchical cloud structure, may sooner or later be connected, yielding the new concept of "Parents of the Land", or rather, an *unlabeled-by-a-single-word concept* that is subsequently named "Parents of the Land". This is how horizontal extension works, i.e. relating things that are considered at the same level by analogy so as to produce new knowledge, new concepts, and sometimes new words or phrases (even though concepts do not need single word denotation to be concepts). In short, horizontal extension is a conjunction that produces conceptual innovation by unification, and thus new concepts that may be vertically extended.[17]

By combining the two kinds of extension, we have another version of the starting analogy model, the one based on four elements. VE is analogy as categorization. In the dynamical process of analogy-building and supposing we do not have two parallel concepts, but we have to find an analogical one, or rather, another concept/domain/situation that is analogous to the initial one, we choose a superordinate category in which we want to include something (a fact, a situation, an element). We see it as a member of that particular category, that is, as analogous to other members of the same category for features that are relevant in the context of analogy we are making. After this first step, features enable other features that guide the search for something parallel (a fact, a situation, an element) in another domain inside the general category we have chosen. The HE give rises to concept innovation or a new concept, which becomes the concrete base for other VE. This is the sense in which HE is an analogy between analogies, an analogy between different things that are categorized in the same way. The HE allows the emerging of a new category core, which becomes the concrete level for new abstractions (new VEs).

So, the process of analogy-making, based on the four-element model, proceeds in a dynamical way with an alternation of VE and HE (VE–HE–VE–HE–VE–HE …), which mirrors the alternation between concrete and abstract elements. Both are required to build the analogical correspondence leading to concept formation or innovation, as every new conceptual correspondence, which is relevant to the analogy-making process, is a new concept, in which the analogy, so to speak, introduces "old", known concepts by categorizing them in a new way. In the dynamical interaction between concreteness and abstraction, abstraction lies on the second level and requires a first level of concreteness. The four elements involved in the process generate an overall model with all the relevant concepts involved, in which some of them are the concrete for the others concepts involved. In the supervised analogy-making process, as occurs in those in scientific discoveries, establishing what is concrete and what is abstract depends on the constraints chosen

[17]The notion of "unification" in language and semantic context has been emphasized, among others, by Jackendoff (1997).

each time for the overall model building. In mundane contexts, the relative concrete and abstract features are an outcome of perceptual and memory retrieval processes.

Although the way in which Hofstadter and Sander develop their theory is not comparable, they provide support for the dynamical explanation of the analogy-making process, especially in scientific discovery. They stress the prominence of concrete/abstract relation and alternation in mathematical progress, especially as regards complex and imaginary numbers: "it would be hard to overstate the importance of geometrical *visualization* in mathematics in general, which is to say of attaching geometrical interpretations to entities whose existence would otherwise seem counterintuitive, if not self-contradictory. The acceptance of abstract mathematical entities is always facilitated if a geometrical way of envisioning them is discovered; any such mapping confers on these entities a *concreteness* that makes them seem much more plausible" (Hofstadter and Sander 2013: 443 [emphasis added]).[18] They extend these remarks to mathematical discovery in general: "The modus operandi of mathematical abstraction is [...]: you begin with a "familiar" idea (that is, familiar to a sophisticated mathematician but most likely totally alien to an outsider), you try to distill its *essence*, and then you try to find, in some other area of mathematics, something that shares this same distilled essence. An alternative pathway towards *abstraction* involves recognizing an analogy between two structures in different domains, which then focuses one's attention on the *abstract structure* that they share. *This new abstraction then becomes a "concrete" concept that one can study*, and this goes on until someone realizes that this is far from the end of the line, and that *one can further generalize the new concept* in one of the two ways just described. And thus it goes..." (Hofstadter and Sander 2013: 449 [emphasis added]). The "two ways just described" can be seen as two ways of regarding the dynamical VE–HE–VE... described above. Thus, the formation of the concrete concept appears to be closely related to a sort of "affordance" of the abstraction process in category alignment, as if the abstraction almost spontaneously emerges from the situation we are faced with when we apply our four-element model of analogy.

This model also guides the choice of level structure that is relevant to the analogical process. HE is on the same level while VE is on two different levels, but the level selection is guided by context pressures and dynamical model building. For example, we may place a leg and an arm on the same level and consider them as playing the same role in an analogical situation (for instance, a diagnosis process); by contrast, a leg and a limb are on a different level because a leg is a member of the category limb, as it is an arm, and the step from a leg to a limb is a VE, an abstraction process due to context pressures. But nothing prevents us from conceiving a situation in which an arm and a leg are on different levels and the leg is a general category to which the arm and other things belong (for instance, a tale about people that move on all fours).

[18]For a similar treatment of this topic from the point of view of conceptual blending see Fauconnier and Turner (2002: 270–274).

Hofstadter and Sander provide many examples of such analogies in physics by trying to reconstruct the Einsteinian analogies in his processes of discovery. For example, in the extension of the Galileian principle of relativity to special relativity, Einstein made vertical and horizontal mental moves (Hofstadter and Sander 2013: 465–468), which can be schematized as follows:

1. Principle of Relativity in Mechanics (from Galileo)
2. Mechanics <==> Electromagnetism (HE)
3. Mechanics Λ Electromagnetism
4. Physics
5. Principle of Relativity in Physics (VE by unification)

where the HE step, the correspondence between mechanics and electromagnetism, leads to their unification, which is equivalent to physics (the step from 3 to 4), which in turn gives rise to the result of extending the principle of special relativity to any kind of physical experiment, for example by asking oneself how optical and electromagnetic phenomena behave in motion (actually, in the same way as in rest). And the analogy was made explicit by Einstein himself.[19] Likewise, the step from special relativity to general relativity can be seen as the outcome of another analogous extension, involving the indistinguishability of an accelerating reference frame from a non-accelerating reference frame as regards, first, any kind of mechanical experiments, and then any kind of physical experiments.[20]Another extension—actually, two VEs from mechanics to physics and from non-accelerating reference frames to every reference frame—gives rise to another analogy as categorization, which forms the basis of a new discovery and a new theory.

6 Conclusion

In this paper, I have put together some ideas about models, analogy and concepts in an attempt to show how the model of analogy works and can be understood in the logic and cognitive fields, in what way models are part of analogy, and how analogy-making and analogical reasoning are consequently a kind of model-based reasoning.

In the first part, I discussed the distinction of analogy in logic and cognitive science, showing that these different fields of research share the same model, which is based on four elements. But while logic deals with analogy from a static viewpoint, cognitive science has become increasingly interested in the dynamical

[19]"That a principle of such broad generality should hold with such exactness in one domain of phenomena, and yet should be invalid for another is a priori not very probable" (Einstein 1920: 17).

[20]Through some thoughts experiments, such as the space lab pulled by a rocket and the ray of light crossing an accelerating lab in a gravitational field. For a discussion see Hofstadter and Sander (2013: 490–495).

processes underlying analogy-making. In the second part, I discussed Nersessian's proposal of hybrid analogies involving incremental model building, which leads to conceptual innovation. Models exploited by analogy in scientific experiments are, in fact, conceptual structures built by means of constraints stemming from different domains involved in the analogy process. In the third part, I debated the idea that analogy-making is based on a construal process, a type of categorization, which has the consequence of unifying, at least partially, analogy and categorization. In the fourth part, I discussed Hofstadter and Sander's theory that analogy is always categorization and vice versa, a very general cognitive process concerning every conceptual cognitive process, from high-level perception to scientific discovery. I have tried to show how this theory is based on concept extension in two different perspectives, VE and HE, each of which is a different form of analogy as categorization. I have also tried to provide a model of dynamical development of relation between concreteness and abstraction, which is involved in the analogy-making process as well as in concept innovation or formation. This dynamical development is not described in the same way in Hofstadter and Sander's theory, though it is consistent with it.

The two kinds of conceptual extensions are consistent with the general four-element model of analogy, involving horizontal and vertical levels of correspondence designed to capture the relationship between concrete and abstract, which is unavoidable in an analogy, even in relative terms. They are also involved in dynamical process of concept innovation and concept creation. This is especially true for HE, which follows the VE process of abstraction and is how new concepts are formed, concepts which in turn yield new abstraction processes. This is why I have claimed that HE produces an analogy between analogies, namely categorizations. Conceptual structures involved in the two kinds of extensions are based on a hybrid model, a conceptual representation built from two different domains involved in the analogy-making process. Model mediation by hybrid models is consequently always categorization (concept innovation or formation) involving analogy between analogies of different domains in the HE–VE–HE… dynamics.

Many things have yet to be understood regarding analogical processes and analogy, the connection between analogy as inference and analogy as dynamical/representational/semantic process, and the way in which concepts and conceptual structures are involved in this capability of reasoning and thought. Difficulties are also encountered when trying to conceive suitable experiments to get an insight into the range of problems raised by analogy. This may, along with the construction of cognitive models and architectures that include the problem of categorization, represent an interesting challenge for further research. Finally, it could be successful dealing with this set of problems in the framework of situated cognition and external representations (Magnani 2009), at the same time trying to explain how we use stored knowledge for producing freshly knowledge, understanding physical world situations and transferring external entities in internal representations to have new abstract models, which nevertheless we use in interaction with external perceptions and representations.

References

Aristotle. (1984). In J. Barnes (Ed.), *The complete works of Aristotle*. Princeton: Princeton University Press.

Atkins, L. J. (2004). Student generated analogies in science: Analogy as categorization phenomenon. *International Conference of the Learning Sciences, June*.

Baronett, S. (2012). *Logic* (2 ed.). Oxford: Oxford University Press.

Barsalou, L. W. (1987). The instability of graded structure: Implications for the nature of concepts. In U. Neisser (Ed.), *Concepts and conceptual development: Ecological and intellectual factors in categorization* (pp. 101–140). Cambridge: Cambridge University Press.

Barsalou, L. W., & Medin, D. M. (1986). Concepts: Static definitions or context-dependent representations? *Cahiers de Psychologie Cognitive, 6*, 187–202.

Bartha, P. (2010). *By parallel reasoning. The construction and evaluation of analogical arguments*. New York: Oxford University Press.

Carey, S. (2009). *The origin of concepts*. New York: Oxford University Press.

Chalmers, D. J., French, R. M., & Hofstadter, D. R. (1992). High-level perception, representation, and analogy: A critique of artificial intelligence methodology. *Journal of Experimental & Theoretical Artificial Intelligence, 4*(3), 185–211.

Dietrich, E. (2010). Analogical insight: Toward unifying categorization and analogy. *Cognitive Processing, 11*(4), 331–345.

Dietrich, E., Markman, A., & Winkley, M. (2003). The prepared mind: The role of representational change in chance discovery. In Y. Ohsawa & P. McBurney (Eds.), *Chance discovery by machines* (pp. 208–230). Berlin: Springer.

Einstein, A. (1920). *Relativity: The special and the general theory*. Reprinted in the 100th Anniversary Edition by Princeton University Press, Princeton, 2015.

Evans, T. G. (1968). A program for the solution of a class of geometric analogy intelligence questions. In M. Minsky (Ed.), *Semantic information processing* (pp. 272–277). Cambridge, Massachusetts: MIT Press.

Falkenhainer, B., Forbus, K. D., & Gentner, D. (1989). The structure-mapping engine: Algorithm and examples. *Artificial Intelligence, 41*, 1–63.

Fauconnier, G., & Turner, M. (2002). *The way we think. Conceptual blendings and the mind's hidden complexities*. New York: Basic Books.

French, R. M. (1995). *The subtlety of sameness. A theory and computer model of analogy-making*. Cambridge, Massachusetts: The MIT Press.

Gentner, D. (1983). Structure-mapping: A theoretical framework for analogy. *Cognitive Science, 7*, 155–170.

Gentner, D., Holyoak, K. J., & Kokinov, B. N. (Eds.). (2001). *The analogical mind. Perspective from cognitive science*, Cambridge, Massachusetts: MIT Press.

Glucksberg, S., & Keysar, B. (1990). Understanding metaphorical comparisons: Beyond similarity. *Psychological Review, 97*, 3–18.

Goldstone, R. (1994). The role of similarity in categorization: Providing a groundwork. *Cognition, 52*, 125–157.

Hall, R. P. (1989). Computational approaches to analogical reasoning: A comparative analysis. *Artificial Intelligence, 39*, 39–120.

Hesse, M. B. (1966). *Models and analogies in science*. Notre Dame: University of Notre Dame Press.

Hofstadter, D. R. (2001). Analogy as the core of cognition. In D. Gentner, K. J. Holyoak, & B. N. Kokinov (Eds.), *The analogical mind. Perspective from cognitive science* (pp. 499–538). Cambridge, Massachusetts: MIT Press.

Hofstadter, D. R., & The Fluid Analogies Research Group. (1995). *Fluid concepts and creative analogies: Computer models of the fundamental mechanisms of thought*. New York: Basic Books.

Hofstadter, D. R., & Sander, E. (2013). *Surfaces and essenced. Analogy as the fuel and fire of thinking*. New York: Basic Book.

Holyoak, K. J., & Thagard, P. (1995). *Mental leaps: Analogy in creative thought*. Cambridge, MA: MIT Press.

Jackendoff, R. (1997). *The architecture of the language faculty*. Cambridge, Massachusetts: MIT Press.

Kokinov, B., & French, R. M. (2003). Computational models of analogy-making. In L. Nadel (Ed.), *Encyclopedia of cognitive science* (Vol. 1, pp. 113–118). London: Nature Publishing Group.

Liberman N., & Förster J. (2009). The effect of psychological distance on perceptual level of construal. *Cognitive Science, 33*(7), 1330–1341.

Liberman, N., & Trope, Y. (2008). The psychology of transcending the here and now. *Science, 322*(21), 1201–1205.

Magnani, L. (2009). *Abductive cognition. The epistemological and eco-cognitive dimensions of hypothetical reasoning*. Berlin, Heidelberg: Springer.

Magnani, L., Nersessian, N. J., & Thagard, P. (1999). *Model-based reasoning in scientific discovery*. New York: Kluwer Academic Publisher.

Marshall, J. (2006). A self-watching model of analogy-making and perception. *Journal of Experimental & Theoretical Artificial Intelligence, 18*(3), 267–307.

Mitchell, M. (1993). *Analogy-making as perception*. Cambridge, Massachusetts: The MIT Press.

Nersessian, N. J. (2008). *Creating scentific concepts*. Cambridge, Massachusetts: MIT Press.

Nersessian, N. J., & Chandrasekharan, S. (2009). Hybrid analogies in conceptual innovation in science. *Cognitive Systems Research Journal, Special Issue: Integrating Cognitive Abilities, 10*, 178–188.

Newell, A. (1990). *Unified theories of cognition*. Cambridge, Massachusetts: Harward University Press.

Ramscar, M., & Yarlett, D. (2003). Semantic grounding in models of analogy: An environmental approach. *Cognitive Science, 27*, 41–71.

Rakison, D. H., & Oakes, L. M. (2008). *Early category and concept development*. New York: Oxford University Press.

Russell, S. J., & Norvig, P. (2010). *Artificial intelligence: A modern approach* (3rd ed.). New Jersey: Prentice Hall.

Thagard, P., Holyoak, K. J., Nelson, G., & Gochfeld, D. (1990). Analog retrieval by constraint satisfaction. *Artificial Intelligence, 46*, 259–310.

Part II
Abduction, Problem Solving, and Practical Reasoning

Abduction, Inference to the Best Explanation, and Scientific Practise: The Case of Newton's Optics

Athanassios Raftopoulos

Abstract Hintikka (1997, 1998) argues that abduction is ignorance-preserving in the sense that the hypothesis that abduction delivers and which attempts to explain a set of phenomena is not, epistemologically speaking, on a firmer ground than the phenomena it purports to explain; knowledge is not enhanced until the hypothesis undergoes a further inductive process that will test it against empirical evidence. Hintikka, therefore, introduces a wedge between the abductive process properly speaking and the inductive process of hypothesis testing. Similarly, Minnameier (2004) argues that abduction differs from the inference to the best explanation (IBE) since the former describes the process of generation of theories, while the latter describes the, inductive, process of their evaluation. As Hintikka so Minnameier traces this view back to Peirce's work on abduction. Recent work on abduction (Gabbay and Wood 2005) goes as far as to draw a distinction between abducting an hypothesis that is considered worth conjecturing and the decision either to use further this hypothesis to do some inferential work in the given domain of enquiry, or to test it experimentally. The latter step, when it takes place, is an inductive mode of inference that should be distinguished from the abductive inference that led to the hypothesis. In this paper, I argue that in real scientific practise both the distinction between a properly speaking abductive phase and an inductive phase of hypothesis testing and evaluation, and the distinction between testing an hypothesis that has been discovered in a preceding abduction and releasing or activating the same hypothesis for further inferential work in the domain of enquiry in which the ignorance problem arose in the first place are blurred because all these processes form an inextricable whole of theory development and elaboration and this defies and any attempt to analyze this intricate process into discrete well defined steps. Thus, my arguments reinforce Magnani's (2014) view on abduction and its function in scientific practise.

Keywords Abduction · Inference to the best explanation · Scientific practise

A. Raftopoulos (✉)
Department of Psychology, University of Cyprus,
P.O. BOX 20537, 1678 Nicosia, Cyprus
e-mail: raftop@ucy.ac.cy

L. Magnani and C. Casadio (eds.), *Model-Based Reasoning in Science and Technology*, Studies in Applied Philosophy, Epistemology and Rational Ethics 27, DOI 10.1007/978-3-319-38983-7_14

List of Abbreviations

ADD It stands for "Additional Manuscript", Cambridge University Library
AT It stands for the edition of Descartes' work by Adam and Tannery (Paris: Leopold Cerf 1897). The Latin numeral indicates the volume of this edition and the Arabic number (s) the page (s)
CSM It stands for the translation of part of Descartes' work in English in three volumes by J. Cottingham, R. Stoothoff, and D. Murdoch (Cambridge: Cambridge University Press, 1985)

Hintikka (1997, 1998) argues that abduction is ignorance-preserving in the sense that the hypothesis that abduction delivers and which attempts to explain a set of phenomena is not, epistemologically speaking, on a firmer ground than the phenomena it purports to explain; ignorance is preserved and, thus, knowledge is not enhanced until the hypothesis undergoes a further inductive process that will test it against empirical evidence. (A set of phenomena are known to occur and, thus, the sentences that state them are known to be true, while the truth-value of the sentence stating the hypothesis is not known. In this sense the phenomena are on a better epistemological ground than the hypothesis. Even though we know that the phenomena occur, however, we do not know their causes and this means, in the venerable philosophical tradition since the pre-Socratic philosophers, that we do not really know them. This is the ignorance related to the phenomena.) Hintikka, therefore, introduces a wedge between the abductive process properly speaking and the inductive process of hypothesis testing, a distinction that, Hintikka thinks, can be traced back to Peirce's views on abduction.

In the same vein, Minnameier (2004) argues that abduction differs from the inference to the best explanation (IBE) since the former describes the process of generation of theories, while the latter describes the inductive process of their evaluation. As Hintikka so Minnameier traces this view back to Peirce's work on abduction. Abduction, properly speaking, should be restricted to describing the processes that lead to the generation of a theory or hypothesis.

Recent work on abduction (Gabbay and Wood 2005) goes as far as to draw a distinction between abducting an hypothesis that is considered worth conjecturing and the decision either to use further this hypothesis to do some inferential work in the given domain of enquiry, or to test it experimentally. The latter step, when it takes place, is an inductive mode of inference that should be distinguished from the abductive inference that led to the hypothesis.

In this paper, I argue that in real scientific practise both the distinction between a properly speaking abductive phase and an inductive phase of hypothesis testing and evaluation, and the distinction between testing an hypothesis that has been discovered in a preceding abduction and releasing or activating the same hypothesis for further inferential work in the domain of enquiry in which the ignorance problem arose in the first place are blurred because all these processes form an inextricable whole of theory development and elaboration, which defies attempts to analyze this intricate process into discrete well-defined steps. Thus, my arguments reinforce Magnani's (2014) view on abduction and its function in scientific practise.

In the first section, I put the discussion in its historical perspective starting with the distinction between the context of discovery and the context of justification and show how the recent considerations on the relation between abduction and IBE are related to the aforementioned original discussion. In the second section I examine Newton's first optical paper to trace the steps of the process of discovery of the non-homogeneous nature of white light. In the third and last section, I use the analysis of Newton's work in optics to argue that the principled distinction between abduction and induction and their consideration as two separate stages of the process of theory generation or discovery fails to take into account the intricacies of actual scientific practise.

The discussion on Newton's first optical paper draws heavily from a previous paper (Raftopoulos 1999) in which I had argued that Newton's argument in support of the thesis concerning the different degrees of refrangibilities of the various colors, as a whole, is an eliminative induction. In the present paper, I pursue further this line of reasoning to argue that Newton's method should be better viewed as an IBE.

1 Clearing the Ground: Abduction and Inference to the Best Explanation

One of the trademarks of the neo-positivist era and its Popperian criticism was the sharp distinction between the context of discovery and the context of justification. Popper's "The logic of Scientific Discovery" aimed, among other things, at showing that the process of discovery of hypotheses and theories is very different from the process of justification or confirmation the reason being that the former is clearly a psychological process that, as such, defies any attempt to explain it rationally by discovering a logic of discovery. The context of justification, on the other hand, covers the whole process of submitting an hypothesis to the test and finding evidence that would support the hypothesis. As such, the context of justification could be the subject matter of enquiry for a logic of discovery, which is inevitably an inductive logic, whence the many attempts to examine the sort of inductive support, that is, the conditions under, and the extent to, which various pieces of evidence support an hypothesis.

A combination of the close study of the history of science and philosophical analysis, however, soon showed that the abovementioned analysis is rather simplistic. As a result, philosophers drew a distinction within the realm of the context of discovery by dividing it into two parts, namely, the context of theory generation and the context of prior plausibility of probability of the hypothesis. The former purports to describe the psychological process of hitting upon an hypothesis, while the latter describes the considerations that lead scientists to think of a hypothesis, which was discovered at the stage of theory generation, as worth pursuing by assessing its prior plausibility or probability, that is, its epistemic and scientific value before the hypothesis be tested by new experiments. Testing the hypothesis by designing new experiments, on the other hand, constitutes the context of theory evaluation and confirmation as the context of justification was now called. The move from a 'context of justification' to a 'context of confirmation' occurred because philosophers realized that justification of a statement entails knowing the statement is true and hypotheses, which usually are universal statements, cannot be known to be true on account of the fact that they are universal statements; it follows that hypotheses are not justified, Hypotheses, however, can be supported by evidence to various extents and this means that they are confirmed to various degrees. These views introduced, on the one hand, a wedge between conceiving of an hypothesis as a possible explanation of a set of phenomena and the initial plausibility or probability of this hypothesis, and, on the other hand, a distinction between these two and the process of theory evaluation.

Inevitably, the context of discovery was related to Aristotle's notion of abduction, construed as the inference in which a series of facts, which are either new, or improbable, or surprising on their own or in conjunction, are used as premises leading to a conclusion that provides an explanation of these facts, as well as to the discussion of the notion of abduction by Peirce.

Even though some philosophers (Harman 1965; Lipton 2004) treat abduction on a par with IBE, Hintikka (1997, 1998) maintained that abduction is ignorance-preserving in the sense that the hypothesis that abduction delivers is not, epistemologically speaking, on a firmer ground than the phenomena it purports to explain; ignorance is preserved and, thus, knowledge is not enhanced until the hypothesis undergoes a further inductive process that will test it against empirical evidence. Hintikka, therefore, distinguishes between the abductive process properly speaking and the inductive process of hypothesis testing, which, Hintikka thinks, can be traced back to Peirce's views on abduction. It follows that, for Hintikka, abduction is different from IBE since the latter presupposes that the abduced hypothesis has been successfully tested because, otherwise, one could not be able to claim that it is the best available explanation.

Similarly, Minnameier (2004) argued that that abduction is different from the IBE since the former describes the process of generation of theories, while the latter describes the inductive process of their evaluation.

> Peirce characterizes abduction as the only type of inference that is creative in the sense that it leads to new knowledge, especially to (possible) theoretical explanations of surprising facts. As opposed to this, IBE is about the acceptance (or rejection) of already established explanatory suggestions. Thus, while abduction marks the process of generating theories or, more generally, concepts, IBE concerns their evaluation. However, if this is so, then both inferential types relate to entirely different steps in the process of knowledge acquisition. (pp. 75–76)

It follows that for these authors abduction should be construed as the logic underlying the context of theory generation, while the logic underlying the context of theory confirmation is, as is traditionally thought, induction. What remains to be seen, if abduction is construed as the logic of theory generation, is whether these philosophers think of the context of theory generation as the traditional context of discovery, or whether they equate theory generation with the context of theory generation as opposed to the context of prior plausibility or probability of the hypothesis, both belonging to the wider context of discovery. Recall that 'theory generation' in the Philosophy of Science was introduced to designate the psychological process of hitting on a hypothesis, a process that, as it was thought, is independent of the prior plausibility/probability of the hypothesis, the estimation of the latter belonging to the context that covers the considerations that lead scientists to think of the hypothesis as worth pursuing.

Gabbay and Wood's (2005) logical scheme of abduction (GW-schema) sheds some light on the distinction between these two contexts, i.e., the context of theory generation and the context of prior plausibility or probability, which taken together cover the traditional context of discovery. The details of the GW-schema need not concern us here.[1] Suffice to say that abduction culminates first in the hypothesis H deemed to be worthy of conjecture—in the GW-schema this conclusion is designated as C(H). Being worthy of conjecture does not entail that H is accepted and used for further scientific purposes, or that H is put to test. The decision to activate or release H for further work in the domain of enquiry to which the attempt to explain the initial set of phenomena gave rise, that is, the decision to start drawing inferences concerning this domain of enquiry, is the second and last conclusion of the abductive process—in the GW-schema this conclusion is designated as H^c. When H is released for inferential work in the original domain of enquiry the abduction is full, whereas, when H is not activated for this purpose, that is, when H is not acted upon, the abduction is partial. Note that in the last case the abduction is considered partial even if H is empirically tested (Woods 2009). There is, thus, a further distinction to be drawn between releasing H for inferential work in the original domain of enquiry and submitting H to experimental test, in addition to the distinction between these two actions and the abductive step of hitting on the hypothesis.

[1]The G-W schema for abduction is as follows:

1. T!α [setting of T as an epistemic target with respect to a proposition α],
2. ¬(R(K, T)) [fact], where R is the attainment relation with respect to T, and K is the knowledge base available to the agent,
3. ¬(R(K*, T)) [fact], where K* is an accessible successor of K in the sense that an agent could construct it in ways that serve to attain targets linked to K,
4. $H \notin K$ [fact], where H is the proposed hypothesis,
5. $H \notin K^*$ [fact],
6. ¬R(H,T) [fact],
7. ¬R(K(H),T) [fact], where K(H) is the knowledge base with the addition of H, which may mean that K has to be somewhat revised,
8. If H * R(K(H),T) [fact], where * is the subjunctive conditional relation,
9. H meets further conditions S1,…, Sn [fact],
10. Therefore, C(H) [sub-conclusion, 1–9].
11. Therefore, Hc [conclusion, 1–10].

At a first glance, it seems that C(H) could be considered as the result of the abductive process that corresponds to the generation of H, the hitting on H, since this step is clearly distinguished from any considerations concerning the prior plausibility/probability of H. H^c, on the other hand, could be considered as the assessment of the prior plausibility or probability of H, since it is reasonable to assume that scientists would decide, under normal circumstances, to release H for inferential work in the relevant domain of enquiry only if they deemed it worth pursuing owing to its significant prior plausibility or probability. This, again, leaves open the possibility that abduction may lead to the generation of an hypothesis that is worthy of conjecture but not worthy of pursuing because it has a low a priori plausibility or probability. If, however, one examines closely the steps of the GW-schema and especially step nine that precedes the conclusion that C(H) is the case and which designates that H should meet a set of further conditions (for example, it satisfies the consistency and minimality constraints (see Magnani 2014, 3), which in essence ensure that H has no plausible rival hypothesis, one tends to conclude that C(H) contains an assessment of the prior plausibility of H. That is, C (H) is the conclusion of an abduction that covers the domain of prior plausibility or probability. This also justifies the widespread view that abduction amounts to guessing reliable hypotheses because, in scientific practice at least, an hypothesis is reliable if it has an acceptable prior plausibility or probability.

A word of caution is needed at this juncture. The above claim should not be meant to exclude the possibility that one may hit on an hypothesis that if true would explain the salient set of phenomena and which, despite the fact that it lacked initial plausibility or probability, was eventually proved to be true; the History of Science has many such examples. There are two possibilities in this scenario. Either the improbable hypothesis is the only one that, in the scientist's view, can account for the phenomena, that is, it is the only game in town, despite its low initial plausibility. In this case, Sherlock Holmes's dictum concerning improbable hypothesis that are the only ones accounting for the phenomena applies. Or, the improbable hypothesis is not the only candidate but the scientist decides that it is worthy of conjecture for some reason or other. Note, however, that in this second case, the condition imposed by step 9 of the GW-schema is not met because the hypothesis in case has more plausible rivals. Thus, if the History of Science reveals cases in which scientists acted in the second way, then either one has to declare that their actions were not rational, according to the GW-schema of abduction, since they did consider worthy of conjecture and even released for inferential work an hypothesis that did not satisfy the conditions set in the ninth step of the abductive inference, or one has to reexamine the reliability of the GW-schema as an account for the actual scientific practise of theory generation.

If, as the above analysis suggests, C(H) includes an assessment of the prior probability/plausibility of H, and in view of the proposed distinction between C(H) and H^c in the GW schema, it would be interesting to examine the conditions under which scientists may decide not to pursue for further inferential work an hypothesis that is deemed worthy of conjecture and has an acceptable prior plausibility of being better than its rival hypotheses, that is, cases in which a scientists does not take the step from C(H) to H^c. I will not pursue this issue here except to note that I think that any such decision would be made on the basis of pragmatical considerations only and not epistemological ones since epistemological considerations are satisfied by the GW-schema of abduction. If this holds true, then this decision is probably taken based on purely pragmatical reasons.

The philosophers discussed thus far assume that the process of theory generation or prior assessment, which they consider to be the abductive, properly speaking, process, differs from the inductive process of hypothesis/theory testing. In this sense, they endorse the old venerable distinction between the process of theory discovery, which they assign to abductive reasoning, and the process of theory justification, which they assign to inductive reasoning, with one notable difference. While traditionally, philosophers of science thought that the context of discovery and, later, the context of theory generation were inherently psychological and, thus, resisted any logical analysis, the recent discussions on abduction attempt to unearth the logical structure of the reasoning underlying theory generation, as the GW-scheme of abduction eloquently shows. This, most likely, results from the fact that the recent accounts of abduction assume that abduction covers the process of the estimation of the prior probability/plausibility of a hypothesis that can be described logically.

There are, however, philosophers who think that the creative process of theory generation cannot really be set apart from the context of testing and confirming theories. In this vein, Magnani (2014, 39) defends a view of abduction according to which

> The proper experimental test involved in the Peircean evaluation phase, which for many researchers reflects in the most acceptable way the idea of abduction as inference to the best explanation, just constitutes a special *subclass* of the process of the adoption of the abductive hypothesis—the one which involves a terminal kind (iv) of actions (experimental tests), and should be considered ancillary to the nature of abductive cognition.

I fully agree with this view and in the next section I discuss Newton's first papers in optics to substantiate this claim.

2 Putting the Distinction into Test: Newton's First Optical Paper

Newton describes in his first optical paper the process of discovery and proof of one of the properties of light, to wit, that different colors have different degrees of refrangibility (Cohen 1958, pp. 47–78). The initial experiment consisted in placing a prism near a small hole in the shutters of the window of a darkened chamber in such a way that sunlight was refracted to the opposite wall. The experiment resulted in the production of a colored pattern on the wall having an oblong form, whereas, according to the laws of refraction, the pattern should have been circular. Newton compared the length of the coloured spectrum with its breadth and found it about five times greater, a result which he called ‘extravagant’. The oblong pattern is a phenomenon, which Newton calls the ‘elongation of the spectrum’ that needs to be explained as is unexpected.

Though Newton may have had from the beginning some reason to believe that the spectrum is due to the various angles of refraction of the different color-producing rays, this did not imply anything regarding the heterogeneity of white light. It is still possible that the different colors could be generated within the prism by means of a modification of homogeneous white light, and then be refracted in different angles, which, as Kuhn reports, was Newton’s first idea when he performed the experiment (Cohen 1958, 34). This is, perhaps, the reason that no mention of the heterogeneity of white light is to be found in in the early optical notes of Newton’s.

The first set of possible causes that occurred to Newton was that “I could scarce think, that the various thickness of the glass, or the termination with shadow or darkness, could have any influence on light to produce such an effect; yet I thought it not amiss, first to examine those circumstances, and so tried, what would happen by transmitting light through parts of the glass of divers thicknesses … (Cohen 1958, 48). None of these circumstances was found “material”, since the fashion of the colors was the same in all cases with different thicknesses, a fact that suggests that the thickness of the glass is not a causal factor of the phenomenon.

The second possible cause is more interesting, since what Newton does is to reject a standard part of any explanation of colors from Anaximenes to Grimaldi, Descartes, and Hooke, namely that colors result from the mixture of light with darkness, which comes from the boundaries of the hole. To test this supposition Newton “tried, what would happen by transmitting light … through holes in the window of divers bigness, or by setting the prism without so, that the light might pass through it, and be refracted before it was terminated by the hole” (Cohen 1958, p. 48). Again, the fashion of the colors did not change, a fact that made Newton reject this hypothesis.

The next possible cause that might explain the phenomenon was the hypothesis that the colors might have been thus dilated because of the unevenness in the glass or some other irregularity. Thinking that if a prism causes an irregular dispersion of the light rays a second prism which refracts light in a contrary way should cancel the regular effects of the first prism while augmenting the irregular ones, he combined two prisms so that they refracted light in contrary ways and performed an experiment whose main feature was that the second prism refracts the light in ‘contrary ways’. The experiment showed that “whatever was the cause of the length, it was not any contingent irregularity”, since “the light which by the first prism was diffused in an oblong form, was by the second reduced into an orbicular one …” (Cohen 1958, p. 48).

The rejection of this hypothesis carried a special weight for Newton because Descartes and Hooke explained the phenomena of colors by appealing to some kind of irregularities. According to Hooke, light is a short vibrating motion in the luminous body. This vibration spreads symmetrically through the surrounding medium, that is, ether. The pulses are at right angle to the beam of light unless they find in their way an interface bounding a different transparent medium. In such cases the pulses are distorted and they cease being normal to the beam. This distortion or irregularity of the pulses constitutes color, blue being the result of an oblique and confused pulse whose weakest part proceeds, and red being the result of a distorted pulse whose strongest part proceeds. Newton's mention to Descartes refers to a non-standard explanation of the color of the tail of comets, put forth in the third part of *The Principles* (AT. IX: 185–88). (Newton's argument, however, is unfair to Hooke. See Raftopoulos 1999 for a discussion.)

Considering another possible cause, Newton writes "Then I began to suspect, whether the rays, after their trajection through the prism, did not move in curved lines, and according to their more or less curvity tend to divers parts of the wall" (Cohen 1958, p. 50).

Newton gives the reason that made him think that the above might be a possible explanation of the phenomenon at issue; this was that he remembered that he had seen a tennis ball, struck with an oblique racket, describing such a curved line.

> For, a circular as well as a progressive motion being communicated to it by that stroke, its parts on that side, where the motion conspires, must press and beat the contiguous air more violently than on the other, and there excite a reluctancy and reaction of the air proportionally greater. And for the same reason, if the rays of light should possibly be globular bodies, and by their oblique passage out of one medium into another acquire a circulating motion, they ought to feel the greater resistance from the ambient aether, on that side, where the motions conspire, and then be continually bowed to the other.... notwithstanding this plausible ground of suspicion, when I came to examine it, I could observe no such curvity in them. (Cohen 1958, p. 50)

Newton's discussion is a cautious attack against an existing hypothesis seeking to explain the phenomena related to the rays of light. This is the hypothesis put forth by Descartes in his *Optics*. Descartes, (AT. VI: 88–9; CSM. 1: 155) in order to explain refraction, reflection and colors appeals to supposed changes in the speed of the small balls that constitute light when they pass from one medium to another with a different density. Newton attacks the account of colors Descartes gave in the *Meteorology* where colors were associated with the various rotational speeds that the particles of light acquire when they pass from one medium to another.

Newton knew that the elongation of the spectrum had been observed before. This elongation, however, was seen in a different light. First, the observed elongation was much smaller than the one reported by Newton owing to the fact that the screen was placed close to the prism and this resulted in a smaller elongation. Second, an explanation of this elongation had been put forward by standard theories of colors, according to which colors result from the modification of light when it interacts with bodies. This elongation was due to the fact that the sun has finite dimensions, so that the light rays falling on the prism were not parallel but, instead, had different angles of incidence. Hooke, for example, believed that the divergence of the rays is caused by the fact that the sun is not a point source of light, and that the resulting divergence could be accounted for by his theory.

Newton, in order to be able to claim that the phenomenon of the elongation of the spectrum of colors supports his explanation of the dispersion of colors against other alternative explanations, had to show that his opponents' explanation of the phenomenon was not satisfactory. Thus, he modified the configuration of the experiment so that the rays coming from opposite parts of the sun's discus were virtually parallel to each other. He found out that the difference between the two configurations could account for 31 or 32 min of divergence (which is the angular size of the sun), and thus the convergence of the beam incident on a small hole. This is much less than the observed elongation of the spectrum, 2° and 49 min. Though this calculation revealed the insufficiency of the other theories of color, Newton proceeded to perform another experiment:

> having placed it [the prism] at my window, as before, I observed, that by turning it a little about its axis to and fro, so as to vary its obliquity to the light, more than an angle of 4 or 5 degrees, the colours were not sensibly translated from their place on the wall, and consequently by the variation of the incidence, the quantity of refraction was not sensibly varied. (Cohen 1958, p. 49)

Newton concludes that this experiment shows that the difference of the incidence of rays flowing from divers parts of the sun could not make them after decussation diverge at a sensibly greater angle than that at which they before converged; there still remained some other cause to be found out to explain the higher divergence.

To understand the role of the series of experiments in the Newtonian method we should turn to Newton's correspondence with Pardies and Lucas because there Newton gives a clear account of the role of experiments. On 21 May 1672, Pardies sent his second letter to Newton, raising the following objection:

> But since I now see that it was in that case that the greater breadth of the colours was observed, on that head I find no further difficulty. I say on that head; for the greater length of the image may be otherwise accounted for, than by the different refrangibility of the rays. For according to that hypothesis, which is explained at large by Grimaldi, and in which it is supposed that light is a certain substance very rapidly moved, there may take place some diffusion of the rays of light after their passage and decussation in the hole. (Cohen 1958, p. 104)

Newton's answer came in the same year (Cohen 1958, emphasis added):

> Hence it has been here thought necessary to lay aside all hypotheses … that the force of the objection should be abstractly considered, and receive a more full and general answer. By light therefore I understand, any being or power of being … which proceeding directly from a lucid body, is apt to excite vision. And by the rays of light I understand its least or indefinitely small parts, which are independent of each other … *This being premised, the whole force of the objection will lie in this*, that colours may be lengthened out by some certain diffusion of light beyond the hole, which does not arise from the unequal refraction of the different rays, or of the independent parts of light. And that the image is no otherwise lengthened, was shown in my letter in Numb. 80 of the Transactions; and to confirm the whole in the strictest manner, I added that experiment now known by the name *Experimentum Crucis*. (pp. 106–107)

Putting aside the issue about hypotheses, let us focus on Newton's response to Pardies' objection. Let us try to reconstruct Newton's argument. Facing an objection as to the cause of refraction, he repeats the way he had used to prove, in his first optical paper, that the rays of different colors have different degrees of refrangibility. This proof consists of two parts: (a) a set of "premises" concerning what Newton understands by "rays of light", and (b) a series of experiments and the *experimentum crucis*. Both parts aim to eliminate possible alternative explanations of the phenomena of refrangibility.

The premises to which Newton refers in his second letter to Pardies are an essential part of Newton's experimental proofs. They include the fact that the rays of light travel in straight lines, that they consist of small parts, and that they are independent of each other. Another assumption used by Newton is found in *Book One*, *Part I*, of the *Opticks* (Newton 1730, 75). After experiment 15 and the discussion that proves proposition *VI*, Newton concludes (Newton 1730)

> And this demonstration being general, without determining what light is, or by what kind of force it is refracted, or assuming any thing farther than that the refracting body acts upon the rays in lines perpendicular to its surface; I take it to be a very convincing argument of the full truth of this proposition. (pp. 81–82)

Newton knows that these assumptions are commonly held by all involved in the study of optical phenomena so it is unlikely that his demonstrations would be contested on these grounds. Some of the premises well entrenched in the background knowledge and Newton feels no need to justify them. For some others, he offers a justification that consists in a kind of rationalization of our experiences. For instance, we see that some part of the light may be refracted or reflected when the rest is not; thus, we may conclude that light consists of independent parts. The conclusion is so obvious in view of our experience that Newton need not add anything else to justify it. Even those premises that are given no justification and, additionally, cannot be grounded directly in experience are accepted by Newton's opponents.

Newton thinks in his letter that the premises block the whole class of alternative explanations that could have arisen from different accounts of the rays of light. Thus, the only possible alternative remaining is (or, as Newton put it "this being premised, the whole force of the objection will lie in this") that "colours may be lengthened out by some certain diffusion of light beyond the hole, which does not arise from the unequal refraction of the different rays." What Newton has in mind are the explanations of colors by Grimaldi, Descartes, and Hooke, which were the main hypotheses to account for the phenomena of colors at that time.

The series of experiments reported in the first optical paper, however, show that such diffusion cannot account for the phenomenon (Cohen 1958, p. 49). Thus, Newton can claim he has shown that the phenomenon cannot be accounted for by these alternatives. Now it is the turn of the *experimentum crucis* to play its part. This experiment confirms in the strictest manner that the lengthening of the image is due exclusively to the different degrees of refrangibility of the rays of light.

The *Experimentum Crucis* is the last step in Newton's proof that the cause of the image could only be that light consists of rays that are differently refrangible (Cohen 1958). The gradual removal of the previous possibilities, says Newton, led him to the *experimentum crucis*. Its configuration consists of two parallel prisms through which the light passes successively. The first prism is slowly turned about its axis and the resulting image of the light is observed on the second board. If Newton's theory was correct, then the second prism should augment the result of the first prism, in the sense that the rays of white light refracted in different degrees and analyzed by the first prism undergo a second refraction which, because of the variation of the degrees of refrangibility, causes an even greater spreading of the beam. The uneven refraction of the rays is observed, confirming that the light consists of rays differently refrangible.

The experiments before the *experimentum crucis* proved the falsity of the main rival theories, but did not confirm Newton's theory. The *experimentum crucis* accomplishes that, by confirming the uneven refraction of the light rays. Since these rival theories were the only alternative ones, given the initial assumptions concerning the light rays, Newton can now claim that his experiments show the falsity of all the other theories, while confirming his. Thus, he claims that his experiments prove his theory. This is why Newton writes to Pardies that the crucial experiment *confirms* in a strict manner the theory.

One might object at this point that Newton could not have known that he had excluded all alternative explanatory hypotheses and, thus, his belief that he his hypothesis has been confirmed to the strictest manner is false. The reply to this objection is threefold. First, Newton does not expect his experiments to eliminate all *possible* causes. These experiments are meant to test all those alternatives that are compatible with certain premises presupposed in the context of enquiry. The different degrees of refraction of the various colors may be caused only by the different refrangibilities of the colors or by certain diffusions of light, given the premise regarding the constitution of the rays of light, namely, that the rays of light consist of indefinitely small parts that are independent of each other. It is clear in his letters that Newton has no doubt that his theory about the cause of the dispersion of colors is true. It is, therefore, evident that, in addition to the aforementioned controlled experiments, his experimental method relies heavily upon certain assumptions that are deemed as unproblematic.

Second, there is another characteristic of Newton's method that restricts further the problem space of possible explanations. This is no other than Newton's famed positivistic attitude towards causal explanations (Hall 1993, p. 59), which led him to distinguish the properties of things from uncertain speculations regarding their causes. Newton searches only for the properties of the rays of light. This reduces significantly the problem space and makes the demand of an exhaustive enumeration of alternative explanations more plausible.

A third factor restricting the search space is Newton's conception of the analogy of nature, as expressed by the two first rules of philosophizing (Newton 1729, pp. 398–400).[2] Since the possible alternatives must be inferred from the

[2]Newton's four rules of reasoning read (Newton 1729, 398–400):

Rule 1: We are to admit no more causes of natural things than such as are both true and sufficient to explain their appearances.

Rule 2: Therefore to the same natural effects we must, as far as possible, assign the same causes.

Rule 3: The qualities of bodies, which admit neither intensification nor remission of degrees, and which are found to belong to all bodies within the reach of our experiments, are to be esteemed the universal qualities of all bodies whatsoever.

phenomena, Newton has to consider only "properties" (and not causal hypotheses) that are known to work in similar cases.

To recapitulate, the series of hypotheses tested are tentative possible explanations of the phenomena. These are the conclusions of weak inferences that do not establish their results with certainty, as opposed to the strong inferences that do yield the maximum certainty that is possible in experimental science. The possible causes of phenomena are inferred by means of inductive, analogical inferences, which allow us to assume that a cause that is known to operate in a certain case also operates in a similar case. The success of an analogical argument hinges heavily upon the extent to which the "target" is similar enough to the "base" to justify the extension of the cause to the new domain. Thus, analogical arguments alone do not yield any significant certainty. In such a situation we have some reason to believe that *X* is the cause of the examined phenomena and, at the same time, we may have reason to believe that a different and incompatible cause operates. The conclusions of weak inferences have the form "*x* may be the cause of *E*", as opposed to strong inferences in which we have strong reasons for believing that "*x* is the cause of *E*."

(Footnote 2 continued)

Rule 4: In experimental philosophy we are to look upon propositions inferred by general induction from phenomena as accurately or very nearly true, notwithstanding any contrary hypotheses that may be imagined, till such time as other phenomena occur, by which they may either be made more accurate, or liable to exceptions.

The first two rules refer to inductive inferences (the term 'inductive inferences' denotes any kind of ampliative inferences, that is, inferences the conclusions of which are not contained in the premises) pertaining to causes. I think that these inferences are instances of causal simplification (or, as they are sometimes called 'analogical inferences') that have one of the following forms:

C_1: Effects *E*1, …, *En* are the same in systems *S*1, … *Sk*. Therefore (By Rule 2) these effects have the same causes in all these systems.

C_2: The cause of effect *E* in system 1 is an *X* with properties *P*1, …, *Pn*. The cause of the same or similar effect *E* in system 2 is an *x* with properties *P*1, …, Pn. Therefore (by Rules 1 and 2) the cause of the effect in system 1 is the same *X* as that which causes the effects in system 2.

C_3: Effects *E*1, …, *En* in system 1 are caused by *C*. Effects *E*1, … *En* are also present in system 2. Therefore (by Rules 1 and 2) effects *E*1, …, *En* are caused by *C* in system 2.

Rule 3 sanctions inferences from properties found to hold for all observed members of a class to the claim that these properties hold for any unobserved members of the class (or for all the members of this class). Furthermore, this Rule justifies inferences to unobserved members of a class and to the unobservable realm.

Rule 4, finally, discloses the method to be used in experimental philosophy In this philosophy propositions are 'inductively' inferred from the phenomena. Though the rule does not specify what kinds of propositions are inferred from the phenomena, Newton states that in natural philosophy we seek to establish the general properties of things and that the method is to be used in our inquiries after the properties of the things. Moreover, the causes of these properties are to be discovered by means of the same method.

A series of weak inferences cedes its place to a strong inference by means of a series of experiments, which seek to test the truth of these inferences, and which culminate in the *experimentum crucis*.

3 Concluding Discussion: Abduction and the Method of Discovery and Proof

The discussion of Newton's first optical paper brings forth the following kinds of inference involved in Newton's experimental proofs:

(a) Inductive inferences from the phenomena to the possible properties of things that could explain them. These inferences are justified by the first two rules of reasoning in the third book of the *Principia* (Newton 1729, pp. 398–400) and they involve 'causal simplification'.
(b) A deduction of consequences from these possible properties, that are tested against the results of controlled experiments.
(c) An elimination of the properties, the consequences of which clash with the phenomena. The surviving property is deemed to be the true cause.
(d) An inductive generalization (justified by the third rule of reasoning) from the studied case to all similar cases.

All these steps form an integrated whole and no part could have existed without the others. Newton's method of discovery and proof is an induction by elimination. This justificatory method exemplifies Nickles' 'generative justification' (Nickles 1988, 40; 1989, 299–304), that is, the methodology according to which a theory is better justified if it can be shown how it was constructed or derived from the background knowledge plus some new experiments. So, Newton does not restrict himself to proposing a theory and submitting it to experimental tests. This trait radically distinguishes Newton's method from the hypothetico-deductive method, according to which the warrant for a theory comes solely from the fact that the experimental evidence can be deduced from it. Instead, Newton shows how this theory was derived from experiments and some background knowledge, namely, his assumptions.

Furthermore, we see that the 'experimental proofs' or '*deductions* from the phenomena', as Newton calls them, involve much more than simple deductive (in the standard sense) inferences. They also contain ampliative inferences, a fact which indicates that Newton used 'deduction' with a wider meaning than it has nowadays. It must be noted here that if Newton considers the list of enumerated alternative explanations to be exhaustive, then the argument by elimination is deductive. We saw, however, that Newton did not think that the experimental proofs are as certain as mathematical demonstrations, and I suggested that this uncertainty might arise from the possibility that the enumeration of alternatives may not be exhaustive.

I claimed that the Newtonian *deduction* includes induction and other ampliative inferences. It may be objected that Newton contrasts *induction* with *deduction* and for this reason one should not include the former in the latter. The motive behind this objection seems to be that Newton uses these two terms differently in the same context, as in the letter to Cotes (Thayer 1953, p. 6) in which Newton states that "principles are deduced from the phenomena and made general by induction."

The answer to this objection is two-fold. First, Newton did not contrast between *induction* and *deduction*. In a letter to Oldenburg (Cohen 1958), we read: "You know the proper method for inquiring after the properties of things is, to deduce them from experiments" (p. 93). Upon the completion of the method propositions pertaining to the properties of the things have been *deduced* from experiments. Now, in *Rule 4* of philosophizing (Newton 1729) Newton writes that: "In experimental philosophy we are to look upon propositions inferred by general induction from phenomena as accurately or very nearly true…" (p. 400). Thus, 'general *induction* from the phenomena' and '*deduction* from the phenomena' are both used to denote the method for deriving the properties of things from the phenomena. This suggests that Newton did not think that they contrasted with one another. This assumption is reinforced by the following passage from the letter to Cotes: "[E]xperimental philosophy proceeds only upon phenomena and deduces general propositions from them only by induction" (Thayer 1953, p. 7). Here it is clear that *induction* is a kind of the *deduction* from the phenomena. Newton's usage of these terms is slippery, but this only shows that Newton shares the same tradition with Descartes, for whom *induction* was not a separate form of inference, but a complicated form of *deduction* (AT XI, Rule VII).

Second, the term *induction* was also used with its modern meaning, meaning an ampliative inference from some members of a set to the entire set. This is how Descartes uses it in *Rule VII* of the *Regulae*, and this is how Newton uses it in the letter to Cotes. The conclusions that have been deduced from the phenomena with respect to a particular experimental set-up can be generalized. Thus, they are the universal properties of things.

All these suggest that Newton does not use the term "discovery" the way philosophers of science did after the distinction between the contexts of discovery and justification. Following the tradition of his time, Newton does not distinguish between "proof" and "discovery". Whenever he refers to his scientific method he starts by stating that this is a method of discovery and proof. This is how Bacon (1990) and Descartes (*Regulae*, AT. X; *Discourse*, AT. VI: AT. VII) conceived of their method. It is worth mentioning that this tradition is maintained until Mill, for whom *induction* is the method of "discovery and proof".

Let us consider Newton's own apocryphal account of how the idea of a universal gravitation was first formed in his mind (the apple, etc.), and his account of his discovery of the theory of colors in the first optical paper. In the former case, Newton describes how he got for the first time an idea, an idea which he elaborated, refined, and tested in order to prove. In the latter case, Newton refers to the process itself of the elaboration of an idea, and calls this whole process the 'discovery' of the different refrangibilities of the rays of light. Thus, we have to conclude that

Newton's 'logic of discovery' should not be taken in the sense of the logic of "theory generation", that is, of the process by which the scientist hits upon a hypothesis, but in the sense of the logic of the process of elaboration of a theory. Newton assumes that this elaboration finally establishes the hypothesis and that the inferences involved, in elaborating the hypothesis, prove it.

If this is what Newton has in mind when he speaks of "discovery", then he is not using the term in the same way as current philosophy of science. Even when philosophers emphasized that the 'logic of discovery' should not be taken to cover the process of theory generation, as logical positivism and Popperianism have traditionally thought, they insisted that it should not be taken to be coextensive with the logic of justification either. Rather, this logic should be deemed to cover the context of prior plausibility or probability of the theory, that is, it should take up the process by which a theory is deemed to be worth pursuing, either because it is plausible, or highly probable, *before* it is tested by new experiments specifically designed to confirm or disconfirm the theory.

This is far from Newton's idea of discovery; in a draft preface for the 1704 *Opticks* he writes (ADD. 3970. 3. Folio 480^v, quoted in McGuire 1970)

> The method of resolution [analysis] consists in trying experiments and considering all the phenomena of nature relating to the subject at hand and (drawing conclusions from them) and examining the truth of these conclusions by new experiments … until you come to the general properties of things. (pp. 184–185)

As this statement shows, experiments testing the propositions drawn from the phenomena are an essential part of the method of discovery and proof.

Let us see the impact of the above considerations, which I have carefully discussed without any mention of abduction lest I distort the impact of the analysis on the notion of abduction, on the accounts of abduction that we discussed in the first section. We have seen that some philosophers (such as Hintikka and Minnameier) view abduction as the process of theory generation, i.e., as the process of hitting upon an hypothesis that, if true, would explain a set of data. It is not clear whether they intend this context to include an assessment of the prior plausibility/probability of the hypothesis. The analysis of the GW-schema of abduction most likely interprets abduction as including the prior assessment of the hypothesis.

According to this view, scientists use abduction to reach at an hypothesis worth considering, which, let us assume, has an acceptable initial plausibility/probability; abduction covers the traditional context of discovery. Then, the decision to test it experimentally is a different, inductive, process; the process of confirmation. Taken together, these two constitute the IBE. Let us now see whether this view of abduction fits Newton's process of discovery. At a first glance, there seems to be a perfect fit. Newton proposes a hypothesis and then tests it to reject it when it conflicts with experience. Finally, after a series of tests, he accepts the hypothesis that withstands the previous tests and also passes the crucial experiment.

This, however, is deceiving. The abductions to the possible causes, the tentative hypotheses, which Newton calls inductions from the phenomena, and the ensuing experimental tests involved in the evaluation phase of these tentative hypotheses constitute a special subclass of the process of the confirmation and eventual adoption of the abductive hypothesis, that is, the hypothesis that Newton finally proposes as the explanation of the phenomena.

The series of tentative hypotheses do not form separate abductive actions. One could not understand Netwon's process of discovery if one analysed this process as a series of separate series of hypotheses formation and empirical testing of these hypotheses because Newton would not have arrived at his final hypothesis, except by chance, had he not undergone the sequence of drawing tentative hypotheses and testing them. His notes and writings show clearly the process that led him from the initial hypotheses he thought could explain the phenomena to the final hypothesis he adopted as he gradually rejected the previously entertained hypotheses. To put it differently, the process of abduction that gives rise to some hypothesis does not function in a vacuum but necessarily takes place in a context and, thus, it is interwoven in a nexus of conflicting alternative hypotheses whose formation and testing plays an in eliminable role in the formation of the abduced final hypothesis. Moreover, even if Newton had hit by accident on this hypothesis, the fact that this hypothesis passes the experimentum crucis does not mean that it is the best game in town unless one has first rejected the other plausible alternative hypotheses, which is exactly what the series of drawing tentative conclusions and testing them aims to accomplish. In fact, one cannot be certain that Newton would have been able to design the experimentum crucis if his mind had not been shaped by the series of experiments that he had already performed. These considerations fully justify Magnani's view quoted in the first section.

This is very important so allow me to dwell upon it a bit further. Attempting to defend the view that hypothesis generation (abduction) and testing of the hypothesis (induction) are two distinct phenomena despite the fact that in conjunction they constitute the backbone of scientific research, one could point out that the analysis of Netwon's work that I have presented does not really clash with this account. This is so because our discussion shows that Newton first hits on an hypothesis and then tests it experimentally. The fact that Newton proposes a series of hypotheses that he subsequently rejects after testing them by taking recourse to experience until he arrives at the final hypothesis only shows that the scientific enquiry is just an interconnecting sequence of abductions and inductions.

Note also that Newton's account renders clear that theory generation, even if viewed as the process of hitting on an hypothesis separate from the subsequent empirical evaluation, is inextricably linked with an initial evaluation of the hypothesis, since one of the factors that determines which hypotheses are arrived at are the first two rules of philosophizing that express the analogy of nature. These tentative hypotheses have inbuilt, as it were, an initial plausibility/probability. In addition, as the quotations from Newton's paper make abundantly clear, the reasons that drive Newton to think of an explanation/hypothesis are also, clearly, reasons that render the hypothesis initially plausible; "I could scarce think, that the various thickness of the glass, or the termination with shadow or darkness, could have any influence on light to produce such an effect; yet I thought it not amiss, first to examine those circumstances …" This means that the distinction between a process of hitting on a hypothesis and a process of initial assessment of its prior plausibility or probability that several philosophers have suggested fails to account for the actual scientific practise.

The examination of Newton's course of research also raises doubts about an important aspect of the GW-schema of abduction. As we saw in the first section, according to that schema a distinction should be drawn between releasing H for inferential work in the original domain of enquiry and submitting H to experimental test, since one could do the latter without undertaking the former task in which case the abduction is still partial. The study of Newton's work shows that this distinction is simplistic. Newton draws inferences from the hypotheses he conceives that are clearly within the original domain of enquiry, namely the study of the degrees of the refrangibilities of the various colors and of the white light; in this sense his abduction is full. However, the purpose of this practise is to use the conclusions of these inferences in order to test experimentally the hypotheses that generated the conclusions. Thus, Newton does not distinguish between engaging in inferential work within the original domain of enquiry and testing these hypotheses; the two are perfectly combined as long as the inferential work produces as conclusions new phenomena. This casts doubt on the validity of the G-W schema as a covering type for all abductions.

References

Bacon, F. (1990). *Novum Organum* (P. Urbach and J. Gibson, Trans.). Open Court: Chicago.

Cohen, I. B. (Ed.). (1958). *Isaak Newton's papers and letters on natural philosophy*. Cambridge, MA: Harvard University Press.

Gabbay, D. M., & Woods, J. (2005). *The reach of abduction*. Amsterdam: North Holland.

Hall, A. R. (1993). *All Was Light: An Introduction to Newton's Opticks*. Clarendon Press: Oxford.

Harman, G. (1965). Enumerative induction as inference to the best explanation. *Journal of Philosophy, 68*(18), 529–533.

Hintikka, J. (1997). The place of C.S Peirce in the history of logical theory. In J. Brunning & P. Forster (Eds.), *The rule of reason*. Toronto: Toronto University Press.

Hintikka, J. (1998). What is abduction? The fundamental problem of contemporary epistemology. *Transactions of the Charles S. Peirce Society, 34*, 503–533.

Lipton, P. (2004). *Inference to the best explanation* (2nd ed.). London/New York: Routledge.

Magnani, L. (2014). Are heuristics knowledge-enhancing? Abduction, models, and fictions in science. In E. Ippoloti (Ed.), *Heuristic reasoning, studies in applied philosophy, epistemology and rational ethics*. Switzerland: Springer.

McGuire, J. E. (1970). Newton's Principles of Philosophy: An Intended Preface for the 1704 Opticks. *The British Journal for the History of Science*, *5*(18), 178–186.

Minnameier, G. (2004). Peirce-suit of truth—Why inference to the best explanation and abduction ought not to be confused. *Erkenntnis, 60*, 75–105.

Newton, I. (1729/1962). *Principia.* Berkeley, CA: University of California Press.
Newton, I. (1730/1952). *Opticks* (4 ed.). New York: Dover.
Nickles, T. (1988). Reconstructing Science: Discovery and Experiment. In D. Batens & J. P. Van Bendegem (Eds.), *Theory and Experiment: Recent Insights and New Perspectives on Their Relation* (pp. 33–53). Dordrecht: Reidel.
Nickles, T. (1989). Justification and experiment. In D. Gooding, T. Pinch, & S. Schaffer (Eds.), *The uses of experiment*. Cambridge: Cambridge University Press.
Raftopoulos, A. (1999). Newton's experimental proofs as eliminative reasoning. *Erkenntnis, 50*(1), 95–125.
Thayer, H. S. (Ed.). (1953). *Newton's philosophy of nature*. New York: Hafner Press.
Woods, J. (2009). Ignorance, inference and proof: abductive logic meets the criminal law. In G. Tuzet & D. Canale (Eds.), *The rules of inference: Inferentialism in law and philosophy* (pp. 151–185). Heidelberg: Egea.

Counterfactuals in Critical Thinking with Application to Morality

Luís Moniz Pereira and Ari Saptawijaya

Abstract Counterfactuals are conjectures about what would have happened, had an alternative event occurred. It provides lessons for the future by virtue of contemplating alternatives; it permits thought debugging; it supports a justification why different alternatives would have been worse or not better. Typical expressions are: "If only I were taller …", "I could have been a winner …", "I would have passed, were it not for …", "Even if ... the same would follow". Counterfactuals have been well studied in Linguistics, Philosophy, Physics, Ethics, Psychology, Anthropology, and Computation, but not much within Critical Thinking. The purpose of this study is to illustrate counterfactual thinking, through logic program abduction and updating, and inspired by Pearl's structural theory of counterfactuals, with an original application to morality, a common concern for critical thinking. In summary, we show counterfactual reasoning to be quite useful for critical thinking, namely about moral issues.

Keywords Critical thinking · Counterfactual reasoning · Abduction · Morality

1 Counterfactual Reasoning

Counterfactual literally means contrary to the facts. Counterfactual reasoning involves thoughts on what could have happened, had some matter—action, outcome, etc.—been different in the past. Counterfactual thinking covers everyday experiences, like regret: "If only I had told her I love her!", "I should have studied harder"; or guilt responsibility, blame, causation: "If only I had said something

L.M. Pereira (✉)
NOVA Laboratory for Computer Science and Informatics,
Universidade Nova de Lisboa, Lisbon, Portugal
e-mail: lmp@fct.unl.pt

A. Saptawijaya
Faculty of Computer Science, Universitas Indonesia, Jakarta, Indonesia
e-mail: saptawijaya@cs.ui.ac.id

L. Magnani and C. Casadio (eds.), *Model-Based Reasoning in Science and Technology*, Studies in Applied Philosophy, Epistemology and Rational Ethics 27, DOI 10.1007/978-3-319-38983-7_15

sooner, then I could have prevented the accident". The general form is: "If the ⟨*Antecedent*⟩ had been true, then the ⟨*Consequent*⟩ would have been true".

Counterfactuals have been well studied in Linguistics, Philosophy, Physics, Ethics, Psychology, Anthropology, and Computation (Baral and Hunsaker 2007; Byrne 2007; Collins et al. 2004; Epstude and Roese 2008; Ginsberg 1986; Halpern and Hitchcock 2015; Lewis 1973; Markman et al. 1993; McCloy and Byrne 2000; Migliore et al. 2014; Pearl 2009; Pereira et al. 1991; Roese 1997; Vennekens et al. 2010), but oddly not much within Critical Thinking. However, people often think how things that matter to them might have turned out differently (Mandel et al. 2005). Researchers from psychology have asked: "Why do people have such a strong tendency to generate counterfactuals?"; "What functions does counterfactual thinking serve?"; "What are the determinants of counterfactual thinking?"; "What are its adaptive and psychological consequences?" Human's ability for the mental time travel required by counterfactual thinking relies on their use of episodic memory. Without this memory humans would be unable to form a stable concept of self along time, consider what might counterfactually have happened instead, and hence human cultures would not have been able to consider evolution paths that took into account past alternatives.

In this paper, counterfactual reasoning is enacted using a three-step logic evaluation procedure (Pereira and Saptawijaya 2016), inspired by the structure-based approach of (Pearl 2009), viz.

1. **Abduction**: to explain past circumstances in the presence of observed evidence, i.e., use the given evidence to determine the unchanging external background circumstances;
2. **Action**: to adjust the logical causal model to comply with the antecedent of the counterfactual, i.e., to impose the truth of the antecedent's hypotheses by means of a forced intervention on the model; and
3. **Prediction**: to predict if the counterfactual's consequent deductively follows, subsequently to steps 1 and 2, i.e., to compute the truth-value of the consequent in the modified intervened model.

The approach is realised by means of logic program abduction and updating. Abduction chooses from available hypotheses (the set A of abducibles) the exogenous variables that constitute the situation's background—i.e., those abducibles or their negations, that best explain the observed given evidence O. An abduced explanation, E, is a subset of A that finds the specific values for exogenous variables, which lend an explanatory support to all currently observed evidence. Note that the abduction procedure guarantees the abduced explanation to be consistent, i.e., disallows both abducible a and its negation a^* to hold in explanation E.[1] Subsequent to abduction, updating modifies those rules to be updated and fixes the initially abduced exogenous background context of the counterfactual statement. That is, updates the knowledge base with some preferred explanation to the current observations and, additionally, the updating also permits causal intervention on the causal knowledge

[1]In the sequel, starred atoms stand for their negations.

model, namely by means of hypothetical updates to the rules, achieved via reserved predicate *make* (illustrated in examples below), so as to render the knowledge base consistently compliant with the antecedent of the counterfactual.

Consider an example (Byrne 2007): *Lightning hits a forest, and a devastating forest fire breaks out. The forest was dry, after a long hot summer*. Let us add more causes for forest fire, i.e., there are two possible alternative causes: *storm*—presuming the lightning—or *barbecue*. The model of this example consists in a set of abducibles

$$A = \{storm, barbecue, storm^*, barbecue^*\}$$

and program P:

$$fire \leftarrow barbecue, dry\_leaves.$$
$$fire \leftarrow barbecue^*, lightning, dry\_leaves.$$
$$lightning \leftarrow storm.$$
$$dry\_leaves.$$

Take counterfactual statement: *If only there had not been lightning, then the forest fire would not have occurred.*

Step 1 Given the observation $O = \{lightning, fire\}$, abduce its explanations E (a subset of A). Note that the observations assure us that both the antecedent and the consequent of the counterfactual were *factually* false. Two possible explanations for O: $E_1 = \{storm, barbecue^*\}$ and $E_2 = \{storm, barbecue\}$. Say E_1 is preferred for consideration, on a criterion of simplicity. Then fix its abduced background context for the counterfactual: i.e., update program P with E_1.

Step 2 Update program P, via an automated transformation, to get a new program T by adding:

$make(lightning^*)$	% Intervention:If there had not been lightning…
$lightning \leftarrow make(lightning)$.	% Note that lightning or otherwise are
$lightning^* \leftarrow make(lightning^*)$.	% now available only by intervention.

where $make/1$ represents an explicit intervention on the model, by forcing its argument true. It corresponds to Pearl's $do/1$ operator. Because the intervention must be made explicit, an implicit default representation would not be adequate. e.g. $make(lightning^*)$ is explicitly imposing on the model that there was no lightning.

Plus, for irrelevancy and consistency, the transformation deletes:

$$lightning \leftarrow storm.$$

Step 3 Verify if the conclusion "the forest fire would not have occurred" is true. Since *fire* is not provable, '*not fire*' holds in the semantics of T for explanation $E_1 = \{storm, barbecue^*\}$ with intervention $make(lightning^*)$. The counterfactual is valid.

2 Counterfactuals in Morality

Typically, people think critically about what they should or should not have done when they examine decisions in moral situations. It is therefore natural for them to engage in counterfactual thoughts of alternatives in such settings. Counterfactual thinking has been investigated in the context of moral reasoning, notably by psychology experimental studies (Byrne 2007), e.g., to understand the kind of critical counterfactual alternatives people tend to think of in contemplating moral behaviours, and the influence of counterfactual thoughts in moral judgment (Mandel et al. 2005; Roese and Olson 2009).

Morality and normality judgments typically correlate. Normality infuses morality with causation and blame judgments. The importance of control, namely the possibility of intervention, is highlighted in theories of blame that presume someone responsible only if they had some control over the outcome (Weiner 1995). The explicit controlled interventions expressed by the counterfactual premises enable to interfere with normality, and hence with blame and cause judgments.

As argued by Epstude and Roese (2008), the function of counterfactual thinking is not just limited to the evaluation process, but occurs also in the reflection one. Through evaluation, counterfactuals help correct wrong behaviour in the past, thus guiding future moral decisions. Moreover, counterfactually thinking about guilt or shame is useful to prevent their future arising, a process of self-cleansing or self-debugging (Niedenthal et al. 1994). Reflection, on the other hand, permits momentary experiential simulation of possible alternatives, thereby allowing careful consideration before a decision is made, and to subsequently justify it.

The investigation in this paper pertains to how moral issues can innovatively be expressed with counterfactual reasoning by resorting to the aforementioned approach. In particular, its application for examining viewpoints on moral permissibility is scrutinized, exemplified by classic moral dilemmas from the literature on the Doctrine of Double Effect (DDE) (McIntyre 2004), and the Doctrine of Triple Effect (DTE) (Kamm 2006).

DDE is often invoked to explain the permissibility of an action that causes a harm, by distinguishing whether this harm is a mere side effect of bringing about a good result, or if this harm is rather the actual means to bringing about the same good end (McIntyre 2004). In Hauser et al. (2007), DDE has been utilized to explain the consistency of judgments, shared by subjects from demographically diverse populations, on a series of moral dilemmas.

Counterfactuals may provide a general way to examine DDE in dilemmas, e.g., the classic trolley problem (Foot 1967), by distinguishing between cause and side effect of performing an action to achieve a goal. This distinction between causes and side effects may explain the permissibility of an action in accordance with DDE. That is, if some morally wrong effect E happens to be a cause for a goal G that one wants to achieve by performing an action A, and E is not a mere side effect of A, then performing A is impermissible. The counterfactual form below, in a setting where action A is performed to achieve goal G, expresses this:

If *not* E had been true, then *not* G would have been true.

The evaluation of this counterfactual form identifies permissibility of action A from its effect E, by identifying whether the latter is a necessary cause for goal G or a mere side effect of action A. That is, if the counterfactual proves valid, then E is instrumental as a cause of G, and not a mere side effect of action A. Since E is morally wrong, achieving G that way, by means of A, is impermissible; otherwise, not.

Note that the evaluation of counterfactuals in this application is considered from the perspective of agents who perform the action, rather than from anothers' (e.g., observers). Moreover, the emphasis on causation in this application focuses on agents deliberate actions, rather than on causation and counterfactuals in general, cf. Collins et al. (2004), Pearl (2009).

In the next examples, the aforementioned general counterfactual method is illustrated by taking off-the-shelf military morality cases (Scanlon 2008). Consider "Terror Bombing", *teb* for short, which means: Bombing a civilian target during a war, thus killing many civilians, in order to terrorise the enemy, and thereby getting them to end the war. DDE affirms *teb* impermissible.

On the other hand, "Tactical bombing" (*tab*) means: Bombing a military target, which will effectively end the war, but with the foreseen consequence of killing the same large number of civilians nearby. DDE affirms *tab* permissible.

Modeling Terror Bombing. Take set of abducibles $A = \{teb, teb^*\}$ and program P:

$$end\_war \leftarrow terror\_civilians. \qquad terror\_civilians \leftarrow kill\_civilian.$$
$$kill\_civilian \leftarrow target\_civilian. \qquad target\_civilian \leftarrow teb.$$

Counterfactual: *If civilians had not been killed, the war would not have ended.* The evaluation follows.

Step 1 Observations $O = \{kill\_civilian, end\_war\}$ with explanation $E = \{teb\}$.
Step 2 Produce program T from P:

$make(kill\_civilians^*)$ % Intervention:If civilians had not been killed
$kill\_civilians \leftarrow make(kill\_civilians).$ % Killing civilians or otherwise
$kill\_civilians^* \leftarrow make(kill\_civilians^*).$ % is now available only by intervention.

Simply deleting *kill_civilians* and adding *kill_civilians**, without employing $make/1$, would throw away the structural information about the intervention, which would hinder the program providing detailed explanations about intervention.

Plus, for irrelevancy and consistency, delete:

$$kill\_civilian \leftarrow target\_civilian.$$

Step 3 The counterfactual is valid since conclusion “the war would not have ended” is true. Indeed, ‘*not end_war*‘ holds in the semantics of updated *T*, added with the abduced, and adopted, unchanging background fact *E*. Hence, the morally wrong action *kill_civilians* is an instrument to achieve the goal *end_war*. It is a cause of *end_war* by performing *teb*, and not a mere side effect of *teb*. Therefore, *teb* is DDE morally impermissible.

Modeling Tactical Bombing. Take set of abducibles $A = \{tab, tab^*\}$ and program *P*:

$$end\_war \leftarrow target\_military.$$
$$kill\_civilian \leftarrow tab. \qquad target\_military \leftarrow tab.$$

The counterfactual is the same as above. The evaluation follows.

Step 1 Observations $O = \{kill\_civilian, end\_war\}$ with explanation $E = \{tab\}$.
Step 2 Produce *T* from *P*, obtaining same *T* as in the terror bombing's model. And, for irrelevancy and consistency, now delete:

$$kill\_civilian \leftarrow tab.$$

Step 3: The counterfactual is not valid, since its conclusion “the war would not have ended” is false. Indeed, *end_war* holds in the semantics of updated *T* plus *E*. Hence, the morally wrong *kill_civilian* is a just side effect of achieving the goal *end_war*. Therefore, *tab* is DDE morally permissible.

A more complex scenario can challenge this application of counterfactuals, to distinguish moral permissibility according to DDE versus DTE. DTE (Kamm 2006) refines DDE particularly on the notion about harming someone as an intended means to harm the person, or instead harming the person only because it is a causal happenstance towards some goal. That is, DTE distinguishes further between doing an action with the intended goal of a harm effect to occur, and doing a action even though a harming effect will instrumentally occur. The latter is a new category of action, which is not accounted for in DDE. Though DTE also classifies the former as impermissible, it is more tolerant to the latter (the third effect), i.e., it treats as permissible those actions performed just because instrumental harm will occur.

Kamm proposed DTE to accommodate a variant of the trolley problem, viz., the Loop Case (Thomson 1985):

A trolley is headed toward five people walking on the track, and they will not be able to get off the track in time. The trolley can be redirected onto a side track, which loops back towards the five. A fat man sits on this looping side track, whose body will by itself stop the trolley. Is it morally permissible to divert the trolley to the looping side track, thereby hitting the man and killing him, but saving the five?

This case strikes most moral philosophers that diverting the trolley is permissible (Otsuka 2008). Referring to a psychology study (Hauser et al. 2007), 56 % of its respondents judged that diverting the trolley in this case is also permissible. To this end, DTE may provide the justification of its permissibility (Kamm 2006). Nonetheless, DDE views diverting the trolley in the Loop case as impermissible.

Modeling Loop Case. Take set of abducibles $A = \{divert, divert^*\}$ and program P, where *save, divert, hit, tst, mst* stand for *save the five, divert the trolley, man hit by the trolley, train on the side track* and *man on the side track*, respectively:

$$save \leftarrow hit. \quad hit \leftarrow tst, mst. \quad tst \leftarrow divert. \quad mst.$$

Counterfactual: *If the man had not been hit by the trolley, the five people would not have been saved.* The evaluation follows.

Step 1 Observations $O = \{hit, save\}$ with explanation $E = \{divert\}$.
Step 2 Produce program T from P:

$make(hit^*)$	% Intervention:If the man had not been hit by the trolley
$hit \leftarrow make(hit).$	% The man being hit by the trolley or otherwise
$hit^* \leftarrow make(hit^*).$	% is now available only by intervention.

And, for irrelevancy and consistency, now delete:

$$hit \leftarrow tst, mst.$$

Step 3 The counterfactual is valid, since its conclusion "the five people would not have been saved" is true. Indeed, *not save* holds in the semantics of updated T plus E.
Hence, *hit*, as a consequence of action *divert*, is instrumental as a cause of goal *save*. Therefore, *divert* is DDE morally impermissible.

DTE considers diverting the trolley as permissible, since the man is already on the side track, without any deliberate action performed in order to place him there. In the above program, we have the fact *mst* ready, without abducing any ancillary action. The validity of the counterfactual "*if the man had not been on the side track, then he would not have been hit by the trolley*", which can easily be verified, ensures that the unfortunate event of the man being hit by the trolley is indeed the consequence of the man being on the side track. The lack of deliberate action (say, by pushing the man—*push* for short) in order to place him on the side track, and whether the absence of this action still causes the unfortunate event (the third effect) is captured by the counterfactual "*if the man had not been pushed, then he would not have been hit by the trolley*". This counterfactual is not valid, because the new observation $O = \{push, hit\}$ has no explanation: *push* is not in the set of abducibles A, and moreover there is no fact *push* either. This means that even without this hypothetical but unexplained deliberate action of pushing, the man would still have been hit by the trolley (just because he is already on the side track). In summary, though *hit* is a consequence of *div* and instrumental in achieving *save*, no deliberate action is required to cause *mst*, in order for *hit* to occur. Hence *divert* is DTE morally permissible.

In order to further distinguish moral permissibility with respect to DDE and DTE, we also consider a variant of the Loop case, viz., the *Loop-Push* case—see also the Extra Push case in (Kamm 2006). Differently from the Loop case, in this Loop-Push case the looping side track is initially empty, and besides the diverting action, an ancillary action of pushing a fat man in order to place him on the side track is additionally performed.

Modeling Loop-Push Case. Take set of abducibles $A = \{divert, push, divert^*, push^*\}$ and program P:

$$save \leftarrow hit. \quad hit \leftarrow tst, mst. \quad tst \leftarrow divert. \quad mst \leftarrow push.$$

Recall the counterfactuals considered in the discussion of DDE and DTE of the Loop case:

- "*If the man had not been hit by the trolley, the five people would not have been saved.*" The same observation $O = \{hit, save\}$ provides an extended explanation $E = \{divert, push\}$. That is, the pushing action needs to be abduced for having the man on the side track, so the trolley can be stopped by hitting him. The same intervention $make(hit^*)$ is applied to the same transform T, resulting in a valid counterfactual: *not save* holds in the semantics of updated T plus E.
- "*If the man had not been pushed, then he would not have been hit by the trolley.*" The relevant observation is $O = \{push, hit\}$, explained by $E = \{divert, push\}$. Whereas this counterfactual is not valid in DTE of the Loop case, it is valid in the Loop-Push case. Given rule $push^* \leftarrow make(push^*)$ in the transform T and intervention $make(push^*)$, we verify that *not hit* holds in the semantics of updated T plus E.

From the validity of these two counterfactuals it can be inferred that, given the diverting action, the ancillary action of pushing the man onto the side track causes him to be hit by the trolley, which in turn causes the five to be saved. In the Loop-Push, DTE agrees with DDE that such a deliberate action (pushing) performed in order to bring about harm (the man hit by the trolley), even for the purpose of a good or greater end (to save the five), is likewise impermissible.

3 Conclusions and Further Work

Computational morality (Anderson and Anderson 2011; Wallach and Allen 2009) is a burgeoning field that emerges from the need of imbuing autonomous agents with the capacity of moral decision making to enable them to function in an ethically responsible manner via their own ethical decisions. It has attracted the artificial intelligence community, and brought together perspectives from various fields: philosophy, anthropology, cognitive science, neuroscience, and evolutionary biology. The overall result of this interdisciplinary research is not just important for equipping agents with some capacity for making moral judgments, but also to help better understand morality, via the creation and testing of computational models of ethical theories.

This paper presented a formulation of counterfactuals evaluation by means of logic program abduction and updating. The approach corresponds to the three- step process in Pearl's structural theory, despite omitting probability to concentrate on a naturalised logic. Furthermore, counterfactual reasoning has been shown quite useful for critical thinking, namely about moral issues, where (non-probabilistic) moral reasoning about permissibility is examined by employing this logic program approach to distinguish between causes and the side effects that are the result of agents actions to achieve goals.

In Pearl's theory, intervention is realised by superficial revision, i.e., by imposing the desired value to the intervened node and cutting it from its parent nodes. This is also the case in the approach presented here, achieved by hypothetical updates via the reserved predicate *make*. Other subtle ways of intervention may involve deep revision, realisable with logic programs (cf. Pereira et al. 2015), and minimal revision (cf. Dietz et al. 2015).

Logic program abduction was used in Kowalski (2011) and Pereira and Saptawijaya (2011) to model moral reasoning in various scenarios of the trolley problem, both from DDE and DTE viewpoints, sans counterfactuals. Abducibles are used to represent decisions, where impermissible actions are ruled out using an integrity constraint, and a posteriori preferences are eventually enacted to come up with a moral decision from the remaining alternatives of action. Subsequent work (Han et al. 2012) refines it with uncertainty of actions and consequences in several scenarios of the trolley problem by resorting to probabilistic logic programming P-log (Baral and Hunsaker 2007).

Side effects in abduction have been investigated in Pereira et al. (2013) through the concept of inspection points; the latter are construed in a procedure by 'meta-abducing' a specific abducible, $abduced(a)$, whose function is only checking that its corresponding abducible a is indeed already abduced elsewhere. Therefore, the consequence of the action that triggers this 'meta-abducing' is merely a side effect. Indeed, inspection points may be employed to distinguish a cause from a mere side effect, and thus may provide an alter- native or supplement to counterfactuals employed for the same purpose.

Counterfactuals may as well be suitable to address moral justification, via 'compound counterfactuals': *Had I known what I know today, then if I were to have done otherwise, something preferred would have followed*. Such counterfactuals, by imagining alternatives with worse effect—the so-called *downward counterfactuals* (Markman et al. 1993)—may provide justification for what was done due to lack of the current knowledge. This is accomplished by evaluating what would have followed if the intent had been otherwise, other things (including present knowledge) being equal. It may justify that what would have followed is no morally better than the actual ensued consequence. We are currently investigating the application of counterfactuals to justify an exception for an action to be permissible (Pereira and Saptawijaya 2015; Saptawijaya and Pereira 2015), which may lead to agents' argumentation following contractualism of Scanlon (1998).

Acknowledgments AS acknowledges the support from Fundação para a Ciência e a Tecnologia (FCT/MEC) Portugal, grant SFRH/BD/72795/2010. LMP acknowledges the support from FCT/MEC NOVA LINCS PEst UID/CEC/04516/2013. We thank Emmanuelle-Anna Dietz for the fruitful discussions.

References

Anderson, M., & Anderson, S. L. (Eds.). (2011). *Machine Ethics*. New York, NY: Cambridge University Press.

Baral, C., & Hunsaker, M. (2007). Using the probabilistic logic programming language P-log for causal and counterfactual reasoning and non-naive conditioning. In *Proceedings of 20th International Joint Conference on Artificial Intelligence (IJCAI)*.

Byrne, R. M. J. (2007). *The rational imagination: How people create alternatives to reality*. Cambridge, MA: MIT Press.

Collins, J., Hall, N., & Paul, L. A. (Eds.). (2004). *Causation and counterfactuals*. Cambridge, MA: MIT Press.

Dietz, E.-A., Hölldobler, S., & Pereira, L. M. (2015). On conditionals. In *Proceedings of Global Conference on Artificial Intelligence (GCAI 2015)*.

Epstude, K., & Roese, N. J. (2008). The functional theory of counterfactual thinking. *Personality and Social Psychology Review, 12*(2), 168–192.

Foot, P. (1967). The problem of abortion and the doctrine of double effect. *Oxford Review, 5*, 5–15.

Ginsberg, M. L. (1986). Counterfactuals. *Artificial Intelligence, 30*(1), 35–79.

Halpern, J. Y., & Hitchcock, C. (2015). Graded causation and defaults. *British Journal for the Philosophy of Science, 66*, 413–457.

Han, T. A., Saptawijaya, A., & Pereira, L. M. (2012). Moral reasoning under uncertainty. In *Proceedings of 18th International Conference on Logic for Programming, Artificial Intelligence and Reasoning (LPAR)* (Vol. 7180 of LNCS, pp. 212–227). Berlin: Springer.

Hauser, M., Cushman, F., Young, L., Jin, R. K., & Mikhail, J. (2007). A dissociation between moral judgments and justifications. *Mind and Language, 22*(1), 1–21.

Kamm, F. M. (2006). *Intricate ethics: rights, responsibilities, and permissible harm*. Oxford, UK: Oxford University Press.

Kowalski, R. (2011). *Computational logic and human thinking: How to be artificially intelligent*. New York, NY: Cambridge University Press.

Lewis, D. (1973). *Counterfactuals*. Cambridge, MA: Harvard University Press.

Mandel, D. R., Hilton, D. J., & Catellani, P. (Eds.). (2005). *The psychology of counterfactual thinking*. New York, NY: Routledge.

Markman, K. D., Gavanski, I., Sherman, S. J., & McMullen, M. N. (1993). The mental simulation of better and worse possible worlds. *Journal of Experimental Social Psychology, 29*, 87–109.

McCloy, R., & Byrne, R. M. J. (2000). Counterfactual thinking about controllable events. *Memory and Cognition, 28*, 1071–1078.

McIntyre, A. (2004). Doctrine of double effect. In E. N. Zalta (Ed.), *The Stanford encyclopedia of philosophy*. Center for the Study of Language and Information, Stanford University, Fall 2011 edition. http://plato.stanford.edu/archives/fall2011/entries/double-effect/

Migliore, S., Curcio, G., Mancini, F., & Cappa, S. F. (2014). Counterfactual thinking in moral judgment: an experimental study. *Frontiers in Psychology, 5*, 451.

Niedenthal, P. M., Tangney, J. P., & Gavanski, I. (1994). "If only I weren't" versus "if only I hadn't": Distinguishing shame and guilt in counterfactual thinking. *Journal of Personality and Social Psychology*, *67*(4), 585–595.

Otsuka, M. (2008). Double effect, triple effect and the trolley problem: Squaring the circle in looping cases. *Utilitas, 20*(1), 92–110.

Pearl, J. (2009). *Causality: Models, reasoning and inference*. Cambridge, MA: Cambridge University Press.

Pereira, L. M., Aparício, J. N., & Alferes, J. J. (1991). Counterfactual reasoning based on revising assumptions. In *Proceedings of International Symposium on Logic Programming (ILPS 1991)* (pp. 566–577). Cambridge: MIT Press.

Pereira, L. M., Dell'Acqua, P., Pinto, A. M., & Lopes, G. (2013). Inspecting and preferring abductive models. In K. Nakamatsu & L. C. Jain (Ed.), *The Handbook on Reasoning-based Intelligent Systems* (pp. 243–274). Singapore: World Scientific Publishers.

Pereira, L. M., Dietz, E.-A., & Hölldobler, S. (2015). *An Abductive Counterfactual Reasoning Approach in Logic Programming*. Draft, Available from http://centria.di.fct.unl.pt/~lmp/publications/online-papers/counterfactuals.pdf

Pereira, L. M., & Saptawijaya, A. (2011). Modelling morality with prospective logic. In M. Anderson & S. L. Anderson (Ed.), *Machine ethics* (pp. 398–421). Cambridge: Cambridge University Press.

Pereira, L. M., & Saptawijaya, A. (2015). Abduction and beyond in logic programming with application to morality. Accepted at *Frontiers of Abduction, a special issue of IfCoLog Journal of Logics and their Applications*. Available from (preprint) http://goo.gl/yhmZzy

Pereira, L. M., & Saptawijaya, A. (2016). Counterfactuals, logic programming and agent morality. In R. Urbaniak & G. Payette (Ed.), *Logic, Argumentation & Reasoning*. Berlin: Springer (forthcoming).

Roese, N. J. (1997). Counterfactual thinking. *Psychological Bulletin, 121*(1), 133–148.

Roese, N. J., & Olson, J. M. (Eds.). (2009). *What might have been: The social psychology of counterfactual thinking*. Hove, UK: Psychology Press.

Saptawijaya, A., & Pereira, L. M. (2015). Logic programming applied to machine ethics. In *Proceedings of 17th Portuguese International Conference on Artificial Intelligence (EPIA)* (Vol. 9273 *of LNAI*). Berlin: Springer.

Scanlon, T. M. (1998). *What we owe to each other*. Cambridge, MA: Harvard University Press.

Scanlon, T. M. (2008). *Moral dimensions: Permissibility, meaning, blame*. Cambridge, MA: Harvard University Press.

Thomson, J. J. (1985). The trolley problem. *The Yale Law Journal, 279*, 1395–1415.

Vennekens, J., Bruynooghe, M., & Denecker, M. (2010). Embracing events in causal modeling: Interventions and counterfactuals in CP-logic. In *JELIA 2010* (Vol. 6341 of LNCS, pp. 313–325). Berlin: Springer.

Wallach, W., & Allen, C. (2009). *Moral machines: Teaching robots right from wrong*. Oxford, UK: Oxford University Press.

Weiner, B. (1995). *Judgments of responsibility: A foundation for a theory of social conduct*. New York, NY: The Guilford Press.

Children's Early Non-referential Uses of Mental Verbs, Practical Knowledge, and Abduction

Lawrence D. Roberts

Abstract Abduction is reasoning which produces explanatory hypotheses. Models are one basis for such reasoning, and language use can function as a model. I treat children's early use of mental verbs as a model for dealing with a problem from developmental psychology, namely, how children's early non-referential use of mental verbs might give children an early grasp of the mental realm. The present paper asks what practical knowledge of mental actions accompanies children's competent use of mental verbs. I begin with examples of non-referential verb uses and some theories from Diessel and Tomasello (Cognit Linguist 12:97–141, 2001) as bases for discussion. I argue that in using mental verbs non-referentially children understand several kinds of relations which people have to situations. Children learn how to use mental verbs to request someone to search for a situation in the past or in the physical surroundings; they learn to express hopes so as to affect their future; they learn to vouch strongly or weakly for the existence of a situation. In all of these cases it appears that the children's main focus is on interactions with people, in which one person's mental action in relation to a situation described in a COMP-clause is intended to have an effect on the other person. Children do not understand the nature or mechanisms of any of these mental actions, but instead focus on practical matters: how to use the verbs to perform certain actions in relation to other people and various situations. It appears that in these early uses children do not view mind as at all separate from the interactional and physical world.

L.D. Roberts (✉)
State University of New York, Binghamton, NY 13901, USA
e-mail: lroberts@binghamton.edu

L. Magnani and C. Casadio (eds.), *Model-Based Reasoning in Science and Technology*, Studies in Applied Philosophy, Epistemology and Rational Ethics 27, DOI 10.1007/978-3-319-38983-7_16

1 Introduction

This paper has two themes, one of content and the other of methodology. The content concerns recent studies[1] of children's early use of mental verbs. These studies conclude that children make both referential and non-referential uses of the verbs, with the latter preceding the former. The priority of the non-referential uses raises the hope that they will illuminate children's very early thoughts about mind. An early paper (Shatz et al. 1983, 319) on such uses suggests they may help us understand "the early development of an ability to distinguish and communicate about the internal world of thoughts, memories, knowledge, and dreams." Nevertheless, the literature has not focused on the question of how the early uses are about mind, probably because of an emphasis on empirical rather than theoretical issues. The literature has also neglected questions about the nature of referentiality, even though the thesis that non-referential uses precede referential ones is generally held. Because there is solid evidence that the uses of mental verbs called "non-referential" precede those called "referential," I agree that the uses called "non-referential" are children's earliest uses of mental verbs. Although I sometimes follow the literature in calling such uses "non-referential," I set aside the question of whether they really are non-referential. My focus is on these earliest uses of mental verbs, about which I ask two questions:

(Q1) What positive semantic and pragmatic features are shared by the various types of early (non-referential) uses of mental verbs?
(Q2) What do these early (non-referential) uses of mental verbs tell us about children's early practical understanding of mind?

There have been treatments of (Q1) in the literature, but my approach is different. Regarding (Q2), I have seen no discussions concerning how early mental verb uses might cast light on children's practical understanding of mind. My interest in the latter question does not imply that children lack any earlier understanding of mind prior to their use of mental verbs. In fact, there is solid evidence that children have a practical understanding of some aspects of mind at one year of age, before they begin to talk (e.g., at one, children know in a practical way that other people have goals (cf. Bruner 1983), and that children share joint attention to things with other people (cf. Carpenter et al. (1998).

The second theme of the paper is methodological, and is based on recent work on C.S. Peirce's notion of abduction (reasoning to explanatory hypotheses). This work suggests that language is a cognitive mediating structure which helps us deal with everyday life. Such practical functioning of language requires its users to have a practical understanding of how to use mental verbs (and related linguistic

[1]The main studies I have in mind are Diessel and Tomasello (2001), Bartsch and Wellman (1995), Bretherton and Beeghly (1982), Furrow et al. (1992), Montgomery (1997, 2002, 2005), Nelson and Kessler Shaw (2002), and Shatz et al. (1983).

practices) as instruments which contribute to their interactions with other people and the world.[2]

Section 2 of the present paper provides data and descriptions from Diessel and Tomasello (2001) for children's three kinds of earliest (non-referential) uses of mental verbs. Section 3 presents the main argument of the paper, which answers the two main questions described earlier:

(a) Linguistic (pragmatic and semantic) accounts of children's early mental verb uses are developed;
(b) The linguistic accounts in conjunction with the data about children's early mental verb use are used in abductive reasoning to develop hypotheses describing the practical knowledge of mind which accompanies children's three kinds of early mental verb uses.

Section 4 characterizes the reasoning used in this paper as involving two types of abduction.

2 Diessel and Tomasello (2001) on Children's Early Non-referential Uses of Mental Verbs

Diessel and Tomasello (2001) takes up (Q1) (concerning the linguistic nature of the early mental verb uses) in regard to syntax, semantics, and pragmatics, with a major focus on syntax, whereas the present paper focuses more on semantics, pragmatics, and children's practical knowledge of mind. Examples and descriptions from Diessel and Tomasello (2001)[3] provide the empirical starting point for my study of children's early uses of mental verbs. In their article, children's early (non-referential) uses of mental matrix verbs are divided into three varieties, which I list here along with some of their examples:

A. Parenthetical epistemic markers: "think," "know," "bet," "mean," and "guess," e.g., I think I'm go in here; It's a crazy bone, I think; I know this piece go; How do you know that a duck?
   The verbs "*think*" *and* "*know*" are primarily characterized in their early uses, according to Diessel and Tomasello (2001), as *epistemic markers* because they indicate the strength of a speaker's belief.
B. Deontic modality markers: "wish" and "hope," e.g., I wish I could play with dis; I wish we can eat; I hope he won't bother you; I hope my cat friends are

[2]The point is not merely that people think about their actions; Magnani (2001, 309) suggests that "Manipulative abduction happens when we are thinking through doing." Cf. also Magnani (1999, 2006).

[3]Diessel and Tomasello (2001) locates the issues within a sophisticated syntactic account, provides a clear and economical array of types of non-referential uses, and makes intriguing suggestions about the presence of illocutionary forces, performatives, and parentheticals in early mental verb uses.

alright; I hope de house won't be on fire. Concerning uses of "I hope," Diessel and Tomasello (2001: 118) says: "they can be seen as deontic modality markers, serving basically the same function as a modal adverb such as hopefully." I agree that "I hope" can be seen as a "modality marker," but I do not understand why the modality is called "deontic," because this term comes from the Greek word for what is obligatory, or what one must do. If "I hope" and "hopefully" indicate a mood of a verb, the appropriate mood would be "optative," a point to which I return later.

C. Discourse directives: "see," "look," and "remember," e.g., See I broke a teeth; See it will work; Look birdie fly; Look (pause) Daddy put it on a wall; You remember I broke my window; Remember you reading de puzzle? These are described in Diessel and Tomasello as attention getters and as indicating a request or a question.

The three early types of mental verb uses are described by Diessel and Tomasello mainly in terms of actions and their functions. These functions result from the use of the entire sentence containing the mental verb. Because each of the three functions is different, they do not by themselves indicate a single common function which might be shared by all three kinds of early mental verb uses.

My approach to studying the early mental verb uses differs from that of Diessel and Tomasello (2001) on three points. First, their approach began with syntax, and on that basis developed pragmatic and semantic accounts. Second, their main semantic/pragmatic accounts appealed to Urmson's parentheticals and Austin's performatives and illocutionary forces. Third, they do not ask what practical knowledge of mind children might have in virtue of their competence in the three types of early mental verb uses. On each point, my approach is mostly different from theirs: First, I begin with pragmatics and semantics. Second, I use Wittgenstein's language games and Austin's view of illocutionary and locutionary acts[4] as bases for explaining the pragmatics and semantics of mental verbs uses. Third, on the basis of the data and linguistic theory, I develop hypotheses for the content of children's practical knowledge of mind as connected to their competence in using mental verbs.

3 Linguistic Mechanisms and Practical Knowledge in the Three Types of Early Mental Verb Uses

Linguistic accounts of children's early mental verb uses: Uses of mental verbs, and the societal practices on which they are based, are instruments by which people deal with others in regard to situations. To discern how such instruments work, one needs to describe the relevant linguistic practices, including especially the relevant pragmatic and semantic mechanisms and structures. For our purposes, we need to

[4]Though Diessel and Tomasello (2001) cites illocutionary force as affecting some early uses of mental verbs, this notion needs some development to cast light on early mental verb use.

understand the mechanisms and structures of Wittgensteinian language games and Austin's distinction between illocutionary and locutionary speech acts. These two accounts together provide an important, though partial, linguistic theory of mental verb use. After using this theory to clarify the linguistic goings on in the three kinds of early mental verb uses, I use these linguistic accounts as foundations for hypotheses about the practical knowledge of mind which children have in virtue of their early mental verb use. My argument begins with a brief introduction to Wittgensteinian language games.

Language Games: Children learn how to use "Look p" and Remember p" in language games, interactions in which actions and words are mixed together. Language games enable a child to learn how to use a word in relationships to other people and the real world. Mental verb uses thereby provide important devices for cooperating with other people in regard to everyday life. Language games embed linguistic and cultural practices of a society. Examples of language games are found in a culture's routines, for instance, those used in caring for infants. A child needs to be fed, washed, have her diapers changed, to have clothes put on her, and to have them taken off. A routine for feeding a small child may run like this: the child is placed in a high chair with a tray, a bib is put on her, a spoonful of cereal is brought to the child's mouth, she eats it, and she is offered more. Talk accompanies this routine, but I focus on just one phrase. After the child eats for a while, and appears to be losing interest, the parent may ask "Do you want some more?" while offering another spoonful of cereal. Perhaps a child may open her mouth and accept the cereal, or may look away, or push the spoon away. Or, in later feedings, a child may nod for yes or no in response to an offer of more food, or may say "more" if the parent is distracted from providing the cereal. Routines in early language games aim at a goal, and assign roles to participants in producing a sequence of actions which lead to the intended result. Although language games enable children to obtain a practical understanding of what a word contributes to the interaction, this practical understanding may be partial, because a child may grasp at first only part of a word's meaning. Another complication is that a child may grasp some words only by understanding their contrast to other words.

An important feature of human language games, but not of the language games in which dogs learn words, is "role reversal imitation." Tomasello (1992, 217–218, cf. 1999, 105–107) assigned this phrase to the fact that when a child learns a new word in an interaction, she is able to use it not only in the role she played (as addressee) in her early interactions involving the word, but also in the role of the speaker. Such imitation is based on cultural learning, in which "the learner participates with the other intersubjectively (in a joint attentional state) and learns about a situation … from the "inside, … " (Tomasello 1992, 217). In such learning, a child understands the other person's perspective and purpose. Carpenter et al. (2005, 275) describe such learning as holistic and involving a "bird's eye view" of the interaction. I agree with these characterizations, and would add that role reversal imitation is a likely outcome of learning a word in a human language game, because the meaning of such a word is its contribution to the game as a whole, and knowing such a game requires knowing not only the word's function in the game, but also the roles of the participants and their contributions to the language game. Role reversal imitation is important for my account of children's practical knowledge connected to mental verb use because it indicates that children understand the roles of both speaker and addressee in a language game even though the data may show them playing only the role of speaker.

C-type uses of mental verbs. I begin with C-type uses because I find their semantic/pragmatic structure more transparent than the other two types. Consider an example of a C-type verb use from Diessel and Tomasello (2001): "Look birdie fly." C-type uses are called "discourse directives" in Diessel and Tomasello (2001), an appropriate name in that these verb uses have the function of directing the discourse toward the described situation ("birdie fly" in the example). Because this focus on the COMP-expressed situation results from the action of using the mental verb, we must investigate the action itself, and not merely its function. By this action, a speaker is asking the addressee to search for and focus on the situation "birdie fly." Therefore, the action is not a matter of asserting "birdie fly," and thereby directing the addressee's attention to the situation, but instead a matter of making a request of the addressee to search for and focus on that situation "birdie fly." Thus a twofold action is involved in the speaker's use of "Look p:" one act is requesting an action by the addressee, and the other act specifies that the addressee's action is to be that of searching the immediate surroundings for "p" (and focusing on it). Children learn from language games that both actions are involved in the meaning of "Look p." A child who has sight would probably also gather that sight is to be used in such searches because it produces the intended result. A blind child would not learn this, because adults would use "see" or "look" to direct such a child to explore the immediate surroundings by touch rather than sight.[5]

The language game model helps one to understand how a child learns to use "Look p" in interactions with others, in order to request them to search for and focus on a situation, and to share attention to it. Although the language game model provides a partial explanation of how children learn their earliest words, it does not explain how the verb use itself works in making its contribution to the interaction. This omission occurs because the language game model by itself provides only the social interactive foundations of a theory of language.[6] It does not explain the distinctive functioning of verbs, which is better handled by Austin's distinction between illocutionary and locutionary speech acts, or at least I will argue this distinction can be developed so as to cast light on mental verb functioning.

On Austin's view, the verb of a main clause is usually used to express both illocutionary and locutionary speech acts. Sometimes a main clause verb is used to express only an illocutionary act, but this is infrequent. The two kinds of speech acts function differently. An illocutionary act is a move in conversation by which a

---

[5]Empirical evidence for this is in Landau and Gleitman (2012, 100).

[6]Montgomery (1997, 2002, 2005) provides clear language–game interpretations of children's early uses of mental verbs, but his Wittgensteinian account does not explain how verbs function as contrasted to adjectives or nouns. Montgomery's account also has the helpful feature of avoiding any discussion of reference.

speaker acts so as to indicate that she is using the verb to relate a proposition to the addressee and the real world[7] (e.g., by asserting that the mentioned situation is actual, or by requesting the addressee to make the situation actual, or by asking whether the situation is actual, or by informing the addressee that the speaker would like the situation to be actual). Because illocutionary acts are the means by which a speaker connects the symbolic functioning of words to the real world and to interacting with other people, they are important for language. In contrast to illocutionary acts, a locutionary act is a move in conversation by which one brings entities, properties, and relations into discourse as content and structures (e.g., "x breaks y," or "x gave y to z," "John," "a cup," "milk," "of"). To combinations of such content, a speaker applies illocutionary acts (of assertion, requesting, etc.). These two types of speech acts are distinct and complementary, even though a single verb use may express both types (for instance, an illocutionary act for requesting, and a locutionary act for supplying some content for the request, e.g., a use of "Stop!"). Content is contributed to discourse not only by locutionary acts but also by contexts.

Although asserting, questioning, and requesting are the most frequently used varieties of illocutionary acts, there are many others. Austin (1962) held that most illocutionary acts not explicit, and instead may be indicated by a variety of factors, including moods of verbs and contexts. In cases of inexplicit illocutionary acts, a speaker in a single use of a verb performs both illocutionary and locutionary speech acts.

In C-type uses of mental verbs, the distinction between illocutionary and locutionary speech acts is easily noticed because a single such use (e.g., "Look p") communicates two distinct actions, each of which has a distinct agent: the speaker performs the illocutionary act of requesting, whereas the addressee is asked to perform the searching for and focusing on situation "p." Both conversers have practical knowledge of these functions of speaker and addressee because they learned the relevant language game for "Look p." In a C-type use of "Look p," the mental verb is in the second person and imperative mood, and in the context, expresses a request (an illocutionary act). Paired with this illocutionary act is a locutionary act expressing the content of the request, namely, searching the immediate physical surroundings for situation "p," and focusing on that situation. Within this locutionary-act meaning of a use of "look p" I distinguish two complementary ingredients: the action of searching for and focusing on a certain situation, and the type of location to be searched, namely, the immediate physical surroundings. I refer to the latter as the "domain" for the search.

Children learn C-type uses of "look" and "remember" from interactions in language games, which fix both the kind of search and the domain to be searched. To learn what domain goes with a particular mental verb, a child must learn in

[7]The account offered here for the contrast of illocutionary and locutionary speech acts is mine, and builds mostly on views of Austin and Searle, who put relatively little effort into a general characterization of the natures and mutual relations of illocutionary and locutionary acts. Such general characterization is the goal of the present section.

language games particular situations which fall under "past shared experiences" or "immediate surroundings." From those particular cases, a child generalizes (probably in a non-verbal way) to the relevant domain for each kind of search. Such generalization results in know-how about where and how to search, but need not involve explicit descriptions of the domain or of the relevant power of mind. Instead the know-how is constituted by a practical skill of making C-type uses in relation to the appropriate power of mind and its domain.

The generalization by which children learn how to use a verb from particular language games involving the verb resembles the process by which children learn early non-mental words. In the latter case, a child learns how to pick out a kind of thing (e.g., a dog or a horse), not by a general description of it, but instead on the basis of her experience of particular individuals in language games. Katherine Nelson[8] has explained such learning to use a word as based on learning the functional contributions which the dog or horse makes to the event which the language game is about. By experiencing samples (or pictures) of dogs in language games, a child may form a practical hypothesis of how to use "dog." Then the child may apply the recently learned word to previously unexperienced members of its extension. In this way, her practical hypothesis goes beyond what the child has experienced, so that the reasoning underlying her application of the word is augmentative. Such use of "dog" by a child indicates her possession of a "practical concept" of dogs, which was formed on the basis of language games. The reasoning from experiencing uses of "dog" in language games to forming such a concept of dogs is, in my opinion, a practical form of abduction.[9] Similar reasoning occurs as a basis for C-type mental verb uses: children learn the functional contributions which sight and memory make to language games in which "Look p" and "Remember p" are used. Children do this mainly on the basis of discerning (without needing to describe) the domains to which the particular situation "p" belongs, which domains are learned from particular examples of situations in language games. Those examples would be of situations with which the child is familiar, and therefore knows that the situation is in the immediate physical surroundings (for "see"), or in past shared experience (for "remember"). Children also learn in a practical way which mental powers are to be used in order to succeed in the relevant language games. Again, no verbal descriptions of the mental powers are needed. In the case of "look," sighted children also learn from language games that the eyes are used, and a gaze is directed at the situation "p," and that the conversers share visual attention to the situation.

The analysis for "Remember p" is parallel to that given for "Look p." "Remember p" is used to question an addressee about a COMP-expressed situation in past shared experience, which is the relevant domain for uses of "remember"

[8]Nelson invented the "functional core hypothesis" which explains children's early grasp of the meaning of nouns as based in what the referents of the nouns contribute to events. Cf. Nelson (1983, 1996).

[9]I argue for this thesis in Roberts (2004).

(as contrasted to the present physical surroundings for "look"). Another difference is that the use of "look" brings in an external sense to do the searching, whereas "remember" does not. Because the domain (past shared experience) for "remember" is the only likely determining factor which children pick up from language games involving the verb, it is especially important for acquiring that verb.

4 Children's Practical Knowledge of Mind

Children have practical knowledge of mind long before they use mental verbs. At age one, they understand that other people have goals and purposes in their actions, and also that other people share with them joint attention to things.[10] When children begin to use mental verbs at around two and a half years of age, they appear to have additional kinds of practical knowledge of mind. This is not explicit knowledge—children cannot describe it. Instead they use the knowledge in their actions, as shown in the data on the early mental verb uses. The hypotheses to be suggested here for the content of such practical knowledge are based on both data and linguistic theories, because children use the mental verbs in accord with the linguistic practices of their culture. All three kinds of early mental verb uses have parallel structures in which conversers relate to each other in regard a certain situation. In this structure, a mental verb is used as an instrument for the function of relating one person to another and to a situation. This structure puts a social dimension in children's early practical knowledge of mind insofar as the verb uses derive from language games, and lead (most of the time) to the speaker and addressee focusing on the same situation "p" and sometimes acting cooperatively in regard to it. Children's earliest uses of mental verbs have a social foundation, and thereby link both conversers to the same situation.

Outline: children's practical knowledge of mind in C-type mental verb uses:

1. As competent users of "See p" and "Look p," children know how to use the appropriate mental power:
   A. Children with sight know how to search, using sight and its appropriate domain, for "p," and to focus on "p."
   B. Blind children also know how to search, using touch and its appropriate domain, for "p," and to focus on "p."
   C. Children learn which power is appropriate to use from language games which provide examples from the domain to which the mental verb uses apply. A practical generalization on the examples enables them to have practical knowledge of the domain.

[10]Cf. Bruner (1983) and Carpenter et al. (1998).

D. The practical knowledge described in (1 A and B) extends not only to sight and touch, but also to a more general level of mind, that of searching for a situation and discerning that one has succeeded in the search.[11]

2. As competent users of "Remember p," children know how to use the appropriate mental power:

    A. They know how to search (using memory and its appropriate domain) for "p," and to focus on "p."
    B. They know in a practical way (from language-game experience) which mental power is to be used, and which domain (past shared experience) is to be searched.
    C. In addition to a practical knowledge of memory, children making competent C-type uses of "Remember p" also have a practical grasp on the more general level of mind described in (1D).

3. Children who are competent in making C-type uses of "see," "look," and "remember" know how to discern whether a search based on such a use is successful or not.[12]
4. In most C-type uses, a speaker who is competent in such use gains practical knowledge that the addressee is attending to the same situation, and each knows the other is attending to the situation (i.e., they are sharing joint attention to it).
5. Children's competence in C-type verb uses reinforces their practical knowledge that engaging in language games is social, generally cooperative, and interpersonally valid. These three features of language games are corroborated by particular features of C-type uses: children can observe in such uses that the seeing or remembering which is requested usually results in joint attention of the conversers to the same situation.

B-type uses of mental verbs. Whereas C-type mental verb uses are cognitive, involving sight, touch, or memory, B-type uses are appetitive, concerning a person's wishes, hopes, or inclinations. Consider an example: "I hope he won't bother you." In this use of "hope," the speaker expresses her inclination toward a future in which a certain male won't bother the addressee. The basic structure of B-type uses has a speaker using a mental verb to relate herself to a COMP-expressed situation, and to inform addressee of that relation. Because the relation is that of expressing her inclination toward having the situation in the future, the correlative domain for B-type uses of "hope" is that of situations which the speaker wants to have in her future. Although B-type verb uses are like C-type ones in introducing connections of a person to a situation, each differs from the other on three (underlined) points:

[11]Landau and Gleitman (2012, 94) make a very similar point when they describe uses of "See p" addressed to blind children as leading to "perceptual exploration and achievement."

[12]I believe children usually succeed in such mental verb uses, but as Millikan (2012) observes, "Conventional coordination patterns need to succeed only often enough to avoid extinction."

(i) B-type uses: the *speaker* asserts an inclination (an *appetitive* relation) toward a certain *future* situation.
(ii) C-type uses: the *addressee* is asked to search for and focus on (*cognitive* relations) a certain *present* (or *past*) situation.

Despite these substantial differences between C and B types of mental verb uses, there are parallels in the practical knowledge that accompanies them. As with C-type verb uses, a child's knowledge of the B-type uses is based on observing and participating in particular language games where the verb is used. Another similarity of B and C type verb uses is that children learn the domains for the relevant COMP-expressed situations from particular examples. In B-type uses, the hoped-for situations are located in the future, and are ones toward which the speaker is attracted. In language games and in everyday life children observe how people act when attracted to a certain situation, and how they act when they obtain, or fail to obtain, the desired situation. In learning both B and C type mental verb uses, children generalize from particular uses of mental verbs in language games to a skill in making such uses themselves.

How do illocutionary and locutionary speech acts fit into B-type mental verb uses? Surprisingly enough, two different accounts (or analyses) of speech acts in such uses have the same semantic content, and seem to yield equally viable accounts of B-type uses. On one analysis, a B-type verb use of "I hope p" expresses a locutionary act which connects the speaker to the COMP-expressed situation, whereas on another analysis, an illocutionary act produces the same connection. On the first analysis, the locutionary act expressed by using "hope" indicates an inclination of a person toward having a future situation "p," and the speaker asserts (by an illocutionary act of assertion) that she is such a person.

The second analysis treats the inclination of the speaker to situation "p" not as the locutionary act content of "hope," but instead as what is expressed by performing a particular kind of illocutionary act, that of expressing a hope for a situation. This kind of illocutionary act is marked by a verb mood (the *optative mood*) in some languages, including classical Greek. On this analysis, the verb would be an explicit illocutionary act verb, in that it functions as an illocutionary act operator, which determines the illocutionary act for the COMP-clause to be the expression of a hope. In such a use of the verb "hope," it has no locutionary-act function; instead its only function is to express a certain illocutionary act which brings in the descriptive content of "hope," namely, the speaker is inclined toward having situation "p." This second analysis is congruent with the characterization in Diessel and Tomasello (2001) of B-type verb uses as "modality markers," because the optative mood is such a marker. In addition, Diessel and Tomasello (2001: 118) states that B-type uses of "hope" serve the same function as "hopefully," an adverb which I take to function in some contexts as an optative mood operator. Both of the analyses for B-type uses of "I hope p" have the same result for successful communications: the conversers share the same practical knowledge, namely, the speaker is inclined toward having situation "p" in her future.[13] Nevertheless, I prefer the illocutionary act analysis because it seems simpler to treat the B-type mental verb use as an illocutionary act operator.

Outline: children's practical knowledge of mind in B-type uses of mental verbs.

1. As a competent B-type user of "I hope p," a child has practical knowledge that the speaker is inclined toward having situation "p" in her future. If the addressee learns this from the B-type verb use, the communication is successful.
2. A child who is competent in B-type uses knows practically that both conversers know the speaker is inclined toward having situation "p." This usually leads to joint attention to the expressed hope, so that each person has practical knowledge that both conversers are sharing attention to the same situation.
3. A child knows practically that telling one's hope to another person may affect the latter's behavior. The speaker is likely to have some goal in telling the addressee about the hope, and the addressee is likely to consider why the speaker is sharing this particular hope.
4. The conversers have practical knowledge that a B-type use differs from a request for the COMP-expressed situation. Such knowledge may be connected with the cooperativeness that is a background feature of language games.

A-type uses of mental verbs. An example of this third type of non-referential use of mental verbs from Diessel and Tomasello (2001) is "*It's a crazy bone, I think.*" This example is an epistemic use of the mental verb in that it is used to express a belief of the speaker and also mark the strength or weakness with which the speaker holds the belief. Non-epistemic uses of "I think p" and "I know p" also exist, but I take them up after treating the more common epistemic uses.

In an epistemic use of "I think p," the speaker informs the addressee that she (the speaker) holds tentatively or weakly that situation "p" actually exists. This communication is equivalent to the speaker making a weak assertion that "p" is actual, because assertions have the function of informing others of what the speaker takes to exist. Contrasting to the tentativeness of "I think p" is the confidence expressed by a use of "I know p." Because this contrast in A-type uses of "think" and "know" concerns the strength of a speaker's commitment to the actuality of situation "p," it is reasonable to view these verb uses as modulating the speaker's illocutionary act of assertion (cf. Shatz et al. 1983). Such modulation is useful because existence claims require evidence, and the strength of an assertion should indicate the strength of the evidence for it. Although speech acts such as modulating an assertion are intended to reflect the speaker's views, they do not represent the actual internal

[13]Diessel and Tomasello (2001, 118) suggests that "hope" in B-type uses does not denote the speaker's desire. The present paper omits discussion of reference and denotation (whose differences are treated in Roberts 1993). Underlying my two analyses for B-type uses of mental verbs is the distinction between locutionary and illocutionary speech acts, rather than notions of reference or denotation.

mental processes by which one produces the speech acts. Instead they represent the claims, requests, questions etc. which the speaker wishes to communicate.[14]

Further evidence for taking A-type uses of "I think p" and "I know p" to function as illocutionary act operators is that "I think p" functions in about the same way as "Maybe p" in most contexts. This use of "maybe" is parallel to that of "hopefully" as described earlier for type-B uses. Both adverbs have a sentential function of a particular type, in that they are used in main clauses to express an illocutionary act operator which affects that main clause.[15] A use of "maybe" usually changes the illocutionary act of the proposition on which it operates from an unmarked assertion to one marked as weak.

Can we describe a domain of situations which is complementary to the modulating action of A-type epistemic uses of mental verbs? I picture such a domain as comprised of situations which the speaker takes to be actual. The A-type use expresses both the actuality of a situation, and the speaker's confidence or hesitancy about her judgment of that actuality.

Do epistemic A-type mental verb uses also have a locutionary act functioning? My view is that they do not. They function only as illocutionary act operators on the associated COMP-clause, and this functioning does not bring in the usual (dictionary) meanings of "think" and "know," which for "think" is "to form or have in the mind" (Merriam-Webster's Collegiate Dictionary, tenth edition, 1996). Children from 2 to 5 years of age lack explicit generic notions such as "mind," but do they have a practical generic notion of "mind? If such is the case, it would be based on their knowledge of more particular types of mental activity. Because uses of "I think p" and "I know p" occur among children's earliest uses of mental verbs, it is unlikely that they know other more specific mental powers which are subcategories of thinking and knowing. Therefore, it is unlikely that these early uses of 'think" and "know" mark a generic notion of mind.

A-type uses of "think" and "know" have illocutionary speech act functions because people have to deal with imperfect knowledge. If one person tells another about the degree of confidence she has in a claim, this information about the speaker's perspective is likely to promote cooperation, and thereby be useful. Why are the mental verbs "think" and "know" used to indicate the strength of one's commitment? Perhaps the origin lies in the strong connection which knowledge has to truth: to know "p" requires that "p" is true, whereas to think "p" does not. Perhaps because "knowing" requires truth, "I know p" came to be used to express confidence in the truth of "p," whereas because "thinking" does not require truth, "I believe p" came to be used to express some lack of confidence in the truth of "p."

The meanings of A-type uses of "know" and "think" differ from the dictionary meanings for the two verbs, which include "forming or having something in mind."

[14]Cf. Tsohatzidis (1994) and Siebel (2003).

[15]Most sentential adverbs have functions of merely making comments on a situation, whereas "maybe" and "hopefully" are illocutionary act operators which have a more foundational role in communication.

I take the latter meanings to be those of locutionary acts, which I have argued are absent from children's A-type uses of the verbs. The priority of children's A-type uses of the verbs for illocutionary acts may indicate that such uses of these verbs are easier or more useful for children than uses involving locutionary act meanings.

Children often acquire one part of a word's meaning before other parts, e.g., in the diary study of Tomasello (1992), the author's child is reported to use the words "sweep" and "work" at first with meanings that include only the physical motions of sweeping and working, but not the goals and results which are included in the standard meanings of these words. These are cases of knowing only part of the locutionary act meanings of the verbs, whereas the partial acquisition involved in children's early uses of "I know p" (or "I think p") is a matter of knowing how to use the verbs to perform certain illocutionary acts, but without picking up the adult locutionary act meanings for the verbs. The basic difference, then, between children's and adults' uses of "think" and "know" is that the children do not grasp the locutionary act meaning of the verbs. This difference may be part of what makes it difficult for three year old children to pass the false belief tests devised by psychologists: such tests require locutionary act notions of knowledge and belief which three-year olds appear not to have.

So far I have supposed that A-type mental verb uses are epistemic, in that they concern the strength of a speaker's assertion that a particular situation really exists. Although this assumption holds for most A-type uses, some appear not to be epistemic: e.g., "*I think I'm go in there*" (Diessel and Tomasello 2001). A child may use this sentence to express an intention to go in there, which it does in a weaker way than would uses of "I'm go in there" or "I know I'm go in there." This appears to be a volitional use of "I think p," quite distinct from epistemic uses. In both types of uses, however, the function of "I think" is to weaken an illocutionary act, whether it is epistemic or volitional. This feature common to both kinds of A-type verb uses provides added support for viewing them as illocutionary act operators, which strengthen or weaken an illocutionary act. Competence in such uses requires one to have a practical understanding of how to weaken (or strengthen) an assertion or an expression of an intention to act.

Although type-A uses of "think" and "know" directly express illocutionary speech acts, rather than mental powers or processes, these speech acts inform the addressee of the strength of a speaker's claim or intent to act. Underlying such personal support for a truth-claim or an intended action are mental processes, but A-type uses of verbs focus on the speaker's personal responsibility for the claim or the intent to act, rather than on the mental processes used by people in evaluating and deciding on truth-claims or courses of action. Perhaps A-type mental verb uses are important because of their effect on personal relationships.

Outline: children's practical knowledge of mind in A-type uses of mental verbs.

1. From epistemic A-type uses of "I think p" and "I know p," an addressee has practical knowledge of the speaker's judgment in regard to whether "p" is actual or not, and also of the speaker's perspective on how strongly or weakly she is

committed to the judgment. Such strength or weakness correlates with the amount of personal responsibility which the speaker takes for her judgment.
2. From volitional A-type uses of "I think p" and "I know p," an addressee has practical knowledge of the speaker's intent to perform the action described in "p," and of the strength of her commitment to that intent. The strength or weakness of this commitment correlates with the amount of personal responsibility which the speaker takes for her expressed intent to act as described in the COMP-clause.
3. Conversers competent in A-type verb uses know how to use the relevant mental powers for having and expressing in a modulated way an assertion (or an intention to act), but this does not imply knowledge about the nature of judgment, volition, the mind, or its mechanisms. As in types C and B uses of mental verbs, children's practical knowledge concerns appropriate uses of particular powers of the mind, rather than their nature or processes.
4. Those who are competent in A-type verb uses know how to share attention to the speaker's modulated judgment or volition, which sharing may provide a basis for planning or action in regard to the COMP-expressed situation.
5. Competence in A-type mental verb uses implies that a speaker has practical knowledge that such uses may affect her relationship to the addressee not only concerning the COMP-expressed situation, but also concerning the addressee's evaluation of the speaker in regard to personal reliability, whether it concerns accuracy of assertions or resoluteness in acting according to one's stated intentions.

5 Three Kinds of Abduction

Abduction is reasoning to explanatory hypotheses which are based on a new concept. "Abduction" is the name for the reasoning which underlies scientific theories, and I propose to call such reasoning the "classical notion of abduction." Two different but related notions of abduction have been used in the present paper. One is practical abduction by which children acquire their early uses of mental verbs, and the second is practical-made-explicit abduction which aims at making explicit the knowledge present in practical abductions—this aim differs from that of providing an explanatory theory of mental processes. I take up practical abduction first. By such abduction, a child generalizes from language games involving the verb use (e.g., "See p") to a skill of using the verb in new situations, but with the same meaning for the word as that experienced by the child in language games. Because the new situations are not exactly like the original ones, mere generalization from earlier uses is insufficient: a child needs a practical hypothesis about the function of the word in language games she has experienced. Such a hypothesis often needs refinement, as shown by children's incorrect extensions of their early vocabulary. Not only is a child's early use of mental vocabulary based on practical hypotheses of how to use the word, but these hypotheses provide the child with a new concept, e.g., uses of "see" or "remember" provide a particular concept of an action which links the conversers to a COMP-expressed situation in the ways described earlier for C-type uses of mental verbs. A child's concept of such an action, however, is practical, in that it concerns how to use the word, which does not require an ability to describe the kind of action. Because the process by which children acquire new words requires both generalization from particular instances and the formation of a practical hypothesis of how to apply the word to new situations, and because this process results in a new concept, I view the reasoning underlying the process as abductive. The skill of using the word in a certain way, according to the practices in language games, amounts to a new practical concept.

In addition to the classical notion of abduction and practical abduction, there seems to be a related third variety, which I call "practical-made-explicit" abduction. Such abduction is central to the present paper, and aims at hypotheses which make explicit the knowledge content implicitly contained in the skill of using a mental verb in accord with the societal practice. Making knowledge explicit is different from explaining the causal mechanisms and structures which produce a phenomenon, but hypothesis making is required for both processes.

This third variety of abduction may seem not to fit traditional descriptions of abduction because it does not seem to introduce a new concept. This is because a practical version of the relevant concept already exists as embedded in the practice of using the word in a particular way. Because the embedded concept is implicit, children need not be able to describe it, or to know much about it or its relations to other things. Because of these deficiencies, practical knowledge seems to be inadequate knowledge. To render it more adequate, one would need to delineate the knowledge explicitly, so that one could inspect its basic nature and its relationships. Such delineation in words of what one knows practically requires one to describe the actions, contexts, and results involved in the practical actions and know-how. Next one must construct hypotheses about the explicit knowledge that would be needed, in the absence of practical knowledge, to produce the same results. Mere intuitions about children's early uses of mental verbs are unlikely to suffice for building such hypotheses for the practical knowledge of mind associated with such verb uses. I found I had to introduce theories from Wittgenstein and Austin in order to figure out what the mental verb uses imply about children's knowledge. This was required because we have to know how language works (or how people use it) in order to figure out the sorts of knowledge that are connected to one's use of mental verbs. Language is a complex instrument, and we need theories of it to understand the implications of its use for knowledge. In forming and refining theories of language, one would use abductive reasoning of the traditional sort. Such theories would then be used to contribute to the practical-made-explicit abduction needed for constructing hypotheses about the practical knowledge of mind that accompanies children's early competence in mental verb use.

Acknowledgements I am grateful to Prof. Michael Tomasello and the Max Planck Institute for Evolutionary Anthropology for opportunities to do research there, and for a pre-publication copy of Diessel and Tomasello (2001). Thanks are also due to Prof. Lorenzo Magnani for encouraging my research on abduction, and to Prof. Marco Buzzoni for pointing out an important clarification needed in the paper.

Bibliography

Austin, J. L. (1962). *How to do things with words* (2nd ed.). Cambridge: Harvard University Press.

Bartsch, K., & Wellman, H. M. (1995). *Children talk about the mind.* New York: Oxford University Press.

Bretherton, I., & Beeghly, M. (1982). Talking about internal states: The acquisition of an explicit theory of mind. *Developmental Psychology, 18*, 906–921.

Bruner, J. (1983). *Child's talk: Learning to use language, New York.* WW: Norton.

Carpenter, M., Nagell, K., & Tomasello, M. (1998). Social cognition, joint attention, and communicative competence from 9 to 15 months of age. *Monographs of the Society for Research in Child Development*, *63*(4), 225.

Carpenter, M., Tomasello, M., & Striano, T. (2005). Role reversal imitation and language in typically developing infants and children with autism. *Infancy, 8*, 253–278.

Diessel, H., & Tomasello, M. (2001). The acquisition of finite complement clauses in english: A corpus-based analysis. *Cognitive Linguistics*, *12*, 97–141.

Furrow, D., Moore, C., Davidge, J., & Chiasson, L. (1992). Mental terms in mothers' and children's speech: Similarities and relationships. *Journal of Child Language, 19*, 616–631.

Gleiitman, L., and Barbara, L. (2012). "Every child an isolate: nature's experiments in language learning," in Rich Languages from Poor Inputs, ed. by Massimo Piatelli-Palmarini and Robert C. Berwick, 91–104, Oxford, Oxford University Press.

Magnani, L. (1999). Model–based creative induction. In L. Magnani, N. J. Nersessian, & P. Thagard (Eds.), *Model-based reasoning in scientific discovery* (pp. 219–238). New York: Kluwer.

Magnani, L. (2001). *Epistemic mediators and model-based discovery in science*. In L. Magnani & N. J. Nersessian (Eds.), *Model-based reasoning, science technology values.* New York: Springer.

Magnani, L. (2006). Symposium on 'cognition and rationalitiy: Part I' the rationality of scientific discovery: Abductive reasoning and epistemic mediators. *Mind and Society*, *5*, 213–228.

Millikan, R. G. (2012). *On meaning, meaning, meaning and meaning in current issues in theoretical philosophy III: Prospects for meaning*, R. Schantz (Ed.), Berlin, New York: de Gruyter.

Merriam-Webster (1996). Merriam-Webster's collegiate dictionary (10th ed.). Springfield, Merriam-Webster.

Montgomery, D. E. (1997). Wittgenstein's private language argument and children's understanding of the mind. *Developmental Review, 17*, 291–320.

Montgomery, D. E. (2002). Mental verbs and semantic development. *Journal of Cognition and Development, 3*, 357–384.

Montgomery, D. E. (2005). The developmental origins of meaning for mental terms. In J. W. Astington & J. A. Baird (Eds.), *Why language matters for theory of mind* (pp. 106–121). Oxford: Oxford UP.

Nelson, K. (1996). *Language in cognitive development: The emergence of the mediated mind.* Cambridge: Cambridge University Press.

Nelson, K., & Shaw, L. K. (2002). Developing a socially shared symbolic system. In E. Amsel & J. P. Byrnes (Eds.), *Language, literacy, and cognitive development* (pp. 27–57). Mahwas: Erlbaum.

Roberts, L. D. (2004). The relation of children's early word acquisition to abduction, *Foundations of Science*, *9*, 307–320.
Roberts, L. D. (1993). How Reference Works: Explanatory Models for Indexicals, Descriptions, and Opacity. Albany: State University of New York Press.
Shatz, M., Wellman, H. M., & Silber, Sharon. (1983). The acquisition of mental verbs: A systematic investigation of the first reference to mental state. *Cognition, 14*, 301–321.
Siebel, M. (2003). Illocutionary acts and attitude expression. *Linguistics and Philosophy, 26*, 351–366.
Tomasello, M. (1992). *First verbs: A case study of early grammatical development*. Cambridge: Cambridge University Press.
Tomasello, M. (1999). *The cultural origins of human cognition*. Cambridge: Harvard University Press.
Tsohatzidis, S. (1994). The gap between speech acts and mental states. In S. Tsohatzidis (Ed.), *Foundations of speech act theory: Philosophical and linguistic perspectives* (pp. 220–233), London, Routledge.

Abduction, Selection, and Selective Abduction

Gerhard Minnameier

Abstract "Selective abduction" is a notion coined by L. Magnani, who contrasts it with the more common notion of "creative abduction". However, selective abduction may easily be confused with inference to the best explanation (IBE). This constitutes a problem, if IBE is reconstructed as an inductive inference. For on the one hand, abduction and induction must be distinct. On the other hand, Gabbay and Woods, but also Hintikka and Kapitan, even include hypothesis selection as part and parcel of the abductive inference per se. Consequently, there seems to be a riddle about what selective abduction clearly means and how it could be distinguished from other forms of reasoning. The contribution tries to solve this problem by explicating selective abduction and embedding it in an overall taxonomy of inferences.

1 Introduction

More than a decade ago, I have argued in "Peirce-Suit of Truth" (Minnameier 2004) that abduction, on the one hand, and inference to the best explanation (IBE), on the other hand, ought not to be confused. On this account, IBE is reconstructed as induction in the Peircean sense, and since abduction, deduction, and induction are the basic inferential types in Peirce's framework, they have to be orthogonal. In "Peirce-Suit of Truth" I also maintain that there is a marked difference between the early and the mature Peirce's inferential triad. Note that Peirce himself has said that "in almost everything I printed before the beginning of this century I more or less mixed up Hypothesis and Induction" (CP 8.221, 1910), where "hypothesis" indicates what he later called "abduction".

G. Minnameier (✉)
Chair of Business Ethics and Business Education, Faculty of Economics and Business Administration, Goethe University Frankfurt am Main, 60629 Frankfurt am Main, Germany
e-mail: minnameier@econ.uni-frankfurt.de

L. Magnani and C. Casadio (eds.), *Model-Based Reasoning in Science and Technology*, Studies in Applied Philosophy, Epistemology and Rational Ethics 27, DOI 10.1007/978-3-319-38983-7_17

The confusion that arises when these two versions of Peircean inferences are not differentiated clearly may be one reason why IBE is often mistaken for abduction. Another reason may be Gilbert Harman's notion of abduction by which he essentially means IBE. All this is well-documented (see Minnameier 2004; Paavola 2006). However, there is still another problem. Look at the following passage from Peirce:

> I say that these three (abduction, deduction, and induction; G.M.) are the only elementary modes of reasoning there are. I am convinced of it both a priori and a posteriori. The a priori reasoning is contained in my paper in the Proceedings of the American Academy of Arts and Sciences for April 9, 1867. I will not repeat it. But I will mention that it turns in part upon the fact that induction is, as Aristotle says, the inference of the truth of the major premiss of a syllogism of which the minor premiss is made to be true and the conclusion is found to be true, while abduction is the inference of the truth of the minor premiss of a syllogism of which the major premiss is selected as known already to be true while the conclusion is found to be true. [CP 8.209 (c. 1905)]

In contrast to the view stated above, Peirce claims here that there was no breach between his early syllogistic version of the three basic inferences and his later account. How can this be, and where would be the line of demarcation between abduction and IBE?

This question is not only relevant for the interpretation of Peirce himself, but also in the context of present-day accounts of abduction, where the notion of "selective abduction" raises the question of whether and to what extent the "selection" of hypotheses includes an IBE-like aspect.

The good news is that this whole issue can be disentangled, if we stick to a few basic principles and differentiations. I will try to show that what Peirce calls "a priori" is equivalent to what Magnani and others call "selective abduction" (Magnani 2001, 2009; Gabbay and Woods 2005; Schurz 2008), while "a posteriori" is equivalent to creative abduction.

In order to reveal this systematically, I will first analyse the three inferences in terms of their specificity and three characteristic subprocesses, of which any inference is composed (Sect. 2). Based on this analysis, selective abduction and IBE can be systematically differentiated (Sect. 3). It turns out that selective abduction is to be reconstructed as the abductive step of knowledge application. However, the application of knowledge is another field where clarification seems necessary. Therefore, the now common differentiation between fact-abduction and theoretical abduction will be addressed in Sect. 4. I shall argue that factual abduction does not mean that facts are (directly) abduced from explanation-seeking phenomena, but rather that laws or other kinds of "theories" abduced in the sense that these are applied to explain the phenomena. If this is true, "fact-abduction" is tantamount to *selective (law-)abduction*. Based on these differentiations, an overall inferential taxonomy will be introduced, in which selective and other kinds of abductions are embedded. However, this taxonomy can merely be sketched in this contribution (for more on this see Minnameier 2016).

2 Three Inferences and Three Inferential Subprocesses

In Minnameier (2004), I suggested the following model of abduction, deduction, and induction, where the three inferences form a recursive triangle, and in Hintikka (1998), I have analysed the criteria for the validity of each type of inference and the cognitive processes of actually making those inferences. This latter aspect deserves our attention in the present context. In one place, Peirce makes clear that any inference is made up of three distinctive elements that are performed in steps. These are called "colligation", "observation", and "judgment":

> The first step of inference usually consists in bringing together certain propositions which we believe to be true, but which, supposing the inference to be a new one, we have hitherto not considered together, or not as united in the same way. This step is called *colligation*. [CP 2.442 (c. 1893)]
>
> The next step of inference to be considered consists in the contemplation of that complex icon … so as to produce a new icon. (…) It thus appears that all knowledge comes to us by observation. [CP 2.443-444 (c. 1893)]
>
> A few mental experiments – or even a single one … – satisfy the mind that the one icon would at all times involve the other, that is, suggest it in a special way… Hence the mind is not only led from believing the premiss to judge the conclusion true, but it further attaches to this judgment another – that *every* proposition *like* the premiss, that is having an icon like it, *would* involve, and compel acceptance of, a proposition related to it as the conclusion then drawn is related to that premiss. [CP 2.444 (c. 1893)]

Thus, we can amend our model by these three distinctive inferential sub-processes (see also Fig. 1). Any inference starts from the *colligation* of certain premises (c). These premises are *observed* so as to produce some result and answer a specific question (o). However, every such result springs to our minds spontaneously in the process of observation, so that observation has to be followed by a *judgement* to make this overall thought process an *inference* (j).

Let us now consider Peirce's famous statement of the abductive inference:

> The surprising fact, C, is observed;
> But if A were true, C would be a matter of course,
> Hence, there is reason to suspect that A is true.
> [CP 5.189 (1903)]

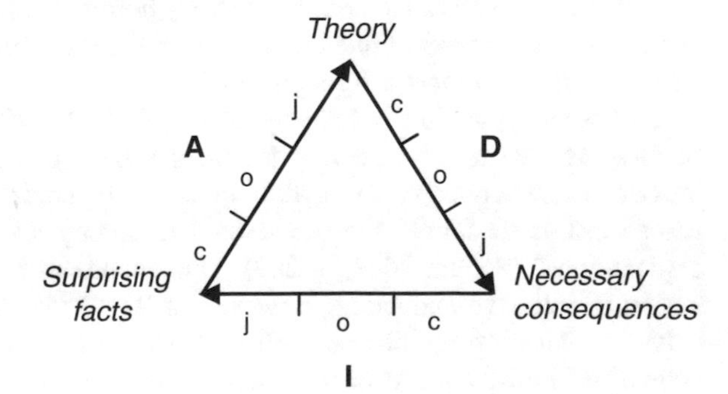

Fig. 1 Abduction, deduction, induction, and their inferential subprocesses

If we look at this statement, we immediately see that it only expresses the *judgemental part*, not the whole process of abductive reasoning. This abductive judgement tells us that the observed hypothesis ("A") actually accommodates the surprising fact ("C").

3 Selective Abduction Versus IBE

Two different views on the selective aspect of abduction can be found in the literature. The first is held by Gabbay and Woods (Gabbay and Woods 2005), who formalise abduction in the following way (p. 47):

1. $T!$ [declaration of T]
2. $\neg(R(K,T))$ [fact]
3. $\neg(R(K^*,T))$ [fact]
4. $R^{pres}(K(H),T)$ [fact]
5. H meets further conditions $S_1,\ldots,S_n$ [fact]
6. Therefore, $C(H)$ [conclusion]
7. Therefore, H^c [conclusion]

On this account, abduction starts with a cognitive target T (e.g. to explain a certain phenomenon) that cannot be met based on the reasoner's background knowledge K (1). R denotes the attainment relation on T, and R^{pres} the presumptive attainment relation on T. If $R(K,T)$ is not possible (2), the reasoner aims at an enhanced successor knowledge base K^*, so that $R(K^*,T)$ holds. According to (3) this is not present, yet. However, the presumptive attainment relation (4) can do the job. H denotes an hypothesis, and $K(H)$ a knowledge-base revised by H. Therefore, (4) essentially establishes that H would, if true, hit the target.

If we stop here and compare this first part of Gabbay and Woods' formalization with my above analysis, we find that statements (1) through (4) already cover the entire abductive inference, where the hypothesis accommodates the initial problem. However, for some reason Gabbay and Woods think this is not enough, and H would have to meet further conditions in order to be accepted, e.g. selected from a set of hypotheses that all satisfy (4). In contrast to their view, my suspicion is that this adds an inductive element to the very notion of abduction and thus conflates abduction with aspects of induction. I would therefore recommend simply to cut the latter part (i.e. numbers 5 through 7) off.

The second position is held by Magnani (2001, 2009) and endorsed also by Schurz, who writes: "Following Magnani (2001, p. 20) I call abductions which introduce new concepts or models *creative*, in contrast to *selective* abductions whose task is to choose the best candidate among a given multitude of possible explanations" (Schurz 2008, p. 202). This sounds as if IBE were included in this notion of selective abduction. However, at least Magnani is careful to distinguish between abduction on the one hand, and IBE on the other. He discusses "selective abduction" in the context of medical diagnosis, where "the task is to 'select' from an encyclopedia of pre-stored diagnostic entities" (Paavola 2006, p. 10; see also Minnameier 2004, p. 19).

This, as well as the mere opposition of *creative* and *selective* abduction, clearly indicates that selective abduction pertains to the *application* of previously established knowledge. Both Schurz and Magnani seem to agree on this. And I, for my part, agree that knowledge application is inferential in the sense that *all* inferences (i.e. abdcution, deduction, and induction) are active either explicitly or implicitly. However, we have to be careful about what "selection" can and cannot mean in the context of abduction. Let us recall one of Peirce's statements on this matter:

> The first starting of a hypothesis and the entertaining of it, whether as a simple interrogation or with any degree of confidence, is an inferential step which I propose to call *abduction*. This will include a preference for any one hypothesis over others which would equally explain the facts, so long as this preference is not based upon any previous knowledge bearing upon the truth of the hypotheses, nor on any testing of any of the hypotheses, after having admitted them on probation. [CP 6.525 (c. 1901)]

Hintikka (1998, p. 503, see also 2007, pp. 44–52) and Kapitan (1997) both read this passage in terms of two different tasks that abduction has to fulfil. Kapitan points out that "(t)he purpose of 'scientific' abduction is both (i) to generate new hypotheses and (ii) to select hypotheses for further examination" (Kapitan 1997, p. 477).

To be sure, however, the last part of Peirce's passage makes clear that the selective aspect be not confused with induction. Furthermore, Peirce stresses that selection docs not mean separating stupid ideas from sensible ones, because whatever the degree of confidence, a hypothesis is abductively valid, if it explains the facts, at least in principle. And elsewhere he notes that "the whole question of what one out of a number of possible hypotheses ought to be entertained becomes purely a question of economy" [CP 6.528 (1901)]. Hence, *this aspect* of selection, i.e. selecting one hypothesis from a set of two or more hypotheses for further examination, concerns abduction only from a practical point of view (i.e. to start with examining the most promising hypotheses), not from a logical one. And it merely marks the transition from abduction to deduction. Thus, the selection that Peirce has in mind does neither pertain to the abductive judgement, nor to IBE.

As it turns out, "selective abduction" in Magnani's sense is nothing else than the *application of previously established knowledge*. In this sense, some suitable background knowledge is activated or "selected" vis-à-vis a certain problem. On this account, selective abduction is to be reconstructed as the abductive step of knowledge application, in particular in the sense that

1. specific (explanatory) concepts or theories are activated (selected) from one's background knowledge, triggered by the initial problem at hand,
2. accepted as the result of the abductive judgement (whereas other spontaneously generated ideas may be rejected as abductively invalid),
3. and, if there are more than one abductively valid ideas, they are ranked according to a priori plausibility (however, only for economical reasons).

To be sure, the latter aspect clearly is the least central one, since it is merely of practical importance. And it should be noted that Magnani, too, does not attribute it to selective abduction when he writes: "Once hypotheses have been selected, they need to be ranked … so as to plan the evaluation phase by first testing a certain preferred hypothesis" (Magnani 2001, p. 73). As also Peirce warns us in [CP 6.525, see above] that it should by no means be confused with inductive reasoning.

This reconstruction of selective abduction as the *abductive step in knowledge application* allows us, finally, so solve the riddle highlighted in the introduction. It concerns what Peirce calls "a priori reasoning" in the passage quoted there, and which he associates with his earlier, syllogistic, concept of abduction (i.e. hypothetical reasoning). This "a priori"-reasoning is equivalent to selective abduction in the sense just described. Hence, when Peirce explains that this *selective* kind of "abduction is the inference of the truth of the minor premiss of a syllogism of which the major premiss is selected as known already to be true while the conclusion is found to be true" [CP 8.209 (c. 1905)],

1. the major premiss to be *selected* is the theory to which one abduces (e.g. $\forall x(Fx \rightarrow Gx)$),
2. based the conclusion (of the syllogism), Ga, which is found to be true and which needs to be explained,
3. and Fa results from the assumption that the occurrence of Ga is a case of $\forall x\,(Fx \rightarrow Gx)$.

With respect to (3), the only question remaining is whether the abduction runs from Ga to the general law $\forall x\,(Fx \rightarrow Gx)$, as I have suggested, or from Ga to Fa, as Schurz (2008) might perhaps argue based on his notion of "factual abduction". This is discussed in the following section.

4 Fact-Abduction and Theoretical Abduction

This is how Schurz formalises the basic form of factual abduction in his "Patterns of Abduction" (Schurz 2008):

Known Law: If Cx, then Ex
Known Evidence: Ea has occurred
==================================
Abduced Conjecture: Ca could be the reason. (Schurz 2008, p. 206)

Following my analysis in the previous section, it can easily be seen that Schurz's syllogism is equivalent to what I have labelled the "abductive judgement". Thus, it is the same as Peirce's famous statement quoted above. The only difference is that Peirce relates to the invention of new concepts, whereas factual abduction relates to the application of existing knowledge, here in terms of a "known law".

Now, the crucial point is that Schurz thinks that "possible facts" (in terms of possible causes) are abduced from actual facts that have to be explained (as effects of some cause). However, if we take abduction as a *cognitive process*, we have to distinguish between colligation, observation, and judgement. This yields a different picture of the inferential process as a whole. In particular, *Ca*, then, is not the result, and the "known law" (If *Ca*, then *Ex*) is not the premise, but the *"theory" abduced to*. The inferential reasoning runs from the *colligation* of *Ea* as the explanation-seeking phenomenon to the *observation* of the known law, which, if true in that situation, yields *Ca* as the possible cause of *Ex,* and the *judgement* that *Ca* could be the reason why *Ea* has occurred. The syllogistic form that is also presented in Peirce's classical beans-example [CP 2.623, 1878], merely expresses the final *judgemental* part; but the core of this abductive inference is the *selection of the known law* in order to explain the fact (i.e. the conclusion of the syllogism). This is how I understand Peirce, when he says that "the major premiss is selected as known already to be true while the conclusion is found to be true" [CP 8.209 (c. 1905)].

Now, if one accepts that factual abduction is basically abduction to known laws and theories, rather than to facts pure and simple, this allows us to unify Schurz's subforms of factual abduction, namely "observable-fact abduction", "first-order existential abduction", and "unobservable-fact abduction" (Schurz 2008, pp. 27–210), because they all reduce to the abduction of known laws and theories. Moreover, it reveals that Schurz's distinction between factual abduction, on the one hand, and law abduction, on the other hand, does not relate to entirely different forms of the abductive inference. The only difference is that law abduction relates to the *creative* abduction of new laws, whereas factual abduction relates to the *selective* abduction as the abductive step of the application of known laws. Schurz explicitly endorses this interpretation (Schurz 2008, p. 212).

Moreover, what is a fact? Laws do not only explain simple facts. The simple law "When it rains, the streets will be wet" does not only explain, in some sense, why the streets may currently be wet, but the law is itself *a fact* about the world we live in. As a consequence, we can think of a cognitive hierarchy moving upwards from the statement of simple facts to laws, and on from simple empirical laws to abstract theories. What's more, we can also construct a downward hierarchy from stated facts to sensual perception and elementary processes of cognitive functioning.[1] Today, the brain as whole and with all its mechanisms is seen as an "active inference machine" (Adams et al. 2013, p. 614) and the activity of the mind "as *intimately* embodied and *profoundly* environmentally embedded" (Clark 2012, p. 275). To my mind, this is of utmost importance also for the *grounding* of Peircean pragmatism.

---

[1]Peirce had reservations against conceiving perceptual judgements as abductions (CP 5.186). However, these concerns do not seem justified (see Minnameier 2016). See also Magnani (2001, 2009).

5 A Sketch of an Overall Taxonomy of Inferences

Finally, let me give you an idea of the overall taxonomy, which is illustrated in Fig. 2 and spelled out in more detail in Minnameier (2016). In one dimension, we have a cognitive hierarchy of levels or reasoning in the sense mentioned in the preceding section. In another dimension, domains are distinguished. Most of the research on abduction has, like Peirce himself, mostly focused on *explanatory* problems, where observed phenomena have to be theoretically explained. However, large parts of academia are not—at least not mainly—concerned with explanatory problems, but rather with *technological* problems (engineering, medical, educational sciences and so forth). Technologies do not have to be true in a strict sense, they have to be effective. Finally, *ethics*—or, to be more precise: normative ethics—aims neither at explanations, nor at technologies, but at principles of justice. As a consequence, we have at least three domains with three different evaluative criteria (or "regulative ideas"): *truth*, *effectivity*, and *justice*.

However, I think that at least one additional type of inferences has to be distinguished. I can only hint at it here, but I do so to make this draft of a comprehensive inferential taxonomy complete. We typically think of abduction, deduction, and induction as *forward-moving*, i.e. from a stated problem to the abduction of a possible solution, the subsequent deductive derivation of necessary consequences, and the final inductive evaluation. However, reasoning in the reverse direction is also possible. In fact, much of our reasoning—in particular scientific reasoning—proceeds in just this way. For example, think of mathematical proofs, i.e. the kind of reasoning that Peirce has called "theorematic deduction" (as opposed to "corollarial deduction") [CP 2.267, 1903]. In my view, this is best understood as an "inverse deduction": We start from the *colligation* of what is the result of a corollarial deduction, and through the *observation* of this premise (i.e., the statement that has to be proved) we look for the base from which we can actually deduce that statement. The final and formal derivation marks the *judgemental* part in this

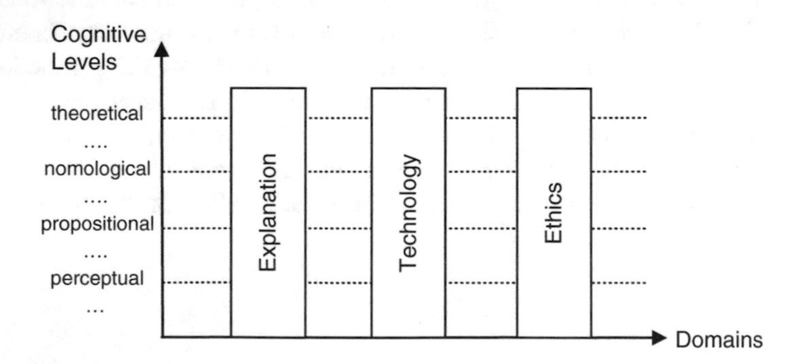

Fig. 2 Cognitive levels and domains. The descriptions of the cognitive levels only represent broad exemplifications. The *three dots* between and bellow the labels indicate that even finer-grained levels may be needed to describe how conceptual levels are built onto one another in a cognitive hierarchy

context. Likewise, *inverse abduction* and *inverse induction* can be constructed, where inverse abduction leads from theories to cases (as examples), inverse induction is the inference from the presumed truth of a theory to a crucial experiment that would confirm it (see Minnameier 2016 for a detailed description).

All in all, we can distinguish forward and backward moving inferences, inferences in different domains and at different cognitive levels. *Selective abduction* can take on all these forms, whenever already established knowledge is merely applied (rather than invented). In other words, the distinction between creative and selective abduction is orthogonal to all the other systematic distinctions made.

6 Conclusion

In the present paper, the notion of "selective abduction" has been analysed, conceptually sharpened and embedded as one form of abduction in a taxonomy of inferential reasoning. Selective abduction comes out as a distinct kind or abductive reasoning, yet broad in scope and covering a range of selective abductions.

As to its distinctness, I have raised objections against Gabbay and Woods' account of abduction which includes hypothesis selection as part and parcel of abduction as such. Magnani's concept of selective abduction, which he distinguishes from IBE, seems more appropriate both in the sense that it is closer to Peirce and that it depicts an important form of abductive reasoning. Accordingly, abductions can be classified into creative abductions, on the one hand, and selective abductions, on the other hand.

As to its breadth in scope, I have argued that selective abduction—or what Peirce calls "a priori" abductive reasoning—is best understood as the abductive application of already established knowledge. Any activation of knowledge vis-à-vis certain explanation-seeking phenomena would then count as a selective abduction. However, not only are such inferences numerous, but there are also different forms of selective abduction, as there are different forms of creative abduction. In particular, I have discussed different cognitive levels, different domains of reasoning, and inverse inferences in this respect.

Selective abductions, therefore, can be manifold and multifarious, but they are at the same time distinct from both creative abduction and IBE.

References

Adams, R. A., Shipp, S., & Friston, K. J. (2013). Predictions not commands: Active inference in the motor system. *Brain Structure and Function, 218*, 611–643.

Clark, A. (2012). Embodied, embedded, and extended cognition. In K. Frankish & W. M. Ramsey (Eds.), The Cambridge handbook of cognitive science (pp. 275–291). Cambridge: Cambridge University Press.

Gabbay, D. M., & Woods, J. (2005). *A practical logic of cognitive systems, Vol. 2: The reach of abduction—Insight and trial*. Amsterdam: Elsevier.
Hintikka, J. (1998). What is abduction? The fundamental problem of contemporary epistemology. *Transactions the Charles S. Peirce Society, 34*, 503–533.
Hintikka, J. (2007). *Socratic epistemology: Explorations of knowledge-seeking by questioning*. Cambridge: Cambridge University Press.
Kapitan, T. (1997). Peirce and the structure of abductive inference. In N. Houser, D. D. Roberts, & J. V. Evra (Eds.), *Studies in the logic of Charles Sanders Peirce* (pp. 477–496). Bloomington: Indiana University Press.
Magnani, L. (2001). *Abduction, reason, and science: Processes of discovery and explanation*. New York: Kluwer.
Magnani, L. (2009). *Abductive cognition: The epistemological and eco-cognitive dimensions of hypothetical reasoning*. Berlin: Springer.
Minnameier, G. (2004). Peirce-Suit of truth: Why inference to the best explanation and abduction ought not to be confused. *Erkenntnis, 60*, 75–105.
Minnameier, G. (2016). Forms of abduction and an inferential taxonomy. In L. Magnani & T. Bertolotti (Eds.), *Springer handbook of model-based reasoning*. Berlin: Springer, (forthcoming).
Paavola, S. (2006). Hansonian and Harmanian abduction as models of discovery. *International Studies in the Philosophy of Science, 20*, 93–108.
Schurz, G. (2008). Patterns of abduction. *Synthese, 164*, 201–234.

Complementing Standard Abduction. Anticipative Approaches to Creativity and Explanation in the Methodology of Natural Sciences

Andrés Rivadulla

Abstract After showing by means of several examples the significant role that standard abduction plays both in observational and in theoretical natural sciences, I introduce in this paper *preduction* as a deductive discovery strategy. I argue that deductive reasoning can be extended to the context of discovery of theoretical natural sciences, such as mathematical physics, and I use the term *theoretical preduction* to denote the way of reasoning that consists in the implementation of deductive reasoning in scientific creativity. Moreover, standard abduction is not always sufficient to provide best explanations in science. It is widely known that during the 1960s, abduction was identified as a form of inference to the best explanation. But when the explanation of many natural phenomena requires more than *spontaneous* acts of creativity, such as those that one might imagine in the cases of Kepler, Darwin or Wegener, much mathematical work has to be done in addition in order to advance justified theoretical explanations. In those cases, the explanation takes place *more preductivo*. I term *sophisticated abduction* the corresponding form of inference to the best explanation.

Keywords Scientific discovery · Standard abduction · Theoretical preduction · Sophisticated abduction

1 Introduction

During the last century the official philosophy of science has to a great extent neglected the relevance of the context of scientific discovery for the methodology of natural sciences. Thus most philosophers have disregarded the creative role played in science by ampliative inferences such as induction and abduction.

A. Rivadulla (✉)
Facultad de Filosofía, Edif. B, Universidad Complutense de Madrid
(Complutense University Madrid), 28040 Madrid, Spain
e-mail: arivadulla@filos.ucm.es

L. Magnani and C. Casadio (eds.), *Model-Based Reasoning in Science and Technology*, Studies in Applied Philosophy, Epistemology and Rational Ethics 27, DOI 10.1007/978-3-319-38983-7_18

Nonetheless abduction has played a significant role for the introduction of highly relevant concepts and hypotheses, both in observational and in theoretical natural sciences: from the postulation of geometrical models of the planetary movements in ancient astronomy to the most recent ideas on dark matter and dark energy in astrocosmology.

My paper is structured in three parts. In Sect. 2, I intend indeed to illustrate the importance of standard abduction by means of examples taken from the history of natural sciences. The distinction between observational and theoretical natural sciences signals that in some kinds of natural sciences the method of hypothesis testing is not predominant. Indeed, sciences grounding on observation, like evolutionary biology, palaeontology and the Earth sciences, are paradigmatic cases of natural sciences that do not apply any Popperian tests. Instead of this they rely basically on abduction as both a discovery and an explanatory practice, and when necessary they resort to so-called additional evidence tests. Theoretical natural sciences, like mathematical physics, rely on their side both on deductive hypothesis testing (context of justification or validity) and on abduction in order to introduce new ideas (context of discovery).

Moreover, theoretical natural sciences make use also of a deductive discovery strategy that I call preduction. Therefore in Sect. 3, I argue that deductive reasoning can be extended to the context of discovery of theoretical natural sciences, such as mathematical physics. I use the term *theoretical preduction*, or simply *preduction*, to denote the way of reasoning which consists in the implementation of deductive reasoning in the context of scientific discovery.

But, since standard abduction is not always sufficient to provide best explanations, the methodology of science sometimes has to proceed *more preductivo*: abduction resorts to preduction. In Sect. 4, I tackle the role of abduction in the context of explanation. It is widely known that during the 1960s, abduction was identified as a form of inference to the best explanation. Nonetheless, the explanation of many natural phenomena requires more than spontaneous acts of creativity such as those that one might imagine in the cases of Kepler, Darwin, Rutherford and many others. These cases require a lot more mathematical work in order to advance justified theoretical explanations. In those cases, the explanation proceeds *more preductivo*. I term *sophisticated abduction* the corresponding inference to the best explanation in theoretical physics.

As I said a few lines above, in this paper I tackle first the role of abduction as a way of reasoning in the methodology of Natural sciences. In particular I present abduction as a form of creativity both in observational and in theoretical sciences. But since this issue is of relevance for the contemporary philosophy of science, and the questions relating to scientific discovery have been neglected for many decades in the official methodology of science, it is necessary to dwell briefly upon the debate on the context of discovery in the philosophy of science.

Is the question of how a new idea is introduced into science, a hypothesis for instance, relevant to the methodology of science? According to Popper and Reichenbach, the methodology of science has focused almost exclusively on systematic aspects of science, the context of justification, and has neglected the issues of conceiving new ideas, the context of discovery. So, for the official philosophy of science of the last decades, the methodology of science consisted nearly exclusively in the application of deductive testing of hypotheses, the so-called Popperian tests. Nonetheless, in some natural sciences, for instance in geology, Dan McKenzie, one of the creators of the plate tectonics model, confesses (2001: 185) that "hypothesis testing in its strict form is not an activity familiar to most earth scientists", and Sclater (2001: 138) affirms that "Earth scientists, in most cases, observe and describe phenomena rather than conducting experiments to test hypotheses." Thus, deductive testing of hypotheses is not the exclusive methodology of natural sciences.

Moreover, the neglect of the context of discovery by the official philosophy of science has since the 1970s been taken over by cognitive scientists and AI researchers, who have developed computational models of scientific creativity; indeed, they have developed a *computational science of scientific discovery* actually (H. Simon, P. Langley etc.).

If we recognize the existence of sciences that do not apply deductive hypothesis testing, sciences for which the context of discovery is methodologically relevant, then we might accept that the distinction between *observational* and *theoretical* sciences in the realm of natural sciences is reasonable indeed. Observational natural sciences are predominantly empirical. They rely on experience, on observation, they apply abduction—inference to the best explanation (IBE)—as a discovery practice, and they do not implement any Popperian deductive testing of hypotheses; only additional evidence tests, when needed.

Theoretical natural sciences, on the other hand, rely both on *abduction* and *preduction* for the introduction of novel hypotheses into science. A typical example of theoretical natural science is mathematical physics. Finally, in theoretical natural sciences abduction can be both: standard and sophisticated.

2 Standard Abduction in the Methodology of Science

Thirty years before Popper and Reichenbach, Charles Peirce had taken up a position on the processes of forming or devising explanatory scientific hypotheses. Peirce gave the name *abduction* to the logical operation of introducing new ideas into science: "All the ideas of science come to it by the way of abduction. Abduction consists in studying facts and devising a theory to explain them" (CP, 5.170).

Thereafter Harman (1965: 88) equated abduction with *inference to the best explanation,* and since the late 1970s abduction has been, for the methodology of science, inference to the best explanation. Thagard (1978: 77), for instance, recognized that "Inference to scientific hypotheses on the basis of what they explain was discussed by such nineteenth-century thinkers as William Whewell and C. S. Peirce, … To put it briefly, inference to the best explanation consists in accepting a hypothesis on the grounds that it provides a better explanation of the evidence than is provided by alternative hypotheses." And Josephson and Josephson (1994: 5) claimed that "*Abduction*, or *inference to the best explanation*, is a form of inference that goes from data describing something to a hypothesis that best explains or accounts for the data."

Abduction has two sides. As Magnani (2001: 17–18), Magnani and Belli (2007: 294) claims, abduction both generates plausible hypotheses and provides best explanations of facts. Indeed, abduction in its both aspects—as creative generation of plausible hypotheses and as IBE—has been widely applied in the natural sciences, as the following cases of the history of science convincingly show:

- Ancient astronomy applied abductive reasoning in order to model planetary movements by means of geometrical models.
- One of the most celebrated abductions was Kepler's postulation of the elliptic character of Mars' orbit (Peirce CP 1.72, 1.73, 1.74; Hanson 1958: 84–85).
- A particularly interesting case of abduction, to be presented later here, is Alfred Wegener's (1880–1930) *continental drift hypothesis* as defended in his *Die Entstehung der Kontinente und Ozeane*, 1915.
- Darwin's evolutionary hypothesis in *On the Origin of Species by Means of Natural Selection*, 1859: "we accept the Darwinian theory of evolution by natural selection as what Peirce called an 'abduction', or what has recently been called an 'inference to the best explanation'" Putnam (1981: 198).
- Ernest Rutherford's *atomic planetary model* was abductively postulated in 1911 on the basis of the alpha particles scattering experiments.
- Etc.

2.1 Abduction in Observational Sciences I. The Postulation of a New Hominin Species: Homo Antecessor

From a methodological viewpoint, palaeoanthropology follows the pattern of a typical empirical science: the recognition of surprising facts, abduction by elimination of mutually exclusive hypotheses, hypothesis revision in the light of additional data, and the beginning of a new cycle. Noonan et al. (2006), in a paper on the comparison of the genomes of Neanderthals and modern humans, claim for instance that "Our knowledge of Neanderthals is based on a limited number of remains and artefacts from which we must make inferences about their biology, behavior, and relationship to ourselves." And Lorenzo (2005: 103) affirms that "Phylogenetic trees are only evolutionary hypotheses built upon a continuously changing empirical basis. It is usual that these hypotheses are tested, and modified, if necessary, on the grounds of new data."

The following steps show summarily how a new hominin species, *Homo antecessor*, was abductively postulated:

1. The surprising fact: The 1995 discovery, at the Sierra de Atapuerca, Spain, of a part of the facial skeleton of a young man—the *Gran Dolina Boy*—of an age of nearly 800 kiloyears.

2. This fossil showed neither the primitive features of *Homo ergaster* nor the derived characters of *Homo heidelbergensis*—500 kiloyears old—inherited by *Homo neanderthalensis*.
3. The skull capacity of the *Gran Dolina Boy*—about 1000 cc, considerably bigger than that of the best preserved skulls of *Homo ergaster*—provided an excellent reason for not considering it to be a specimen of *Homo ergaster*.
4. Neither could the *Gran Dolina Boy* be *Homo erectus*, since this one is mainly distributed throughout Asia (as well as in Israel and in Georgia), and the *Homo erectus* fossils are considerably older.
5. Abductive conclusion: The *Gran Dolina Boy* could only be a specimen of a new species, which was called *Homo antecessor*: "a common predecessor of both the evolutionary line that in Europe led to *Homo neanderthalensis* and of the evolutionary line that in Africa led to the modern populations of *Homo sapiens*." (Bermúdez de Castro 2002: 35).

2.2 Abduction in Observational Sciences II. The Case of the Continental Drift Hypothesis

The postulation of the continental drift hypothesis has proceeded—confesses Wegener (1966: 167)—in a purely empirical way: "by means of the totality of geodetic, geophysical, geological, biological and palaeoclimatic data. … This is the *inductive method*, one which the natural sciences are forced to employ in the vast majority of cases." (Of course *abduction* was completely unknown in 1915, the year of publication by Wegener of his *The Origin of Continents and Oceans*.) The observations supporting Wegener's hypothesis were the following (Cf. Rivadulla 2010):

1. *Geodetic data*: Observation of the continuous separation of Europe and America on the basis of astronomical, radiotelegraphic and radio-emission measures,
2. *Geophysical data*: Compatibility of both the *Fennoscandian rebound* and the *isostasy* hypothesis with lateral continental displacements.
3. *Geological data*: Affinities between the plateaus of Brazil and Africa and the mountains of Buenos Aires and the Cape region.
4. *Palaeontological data*: Distribution of the Glossopteris flora—a fern fossil register—in Australia, South India, Central Africa and Patagonia, and of Mesosaurus in Africa and South America.
5. *Palaeoclimatical data*: The Spitzberg Islands, currently under polar climate, must have enjoyed a much warmer climate in the Mesozoic and in the Palaeozoic.

2.3 Abduction in Theoretical Natural Sciences I. The Case of Dark Matter

Between the years 1932–1933, Jan Hendrik Oort (1900–1992) and Fritz Zwicky, working independently from each other, found out that stars orbiting at remote distances around their galaxies' centres do move more quickly than expected.

According to Kepler's Third Law in Newtonian form, $P^2 = \left(\frac{4\pi^2}{G_N M}\right) D^3$, in the idealized case of circular orbit, the orbital velocity of the star should behave like $v_{orb} \propto \sqrt{1/D}$, i.e.: $v_{orb} \propto D^{-1}$. This means that it is inversely related to the star's distance: the larger the distance, the smaller the star orbital velocity.

Instead of concluding that Kepler's Law is wrong, physicists abductively hypothesized that a much greater quantity of matter must exist than that which is being directly observed, and which is responsible for the rapid orbit of faraway stars. They called it *dark matter*.

2.4 Abduction in Theoretical Natural Sciences II. The Case of Dark Energy

In 1998, observations of supernovae of type Ia, situated 4300 Mpc (mega parsecs [1 parsec = 3.26 light years]) away from us, show that they are more distant than would be expected if the Universe were to contain only matter—both ordinary, i.e. baryonic, and dark matter—since the gravitational attraction would slow down its expansion. These observations suggest that the Universe is not only expanding, but that it is also accelerating.

In order to provide an explanation for this unexpected phenomenon, physicists abductively proposed the hypothesis of the existence of some *dark energy*. For this discovery Saul Perlmutter, Brian Schmidt and Adam Riess were awarded the Nobel Physics Prize in 2011.

3 Complementing Abduction. The Role of Preductive Reasoning in the Creative Processes of Theoretical Natural Sciences

According to Peirce (CP, 5.145), "Induction can never originate any idea whatever. No more can deduction." Indeed, "deduction merely evolves the necessary consequences of a pure hypothesis" (CP, 5.171).

My main concern now is: *Can deductive reasoning be used in the context of scientific discovery*? My answer will be: Yes, it can. Indeed I maintain that in the methodology of theoretical physics, we can implement deductive reasoning in the context of discovery, beyond its ordinary uses in the context of justification and that a new form of reasoning in scientific methodology, which I call *theoretical preduction* or simply *preduction*, can be identified.

Let me start with following quotation by Eddington (1926: 1) on the knowability of stellar interiors: "At first sight it would seem that the deep interior of the sun and stars is less accessible to scientific investigation than any other region of the universe…, the interior of a star is not wholly off from such communication. A gravitational field emanates from it …; further, radiant energy from the hot interior after many deflections and transformations manages to struggle to the surface and begin its journey across space. From these two clues alone a chain of deduction can start, which is perhaps the more trustworthy because it is only possible to employ in it the most universal rules of nature—the conservation of energy and momentum, the laws of chance and averages, the second law of thermodynamics, the fundamental properties of the atom, and so on. There is no more essential uncertainty in the knowledge so reached than there is in most scientific inferences." In my opinion in this quotation Eddington pre-announces this form of deductive discovery in theoretical astrophysics that I call preduction and can be identified as follows:

1. *Preduction* consists in resorting to the available results of theoretical physics as a whole, in order to *anticipate* new ideas by mathematical *combination* and manipulation of these results, in a form which is compatible with dimensional analysis—although not every combination need be heuristically fruitful.
2. The results postulated *methodologically* as premises proceed from differing theories, and any accepted result can serve as a premise—on the understanding that *accepted* does not imply *accepted as true.*
3. This suggests the notion of a hypothetical-deductive method. Indeed, *preduction* is an implementation of the deductive way of reasoning in the context of scientific discovery.
4. Since the results, which are the premises of the preductive way of reasoning, derive from different theories, preduction is *transversal* or *inter-theoretical* deduction.
5. This is what makes it possible to anticipate new ideas in physics.
6. Physicists apply preductive reasoning in a spontaneous way in order to anticipate as yet unavailable ideas, hypotheses or theoretical results.
7. Preduction differs from abduction by the fact that the results of preductive reasoning do not come from empirical data but they are deductively derived from the available theoretical background taken as a whole.

3.1 The Preductive Discovery of the Stellar Interior Model

The *discovery* of the interior structure of main sequence stars amounts to preducing a theoretical model of the stellar interiors. This model consists of five basic differential equations (five gradients): Hydrostatic equilibrium, mass conservation, interior luminosity, temperature gradient for radiative transport and temperature gradient in case of adiabatic convection (Cfr. Ostlie and Carroll 1996: 365).

The idealisations needed for hydrostatic equilibrium are those of a spherically symmetric and static star. The corresponding *preductive* procedure consists in the *combination of three theories*: Newtonian mechanics (second and third laws, and the universal gravitation law), classical statistics mechanics (the Maxwell-Boltzmann distribution of the ideal gas pressure), and quantum physics (Planck's radiation law of the radiative pressure of a black body), being the total pressure the combination of both the ideal gas pressure ($P_{ig} = \rho kT/\bar{m}$) and the radiation pressure ($P_{rad} = (1/3)aT^4$).

The idealisations assumed for the obtaining of the temperature gradient are also that of a static sphere with black body conditions plus the conditions of adiabatic expansion. The corresponding *preductive* procedure consists in the *combination of following disciplines*: classical physics and quantum physics for the temperature gradient of radiative transfer (combination of the equation of radiative transfer with the equation of the black body radiation pressure), and classical statistical mechanics and thermodynamics of adiabatic processes (for the obtaining of the temperature gradient of a monoatomic ideal gas expanding adiabatically).

The full account of the theoretical model of stellar interiors is completed with the mass conservation equation and the equation of the luminosity gradient, with the latter depending on the energy generated by both nuclear and gravitational processes.

The preduction of the stellar interior model presented here follows the basic lines anticipated by Eddington in (1926): combination of accepted results of different physical theories and disciplines that allow to anticipate as-yet unavailable knowledge on the deep interior of main sequence stars.

4 Sophisticated Abduction

In Sect. 2, I have focused on the creative aspect of abduction. But when concerned with *inferences to the best explanations,* sometimes the available empirical data do not directly suggest an attractive explanation, as we might imagine they would for Rutherford's postulation of the planetary atomic model or Wegener's continental drift hypothesis. Very frequently, hard (mathematical) work is needed for instance in theoretical physics. In these cases, preductive reasoning serves abduction for the purposes of providing satisfactory explanations for as-yet unexplained constructs. In such cases, the theoretical explanation takes place *more preductivo*. Since the inference to the best explanation depends on the implementation of preductive reasoning on the context of theoretical explanation, it is not standard abduction which is being applied here. I name this procedure *sophisticated abduction*.

4.1 *The Case of Planck's Radiation Law*

The search for a theoretical explanation of the black body radiation (Kirchhoff) was one the most urgent tasks theoretical physicists were faced with at the end of the nineteenth century.

With the contributions by Stefan, Boltzmann and Wien, empirically confirmed by Lumer and Pringsheim in 1897, classical thermodynamics had exhausted its explanatory possibilities. Further contributions by Rayleigh and Jeans combining statistical mechanics and electrodynamics were only partially satisfactory (ultraviolet catastrophe).

Finally, in December 1900, Max Planck, *combining* electrodynamics with Boltzmann's statistical mechanics and with Wien's first law (classical thermodynamics), plus the assumption $E = h\nu$, produced—preduced—his famous *Black Body Radiation Law*.

Planck's procedure is unquestionably preductive. But since his purpose was to offer a theoretical explanation of a challenging empirical result, Planck provided a Best Explanation of the black body radiation, thus performing a characteristic abductive inference.

4.2 *Some Further Examples of Sophisticated Abduction*

Here is a short list of further cases of sophisticated abduction that can be found in theoretical physics. In all these cases the explanation proceeds via *preductiva*:

- Einstein's explanation of the photoelectric effect.
- Planck's Radiation Law, as it obtains in the framework of Bose-Einstein quantum statistical physics.
- The explanation of the variability of *Cepheids* or *pulsating stars.*
- The construction of theoretical models for *cataclysmic explosive variables: novae* and *supernovae* (for instance *SN*1987*A*).
- The theoretical explanation of neutron stars.
- The theoretical explanation of *p*ulsating *s*ources of *r*adio waves—pulsars—(for instance *PSR* 1929+21).
- Etc.

5 Conclusion

In Sect. 2, I have shown the fruitfulness of standard abduction in the methodology of both observational and theoretical natural sciences. In Sect. 3, I have crossed the frontiers of abduction for the postulation of theoretical preduction as a deductive practice of creativity in theoretical natural sciences. Finally, in Sect. 4, I have completed the view on Inference to the Best Explanation by the assessment of sophisticated abduction as a specific form of abductive reasoning that relies on theoretical preduction.

To sum up: I join the very many contemporary philosophers who have tried to overcome this major error of methodologists over recent decades, that of neglecting the relevance of the context of discovery for the philosophy of science.

Acknowledgment Complutense Research Group 930174 and Research Project FFI2014-52224-P supported by the Ministry of Economy and Competitiveness of the Government of the Kingdom of Spain.

I very much thank one anonymous referee for helpful comments on an earlier version of this paper.

References

Bermúdez de Castro, J. M. (2002). *El chico de la Gran Dolina. En los orígenes de lo humano*. Barcelona: Crítica.

Eddington, A. (1926). *The internal constitution of stars*. Cambridge: Cambridge University Press.

Hanson, N. R. (1958). *Patterns of discovery. An inquiry into the conceptual foundations of science*. Cambridge: Cambridge University Press.

Harman, G. H. (1965). The inference to the best explanation. *The Philosophical Review, 74*(1), 88–95.

Josephson, J. R., & Josephson, S. G. (Eds.). (1994). *Abductive inference. Computation, philosophy, technology*. New York: Cambridge University Press.

Lorenzo, C. (2005). Primeros homínidos. Géneros y especies. In E. Carbonell (Ed.), *Homínidos: Las primeras ocupaciones de los continentes*. Barcelona: Ariel.

Magnani, L. (2001). *Abduction, reason and science. Processes of discovery and explanation*. New York: Kluwer Academic/Plenum Publishers.

Magnani, L., & Belli, E. (2007). Abduction, fallacies and rationality in agent-based reasoning. In O. Pombo & A. Gerner (Eds.), *Abduction and the process of scientific discovery* (pp. 283–302). Lisboa: Colecçao Documenta, Centro de Filosofia das Ciências da Universidade de Lisboa.

McKenzie, D. (2001). Plate tectonics: A surprising way to start a scientific career. In N. Oreskes (Ed.), *Plate tectonic*. Boulder, Colorado: Westview Press.

Noonan, J., et al. (2006). Sequencing and analysis of neanderthal genomic DNA. *Science*, *314*.

Ostlie, D. A., & Carroll, B. W. (1996). *An introduction to modern stellar astrophysics*. Reading, MA: Addison-Wesley Publ. Co., Inc.

Peirce, C. S. (1965). *Collected papers, CP*. Cambridge, MA: Harvard University Press.

Putnam, H. (1981). *Reason, truth and history*. Cambridge: Cambridge University Press.

Rivadulla, A. (2010). Complementary strategies in scientific discovery: Abduction and preduction. In M. Bergman, S. Paavola, V. Pietarinen, & H. Rydenfelt (Eds.), *Ideas in action: Proceedings of the Applying Peirce Conference. Nordic Studies in Pragmatism 1* (pp. 264–276). Helsinki: Nordic Pragmatism Network.

Sclater, J. G. (2001). Heat flow under the oceans. In N. Oreskes (Ed.), *Plate tectonics.*

Thagard, P. (1978). The best explanation: Criteria for theory choice. *The Journal of Philosophy, 75*(2), 76–92.

Wegener, A. L. (1966). *The origin of continents and oceans*. New York: Dover.

The Ontogeny of Retroactive Inference: Piagetian and Peircean Accounts

Donna E. West

Abstract Like Peirce, Piaget posits that the type of problem-solving truly retrospective in nature is embedded in action schemes (Empirical and Pseudoempirical Abstraction), inducing increasingly diverse courses of action, and spontaneous explanations. Piaget insists that working from the consequence to determine the premises (Retroactive Reasoning) represents a formidable means to develop viable hypotheses which rest upon the means to reverse and compensate in novel ways. He accounts for amplified social and logical reasoning in the face of unexpected consequences (Reflecting/Reflected Abstraction), when reasoning extends beyond present appearances to incorporate diverse orientations, e.g., changes in event participants' location/orientation and object motility/dimension modifications, e.g., changes in mass do not automatically result in form alterations. Children propose arguments (identity, reversibility, compensation), illustrating objective explanations for changes in appearances. Afterward, children assert others' epistemic, and deontic idiosyncrasies (as bystanders). This form of Reflected Abstraction unequivocally demonstrates modal logic; it is free from perceptual constraints. These perspective-taking competencies ultimately trigger well-founded recommendations for diverse courses of action in would-be events (West 2014a; West 2014b), representing Piaget's commitment to germinating plausible hypotheses. For Peirce, recommending courses of action independent of experiencing them (MS 637: 1909) embodies his pragmatic maxim.

Keywords Piaget · Peirce · Action schemes · Retroactive inference

D.E. West (✉)
State University of New York at Cortland, PO Box 2000, Cortland, NY 13045, USA
e-mail: donna.west@cortland.edu; westsimon@twcny.rr.com

L. Magnani and C. Casadio (eds.), *Model-Based Reasoning in Science and Technology*, Studies in Applied Philosophy, Epistemology and Rational Ethics 27, DOI 10.1007/978-3-319-38983-7_19

1 Introduction

This inquiry demonstrates the existence of compelling similarities (heretofore unaddressed) between Piaget's and Peirce's models of hypothesis-making, and dispels assumptions which have heretofore mischaracterized both. Three critical assumptions will be called into question, although attention will be given to the latter two: that Peirce's commitment to realism rejects Piaget's constructivism altogether; that language alone (not spatial/temporal primitives) drives logic; and finally that retroactive inferencing (abductive reasoning) always needs to be expressed-articulating explicit rationale for the legitimacy of the novel inference. These unfounded assumptions have obscured any serious comparison between the two models. They ignore the existence of universal and a priori event structures, and particular propensities to regulate (via action and mental objects) shifting components central to logic which are pregnant in ontological constructions. The existence of spatial primitives such as: motion, force and blocked path (Mandler and Cánovas 2014: 514–515) serves as the foundation for constructing event profiles, and in turn, maps how children "instinctually" integrate events into episodes by exploiting affordances (cf. West 2014b for further discussion of events as affordances), which result in insightful inferences. "…On the basis of what is known to date about concepts in the first months of life, …all the information being conceptualized appears to be spatial in nature, either describing what something looks like and how it moves or what happens in the events in which it participates"(Mandler and Cánovas 2014: 512). Accordingly, image schemas provide the common framework for prelinguistic and linguistic representations of event profiles (Mandler and Cánovas 2014: 513). Despite the fact that the existence of spatial primitives was not an argument to which Piaget or Peirce were exposed, they would not have rejected it out-of-hand. In fact, Piaget (1977/2001: 300) ascribes to the primacy of spatial factors as follows: "At the beginning, then, space is a junction point between the properties of the object and the operations of the subject, but there is much to be learned from its development." Shortly thereafter Piaget (1977/2001: 301) augments the conviction that spatial concepts are necessary to perceive relations across all developmental levels as follows: "…space conserves more than ever its role as mediator between subject and objects, allowing the subject to assimilate the diversity of those objects' manifestations in an intelligible manner."

The fact that both Piaget and Peirce accorded great weight to internal factors, from within the organism, demonstrates their appreciation for more constructivist viewpoints—that generation of mental schemes/hypotheses is not primarily dependent upon external factors, including language input. At the same time, the fact that Piaget and Peirce were adamant opponents of psychologism (Smith 1999: 87) demonstrates their commitment to the influence of the individual, as well as phenomena apart from the subject, himself, to obviate relations in the physical world, especially those which propel deictic regenerative event structures and explanations for their states of affairs. Whereas Peirce attributes the latter to the dialogic interaction between the continuum and insight of the individual, Piaget limits this factor to the universal need to regulate external and internal systems through equilibration and disequilibration (Piaget 1975/1985: 142). Because equilibration entails internal processes—establishing a balance between accommodating and assimilating mental schemes—it likewise requires reconciling contradictions of external regularities with children's previous schemes—implying some adherence to a continua within which a balance with nature at large becomes relevant.

It is evident then, that constructivist accounts, like that of Piaget, do not dismiss altogether factors beyond empirical ones; nor does Peirce's realism or phenomenology (in their emphasis on extra-sensory influences on conscious and unconscious processes) discount a constructivist approach. The very fact that Peirce makes a significant and primary place for Firstness-based factors (unbidden, internal foci) in the process of knowing testifies to his appreciation for internally driven, subjective and creative endeavors. Although constructivism does not directly embrace learning from internal insight within a realism paradigm (instinct), its primary tenet—that organisms have an active role in picking up knowledge—does not directly oppose an account which charges that at junctures in development, a flash of insight (often informed by Firstness-based sources) may assert itself, lending significant innovation to the process. More obviously, constructivism supports the development of later skills inherent in the revisionary nature of abductive processes, i.e., consciously using imperatives to determine which premise is more plausible from among the contenders (cf. West 2015a for further expansion of imperatives driven which underlie abductions). In fact, the source and composition of revisionary abductions are of a quite distinctive character from those which derive from instinct at early developmental stages. The distinction derives from the type of knowledge and explanation underpinning the hypothesis—either logica utens typically applying in the case of the latter, and logica docens in the former. Nonetheless, abductions can arise from both sources, especially if they are revisionary in nature. Novel hypotheses derive from instinct/insight alone (independent of explicit knowledge) arise primarily from practical, action-based pre-linguistic schemes which are unconscious (Gopnik 2009: 27).[1] Such hypotheses ordinarily emanate from what Peirce refers to (1901: MS 692: 5) as logica utens—implicit knowledge, "logic in possession" which is anchored in either innate dispositions or in universal sensorimotor intelligence. Conversely, hypotheses which surface largely from explicit, learned knowledge, arise from logica docens, …"legitimate

[1]According to Gopnik (2009: 27) even prior to using language children (before 1;6) engage in pretend behaviors, imagining "the ways things might be different." Gopnik cites to such displays of conduct as, combing hair with a pencil; or substituting a block for a car in transit. These behaviors demonstrate implicit, novel hypotheses about similar functions across objects, suggesting that one object might be substituted for another in a similar context. Here children infer, prior to the onset of language, that one object can be employed in similar fashion to another. Although explicit rationale is not offered to explain what contributes to the effectiveness of the comb, for example, to tidy the hair, the inference, nonetheless, qualifies as an abduction—a reasonable hunch about what might contribute to or produce a surprising consequence, tidy hair via pencil strokes.

doctrine learned by study" (1901: MS 692: 5). These hypotheses are fueled by more conscious operations which consider the internal congruity of external systems. In logica docens, assumptions which culminate in abductions are more endoporeutic in nature—they operate from the outside in, grounding themselves in external principles; and they often underscore the need to ultimately modify internal, predetermined systems (cf. Pietarinen 2006: 183). Accordingly, although pragmatic, action driven inferences more often underlie abductions grounded in logica utens, such practical knowledge can likewise inform (but not define) revisionary hypotheses which rely additionally upon other types of acquired knowledge, i.e., those arrived at more indirectly and more consciously. In fact, abductions deriving from logica docens and the hybrid type transcend Woods' (2013: 367–368) ignorance preserving kind because although they initially may emanate from practical inferences, they draw more heavily upon objective, more endoporeutic principles.

2 Epistemological Considerations

Although superficial differences exist between the epistemological models of Jean Piaget and C.S. Peirce, their primary tenets are aligned. Four overarching assumptions obviate commonalities between Peirce and Piaget: the semiotic basis for building a logical system,[2] the foundational place of indexical devices when transitioning from mechanical to more advanced thought, the revisionary nature of implicit explanatory inferences, and the facilitating role of inter/intra-subjective dialogue in refining logical systems.

The point of departure of the two models is that truth-seeking and scope (illative determinations) are inextricably bound because they are both quantified in Peirce's existential graphs, in line with Pietarinen and Bellucci's (2015: 10) claim. Peirce unites illative (inferencing) and truth value issues in his existential graphs (1893: MS 559: 8); while Piaget makes this claim implicitly with repeated assertions that via disequilibration and equilibration, children reconcile their logical systems to incorporate increasingly more objective assumptions without dispensing altogether with less advanced approaches to knowledge-seeking (Piaget 1975/1985: 142). As a consequence, hypotheses need to be revised to explain what appear to be novel phenomena or puzzling events. Truth values change consequent to their amplification into further contexts or to their limitation to discrete contexts. This process of widening/narrowing the field of truth (establishing its scope), is exemplified in what Piaget refers to as a cyclic process in which early schemes are not discarded, but are enriched by more abstract, symbolic representations, a process

[2]It is obvious that Peirce did not intend his model to derive primarily from logic but from semiotic and pragmatic principles, in that he explicitly dispelled the fact that his model is "in the world of formal logic" (1905: MS 1134: 3). This claim is supported by a recent article by Bellucci: "That the science of logic is better considered as Semeiotic…is indeed one of the most fundamental tenets of Peirce's mature philosophy of logic" (2014: 524).

which is revisionary in nature. Ultimately, invariant principles are invoked implicitly by proposing internally derived discoveries (akin to Peirce's "guessing instinct"), housed in objective hypotheses to explain anomalous consequences. As such, their models rest upon constructivist processes toward that end, while not precluding altogether reliance on universal proclivities (force, movement, blocked path) to uncover explanations for spatial and temporal primitives (cf. Mandler 2010; Mandler and Cánovas 2014 for further discussion of spatial universals).

Piaget's invariant stage theory supplies the key to which type of abductive reasoning is operational. Because the purpose of proposing novel inferences changes consequent to the logical level which children adopt at the moment of proposing the abduction, the nature of the explanation to account for the relationship between an unexpected result and a premise for its actualization, is likewise qualitatively distinct. As such, abductions which preclude the child from operating at higher levels of logic represent the ignorance-preserving type, in that they ignore facts (objective principles, special conditions) which would disconfirm the hypothesis. In particular, children ignorance-preserve when the explanations for their hypotheses (however novel and plausible in their system) conform to a parochial, internally constructed system.

3 Empirical Abstraction

Piaget describes reasoning as a primarily empirical process, conducted to satisfy practical means to practical ends. At the outset, novel inferences are carved out as what Stjernfelt (2014: 118–119), drawing on NEM III: 493–494 (1907), refers to "action habits," when working from puzzling ends to the reproduction of those ends. Here Stjernfelt concludes that, in his later writings (1907) Peirce regarded that inference may surface from unconscious, not merely from conscious deliberation: "Central to the pragmatist doctrine …is that the conclusions of inferences are primarily action habits rather than psychic or mental representations only… This leads Peirce to the important step of conceptually separating inference from consciousness." During the practical, prelinguistic period novel inferences do not contain explicit explanations; rather rationale for their adoption is implicit—housed in the series of manipulations ascribed to reproduce what had been a puzzling consequence.

These practical, action habits are characteristic of Piaget's "Empirical Abstraction" stage (1977/2001: 303), encompassing his Sensorimotor stage (0;0–1;6). At the outset, behavior primarily consists of reflexive responses (e.g., Palmer/rooting reflex) —biologically predisposed conduct: "First let us point out that the assimilatory schemes a subject starts with are innate, few in number, and very general in terms of what they can assimilate. Sucking (a scheme that quickly extends beyond nursing), looking, listening, and touching (which begins with the palmar reflex and subsequently extends into intentional prehension) are all examples" (Piaget 1975/1985: 69).

These reflexes do not approach a scheme in which sense data are coordinated with action. Rather, each targeted conduct is perceived separately from an organized problem-solving schema. Later in the sensorimotor stage motor conduct is guided by indexically driven sensory systems; and action is targeted, rising to the level of intentional and goal directed schemes. Although coordinating two indexes (vision and reach) is a quintessential illustration of sensorimotor schemes (at 0;4) indexically driven event coordination does not emerge until the second year. Piaget describes the process in the sensorimotor stage of observing single objects in associated (ordinarily intrinsic places) as "operating on observables:" "*Empirical Abstraction* draws its information from objects as such, or from material characteristics of the subject's actions, thus in general from observables" (1977/2001: 303).

Like Peirce, Piaget explicitly ascribes a special place to indexes in the course of development: "On each new level, what we have called "reflexion" gives rise to new equilibrations through regulation of indexes, and so forth" (Piaget 1975/1985: 30). He holds that indexes hasten coordination of sensorimotor skills and later integrate mental reflections—organizing events into a logical structure, relating or distinguishing them (West 2014a). In this way, indexes (gaze trajectories, reach) as action habits make apparent spatial contingencies among agents and externals. They facilitate the inclusion of objects and actions into events, by drawing physical paths between co-existent objects, establishing a syntax of events (West 2015b). This structured but tacit coalescence of objects and participants into a single scheme can only be orchestrated by indexical signs—capitalizing on the relevance of co-occurring objects in the participant pool. Implicit inferences are born when participants draw unconscious practical connections between objects across similar contexts, using a semiotic device which "asserts nothing" (1885: EP 1: 226),[3] but which as a consequence of its semantic emptiness, endows it with foundational, prelinguistic and pragmatic force. Stjernfelt (2015: 1032) supports this determination: "…colocalization seems to form a primitive prelinguistic syntax sufficient to connecting the subject [index] and the predicate [icon] tokens as a sign of a combination of the subjects and predicates themselves in a proposition." Initially, these indexes are visual in nature; and objects are largely concrete and immediate, not contained within abstract conceptual aggregates. For example, gaze and reach as initial indexes, are employed to establish a focus on object's appearance, which includes location and material properties; later, in Piaget's pseudo-empirical stage, indexes, such as pointing[4] and joint gaze, cement objects to particular events and establish a sequence among events to create an episode: "When the object has been modified by the subject's actions and enriched with properties taken from their coordinations (for instance, when the elements in a set are ordered), abstraction that

[3]"The index asserts nothing; it only says 'There!' It takes hold of our eyes, as it were, and forcibly directs them to a particular object, and there it stops" (1885: EP 1: 226).

[4]"Of this nature are all natural signs and physical symptoms. I call such a sign an *index*, a pointing finger being the type of the class" (1885: EP 1: 226).

ranges over these properties is called *pseudo-empirical*" (Piaget 1977/2001: 303). The actions which modify and enrich objects and which are responsible for coordinating them are nothing short of static and motion indexes (gaze, reach)—the very elementary semiotic tools which implicitly unite substance with function and with alterations in function.

In Piaget's stage of Empirical Abstraction, children restrict implicit inferences to directly observable contiguous events whose components have not yet been coordinated into episodes, namely transductions (Piaget 1924/1959: 197). Transductive thought entails connecting events by virtue of action schemes on co-existing objects: "…the succession of relations constructed by the sum of movements—whether performed, begun, or imagined—does present something that is equivalent to a reasoning process, but …these actions are not reversible…" (Piaget 1924/1959: 197–198). Later, in the Pseudo-empirical stage (advent of preoperational thought) however, precedent events coordinated with other present factors, implicitly explain anomalous consequences. At the concrete operational stage, however, explanations factor in events/conditions which may defy what is perceptually apparent. In cause-effect scenarios, this same chain of advancement operates, demonstrating the revisionary nature of inferencing even at early ages.

Findings from Cohen and Amsel (1998) support the existence of implicit abductions in Piaget's Pseudo-empirical Abstraction stage. Prior to 2;0, what children use to infer causality is spatial contiguity between objects/events, and especially those which include direct contact with one another, e.g., those involving collision. Afterward, children can infer causal relationships between non-contact co-present objects/events (Sobel and Buchanan 2009). Still later, at 4;0 and thereafter irrelevant co-present events are distinguished in favor of using relevant co-present events to determine cause (Buchanan and Sobel 2011: 2063).

In any case, the action schemes which underlie both instinctive and more revisionary abductions are, without question, not wholly dependent upon linguistic competencies; instead, they are grounded in practical and empirical systems of explanation, consonant with Magnani's claim (Magnani 2001: 54–55, 60–62, 2009: 374) that even the most revisionary abductions are informed by manipulative schemas. Magnani argues: "The various procedures for manipulating objects, instruments, and experiences will be in their turn reinterpreted in terms of procedures for manipulating concepts, models, propositions, and formalisms" (2001: 55). Magnani's claim resonates with Piaget's repeated assertion that sensori-motor manipulations serve as a basis for preoperational and concrete operational intelligence. Magnani (2001: 60–62) sets forth two types of manipulative mediators (used in early abductions) which begin to operate at Piaget's sensorimotor stage and continue in the Pseudo-empirical stage: bodily epistemic mediators, e.g., fingers to count, and external epistemic mediators, e.g., narratives, construction of diagrams. According to Magnani (2001: 60) in these cases, "action performs an epistemic and not merely performatory role…".

Additional illustrations of manipulative schemas leading to revisionary hypotheses which materialize early on (at Piaget's sensorimotor stage—before 1;6), arise when modifications are made to children's object concept (object permanence). These abductions entail recommending the most successful action schema to recover a hidden object. The surprising consequence is failure to find the object in question under the original hiding place. The course of action adopted here flowing from the insight and the hypothesis, is not to search for the object in other locations, but to look under the original hiding place, despite observations indicating a change in hiding place. Mandler (2004: 225) describes children's adherence to an originary (faulty) inference in this hiding place paradigm as a lack of means to "inhibit an already planned reaching response." Diamond (1985: 875, 880) proposes that looking under the original hiding place is merely a consequence of an inability to transcend a well-entrenched behavior–response "perseveration" or "failure to resist the habit." While these accounts acknowledge the effect of action habits on object concepts (identified by Stjernfelt's (2014: 118–119) interpretation of Peirce), they fail to note the increased import of embodied action verses observed action. Were children actually the agents of the action—themselves carrying the object to the non-original place and hiding it, their indexical memory for the change in location is unlikely to have experienced interference from previous action habits, as Mandler and Diamond assume. In fact, Mandler and Diamond assume that children's conduct is reducible to automatic behaviors, which fails to attribute unconscious inferencing capabilities before 1;0. Later (toward the close of the sensorimotor stage), revisionary abductions propose that investigation in the last hiding place only is required to retrieve the hidden object. In short, after deployment of a number of unsuccessful sensorimotor based manipulative inferences—to retrieve the hidden object from the original place, children revise their abductions (their action habits)—tacitly noting the factual implausibility of previous inferences. In contradistinction to Mandler's and Diamond's interpretations, children's own manipulation of indexes (gaze, arm extension) to displace the object to the second hiding place may, in fact, hasten apprehension of the displacement.

4 Pseudoempirical Abstraction

These early practical abductions (driven by empirical/pseudo-empirical abstraction) are grounded in Peirce's category of Secondness and his Dynamical Object. Secondness constitutes the most elementary of the categories (when compared with Firstness and Thirdness): "…Secondness is the easiest to comprehend, being the element that the rough-and-tumble of this world renders most prominent" (1903: EP2:268). This is so given the prominence of external factors consequent to limited means to hold representations in working memory (WM). The state of neurological development dictates this adherence to observables, in view of limited WM resources and capacity—two slots only to be filled with representational material. As a consequence, the prominence of Secondness—its brute insistence on attention to the here and now—compels children to draw upon instantiations of co-localized objects and targeted action schemes to make sense of externals in the physical world (West 2015c: 7). Accordingly, Peirce's Dynamical Object, that which is ordinarily present to the senses, but can likewise be unreal or fictional (1909: 8.314), and not yet his Immediate Object, the instantiations of similar objects "as represented in the sign," i.e., the part of the sign which represents the Dynamical Object (1909: 8.314), qualifies as the primary material upon which children operate. Secondness supplies the impetus or as Peirce terms it, "compulsion"[5] to use action and reaction schemes,[6] and to apprehend spatial/temporal relations across contexts of co-present Objects. In short, Peirce's category of Secondness provides the venue to develop sensorimotor schemes (West 2015c: 2); and index is the primary semiotic device that unites actions with particular Dynamical Objects—the raw material to construct schemes and ultimately hierarchical event structures.

In fact, it is the effect of puzzling/surprising consequences played out in the arena of Secondness that serves as a significant impetus to organize (order) and coordinate event structures into episodes (dynamic bundles of events which adhere to Peirce's concept of habit because they cohere logically). A particular consequence is deemed to be puzzling when it does not conform to expectations about what ordinarily is temporally contiguous or co-occurring with it. With the advent of Peirce's immediate Object and use of joint indexes, emerges some refinement of how event slots are filled with different participant roles of agent and receivership, which, although social, provide a living template of the concept of contributoriness to affect a consequence. At this juncture, novel inferences regarding how a consequence materializes are enhanced by dynamic, illative displays of how self and other alter a consequence by changing emotions, actions, or perceptions. This joint enterprise is markedly distinct from merely moving objects or observing what Piaget refers to as "observables."

Piaget's pseudoempirical abstraction is characterized by self actions which, at first, are perceived as separate from each other; as such, they are not coordinated to bring about a planned end (Piaget 1977/2001: 303). "When the object has been modified by the subject's actions and enriched with properties taken from their coordinations (for instance, when the elements in a set are ordered), abstraction that ranges over these properties is called *pseudo-empirical.*" Even when actions are performed in sequence, apprehension of the contributory nature of each action is not initially apparent. Children evidence this uncoordinated action performance when they engage in sequences which they do not perceive to be alterable, e.g., in deferred imitation paradigms, where children reenact a behavior sequence exactly as they saw it after a twenty-four hour interval to exact the same consequence. Such does not qualify as an abduction, since actions are performed as a whole, ascertaining an end but without apprehending the contributorial effect of each conduct and without the means to truncate or delete any portion of the event sequence.

[5]"That hardness, that compulsiveness of experience, is Secondness" (1903: EP 2: 268).

[6]"A door is slightly ajar. You try to open it. Something prevents. You put your shoulder against it, and experience a sense of effort and a sense of resistance. These are not two forms of consciousness; they are two aspects of one two-sided consciousness. It is inconceivable that there should be any effort without resistance, or any resistance without a contrary effort" (1903: EP 2: 268).

At this stage, objects, properties and locations can be modified, but not their sequence (Piaget 1977/2001: 303). This discrepancy makes apparent that within the Pseudoempirical Abstraction stage, spatial relations (unlike temporal ones) can be modified, such that children use their own body to orchestrate particular orientational and locational displacements and refer to such with indexical signs. Temporal alterations are far less subject to attenuation from present events at this stage, since index can not be conveniently called upon to transpose children from one moment to another.

This is supported by Tulving's (2005: 7) finding that children do not "time travel," referring to a coherent series of past events in which they have participated, viz., an episode, until after 4;0. This is the case largely because remembering beyond the present and organizing/coordinating events into sequences based upon which event affects which and how they do so depends upon more advanced representational/memory skills than do spatial integrations. Spatial representations are fashioned upon iconic imaging of single events, absent reconstruction into a coordinated/coherent structure with functionally related events; whereas temporal representations depend upon an increased ability to make implicit inferences—which events fit and where—bundling them into episodic structures (cf. Baddeley 2007: 12–13 for elaboration on the episodic buffer in working memory). In short, inferring relations between/among events (intrinsic to temporal representations) constitutes far more challenging mental integrations, given the increased need to rely upon novel inferencing as the logical glue, coordinating premises into coherent conclusions and the reverse. Because spatial relations are more elementary than are temporal ones, given their dependence upon analogous featural representations (encoding co-present relations via perceptual comparisons of location, color, shape), these types of situational relations require the facilitating presence of Index to call attention to and obviate the iconic features which define classes. Use of index draws attentional paths between objects, hastening notice of similar perceptual features. This compulsive path-finding function is unique to index; it leaves a discernable footprint within each context, which transcends mere locational data, given iconic coupling. Later in development at Piaget's Reflecting Abstraction and Reflected Abstraction stages, when this double sign (Dicisign) supplies a template of information (cf. Stjernfelt 2015: 1023), the meaning which it imports becomes symbolic in nature (1906: MS 293). Its more symbolic nature shepherds in a new function of index—the means to appreciate the practical possibilities inherent to shifting agency-receivership roles encoded in pronoun use (for further elaboration, cf. West 2013: Chap. 2).

As such, the presence of Peirce's Dicisign, double sign in which index enriches iconic properties (1905: MS 284: 43), particularly facilitates early spatial relations, supplying a co-present marker of the original location. The functionality of the Dicisign in early development is formidable: "Thus, every proposition is a compound of two signs, of which one functions significantly, the other denotively. The former is intended to create something like a picture in the mind of the interlocutor, the latter to point to what he is to think of that picture as being a picture of" (1905: MS 284: 43). The power of Peirce's index in the Dicisign to situate and create propositions from icons (pictures in the mind) is undeniable—it affixes a symbolic meaning (interpretant) with the object of the sign in each application. Accordingly, it demonstrates underlying inferences, not yet made explicit, but which linger in the mind from applications of other uses of indexes with similar icons. For example, using two indexes (pointing to another then to an object, while gazing toward it) establishes an action habit, a chain of command either an imperative or a declarative between the child-signer, the sign receiver, and the object (West 2015b). Implied inferences include: index aggregates which command the sign receiver to access the object for the child-signer, or declarations identifying the kind/class of the object in question.

During Piaget's Pseudo-Empirical stage, Peirce's Dicisign materializes as joint Sinsigns (signs operating in the here-and-now of existence to share focus on an object via gaze/pointing; it hastens apprehension of perspective-taking templates, and highlights differences in modal logic across distinct players taking the same role (West 2015b). Index constitutes the orientation of the agent's body and gaze direction; while the icon inhabits the shape of the full spatial array perceived by the gaze initiator at that moment. This coupling of index with icon establishes Index as the ultimate path-finder. Index draws a parameter around potential objects—making plain the shape of action trajectories (icon). What is salient at this developmental stage is that the object of the agent's focus is determined by the semiotic device—the individuating character of index.

The indexical sign/s, obviates for the interlocutor not merely the Dynamical Object (the real object), but likewise the emergence of the Immediate Object: "the idea the sign is built upon" (1907: EP 2:407). But, the Immediate Object does not surface until Piaget's Reflecting Abstraction stage, when children begin to employ informational indexes to settle upon some invariant meaning common to interlocutors, superseding the Dynamical Object (1903: EP 2: 276; Stjernfelt 2015: 1023). At this juncture, meanings express classificatory, not merely capricious attentional impulses, marking the genesis of double object use (Primary and Secondary) (1908: SS 83). The former entails the object outside of the sign (in the real world) while the latter refers to the object as contained within the sign—inclusive of discourse matters. The Secondary Object of the Dicisign can be illustrated as a contour between neighboring objects (the functional relations holding among objects to eventually construct coordinated event templates). The emergence of the Secondary Object illustrates the relevance of possible/potential objects were they to materialize in the context under focus. As Stjernfelt (2015: 1028) comments: "thus, the Dicisign gives flexibility where implicit information is agreed upon the interlocutors and the specific universe of discourse they address." The Dicisign helps to draw relations between objects in their universe, and the implicit discourse focus which is maintained and shifted by the sign users. The implicit way in which Dicisigns refer to specific foci via attentional devices, together with their natural means to make relevant the host of discourse shifts, obviates their indispensability to building inferential reasoning. While enhancements capture novel/analogous uses of the sign (metaphoric ones), presuppositional functions of the Dicisign promote perspectival skills (1908: 8.179, SS 70). In short,

Peirce's Dicisign presupposes and enhances knowledge of other's focus, drawing diverse perspectival templates between persons and objects. These perspectival skills serve as the bedrock for abductive rationality, because the relations which they suggest serve as the raw material for referring to would-be objects and recommending a fitting course of action for others, a rudimentary competency underlying novel inferencing (1909: MS637: 12).

5 Reflecting Abstraction

When sequences of self action are coordinated, such that the relevance of each action is attached to the respective consequence, children are considered to have reached Piaget's Reflecting Abstraction stage (Piaget 1977/2001: 289–295). Smith (2005: 519) underscores Piaget's insistence that action underlies more advanced mental coordinations by situating Reflecting Abstraction within the bounds of "action-logic." At this period, action sequences are goal directed; they no longer constitute mere automatic, indivisible wholes, but reasoned event aggregates. Despite their unconscious nature as unexpressed premises whose materialization results in the initial unexpected consequence, they, nonetheless, qualify as abductions. In fact, these early abductions do not primarily rely upon linguistic competencies, since performance at this age (1;9) does not ordinarily exceed single word utterances—syntactic and semantic structures which demonstrate logical relations.

The existence of these early implicit assertions from puzzling consequences is consonant with Magnani's claim that inferences based on tacit knowledge (2001: 8), especially those which derive from manipulating physical and mental schemes are candidates for abductive processes. These implicit inferences consist in behaviors which rise to the level of propositions/assertions, but which do not offer explanatory rationale, e.g., imperative pointing to gain access to an object. As such, inferences can qualify as abductions independent of whether the underlying hypothesis is a conscious affair, and independent of whether explicit explanations are articulated. Accordingly, plausible hunches (good guesses) regarding whether canine are classified separately from humans consequent to the sudden realization of their inability to retrieve a toy from a new hiding-place, constitutes an abduction, however tacit it may be. The existence of this type of novel inference qualifies as an abduction, if evidence suggests that children have, in fact, constructed a new class of entities, differentiating those which had been previously associated/included in the same class. In short, Piaget's Reflecting Abstraction stage is characteristic of just this kind of tacit construction—novel inferences which indicate, however implicitly, internally reconstructed mental categories which rest upon modified assumptions. The exercise of distinctive action habits implies alterations in children's previously conceived of hunches, sufficient to implement different action schemes on the respective objects, e.g., sorting objects according to color rather than shape.

Piaget's rotation experiment exemplifies how children gradually implement new action habits which derive from novel mental coordinations, and which eventually qualify as plausible problem-solving strategies. In this experiment, the mid-point of a bar was fastened to a horizontal surface, such that it could be manipulated to circulate (clockwise/counter-clockwise) around the table surface. Children between 0;10 and 3;0 were individually instructed to employ a bar to access an object which was at different locations (in different trials) on a table surface. They were advised not to walk to another side of the table, nor to climb on the table's surface to access the toy. At no point in the experiment were Piaget's subjects explicitly instructed to employ the bar to access the toy, in order to encourage the use of natural constructive problem-solving competencies from novel consequences.

Findings reveal that before 1;9 none of Piaget's subjects were able to utilize the bar to successfully bring the object into their field of grasp without altering their original location (side of table). At 0;10 infants did not have an intentional plan in place to access the toy: they merely attempted to discover physical linkages to the bar, pulling on its support: "While pulling on a support in different ways, the subject unintentionally makes it turn a little, records this observable along with the displacements imparted to the desired object, and all of this is a matter of empirical abstraction along with simple repetitions of what he has just seen. But to draw an intentional pattern of behavior from rotations, even partial rotations, the observables must also be assimilated to action schemes that are more or less well coordinated" (Piaget 1977/2001: 293). Even when randomly moving the bar, the youngest subjects were unable to apprehend its culminating function—to ascertain their goal (Piaget 1977/2001: 290). Prior to 1;9, children were unable to systematically employ the bar to bring the object around to their location, independent of the beginning point or directionality of the bar with respect to the toy. By 1;9, however, with significant advances in use of index: "…the subject now rotates the wooden bar completely. Consequently, he can bring the toy to himself without having to change positions" (Piaget 1977/2001: 291). With the use of an indexical device (pointing and circulating the bar toward the child to displace the toy), children begin to employ mediational indexes, apart from immediate, corporal ones, to access the desired object. Beyond 2;0, children performed at the highest levels, given the realization that circulation of the bar in either direction could bring about the desired end, i.e., the "inversion" of the indexical device (rotating the bar in the reverse direction). Here the bar is utilized intentionally, and in a planned manner to bring the toy to the subject. Essentially, at Piaget's Reflecting Abstraction stage, the bar serves as an instrument or intermediary to reach a particular end. At this juncture, the path of the bar is not fixed (its direction can be reversed); and the means (direction of the bar's rotation) to the end alters according to the varied location of the toy at the point the bar pushes the toy.

What Piaget refers to as an "intermediary" (device other than one's own body) can bring about the same ultimate goal of effectuating access to the toy. "Finally, while we must remind ourselves that empirical abstraction (contact with the observables) never stops being indispensable, Level 5 [the highest level of performance on the rotation task] exemplifies inversion of the direction of actions at its maximum" (Piaget 1977/2001: 294). The means to reverse the action to contribute to the result reveals some attenuation of events from their original order—a skill useful in developing well-formed inferences. Accordingly, settling upon alternative ways to reach the same effect, essential competencies mastered in reflecting abstraction, represents a transition from practical to higher levels of mental consideration.

What is distinctive at the Reflecting Abstraction period is the means to keep various stages of progression toward a goal in the mind simultaneously—to recognize class inclusion and to orchestrate event sequencing toward a combinatorial effect. "… objects do not constitute operators independent of the subject but are coordinated insofar as they are endowed with properties such as order and class and class membership that the subject's operations confer upon them" (Piaget 1975/1985: 51). Making event consequences topics of inquiry (applying agent and receiver slots to others) rather than operating on random self experiences, effectuates more objectified foundations to generate more plausible abductions for particular event participants. In fact, Peirce's injunction that abductions/retroductions are exemplified by "recommending a course of action" (1909: MS637:12) underscores the indispensability of pragmaticism in generating abductions, and makes plain that action habits constitute the hallmark of abductive reasoning. This is critical when conceiving practical abductions—recommending plausible courses of action, in that it increases the likelihood of success not merely for self but for others. Recommending strategies/courses of action based on good guesses (consonant with Peirce's notion of insight by way of instinct) and subjecting guesses to some cursory comparison with alternative approaches can likewise prevent unwelcome eventualities.

Peirce's concept of abduction as instinct (although operational at earlier stages in development) acquires a new complexion at the Reflecting Abstraction stage. It is no longer a "flash" entirely from feeling (Firstness), but constitutes "an act of insight" (1903: 5.181). At this period, acts of insight are recognized to be "fallible" (1903: 5.181). Foundational to the apprehension that inferences are fallible is the development of intersubjective competencies—appreciation for the legitimacy of diverse vantage points, particularly salient in dialogue exchanges. Fallibility is obviated when one course of action recommendation for one party is determined not to be effectual, while, for another in the same context, it is viable. Suggesting courses of action via one's own engagement in action schemes is but one way (although implicit) that these recommendations can be made. Nonetheless, Peirce does not limit courses of action to manipulation of objects, but incorporates linguistic avenues for abductors to recommend human-to-human treatment through inter subjective or intrasubjective dialogue.

In fact, both Peirce and Piaget explicitly accord dialogue a special place in the emergence and implementation of model-based reasoning. Piaget insists that expressing thoughts to others is rudimentary to clarifying logical connections such as causal inferences, and in turn to generating plausible courses of action. He indicates that articulating one's thoughts about how events are coordinated (ordinarily once the Reflecting Abstraction stage is well underway), precludes manifold errors in reasoning. "What then gives rise to the need for verification? Surely it must be the shock of our thought coming into contact with that of others, which produces doubt and the desire to prove. If there were not other people, the disappointments of experience would lead to over-compensation and dementia. We are constantly hatching an enormous number of false ideas, conceits, Utopias, mystical explanations, suspicions, megalomaniacal fantasies, which disappear when brought into contact with other people. The social need to share the thought of others and to communicate our own with success is at the root of our need for verification" (Piaget 1924/1959: 204). Here, Piaget asserts that the necessary component for the ability to guess right (generating abductions) is the means to revise our own assumptions such that they are comprehensible to others. In other words, framing assumptions to others provides the impetus to formulate revisions to originally unexpressed hunches. Dialogue for Piaget serves as a forum to recognize doubts, and to modify them to mediate false suspicions, fantasies and exaggerated utopia, before they lead us into error.

The place of error and ignorance in Peirce's model is more radical still; it stands at the threshold of self-identity, in that recognition of the self as fallible in ontogeny illustrates the emergence of the subject as a separate sentient being (1867–1871: 169).[7] Peirce identifies two forms of dialogue which naturally proceed from the intersubjective to the intrasubjective (cf. West in press for a more detailed account of Peirce's use of dialogue as habit). The former is exemplary of Piaget's Reflecting Abstraction stage, while the latter relies upon metacognitive skills developed at Piaget's Reflected Abstraction stage. The first form of dialogue which Peirce identifies entails interaction between separate individuals. Using signs to interact with another requires tacit agreement (or as Peirce terms it "common ground") between the interlocutors to share focus on objects and to assign similar interpretants to signs. "No man can communicate the smallest item of information to his brother-man unless they have που στωσι [a place to stand] of common familiar knowledge; where the word 'familiar' refers less to how well the object is known than to the manner of knowing" (1908: MS 614). Peirce convinces us that knowing the object well (its perceptual characteristics in narrow contexts) pales in comparison to knowing its functionality in diverse contexts while utilizing higher levels of mental powers (inferencing) to seek out new possibilities of use. This latter form of knowledge is superior consequent to the profound effect of particular semiotic tools to reconstitute representations of objects. The use of semiotic tools (particularly legisigns which direct events internally and externally) illustrates manipulative abduction on still higher planes—from practical manipulation of objects, to affecting the self and others via mental and linguistic tools. The latter qualifies as a manner of knowing, a revisionary instrument for abduction, in that interlocutors' signs serve a mutual purpose—to transmit new platforms for causal determination. The manner in which interlocutors choose to show an object of focus to another

[7]"Error and ignorance, I may remark, are all that distinguish our private selves from the absolute ego" (1867–1871: 169).

(the signs they select to represent the object/meaning) reveals something substantial about what the signer presumes is foremost in the interlocutor's mind.

In fact, the type of sign which the signer selects indicates what he/she assumes the interlocutor's interpretants to be. If the signs are primarily iconic and/or indexical, such that causal inferences are founded upon spatial contiguity, co-existence between plausible causal factors and their surprising consequences, interpretants are likely to be Dynamical, and of the Emotional or Energetic type. Obviously, the manner of knowing the object (so critical for Peirce)—which often entails the sensory modality responsible for knowledge construction, demonstrates the level of object knowledge. For example, the choice of index alone as the only/primary sign vehicle ordinarily illustrates superficial familiarity or unforeseen relevance of a particular object. This sign is employed early on in ontogeny; and especially intrinsic to its use is its percussive nature.[8] Hence, children's early sign selection, in particular, often occurs by default—sign vehicle is ordinarily index because it needs to operate in line with sensorimotor intelligence to effectively showcase objects' novelty/uniqueness (West 2014b: 171). At this stage, the character of knowledge of the object as a dynamical Object, makes obvious the manner of knowing the object—merely as real and manipulatable.

Later, when children wish to make a declaration that a particular object belongs to a class or that it is a prototype of a class, they choose an indexical legisign to do so, which implies the habitation of the common features of that class (West 2013: Chap. 2). In this way, the sign vehicle is altered to show the partner in the dialogue the signer's knowledge state—the manner of knowing the object. Here children use nouns as indexical legisigns to represent propositions—asserting a class of objects into which the object in question is to fit. In this way, names indicate inferences about which hierarchy is most relevant to the child-signer. Novel inferences can be evidenced at or before 2;0, in children's lexical overgeneralizations of "doggie" to refer to all quadrupeds. These generalizations are not mere errors in judgment as is often assumed, but are novel inferences—implicitly incorporating assumptions about the responsiveness or ferocity of other animals. These single word utterances may even serve as truncated arguments in warning others to stay away or in encouraging them to caress the animal. In short, within dialogical contexts nouns are deliberately (but often unconsciously) chosen to communicate implicit inferences about the objects' identity and functionality. As such, they serve to introduce implicit inferences from one signer to another; and as Piaget concludes, the existence of another mind to process these novel propositions/arguments serve a regulatory function—guarding against asserting anomalous/outrageous claims.

A still higher level of abductive reasoning is present when children choose to use demonstrative and personal pronouns to imply shifting spatial and conversational roles (cf. West 2011: 96–97; West 2013: Chap. 2 for further explanation of how

[8]In his 1908 draft letter to Victoria, Lady Welby, Peirce identifies the "Percussive" interpretant as bearing the element of Secondness in his sixth trichotomy—"Of the Nature of the Dynamic Interpretant" (1908: EP 2: 490). West (2013: 114) elaborates on this with the assertion that, "the Percussive [interpretant] gives rise to a sudden, single, emotional experience."

pronouns facilitate semiosis). It is in these more diversified event templates (when roles/slots can be filled with different objects/person participants) that children begin internalizing action structures into objective event avenues. In fact, using the same pronoun for self in distinctive event roles, and for others in those roles serves as the transition to internal dialogue, an increasingly more internalized, self-regulatory system.

6 Reflected Abstraction

When arguments become conscious tools for changing habits or regulating problem-solving approaches, children are considered to have reached Piaget's Reflected Abstraction stage. "Finally, we call the result of reflecting abstraction *reflected* when it has become conscious, regardless of its developmental level" (1977/2001: 303). In Piaget's Reflecting Abstraction Stage thought is reflective in nature, in that it is conscious, coordinating events and states into organized structures; but, it does not reflect back upon itself as is the case in the Reflected Abstraction stage. It is not thought only which ultimately must be conscious, but the knowledge of the effects which novel arguments can produce: "Reflected Abstraction lags rather systematically behind the reflecting process until the point… at which it becomes the necessary instrument of reflections on prior reflection; eventually it allows the formation of metareflection, or reflective thought, which then makes it possible to constitute logico-mathematical systems of a scientific nature" (Piaget 1977/2001: 318). As such, metacognitive skills are paramount at the Reflected Abstraction stage, since anticipation of the effect of conduct on states of affairs is indispensible to strategy-making. Abductive reasoning at this stage must incorporate inventing strategies which culminate in successful argumentation for the self and for others. Moshman (1996: 409) elaborates on the relevance of Reflected Abstractive metaskills to abductive reasoning, claiming that it is not until 6;0 that these skills can begin to be mastered (Moshman 2015: 76). Moshman (1996: 409) attributes the complexity of metacognitive skills underlying explicit inferences to two mutually integrated processes: "(construction of a new subject at a higher level of abstraction and (b) reconstruction of the old subject as an object of understanding." Moshman (2015: 74) further demonstrates the relevance of metaskills to logical reasoning: "Metalogical knowledge understanding is conceptual knowledge about logic, especially with regard to the nature of logical justification and truth." He claims that metalogical understanding entails apprehension of the properties of inferences, in the form of propositions/arguments, and acknowledges that children's justification for them may be strong or weak. But, what Moshman fails to address is Piaget's distinction between proactive and retroactive inferences.

Like Peirce, Piaget insists that working from the consequence to determine the premises (amounting to abductive reasoning) represents a formidable means to develop viable hypotheses which rest upon retroactive implications. "…A second group of implications binds an element E not to its later consequences but to its antecedents or 'prior conditions' that can be multiple but for all that sufficient. Here we shall speak of 'retroactive' implications, each proactive discovery capable of leading to retroactive recastings" (Piaget 1980/2004/2006: 9). For Piaget, proactive implications necessarily entail engaging in retroactive ones. Piaget recognizes here the ampliative effect of retroactive implications on all inferences. Piaget's claim transcends Peirce's notion of retroduction[9] (reasoning from consequences to antecedents), in that pregnant in every proactive implication (reasoning from premises to consequences) is the ability to logically reverse the sequence. Otherwise, inferences of any type would be nothing short of contained wholes absent the means to alter and analyze event relationships within them; and abductions would be cut off altogether. In short, for Piaget mentally reversing[10] components of event sequences is what allows them to be disassembled and reassembled as novel logical episodes.

In fact, the means to engage in retroactive reasoning is the hallmark for thinking at the Reflected Abstraction stage; it underscores the importance of event sequences and their reversals to arrive at reasonable but plastic argument structures. Nonetheless, with his emphasis on the purpose of internal dialogue, Peirce extends Piaget's emphasis on retroactive reasoning still further. The internal process of structuring and restructuring events into episodes is foundational to retroductive reasoning; and as such, arguing with the self through inner dialogue (from consequence to antecedents) is vital. Having the means not merely to influence others in issuing novel arguments, but presenting such to the self, provides the tools to eventually invent scientifically sound models (utilized abductively in this context, although other forms of reasoning may benefit from dialogic iterations).

Peirce further testifies to this powerful tool to refine abductive reasoning: "…it is…a necessity of logic that every logical evolution of thought should be dialogic" (1906: CP4.551). Posing two plausible arguments to one's own person, such that conflicts of logic are exposed demonstrates a heightened command of world

[9]The following passage from Peirce from his later writings reveals the foundational place of retroduction in developing logic and representational thinking: "I consider Retroduction…to be the most important kind of reasoning, notwithstanding its very unreliable nature, because it is the only kind of reasoning that opens up new ground. [—] Retroduction gives hints that come straight from our dear and adorable Creator. We ought to labour to cultivate this Divine privilege. It is the side of human intellect that is exposed to influence from on high. With this investigation starts. Having once formed a conjecture, the first thing to be done is to draw Deductions from it and compare them with observations 1911: NEM 3:206.

[10]In making the argument regarding reversibility, Piaget refers directly to Peirce's notion of retroduction as one form (albeit primary) of retroductive inference/implication: "…action implications, just as implications between statements, may take three forms: (1) a "proactive" form (which Peirce called "predictive"), in which case $A \rightarrow B$ means that B is a new consequence derived from A; (2) a retroactive form (which Peirce called "retrodictive"[sic]), according to which B implies A as a preliminary condition; and (3) a justifying form, which relates (1) and (2) through necessary connections that thus attain the status of "reasons" (Piaget and Garcia 1980/1987/1991: 121). Piaget (1981/1986: 57) likewise connects the retroactive with the proactive in view of "possibilities already realized before the task."

knowledge, and reveals an intimate familiarity with the state of one's own logical system. By "every logical evolution of thought" Peirce refers to the natural unfolding of abductive reasoning. This is so given the natural way in which good guesses emerge—they surface unbidden; and constitute, even in raw form, a "tendency to guess right" (1909: MS637). In fact, Peirce's concept of guessing right becomes more refined at Piaget's Reflected Abstraction stage, in that self-control over structuring events is ultimately assimilated into children's own logical system by their own conscious deliberation.

Piaget expresses the import of self-control on abductive reasoning as follows "It is a matter of course that the higher system constitutes a regulator exercising control over the regulations of lower levels. That is the case wherever reflexion occurs, because reflexion is a reflexion "on" what has been acquired previously. Reflexion thus represents the prototype of a regulation of regulations, since it is itself a regulator and takes control over whatever is controlled by previous regulations" (Piaget 1975/1985: 30). Similarly, Peirce demonstrates the critical place of self control to abductive rationality as follows: "As a process, abduction moves from the category of uncontrolled to controlled thought in the transition from perceptual judgment to guessing" (1903: CP5.181). It is a form of consciously taking a habit," recognizing the plasticity of knowledge, while relying upon the knowledge/conduct to produce a particular consequence: "A reasoning must be conscious; and this consciousness is…a sense of taking habit, or disposition to respond to a given kind of stimulus in a given kind of way" (1905: CP5.440). Increased self-control ensures greater plausibility of the abduction, since exercising self-control over arguments entails checking the validity of competing arguments, and precluding conflicts within the system. Peirce elaborates on the process by which internal self-control results in revisionary abductions, namely, the process of intrasubjective dialogue: "Thinking always proceeds in the form of a dialogue, between different phases of the ego" (1898: CP4.6). In 1908, Peirce reiterates this conviction, describing "All thinking is necessarily a sort of dialogue, an appeal from the momentary self to the better considered self of the immediate and of the general future" (SS 195). Self to self-appeal is a quintessential form of retroduction for Peirce, in that previous abductions compete with accommodated ones toward the enterprise of making the best guess. In fact, Peirce's selection of "retro-"demonstrates the need to harken back to already conceived of hypotheses in the process of establishing more fitting ones.

Given its reliance upon metaskills which emerge during Piaget's Reflected Abstraction stage, this intrasubjective exercise is a rather late and protracted development, beginning beyond four years of age and continuing throughout Piaget's Concrete Operational stage (Piaget 1975/1985: 30). To effectively convince the self of the increased veridicality of an hypothesis over another which has previously been held, the same mind must have the means to exercise self-control to preclude non-adherence to ignorance preservation by relaxing or tightening assumptions. Piaget acknowledges the import of using objective perspectives to generate plausible hypotheses. In Piaget's Reflected Abstraction stage, logic is treated as an object—an objective inquiry in which the legitimacy of hypotheses is in their effects on reality at large. To use logic intrasubjectively as a tool to ascertain objective truths, children need to have already exercised at least rudimentary perspective-taking competencies wrought by advancing from topological to projective space (Piaget and Inhelder 1948/1967: 153–154)—the means to imagine one's self in another place (physically and/or psychologically).[11]

According to Peirce, effects which flow from projective relations qualify as Logical and Final Interpretants. This self-talk leads to what Peirce refers to as the "twigging of ideas" (1913d: MS930). It presumes the power to perspective-take, as Piaget and Inhelder (1948/1967: 241) describe it—switching perspectival roles. Perspective shifting affords both hypothesis rumination and revision prior to becoming viable turns-of-action/argumentation. This internal back-and-forth (in Piagetian terms, reversibility of roles) allows greater autonomy for reasoning, without relying upon another's logical and representational system (or their approval) to carve out/twig ideas. Once attaining Piaget's Reflected Abstraction stage, children can draw upon their own perspective-shifting abilities to pose novel remedies (objectively-based courses of action), by independently evaluating the relative authenticity/effectiveness of each premise. As such, ignorance preservation is minimized, while unbidden well-formed hunches make their mark.

7 Conclusion

The consonance that permeates Piaget's and Peirce's models of abductive reasoning is remarkable. The constructivistic and developmental nature of the former, ratifies the integration of pragmaticism and realism of the latter. Both models begin with the assumption that empirical factors do not fully account for plausible discoveries, and that such discoveries ultimately constitute habits/dispositions of mind and action, accounting for how action and logic meet. Both models attribute consciousness and self control a primary place in the abductive turn—when inferring plausibly from unexpected consequences. They further maintain that social dialogue and later intrasubjective dialogue foster well-formed explanatory inferences, given their indexical and iconic nature. Inferences using index with icon trace not merely reversible paths from one perspective to another, but predict the actualization of social/conversational role-shifts. In short, both Piaget and Peirce predicate that mental reversibility (premises to an argument) is foundational to critical advances in hypothesis-making.

[11]Perspective-taking for Piaget presumes projective relations in that: "Projective space…begins psychologically at the point when the object or pattern is no longer viewed in isolation, but begins to be considered in relation to a 'point of view.' This is either the viewpoint of the subject, in which case a perspective relationship is involved, or else that of other objects on which the first is projected" (Piaget and Inhelder 1948/1967: 153–154).

References

Baddeley, A. (2007). *Working memory, thought, and action*. Oxford: Oxford University Press.

Bellucci, F. (2014). Logic, considered as semeiotic: On Peirce's philosophy of logic. *Transactions of the Charles S. Peirce Society, 50*(4), 523–547.

Buchanan, D., & Sobel, D. (2011). Mechanism-based causal reasoning in young children. *Child Development, 82*(6), 2053–2066.

Cohen, L. B., & Amsel, G. (1998). Precursors to infants' perception of causality. *Infant Behavior & Development, 21*(4), 713–732.

Diamond, A. (1985). The development of the ability to use recall to guide action, as indicated by infants' performance on AB. *Child Development, 56*, 868–883.

Gopnik, A. (2009). *The philosophical baby: What children's minds tell us about truth, love, and the meaning of life*. New York: Farrar, Straus and Giroux.

Magnani, L. (2001). *Abduction, reason, and science: Process of discovery and explanation*. New York: Kluwer Academic/Plenum Publishers.

Magnani, L. (2009). *Abductive cognition: The epistemological and eco-cognitive dimensions of hypothetical reasoning*. Berlin: Springer.

Mandler, J. (2004). *The foundations of mind: Origins of conceptual thought*. Oxford: Oxford University Press.

Mandler, J. (2010). The spatial foundations of the conceptual system. *Language and Cognition, 2* (1), 21–44.

Mandler, J., & Cánovas, C. P. (2014). On defining image schemas. *Language and Cognition, 6*(4), 510–532.

Moshman, D. (1996). The development of metalogical understanding. In L. Smith (Ed.), *Critical readings on Piaget* (pp. 396–415). London: Routledge.

Moshman, D. (2015). *Epistemic cognition and development: The psychology of justification and truth*. London: Psychology Press.

Peirce, C. S. (i. 1866–1913a). In C. Hartshorne & P. Weiss (Eds.), *The collected papers of Charles Sanders Peirce* (Vol. I–VI). Cambridge, Massachusetts: Harvard University Press (1931–1935). A. Burks (Ed.), (Vol. VII–VIII). Same publisher (1958).

Peirce, C. S. (i. 1866–1913b). In N. Houser & C. Kloesel (Eds.), *The essential Peirce: Selected philosophical writings* (Vol. 1). Peirce Edition Project (Eds.), (Vol. 2). Bloomington: University of Indiana Press (1992–1998).

Peirce, C. S. (i. 1866–1913c). In C. Eisele (Ed.), *The new elements of mathematics* (Vol. III). The Hague: Mouton Press (1976).

Peirce, C. S. (i. 1866–1913d). In R. Robin (Ed.), Unpublished manuscripts are dated according to the *Annotated Catalogue of the Papers of Charles S. Peirce*. Amherst: University of Massachusetts Press (1967), and cited according to the convention of the Peirce Edition Project, using the numeral "0" as a place holder.

Peirce, C. S. (i. 1867–1871). In Peirce Edition Project (Eds.), *Writings of Charles S. Peirce: A chronological edition* (Vol. 2). Bloomington: Indiana University Press, 1984.

Peirce, C. S., & Welby, V. (i. 1898–1912). In C. Hardwick & J. Cook (Eds.), *Semiotic and significs: The correspondence between Charles S. Peirce and Victoria, Lady Welby*. Bloomington: University of Indiana Press (1977).

Piaget, J. (1924/1959). *Judgment and reasoning in the child* (M. Warden, Trans.). Paterson, NJ: Littlefield, Adams, & Co.

Piaget, J. (1975/1985). *The equilibration of cognitive structures* (T. Brown & K. J. Thampy, Trans.). Chicago: University of Chicago Press.

Piaget, J. (1977/2001). *Studies in reflecting abstractions* (R. L. Campbell, Trans.). London: Routledge.

Piaget, J. (1980/2004/2006). Reason. (L. Smith, Trans.) *New Ideas in Psychology 24*(1), 1–29. *Dates correspond to initial composition, publication in French, and subsequent publication in English.*

Piaget, J. (1981/1986). *Possibility and Necessity, Vol. 1: The Role of Possibility in Cognitive Development* (H. Feider, Trans.). Minneapolis: University of Minnesota Press.
Piaget, J., & Garcia, R. (1980/1987/1991). In P. M. Davidson & J. Easley (Eds.), *Toward a logic of meanings*. Hilldale, NJ: Lawrence Erlbaum Associates. *Dates correspond to initial composition, publication in French, and subsequent publication in English.*
Piaget, J., & Inhelder, B. (1948/1967). *The child's conception of space* (F. J. Langdon & J. L. Lunzer, Trans.). New York: W.W. Norton and Company.
Pietarinen, A.-V. (2006). *Signs of logic: Peircean themes of the philosophy of language, games, and communication*. Heidelberg: Springer.
Pietarinen, A.-V., Bellucci, F. (2015). What is so special about logical diagrams? 1–15.
Smith, L. (1999). Epistemological principles for developmental psychology in Frege and Piaget. *New Ideas in Psychology, 17*(2), 83–117.
Smith, L. (2005). Studies in reflecting abstraction. *British Journal of Educational Psychology, 75*, 518–519.
Sobel, D., & Buchanan, D. W. (2009). Bridging the gap: Causality at-a-distance in children's categorization and inferences about internal properties. *Cognitive Development, 24*, 274–283.
Stjernfelt, F. (2014). *Natural propositions: The actuality of Peirce's doctrine of dicisigns*. Boston: Docent Press.
Stjernfelt, F. (2015). Dicisigns: Peirce's semiotic doctrine of propositions. *Synthese, 192*(4), 1019–1054.
Tulving, E. (2005). Episodic memory and autonoesis: Uniquely human? In H. S. Terrace & J. Metcalfe (Eds.), *The missing link in cognition: Origins of self-reflective consciousness* (pp. 3–56). Oxford: Oxford University Press.
West, D. (2011). Deixis as a symbolic phenomenon. *Linguistik Online, 50*(6), 89–100.
West, D. (2013). *Deictic imaginings: Semiosis at work and at play*. Heidelberg: Springer.
West, D. (2014a). Piaget's system of spatial logic: The semiosis of index. *Semiotica, 202*, 459–480.
West, D. (2014b). Perspective switching as event affordance: The ontogeny of abductive reasoning. *Cognitive Semiotics, 7*(2), 149–175.
West, D. (2015a). Recommendations as imperative propositions in the operation of abductive reasoning: Peirce and beyond. In M. Bergman & J. Queiroz (Eds.), *The commens encyclopedia: The digital encyclopedia of Peirce studies* (New ed.). Pub. 150526–0100a.
West, D. (2015b, August). *The semiosis of Peirce's dicisign in early habit-formation*. Paper presented at the 9th conference of the Nordic Association for semiotic studies, Tartu, Estonia.
West, D. (2015c). The work of as habit in the development of early schemes. *Public Journal of Semiotics, 6*(2), 1–13.
West, D. (2015d). Dialogue as habit-taking in Peirce's continuum: The call to absolute chance. *Dialogue (Canadian Review of Philosophy), 54*(4), 685–702.
Woods, J. (2013). *Errors of reasoning: Naturalizing the logic of inference*. London: College Publications.

Abduction and Model-Based Reasoning in Plato's *Republic*

Priyedarshi Jetli

Abstract I begin with a typology of reasoning and cross it with types of processes. I demonstrate that the thrust of Plato's *Republic* is theory-building. This involves the critical and dialectic processes which are paradigms of Platonic methodology. Book I displays abductive analogical reasoning joined by an induction that is embedded in a deduction; hence there is a deduction–induction–abduction chain. In Book VI, Plato constructs a visual model of the divided line, which also displays model-based and abductive hypothesis generation that is essential to theory building. Book VII provides an abductive metaphor model of the allegory of the cave. Both models depict degrees of reality and the ascendency of knowledge. The multimodal model-based allegory has far reaching applications from criminal justice to information systems. I conclude by capturing the narrative of the *Republic* as a critical and dialectic process of theory building (of justice) using deductive–inductive–abductive chains, an abductive visual model and an abductive metaphor model. Hence, the *Republic* is simultaneously a masterpiece of deductive reasoning and a marvel of complex model-based abduction, involving visual models, analogies and metaphors.

Keywords Model-based reasoning · Deduction · Induction · Abduction · Theory · Divided line · Allegory of the cave · Analogy · Visual model · Metaphor · Plato · Peirce

Critical still and Dialectic

> From metaphor thou springest—
> Analogy and visual model;
> Theory thou abductest
> On model-basis still dost soar, and reasoning ever abductive[1]
> —Adapted from Percy Bysshe Shelley, 'To a Skylark' (1820)

---

[1]Some of the words of Shelley's poem 'To a Skylark' (1820) have been changed, the spirit is maintained.

P. Jetli (✉)
Department of Philosophy, University of Mumbai,
Vidyanagari, Kalina, Santacruz (E), Mumbai 400 098, India
e-mail: priyedarshij@yahoo.com; pjetli@gmail.com

L. Magnani and C. Casadio (eds.), *Model-Based Reasoning in Science and Technology*, Studies in Applied Philosophy, Epistemology and Rational Ethics 27, DOI 10.1007/978-3-319-38983-7_20

1 Introduction

Plato (c.427–c.347 BCE) is recognized as the master of deduction. His use of induction and abduction is most often overlooked but it is methodologically embedded in most of his deductions. There is an abundance of model-based abduction, analogical reasoning and case-based induction in Plato. I demonstrate that most of Plato's compact reasoning provides paradigm examples of Magnani's 'abduction–deduction–induction cycle' (Magnani 2000, p. 25). As Olsen states, 'In general Plato presents puzzles, problems, and incomplete analysis, from which the reader may infer (abduct) the solutions (or adequate hypotheses)' (Olsen 2002, p. 86).

In this paper I trace various types of model-based abductive reasoning in the *Republic*. In Book I Plato uses integrated deductive–abductive–inductive chains. To discard inadequate definitions of 'justice', such as the one given by Thrasymachus, Plato provides a deductive argument, embedded in which is an abductive analogy combined with an induction. In Book VI, Plato employs an abductive (visual) model in the famous divided line. In Book VII the abductive metaphor model of the allegory of the cave depicts the degrees of reality as well as the ascendency of knowledge.

The theoretical purpose of the *Republic* is to establish the Platonic definition of justice. The three models within the narrative hence become: theoretical deductive–abductive (analogical)–inductive; theoretical abductive (visual), and theoretical abductive (metaphor) models respectively. Therefore, the *Republic* is not simply a masterpiece of deductive reasoning but it employs, in its most famous passages, model-based abduction of the visual and metaphor varieties. The dialogue then is a multi-modal integration of sentential, model-based, and case-based reasoning incorporating deduction–induction–abduction chains.

In Sect. 2, I provide a heuristic classification of reasoning and processes. This is a tentative and conjectural classification with the purpose of providing some insights into understanding the uses of model-based reasoning and abduction in the *Republic*. My classifications are neither alternatives to nor complements of contemporary classifications of this sort.

In Sect. 3, I trace the origins of deduction, abduction and induction from Homer to Plato. We find the origins of abductive metaphor as a form of reasoning in Homer. Then, in the Ionians: Thales, Anaximander and Anaximenes; we find abductive conjectures in the idea of there being an *urstoff* of the world, close to inference to the best explanation. Thales, traditionally also marks the beginning of deduction with his proofs in geometry. In Parmenides and Zeno we have the advent of *reduction ad absurdum* arguments. And finally in Anaxagoras and Empedocles abduction is used to posit counteracting forces as cosmological principles.

In Sect. 4, I display Magnani's deduction–abduction–induction cycle in Book I of the *Republic*, an abductive analogy in conjunction with a case-based induction is embedded inside a deductive argument that demonstrates the inadequacy of the definition of 'justice' provided by Thrasymachus. So, what is often thought of as a paradigm of deduction in Plato turns out to be a good example of an integrated multi-modal deduction–abduction–induction reasoning.

In Sect. 5, I demonstrate how the divided line model in Book VI is a paradigm of a visual abductive model which depicts reality as well as the ascendency of knowledge. If Plato were asked where in his writings is his metaphysics and epistemology captured in a nutshell, he would most likely have pinpointed the divided line. The divided line and Plato's discussion of it in the dialogue remarkably coincides with features of visual abductive models that contemporary workers in this area point out.

In Sect. 6, I discuss the allegory of the cave in Book VII as an abductive model of the metaphor variety. This is perhaps the single most famous piece of reasoning in the history of philosophy. The metaphor Plato uses here for the ascendency of knowledge satisfy the three stages of recognition, analysis and interpretation as pointed by D'Harris, a contemporary worker on metaphor. It also satisfies the criterion of a strong metaphor as laid out by Max Black. So, again Plato's use of the allegory of the cave as in the case of the divided line is visionary.

In Sect. 7, I compare the visual abductive model of the divided line with the metaphor abductive model of the allegory of the cave. An isomorphism between the two can be established, perhaps only because both are abductive models not pieces of deductive reasoning. This establishes the primacy of abduction over deduction in the heart of Plato's most highly regarded work.

In Sect. 8, I provide a narrative sketch of the whole dialogue. I conclude that the types of processes and reasoning in the *Republic* can be best captured by the following: ≺Critical—Dialectical [theoretical {sentential (deductive)—case-based (inductive)—model-based (abductive ⟨analogy—visual—metaphor⟩)}]≻. That is, the dialogue integrates the processes of critical thinking with a dialectic, which, in turn are carried out by an overarching theoretical reasoning, the main purpose of which is to develop the correct definition of 'justice'. This proceeds first by a polemic stage in Book I, which, though it is mainly deductive in a sentential argument form, it does contain embedded in it an abductive analogy and a case-based induction. Then, the divided line employs a visual abductive model and the allegory of the cave a metaphor abductive model. Hence, the *Republic* is not a solely deduction centered dialogue. Rather, it employs exemplary deduction–abduction–induction cycles, and it is densely model-based abductive in its most crucial parts.

2 Classification of Reasoning and Processes

I begin with classification of different types of reasoning. This is an open-ended classification. Following Peirce, I could begin with deduction, induction and abduction as three types of reasoning. Peirce emphasized that a crucial activity of science, that of hypothesis formulation, conjecture, discovery and invention was not captured by deduction and induction and thereby abduction was required. I sustain Peircean insights and contributions but extend 'reasoning' to all types inside and outside of science. The order in which I display the classification is a convenient heuristic device to understand the types of reasoning displayed in Plato's dialogues. The classification is neither an alternative to nor a complement to any such classifications being discussed today.

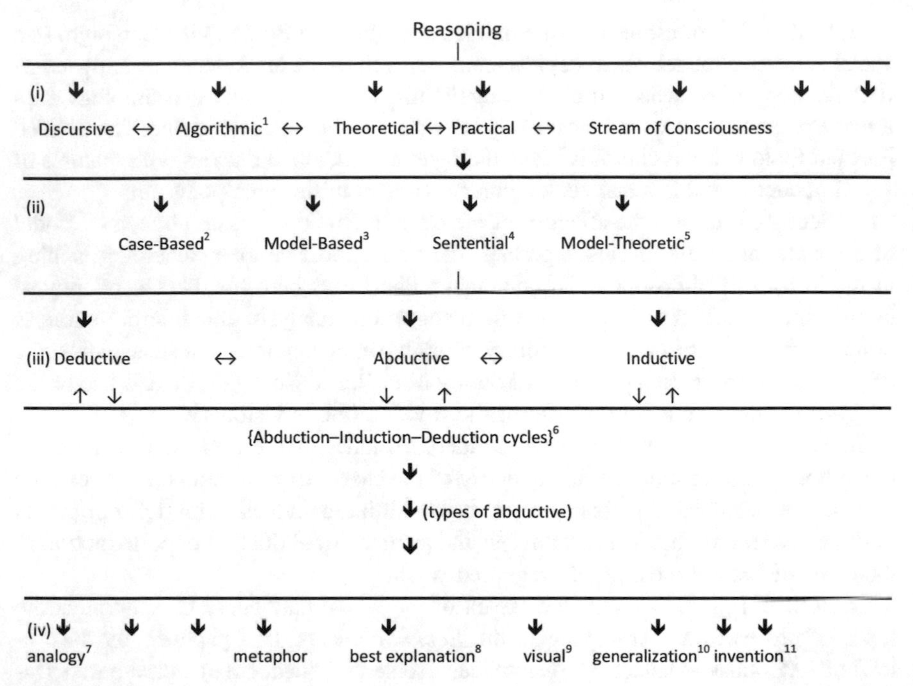

Fig. 1 Reasoning. *1* Algorithms are normally thought to be deductive. In my earlier paper on Plato's *Meno* I contend that algorithms may involve deduction when used but an algorithm itself, is not deductive, and it is a different sort of thinking involving more intuition and even visualization than in deductive inferences (Jetli 2014). *2* By case-based I mean '[…] process […] which helps store, organise, retrieve, and reuse experiential knowledge […].' (Shuguang et al. 2000, abstract). Also see Bergmann and Wilke (1998). *3* Though model-based reasoning is most commonly thought of in terms of physical models (Lidz 1995), I take it to be broader and best characterized by '[…] the knowledge base is represented as a set of models […] of the world rather than a logical formula describing it'(Khardon and Roth 1997, p. 171). Also, the notion of model can be extended to mental models: '[…] forms of model-based reasoning: analogical, visual, and simulative modeling. Further, the psychological theory of mental modeling provides a basis in human cognition for taking the external traces of modeling […]' (Nersessian 1999, p. 14). I prefer the even broader conception of Brenner as reasoning with the aid of models, about models and model-determined (Aguayo 2011, p. 34). *4* 'related to logic and to verbal/symbolic inferences' (Magnani 2006, p. 108, 2009, p. 5). *5* I mean 'model-theoretic' in the sense Giere uses it as 'a set theoretical structure' (Giere 1999, p. 42). *6* See Magnani (2000, p. 25). *7* See Clement and Núñez Oviedo (2003, p. 5). *8* Abduction is often identified with inference to the best explanation. See Bylander et al. (1991, p. 25), Flach (1996, p. 31), Josephson and Josephson (1994, p. 5). Some however think that a distinction should be maintained between abduction and inference to the best explanation (see Minnameier 2004, p. 75, Paavola 2006, p. 96). However I consider inference to the best explanation to be a type of abduction. *9* See Thagard and Shelley (1997). *10* See Tiercelin (2005, pp. 395–396). *11* See Clement and Núñez Oviedo (2003, p. 11)

In Fig. 1, the two-sided arrows at each level mean that they interact with each other as a particular piece of reasoning could be a synthesis of two or more of these classifications. Level (ii) is a classification into which each of the level (i) types can be partitioned. The same is the case for level (iii) in relation to level (ii). At level (iv) I have only provided types of abductive reasoning, leaving out types of deductive and inductive reasonings for the purposes of this paper. Each level could have other species as well.

These levels express the bias of a philosopher. Going back to Aristotle, the main division in philosophy is between theoretical and practical thinking, so I begin at level (i) with the distinction between theoretical and practical. Level (ii) I place next in the light of a wider perspective of 'science' that stretches beyond the physical, natural and life sciences. It includes medical science, social sciences, cognitive science and computer science. I then bring the Peircean distinction at level (iii).

These levels also work beautifully for understanding the methodology of Plato. The purpose of the *Republic* is to find the correct definition of 'justice'. This is a search for a theory of justice. In order to construct the theory Plato begins first with arguments against the existing alternative definition of 'justice'. This involves formulating arguments in sentential format. Hence it is at level (ii). The arguments themselves are deductive arguments and at level (iii).

The main bone of contention here may be the relation between model-based reasoning and abduction. Is model-based reasoning a species of abduction or is abduction a species of model-based reasoning? I hope to argue for the latter in a future paper, here, I simply assume it. I am taking a Hansonian line of the primacy of the context of discovery over the context of justification rather than the Popperian line of the primacy of the context of justification over the context of discovery. The trichotomy of deduction, induction and abduction is generally presented as three types of inferences. And the word 'inference' suggests that these have to do with the context of justification.

I am adopting another unorthodox line for which I will need a lot of work against the current of Plato scholars. Most of Plato's dialogues are hugely polemic and stay at the context of justification. When some theory has to be presented Plato often resorts to some myth or visual models and allegory as in the case of the *Republic*. This is the context of discovery. Knowledge for Plato comes at the point of recollection, and when we recollect, even though at the end of a long dialectic process, it is a matter of discovery rather than justification.

I use 'abduction' in a wide sense motivated by insights of Lorenzo Magnani:

> Many commentators always criticized the Peircean ambiguity in treating abduction in the same time as inference and perception. [...] perception and imagery are kinds of that model-based cognition which we are exploiting to explain abduction: [...] we can render consistent the two views, beyond Peirce, but perhaps also within the Peircean texts, taking advantage of the concept of *multimodal* abduction, which depicts hybrid aspects of abductive reasoning (Magnani 2006, p. 120).

Hence abductive analogy is hybrid since analogy is not a subspecies of abduction, but a kind of reasoning that is prevalently abductive.

Peirce divided inferential reasoning into deductive, inductive and abductive:

> If we are to give the names of Deduction, Induction, and Abduction to the three grand classes of inference, then deduction must include every attempt at mathematical demonstration, […]; Induction must mean the operation that induces an assent, […]; while Abduction must cover all the operations by which theories and conceptions are engendered (Peirce 1957, p. 237).[2]

'Inference' should not be taken literally. Perhaps deduction and induction can still be thought of as inferences. But if we are to accept Peirce's last sentence here, then some operations by which theories are engendered may not be inferences as is the case in visual and metaphor abduction. So, it is best to interpret 'inference' here as reasoning.

Peirce later writes in 1905: '[…] there are but three elementary kinds of reasoning. The first, which I call *abduction* […]. The second kind of reasoning is *deduction*, […]. The third way of reasoning is *induction* (Peirce 1931–1958, CP 8.209).[3] And in an even later work, the word 'elementary' is also removed: 'Th(e) three kinds of reasoning are Abduction, Induction and Deduction' (Peirce 1931–1958, CP 5.145).[4] I take reasoning to be broader than inference following a welcome transition that Peirce makes from types of inferences to types of reasoning from 1903 to 1905. Sami Paavola has distinguished between abductive inference and abductive instinct within Peirce (Paavola 2005). With my typology it is possible that model-based reasoning could have the varieties of deduction, induction or abduction.

'Model-based reasoning' is generally taken to be synonymous to 'abduction' or a species of abduction. For example:

> Many animals […] make up a series of signs and are engaged in […] semiotic activity — which is fundamentally *model-based* — they are at the same time engaged in "being cognitive agents" and therefore in thinking intelligently. […] An important effect of this semiotic activity is a continuous process of "hypothesis generation" […]. This activity is at the root of a variety of *abductive* performances, which are also analyzed in the light of the concept of affordance […] (Magnani 2007, p. 3).

As 'hypothesis generation' is most commonly identified with abduction since Peirce, I read Magnani's passage as an equivocation of 'model-based reasoning' with 'hypothesis generation' and thereby with 'abduction'. However, I take 'model-based reasoning' to be broader in that it includes induction and deduction as we do talk about deductive and inductive models. Magnani has demonstrated that non-human animals use model-based abductive reasoning, and others have demonstrated that non-human animals use model-based inductive reasoning, only the presence of model-based deductive reasoning is lacking in them.

---

[2]For further documentation of Peirce's classification of deduction, induction and abduction as three types of inferences, see Aguayo (2011, p. 33), Aliseda (2000, p. 47), Angué (2009, pp. 65–66), Paavola (2006, p. 95), Pape (1997, p. 199); Svennevig (2001, p. 1), Upshur (1997, p. 205).

[3]Accessed from Hoffman (1999, pp. 271–272).

[4]Accessed from Koschmann (2003, p. 4).

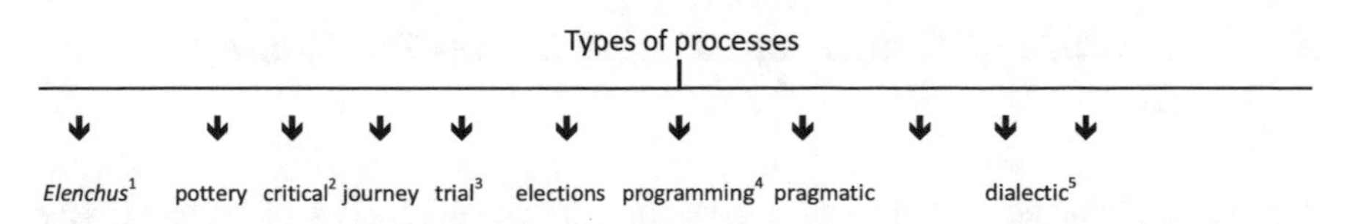

Fig. 2 Types of processes. *1* This is the Socratic *elenchus* and the dialectic is a sub-process of it. *2* Plato's methodology consists of the following stages: first, the polemic stage, where alternative theories are considered revised and rejected; second, the constructive stage, where Plato's theory is constructed as a synthesis of and transcending the rejected theories; third, the critical stage, where the constructed theory is criticized and put to test, fourth, the revision stage. *3* 'The entire course of a judicial proceeding' (FARLEX 2012). *4* 'a. A running software program or other computing operation. b. A part of a running software program or other computing operation that does a single task' (FARLEX 2012). *5* For Plato dialectic is the process which leads to the final recollection of Forms and it is usually embedded in the Socratic *elenchus*. Here by 'dialectic' I mean in the wider sense as used by Heraclitus and Parmenides, and later by Hegel. In a sense in which everything is thrown in for debate without any possible resolution on the horizon, which in fact leaves room for creative and innovate resolutions to the problem posed

In normal human activity of any kind, especially in science and technology, the types of reasoning intersect with types of processes (Fig. 2).

These are examples of processes and arbitrarily chosen as there are thousands of such processes. I take the operating definition of 'process' as: '1. A series of actions, changes, or functions bringing about a result [...]; 2. A series of operations performed in the making or treatment of a product' (FARLEX 2012).

Though processes are normally thought of as concrete and actual, there are also philosophical processes, and this characterizes all of Plato's dialogues, where the processes are normative or thought experiments rather than actual. Such a distinction is found in software engineering: 'One can distinguish two main types of process models: *prescriptive* and *descriptive* process models' (Münch et al. 2012, p. 20). The three main processes that guide most of Plato's dialogues are the Socratic *elenchus* (or pedagogy), the dialectic (the road to recollection) and critical, which is comprehensive philosophical activity involving polemic, constructive (theory-building) and revision stages. The intersection of processes and reasoning is hybrid.

I use the following notational scheme to classify different types or reasoning integrated with processes: ≺Process [level (i) reasoning {level (ii) reasoning (level (iii) reasoning ⟨level (iv) reasoning⟩)}]≻. Any n-dashes would indicate more than one type of reasoning being used at any level. The whole of the *Republic* can be summarized cumulatively as: ≺Critical–Dialectic [theoretical {sentential (deductive)—case-based (inductive)—model-based (abductive ⟨analogy–visual–metaphor⟩)}]≻.

3 From Homer to Plato

I now trace the history of different types of reasoning from Homer to Plato.

3.1 Literature of Homer and Hesiod in 8th–7th Century BCE

The basic syntactic structure of a reasoning in verse consists of meter, rhyme, alliteration, and so on, as in the opening of the *Iliad*:

> His bow and quiver both behind him hang,
> The arrows chink as often as he jogs, 50
> And as he shot the bow was heard to twang,
> And first his arrows flew at mules and dogs.
> But when the plague into the army came,
> Perpetual was the fire of funerals;
> And so nine days continued the same. 56 (Homer 1975)

The meter is dactylic hexameter (Wikipedia 2012). The rhymes have been preserved in the English translation by Hobbes, though this particular verse does not display alliteration.

The semantic structure of reasoning in literature consists of metaphor, simile, allegory, and so on. Consider Hesiod's *Shield of Heracles*:

> So he arose from Olympus by night pondering guile in the deep of his heart, and yearned for the love of the well-girded woman. Quickly he came to Typhaonium, and from there again wise Zeus went on and trod the highest peak of Phicium: there he sat and planned marvelous things in his heart. So in one night Zeus shared the bed and love of the neat-ankled daughter of Electryon and fulfilled his desire; and in the same night Amphitryon, gatherer of the people, the glorious hero, came to his house when he had ended his great task (Hesiod 2012, pp. 30–38).

One could read this as a literal story but since the characters are mythological it is best to read it as metaphorical. One metaphoric meaning of 'night' is that of darkness (Kirk and Raven 1957, p. 22), so that both of these acts were committed in ignorance.

3.2 The Origins of Deduction, Abduction and Induction in Thales and the Ionians of 6th to 5th Century BCE

Thales (c. 624–c. 546 BCE) is generally credited to be the first to provide deductive proofs in geometry: 'Thales […] began the process of deriving theorems from first principles that we still use today' (Experiment-Resources.com 2012). Deduction as a form of reasoning is partitioned into syntactical and semantical. This is the basic model of all deductive reasoning.

Thales conjectured that water was the *urstoff* of the world. How did Thales reach this conclusion? That water was the first principle with which all plurality and change could be explained was a hypothesis, perhaps an inference to the best explanation: 'Thales is offering a "hypothesis" regarding how water changes to form the plurality of reality. In doing so he *abduces* an explanation'.[5] This is supported by Kirk and Raven: 'From the analogy of his immediate successors we might have expected Thales to have abduced meteorological reasons, more conspicuously, in support of the cosmic importance of water' (Kirk and Raven 1957, p. 89). This is abduction of the variety of inference to the best explanation.

After Thales began the formation of hypotheses, others after him formed other hypothesis: Anaximander (c. 610–c. 546 BCE) posited the unlimited as the *urstoff* and first principle; Anaximenes (c. 585–c. 528 BCE) posited air and Heraclitus (c. 535–c. 475 BCE) posited fire. So, abduction as conjectures through inference to the best explanation was well under way at the beginnings of Western science.

In response to Thales' conjecture that 'everything is water' someone objected:

If everything is water, then everything is wet.
Some things are not wet.
Therefore, Not everything is water.

'Some things are not wet' is the same as 'not all things are wet' or 'it is not the case that everything is wet'. Once we replace the second premise with the last of these statements we get the following form:

If p, then q
not-q
Therefore, not-p

This is *modus tollens*. Though *modus ponens* is the most common rule of inference used in axiomatic formal systems, *modus tollens* has had a great significance in the history of science, in the counterexample technique of disproving hypotheses. It is no surprise that any documented use of *modus tollens* may precede that of *modus ponens*. When one offers a hypothesis, not reached by deductive argument, and an opponent wants to immediately react because she is skeptic about the hypothesis, then she will immediately posit a counterexample and use *modus tollens*.

Both *modus ponens* and *modus tollens* are rules of inference that the human brain seems to be hard wired with and they have hardly ever been challenged. No doubt we use rules of inference in day to day reasoning, but how do we acquire knowledge that these are rules of inference? Aren't these conjectures which we cannot arrive at deductively? Perhaps all rules of inference can be derived from *modus ponens*, but then *modus ponens* would be a conjecture not arrived at through deduction, nor through induction, hence the discovery of *modus ponens* as a rule of inference must be abductive.

---

[5]McMahon, K.: Thales of Meletus. http://www.kevindmcmahon.com/Reseda/philosophy/philosophy%20web%20pages/Greek%20and%20Roman/Thales/thales6.htm.

3.3 *Deductive* Reductio Ad Absurdum *Arguments in Zeno in 5th Century BCE*

Zeno's (c. 490–c. 430 BCE) arguments were *reductio ad absurdum* deductive arguments.

Arrow argument: Suppose a moving arrow. According to the Pythagorean theory the arrow should occupy a given position in space. But to occupy a given position in space is to be at rest. Therefore the flying arrow is at rest, which is a contradiction (Copleston 1962, p. 57).

Stadium argument: Suppose I want to travel from point A to point B on a straight line, I must first reach the half-way point, which will take some time, then I must reach the half-way point between the half-way point and the goal, which will take some more time, and so on, so that it would take infinite time to reach from A to B so that I will never reach there, hence there is no motion.

The argument, as Aristotle states: '[…] *asserts the non-existence of motion on the ground that that which is in locomotion must arrive at the half-way stage before it arrives at the goal. […] many arguments against common opinion, such as Zeno's that motion is impossible and that you cannot traverse the stadium*' (Kirk and Raven 1957, p. 292). Hence, we have two variations of the deductive *reductio ad absurdum* argument against motion.

For each variation there are several nominalized instances of the argument. In the stadium argument, one instance could be where point A is the colosseum and point B is Il Vittoriano in Rome; another instance is where point A is India Gate and point B is Rashtrapati Bhavan in New Delhi; and another instance where point A is the Capitol building and point B is the Washington Monument in Washington D.C.; and so on.

Since Zeno began with assumptions such as there is plurality or there is motion his arguments are deductive *reductio ad absurdum* arguments inspired by Pythagorean mathematics (Kneale and Kneale 1962, p. 8).

3.4 *Abductive Conjecture of Opposing Forces in Anaxagoras and Empedocles in 5th Century BCE*

Empedocles (c. 490–c. 430 BCE) posited all the natural elements together as being the *urstoff* (Copleston 1962, p. 62; Kirk and Raven 1957, p. 328), perhaps because we observe these elements everywhere and inductively generalize from them that together they must constitute the *urstoff* of the world. Empedocles also adopts from Heraclitus the opposing forces of Love and Hate.

Anaxagoras (c. 510–428 BCE) was a precursor to alchemy and chemistry as he posited about thirty basic elements which constituted the *urstoff* of the world. Each thing was a mixture in different proportions of all of these elements. Love was the force that accounted for the bonding of these elements to form ordinary objects and

Hate was the opposing force that accounted for decomposition. Both the number of elements as well as the opposing forces of Love and Strife are abductive conjectures.

3.5 Presocratic Conclusions: Origins of Deduction and Abduction

In the Presocratics we find plenty of deduction, but we also find abductive reasoning in terms of conjectures and hypotheses as well as for the justification of deduction. We may ask, why does the *reduction ad absurdum* form of deductive argument work? The answer would have to involve an abductive inference. There is then evidence of abductive–deductive chains as far back as Thales. However, there does not seem to be much use of induction. Kirk and Raven posit that Thales could have used induction in his argument for souls but they reject it immediately (Kirk and Raven 1957, p. 95). I have in this survey of the Presocratics attempted to lay out all the forms of reasoning used by them. Deduction was surely dominant from Thales to Zeno, but as I have shown abduction in various forms was very much around, and with the near absence of induction; we could summarize Presocratic reasoning as a multimodal deduction–abduction cycle.

4 Deduction–Abduction–Induction Cycle in Book I of the *Republic*

In Book I the purpose is to go through the available proposed definitions of 'justice'. The apex is Socrates' construction of a substantial argument against Thrasymachus's definition that justice is whatever is to the advantage of the stronger [338c]:

> physician is a healer of the sick;[6] […] pilot is a ruler of sailors; […] an art seeks the advantage of its object; […] medicine does not consider the advantage of medicine but of the body; […] nor horsemanship of horsemanship but of horses; […] it has no need–but for that for which it is the art; […] Then no art considers or enjoins the advantage of the stronger but every art of the weaker which is ruled by it; (Plato 1989, 341c–342d).

By analogy then ruling as an art and the ruler serves the interest of the subjects and not his own interest. The conclusion seems to be an inductive generalization from the three cases. Are three cases enough for an inductive generalization? Plato leaves as an exercise for the readers to provide multiple other cases such as:

[6]Medical analogies are very common throughout Plato's dialogues as Lidz claims: 'such as comparing justice and injustice to health and illness' (Lidz 1995, p. 529).

teaching does not consider the advantage of teaching but of the students; a midwife does not consider the benefit of midwifery but of the mother and child; and so on.

Socrates concludes:

> [...] neither does anyone in any office of rule in so far as he is a ruler consider and enjoin his own advantage but that of the one whom he rules and for whom he exercises his craft, and he keeps his eyes fixed on that and on what is advantageous and suitable to that in all that he says and does (Plato 1989, 342e).

Both induction and analogical reasoning are used completing 'Peirce's remark that analogical reasoning be a compound inference made up of abduction and induction, [...] abduction, induction, and deduction stand in a dynamic relation whereby abduction leads to new explanatory or technological hypotheses [...]' (Minnameier 2010, p. 109).

Three instances are hardly sufficient for an adequate induction so it is analogical reasoning rather than induction that is dominant in this argument. The analogical reasoning used here is not for explanation but for thinking as was the case with Kepler who 'used analogies to think with and not simply to explain' (Gentner 2002, p. 33). So, inductive reasoning is embedded in the deductive argument here, but without the analogical step, the conclusion cannot be reached.

The fulcrum of the argument is the analogy among the examples offered and the case of the ruler. This requires understanding the structural similarity among these cases. This is vintage analogical reasoning embedded in an induction which is further embedded in a deduction. Once the analogy is established, one can generalize by induction to the hypothesis that in every art which is analogical to the examples the art serves the interest of the subjects for whom the art is administered. Now, by deduction from this generalization we derive the instance that in the art of ruling the ruler serves the interest of his subjects for whom the art of governing is administered. Induction and abduction complement each other here as they do in current computer science: 'The complementarity between abduction and induction [...] abduction providing explanations from the theory while induction generalizes to form new parts of the theory' (Flach and Kakas 2000, p. 24). As Josephson (2000, p. 40) claims we can treat 'inductive generalization as an instance of abduction'.

This argument has the surface structure of an inductive generalization from inductive instances, and then a universal instantiation from the generalized proposition as the subsidiary conclusion. However, the deep structure is analogical reasoning. Not every art will have the structure of its being for the object of the art where the object is other than the art itself such as in painting. Hence, clear examples of arts with very clear objects other than the arts themselves are picked, and ruling is considered to be such an art with the clear object being the ruled, the subjects, for whose advantage the art of ruling is performed.

5 Visual Model of the Divided Line in Book VI

The divided line is constructed to display the distinction between the visible and the intelligible (*509d*). It is an *abductive visual* model:

> […] You surely apprehend the two types, the visible and the intelligible.
> I do.
> Represent them then, as it were, by a line divided into two unequal sections and cut each section again in the same ratio—the section, that is, of the visible and that of the intelligible order—
> and then as an expression of the ratio of their comparative clearness and obscurity you will have, as one of the sections of the visible world, images. By images I mean, first, shadows, and then reflections in water and on surfaces of dense, smooth, and bright texture, and everything of that kind, if you apprehend.
> I do.
> As a second section assume that of which this is a likeness or an image, that is, the animals about us and all plants and the whole class of objects made by man.
> I so assume, he said (Plato 1989, 509d–510a) (Fig. 3).

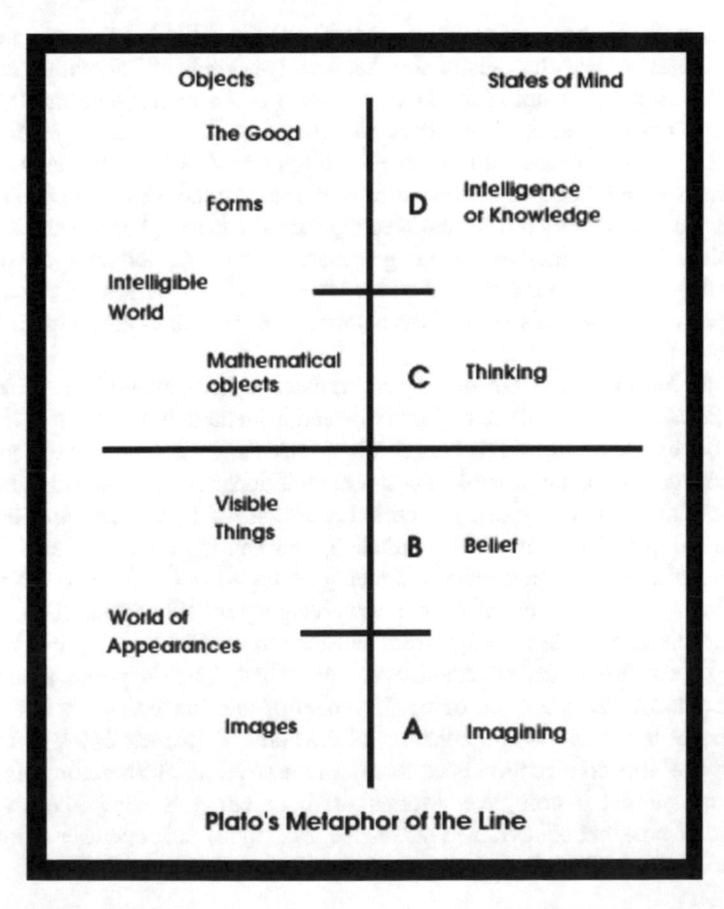

Fig. 3 The divided line. *Source* https://aquileana.wordpress.com/2014/04/03/platos-republic-the-allegory-of-the-cave-and-the-analogy-of-the-divided-line/

A is the lowest of the lower section, that of imagination and B, the upper of the lower is the perceptual. C is the lower of the upper, that of mathematical objects and D is the upper of the upper, that of Forms.

The divided line as a model of reality is now laid out:

> Would you be willing to say, said I, that the division in respect of reality and truth or the opposite is expressed by the proportion—as is the opinable to the knowable so is the likeness to that of which it is likeness?
> I certainly would.
> Consider again the way in which we are to make the division of the intelligible section.
> In what way?
> *By the distinction that there is one section of it which the soul is compelled to investigate by treating as images the things imitated in the former division, and by means of assumptions from which it proceeds not up to a first principle but down to a conclusion, while there is another section in which it advances from its assumption to a beginning or principle that transcends assumption, and in which it makes no use of the images employed by the other section, relying on ideas only and progressing systematically through ideas.*
> I don't fully understand what you mean by this, he said. (my *emphasis*) (Plato 1989, 510a-b).

This and the further explanation to follow in *510c-511e* is Plato in a nutshell.

Plato makes a very clear distinction between two kinds of inferential reasoning. The first, that from assumptions to conclusion ('*down to a conclusion*'). This is deduction. The second is from ideas to first principles ('*to a ... beginning or principle*'). This is classic inference to the hypothesis or abduction. Deductive reasoning is mainly used in mathematics as proofs. Abductive reasoning is used in the dialectic and the process for recollecting forms, wherein, lies the acquisition of knowledge. While deductive reasoning operates at level C, abductive reasoning is used at level D. Furthermore the classification and discussion of the different features of varying models would also belong, as all pure theories would belong, to level D.

This model shows the ascendancy of knowledge from imagination to veridical perception at the bottom of the major divide and from thought to reason in the upper half of the major divide. Even though Plato appeals to mathematical proportions, the reasoning is minimally deductive, not at all inductive, but is heavily abductive. The role of abduction in visual perception is best stated by Shanahan: 'The bridge between the quantitative and the qualitative, between the numerical and the symbolic, is made by abduction, which, drawing on high-level knowledge, is the final arbiter in the fixation of belief for the perceiving agent' (Shanahan 2005, p. 130).

The ascendancy of knowledge from illusions to veridical perception to mathematical objects to Forms is exhaustively described. This is prime model-based reasoning. In model-based reasoning, 'a model of the domain is first constructed, consisting of the structure and behavior of the domain' (Kumar 2002, p. 1). In the 'divided line' this construction is carried out to perfection. Furthermore, 'Insofar as the domain model is complete, Model-based reasoning is comprehensive in its coverage of possible behavioral (and hence, structural) discrepancies. This is not necessarily true of Rule-Based systems […]' (Kumar 2002, p. 2). The rejection of the proposed definitions of 'justice' in Book I is a rule-based system in which the proposed definitions are rejected because they violate the rules of adequate definitions, leaving not much room for flexibility. Here however, the structure of the models provides, actually accommodates all variations in what people perceive and what they know, yet the ascendency of knowledge depicted by the divided line remains.

6 Metaphor as Model-Based Abduction in the Allegory of the Cave in Book VII

Book VII presents the most famous of Plato's models:

> Next, said I, compare our nature in respect of education and its lack to such an experience as this. Picture men dwelling in a sort of subterranean cavern with a long entrance open to the light on its entire width. Conceive them as having their legs and necks fettered from childhood, so that they remain in the same spot, able to look forward only, and prevented by the fetters from turning their heads. Picture further the light from a fire burning higher up and at a distance behind them, and between the fire and the prisoners and above them a road along which a low wall has been built, as the exhibitors of puppet shows have partitions before the men themselves, above which they show the puppets.
> All that I see, he said (Plato 1989, 514a-b) (Fig. 4)

Since this is a literary model, whether the reasoning is allegorical or metaphorical or one using simile, I may use the relative freedom of literature to posit an interpretation that is not normally pointed out by Plato scholars.

'[…] they remain in the same spot, able to look forward only, […]' may be interpreted as forward reasoning only, proceeding from premises to conclusion. Hence, what we traditionally call 'deduction' comes down to the lowest level in the Platonic hierarchy as in the allegory of the cave this is the lowest epistemic stage.

> See also, then, men carrying past the wall implements of all kinds that arise above the wall, and human images and shapes of animals as well, wrought in stone and wood and every material, some of these bearers presumably speaking and others silent.
> A strange image you speak of, he said, and strange prisoners.
> Like to us, I said. For to begin with, tell me do you think that these men would have seen anything of themselves or of one another except the shadows cast from the fire on the wall of the cave that fronted them?
> How could they, he said, if they were compelled to hold their heads unmoved through life? (Plato 1989, 515a-b).

'Like to us' is the main metaphor. When we are focused on an argument geared towards a certain conclusion, the relation to reality is lost. Whether the argument is about propositions that refer to facts or about propositions about illusions does not seem to be important as the argument in itself is either valid or invalid.

> And again, would not the same be true of the objects carried past them?
> Surely.

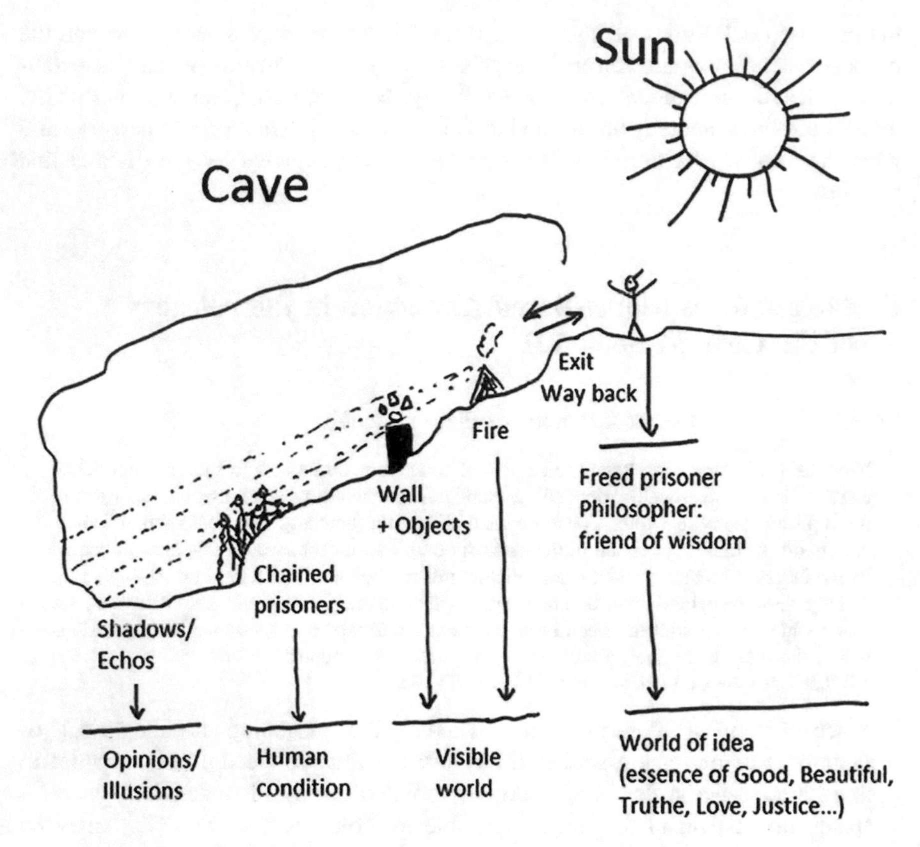

Fig. 4 The allegory of the cave. *Source* https://in.pinterest.com/pin/495958977687405517/

If then they were able to talk to one another, do you think that they would suppose that in naming the things that they saw they were naming the passing objects?
Necessarily.
And if their prison had an echo from the wall opposite them, when one of the passers-by uttered a sound, do you think that they would suppose anything else than the passing shadow to be the speaker?
By Zeus, I do not, said he. (Plato 1989, 515b)
Then in every way such prisoners would deem reality to be nothing else than the shadows of the artificial objects.
Quite inevitable, he said.
Consider, then, what would be the manner of the release and healing from these bonds and this folly if in the course of nature something of this sort should happen to them. When one was freed from his fetters and compelled to stand up suddenly and turn his head around and walk and to lift up his eyes to the light, and in doing all this felt pain and, because of the dazzle and glitter of the light, was unable to discern the objects whose shadows he formerly saw, what do you suppose would be his answer if someone told him that what he had seen before was all a cheat and an illusion, but that now, being nearer to reality and turned toward more real things, he saw more truly? And if also one should point out to him each of the passing objects and constrain him by questions to say what it is, do you not think that he

> would be at a loss and that he would regard what he formerly saw as more real than the things now pointed out to him?
> Far more real, he said (Plato 1989, 515c-d).

We have now moved from stage A of the divided line to stage B. Both stages have to do with perception, though at stage A one perceives illusions and at stage B one perceives real physical objects, the objects which are ultimately causally responsible for the illusions.

The allegory shows more clearly how we move from knowledge of immediate perceptions to the knowledge of the existence of physical objects and the physical world.

The divided line and the allegory of the cave complement each other:

> And if he were compelled to look at the light itself, would not that that pain his eyes, and would he not turn away and flee to those things which he is able to discern and regard them as in very deed more clear and exact than the objects pointed out?
> It is so, he said.
> And if, said I, someone should drag him thence by force up the ascent which is rough and steep, and not let him go before he had drawn him out into the light of the sun, do you not think that he would find it painful to be so haled along, and would chafe at it, and when he came out into the light, that his eyes would be filled with its beams so that he would find it painful to be so haled along, and would chafe at it, and when he came out into the light, that his eyes would be filled with its beams so that he would not be able to see even one of the things that we call real?
> Why, no, not immediately, he said.
> Then there would be need for habituation, I take it, to enable him to see things higher up. And at first he would most easily discern the shadows and, after that, the likenesses or reflections in water of men and other things, and later, the things themselves, and from these he would go on to contemplate the appearances in the heavens and heaven itself, more easily by night. Looking at the light of the stars and the moon, than by day the sun and the sun's light.
> Of course.
> And so, finally, I suppose, he would be able to look upon the sun itself and see its true nature, not by reflections in water or phantasms of it in an alien setting, but in and by itself in its own place.
> Necessarily, he said (Plato 1989, 515e-516b).

We have now moved from stage B to stages C and D, the world of Forms. Just as men who first stepped out of the cave would be blinded by the light of the sun, moving from the world of vision to the world of Forms would be blinding, but with gradual training one could progress from grasping mathematical objects and lower Forms to the grasping of higher Forms.

> And at this point he would infer and conclude that this it is that provides the seasons and the courses of the year and presides over all things in the visible region, and is in some sort the cause of all these things that they had seen. (Plato 1989, 516b-c)
> […]
> And consider this also, said I. If such a one should go down again and take his old place would he not get his eyes full of darkness, thus suddenly coming out of the sunlight? [516e]
> […]

> [...] and would it not be said of him that he had returned from his journey aloft with his eyes ruined and that it was not worth while even to attempt the ascent? [...] (Plato 1989, 517a).

This is the essential Platonic epistemology and pedagogy. The ascendency to knowledge must be undertaken by each person on her/his own and not guided by any enlightened person who has already gone through the process.

The reasoning used here is neither deduction nor induction, but model-based abduction of the analogical and metaphoric kind. This type of abduction is labeled by Gary Shank and Donald Cunningham as 'metaphor/analogy': 'The third mode leads to what Peirce identified as the Open Iconic Type (or Metaphor/Analogy). This type of inference deals with the manipulation of resemblance to create or discover a possible rule' (Shank and Cunningham 2012, p. 5).

The allegory of the cave satisfies the basic stages of reasoning by metaphor, that is, 'first stage [...] *recognition* [...] what the expression is about, second stage [...] *analysis* [...] transfer the information we have on the secondary subject to the primary subject, third stage [...] *interpretation* [...] the information we have of the primary subject is extended' (D'Harris 2002, p. 26). It is what Max Black calls a 'strong metaphor': 'I propose to call a metaphor that is both markedly emphatic and resonant a *strong metaphor*' (Black 1993, p. 26). The vividness of the metaphor is definitely emphatic and seems to draw one in only one direction not allowing for various interpretations in the ascendency of knowledge.

The process of the allegory of the cave, a prescriptive process, conforms to what is seen by some as a requirement for software processes:

> The development process used in a project should be well-defined and documented so it can be understood by all developers and its application can be monitored and evaluated. Software development processes can be described using natural language, but also modeling languages that are specific to the task such as SPEM. Since models are supposed to raise the abstraction level of information processing, bring it near to human understanding, and be less ambiguous than natural language, we prefer the last approach to describe a software development process. Furthermore, the development of the software process facilitates reuse of the process by instantiation and *execution of the model into multiple projects*. (my *emphasis*) (Porres and Valiante 2006, p. 129).

In the history of ideas the allegory of the cave has been used for multiple projects in almost every discipline. A quick search on Google scholar shows that there are 52 entries for 'allegory of the cave' (in the exact phrase) in titles of articles and 5090 entries for it appearing anywhere in the article.[7] This is all restricted to the 21st century.

The variety of references to the allegory of the cave range from an anthropologist talking about the 'virtual world' (Boellstorff 2008, p. 33) to an Australian police academy article on the 'outsider–insider' dichotomy (Vickers 2000, p. 506); to

[7] https://scholar.google.co.in/scholar?as_q=&as_epq=allegory+of+the+cave&as_oq=&as_eq=&as_occt=title&as_sauthors=&as_publication=&as_ylo=2001&as_yhi=2015&btnG=&hl=en&as_sdt=0%2C5.

teaching ethics in the police academy (Conti and Nolan 2005, p. 183); to seek the truth for its own sake in criminal justice theory (Souryal 2011, pp. 13, 42).

The relevance of the allegory of the cave to information science is stated as:

> The allegory of the cave (Plato 1967, Rep. 514-518) can be seen as an inverted information utopia of what a philosophical view of the 'unchanging' and 'supra-sensible' world brings about. Instead of the multiplicity of forms or messages reproduced in front of the cave, of which the prisoners can only see the shadows and talk about them, the Platonic dialectic presents a world where there is no more need for information because the forms themselves are the permanent subject of an eternal communication structure. [...] Learning to see the sensible world under the perspective of the 'world' of mathematical structures and of the 'ideal forms' means nothing more and nothing less than finding the 'utopian' place, i.e. the place or the perspective from where it is possible to see it as forever 'in-formed'. Plato's information utopia is a communication utopia. [...] like a networked cave, a surrogate of the 'hyper reality' of the divine 'intellectual place' (*tópos noetós*) of pure 'in-formation' or pure communication (Capurro 1996, p. 265).

7 Comparing the Two Models: The Divided Line and the Allegory of the Cave

The allegory has a wider appeal than the divided line. It is more generalized and more universal since it latches on to the structure of a linear narrative, which is common to almost all literature and story-telling, which also underlies the divided line but not so explicitly as in the allegory of the cave. In the divided line one can start from the top and go down, whereas in the allegory we are bound by the arrow of time and go only from the beginning of the prisoner looking at the wall of the cave to the sun. Hence, the allegory depicts the ascendancy of knowledge more explicitly.

Plato had a covert purpose in using the allegory as an afterthought after the divided line. One of the aims of the *Republic* is to establish that the Form of the Good is the highest of all the Forms. This cannot be done with the divided line model, but is accomplished through the allegory in which the sun under the light of which the other Forms are shining in the world outside the cave, symbolizes the Form of the Good.

These are classic cases of model-based reasoning defined as 'Model Based Reasoning is an idea of using a knowledge base or a form of reasoning and argument through the use of a physical system or object. The model then becomes a universal problem solving engine [...]'.[8] The models also enable us to 'derive relevant conclusions from these models' (Bosse et al. 2008, p. 352). Both models are metaphors of the basic type, called 'spotlight metaphor' where the source domain of the divided line and the allegory of the cave 'illuminates' the target domain of the ascendency of knowledge and the degrees of reality in Plato (Johnson 2002, pp. 2–3).

[8]Healthinformatics, http://healthinformatics.wikispaces.com/Model+Based+Reasoning.

The two models are structurally isomorphic. Isomorphisms may or may not be reducible to pure deductive reasoning, but the two models themselves cannot be established solely through deductive or inductive reasoning, rather they are loaded with abduction. We may nonetheless place them under the genus of abduction–deduction–induction cycle.

The *Republic* on the whole provides a theoretical framework of the human psyche which is fundamental for defining 'justice in the individual' and the anatomy of the state which is fundamental for defining 'justice in the state'. However, it is not a theoretical framework geared towards causal explanations but towards functional explanations, and this should be clear on any reading of the *Republic*. Functional explanations are characteristically abductive: 'A pragmatic, abductory theory of science involves functional rather than causal explanations, such that behavior is *accounted for* rather than *predicted*' (Svennevig 2001, p. 13). Both the processes provided in the two models are prescriptive rather than descriptive (Münch et al. 2012, p. 22).

8 Contextualizing the Models Considered in the Context of the *Republic* Narrative

The *Republic* is considered by some polls to be the best philosophy book ever.[9] It is used by philosophers, literary critics, historians, political scientists and economists. The narrative is sketched in Fig. 5.

As the main purpose of the *Republic* is a theoretical construction of the definition of 'justice' and a theory of justice to follow it, I take the theoretic model as the overarching model.

We can hence depict the Book I refutation of Thrasymachus's definition of 'justice' as: [theoretical {sentential (deductive)—case-based (inductive)—model-based (abductive ⟨analogy⟩)}]. The divided line is: [theoretical {model-based (abductive ⟨visual⟩)}]. The allegory of the cave is [theoretical {abductive (metaphor)}]. Bringing in the critical and dialectic processes that are employed in the *Republic*, we can capture the structure of the narrative in terms of processes and reasoning as: ≺Critical—Dialectical [theoretical {sentential (deductive)—case-based (inductive)—model-based (abductive ⟨analogy—visual—metaphor⟩)}]≻. Hence, the *Republic* is simultaneously a masterpiece of deductive reasoning and a marvel of complex model-based abductive reasoning, involving visual models, analogies and metaphors.

[9]See http://www.guardian.co.uk/uk/2001/sep/07/books.humanities.

Narrative of the *Republic*

↓

Plato is in search for the definition of 'justice' and he means justice in the individual, that is, what makes a person a just person, or a morally good person

↓

Alternative definitions of 'justice' available at the time are considered and rejected one by one.

↓

Plato now begins the construction of his definition of justice.

↓

Plato claims that sometimes it is easier to see at the macro level than at the microlevel.

↓

Hence Plato begins to construct the definition of 'justice' in the state, the state being the macro enlargement of the micro individual and it is an enlargement because they are structurally similar both having three parts

↓

Plato constructs an Ideal State based on a division of labor type of economy

↓

The guardians are to be the rulers of the state hence they must possess wisdom

↓

Hence, there is a need to discuss 'what is knowledge' and 'what is reality'?, i.e. epistemology and ontology

↓

The divided line, analogy of the sun and allegory of the cave are used as epistemological and ontological models

↓

Justice in the state is defined as guardians ruling with wisdom, the soldiers defending with courage and the workers working with temperance, that is, justice is a balance created by each class doing their own work

↓

Analagously, justice in the individual justice is defined as a balanced individual in whom the virtue of wisdom is functioning to acquire knowledge, the virtue of courage to be brave and the virtue of temperance to keep desires under control

↓

Plato now considers the question 'Why should one be just?' or 'does justice pay?'

↓

Plato answers this question by considering long term happiness as being central to human living

Fig. 5 Narrative of the *Republic*

References

Aguayo W. P. (2011). La theoría de la abducción de Peirce: Lógica, methodológia e instinto. *Ideas y Valores, 60*, 33–53.

Aliseda, A. (2000). Abduction as epistemic change: A Peircean model in artificial intelligence. In P. A. Flach & A. C. Kakas (Eds.), *Abduction and induction: Essays on their relation and integration* (pp. 45–58). Dordrecht: Kluwer Academic Publishers.

Angué, K. (2009). Rôle et place de l'abduction dans la creation de connaissances et dans la méthode scientifique Peircienne. *Récherches Qualitatives, 28*(2), 65–94.

Bergmann, R., & Wilke, W. (1998). Toward a new formal model of transformational adaptation in case-based reasoning. In: H. Prade (Ed.), *Proceedings of the 13th European Conference on Artificial Intelligence (ECAI-98)* (pp. 53–57). London: Wiley

Black, M. (1993). More about metaphor. In A. Ortony (Ed.), *Metaphor and thought* (pp. 19–41). Cambridge: Cambridge University Press.

Boellstorff, T. (2008). *Coming of age in second life: An anthropologist explores the virtually human*. Princeton, NJ: Princeton University Press.

Bosse, T., Both, F., Gerritsen, C., Hoogendoorn, M., & Treur, J. (2008). Model-based reasoning methods within an Ambient Intelligent Agent Model. In R. Bergmann, K.-D. Althoff, U. Furbach & K. Schmid (Eds.), *Constucting Ambient Intelligence, AML 2007 Workshop Proceedings, LNCS* (Vol. 11, pp. 352–370), Heidelberg: Springer.

Bylander, T., Allemang, D., Tanner, M., & Josephson, J. (1991). The computational complexity of abduction. *Artificial Intelligence, 49*(1–3), 25–60.

Capurro, R. (1996). On the genealogy of information. In K. Kornwachs & K. Jacoby (Eds.), *New questions to a multidisciplinary concept* (pp. 259–270). Berlin: Akademie Verlag.

Clement, J., & Núñez Oviedo, M. C. (2003). Abduction and analogy in scientific model construction. In *Proceedings of NARST*, Philadelphia, PA, March 23–26. http://people.umass.edu/~clement/pdf/clement_nunez_paper.pdf

Conti, N., & Nolan, J. J. (2005). Policing the Platonic cave: Ethics and efficacy in police training. *Policing and Society, 15*(2), 166–186.

Copleston, F. (1962). A history of philosophy, volume I: Greece and Rome. New York: Image Books, Doubleday.

D'Harris, I. (2002). A logical approach to the analysis of metaphors. In L. Magnani, N. J. Nersessian, & C. Pizzi (Eds.), *Logical and computational aspects of model-based reasoning* (pp. 21–38). Dordrecht: Kluwer Academic Publishers.

Experiment-Resources.com. (2012). Thales and the deductive method. http://www.experiment-resources.com/thales.html

FARLEX. (2012). The free dictionary. http://www.thefreedictionary.com/process

Flach, P. A. (1996). Abduction and induction: Syllogistic and inferential perspectives. In P. A. Flach, A. C. Kakas (Eds.), *Proceedings of the ECAI'96 Workshop on Abductive and Inductive Reasoning, Budapest, Hungary* (pp. 31–35). London: Wiley

Flach, P. A., & Kakas, A. C. (2000). Abductive and inductive reasoning: Background and issues. In P. A. Flach & A. C. Kakas (Eds.), *Abduction and Induction: Essays on their relation and integration* (pp. 1–30). Dordrecht: Kluwer Academic Publishers.

Gentner, D. (2002). Analogy in scientific discovery: The case of Johannes Kepler. In L. Magnani & N. J. Nersessian (Eds.), *Model-based reasoning: Science, technology, values* (pp. 21–40). New York: Kluwer Academic/Plenum Publishers.

Giere, R. (1999). Using models to represent reality. In L. Magnani, N. J. Nersession, & P. Thagard (Eds.), *Model-Based reasoning in scientific discovery* (pp. 41–58). New York: Kluwer Academic/Plenum Publishers.

Hesiod. (2012). *Shield of Heracles* (H. G. Evelyn-White, Trans.). http://www.perseus.tufts.edu/hopper/text?doc=Perseus%3Atext%3A1999.01.0128%3Acard%3D402

Hoffman, M. (1999). Problems with Peirce's concept of abduction. *Foundations of Science, 4*, 271–305.

Homer. (1975). *Iliad* (ca. 8th Century BCE) (T. Hobbes, Trans). http://records.viu.ca/~johnstoi/homer/hobbesiliad.htm, London.

Jetli, P. (2014). Abduction and model based reasoning in Plato's *Meno*. In L. Magnani (Ed.), *Model-based reasoning in science and technology: Theoretical and cognitive issues* (pp. 221–246). Berlin, Springer.

Johnson, M. (2002). Metaphor-based values in scientific models. In L. Magnani & N. J. Nersessian (Eds.), *Model-based reasoning: Science, technology, values* (pp. 1–20). New York: Kluwer Academic/Plenum Publishers.

Josephson, J. R. (2000). The philosophy of abduction and induction. In P. A. Flach & A. C. Kakas (Eds.), *Abduction and induction: Essays on their relation and integration* (pp. 31–44). Dordrecht: Kluwer Academic Publishers.

Josephson, J. R., & Josephson, S. G. (1994). *Abductive inference: Computation, philosophy, technology*. Cambridge: Cambridge University Press.

Khardon, R., & Roth, D. (1997). Defaults and relevance in model based reasoning. *Artificial Intelligence, 97*, 169–193.

Kirk, G. S., & Raven, G. E. (1957). *The Presocratic philosophers: A critical history with a selection of texts*. Cambridge: Cambridge University Press.

Kneale, W., & Kneale, M. (1962). *The development of logic*. Oxford: Oxford University Press.

Koschmann, T. (2003). Peirce's notion of abduction and Deweyan inquiry. In *Conference Proceedings, Paper 2*. American Research Association, Chicago IL. http://opensiuc.lib.siu.edu/meded_confs/2

Kumar, A. N. (2002). Model-based reasoning for domain modeling in a web-based intelligent tutoring system to help students learn C++ Programs. In *6th International ITS Conference*,

Biarritz, France and San Sebastian, Spain, June. http://citeseerx.ist.psu.edu/viewdoc/download?doi=10.1.1.84...rep

Lidz, J. W. (1995). Medicine as metaphor in Plato. *The Journal of Medicine and Philosophy, 20*, 527–541.

Magnani, L. (2000). *Abduction, reason and science: process of discovery and explanation*. New York: Kluwer Academic Publishers.

Magnani, L. (2006). Multimodal abduction: External semiotic anchors and hybrid representations. *Logic Journal of the IGPL, 14*(2), 107–136.

Magnani, L. (2007). Animal abduction: from mindless organisms to artificial mediators. In L. Magnani & P. Li (Eds.), Model-based reasoning in science, technology, and medicine, series studies in computational intelligence 64 (pp. 3–38). Berlin: Springer.

Magnani, L. (2009). Abductive cognition: The epistemological and eco-cognitive dimensions of hypothetical reasoning. Cognitive Systems Monographs 3, Berlin, Springer.

Minnameier, G. (2004). Peirce-suit of truth—Why inference to the best explanation and abduction ought not to be confused. *Erkenntnis, 60*, 75–104.

Minnameier, G. (2010). Abduction, induction and analogy: On the compound character of analogical inferences. In: L. Magnani, W. Carnielli & C. Pizzi (Eds.), *Model based reasoning in science and technology: Abduction, logic, and computational discovery. Studies in computational intelligence 314* (pp. 107–120). Berlin: Springer.

Münch, J., Armbrust, O., Kowalczyk, L., & Soto, M. (2012). *Software process definition and management. The Fraunhofer IESE series on software and systems engineering*. Berlin: Springer.

Nersessian, N. J. (1999). Model-based reasoning in conceptual change. In L. Magnani, N. J. Nersession, & P. Thagard (Eds.), *Model-based reasoning in scientific discovery* (pp. 5–22). New York: Kluwer Academic/Plenum Publishers.

Olsen, S. (2002). Plato, Proclus and Peirce: Abduction and the foundations of the logic of discovery. In R. B. Harris (Ed.), *Neoplatonism and contemporary thought part one* (pp. 85–102). Albany, NY: State University of New York Press.

Paavola, S. (2005). Peircean abduction: Instinct or inference? *Semiotica, 153*(1–4), 131–154.

Paavola, S. (2006). Hansonian and Harmanian abduction as models of discovery. *International Studies in Philosophy of Science, 20*(1), 93–108.

Pape, H. (1997). Abduction and the topology of human cognition, review of Ansgar Richter, Der begriff der abduction bei Charles S Peirce. *Modern Logic, 7*(2), 199–221.

Peirce, C. S. (1957/1903). The logic of abduction. In Thomas, V. (Ed.), Essays in philosophy of science (pp. 235–255). New York: Liberal Arts Press.

Peirce, C. S. (1931–1958). The collected papers of Charles Sanders Peirce. In C. Harthshorne, P. Weiss & A. Burks (Eds.), Cambridge: Harvard University Press.

Plato. (1989). Republic (P. Shorey, Trans.). In E. Hamilton, H. Cairns (Eds.), *The collected dialogues of Plato. Bollingen series* (pp. 575–844). Princeton: Princeton University Press.

Porres, I., & Valiante, M. C. (2006) Process definition and project tracking in model driven engineering. In J. Münch & M. Vierimma (Eds.), *Product-focused software process improvement. Proceedings of 7th International Conference on PROFES 2006*, Amsterdam, The Netherlands, LNCS, 12–14 June. (Vol. 4034, pp. 127–141). Berlin: Springer.

Shanahan, M. (2005). Perception as abduction: Turning sensor data into meaningful representation. *Cognitive Science, 29*, 103–134.

Shank, G., & Cunningham, D. J. (2012). Modelling the six modes of Peircean abduction for educational purposes. http://www.cs.indiana.edu/event/maics96/Proceedings/shank.html

Shelley, P.B. (1820). To a skylark. http://www.bartleby.com/101/608.html.

Shuguang, L., Qing, J., & George, C. (2000). Combining case-based and model-based reasoning: A formal specification. UNU/IIST Report No. 188.

Souryal, S. S. (2011). *Ethics in criminal justice: In search of the truth* (5th ed.). Burlington, MA: Elsevir.

Svennevig, J. (2001). Abduction as a methodological approach to the study of open interaction. *Norskrift, 103*, 1–22.

Thagard, P., & Shelley, C. (1997). Abductive reasoning: Logic, visual thinking, and coherence. http://cogsci.uwaterloo.ca/Articles/Pages/%7FAbductive.html

Tiercelin, C. (2005). Abduction and the semiotics of perception. *Semiotica, 153*(1–4), 389–412.

Upshur, R. (1997). Certainty, probability and abduction: Why we should look to C.S. Peirce rather than Gödel for a theory of clinical reasoning. *Journal of Evaluation in Clinical Practice, 3*(3), 201–206.

Vickers, M. H. (2000). Australian police management and research education: a comment from "outside the cave". *Policing: An International Journal of Police Strategies and Management, 23*(4), 506–524.

Wikipedia. (2012). http://en.wikipedia.org/wiki/Dactylic_hexameter

Defining Peirce's Reasoning Processes Against the Background of the Mathematical Reasoning of Computability Theory

Antonino Drago

Abstract In the present paper Peirce's inferential processes are accurately defined against the background of the four ways of reasoning in Computability theory, i.e. general recursion, unbounded minimalization, oracle and undecidabilities. It is shown that Peirce anticipated almost all them.

1 Beyond the Monist Interpretations of Peirce's Abduction

Peirce was primarily concerned with methods of inquiry and the growth of knowledge. He regarded his philosophy, pragmatism, as linked to scientific research. However, a severe appraisal of his philosophical works has also been made:

> Peirce scholarships is a painstaking business. His mind was labyrinthine, his terminology intricate, and his writings are, as he himself confessed, "a snarl of twine". (Davis 1972, p. vii)[1]

One of Peirce's main aims was to define all inferential processes. He stressed that

> I have constantly since 1860, or 50 years, had this question (classification of types of reasoning) prominently in mind…(7.98)[2]

---

[1]However, the author adds on the same page: "The reader may be sure that any time he spends with Peirce…. [he] will splendidly widen his intellectual horizons.".

[2]I recall that two numbers in round brackets refer to (Peirce 1931), the first number means the volume and the second number the section.

A. Drago (✉)
Dipartimento di Scienze fisiche, Università di Napoli "Federico II",
Via P.C. Benvenuti 3, 56011 Pisa, Calci, Italy
e-mail: drago@unina.it

L. Magnani and C. Casadio (eds.), *Model-Based Reasoning in Science and Technology*, Studies in Applied Philosophy, Epistemology and Rational Ethics 27, DOI 10.1007/978-3-319-38983-7_21

Beyond deduction and induction he suggested a new way of reasoning, his celebrated "abduction".

Yet, it is well-known that during his life he changed his views on these notions.

> Over many years Peirce modified his views on the three types of argument [deduction, induction, abduction], sometimes changing his views but mostly extending them by expanding his commentary upon the original trichotomy. (Burch 2012, Sect. "Induction, Abduction, Deduction")[3]

In 1910 Peirce himself admitted that concerning the distinction between abduction and induction:

> … in almost everything I printed before the beginning of this century I more or less mixed up hypotheses [abduction] and induction. (8.227)

He changed so much that the different definitions he suggested are not mutually compatible. (Fann 1970, pp. 9–10)

At least two periods have to be distinguished in his reflection on abduction, spanning more than fifty years. (Hilpinen 2004, p. 648ff.)

McMullin remarked that even the study of abduction alone is laborious:

> It is not easy to disentangle the theme abduction… from the enormously complex and sometimes idiosyncratic metaphysical and psychological system Peirce laboured to build and rebuild. (McMullin 1992, p. 89)

Moreover, while the deduction process has long since been defined and at present is commonly assumed to be an almost unquestionable process, the other kinds of reasoning have been accurately explored in last decades and common agreements on their definitions are still lacking.

In particular, the abduction process has been interpreted in a variety of ways[4]:

> Most philosophers agree that this type of inference is frequently employed, in some form or other, both in everyday and in scientific reasoning. However, the exact form as well as the normative status of abduction are still matters of controversy.… Precise statements of what abduction amounts to are rare in the literature on abduction. (Peirce did propose an at least fairly precise statement but… it does not capture what most nowadays understand by abduction). Its core idea is often said to be that explanatory considerations have confirmation-theoretic import, or that explanatory success is a (not necessarily unfailing) mark of truth. (Douven 2011, "1.2 The ubiquity of abduction").[5]

In addition, Peirce wanted to reduce all his kinds of reasoning to one only (5.278); but—as in an analogous case one scholar put it:

---

[3]For an introduction to Peirce's thinking on inference processes, see Fann (1970).

[4]In particular, two different classifications of the several meanings that the notion abduction may assume have been suggested by two authors (Schurz 2008; Hoffmann 2010). The latter author obtains fifteen forms of abductions, yet he classifies them according to the 'outside', i.e. by means of external issues and the resulting effects; in my opinion these are superficial features. The classification of the former author will be analysed later.

[5]See also Stalker (1994), McKaughan (2008, p. 448–449), Chiasson (2001).

> it remains to be established precisely how clear Peirce himself was about this matter. (Hintikka 1998, p. 531).

By following however Peirce's intent, almost all scholars tried to explain the notion of abduction as a single rule for all the instances of the kind of reasoning presented by him. Yet, recently a new viewpoint on Peirce's abduction was presented, no longer attributing to Peirce a monist, but an essentially pluralist, view of this inference process. Above all a paper by Schurz (2008) was decisive in abandoning the monist interpretation of abduction.

> The majority of the literature on Abduction has aimed at *one of most general* schema of abduction (for example the best hypothesis) which matches every particular case. I do not think that good heuristic rules for generating explanatory hypotheses can be found along this route, because these rules are dependent of the *specific type* of abduction scenario, for example, on whether the abduction is mainly selective or creative [see Magnani 2001] (etc.).… I will rather pursue a *new route to abduction* which consists in modelling various particular schemata of abduction, each fitting to particular kinds of conjectural situation. (Schurz 2008, p. 205)

Schurz supports, by means of a series of several instances of inference processes, three theses, the first of which is the following:

> Thesis 1 (Induction versus abduction): … induction and abduction are two distinct families of ampliative reasoning kinds which are not reducible one to another. (Schurz 2008, p. 202)[6]

In the following I will interpret Peirce's writings on inference processes in correspondence to four processes of reasoning in Computability theory (CT): recursion, minimalization, oracle and undecidabilities. They correspond also to the four prime principles of reasoning that one can detect in the several physical theories, respectively: causality, extremants, existence of a mathematical being, limitation (Drago 2015; Drago 2016a, b).[7] My results will agree and support Schurz's above thesis.

At first glance the comparison between the set of Peirce's three inference processes with the set of preceding four CT's processes of reasoning meets a basic difficulty. It is defective since a fourth Peirce's reasoning kind seems to be lacking. But he conceived a computing machine through "impotencies". Moreover, three of Peirce's writings on a philosophical subject present a reasoning kind concluding the

[6]This thesis is the same as Hintikka's (1998, p. 507).

[7]Peirce implicitly referred his reflections to his wide knowledge of scientific theories, mainly chemistry; (Drago 2015) but his scholars rarely referred his reasoning to some inductive processes in the history of classical chemistry. Among the philosophers of chemistry, only van Brakel (2000, pp. 21–22) devoted a page to Peirce. Rather two other instances of inductive processes have been studied: (1) Kepler's theory of the orbit of Mars (1.73); which was further illustrated by (Hanson, ch. IV). Nickles made valid criticisms of this analysis (1980, pp. 22–23). In any case, Kepler's theory belongs to a too informal context for suggesting certain conclusions on his underlying logic. (2) The kinetic theory of gas (2.639) which however, is too loosely treated by Peirce; moreover, this instance was contested as inappropriate to an investigation of his suggestions (Achinstein 1987).

existence of "incapacities" in human reasoning. Both impotencies and incapacities may correspond to undecidabilities.

One more difficulty we will meet in the following investigation of the meanings that Peirce attributed to his trichotomy will be the above-mentioned obscurities of Peirce's writings, in particular the lack of clear definitions of either abduction or induction. Owing to these difficulties my collection of quotations from Peirce's writings necessarily represents no more than a particular selection of them.

Notice that the four CT's reasoning processes are formally defined in *mathematical* terms, whereas Peirce considerations are of a philosophical nature. Yet, we are allowed to associate philosophical descriptions of reasoning processes with mathematical techniques of calculation because each mathematical calculation is the material expression of a reasoning process.

This comparison provides a new viewpoint for analysing Peirce's suggestions not by means of intuitive ideas, but a formal language used throughout the millennia, which eventually produced four distinct mathematical techniques of CT. Their formal distinctions will suggest an accurate classification of Peirce's inferential processes.

A comparison of intuitive, philosophical ideas with formally mathematical ideas does not allow, of course, any formal conclusion—except in exceptional cases; indeed, in the case of deduction, Peirce, who was among the first, suggested a mathematical correspondent process of arguing, recursion. However, the established correspondences will be accurate in the spirit of CT, which basically links together—through Turing-Church's thesis—an intuitive notion of computability with its formal notion.

Moreover, help is offered by one more instantiation of the possible reasoning processes, i.e. the experience accumulated over the centuries by several physical theories. These theories are also formalized in mathematical terms, but their principles link together this mathematical formalization with the operative, intuitive reality. The common reasoning processes of these principles can be considered according to four kinds—respectively, deduction, minimalization, extremants, impossibilities—which will prove to correspond with the previous ones.

Conversely, this clarification of Peirce's trichotomy (plus his impotencies) will suggest that his *philosophical* effort has anticipated CT's four process of reasoning.

2 Comparing Peirce's 'Incapacities' to CT's *Undecidabilities*

Let us as first consider a kind of reasoning which Peirce made use of, but which he did not rationalize and thus did not include in his classification of inferential processes.

Being a pragmatic philosopher, Peirce maintains that the process of building knowledge starts from observations or experimental data.

We can say that our knowledge may be said to rest upon observed facts (Peirce 1965, Vol. 6, p. 356).

The same holds true for CT, a theory which basically refers to calculations performed by 'mechanical' machines.

However, to such 'facts' Peirce adds hypotheses (also considered by him as questions). So that:

> If the pragmatism is the doctrine that every conception is a conception of conceivable practical effects, it makes conception reach far beyond practical. It allows any practical imagination, provided this imagination ultimately alights upon a practical effect; and thus many hypotheses may seem at first glance be excluded by the pragmatical maxim that are not really so excluded. (Peirce 1965, V, p. 122, 5.196)

Let us remark that also CT's calculations are subject to programs which may be very abstract from the calculation operations, in a parallel way in a physical theory some ideal hypotheses are added to experimental data and mathematical techniques.

But not all idealizations are allowed. Many of them have to be excluded because in the end they do not produce "tangible results". Moreover, some others have to be excluded for theoretical reasons only.

This point is stressed in his remarkable reflections on "reasoning machines". In fact, Peirce was one of the first philosophers to have pondered on them. According to him, the "secret of all reasoning machines" is the following:

> … whatever relation among the objects reasoned about is destined to be the thing of a ratiocination, that same relation must be capable of being introduced between certain parts of the machines (Peirce 1887, p. 168)

According to this "secret"

> … a man may be regarded as a machine which turns out, let us say, a written sentence expressing a conclusion, the man-machine been fed with a written statement of fact. Since this performance is no more than a machine might go through, it has no essential relation to the circumstance that a machine happens to work by geared wheels, while a man happens to work by an ill-understood arrangement of brain cells. (Peirce 1933, 2, p. 33)

He concluded by stating that a machine has two "inherent impotencies".

> [First,]… it is destituted of all originality, of all initiative [= not previously programmed behavior].[8] It cannot find its own problems… It cannot direct itself between different possible procedures.
>
> [Second,]… the capacity of the machine has absolute limitations; it has been contrived to do a certain thing and it cannot do anything else. (Peirce 1887, pp. 168–169)

The above propositions define in philosophical terms some general limitations of the computers.

---

[8]In the following, the negated words belonging to doubly negated proposition will be underlined for an easy inspection by the reader. Notice that even a single word may be equivalent to such a kind of proposition, in particular a **modal** word; e.g. **possible** = it is not the case that is not (more in general, it is well-known that the modal logic is translated into intuitionist logic by means of S4 model) such a kind of word will be bold.

Let us notice that these propositions all are doubly negated propositions; moreover, they are not equivalent to the corresponding affirmative propositions for lack of pragmatic, operative evidence (DNPs); for instance the first DNP is not equivalent to "Machine's behaviours are all pre-established", because no one can be sure that the hidden internal computation process of a complex "reasoning machine" behaves in the exact way it was programmed.[9]

This point is very relevant, since, according to mathematical logic, the double negation law ($\neg\neg A \rightarrow A$) fails in almost all non-classical logic (whereas the opposite logical law, $A \rightarrow \neg\neg A$, is always valid). Hence, the failure of the double negation law in a proposition qualifies this proposition as belonging to non-classical logic—in particular, intuitionist logic. (Prawitz and Melmnaas 1968; Dummett 1977) Hence, previous Peirce's conclusions, being all DNPs, show that *he is reasoning in non-classical logic*.

Let us add that in classical logic one draws from certain axioms assured consequences thanks to its law of the excluded middle (whose validity is equivalent to that of previous logical law). Yet, being defined by two negations with respect to an unknown totality, the content of a DNP is not circumscribed, hence a DNP cannot play the role of an axiom. Moreover, each DNP, which is not equivalent to its corresponding affirmative proposition, cannot be derived as a consequence of axioms, because the latter are affirmative propositions, which are all equivalent to their corresponding doubly negated propositions.

In non-classical logic productive reasoning is still possible. One proceeds by accumulating, through ad absurdum proofs composed of DNPs, more and more evidence supporting a new hypothesis.[10] Carnot's theorem of thermodynamics

---

[9]This point deserves particular attention, because the current usage of the English language exorcises DNPs as representing a characteristic feature of a primitive language, English linguists are dominated by a long tradition which L. Horn called a "dogma" (Horn 2001, pp. 79ff.; Horn 2008). This linguistic dogma asserts the absolute validity of the double negation law: whenever a DNP is found in a text, it has to be changed into the corresponding affirmative statement, because those who speak by means of DNPs want to be, for instance, unclear. Evidence for this "dogma" is the small number of studies on double negations in comparison with the innumerable studies on a single negation. The following three well-known DNPs belonging to mathematics, physics and classical chemistry show that this logical-linguistic feature pertains to scientific research since its origin. In Mathematics it is usual to assure that a theory is "without contradictions"; to state the corresponding affirmative statement, i.e. the consistency of the theory, is impossible, owing to Goedel's theorems. In theoretical physics it is usual to study the in-variant magnitudes. The proposition does not mean that the magnitudes stay fixed. In order to solve the problem of what the elements of matter are, Lavoisier suggested defining these unknown entities by means of a DNS: "If… we link to the name of elements or principles of corps the idea of last term to which [through chemical reactions] arrive at the analysis, all the substances which we were not capable to decompose through any tool are for us, elements", where the word 'decomposable' naturally carries a negative meaning and stands for 'non-ultimate' or 'non-simple' (Lavoisier 1862–92, p. 7). As a matter of fact, Grzegorczyk (1964) independently proved that scientific research may be formalized through propositions belonging to intuitionist logic, hence through DNPs.

[10]Notice that it is commonly maintained that *ad absurdum* proofs can be all translated into direct proofs (Gardiès 1991). Unfortunately, it is rarely remarked that this translation is possible only by applying classical logic to its conclusion, i.e. the absurdity of the negated thesis $\neg Ts$, that one

represents the most celebrated ad absurdum theorem which supports a very productive reasoning.

Notice that a book is recognized at glance as belonging to CT by its inclusion of a great number of ad absurdum theorems; no other branch of modern Mathematics appears thus.

Although lacking ad absurdum proofs, Peirce's previous remarks have to be considered as remarkable hints for a future CS.

The previous way of reasoning according to limitations played a decisive role in Peirce's mind. Indeed, this same way of reasoning is followed by him in the more general context of his crucial philosophical polemic against Descartes' philosophy. Three writings (Peirce 1868a, b, c) have stressed the idea of limitation as concerning also more general human reasoning.

In the celebrated paper "Some Consequences of Four Incapacities" (Peirce 1968b) he illustrates a short theory aimed to refute what Descartes had suggested as basic for his philosophy of knowledge, the methodical doubt. Peirce begins his argument by stating the problem; he opposes to Descartes' thesis the following DNPs: "We cannot begin with complete doubt." (p. 212)

Then he proceeds to disprove it by means of an ad absurdum argument. If Descartes' maxim was true:

> … no one who follows the Cartesian method will ever be satisfied until he has formally recovered [if he will do not formally recovered] all those beliefs which in form he has given up. It [= this procedure of giving up] is therefore, as useless [or absurd] a preliminary as going to the North Pole would be in order to get to Constantinople by coming down regularly upon a meridian. (p. 212)

Then he considers the corresponding affirmative proposition of Descartes' methodical doubt—"Whatever I am clearly convinced of, is sure"—and he refutes it by means of one more ad absurdum proof:

> If I were really convinced, I should have done with reasoning, and [, absurd, I] should require no [more a] test of certainty.

The conclusion is the following DNP, which I call (A):

> Every un-idealistic philosophy supposes some absolutely in-explicable, un-analyzable ultimate; in short, something resulting from mediation itself not susceptible of mediation. (Peirce 1984, p. 213)

Let us notice that at the beginning of this paper he had put the following proposition about un-analyzable ultimates; I call this proposition (B):

> We must begin with all the prejudices which we actually have when we enter upon the study of philosophy…. These prejudices are not to be dispelled by a maxim [i.e. the

(Footnote 10 continued)

wants to prove, $\neg\neg Ts$. But this last step, the translation of $\neg\neg Ts$ in Ts, is interdicted by intuitionist logic (otherwise a direct proof would be more appropriate). Hence, an essentially ad absurdum proof is a characteristic argument of intuitionist logic.

Cartesian one], for they are things which it does not occur to us *can* be questioned. (Peirce 1984, p. 212)

Owing to its initial location, this proposition (B) seems an a priori tenet. Actually, this proposition of Peirce's derives from the previously quoted proposition (A)—in which he speaks about un-analyzable prejudices; he implicitly considers (A) as sufficiently supported by the evidence previously accumulated, for applying to it the principle of sufficient reason and hence to obtain the corresponding affirmative proposition (B); afterwards, this result is considered a hypothesis from which one deductively derives all consequences, to be tested against the experiments.[11]

In conclusion, he obtained, by means of two ad absurdum arguments, that the human mind cannot decide the truth of some propositions, which hence have to be considered "legitimate prejudices".

At this point Peirce (1868a) summarises a previous paper, where he presented seven theses by means of seven ad absurdum proofs (three of them—in the discussions of the theses 1, 3, 5—rely on the *regressus* ad infinitum). Since their demonstrations are of a non-apodictic nature, their conclusions cannot be presented as assured truths, hence also they are presented by means of DNPs (e.g., there is no evidence for the affirmative proposition corresponding to the DNP no. 3: "Each thinking process enjoys a sign"):

1. We have no power of Introspection, but all knowledge of the internal world is derived by hypothetical reasoning from our knowledge of external facts.
2. We have no power of Intuition, but every cognition is determined logically by previous cognition.
3. We have no power of thinking without signs.
4. We have no conception of the absolutely incognizable (Peirce 1984, p. 213).[12]

These propositions all concern limitations of the human mind. He does not regard them as axioms from which to derive certain conclusions; rather,

---

[11]As it will be illustrated in Sect. 6 this principle (itself a DNP: "Nothing is without reason") allows us to change a DNP into its corresponding affirmative proposition (in our case in the last quotation).

[12]More versions of the same theses occur on pp. 223, 228, 240, 241. Two more incapacities are presented by Peirce in 5.200 "…because all our ideas being more or less vague and approximate, what we mean by saying that a theory is true can only be that a theory is very near true. But they [some logicians] do not allow to say that anything they put forth as an anticipation of experience should assert exactitude, because exactitude in experience would imply experience in endless series, which is impossible…. We therefore have a right, they [the logicians] will say, to infer that something *never* will happen, provided it be of such a nature that it would not occur without being detected." (Peirce 1965, p. 124, 5.199 and 5.200). These propositions are also DNPs. By incidence, let us remark that these propositions anticipated the great debate among mathematicians that took place in the middle of the last century. Almost to defy such statements by Peirce, Hilbert had solemnly affirmed: "In mathematics there is no *Ignorabimus*" (Hilbert 1902, p. 445). But Goedel's theorems subsequently disproved Hilbert's thesis, leading mathematicians to accept innumerable undecidabilities.

> These propositions cannot be regarded as [absolutely] certain [as the axioms are],... it is now proposed to trace them [his "prejudices"] out to their consequences (Peirce 1868b, p. 213)

This is just what one does when one manages rather than an absolute truth a merely probable proposition, i.e. a hypothesis as a DNS is, which is then tested against the consequences of its affirmative version. In fact, Perice's reasoning conforms to an alternative model of a systematic organization of a theory to the usual, apodictic one (AO). Its development is represented by the following sequence: to put a crucial problem (in Peirce's philosophical papers the problem is the following: Is Descartes' methodical doubt valid?), to accumulate ad absurdum arguments in order to prove a DNP which is a universal predicate, to translate it into an affirmative predicate to be used as an hypothesis from which to obtain propositions to be tested with reality. I recognized this alternative model by analysing the original texts of several mathematical and physical theories; I call this kind of organization a problem-based one (PO) (Drago 2012).

Notice that CT is a PO theory; its basic problem is to define a computation; or rather, whether the intuitive notion of a computation is the same as the formal notions of computation.

In conclusion, Peirce stressed that one can reason philosophically even when the reasoning includes some "prejudices", i.e. propositions whose truths are undecidable using the tools of human mind. Indeed, in CT the undecidabilities of Turing machines do not obstruct the building of a general and well-founded theory. Also in physical theories a limitation does not obstruct the building of a theory. For instance, thermodynamic theory was founded on the two impossibilities of the perpetual motion of the first and the second species; the theory obtains a plausible DNP on all heat machines; afterwards its affirmative versione is exploited for deducing all the laws of these machines.[13]

Peirce has the merit of having anticipated through his "incapacities" the reasoning of CS through undecidabilities, recognized in formal terms by mathematicians only fifty years later (Goedel's theorems of 1931).[14]

[13]Let us recall that thermodynamics' principle of the impossibility of perpetual motion suggested to Einstein the trigger idea for building special relativity; he was led by it to think that a body's velocity greater than or equal to the speed of light is impossible.

[14]Why did he not regard this reasoning about incapacities as a kind of reasoning that was independent of the others? One may suggest two reasons. First, his above-mentioned three writings were a polemic against Descartes, rather than an investigation of kinds of reasoning. Second, the strategy of the kind of reasoning on incapacities is the opposite of his usual strategy; he was interested in the ampliative, rather than the restrictive, kinds of reasoning -, which is what incapabilities are. (About the notion of strategy in Peirce, see Hintikka (1998, pp. 51ff). As such this kind of reasoning is quite different from the usual ones. In particular, since they refer to totalities undecidabilities are essentially different from deductions and inductions.

3 Comparing Peirce's Deduction to *Recursion*. His Priority on the Latter Notion

Let us consider in this section and in the following three sections the trichotomy of kinds of reasoning which have been *declared* by Peirce.

The following quotation illuminates the meanings that Peirce attributes to the various processes of inference:

> If we are to give the names of Deduction, Induction and Abduction to the three grand classes of inference, then Deduction must cover every attempt at mathematical demonstration, whether it is to relate to single concurrences or to "probabilities", that is to statistical ratios; Induction must mean the operation that induces an assent, with or without quantitative modifications, to a proposition already put forward, this assent or modified assent being regarded as the provisional result of a method that must ultimately bring the truth to light; while abduction must cover all the operations by which theories and conceptions are engendered. (5.590)

As summarized by one scholar, Peirce attributes to all of them the following features:

> … deductions [and we can add, limitations] are *non-ampliative* and *certain*… In contrast, inductions and abductions are *ampliative* and *uncertain*. (Schurz 2008, p. 202)

The previous quotation makes apparent that Peirce attributes a common meaning to deduction (the meanings that he attributes to the other two notions will be considered in the three following sections).

Peirce analyses this logical process also in terms of the ancient formal kind of reasoning, the classical syllogism. Since the time of Euclid's geometry, Mathematics presented it as the best example of a theory relying on the inference process of deduction, according to this ideal model of Aristotle's apodictic science (AO), (Beth 1959, Sect. 1.2). Since the time of Newton, physical theories had also appeared to conform to this model, given that Newton's celebrated theory of mechanics was founded as a system of laws deductively drawn from the celebrated three principles

The following quotation presents a more developed definition of deduction by Peirce:

> Deduction is reasoning which proposes to pursue such a method that if the premises are true the conclusions will in every case be true. (Eisele, ed. *The New Elements*, vol. 4, p. 37)

After the birth of mathematical logic, we can state without doubt that Peirce's deduction is the logical process governed by formal classical logic.

Moreover, Peirce has regarded as distinct two kinds of deductive reasoning, theorematic, in which the conclusion follows mediately or non-trivially from some modification of the premises, and corollalial, in which the deduction follows immediately or trivially from some modification of the premises:

> Corollalial deduction is where it is only necessary to imagine any case in which the premises are true in order to perceive immediately that the conclusion holds in this case. (Peirce, MS, I, 73).

In Mathematics—and in particular in CT—recursion is a particular way of deducing new numbers or functions from the previous ones by arithmetical calculation. It is an instance of deduction which is shaped in arithmetical terms according to a purely 'mechanical' method; for this reason recursion is deductive reasoning of the kind that Peirce called corolliałal.

In addition, the formal process of recursion belongs to Peirce's thinking since he founded the theory of arithmetic on it. He is one of the historical fathers of recursion.[15]

In conclusion, we can say that Peirce invented this inferential process—the most accurately defined and developed by him. In other words, his philosophical reflection was so accurate that a mathematical technique, recursion. Fifty years later this mathematical technique played the role of the basic mathematics of CT.[16]

4 Comparing Peirce's Abduction with CT's *Oracle*

Let us consider the two remaining kinds of Peirce's inferential process. He attributed to them the same importance as to deduction. Indeed, Peirce maintained that logicians are too concerned with "[the] *security* (approach to certainty) of each kind of reasoning" (8.384), to devote due attention to non-deductive reasoning, which is of the greatest importance notwithstanding its uncertainty, because it constitutes a process of inference of an ampliative kind.

It is a great merit of Peirce to have stressed that beyond the induction, one has to add one more inferential process, abduction.

> I call all such inference by the peculiar name, *abduction*, because its legitimacy depends upon altogether different principles from those of other kinds of inference. (6.525)

Peirce staunchly emphasised that abduction is not a fancy, but reasoning, and indeed, formal reasoning.

[15]Peirce (1881), "On the Logic of Numbers" (3.252–288). Notice that the Dedekind's most celebrated work on the same subject was edited seven years later (Dedekind 1888).

[16]Notice that recursion, although apparently a very simple technique, may make use of the actual infinity. Indeed, the general recursive functions are defined e.g. by a diagonalization process on all the elementary ones (see Davis et al. 1994, p. 105ff.). Also unbounded minimalization introduces non constructive elements (see Davis et al. 1994, pp. 57–58). The oracle manifestly constitutes a non-constructive move. It is roughly defined as a black box which is able to decide certain decision-making problems, otherwise unsolvable, through a single operation. It is more precisely defined as follows: "A number m that is replaced by $G(m)$ in the course of a G-computation… is called an *oracle to query to the G-Computation*" (where a G-computation is a computation of a partial recursive function G under specific conditions; Davis et al. 1994, p. 197).

> Although it is very little hampered by logical rules, nevertheless is logical inference, asserting its conclusion only problematically or conjecturally, it is true, but nevertheless having a perfect definite logical form. (5.188)

According to Peirce abduction

> … is reasoning and though *security* is low, its *uberty* is high. (8.388)

However, abduction was disregarded for one century also because Peirce left a questionable legacy:

> He had to try different paths at different times and as a consequence his terminology varied from time to time. Also, Peirce's discussions on abduction are often connected with his discussion on other topics. (Fann 1970, p. 6)

Some scholars have regarded Peirce's abduction as a novelty to the extent that it suggested the process of choosing a hypothesis among several other hypotheses. By stating that his philosophy is oriented by a principle of economy Peirce gave considerable support to this interpretation:

> … the leading consideration in Abduction is the question of Economy - Economy of money, time, thought and energy. (5.600)

This kind of reasoning was analysed, accurately defined and improved by several scholars, in particular (Josephson and Josephson 1994), so much so that at present it represents the most widely accepted interpretation of abduction. According to Hintikka, this interpretation "…. has a great deal of initial plausibility", yet, it does not include the core-meaning of Peirce's new notion: "This view on Peirce's abduction is seriously oversimplified at best." (Hintikka 1998, pp. 506–507).

In order to interpret the notion of an abduction let us start by stressing the independence of an abduction from both the operative context and calculations. According to Peirce:

> [An induction] infers the existence of phenomena such as we have observed in cases that are *similar* [, while an abduction] supposes something of a different kind from what we have directly observed, and frequently something which it would be impossible for us to observe directly. (2.640).

Of course, this inference process has to be considered a creative process obtaining an element by non-operative means. The following definition by Peirce suggests some characteristic features of this process:

> The validity of a presumptive adoption of a hypothesis [= abduction] being such that its consequences are capable of being tested by experimentation, and being such that the observed facts would follow from it as necessary conclusions, that hypothesis is selected according to a method which must ultimately lead to the discovery of the truth. (2.781)

Let us add one more qualification:

> Its only justification is that its method [read: introduction] is the only way in which there can be any hope of attaining a rational explanation. (2.777)

For a long time Peirce tried to translate his disparate ideas on abduction into some syllogisms of a similar kind of those of Aristotle. According to Burch, the first instance of Peirce's presentation of abduction as a new kind of syllogism was the following one:

> [Rule] All balls in this urn are red; [Case] all balls in this particular random sample are red; [Result] therefore, all balls in this particular random sample are taken from this urn...
> It should be clear that abduction is never necessary inference. (Burch, Sect. 3)

This kind of abductive syllogism manifestly produces a verdict of an oracle, when we mean the notion of an oracle in intuitive terms. By accurately inspecting the definition in the penultimate quotation ("The validity...") we remark that Peirce refers not to a process which provides a basis for obtaining something new, but rather to a logical process which is consequent on having obtained the new element ("consequence", "rational explanation"), possibly validating it.

Cellucci has captured this logical meaning of abduction by defining it in the following way:

> Given a set of sentences Γ and a sentence C not derivable from Γ, to find a sentence A such that C is derivable from Γ + A. The set Γ consists of the already available hypotheses, or background hypotheses, C is the problem to be solved, A is the new hypothesis sought for. (Cellucci 1998, p. 235)

Hence, I interpret an abduction in CT as an inference process, attributing hypothetical existence to a mathematical object which is not discovered or validated by means of a method for obtaining it (e.g. an approximation process), yet it is fruitful for its logical consequences—e.g. Zermelo's axiom.[17]

In particular, I see an abduction in the application of the classical law of the excluded middle to the search for a solution of a problem. For instance, let us consider the question whether there exist two irrational numbers, a and b, such that a^b is a rational number. Abduction: Let us consider $\sqrt{2}^{\sqrt{2}}$. This number is either rational or irrational. If it is rational, we have the desired answer: $a = b = \sqrt{2}$. Otherwise, the power $\sqrt{2}$ of $\sqrt{2}^{\sqrt{2}}$ gives 2, which is a rational number; in which case the answer is $a = \sqrt{2}^{\sqrt{2}}$ and $b = \sqrt{2}$. We obtain as a certain answer a mathematical element, although it is an ideal one because we ignore which specific number it is.

---

[17]Notice that in the history of science an abduction was an elusive notion. In the case of the Kinetic Theory of Gases, the atomic hypothesis started as a mere guess in Galilei and Boyle, became an abduction-oracle in the first attempts at building a theory through its consequences (Newton and D. Bernoulli attributed reality to atoms without any experimental evidence, but only to obtain new results in a deductive development from this hypothesis) and subsequently became an extremant when Avogadro calculated his celebrated number; eventually, in the 20th century it received experimental evidence. It is not possible to study it with the expedient of the historical case of D. Bernoulli and Newton, because these authors communicated their logical inferences in the language of mathematics, without references to their logical processes. In other words, in the history of science abduction pertains to a genetic stage of the historical development of a theory, without certain documents that need to be analyzed.

Here we have extended—in an ideal way—the classical law of the excluded middle, imposing a sharp dichotomy, from a finite domain of elements to an infinite domain.

From the above we conclude that the link which in the above has been presented in intuitive terms, is true in general: an inference process of an abduction corresponds in CT to the *oracle function*, being by definition an oracle capable of giving an operative, practical answer, whose validity is verified by following its consequences. Hence, it essentially refers to an AO context.

Since the abducted hypothesis has to be crucially tested for validity by a subsequent verification, Peirce considered it sufficient for generating this hypothesis also the observation of a single instantiation of it, although a single instantiation is of course quite insufficient for validating the hypothesis. In order to underline this exceptional way of generating hypothesis from a single instance, Peirce called abduction also retroduction.

5 One More of Peirce's Kinds of Abduction: A Crucial Logical Principle for a Problem-Based Theory

One scholar of Peirce suggested an important distinction between an abduction concerning a single element of a theory—either a number or a hypothesis—and a theory in its entirety, i.e. a complex of several laws, theoretical terms, principles and mathematical techniques. (McMullin 1992, p. 90ff.) One may suppose that Peirce wanted to include this kind of reasoning when he wrote as follows:

> Scientific abduction includes all the [logical] operations whereby theories are engendered. (5.590)

As a consequence, I suggest a distinction between two kinds of abduction, according to whether it produces a hypothesis or a theory.

Niiniluoto (1999, pp. 436–439) has cleverly re-constructed Peirce's various kinds of syllogisms. In his opinion the following instance represents the more general schema of Peirce's abduction:

> [Rule]The surprising fact C is observed,
> [Case] But if A is true, C would be a matter of course;
> [Result] Hence there is a reason to suspect that A is true. (5.189)

A previous paper (Drago 2013) suggested that the word "suspect" is equivalent to the DNP: "We can<u>not</u> <u>exclude</u> that…". This DNP is not logically equivalent to the corresponding affirmative proposition—"There is a reason to affirm that A is true"—since the latter is manifestly unsupported by scientific evidence.[18]

[18]Also the conclusions of the forms 2.702 and 2.706 are non-affirmative statements: "… *probably* and *approximately*….".

This interpretation is confirmed by Peirce's unpublished draft of a letter to Victoria Welbey (16 July 1905); he illustrates abduction by reiterating the scheme of *modus tollens*; the last sentence first ends with a question mark and then it is developed till up to conclude by similar words to "suspect":

> [Result] Therefore A is not true?
> Instead of'interrogatory', the mood of the conclusion might more accurately be called 'investigand', and be expressed as follows:
> 'It is to be inquired whether A is not true.'(Peirce MS L 463)

Peirce's conclusive words above all represent a problem which may be formulated by means of the same DNP as in the above ("We cannot exclude that…"). I conclude that Pearce's previous two syllogisms, both concluding the above DNP belong to intuitionist logic.

Of course, Peirce's aim was to change the above predicate into the corresponding affirmative predicate, in order to derive from it all possible implications to be tested by experiments. This attitude is the same as the previous inferential process which was qualified as an abduction; in particular, Cellucci's definition applies again; moreover, it seems even more appropriate to this latter case. Hence, I also call the present one an abduction, although at the level of an entire theory.

In logical terms we write the above change as follows:

$$\exists x \neg\neg f(x) = > \exists x f(x),$$

where the sign => does not mean a logical implication because this change from a DNP, which is an intuitionist predicate, to the corresponding affirmative predicate, belonging to classical logic, is not justified by a law of any specific logic. This change is rather allowed by the principle of sufficient reason, to which Peirce intuitively alluded when he suggested that one has to infer by relying also on the natural instinct of the human mind. (Drago 2013, p. 334). Since this change of an intuitionist predicate into the corresponding classical one is not a logical deduction, I call it a *principle*, i.e. Peirce's principle. It is a principle also due to its level of generality; any result of its application, i.e. $f(x)$, is validated by the logical consequences drawn from $f(x)$ deductively.

This inferential process can be regarded as representing a new kind of abduction, concerning the last predicate of a PO theory. With respect to the initial problem—i.e., to obtain a hypothesis capable of beginning a theory of a given field of phenomena, this principle, by obtaining a classical predicate as a change of—not an implication from—intuitionist arguments, constitutes an oracle answer. Its result is an ideal element with respect to the intuitionist context; just as the answer of a CT'S *oracle* comes from outside the discussion of a given problem.[19]

[19]I took advantage of Schurz's comprehensive appraisal of the several meanings of an abduction processes: "I will classify patterns of abduction along three dimensions: (1) along the kind of *hypothesis* which is abduced, i.e. which is produced as a conjecture. (2) along the kind of *evidence* which the abduction intends to explain, and (3) according to the *beliefs* or *cognitive mechanisms*

Summarizing, *Peirce's latter instance of a syllogism illustrating an abduction actually represents a change from the conclusive DNP to the corresponding affirmative predicate; the change is obtained by a logical translation—here called Peirce's principle—from a particular intuitionist predicate to its corresponding classical predicate; from it, now considered a new hypothesis axiom, a new theory is developed according to the deductive method. Since it concerns two kinds of logic at the same time, this translation of a predicate—playing a crucial role in a PO theory—represents the most general form of abduction.*

In conclusion, in Peirce's writings we find two kinds of abduction:

1. a hypothesis for a single datum or hypothesis, which cannot be obtained by an approximation process and not even by an idealization of this process; rather it is justified by the classical derivations from it.
2. a decisive hypothesis for building a new theory, i.e. a translation of a non-classical predicate into the corresponding classical predicate, which is called "Peirce's principle of abduction".

When the abduction is interpreted as before—i.e. either as (in previous section) an oracle producing either a single notion or (in present section) a decisive hypothesis concluding a PO theory—, it corresponds to Peirce's previous quotation:

> all operations by which theories and [ideal] conceptions are engendered. (5.590)

Yet, Peirce's illustration of this inferential process is inadequate. Let us remark that an abduction essentially belongs to a PO theory (in an AO theory no creative inferences are needed). In such a kind of theory the conclusive predicate is obtained by some ad absurdum arguments. Although in his philosophical reflections—as we saw in the above—, he made an essential use of this kind of argument, Peirce ignored it and he considered an abduction an isolated act of the human mind. Peirce did not know that abduction, although it is verified through its implications in a deductive context,—as stressed in previous quotations in Sect. 4—is generated in a PO context; and hence that the logic of this kind of abduction is not classical logic plus some partial modifications—like his inventions of new versions of classical syllogisms—, but intuitionist logic, to which his resulting syllogisms actually belong. As a consequence of his shortening the entire process, Peirce merged the two kinds of abduction into only one, to which he attributed a lot of features.

(Footnote 19 continued)

which *drive* the abduction." (Schurz 2008, p. 205). His decisive "Result 3" seems to summarize the previous dimensions: "In all cases the crucial function of a pattern of abduction… consists in its function as a *research strategy* which leads us, for a given kind of *scenario*,… to a most promising explanatory conjecture which is then subject to further test." (Schurz 2008, p. 205). I agree with both dimensions and Result 3 obtained by Schurz' since in my opinion he intuitively referred to the two above illustrated dichotomies on the kind of organization of a theory ("scenario") and the kind of infinity ("cognitive mechanics which drive the abduction"). In particular, the word "belief" in Schurz's third dimension (Schurz 2008, p. 205) corresponds to the logical step which plays the role of the conclusion of a PO theory.

In CT the notion of an oracle also corresponds to the second kind of abduction. CT being organized as a PO, the concluding predicate is the so-called Turing-Church's thesis, which correctly is not called a 'theorem' because cannot be obtained in an AO theory, indeed, it is a DNP, as the strange words of his enunciation reveals.[20] In CT the change of the conclusive predicate into the corresponding affirmative predicate is not assured by Peirce's principle, but by Leibniz' principle of sufficient reason, because its logical formula is the following one:

$$\neg\exists x \neg f(x) \neg\neg\forall x f(x) = > \exists x f(x),$$

6 Comparing Peirce's Induction to CT's *Minimalization*

Let us start by interpreting Peirce's induction through some steps. As a first step, I agree with McMullin's appraisal about what Peirce meant by an induction:

[Peirce's] Induction is *strictly* limited to the observable domain. And it is only in a very weak sense explanatory (McMullin 1992, p. 90).

I interpret this statement as stressing that the reference domain of definition of induction is for the main part given by operative, constructive tools. I add a definition of it in philosophical terms; it is a regularity which is transferred even idealistically from observed to possibly unobserved.[21]

Both the quotation and the definition characterize an induction by means of a context not including ideal elements; in mathematical terms, its basics appeal to the potential infinity only (PI). But its entire process may include AI.

As a second step, in order to make homogeneous (*si può usare 'homogeneous' riferito ad una comparazione?*) the comparison of Peirce's informal induction with CT's mathematical techniques, I consider mathematically parameterized the data, hypotheses, functions involved in an induction, so that the following discussion concerns mathematical sets of numbers and functions.

As a third step, in order to find out a correspondence between Peirce's induction with a complete mathematical process, let us take an orientation from a particular case; I associate the creative inferential process of an induction—which obtains "an

[20]For instance, Kleene states Church's thesis as follows: "… it cannot conflict…" (Kleene 1952, pp. 318–319). Although ignoring the exact linguistic expression of a DNP, several authors presented Turing-Church's thesis through similar words: Goedel called it a "**heuristic** principle", (Davis, p. 44). Post a "working **hypothesis**" (Davis 1965, p. 291). The same Church presented it through the words: "… it is **thought** to correspond satisfactorily…" (Davis 1965, p. 90) Remarkable is the masked proposition in Davis et al. (1994, pp. 68–69): "we have reason to believe that…", where the word "believe" is a subjective word which actually is the DNP: "it is not false that it is …".

[21]This definition is a modification of Schurz' definition (Schurz 2008, p. 202).

assent", as Peirce calls the final element in the first quotation of Sect. 3—with the following very common inference process.

The following quotation leads to stress a Peirce's inference process, an approximation:

> approximation must be the fabric out of which our philosophy has to be built. (1.404)

Here he really meant an "abduction", but I consider the case of an approximation to be too common to be related to the notion of abduction, which Peirce claimed to be an absolutely original notion. Indeed, the following proposition suggests that Peirce regarded an approximation as the first meaning of his notion of induction:

> Induction is reasoning which professes to pursue such a method that, being persistent in it, each special application of it.... must at least indefinitely approximate to the truth, about the subject in hand, in the long run. (Eisele 1985, vol. 4, p. 37)

This definition agrees with this Peirce's suggestion of induction as an approximation process, provided that "in the long run" is intended as "at infinity", be it either PI or AI. In fact, he speculated upon several notions, e.g. hardness (5.207), which result from approximation processes. In its best form an approximation is a limit process.

Yet, this kind of calculation has to be considered in the light of the dichotomy regarding the kind of infinity. Indeed, in classical mathematics, which makes use of actual infinity (AI), e.g. Zermelo's axiom, a series (possibly including ideal numbers) obtains in all cases a single final result; indeed, the classical mathematician, in virtue of his Platonist philosophy of mathematics, claims that the result does exist, provided that no contradiction will result. Notice that Cauchy's ε-δ technique of approximation of a limit appeals to AI to the extent that it abruptly reduces to a single point representing the final element the two-points interval defined by the approximation process (Kogbetliantz 1968, App. 2).

On the other hand, in constructive mathematics—the mathematics that makes use of potential infinity (PI; Bishop 1967) only—, an approximation process, constituted by a converging series of operative or constructive numbers, obtains no more than approximations to a final element (which however is achieved if some conditions are added—e.g. it is independently known that the final element is a natural number). Hence, it does not correspond to the usual meaning of induction, which is commonly thought of as only in the favourable case of the achievement of the final result. Notice that the ancient method of e.g. approximating the length of the circumference by means of the perimeters of a series of polygons inscribed in it, added an ad absurdum argument for equating the final element to the limit number. According to the ancient method, only by adding the ad absurdum proof may the final element be taken as the result of an induction process.

However, from the knowledge of the final element one cannot obtain such an element recursively, because the attribution of the final element implies the use of the AI. In other words, this case an induction is of a creative kind.

Owing to the great capacity of calculation of a computer this mathematical operation does not require what the human mind needs to manage a large amount of data, that is, an ordering of them. In the latter case—i.e. a set of ordered data—an induction corresponds to a mathematical process of a limit process of a convergent series of numbers.

Moreover, the case of a limitation to PI in CT corresponds to a *bounded minimalization* process, whose definitory set is finite and whose result—the calculation being performed by a primitive recursive function—is also finite.[22] Hence, it is not a process producing an essentially new element, rather it is a mere calculation from finite elements to finite elements by means of finite tools.

Now let us consider the case in which the mathematized induction process appeals to AI. In such a case the mathematical technique may present different aspects: the method of the least squares, search for a best hypothesis under some constraints, calculus of variations, etc. Some other mathematical techniques of attributing an ideal element, approached by a mathematical technique of calculation, can be considered instances of this case of induction; e.g. the process of interpolating from a list of observational data a hypothesis-mathematical function including all of them as particular cases.[23] This subject was for long time investigated and clarified by statistical analysis. The extrapolation is a similar process of inference.

[22]For instance in the history of theoretical physics the approximating series mattered. The case of an ideal body for building the theory of impacts was very controversial. Wallis and Newton have suggested a perfectly hard body; it does not change its shape however violent the impact; of course the conservation of energy is no longer allowed. This ideal body cannot be approximated by a series of ever harder bodies because its definition refers only to the behaviour of the final body of any possible approximation series. Instead Leibniz has suggested as an ideal body the perfectly elastic body. The approximating instances of elasticity, by pertaining to the observable domain, may be represented by a series of values of a parameter characterizing this specific feature; in such a way one obtains a mathematical converging series of the values of the elasticity index of the material objects. This process may be defined also—in physical terms—as a process that obtains a final element by operative means. It was not until 1850 that the hard body model was dismissed and the conservation of energy was considered a general law. In conclusion, since lacking any approximating series, the abstraction of a perfectly hard body was a false induction, a mere guess. In theoretical physics the inertia principle, the door to modern physics, is not the a result of an approximation process; it holds true only when one considers a body without friction and gravity (a situation which cannot be approximated—as Galileo put it—by a variable incline). For this reason it is a theoretical principle, not derived from other principles.

[23]Being a merely probable proposition, it has to be correctly stated as "It is not the case that it is not…", i.e. a DNP. The lack of this accurate logical version of his notion surely made Peirce's logical elaboration of these inference processes difficult.

According to the previous analysis, the selection of the best hypothesis among a set of hypotheses under some constraints is an "extremization process" performed by means of a mathematical calculations; hence, it has to be considered a case of induction, rather than abduction.

In CT there exists a similar process, unbounded minimalization, whose calculation allows the dominion of definition to be infinite. Here we have a difference between Peirce's (mathematized) induction and the possibly corresponding CT process; the former concerns the final result, the latter concerns the dominion of definition. Another CT version of creative induction is recognized in the *generalized recursion*; for instance the definition of a generalized recursive function by means of the a diagonalization process does not correspond to any of the previous mathematical processes; its result—"the assent"—comes from an appeal to AI (it allows us to deal with the set of *all* primitive recursive functions). Its idealization process suggests rather than a single point, a single function by considering the totality of (the primitive recursive) functions.

However, all these processes are included in the category of "extremants", a category that in theoretical physics includes all kinds of appeals to AI aimed at obtaining a result outside a given mathematical context, be it an approximating series, or the dominion of definition of the process, or other issues.

For instance Maupertuis' principle of minimal action, Fermat's principle of the minimal path, etc. Notice that the calculus of these extremants concerns not an ordered set, but a set of continuous functions. All these physical principles claim to obtain more than what operative means allow; they transfer regularities in the physical realm to idealized final results.

In conclusion, Peirce's inferential process of an induction corresponds to CT's two mathematical techniques:

1. *Unbounded minimalization*; in classical mathematics it partly corresponds to the process of attributing a final element to approximations; in theoretical physics it corresponds to the extremant techniques.
2. *General recursion*, taking into account idealistically the complete set of the primitive recursive functions.

Inappropriately one more process may be added, the bounded minimalization; the great capacity of calculation of a computer appears to the human mind to be performing an induction when it computes a kind of process that is in practice impossible by humans.

In retrospect, the first difficulty met by scholars attempting to disentangle Peirce's notion of induction came from having disregarded the formal distinction between PI and AI, which is manifestly relevant in such a kind of inferential process. The second difficulty came from the ignorance of the alternative organization of a theory to AO, i.e. PO; if located in an AO context the verification process of an induction misses any meaning.

7 Conclusions

Concerning Peirce's treatment of abduction I conclude that:

(A) Peirce cleverly introduced a new kind of inference process, abduction.
(B) He did not clearly distinguish among the inductive processes, mainly between induction and abduction.
(C) He coalesced many conceptual and mathematical notions into one philosophical notion, abduction.

Moreover, through a comparison with CT's principles[24] I obtained a complete characterization of Peirce's inferential processes; this comparison led to the addition of one more process which he unwarily made use of, limitation. The other term of this comparison belongs to Mathematics and/or Mathematical logic; these formal sciences may be a little crude in reducing the richness of Peirce's reflections; yet, they offer certain points of reference for all other interpretations of Peirce's philosophy. For instance, for the first time we obtained that the mathematical notion of induction is essentially distinct from the logical notion of abduction owing to the context, respectively the mathematical and/0r the logical context; this constitutes a formal distinction which characterizes the structure of Peirce's reflections.

A Table 1 summarizes the results (recall that the column headed AI&PO gives less tight relations than the others).

By the way, we have recognised CT's basic choices on the kind of organization and on the kinds of infinity, and consequently its kind of theoretical development, including DNPs, ad absurdum arguments, conclusive DNP and its change into an affirmative predicate. This development is the same as the physical theories whose organization is a PO.

Peirce was one of the first philosophers to ponder on "reasoning machine". Moreover, we have shown that no other philosopher's reflection was so comprehensive as Peirce' in considering all possible inference processes of CT, so that Peirce anticipated the ways of arguing in CS; i.e., he anticipated the results obtained by the effort of computer scientists for over half a century to discover all possible kinds of reasoning which can be represented by operative (pragmatical is Peirce's word) tools, which is what computer operations are.

For this reason I suggest that Peirce should be considered the philosophical father of CT's reasoning.

[24]Why was it possible to compare Peirce's reflection on inferential processes with those of CT rather than with those of physical theories? Because each physical theory at most exploited one inferential process only. Only Mendeleev may have exploited all these processes together in his obscure way in order to build the table of chemical elements (Drago 2015).

Table 1 Correspondence between the basic choices, Peirce's kinds of reasonings and computability theory's principles

MST: ideal elements (AI) or not (PI) classical logic (AO) or non-classical logic (PO)	AI&AO	AI&PO	PI&AO	PI&PO
Theory: Prime principles in Theoretical Physics	Causality	Extremants	Existence of a mathematical object	Constraints, limitation
Computability theory's ways of reasoning	Generalized Recursion	Unbounded minimalization	Oracle	Undecidabilities
Peirce's four ways of reasoning	Deduction	Induction	Abduction	Incapacities
Nature of his processes of reasoning	Conservative	Ampliative		Limitative
Interpreters of Peirce's kinds of reasoning	*All scholars*: deduction	*Drago*: constructive approximation process, with or without final point	*Josepheson:* the best hypothesis. *Hintikka:* an interrogative inquiry. *Drago*: Oracle of an element or a hypothesis	*Drago*: undecidabilities. abduction of Peirce's principle
General logical principle	Principle of non-contradiction		Principle of sufficient reason	

Acknowledgements I acknowledge Prof. David Braithwaite who corrected my poor English.

Bibliography

Achinstein, P. (1987). Scientific discovery and Maxwell's kinetic theory. *Philosophy of Science, 54*(3), 409–434.

Beth, W. E. (1959). *Foundations of mathematics*. Amsterdam: North-Holland.

Bishop, E. (1967). *Foundations of constructive analysis*. New York: Mc Graw-Hill.

Burch, R. (2012). Charles Saunders Peirce. In E. N. Zalta (Ed.), *Stanford encyclopedia of philosophy*. http://plato.stanford.edu/entries/peirce-benjamin/

Cellucci, C. (1998). The scope of logic: Deduction, abduction, analogy. *Theoria, 64*, 217–241.

Chiasson, P. (2001). Abduction as an aspect of retroduction. *Digital Encyclopedy of Charles S, Peirce*. http://www.digitalpeirce.feeunicamp.br/p-abachi.htm

Dauben, J. W. (1977). C.S. Peirce's philosophy of inifnite sets. *Mathematics Magazine, 50*, 125–135.

Davis, W. H. (1972). *Peirce's epistemology*. The Hague: Martinus Nijhoff.

Davis, M., et al. (1994). *Computability, complexity and languages, fundamental of theoretical computer science*. New York: Academic Press.

Douven, J. (2011). Abduction. In: E. N. Zalta (Ed.), *Stanford encyclopedia of philosophy*. http://plato.stanford.edu/entries/abduction/

Drago, A. (1990). I quattro modelli della realtà fisica. *Epistemologia, 13*(1990), 303–324.

Drago, A. (1997). New Interpretation of Cavalieri's and Torricelli's method of indivisibles. In J. Folta (Ed.), *Science and technology of Rudolfinian time* (pp. 150–167). Praha: Nat. Technical Museum.

Drago, A. (2003). The introduction of actual infinity in modern science: mathematics and physics in both Cavalieri and Torricelli, *Ganita Bharati. Bulletin Society Mathematical India, 25*, 79–98.

Drago, A. (2012). Pluralism in logic: The square of opposition, Leibniz' principle of sufficient reason and Markov's principle. In J.-Y. Béziau & D. Jacquette (Eds.), *Around and beyond the square of opposition* (pp. 175–189). Basel: Birkhaueser.

Drago, A. (2013). A logical model of Peirce's abduction as suggested by various theories concerning unknown entities. In L. Magnani (Ed.), *Model-based reasoning in science and technology. Theoretical and cognitive issues* (pp. 315–338). Berlin: Springer.

Drago, A. (2015). Il chimico-filosofo Charles S. Peirce sulla tabella di Mendeleieff e sui tipi di inferenza per costrurla. In M. Taddia (Ed.), *Atti XVI Conference of GNSFC 2015* (in press).

Drago, A. (2016). The four prime principles of theoretical physics and their roles in the history. *Atti Fondazione Ronchi, 70*(6), 657–667.

Drago, A. (2016). I quattro principi primi della fisica teorica. In L. Fregonese (Ed.), *Atti Sisfa 2011* (in press).

Dummett, M. (1977). *Principles of intuitionism*. Oxford: Clarendon Press.

Eisele, C. (1985). *Historical perspectives on Peirce's logic of science: A history of science*. Amsterdam: Mouton.

Fann, K. T. (1970). *Peirce's theory of abduction*. The Hague: Nijhoff.

Feferman, S. (1998). *In the light of logic* (pp. 77–93). Oxford: Oxford U.P.

Fisch, M. H. (1972). Peirce and Leibniz. *Journal History of Ideas, 33*, 485–496.

Flach, P. A., & Kakas, A. (2000). Abductive and inductive reasoning: Background and issues. In P. A Flach & A. Kakas (Eds.), *Induction and abduction. Essays on their relations and integration* (1–27). Boston: Kluwer Academic Press.

Gardiès, J.-L. (1991). *Le raisonnement par l'absurde*. Paris: PUF.

Grzegorczyk, A. (1964). Philosophical plausible formal interpretation of intuitionist logic. *Indagationes Mathematicae, 26*, 596–601.

Hanson, R. N. (1958). *Pattern of discovery*. Cambridge: Cambridge U.P.

Hilbert, D. (1902). Mathematical problems. *Bulletin of the American Mathematical Society, 8*(10), 437–479.

Hilpinen, R. (2004). Peirce's Logic. In D. M. Gabbay & J. Woods (Eds.), *Handbook of history of logic* (Vol. 3, pp. 611–658). New York: Elsevier.

Hintikka, J. (1998). "What is abduction? The fundamental problem of contemporary epistemology. *Transactions of the Charles S. Peirce Society, 34*, 503–533.

Hoffmann, M. H. G. (2010). "Theoretic Transformations" and a new classification of abductive inference. *Transactions of the Charles S Peirce Society, 46*(4), 570–590.

Horn, L. (2001). The logic of logical double negation. In *Proceedings Sophia Symposium on Negation*, Tokyo, University of Sophia, pp. 79–112.

Horn, L. (2008). On the contrary. In *Proceedings Conference Logic Now and Then*. Brussels (in press).

Josephson, J. R., & Josephson, S. G. (1994). *Abductive inference: Computation, philosophy and technology*. Cambridge: Cambridge University Press.

Kakas, A. C., & Mancarella, P. (1990). Generalized stable models: A semantics for abduction. In *Proceedings E-CAI-90*, pp. 385–391.

Klein, M. J. (1982). Some turns of phrase in Einstein's early papers. In A. Shimony (Ed.), *Physics as natural philosophy* (pp. 364–373). Cambridge, MA.: MIT Press.

Lavoisier, A.-L. (1862–92). *Oeuvres de Lavoisier*, Paris, t. 1.
Magnani, L. (2001). *Abduction, reason, and science. Process of discovery and explanation*. Dordrecht: Kluwer Academic Press.
McKaughan, J. (2008). From Ugly Duckling to Swan: C.S. Peirce, Abduction, and the Pursuit of Scientific Theories. *Transactions of the Charles S. Peirce Society, 44*, 446–468.
McMullin, E. (1992). *The inference that makes science*. Milwaukee: Marquette University Press.
Nickles, T. (Ed.). (1980). *Scientific discovery, logic and rationality*. Boston: Reidel.
Niiniluoto, I. (1999). Defending abduction. *Philosophy of Science, 66*, 346–351.
Peirce, C. S. (1965). *Collected papers of Charles Sanders Peirce*. In C. Hartshorne & P. Weiss (Eds.), Harvard University: Cambridge, Massachusetts.
Peirce, C. S. (1984). Some consequences of four incapacities (orig. 1868). In *Writings of Charles Spencer Peirce* (Vol. 2, pp. 211–242). Bloomington: Indiana University Press.
Peirce, C. S. (1968). Questions concerning certain claimed human faculties. *Journal of Speculative Philosophy*, *2*, 103–114; (1868). Some consequences of four incapacities. *Journal of Speculative Philosophy*, *2*, 140–157; (1869). Grounds of validity of the laws of logic: Further consequences of four incapacities. *Journal of Speculative Philosophy*, *2*, 193–208. All collected in (Peirce 1984, Vol. 2) C. S. Peirce (1931): *Collected paper by Charles S. Peirce* (Vol. 1–8). Cambridge MA: Harvard (1931–1958).
Prawitz, D., & Melmnaas, P.-E. (1968). A survey of some connections between classical intuitionistic and minimal logic. In H. A. Schmidt, K. Schütte, & H.-J. Thiele (Eds.), *Contributions to mathematical logic* (pp. 215–229). Amsterdam: North-Holland.
Schurz, G. (2008). Patterrn of abduction. *Synthese, 164*(2), 201–234.
Stalker, D. (Ed.). (1990). *Grue, the new riddle of induction*. Open Court, Chicago and La Salle, 1994.
van Brakel, J. (2000). *Philosophy of Chemistry*. Leuven: Leuven University Press.
Webb, J. (1980). *Mechanism, mentalism and metamathematics*. Dordrecht: Reidel.
Weyl, H. (1987). *The continuum: A critical examination of the foundation of analysis* (orig. 1918). Kirksville MO.: T. Jefferson University Press.

Perception, Abduction, and Tacit Inference

Rico Hermkes

Abstract The aim of this paper is to develop an inferential conception of perception. For establishing such a conception, we turned to Peirce, who had been first in systematically linking perception to abductive inferences. However, he did not conceive perception as abduction, because of the impossibility for conscious control. Thus, Peirce spoke of an "extreme case of abduction". An essential question is, if we can succeed in developing a conception that incorporates tacit inferences. Peirce's semiotic conception as well as Polanyi's theory of implicit knowledge work fairly well as a theoretical starting point. By appointing equifunctionality between non-symbolic sign activity and applying logical rules it is possible that inferences can be also realized unconsciously. Furthermore, we point out that feelings may be regarded as indices and illustrate how affirmative judgements can be realized using such indices for judging the validity of an inference. The "Predictive Processing-Approach" seems suitable for finalizing the picture. According to this approach perceptions may be conceived as an inferential triad rather than mere abductive inferences. The upshot is that it enables us to extend it to other forms of unconscious cognition and intuition.

1 Introduction

Perception is generally understood as unconscious cognition leaving us mostly unaware of its processing. Nonetheless, in the pragmatist tradition dating back to Charles Sanders Peirce, perceptions consist of generating hypotheses and presumptions about the world. The essential question is, if there is an internal "logic" at work giving rise to such hypotheses, i.e. whether it is possible to reconstruct perceptions as abductive inferences. Peirce himself seriously doubts that perception can

R. Hermkes (✉)
Faculty of Economics and Business Administration, Business Ethics and Business Education, Goethe University Frankfurt am Main, D-60629 Frankfurt am Main, Germany
e-mail: hermkes@econ.uni-frankfurt.de

L. Magnani and C. Casadio (eds.), *Model-Based Reasoning in Science and Technology*, Studies in Applied Philosophy, Epistemology and Rational Ethics 27, DOI 10.1007/978-3-319-38983-7_22

be understood as inference, as they are not controlled by the mind. He therefore dismisses perception from his system of inferential cognition. However, he also holds that nothing can be conceived in the human mind, which has not been subject to the senses before (CP 5.181). However, since Peirce dismisses perception from systematic knowledge acquisition, he finds himself confronted with the problem that the initial premises of deliberate (inferential) thinking are, as a matter of fact, unsecure and that consequently all human knowledge would be grounded on sand. Peirce tries to cope with this unwieldy issue by taking perceptions as an inferential boundary case, an "extreme case of abductive inferences" (EP, p. 227), thereby at least assigning equifinality to inferences for perception.[1]

Conversely, by adopting Peirce's semiotics and a more recently published neurobiological approach to perception, I shall argue that perceptions can be considered as inferences (abductive inferences in particular) that can be incorporated into an inferential taxonomy. In Chap. 2, I shall start with presenting Peirce's conception of perception and continue by sketching an inferential conception for perception in six propositions based on Peirce. In Chap. 3 I will explicate these propositions in more detail. In the course of our endeavor, Polanyi's theory of implicit knowledge will be of further support. Polanyi contributes three aspects to it: Firstly, his considerations are based on the assumption that tacit inferences occur during perception. Secondly, he considers tacit inference as an active process rather than a passive one. Thirdly, Polanyi incorporates affirmations into his inferential approach even though he does not elaborate this idea thoroughly. In this respect we may consider Polanyi's approach as a model for an inferential theory of perception, not only comprising conscious but also unconscious inferences. That is why I shall dedicate Chap. 4 to Polanyi's theory of implicit knowledge.

The upshot of establishing an inferential taxonomy of perception is that it enables us to extend it to other forms of intuitive cognition.

2 An Inferential Conception of Perception Sketched in Six Propositions

2.1 Starting Point: Peirce About Perception and Abduction

In Peircean pragmatism, perceptions are generally comprehended as primordial cognitive processes (CP 5.55; cf. Nesher 2001). Peirce describes perceptual judgments—the results of perceptions—as "the starting point or first premise of all critical and controlled thinking" (CP 5.181). He states: "But the content of the perceptual judgment cannot be sensibly controlled now, nor is there any rational hope that it ever can be" (CP 5.212). Thus, there are cognitive processes at work to

[1]Peirce about intuition: "Though inferential in their nature, they are not exactly inferences" (EP, p. 11).

which we have no conscious access or control. At the same time, Peirce conjectures that a perceptual judgment entails a classification: "(T)he very decided preference of our perception for one mode of classing the percept shows that this classification is contained in the perceptual judgment" (CP 5.183).

Apparently, Peirce is not totally clear about the inferential status of perceptions. On the one hand, he conceives perception as an uncontrolled, unconscious and involuntary process, on the other hand as an "extreme case of an abductive inference" (CP 5.180). To clarify this issue let us recall Peirce's abduction schema:

(P1) The surprising fact, C, is observed,
(P2) But if A were true, C would be a matter of course,
(K) Hence, there is reason to suspect that A is true (CP 5.189).

Following Minnameier (2016) this schema does not describe the entire process of abductive inference, but only the final judgmental step. According to Peirce, every inference consists of three inferential steps. The first one is colligation (the collection of premises), the second one observation (contemplation and manipulation of premises, arriving at a conclusion)[2] and the third one is judgement (validity check of the conclusion). Colligation comprises the registration of the surprising fact C (the data). The observation step accounts for merging the data and converging them to hypothesis A (corresponding to the explanation of data). In the judgemental step the transformation into a propositional format finally takes place. Relating to perception, Peirce writes: "By a perceptual judgment, I mean a judgment asserting in propositional form what a character of a percept directly present to the mind is" (CP 5.54).

From this we may conclude that perception in principle can be constructed as an abductive inference. Peirce however, does no fully comply with this idea as it seems basically impossible for him to exert control over perceptual processes. However, if succeeded in developing an inferential conception where unconscious abductions were systematically incorporated, such a conception would allow us to include perceptions without difficulties.

Admittedly, there are two other issues left to be clarified. Where does hypothesis A (as a percept or perceptual judgement) arise from? And how are the percept and the perceptual judgement related to each other? For closer examination of the first question we are inclined to turn back to Peirce who offers us clues for revealing what happens during the generation of the hypothesis. He writes: "The abductive suggestion comes to us like a flash [...]. It is true that the different elements of the hypothesis were in our minds before; but it is the idea of putting together what we had never before dreamed of putting together which flashes the new suggestion

[2]The observation step does not only comprise conscious and controlled cognitive activity. Furthermore, it is contemplative and might be driven by a "force majeure", as Peirce puts it (CP 5.581; cf. Minnameier 2016). The subject's role consists in following its direction and course mindfully. The observation step terminates with the conclusion, i.e. the unfolding of a regularity.

before our contemplation" (CP 5.180). This explanation already points to an involvement of a process that we can hardly imagine as a purely logical one. In fact, it seems that this description of the observational step indicates a form of tacit process, even though Peirce does not consider tacit inferences explicitly.

For answering the second question, we shall take a closer look on Peirce's concepts of *percept* and *perceptual judgement*. It is striking that Peirce uses perceptual judgement ambiguously. He simultaneously refers to it as the process (or better: procedure) of judgement and to its result. Peirce writes: "A judgment is an act of formation of a mental proposition combined with an adoption of it or act of assent to it" (EP, p. 191; see also Magnani 2009, p. 269). Herein he regards perceptual judgement as a procedure. Elsewhere Peirce adheres strictly to the distinction between percept und perceptual judgement referring to the judgemental step as a result of abduction: "The percept of course is not itself a judgment, nor can a judgment in any degree resemble a percept" (CP 5.54). And: "Even after the percept is formed there is an operation, which seems to me to be quite uncontrollable" (EP, p. 191). Due to these statements percept and perceptual judgement can be considered as results of two different procedures. Both are unconscious, whereas the procedure of establishing a percept is followed by the procedure of judging (arriving at the perceptual judgement).

Therefore, it is possible that Peirce allocated the generation of the percept to the observational step and the perceptual judgement to the judgemental step. Another interpretation would result in modelling both as two separate inferences: the first in establishing the percept, the second in generating the perceptual judgement (as a proposition), which is, as Peirce puts it, the "starting point or first premise of all critical and controlled thinking" (CP 5.181). In this respect, it seems reasonable to reconstruct perception as more than just one inference.

At this point we may conclude in a nutshell:

- If we can justify the assumption of unconscious inferences, it will pave our way for conceiving perception as abduction.
- Peirce's considerations of the observational step point to an involvement of "tacit elements".
- Peirce's application of the terms *percept* and *perceptual judgement* suggests the following interpretation: Perception is not limited to just one process. We can imagine different forms of perception (at least two).

I will now present six propositions emanating from these preliminary considerations.

2.2 *Perception as Inference—Six Propositions*

The first proposition deals with the taxonomy of inferences (especially abduction) and integrates perception into this taxonomy.

(1) *According to Minnameier* (2016), *abduction can be classified in two dimensions, (A) the domain of reasoning and (B) the cognitive level of understanding. Perception can be conceived as abductive inference belonging to the explanatory domain and localized at lower cognitive levels of understanding.*

Dimension A can be subdivided into the explanatory, the technological and the ethical domain. Perceptions may be allocated to the explanatory domain as they lead to hypotheses explaining the given sensory data. Concerning dimension B, and as already stated in Chap. 2, we can distinguish perceptions at different cognitive levels of understanding. Thus,

(2) *We can discriminate (at least) two forms of perception corresponding to specific levels of cognitive understanding.*

At Level 1 (the lower level) understanding is limited to a registration of a regularity (see Minnameier 2016). At Level 2 (the upper level) individuals realize understanding by questioning how this regularity can be reasonably explained and classified, i.e. either by integrating the regularity into an appropriate category or by creating a new category. In addition, perceptions can be analyzed using a finer grained resolution, which Peirce already employed by conceiving three inferential sub-steps:

(3) *Perceptions comprise colligation, observation, and judgement. Colligation accounts for the availability of data, observation refers to the treatment of data (e.g. integration), and judgement refers to the adoption or rejection of the result of the observation.*[3]

This leads us to the question: What are the kinds of data to be processed inferentially?

(4) *Abductions are not only realized in symbolic formats but also in other formats, e.g. icons and indices.*

With this proposition we are tying up with Magnani's conception (Magnani 2009, 2015) where inferences are no longer bound to propositions (a symbolic format), but considered as a form of sign activity more generally. Magnani coins this idea by stating "iconicity hybridates logicality" (Magnani 2009, p. 265). According to this idea, operating in sign formats can be considered as equifunctional to applying logical rules. Furthermore, non-symbolic signs are also involved in judgements. In Peircean systematics, the determination of validity occurs during the judgemental step. According to El Kachab (2013) and Minnameier (2016) an abduction is valid, if it (i) explains the facts, and (ii) if it is susceptible to empirical verification.

[3]Thereby, we define an inference as the generation of a new item (data) from a set of already known items (data) and its adoption or rejection. In this respect, abductive inferences can be considered as updating a belief.

(5) *Due to the first criterion (i) an abductive inference would be valid, if there was a sign signaling that the inferred conclusion explains the surprising fact.*

Damasio (2006, 2012) proposes an embodied emotion-related signaling system introducing such indices as somatic markers. Peirce highlights the importance of feelings in the context of inferential judgements agreeing that feelings can be integrated into the sign repertoire (see CP 5.375, EP p. 21ff).

Coping with the second criterion (susceptibility to empirical verification), takes us beyond abductive inferences. This leads us to a recently published inferential approach of perception called "Predictive Processing" also named "Predictive Coding" (see Friston 2011; Clark 2015).

(6) *According to the theory of "Predictive Processing" (PP) perception is not restricted to abductive inferences; it can rather be conceived as an inferential triad.*

The PP-approach can be summarized by the fact that the brain operates as an ever-active inference "engine" (Clark 2015, p. 2), realizing predictions about the upcoming sensory input. This idea accounts for all sensory modalities. If the predictions do not meet the sensory data, a prediction error results, which is considered as unpleasant *surprise* (in the Peircean sense; cf. abduction schema, Chap. 2) by the system. The whole game of the PP therefore centers in minimizing surprise, i.e. the reduction or elimination of the prediction error (Friston 2011, p. 488). From the Peircean point of view (a) such predictions can be regarded as deductions, (b) proving these predictions on the basis of incoming sensory data as inductions and (c) the minimization of the prediction error as abduction that leads to new hypotheses about the surprising data.

Following this brief presentation of the six propositions I will illustrate the fundamental aspects of this inferential conception in Chap. 3 in more detail.

3 Elaboration of the Inferential Conception for Perception

Our agenda comprises 6 topics: (1) localizing perception within the inferential taxonomy (reviewing a current state of perceptual research which opens out into our inferential taxonomy), (2) two forms of perception, (3) inferential sub-processes and the tacit nature of perception, (4) semioticity: inferences in non-symbolic formats, (5) feelings as indices and their role for inferential judgements, (6) "Predictive Processing" and the inferential triad.

These topics refer to the six propositions sketched in Chap. 2. I will by deal with these topics addressing them in this consecutive order.

3.1 Localizing Perception Within an Inferential Taxonomy

The first proposition concerns the two-dimensional inferential taxonomy. Dimension A comprises three domains: the explanatory, the technological and the ethical domain.[4] An abduction in the explanatory domain leads to an explanation, which can be finally verified or falsified by induction (for the inferential triad see Minnameier 2004, 2010). Correspondingly, abduction in the technological domain leads to new technical or technological inventions. Its effectivity can be proved by the following induction step. In the ethical domain abduction selects a moral principle or rule for its application in a situation involving moral relevance.[5] The inductive evaluation finally assesses the appropriateness of the selected principle. We can therefore differentiate the three dimensions by (i) the way abductive hypotheses can be generated (explanation, technological invention or technical tool creation or selection, moral principle selection) and by (ii) the criteria of validity applied for the following inductive inference: truth in the explanatory domain, effectivity in the technological domain and justice in the ethical domain, respectively.

Dimension B points to the hierarchy of different cognitive levels of understanding. Minnameier assumes, that "concepts and theories are built upon one another across cognitive levels, from elementary perceptions and actions to high-level scientific theories" (Minnameier 2016). In this respect, inferences take place already at very low sensory levels, where understanding is enveloped in embodied formats (for bodily rooted concepts cf. Johnson 1987; Lakoff 1988; Glenberg and Kaschak 2002; Nunez 2008; Goldman 2012). This understanding can potentially evolve to abstract conceptual understanding at higher levels.[6]

Schurz also conceptualizes a hierarchy of cognitive understanding, where the understanding at lower levels may emerge to higher levels by means of abductive

---

[4]Thaggard conceives five domains in his taxonomy, specified by what needs to be explained via abduction and the kinds of hypotheses that provides explanations (see Thaggard 2007).

[5]The classification of abduction in these three domains does not imply that each problem solving procedure can be precisely allocated to a single dimension. According to the investigations of the Gestalt psychologist Karl Duncker, solutions for real-world problems require at least a sequence of two subsequent abductions that are located in different dimensions. Initially, a basic principle is abduced for solving the issue, which also accounts for its explanation. Duncker calls this penetration of the problem. Subsequently, a technological realization for this problem is to be abduced, that Duncker names realization. Therefore, every creative problem solving consists of two separate abductions, one in the explanatory domain and the other in the technological domain (see Duncker 1926, 1935, 1945).

[6]At least four kinds of taxonomies of cognitive hierarchies can be distinguished: (1) hierarchy of cognitive levels of understanding (see Minnameier 2016), (2) hierarchy of ontic magnitudes (neurons, neuronal networks, cognitive microstates, etc.; see Alisch 1995; Alisch 2010), (3) hierarchy of cognitive levels of processing (retina, thalamus, visual cortex, higher cortical areals; see Clark 2015), (4) hierarchy of cognitive and meta-cognitive levels (see Nelson and Narens 1994). Note that Propositions 1 and 2 relate to (1) the cognitive level of understanding. Talking about ontic magnitudes, I shall use the term *scales* hereinafter.

processing (see Schurz 2008). He defines creative abduction as an integration of (empirically observed) correlated regularities to a common cause. The common cause constitutes a higher order concept unifying these regularities. Schurz' framework comprises three cognitive levels of understanding, starting with observational concepts, followed by first order dispositional concepts and finishing with second order dispositional concepts.

Both conceptions settle in a mutual agreeable description, in that they start off at a very fundamental embodied level evolving to an understanding at higher levels via inferences. The function of abduction consists in achieving a higher level of understanding. Thus, perceptions may also be conceptualized as inferences.

There has been a long tradition, beginning with Helmholtz in (1867), of spelling out perception in inferential terms. One of Helmholtz's basic convictions is that sensory data function as signs and that their meaning can be learnt by the conscious mind. In accordance with J.S. Mill, Helmholtz assumes that perceptions are inductive inferences (Helmholtz 1867, p. 447ff) thereby providing meaning to these signs. Helmholtz is quite aware of the problem that goes along with unconscious inferences. He states, that even though perceptions are unconscious processes, they obviously perform the same work as conscious inferences and providing the same outcomes (Helmholtz 1867, p. 449). In this respect, Helmholtz supposes equifunctionality and equifinality between unconscious perceptions and conscious inferential acts.

Brunswik (1934) stresses the hypothetical (and therefore abductive) character of perceptions presenting his lens model of perception. One of his basic assumptions concerns the existence of a causal relationship between the objects of the environment which is contained in the sensory data. But this information is equivocal and therefore hypotheses need to be generated for the disclosure of the inherent causal relationship (Tolman and Brunswik 1935). Nevertheless, Brunswik, does not construe the generation of deliberate hypotheses and perceptions to be similar to acts of cognition (Brunswik 1948).

This mental leap was finally taken by Bruner (1957). To his mind inferences involved in perception do not substantially differ from those at higher cognitive levels. He states: "it is evident that one of the principal characteristics of perceiving is a characteristic of cognition generally. There is no reason to assume that the laws governing inferences of this kind are discontinuous as one moves from perceptual to more conceptual activities" (Bruner 1957, p. 124). What are the inferences that Bruner speaks of? He states: "In learning to perceive, we are learning […] appropriate categories and category systems, learning to predict and to check what goes with what" (Bruner 1957, p. 126). Interestingly, talking about categorization Bruner interchangeably uses the term model building (Bruner 1957, p. 126). These models can be realized in different representation formats. Bruner mentions enactive, iconic and symbolic representations that evolve ontogenetically in this order (Bruner 1974). This is of great importance as Bruner's conception points to the fact that inferences are no longer restricted to a symbolic format, but can be also realized in embodied and iconic formats.

3.2 Two Forms of Perception

A further question concerns the number of assumed processes; that is why I referred to *lower* levels of cognitive understanding (plural) in proposition 2 and not to just one *low* level (singular). As pointed out in Chap. 2 there are two possible interpretations of Peirce's presumptions concerning perception. According to the first alterative, the generation of the percept and the perceptual judgement is performed in two separate steps, but within one single inference. The second alternative displays separate inferences for each one.

Pursuing the first alternative we are confronted with the issue that the formation of the percept finishes in a sensory format. If so, there is an open account that needs to be settled: its transformation into a propositional format. This needs to happen at the beginning of the judgemental step and is performed unconsciously. Advocating the second alternative we may model perceptions at several cognitive levels of understanding (at least two). At Level 1 (the lower level) understanding is limited to registration of a regularity. At Level 2 (the upper level) individuals realize understanding by questioning how this regularity can be reasonably explained.

Localizing perception at two separate levels conforms with Herbart's (1892) early idea of distinguishing "Perzeption" and "Apperzeption".[7] The term *Perzeption* applies to the generation of an immediate percept, *Apperzeption* to the individuation of this percept allocating it to a certain category.

Magnani's analysis of the syllogistic character of "Peircean perceptions" is consistent with the above-quoted perspective of *Apperzeption* as "the act of subsuming sense data or 'percepts' under concepts or ideas to give rise to perceptual judgements" (Magnani 2009, p. 274). *Apperzeption* therefore can easily be assigned to Peirce's syllogistic schema:

> (P1) A well-recognized kind of object, M, has for its ordinary predicates P [1], P [2], P [3], etc.
> (P2) The suggesting object, S, has these predicates P [1], P [2], P [3], etc.
> (C) Hence, S is of the kind M. (CP, 8.64; see Magnani 2009, p. 274).

Once again, please note, that this schema merely comprises the judgemental step and not the entire inference.

Raftopoulos (2001a, b) proposed a similar structure for perceptual processing. He presumes three processes and calls them "sensation", "perception" and "observation".[8] Sensation relates to the constitution of the retinal image, perception to the processing of sensory data constituting a structured visual representation (e.g. generating a percept). The final process of observation reflects the perceptive hypothesis (applying cognitive categories), and seems to be congruent with Herbart's idea of *Apperzeption*, or in Peircean terms: a perceptual judgement.

---

[7]I will stay with the German wording to avoid confusions with the much broader concept of perception sensu Peirce.

[8]This is not to be mistaken for the Peircean observation concept as the second inferential step.

However, the constitution of the retinal image, i.e. the physiological process of sensory transduction, should be ruled out as a separate inference (see also Clark 2012, p. 759).

However, the relationship between *Perzeption* and *Apperzeption* shall not be misconceived in that *Apperzeption* is always taking place—in fact *Perzeption* also occurs without *Apperzeption*. Karl Lange elucidates this argument by citing an excerpt of Jean Paul's reminiscences, where Jean Paul compares his travelling experiences to Goethe's. He writes:

> "While travelling Goethe perceives everything definitively, whereas I am experiencing myself as totally deliquescent. Everything I perceive is melted in a romantic way. I travel through cities, without recognizing the things underway; what merely stimulate me are scenic landscapes. [...] Even I am aware of all individualities of life; I do not question them and forget about them quite soon." (cited from Lange 1912, p. 17, translation by the author).

It is worth taking his statement literally, not only because it highlights the differences between the epochs of German classicism and romanticism (Goethe was a vehement critic of romanticism). Moreover, it relates to an inner credo of comprehending the world, by experiencing it in pure perceptive way, i.e. by the immediate experience—and one might add, in the denial of *Apperzeption* as an act of conceptual comprehension. The point, where Jean Paul talks of individuating without questioning, seems particularly remarkable. He experiences these impressions, but has no interest in its conceptual integration at the same time. This position clearly differentiates *Perzeption* from *Apperzeption*. Of course, the same distinction applies to hearing and the auditory modality. For example, listening to music (abducing a hidden rhythm or a *Sound-Gestalt*) can be considered as *Perzeption*, whereas spoken language comprehension as *Apperzeption*. Under certain circumstances listening to music may also be considered as *Apperzeption*, that is if we intend to infer an explicit meaning of a tonal sequence, e.g. as a flowing river (Smetana's Vltava) or a sword stroke (Beethoven's Egmont) or if we abduce an already known song or a piece of music.

3.3 Inferential Sub-processes and the Tacit Nature of Perception

In Proposition 3 we have classified perceptions more explicitly into colligation, observation und judgement. Remember Chap. 2, where Peirce describes the observational step of an inference. He talks about "contemplation", the magic appearance of a "force majeure", of inspiration that "comes to us like a flash". These quite metaphorical terms point at implicit knowledge approaches, mainly advocated by Gilbert Ryle and Michael Polanyi. In fact, Peirce's description of the observational step reminds us of Ryle's analysis of human reasoning, when he asks, "via what interim paces had he (the thinking subject) marched from where he had been to where he is now" (Ryle 1976, p. 71). Or how we may illuminate cognition in utterances like "Oh, it just came to me"? Polanyi developed an approach to solving the issue. He presumes tacit inferences and localizes the process of tacit integration within it. Tacit integration focuses exactly the "black box" (Starting from the premises, how to attain a conclusion?). Furthermore, Polanyi adds another process called tacit confirmation that persists in the affirmative adoption or rejection of the conclusion.

In Chap. 4 I will provide a concise introduction of Polanyi's theory and discuss it as a model of an inferential conception of perception.

3.4 Semioticity: Inferences in Non-symbolic Formats

In proposition 4 I claimed that inferences are not only realized in symbolic formats but also in other formats e.g. icons and indices.

In his semiotic approach Peirce discriminates three types of signs: indices, icons and symbols. Symbols are arbitrary and conventional signs (CP 2.299), whereas icons resemble the objects that they denote (CP 2.276, CP 2.247). Indices are signs that are connected to the objects by an "existential relation" (CP 6.318). Its relationship can be causal, e.g. smoke may be an index for fire; footprints in the snow may be indices for people that have passed. Demonstrative pronouns are another example for indices pointing to an object.

The extension of the previous set of signs widens our possibilities and allows us to look at inferences from an alternate point of view. Magnani (2009) interprets inferences almost in the same manner as Peirce (as "a form of sign activity"; Magnani 2009, p. 265), by shifting the focus from logicality to semiotics. He states that "iconicity hybridates logicality" (Magnani 2009, p. 265) and concludes that sign characteristics (e.g. spatial configurations) may be equifunctional to logical rules. According to Magnani visual abduction represents a typical iconic abduction. Magnani consolidates iconic abductions under the term of model-based reasoning occurring "when hypotheses are instantly derived from a stored series of previous experiences" (Magnani 2009, p. 268).

Pape (1997) analyzes the Peircean approach and supposes isomorphisms between thinking and perception, implying that the ontic reality of visual experiences is also capable of expressing logical relations.

Now we have analyzed icons as non-symbolic formats, I would like to shift our focus to indexical signs that Peirce proposed as a third sort of signs. Indices play a central role in Pylyshyn's theory of perception (see Pylyshyn 2009). Pylyshyn addresses indices as mental signs that allow us to separate objects from the environment. He writes: "in the presence of a visual stimulus, we can think thoughts that involve individual things by using a term such as 'that' [...] where the term 'that' [...] refers to something we have picked out in our field of view without reference to what conceptual category it falls under or what properties it has" (Pylyshyn 2009, p. 7). What are the differences between such indexical signs and other signs like icons or symbols? Symbols and icons exemplify the classical idea of representation i.e. mirroring the world in the mind. Indices, however, "function without an encoding of objects' properties" and can therefore be conceived as pointers to the objects (Pylyshyn 2009, p. 5). Pylyshyn specifies indices as "something like a finger that would stay attached to a particular element and could be used to maintain a correspondence between the individual element that was just noticed now and one that had been represented in some fashion at an earlier time" (Pylyshyn 2009, p. 3). In this context we should be aware that Pylyshyn is talking about mental indices here and not about gestures. An interesting point about indexing is that the perceiving subject thereby attains access to a visual object, which Pylyshyn also calls proto object, as it is blank in its categorial properties. This proto object has just been tracked by the subject. Bruner defines such an act of indexing as "primitive categorization", referring to it as a "'silent' process that results from the perceptual isolation of an object or event" (Bruner 1957, p. 130f) (cf. Glenbergs "Indexical hypothesis" in his approach of language comprehension; see Glenberg 2008).

Summing up the possibilities of reconstructing inferences in unconscious and non-symbolic formats mentioned above, it seems evident that inferences are not restricted to *Apperzeption*, but apply to *Perzeption*, too.

Such an inferential conception is suitable for the explanation of phenomena like tilted images, e.g. the Necker cube. Holding the cube as a visual object in the frontal perspective is only of short duration, as the binding to the object gets lost soon after and another object gets established (cube this time seen from above). But again the new binding to this object does not persist. As a result we attain two alternating objects (as perceived), the cube in frontal perspective vs. cube seen from above. The instable binding is probably due to the ambiguity of the incoming sensory data. This ambiguity leads to different abductive hypotheses, but none of these hypotheses can be sustained permanently. Both hypotheses are equally valid, so that none is finally confirmed.

By addressing the issue of validity and verification, we arrive at the next point. How can we make sure that an abductive inference is valid?

3.5 Feelings as Indices and Its Role in Inferential Judgements

In proposition 5 we agreed that an abduction is valid, if it (i) explains the facts, and (ii) if it is susceptible to empirical verification. In proposition 4 we agreed, that non-symbolic formats are equifunctional to applying logical rules in the generation of a conclusion. Similarly, we may broaden our view regarding the validation during the judgement step by integrating indices.

According to these considerations, an abductive inference would be valid, if there is an index at place, signaling that the inferred conclusion is explaining the surprising fact. Thus, Damasio (2006, 2012) came into play who proposed an embodied emotion-related signaling system by calling such indices somatic markers. In his somatic marker theory these indices serve to validate decisions as correct or incorrect (from subject's perspective).

Peirce likewise highlights the importance of feelings for inferential judgements. He states: "The feeling of believing is a more or less sure indication of there being established in our nature some habit which will determine our actions. Let us recall the nature of a sign and ask ourselves how we can know that a feeling of any sort is a sign that we have a habit implanted within us." (CP 5.371).[9]

What does this mean for the concept of truth in an inferential context? Peirce continues: "The most that can be maintained is, that we seek for a belief that we shall think to be true. But we think each one of our beliefs to be true, and, indeed, it is mere tautology to say so. For truth is neither more nor less than that character of a proposition which consists in this, that belief in the proposition would, with sufficient experience and reflection, lead us to such conduct as would tend to satisfy the desires we should then have. To say that truth means more than this is to say that it has no meaning at all." (CP 5.375).

From these statements we may conjecture that the validity of an abductive inference can also be conveyed by a feeling. It seems worth considering Ciompi (1993, 1998, 2011) in this context who introduces the notion of *affect logic*. Hence, we might enhance our inferential conception by adding emotional components to it.

3.6 Predictive Processing and the Inferential Triad

We now arrive at the second aspect of validity concerning abductive inferences: the susceptibility for empirical verification. By this we address a criterion that reaches beyond the abductive inference, necessitating another expansion of our current inferential conception for perception. Following this demand I shall refer to a recently published inferential approach called "Predictive Processing" or "Predictive Coding" (Friston 2011; Clark 2012, 2013, 2015; Friston et al. 2012; Bastos et al. 2012; Hohwy 2013; Pickering and Clark 2014). According to the approach of *Predictive Processing* perception is not restricted to abductive inference, but can be conceived as an inferential triad.

Briefly, in this approach, predictions about the world are constantly generated by the sensory system. These predictions are derived from a hierarchical neurobiological "generative model" (Clark 2015; Pickering and Clark 2014) comprising different levels of neuronal processing. In this generative model predictions are unfolded as sensory patterns in a top-down direction. From an inferential point of

[9]Please note, that Peirce refers to a feeling as subjective intensity, "which is supposed to be immediately, and at one instant present to consciousness" (EP, p. 22). This quotation does not mean that believing is a feeling, but there is a feeling, which sustains a belief to be true.

view these a prediction can therefore be regarded as deduction, and proving the prediction on the basis of the successive sensory data, as induction. If the incoming sensory data do not meet the current prediction, the resulting deviation is called "prediction error". For the system, a *prediction error* would be *surprising* (in a Peircean sense). Clark considers the entire Predictive Coding theory as a game centered on minimizing the prediction error. One interesting point of this approach is that the ability for realizing inferences is already embodied in the biological system. Clark relates to the brain as an ever-active inference "engine" (see Clark 2015, p. 2). Büchel et al. adapt this framework for explaining the mechanisms of pain (placebo effects, chronic pain) and hypothesize "that the brain is not passively waiting for nociceptive stimuli to impinge on it but is actively making inferences based on prior experience and expectations" (Büchel et al. 2014, p. 1223).

A similar constructivist view towards perception was published by Bruner, entitled "Perceptual readiness" (Bruner 1957), which accentuates a triadic inferential conception regarding to cognitive scales (see proposition 2). Perceptual readiness means that individuals already expect the objects the environment is about to offer. Bruner models such expectations as accessibility to certain categories. One function of these predictions is "to minimize the surprise value of the environment" (Bruner 1957, p. 133), which sounds very much like an idea of "mental Predictive Coding". His presumption of a final confirmation step within the perceptual inferences (Bruner 1957, p. 131) points to an involvement of induction, besides abduction (see proposition 1) and prediction (deduction).

An adjacent question concerns the interplay of the three inferences and the underlying logic for this inferential system. One option would be assuming inferential modules, which receive an input and produce an output. The output of such a (abductive) module would then function as input for the subsequent (deductive) module. An alternative would be assuming a unifying logic comprising all three inferences. A candidate could be a default reasoning approach that allows using defeasible presumptions for deductive inferences. That connects abductions and deductions (Rescher 2007). Another choice consists in an intuitionistic logical approach (van Dalen 2001), which especially accounts for the constructivist view towards human cognition.

4 Tacit Knowing as a Model for the Inferential Approach of Perception

4.1 Polanyi's Theory of Implicit Knowledge

An overall conception of intuitive cognition is presented by Polanyi's theory of implicit knowledge (Polanyi 1962a, b; 1965; 1967a, b, 1968a). In this theory, perception, decision making, language comprehension, motor skills and the utilization of heuristics are put under one roof strongly emphasizing the necessity and relevance of unconscious cognition. Furthermore, conscious control might have a deteriorating impact on "tacit knowing" (Polanyi 1962b, p. 601).

Polanyi conceives implicit knowledge—perception being a "prototype" of it—as a mental act of tacit (or implicit) integration. By tacit integration he means that two terms—the proximal and the distal term—are being related to each other.

I would like to give an example dealing with tactile perception: Let us imagine that we are using a stick for orientation in the dark. In this act the subsidiaries which constitute the proximal term are (i) the sensation of holding the stick in the hand which is generated by tactile receptor information, (ii) specific muscle contractions of the fingers, the hand and the arm, as well as (iii) the subjective sensation of these muscle contractions. The meaning of the perceived object (the distal term) is conferred through these subsidiaries. If we touch an obstacle with the stick in the dark and come to the conclusion: "This is a soft object", than this softness is due to the specific muscle configuration and its resulting sensation for the subject. Another muscle configuration and sensation would constitute another distal object; we would for instance say: "This is a stiff object". In Polanyi's theory subsidiaries serve as instruments in the sensory orchestra for composing the perceptual object (distal term). They all play an instrumental role within the process of tacit integration.

4.2 *Tacit Knowing and Abductive Inferences*

In Chaps. 2 and 3 we already indicated that Peirce does not assume tacit inferences. Nevertheless, we may interpret his remarks as indications pointing to tacit "elements", when talking about the generation of a conclusion from the premises: the appearance of a force majeure, the inspiration that come to us like a flash. To my mind this can be interpreted as a tacit process. I therefore think, that Polanyi's "tacit integration" is an appropriate concept for characterizing the path from the premises to the conclusion.

However, we can interpret Polanyi's tacit integration in two different fashions. One is to conceive it in an active way, the other would be interpreting it in a passive way. If we agree on interpreting it in the latter manner, we find ourselves in cybernetics. From this perspective Ashby (1981) supposes, it is the environment, which imposes its hidden structures on the system (thereby referring to the brain). There are laws of transition in the environment (environmental patterns). These environmental patterns are adopted by the system. Ashby calls this an inductive inference and describes it as a process, where the system is considered to be passive: "Thus pattern in the environment inevitably tends to 'diffuse' into the system (Mullins 2002, p. 317) [...] it tends to force its way in" (Ashby 1981, p. 318). Ashby illustrates this driving force by adopting a metaphor: "it occurs when one stirs a dish of sand with one's fingers; if the finger makes only circular motions one will find afterwards clear marks of circularity in the sand" (Ashby 1981, p. 318). Likewise, the brain "mirrors" the causal laws of the environment, as the patterns in the sand resemble the circular finger movements. Hence, the brain represents the causal structure of the world.

Taking it the other way we may consider tacit integration as an active act. In this case, we would be talking about the creative power of the mind and not just about environmental forces at work. I assume that Polanyi favors the second position, because he conceives tacit integration as a "tacit triad" (Polanyi 1968a, p. 30) consisting of (1) the subsidiaries (proximal term), (2) the focal (distal) term, and (3) the "person, a knower, [who] sustains this integration" (Polanyi 1968a, p. 30).

Mullins puts it aptly by describing tacit integration as a "coordination or coalescence of subsidiaries" (Mullins 2002, p. 207). This refers to the different scales involved in tacit integration (cf. Neuweg's systematic overview about the constituents of the proximal term (Neuweg 2004, p. 189ff)). Thus, the proximal term is not only a mere receptacle for a number of various elements, but also constitutes a collective for all accessible information, which is (i) stored in different scales, (ii) in various formats and representational modalities, (iii) available for the process of tacit integration. However, certain questions remain open, also unanswered by Polanyi. What scales are relevant for tacit integration? What is the lowermost scale to be considered; e.g. is it necessary to include quantum effects for conceiving tacit integration (see Bohm 1973, 1990; Stapp 2009), and if the proximal term (subsidiaries) contains accessible information, what kind of information would that be? Stonier (1992, 1997) distinguishes between structural information and processible information. Structural information concerns the grade of organization and order of the organism (cf. Alisch 1995, p. 49). This describes incorporation rather than a mere memory function. From this, we get a notion for the meaning of the term "body memory". According to Fuchs such body memory is an "implicit memory [...] based on the habitual structure of the living body" (Fuchs 2012, p. 9). Polanyi considers such incorporated information of the body memory as being part of the proximal term (Polanyi 1968b, 1969). But how can we convert structural information into processible information, and in what format will this converted information then be presented? Finally, how is the integration into other formats being effected?

To gain a more detailed understanding of tacit inferences, it is necessary to follow up with these questions. Polanyi also offers a tacit version for inferential judgement. In his article "Problem solving" Polanyi questions the validity of an inference and arrives at the conclusion that it might be a matter of emotional qualities (Polanyi 1957, p. 91). He presumes a process for the judgement step that he calls "tacit confirmation" (Polanyi 1957, p. 102), serving as an affirmative approval of the conclusion. However, he does not pursue the idea of tacit confirmation in his later writings.

Concerning validity, the idea of an affirmative adoption or rejection of the conclusion can serve as a promising starting point for further considerations (cf. Proposition 4). We can further speculate about affirmative adoption and rejection as a basic principle of human reasoning that also accounts for (unconscious) deductive and inductive inferences. This question, however, would have to be dealt with in a separate paper.

5 Conclusion

The aim of our considerations was to develop an inferential conception of perception. For establishing such a conception, we turned to Peirce, who had been first in systematically linking perception to abductive inferences. However, he did not conceive perception as abduction, because of the impossibility for conscious control. Thus, Peirce spoke of an "extreme case of abduction". An essential question was, if we can succeed in developing a conception that incorporates tacit inferences. Peirce's semiotic conception as well as Polanyi's theory of implicit knowing worked fairly well as a theoretical starting point. By appointing equifunctionality between non-symbolic sign activity and applying logical rules it is possible that inferences can be also realized unconsciously. Furthermore, we pointed out that feelings may be regarded as indices. We illustrated how affirmative judgements can be realized using such indices for judging the validity of an inference. By assuming tacit integration and tacit confirmation Polanyi offers us a case of tacit inferences. A closer analysis revealed that such tacit inferences can be conceived as active being contrary to the cybernetic approach sensu Ashby. Polanyi also employs affirmative aspects for judging the validity of an inference.

The Predictive Processing-Approach seems suitable for finalizing the picture. According to this approach perceptions may be conceived as an inferential triad rather than mere abductive inferences. This approach originating from cognitive brain research accounts for neuronal processing. Bruner's constructivist approach might be useful in completing this view assuming an inferential triad at cognitive scales.

The upshot is setting up an inferential conception for perception would allow us to widen it to other forms of unconscious cognition. That implies, that e.g. gut feelings (cf. Gigerenzer 2008) and intuitive decisions may no longer be considered as random or being reduced to merely biological programs. From this inferential perspective we may now relate to intuitive cognition as skills that are subject for improvement by learning.

An inferential conception encompassing tacit inferences gives way beyond the issues raised in this article. There is e.g. evidence that tacit "elements" are involved in deliberative thinking as well (cf. Nisbett 1977; Suppes 2003; Gigerenzer and Brighton 2009; Dreyfus and Dreyfus 1986; Dreyfus 2006). Starting from perceptions and instinctive cognition as gut feelings this conception may also comprise deliberative thinking and applies for modelling skilled performance too. This article hopefully contributes to the understanding of intuitive cognition reaching from perception to high levels of skilled performance.

References

Alisch, L.-M. (1995). *Pädagogische Wissenschaftslehre. Zum Verhältnis von Ethik, Psychologie und Erziehung*. Münster: Waxmann.

Alisch, L.-M. (2010). Distanz zu groß, Komplexität zu hoch: Naturalisierung vertagt. In S. Schlüter & A. Langewand (Eds.), *Neurobiologie und Erziehungswissenschaft* (pp. 186–208). Heilbronn: Klinkhardt.

Ashby, W. R. (1981). Induction, prediction, and decision-making in cybernetic systems. In R. Conant (Ed.), Mechanisms of intelligence: Ashby's Writings on cybernetics (pp. 313–324). Intersystems Publications.

Bastos, A. M., Usrey, W. M., Adams, R. A., Mangun, G. R., Fries, P., & Friston, K. J. (2012). Canonical microcircuits for predictive coding. *Neuron, 76*, 695–711.

Bohm, D. (1973). Quantum theory as an indication of a new order in physics. B. Implicate and explicate order in physical law. *Foundations of Physics, 3*, 139–168.

Bohm, D. (1990). A new theory of the relationship of mind and matter. *Philosophical Psychology, 3*, 271–286.

Bruner, J. (1957). On perceptual readiness. *Psychological Review, 64*, 123–152.

Bruner, J. (1974). *Toward a theory of instruction*. Cambridge: Harvard University Press.

Brunswik, E. (1934). *Wahrnehmung und Gegenstandswelt. Grundlegung einer Psychologie vom Gegenstand her*. Leipzig: Deuticke.

Brunswik, E. (1948). Statistical separation of perception, thinking, and attitudes. *American Psychologist, 3*, 342.

Büchel, C., Geuter, S., Sprenger, C., & Eippert, F. (2014). Placebo analgesia: A predictive coding perspective. *Neuron, 81*, 1223–1239.

Ciompi, L. (1993). *Die emotionalen Grundlagen des Denkens. Entwurf einer fraktalen Affektlogik*. Göttingen: Vandenhoek & Ruprecht.

Ciompi, L. (1998). *Affektlogik*. Stuttgart: Klett-Cotta.

Ciompi, L. (2011). *Gefühle, Affekte, Affektlogik*. Wien: Picus.

Clark, A. (2012). Dreaming the whole cat: Generative models, predictive processing, and enactivist conception of perceptual experience. *Mind, 121*, 753–771.

Clark, A. (2013). Whatever next? Predictive brains, situated agents, and the future of cognitive science. *Brain Sciences, 36*, 181–253.

Clark, A. (2015). Embodied prediction. In T. Metzinger & J. M. Windt (Eds.), Open MIND: 7(T) (pp. 1–21). Frankfurt: Main.

Damasio, A. (2006). *Descartes' error: Emotion, reason and the human brain*. New York: Harper Collins.

Damasio, A. (2012). *Self comes to mind*. London: Vintage.

Dreyfus, H. L. (2006). Overcoming the myth of the mental. *Topoi, 25*, 43–49.

Dreyfus, H. L., & Dreyfus, S. E. (1986). *Mind over machine: The power of human intuition and expertise in the era of computer*. New York: Free Press.

Duncker, K. (1926). A qualitative (experimental and theoretical) study of productive thinking (solving of comprehensible problems). *Pedagogical Seminary and Journal of Genetic Psychology, 33*, 642–708.

Duncker, K. (1935). *Zur Psychologie produktiven Denkens*. Berlin: Springer.

Duncker, K. (1945). On problem solving. *Psychological Monographs 58* (i—113).

El Kachab, C. (2013). The logical goodness of abduction in C.S. Peirce's thought. *Transactions of the Charles S. Peirce Society, 49*, 157–177.

Friston, K. (2011). What is optimal about motor control? *Neuron, 72*, 488–498.

Friston, K., Adams, R. A., Perrinet, L., & Breakspear, M. (2012). Perceptions as hypotheses: Saccades as experiments. *Frontiers in Psychology, 3*, 1–20.

Fuchs, T. (2012), The phenomenology of body memory. In S. C. Koch, T. Fuchs, M. Summa & C. Müller (Eds.), *Body memory, metaphor, and movement* (pp. 9–22). Amsterdam: John Benjamins Publishing Company.

Gigerenzer, G. (2008). *Gut feelings: The intelligence of the unconscious, reprint*. London: Penguin Books.

Gigerenzer, G., & Brighton, H. (2009). Homo heuristicus: Why biased minds make better inferences. *Topics in Cognitive Science, 1*, 107–143.

Glenberg, A. M. (2008). Embodiment for education. In P. Calvo & A. Gomila (Eds.), *Handbook of cognitive science: An embodied approach* (pp. 355–372). Amsterdam: Elsevier.

Glenberg, A. M., & Kaschak, M. F. (2002). Grounding language in action. *Psychonomic Bulletin and Review, 9*, 558–565.

Goldman, A. (2012). A Moderate approach to embodied cognitive science. *Review of Philosophy and Psychology, 3*, 71–88.

Helmholtz, H. (1867). *Handbuch von der physiologischen Optik*. Leipzig: Voss.

Herbart, J. F. (1892). Von der apperzeption, dem inneren Sinne, und der Aufmerksamkeit. In K. Kehrbach (Ed.), Johann Friedrich Herbarts Sämtliche Werke (Vol. 6, pp. 140–151). Langensalza: Verlag Hermann Beyer & Söhne.

Hohwy, J. (2013). *The predictive mind*. Oxford: Oxford University Press.

Johnson, M. (1987). *The body in the mind: The bodily basis of meaning, imagination, and reason*. Chigaco: University of Chicago Press.

Lakoff, G. (1988). Cognitive semantics. In U. Eco, M. Santambrogio & P. Violi (Eds.), *Meaning and mental representations* (pp. 119–154). Bloomington: Indiana University Press.

Lange, K. (1912). *Apperzeption. Eine psychologisch-pädagogische Monographie. Elfte und zwölfte (Doppel-)Auflage*. Leipzig: Voigtländer.

Magnani, L. (2009). *Abductive Cognition. The epistemological and eco-cognitive dimension of hypothetical reasoning*. Berlin: Springer.

Magnani, L. (2015). Understanding visual abduction. The need of the eco-cognitive model. In L. Magnani & P. Li (Eds.), *Philosophy and cognitive science II, western and eastern studies* (pp. 117–139). Heidelberg: Springer.

Minnameier, G. (2004). Peirce-suit of Truth: Why inference to the best explanation and abduction ought not to be confused. *Erkenntnis, 60*, 75–105.

Minnameier, G. (2010). The logicality of abduction, deduction and induction. In M. Bergman, S. Paavola, A.-V. Pietarinen & H. Rydenfelt (Eds.), *Ideas in Action: Proceedings of the Applying Peirce Conference, Nordic Studies in Pragmatism 1* (pp. 239–251). Helsinki: Nordic Pragmatism Network.

Minnameier, G. (2016). Forms of abduction and an inferential taxonomy. In L. Magnani & T. Bertolotti (Eds.): *Springer handbook of model-based reasoning*. Berlin: Springer (forthcoming).

Mullins, P. (2002). Peirce's abduction and Polanyi's tacit knowing. *The Journal of Speculative Philosophy, 16*, 198–224.

Nelson, T. O., & Narens, L. (1994). Why investigate metacognition?. In J. Metcalfe & A. P. Shimamura (Eds.), Metacognition: Knowing about knowing (pp. 1–25). Cambridge: MIT Press.

Nesher, D. (2001). Peircean epistemology of learning and the function of abduction as the logic of discovery. *Transactions of the Charles S. Peirce Society, 37*, 23–57.

Neuweg, G. H. (2004). *Könnerschaft und implizites Wissen. Zur lehr-lerntheoretischen Bedeutung der Erkenntnis—und Wissenstheorie Michael Polanyis*. Münster: Waxmann.

Nisbett, R. E., & DeCamp Wilson, T. (1977). Telling more than we can know: Verbal reports on mental processes, *Psychological Review 44*, 231–259.

Nunez, R. E. (2008). Mathematics, the ultimate challenge to embodiment: truth and the grounding of axiomatic systems. In P. Calvo & A. Gomila (Eds.), Handbook of cognitive science: An embodied approach (1st ed., pp. 333–353). San Diego: Elsevier.

Pape, H. (1997). Die Sichtbarkeit der Logik und die visuelle Topologie des Geistes. In H. Pape (Ed.) *Die Unsichtbarkeit der Welt* (pp. 378–460). Frankfurt am Main: Suhrkamp.

Pickering, M. J., & Clark, A. (2014). Getting ahead: Forward models and their place in cognitive architecture. *Trends in Cognitive Science, 18*, 451–456.

Polanyi, M. (1957). Problem solving. *The British Journal for the Philosophy of Science, 8*, 89–103.
Polanyi, M. (1962a). *Personal knowledge. Toward a post-critical philosophy*. Routledge: London (corrected edition).
Polanyi, M. (1962b). Tacit knowing: Its bearing on some problems in philosophy. *Reviews of Modern Physics, 34*, 601–616.
Polanyi, M. (1965). The Structure of consciousness. *Brain, LXXXVIII*, 799–810.
Polanyi, M. (1967a). The logic of tacit inference. *Philosophy, 41*, 1–18.
Polanyi, M. (1967b). Sense-giving and sense-reading. *Philosophy, 42*, 301–325.
Polanyi, M. (1968a). Logic and psychology. *American Psychologist 23*, 27–43.
Polanyi, M. (1968b). The Body-mind relation. In W. R. Coulson & C. R. Rogers (Eds.), *Man and the sciences of man* (pp. 85–102). Columbus: Merrill Publishing Company.
Polanyi, M. (1969). On body and mind. *The New Scholasticism, 43*, 195–204.
Pylyshyn, Z. W. (2009). Perception, Representation, and the World: The FINST that Binds, in D. Dedrick & L. Trick (Eds.): *Computation, cognition, and Pylyshyn* (pp. 3–48). Cambridge: MIT Press.
Raftopoulos, A. (2001a) Is perception informationally encapsulated? The issue of the theory-ladenness of perception. *Cognitive Science 25*, 423–451.
Raftopoulos, A. (2001b). Reentrant neural pathways and the theory-ladenness of perception. In *Supplement: Proceedings of the 2000 Biennial Meeting of the Philosophy of Science Association. Part I: Contributed Papers (Sep. 2001). Philosophy of Science 68,* S187–S199.
Rescher, N. (2007). Default reasoning, In D. Jacquette *Handbook of the philosophy of science, philosophy of logic* (pp. 1163–1171). Amsterdam: Elsevier.
Ryle, G. (1976). Improvisation. *Mind, 85*, 69–83.
Schurz, G. (2008). Patterns of abduction. *Synthese, 164*, 201–234.
Stapp, H. (2009). *Mind, matter and quantum mechanics* (3rd ed.). Berlin: Springer.
Stonier, T. (1992). *Beyond information. A natural history of intelligence*. London: Springer.
Stonier, T. (1997). *Information and meaning. An evolutionary perspective*. Berlin: Springer.
Suppes, P. (2003). Rationality, Habits and freedom. In N. Dimitri, M. Basili & I. Gilboa (Eds.), *Cognitive processes and economic behavior* (pp. 137–167). New York: Routledge.
Thaggard, P. (2007). Abductive inference: From Philosophical analysis to neural mechanisms. In A. Feeney & E. Heit (Eds.), *Inductive reasoning: Experimental, developmental and computational approaches* (pp. 226–247). Cambridge: Cambridge University Press.
Tolman, E. C., & Brunswik, E. (1935). The organism and the causal texture of the environment. *Psychological Review, 42*, 43–77.
van Dalen, D. (2001). Intuitionistic logic. In L. Gobble (Ed.), *The Blackwell guide to philosophical logic* (pp. 224–257). Malden: Blackwell.

Abduction in One Intelligence Test. Types of Reasoning Involved in Solving Raven's Advanced Progressive Matrices

Małgorzata Kisielewska, Mariusz Urbański and Katarzyna Paluszkiewicz

Abstract Given that Raven's Advanced Progressive Matrices (APM) as an intelligence test with robust psychometric properties is considered to be a good measure of reasoning ability component of general intelligence, particularly its fluid factor, one would expect that uncovering the determinants of APM performance, especially reasoning patterns, could significantly contribute to understanding of intelligence. Our aim in this study was to identify types of reasoning processes involved in solving Raven's Advanced Progressive Matrices test. To this end we carried out two studies: one involving eliciting verbal protocols in the form of Socratic tutorial dialogues and one involving controlling eye-fixation patterns. Results suggest that hypotheses generation and testing, involved in solving APM tasks, essentially amounts to abductive reasoning.

1 Introduction

Marr and Vision (1982), describing levels of analysis of information processing system, distinguished its three different levels: computational, algorithmic and implementational. Computational level is responsible for describing what does the system do, algorithmic—how to achieve a solution, implementation level—what its' real neural activity is. This distinction is widely applied in the study of cognition. As an inspiration for researchers it can somehow explain a cognitive, or practical, turn in logic (Gabbay and Woods 2005), where algorithmic level becomes crucial, both from descriptive and prescriptive perspective (Stanovich 1999), and implementation level becomes of interest. This new paradigm can be described as a

M. Kisielewska · M. Urbański (✉) · K. Paluszkiewicz
Institute of Psychology, Adam Mickiewicz University,
A. Szamarzewskiego 89AB, 60-568 Poznań, Poland
e-mail: Mariusz.Urbanski@amu.edu.pl

M. Kisielewska
e-mail: Malgorzata.Kisielewska@amu.edu.pl

K. Paluszkiewicz
e-mail: k.paluszkiewicz@amu.edu.pl

L. Magnani and C. Casadio (eds.), *Model-Based Reasoning in Science and Technology*, Studies in Applied Philosophy, Epistemology and Rational Ethics 27, DOI 10.1007/978-3-319-38983-7_23

shift from evaluating reasoning from a normative deductive point of view to a more everyday-life perspective, where reasoning, problem solving and decision making are interpreted as involving similar processes. Our study, aiming at identifying types of reasoning processes involved in solving Raven's Advanced Progressive Matrices test tasks, contributes to this new perspective.

The paper is organized as follows. In the first section we characterize basic differences between well-defined vs ill-defined problems, as the tasks in question are exemplary of the second ones. In the second section we describe Raven's Progressive Matrices test family and outline previous research aimed at revealing cognitive processes involved in solving this kind of tasks, while in the third section we give details on our own two studies. In the fourth section we present our general findings, which are followed by short concluding remarks.

2 Well-Defined Versus Ill-Defined Problems

One of the most popular classification of problems with respect to their characteristics and structure is distinction between so-called well-defined and ill-defined oncs (Reitman 1965) or well-structured and ill-structured ones (Simon 1973). Well-structured problems are a class of tasks which contain all the information that is sufficient to solve them, have convergent answer and single process leading to proper, final solution (Simon 1973, p. 183). The subject solving such a problem has a clearly specified goal, knows all the rules and the criterion for deciding whether the goal has been achievcd or not; moreover, all the pieces of information that are necessary to solve the task can be obtained from the formulation of the problem (Orzechowski et al. 2008, p. 488). Distinction between well-structured and ill-structured problems proposed by Simon (Simon 1973) is determined relatively to the agent of problem solving process, his available knowledge and particular problem to be solved. Problems may be well or ill-structured depending on individual's knowledge and solving abilities. Ill-structured problems differ from ill-defined ones in that ill-defined problem may not allow a clear solution strategy, but may allow single, correct answer (i.e. accepted by qualified experts) (Hayes 1978).

Ill-defined problems, as opposed to well-defined ones, are a class of tasks that are incomplete in terms of specification of components of a problem space—they present a dilemma for planning the solving process (Hayes 1978). It is not obvious for the subject whether information given is relevant to the task or to the goal definition; also, it is not obvious when the goal state conforms to the requirements of the proper solution if it ever does so. Strategies of solving this kind of problems depend on different approaches, but their subjective soundness may vary: addition of some kind of constraints enable subject to clarify at least one component of a problem space (for example by adding a set of rules that can be defined as legal moves, operationally define the goal or dividing initial problem into a few less complicated or better structured sub-problems), but none of those 'moves' guarantee success in finding satisfying solution.

Based on specification of components of the problem space, distinction between well-defined and ill-defined problems lies in the assumption, that solving a well-defined problem can be described as a task of moving from start-state to (known) goal-state by applying appropriate operations under given constraints (Hayes 1978). Such transition can be optimised by analyzing different sequences of moves: planning process within the process of solving a well-defined problem involves evaluation of moves conducted prior to their selection.

Ill-defined problems are often described as a kind of tasks that involve some kind of creativity on the side of a subject, who actively searches for a solution that can be accepted. This active and creative performance is commonly observed in different 'creativity-related' domains and tasks. For example, design activity can be described as one that

> involves the processes of making hypothetical statements and evaluations from beginning to the end of the course of a design (Chan 2014, p. 52),

which corresponds well with Magnani's account of abduction, understood as

> the process of inferring certain facts and/or laws and hypotheses that render some sentences plausible, that explain (and sometimes discover) some (eventually new) phenomenon; it is the process of reasoning in which explanatory hypotheses are formed and evaluated (Magnani 2009, p. 8).

2.1 *Mind Maze*

In one of our previous studies on reasoning and problem-solving (Urbański et al. 2016) we used a game called 'Mind Maze' by Igrology—a card game based on factual stories representative for ill-defined problems. The mechanism of this game can be concisely described as making sense of puzzling information. Players are told a strange or surprising story by a gamemaster. Their main task is to find out why and how the story with twist in the plot happened by asking as few questions as possible. Players have to discover all prescribed key pieces of information only by asking polar questions in turns. The only responses that gamemaster gives to players is 'Yes' and 'No'. Eventually during the gameplay gamemaster can choose to answer with 'Not important' when players ask questions that throw them off the track. After each finished gameplay subjects were asked about the details of their process of discovering explanations for the presented stories. All subjects tried to recreate the path of finding arguments and questions that helped them in the process of discovering key pieces of information leading to the explanation for the story.

The game is somewhat similar to the game of 20 questions, with one important difference. In case of 'Mind Maze' the crucial initial part of each gameplay is reasoning to an interpretation (Stenning and Van Lambalgen 2008), aimed at structuring epistemic goals of the subject; the subject needs to determine what are key pieces of information he or she is to establish (Urbański et al. 2016). In case of the game of 20 questions the structure of an epistemic goal is just given.

One of the main conclusions derived from this study was that during the gameplay in the process of finding explanations for all surprising stories subjects were using abduction as a type of reasoning that was significant for the whole process. Subjects were describing post factum their way of reasoning in terms of justifying their steps of discovering the story plot. In many cases it was impossible to identify which piece of information implied another in the context of discovering the explanation, because subjects were not able to reconstruct the process clearly. Describing the guessing process post factum was somewhat confusing for subjects. They were often fixating on the outcome and asked about the beginnings of the whole process tended to skip describing a few steps of acquiring hints and key information. Nevertheless, we were able to identify two mechanisms on which justification of posing of consecutive questions was based: sifting and funneling of relevant information (Urbański and Żyluk 2016). Both mechanisms may be modeled formally in terms of semantic relations between questions, within the framework of Inferential Erotetic Logic (Wisniewski 2013).

3 Raven's Progressive Matrices

Results of Raven's Progressive Matrices (RPM) tests family, consisting of Standard (SPM), Advanced (APM), and Coloured versions, are generally thought of as a good measure of reasoning ability component of general intelligence, especially its fluid factor, with low level of culture-loading (Blair 2010). In each task (test matrix) subject is asked to identify a missing element that completes the pattern. Each version of Raven's tests was designed for a group of participants of different ability level. The Advanced version used in our research was developed for measuring intellectual ability between adults and adolescents at the high end of intellectual ability. When administered under timed conditions, the APM can also be used to assess intellectual efficiency (Raven et al. 2003).

The structure of tasks in RPM corresponds to the one of ill-defined problems. Test-takers solving this nonverbal tasks know only that they need to find patterns to define rules for each matrix in order to finally come up with the proper solution and identify the piece that completes the martix. Only one piece from eight given possible answers fits the matrix and fully corresponds with other elements.

3.1 Raven's Advanced Progressive Matrices (APM)

Raven's Advanced Progressive Matrices (APM) contains 48 test items: presented as a training set of 12 items (Set I), and another of 36 items (Set II). In each test item, the subject is asked to identify the missing element that completes a pattern. APM consists of test items that are 3×3 matrix of figural stimuli organized according to latent rules. Subject has to choose one piece that completes the matrix from eight possible answers. Items become increasingly difficult as progress is made through each set.

Raven's Advanced Progressive Matrices test demonstrates very robust psychometric properties. As for internal consistency reliability, Cronbach's alpha coefficient for APM typically fits in the range 0.75–0.95. As for the content validity, APM has been used as a measure of eductive ability:

> the ability to evolve high-level constructs which make it easier to think about complex situations and events (Raven et al. 2003, p. G8).

It has been also described as a test of analytic intelligence:

> the ability to reason and solve problems involving new information, without relying extensively on an explicit base of declarative knowledge derived from either schooling or previous knowledge (Carpenter et al. 1990, p. 404).

It has been shown that there is medium to strong correlation between APM results and results of tests of deductive reasoning of different difficulties, with no collinearity (Urbański et al. 2013).

3.2 Cognitive Processes in Solving Raven's Matrices

There were some attempts to reveal the cognitive processes underlying RPM's solutions. Findings based on both behavioral and neuroimaging studies suggest presence of two qualitatively different strategies (Kunda et al. 2012). It is claimed that the main difference in performance between different test items is related to the type of mental representations involved in their solutions—the superiority of visual strategy (based on iconic representations) versus verbal strategy (based on propositional representations). Evidence from factor analyses of SPM and APM shows that test items combined in RPM test family can be divided into various categories that differentiate test-takers well in terms of preference for either verbal or visual strategy. Moreover this kind of classification divides most of RPM test items into groups that can be described as ones favouring one of the two strategies, or 'favouring both' (DeShon et al. 1995).

Despite the evidence for accuracy of classification of RPM test items based on type of processing, most computational accounts on RPM test family solutions rely on translating visual data into propositional representations that are used as an input for computational algorithmic reasoning conducted on RPM test items (Kunda et al. 2012). Some of the most significant computational models of cognitive processes involved in solving APM tasks are:

- model based on hand-coded symbolic description of tasks with predefined rules, by Carpenter et al. (1990),
- spiking neuron model based on translated propositional representations, by Rasmussen and Eliasmith (2011),
- model based on image transformations with no usage of propositional form, by Kunda and colleagues (affine and fractal model of solving scanned images of APM test items with no hand-coded propositions) (Kunda et al. 2012, 2013).

Carpenter et al. (1990) devised two computer simulations, FairRaven and BetterRaven (based on hand-coded propositions), which express the difference between good and extremely good performance on APM. Kunda et al. (2010, 2013) described the Affine model and Fractal model that simulate modal reasoning by using only iconic visual representations together with affine and set transformations over these representations to solve a given RPM and APM problems. Fractal model performs at levels equivalent to the 95th percentile for set I (solved all 12 test items correctly) and 75th percentile for set II (26 out of 36 items were correct) for test takers aged from 20 to 62 years-old for APM, which

> align strongly with evidence from typical human behaviour suggesting that multiple cognitive factors underline problem solving on the APM, and in particular, that some of these factors appear based on visual operations (Kunda et al. 2012, p. 1832).

Another aspect of cognitive processing that is commonly taken into account in research conducted on APM task solutions is a question of task management combined with other abilities. Most notably, Carpenter and colleagues showed that APM measures

> the common ability to decompose a problem into manageable segments (…), the differential ability to manage the hierarchy of goals and subgoals (…), and the differential ability to form higher level abstractions (Carpenter et al. 1990, p. 429).

Based on this evidence it can be assumed that high-end scoring subjects are skilled in combining the execution of all of those three abilities within the process of solving APM test items.

4 Two Short Studies

We conducted two studies based on VA-APM (Visual-Analytical APM)—shortened 14-item version of APM Set II proposed by Vigneau and Bors (2008). VA-APM is based mainly on the taxonomy and results derived from DeSchons' work (DeSchons et al. 1995) that is taking into account item processing type aspect of cognitive processing in APM. VA-APM subtest proposed by Vigneau and Bors consists of even number of test matrices classified by the authors as visual and analytical stimuli. Test items classified by DeSchon and colleagues as favoring both strategies were not included in VA-APM. VA-APM as a set of selected items

> represents various types of items and a broad range of difficulties (Vigneau and Bors 2008, p. 264)

comparable to full version of APM. Such short forms of APM retain psychometric properties of original version of the test (Vigneau and Bors 2008), also its predictive validity (Bors and Stokes 1998).

4.1 *Study 1*

4.1.1 Method

First study was conducted with APM in nonverbal condition (that is, a procedure identical to the standard procedure for APM) and VA-APM in verbal protocols condition. Experimental session consisted of three parts—completion of APM with unlimited time and (after a small break) completion of shortened version of APM—VA-APM with recorded verbal report (also without time limitation) preceded by verbal protocol training. Experimental session ended with short interview with participant concerning the process of solving VA-APM with verbal report.

4.1.2 Participants

10 volunteers took part in the first study. Participants were students and graduates from A. Mickiewicz University in Poznań ranged from 21 to 28 years in age (M = 25; SD = 1.77), 6 of them were women.

4.1.3 Procedure

Session was conducted individually with each participant during one session. Participants were instructed that experimental session consists of three parts: APM test administered with standard procedure, verbal report training and VA-APM with recorded verbal protocol with interview. Small break was included between completing APM and second part with verbal protocol condition. Participants had unlimited time for completing each part of the study.

First subjects had to complete the Raven's Advanced Progressive Matrices test (Set I & Set II) in unlimited time condition administered with instructions that were read by the experimenter.

In the second part of the study we employed verbal protocols in the form of Socratic tutorial dialogues, as proposed by Stenning and Van Lambalgen (2008). First, for training and instructional purposes, participants were asked to solve two simple tasks with additional usage of verbal protocol and, as proposed by Bors and Strokes (1998), completed two test-items selected from second part of Set I APM. After completing this step, experimenter informed that from this point on session will be recorded. Participants were instructed to report the whole process of solving each of test-items from VA-APM. After completing VA-APM test participants were asked a few additional questions concerning methods and strategies used during solving those test-items and any difficulties they encountered completing this part of the study.

4.1.4 Results

The differences in solutions to the very same items in the standard condition vs. in the verbal protocol conditions confirmed Lane and Schooler (2004) findings that describing one's thought processes or analyzing a judgment may impair performance. Also, solution times were longer in the case of verbal protocol condition.

Verbal reports obtained from higher scoring participants were brief and less specific, especially for easier test-items (first half of VA-APM), as compared to lower scoring participants. Reports of higher scoring participants did not contain many details describing properties of elements of matrices or rules that were generated during the solving process for rows or columns. Higher scoring participants were describing only general relations between elements in the matrix in an abstract matter, for example:

> The second figure [in the row] is subtracted from the first figure [from this row] and that what is left is the answer, so the answer is number eight' (subject 05, complete verbal report for 3rd test-item from VA-APM).

Higher scoring participants reported the solution process in 2 or 3 sentences (including reporting the chosen answer). The process of finding solution and picking the answer was somewhat 'faster' than thinking-aloud and reporting it.

Lower scoring participants tended to report more specific item characteristics and rules describing elements in columns and rows even for test-items from the first half of VA-AMP. Their reports contain neatly described relations noticed between adjacent elements in the matrix. Reports are still brief, but not so general as verbal reports from higher scoring subjects.

Verbal reports from both higher scoring and lower scoring participants confirmed that subjects kept on focusing on finding corresponding regularities between two adjacent elements in the matrix at a time to identify a plausible rule, then jumping to third element to test correctness of the possible rule. This subprocess of rule generation and evaluation is an integral part of constructive matching strategy (Wiley et al. 2015). Participants were evidently aware that they were employing this strategy and that it has some virtues. When asked about stages of solving process in VA-APM during the interview part of the session, participants pointed at matrix elements comparison in pairs as most basic step for the whole process. There were a few cases where a test-item was simple enough for participants to point at the answer almost immediately. In this situation participants reported that after a brief glimpse on the matrix, crucial rule 'popped up' so that they could point at the answer really quickly.

Participants used also falsificatorial exclusion strategy (Gittler and Wuerfeltest 1990), but only as a support for constructive matching one. Both verbal reports and interviews showed that falsificatorial exclusion strategy is a supporting strategy for solving APM test-items. Participants used it only in situations where 'pure' constructive matching strategy was not sufficient. Subjects generated some rules that enabled exclusion of possible answers, narrowing the answer pool to 2 or 3 promising answers, then used falsificatorial exclusion strategy only for evaluation of these cases as possible solutions. Falsificatorial exclusion strategy was reported only for test-items from second half of the VA-AMP.

4.2 *Study 2*

4.2.1 Method

In the second study we controlled eye-fixation patterns using eye-tracker RED SMI 60 Hz. Sessions were conducted individually at the Action and Cognition Laboratory, Institute of Psychology, Adam Mickiewicz University. Participants solved VA-APM in nonverbal condition. Items presented in this study were obtained from paper version of Set II APM and adjusted for the experimental requirements with standard RED SMI software. Based on Williams and McCord (2006) research we assumed that digitalized form of VA-APM is characterized with the same psychometric properties as paper equivalent.

4.2.2 Participants

Study was administered to a group of 10 volunteers. Participants were students and graduates from A. Mickiewicz University in Poznań ranged from 19 to 28 years in age ($M = 24.7$; $SD = 3.13$), 4 of them were women.

4.2.3 Procedure

Participants were instructed by the experimenter that the session consists of three parts: calibration of the eye-tracker, short training part and experimental procedure that contains 14 test-items analogical to the items from the training part. All instructions and information concerning completing each part were provided on screen. We modified standard instructions for APM to eliminate any specific hints suggesting method of solving test-items (original instruction suggests a 'way of looking' at the matrix) and reformulated sentences so that instruction did not suggest any starting point. We wanted to avoid any suggestions so that participants could solve each test-item in most natural, undisturbed way which was crucial for this study.

For instructional purposes, as suggested by Bors and Strokes (1998), participants completed two items from APM Set I (matrices 6th and 9th, which were most suitable in terms of elements salience and difficulty level). Additionally, participants were instructed to report aloud the number of selected answer for each item and to choose the same number from a list by clicking on it (the list appeared on the next screen seen after they pressed a key on the keyboard).

After completing the trial part, participants were administered the VA-APM version of Set II APM with the modified instruction. The answering procedure remained the same as in the trial part. Participants had unlimited time for completing each part of the session.

4.2.4 Results

The results were analyzed based on the video recordings of eye gaze plots (eye-fixation paths recorded in real time for each matrix and each participant individually) and heat maps. Heat maps contain eye fixation length averaged for each test-item divided into two groups for each matrix—separating data between correct and incorrect solutions. The 'warmer' the colour, the longer participants were focusing in this area (see: Fig. 3). While eye gaze plots (see: Figs. 1 and 2) allow to analyze the solving process step by step, heat maps show the overall focus tendencies; averaged for participants, they show regularities in 'attractiveness' of specific areas in the matrix.

Eye gaze plots analyses allowed for the conclusion that higher scoring participants were looking at the matrices mostly in a specific order—by jumping between two adjacent items in the matrix focusing on identification of item properties that are potentially meaningful and searching for regularities. Most participants started from top left corner and (eventually after brief look through a whole row or column) kept jumping between two elements at a time. We could easily observe subprocesses within construction of a potentially meaningful rule. Subjects were focusing on two elements then jumping to third in a group (if the elements were in one row

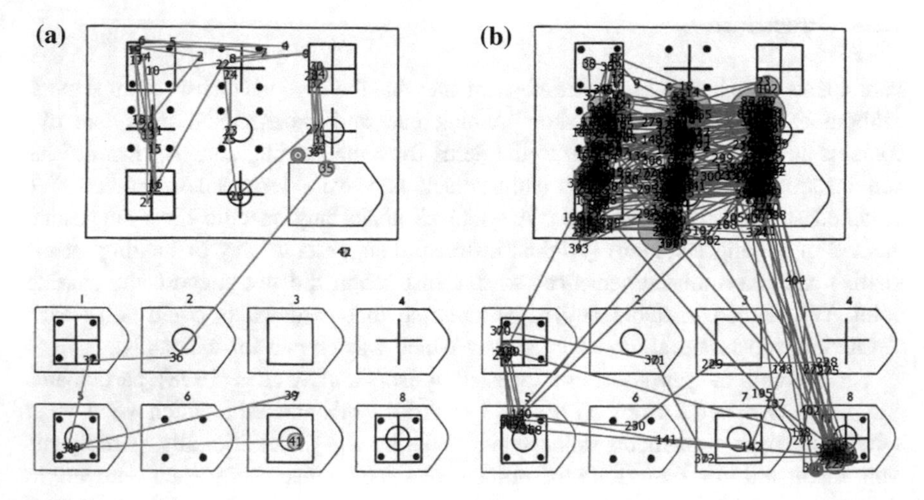

Fig. 1 Correct versus incorrect solution—a comparison of effects at the final stage of the problem solving process between a skillful, high-scoring participant (**a**) and a lower-scoring one (**b**), who overdid at the rules generation stage [based on performance of subject 04 (correct) and subject 10 (incorrect), task no. 22]

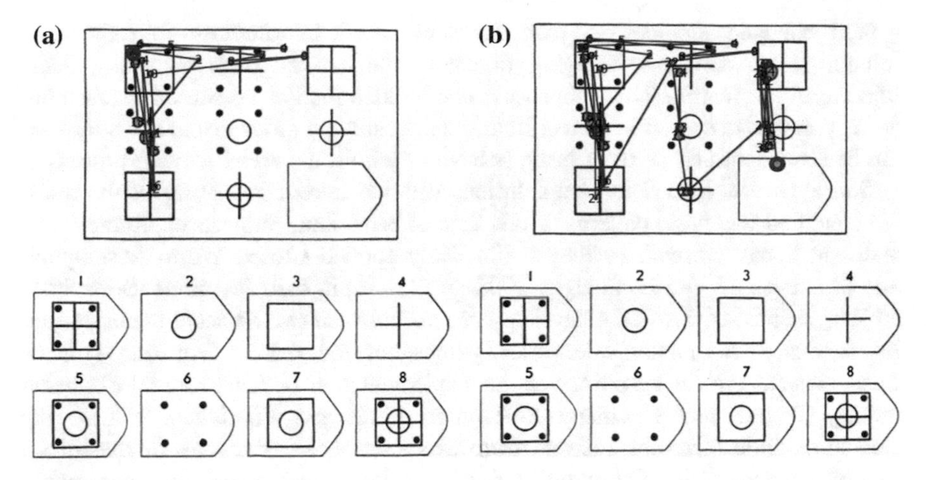

Fig. 2 Screenshots show how high-scoring subjects analyse one of the matrices (based on subject 04 performance, task no. 22). Starting from *top left corner* he jumped methodically between the first element and the one below, then tested generated rule with third element in the column (**a**). After setting up the first rule, he returned to the second analyzed pair of items (**b**)—*top row*, then analyzed another pair in the second column and confirmed rule jumping to the third item in this column, next moved to the last column and analyzed items in pairs again

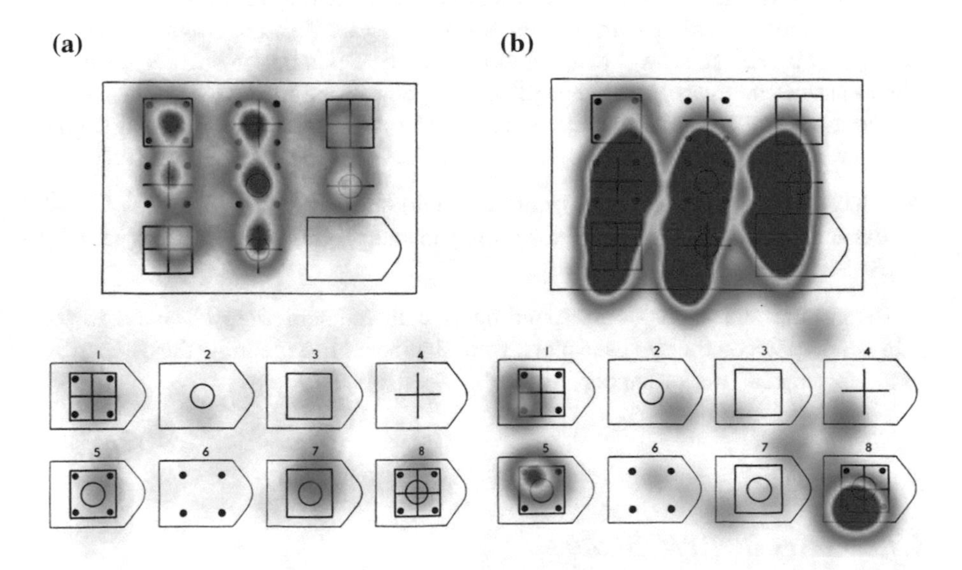

Fig. 3 Heat maps: averaged eye fixation length for correct solutions (**a**) versus incorrect solutions (**b**). Both shows that matrix was analysed 'in columns', but with slightly different patterns and the outcome in answers area (task no. 22)

—third ‘checked’ element was from the same row, if in columns—from the same column). Last step was verifying potential rule derived from analyzing items adjacent in the matrix. This subprocess can be identified as constructive matching strategy. It was observed in all test-item problem solving processes in this study—it can be also identified as most basic behaviour within the APM solving process.

Some subjects tended to give solutions without thorough analyses of the tasks. We identified two basic patterns of this kind of behaviour. Subjects exhibiting it we shall call ‘hasty closers’, as they ineffectively applied Closed World Assumption (see Stenning and Van Lambalgen 2008, pp. 33–34) in their reasoning. Sometimes subjects, instead of detailed analysis proceeded between two adjacent items, peeped the answer area quickly after brief glimpse at the matrix. For one type of ‘hasty-closers’ typical behavior was an implementation of falsificatorial exclusion strategy (Preckel and Thiemann 2003, after: Gittler and Wuerfeltest 1990)—they tried to exclude incorrect answers from available pool rather than to construct a possible correct one in the first place. Falsificatorial exclusion strategy is commonly reported as less effective than constructive matching strategy (Hayes 1978). Another type of ‘hasty-closers’ were participants who had problems with rule generation processes, evaluation of rules or simply were not so skilled in managing goals and subprocesses (or they overdid at the stage of properties identification, rules generation or didn’t manage to maintain enough accurate ‘threads’). This group used falsification exclusion strategy as an alternative to constructive matching strategy which was their first choice. Most commonly, after failing at some point of the process of problem solving (in the outcome of ‘closest fit’ to not-so-good educated rules), they picked an answer that was somehow corresponding to already generated rules—answer that was ‘good enough’ at this point of the process.

Carpenter et al. (1990) indicates two subprocesses that distinguish among higher and lower scoring subjects:

- eduction of abstract relations (rule induction) and
- dynamic goal management (in working memory) [see also (Verguts and De Boeck 2002)].

Behaviours and actions corresponding to both of them were observed in our study and, based on the eye gaze plots, were significant in explaining the differences between subjects’ performance.

5 Discussion

5.1 Biases in RPM Studies

Despite the wide variety of studies concerning strategies, methods and different aspects of solving RPM and RPM-like tasks, many of those studies remain biased. The main problem stems from typical methods used in research concerning this type of test-items.

As for computer simulations used in RPM test family research one of the most popular approaches is based on employing some kind of propositional representations [hand-coded or derived from automatically computed visual data, like CogSketch generating spatial representations (Forbus et al. 2010)], as opposed to less popular iconic visual representations-based models.

Similar to computer model-based studies most experiments with human test-takers have also their own biases. Studies involving participants solving RPM and RPM-like test items are favouring verbal reports as a method for obtaining data. Common problem with verbal report protocols in RPM-based studies is verbal overshadowing effect known as

> the act of verbal reporting (...)[that] biases individuals towards using verbal strategies and/or impairs their use of visual strategies (Kunda et al. 2012, p. 1829).

DeSchon et al. (1995) found that verbal report protocols method impairs accuracy of execution in solving APM tasks in about half of the test items—specifically those that are classified as favouring visual strategies. Verbal overshadowing causes often problems in analogical retrieval [noted even in experiments based only on verbal material (Lane and Schooler 2004)], impairs noticing deeper connections and similarities between artefacts and causes skimming over task surface instead of deeper problem processing. This phenomenon is crucial in analysing processes involved in solving RPM and RPM-like items. Test-takers who did not verbalize thoughts during solving RPM tasks are more likely to use different strategies (suitable for particular task or more convenient for them as a general strategy) instead of being focused on using mostly verbal strategies derived from verbalising the whole solving process at the expense of working on information that is less easily verbalized.

Verbal overshadowing effect was observed mostly in lower scoring participants (Carpenter et al. 1990). Carpenter and colleagues noted also that higher error rates which were observed in group of participants who were reporting verbally the process were correlated with higher response times and higher number of rules that were involved by the subjects in the process.

Both studies reported in this article were aimed at obtaining insights into types of reasoning processes involved in solving APM test tasks. Verbal protocols and eye-tracking used as research methods complemented each other well, and employing both of them was aimed at compensating for possible biases that are commonly present in similar studies.

5.2 *Closed World Assumption in APM Solutions*

Our main finding is that Closed-World Assumption (CWA) was crucial for the whole process of solving test items in APM. The process of identifying properties, generating enough and accurate rules that account for the particular problem was fundamental for finding solutions. That this process was not always monotonic is witnessed by the 'But wait' moments of the subjects, identifiable in the verbal protocols mostly. Quite often they postponed tentative solution in favour of another one because of previously unnoticed property of test items. Higher scoring subjects applied CWA in a way which enabled deductive reasoning as an effective mean of hypotheses testing and proving (typically they used reasoning based on modus tollens and modus ponens). Those participants who scored lower as opposed to high-scoring subjects were mostly hasty closers, in the sense we described above (see Sect. 3.2.4). However, some of them did the opposite and overdid rule generation. With too many of possible rules they were unable to establish sound criteria of choice between them. Lower scoring subjects also tried to employ deduction in the evaluation process but ended up accepting solutions which were 'closest' to not-so-good educted rules. This was especially visible for test-items from the second half of Set II APM, where the number of rules and elements in a test-item (sometimes also juxtaposed with element salience) increases, raising the difficulty level and undermining the process of finding correspondence between elements in the matrix (Roberts et al. 2007). The ability to apply CWA in an effective way may be interpreted as a crucial part of an ability that is necessary for solving APM (and RPM in general) test items seen as analogous to ability of problem-solvers described by Simon (1973), which contributes to identify particular problem as ill-structured or well-structured one. We can also stipulate that application of CWA can be perceived as a form of epistemic self-regulation [by analogy to Competence ← – → Procedural processing model; See Ricco and Overton (2011)], but simple identification of epistemic self-regulation with goal management would be a bit too trivial.

5.3 *Solving APM as a Case of Abduction*

Combining interpretation from previous studies with our own results from both studies we may suggest that hypotheses generation and testing, involved in solving APM tasks, essentially amounts to abductive reasoning. Reasoning performed by the subjects fits well into the explanatory-coherentist model of abduction (Urbański 2009), and can also be interpreted as a case of model-based creative abduction

> related to the exploitation of internalized models of diagrams, pictures, etc. (Magnani 2009, p. 11)

The problem-solving processes observed in both studies can be also summarized as a kind of reasoning analogical to creative abductions described by Schurz as

> conceptual abstraction based on isomorphic or homomorphic mapping (Schurz 2008, p. 217).

It is doubtful whether observed processes can be interpreted as an example of explanatory abduction. Rules generated within a problem solving processes in APM

> are secured by instrumental considerations and accepted because doing so enables one's target to be hit (Magnani 2009, p. 77)

which indicates them as instrumental abduction example.

The process of evaluation of rules (or sets of rules), which were generated on the basis of observed regularities in order to find a prospective one, is based mainly on different forms of coherence (including entailment as an evaluation tool). All abductive hypotheses that came into being in the process were evaluated in terms of their coherence—the one that maximizes constraints satisfaction becomes an accurate rule (see Thagard 2000; Thagard and Verbeurgt 1998). Maximization of coherence is based on consistency of the group of rules with all elements in the matrix and its delivering the best possible fit for one particular element from an answer pool that completes the matrix.

Previous research concerning reasoning processes involved in solving RPM tasks (APM included) combined with our study suggest that those processes are compound, that is, subsuming a few particular forms of inference (Ajdukiewicz 1974). Abduction understood as a compound form of reasoning (Urbański 2009, pp. 161–169) accounts well for the set of their building elements, which consists of the process of identification of meaningful item properties that leads to regularity search, the generation of rules and testing based on available data (derived both from matrix elements and pool of possible answers). One striking feature of APM tasks solving is demonstrably incremental nature of the processing.

It should be noted however, that, although we found all this elements in all of the subjects' solutions, the exact connections between them (order of execution, prioritization of information obtained, etc.), and, as a result, the exact structure of reasoning performed, differed from subject to subject.

6 Conclusion

Overall, our results are consistent with studies concerning processes involved in solving RPM tests family tasks. Awareness of possible biases related to applications of particular research methods is crucial in identifying types of reasoning processes involved in solving RPM tasks. Combining results from verbal protocol analysis and controlling eye-fixation patterns turned out to be an effective way of obtaining data that deliver insightful observations and meaningful examples for reconstructing complexities of reasoning patterns of problem solving processes in APM. Our claim that hypotheses generation and testing, involved in solving APM tasks, essentially amounts to abductive reasoning is supported both by eye-tracking and verbal protocols data. In making sense of puzzling APM task the subjects employed Closed World Assumption, with coherence as the main criterion against which the hypotheses were evaluated. Nevertheless, as usual, further research are needed, in particular in order to obtain more comprehensive insights into the problem of individual differences not only in the level of performance in APM tasks, but also in the structure of performed reasoning.

Acknowledgments Research reported in this paper were supported by the National Science Centre, Poland (DEC-2013/10/E/HS1/00172).

References

Ajdukiewicz, K. (1974). *Pragmatic logic* (O. Wojtasiewicz, trans.). Dordrecht: D. Reidel.

Blair, C. (2010). Fluid cognitive abilities and general intelligence. In R. M. Lerner, W. F. Overton, A. M. Freund & M. E. Lamb (Eds.), *The handbook of life-span development*. Wiley.

Bors, D., & Stokes, T. (1998). Raven's Advanced Progressive Matrices: Norms for first-year university students and the development of a short form. *Educational and Psychological Measurement, 58*(3), 382–398.

Carpenter, P., Just, M., & Shell, P. (1990). What one intelligence test measures: a theoretical account of the processing in the raven progressive matrices test. *Psychological Review, 97*(3), 404–431.

Chan, C. S. (2014). *Style and creativity in design* (Vol. 17). Berlin: Springer.

DeShon, R., Chan, D., & Weissbein, D. (1995). Verbal overshadowing effects on Raven's Advanced Progressive Matrices: Evidence for multidimensional performance determinants. *Intelligence, 21*(2), 135–155.

Forbus, K., Lovett, A., & Usher, J. (2010). A structure-mapping model of Raven's Progressive Matrices. *Proceedings of CogSci, 10*, 2761–2766.

Gabbay, D., & Woods, J. (2005). The practical turn in logic. In G. M. Gabbay & F. Guenthner (Eds.), *Handbook of Philosophical Logic* (2nd Ed., pp. 15–122). Berlin: Springer.

Gittler, G., & Wuerfeltest D. (1990). Ein rasch-skalierter test zur messung des raeumlichen vorstellungsvermoegens. Theoretische grundlagen und manual. Beltz Test, Weinheim.

Hayes, J. (1978). *Cognitive psychology: Thinking and creating*. Belmont: Dorsey.

Kunda, M., McGreggor, K., & Goel, A. (2010). Taking a look (literally!) at the raven's intelligence test: Two visual solution strategies. In *Proceedings of 32nd Annual Meeting of the Cognitive Science Society, Portland.*

Kunda, M., McGreggor, K., & Goel, A. (2012). Reasoning on the Raven's Advanced Progressive Matrices test with iconic visual representations. In *34th Annual Conference of the Cognitive Science Society* (pp. 1828–1833).

Kunda, M., McGreggor, K., & Goel, A. (2013). A computational model for solving problems from the raven's progressive matrices intelligence test using iconic visual representations. *Cognitive Systems Research, 22*, 47–66.

Lane, S., & Schooler, J. (2004). Skimming the surface verbal overshadowing of analogical retrieval. *Psychological Science, 15*(11), 715–719.

Magnani, L. (2009). *Abductive cognition: The epistemological and eco-cognitive dimensions of hypothetical reasoning* (Vol. 3). Berlin: Springer Science and Business Media.

Marr, D., & Vision, A. (1982). *A computational investigation into the human representation and processing of visual information*. WH San Francisco: Freeman and Company.

Orzechowski, J., Nęcka, E., & Szymura, B. (2008). *Psychologia poznawcza*. Wydawnictwo Szkoly Wyższej Psychologii Spolecznej "Academica".

Preckel, F., & Thiemann, H. (2003). Online-versus paper-pencil version of a high potential intelligence test. *Swiss Journal of Psychology/Schweizerische Zeitschrift für Psychologie/ Revue Suisse de Psychologie, 62*(2), 131.

Rasmussen, D., & Eliasmith, Ch. (2011). A neural model of rule generation in inductive reasoning. *Topics in Cognitive Science, 3*(1), 140–153.

Raven, J., Raven, J., & Court, J. (2003a). *Manual for Raven's Progressive Matrices and vocabulary scales (Section 1: General overview)*. San Antonio, TX: Harcourt Assessment.

Raven, J., Raven, J., & Court, J. (2003b). *Manual for Raven's Progressive Matrices and vocabulary scales (Section 4: Advanced progressive matrices)*. San Antonio, TX: Harcourt Assessment.

Reitman, W. (1965). *Cognition and thought: An information processing approach*. New York: John Wiley & Sons, Inc.

Ricco, R., & Overton, W. (2011). Dual systems competence–procedural processing: A relational developmental systems approach to reasoning. *Developmental Review, 31*(2), 119–150.

Roberts, M., Meo, M., & Marucci, F. (2007). Element salience as a predictor of item difficulty for Raven's Progressive Matrices. *Intelligence, 35*(4), 359–368.

Schurz, G. (2008). Patterns of abduction. *Synthese, 164*(2), 201–234.

Simon, H. A. (1973). The structure of ill structured problems. *Artificial Intelligence, 4*(3–4), 181–201.

Stanovich, K. (1999). *Who is rational? Studies of individual differences in reasoning*. Park Drive: Psychology Press.

Stenning, K., & Van Lambalgen, M. (2008). *Human reasoning and cognitive science*. Cambridge: MIT Press.

Thagard, P. (2000). *How scientists explain disease*. Princeton: Princeton University Press.

Thagard, P., & Verbeurgt, K. (1998). Coherence as constraint satisfaction. *Cognitive Science, 22* (1), 1–24.

Urbański, M. (2009). Rozumowania abdukcyjne. Modele i procedury. Adam Mickiewicz University Press, Poznań.

Urbański, M., Paluszkiewicz, K. & Urbańska, J. (2013). *Deductive reasoning and learning: A cross-curricular study*. Research report. Institute of Psychology, Adam Mickiewicz University.

Urbański, M. & Żyluk, N. (2016). *Sets of situations, topics, and question relevance*. Research report. Institute of Psychology, Adam Mickiewicz University.

Urbański, M., Żyluk, N., Paluszkiewicz, K., Urbańska, J. (2016). A formal model of erotetic reasoning in solving somewhat ill-defined problems. In D. Mohammed & M. Lewiński (Eds.), *Argumentation and Reasoned Action. Proceedings of the 1st European Conference on Argumentation*. London: College Publications (in print).

Verguts, T., & De Boeck, P. (2002). The induction of solution rules in Raven's Progressive Matrices test. *European Journal of Cognitive Psychology, 14*(4), 521–547.

Vigneau, F., & Bors, D. (2008). The quest for item types based on information processing: An analysis of Raven's Advanced Progressive Matrices, with a consideration of gender differences. *Intelligence, 36*(6), 702–710.

Wiley, J., Loesche, P., & Hasselhorn, M. (2015). How knowing the rules affects solving the Raven Advanced Progressive Matrices test. *Intelligence, 48*, 58–75.

Williams, J., & McCord, D. (2006). Equivalence of standard and computerized versions of the Raven Progressive Matrices test. *Computers in Human Behavior, 22*(5), 791–800.

Wiśniewski, A. (2013). *Questions, inferences, and scenarios*. London: College Publications.

Thought Experiments as Model-Based Abductions

Selene Arfini

Abstract In this paper we address the classical but still pending question regarding Thought Experiments: how can an imagined scenario bring new information or insight about the actual world? Our claim is that this general problem actually embraces two distinct questions: (a) how can the creation of a just imagined scenario become functional to either a scientific or a philosophical research? and (b) how can Thought Experiments hold a strong inferential power if their structures "do not seem to translate easily into standard forms of deduction or induction"? (Bishop in Philos Sci 66(4):534–541, 1999). We contend that, in order to answer both questions, we should consider the relation between the creation of the imagined scenario and the inferential power of Thought Experiments. Specifically, we will analyze Thought Experiments from an eco-cognitive point of view as goal-oriented objects, explaining their inferential power considering their generation as the result of abductive cognition and the construction of an imagined scenario as a process of scientific modeling. This will lead us to consider the creation of a Thought Experiment as a case of sophisticated model-based abduction.

1 Introduction

How can an imagined scenario bring new information or insight about the actual world? Variously and frequently re-addressed, this question encompasses the main challenges faced by philosophers in the analysis of Thought Experiments (hereafter TEs). Despite its apparent clear focus, this question (which could be considered the *original question*) embraces two secondary but still relevant issues regarding TEs:

(a) how can a just imagined scenario become highly functional to either a scientific or a philosophical investigation?

S. Arfini (✉)
Department of Philosophy, Education and Economical-Quantitative Sciences,
University of Chieti and Pescara, Chieti, Italy
e-mail: selene.arfini@unich.it

L. Magnani and C. Casadio (eds.), *Model-Based Reasoning in Science and Technology*, Studies in Applied Philosophy, Epistemology and Rational Ethics 27, DOI 10.1007/978-3-319-38983-7_24

(b) how can thought experiments hold a strong inferential power if their structures "do not seem to translate easily into standard forms of deduction or induction"? (Bishop 1999).

Many philosophical investigations have been divided over either one of these two focal points regarding TEs.

In order to solve the problem presented by the issue (a), numerous philosophers have investigated the cognitive modalities within which the scenario is created. Brown and others have considered the scenario a result of an a priori reasoning (Brown 1999a), some have taken the chance to discuss the properties of visual models (e.g. Gendler 2000 and Bishop 1999), and others have focused on whether the scenario can or cannot be constructively reenacted using computational simulations (Nersessian et al. 2012; Di Paolo et al. 2000; Skaf and Imbert 2013). Addressing the issue (b), many other authors considered the analysis of TEs inferential structure the principal direction of inquiry in order to prove or attack TEs reliability. For example, according to the proponents of the "intuition account", the inferential power of TEs is partially explained by the fact they are triggers of (or vessels for) plain argumentations and they can be defined as "intuition pumps", (Dennett 1984; Bealer 1998; Thagard 2014); a similar consideration prodded Norton to label TEs as "just" disguised deductive arguments, (Norton 2004) and drove Häggqvist and Wilkes to focus on the logical properties of the narrative shape of TEs (Häggqvist 1996; Wilkes 1999). A third category of authors mainly consider the comparison between TEs and empirical or "real" experiments and from there they deal with either the issue *a* or *b*. In our opinion, also those who follow this path end to focus on either the creation of the scenario or the inferential activity enabled by a thought or a real experiment. The first group generally looks at the possibility of the *recreation* of the scenario in the empirical environment rather than simply imaging it, for instance Buzzoni (2013). The second group, instead, proposes to already consider the logical structure of a TEs the abstract reenactment of a real experiment, e.g. Lennox (1991), Sorensen (1992).

In this paper we aim at reflecting again over the *original question* considering the *a* and *b* issues as essentially intertwined. We will consider the construction of an imagined scenario as deeply connected to the particular inferential process which underlies the creation of a TE, and at the same time the inferential structure of a TE as depending on the development of the scenario. In order to clearly present the perspective from which our investigation begins, we shall see the *original question*: "How can an imagined scenario bring new information or insight about the actual world?" as an abbreviation of a more specific one: "what is the relation between the creation of imagined scenarios and the inferential reasoning performed in the construction of TEs that constitutes them as effective and highly functional tools in the scientific and philosophical research?"

Hence, to consider the newly presented *original question*, in the first section we will examine the reasons why many epistemologists saw TEs as "exceptional tools" in the history of thought. Then we will present our different evaluation, motivated by the employment of an eco-cognitive and a bottom-up perspective, from which we will speak about TEs as goal-oriented objects and highly functional problem-solving methods. In the second section we will analyze the process of creation of TEs as a case of *abductive cognition*, employed in order to solve or better understand intricate issues. Then, in the third section, we will proceed to describe the construction of the imagined scenario as a *mental modeling* of theoretical possibilities. Finally, in the fourth section, we will explain the relation between the inferential pattern from which TEs emerge and the development of the imagined scenario as *converging* on the construction of TEs as *model-based abductions*.

Shall we start, then, speaking about the controversial exceptionality of TEs and the unexplored advantages of adopting a bottom-up and eco-cognitive point of view in their analysis.

2 Thought Experiments and Ordinary Reasoning

2.1 The Mystique of Thought Experiments: Exceptional Tools of Armchair Philosophy

Despite TEs played an important role in the history of thought, many modern epistemologists consider them exceptional—both unusual and extraordinary—tools in the development of science and philosophy. Indeed, TEs have been defined with words of wonder, even enthusiasm, when appraised as much as with terms of suspect and irritation when attacked. As David Gooding wrote, therc is a sort of "thought experimental mystique" that maintained through the years of analysis of their methodology (Gooding 2002),[1] and the mystique regarding TEs is actually understandable. TEs do not require particular instrumentations nor a specialized environment to be performed. Scientists and philosophers have used different TEs in various fields of study in order to prove theories, in order to argument against theses and to find easy ways to comprehend and express particular issues. As well-described by Rowbottom, "they appear, on the surface, to be means by which to delimit ways the world might be from the armchair" (Rowbottom 2012). Even Norton, from his not-enthusiastic point of view, repeatedly admitted how TEs can bring astonishment when confronted.

> A scientist – Galileo, Newton, Darwin or Einstein – presents us with some vexing problem. We are perplexed. In a few words of simple prose, the scientist then conjures up an experiment, purely in thought. We follow, replicating its falling bodies or spinning buckets in our minds, and our uncertainty evaporates. (Norton 2004, p. 1139)

[1]Actually, the reactions of appraisal and wonder are so common that who does not comprehends nor approves the definition of TEs as an extraordinary tool—as Gooding, for one—has to extensively justify his lack of enthusiasm (Gooding 2002).

Considering TEs from this perspective, it actually seems that they are amazing tools which permit to glimpse "the relevant laws—not the regularities, but the universals themselves" (Brown 1991b, p. 127).

First, we should admit that the wonder or the rejection about TEs came often from scientists and philosophers who were not TEs-users. The authors who employed TEs rarely expressed amazement for their functionality and, actually, sometimes they did not even affirm that their argumentations were based on TEs [for example, Einstein never claimed that he was using such cognitive devices (Galili 2007)].

Second, we can recall the balanced kind of wonder expressed by the physicist Ernst Mach, one of the firsts who paid attention to TEs as a recurring methodology in science and tried to explain their success. In his paper "On Thought Experiments" he expressed wonder about the use of TEs as the instantiation of what Price and Krimsky (the two translators of his paper) defined as "the propensity of men to experiment […] and while all experiments are guided by theory, not all experiments require a laboratory" (Mach 1976, p. 450). Indeed, Mach, speaking of TEs, referred to the capacity of a child to experiment the possibilities of his body through its manipulation, the ability of dreamers, poets and "builders of castles in the air" to thought experiment the unknown possibilities of reality. Before naming the scientific TE of *Newton's Bucket*, Mach referred to the *ordinary reasoning*, the capacity and propensity of ordinary men to manipulate an imagined settings in order to answer to more or less practical needs.

The connection between the use of TEs in philosophy and science and the employment of simulative reasoning in ordinary circumstances was not just an ingenuous consideration over a (once) newly analyzed topic. In more recent times, the same reflection has been also redeemed by the proponents of a naturalistic view over TEs as "mental models" (Nersessian 1992, Bishop 1999, Gendler 2000). Especially, this reflection was highlighted by Nersessian, who described Mach's seminal ideas on *Gedankenexperiment*, tracing a connection between scientific and philosophical TEs and the model-based reasoning performed by ordinary agents in daily situations:

> While thought experimenting is a truly creative part of scientific practice, the basic ability to construct and execute a thought experiment is not exceptional. The practice is highly refined extension of a common form of reasoning. It is rooted in our ability to anticipate, imagine, visualize, and re-experience from memory. That is, it belongs to a species of thinking by which we grasp alternatives, make predictions, and draw conclusions about potential real-world situations (Nersessian 1992, p. 292).

These considerations are not intended to shape a deflationary picture of TEs in scientific contexts. Instead, they help remind that even an extraordinary scientific or philosophical reasoning is just a refined extension of ordinary reasoning. Hence, even TEs can be described as instances of ordinary simulative or analogical reasoning used in extraordinary circumstances—as in scientific or philosophical research.

Thus, in order to consider TEs as the results of ordinary reasoning employed with epistemological value in scientific practice, we should analyze their generation as goal-oriented objects created to answer to specific dilemmas. In order to elaborate this task, in the next subsection we will adopt an eco-cognitive perspective, which will consent us to connect the dots from the reasoning behind the capacity of the ordinary agent to say "If I were in your shoes…"? and Schrödinger's ability to display the concept of quantum superposition through the paradox of a cat simultaneously alive and dead in a box.

2.2 From an Eco-cognitive Perspective: Thought Experiments as Problem-Solving Methods

First of all, we can define the eco-cognitive point of view as a naturalistic approach in epistemology and logic, that has been introduced by Magnani in the last decade (Magnani 2009, 2015a). As Magnani proposed, a research which is set up from an eco-cognitive point of view is meant to investigate the cognitive resources and heuristics employed by the "practical agent", that is an individual agent operating "on the ground", in the circumstances of real life (Magnani 2009). In this context, the practical agent employs a TE as a specific goal-oriented object, as a *problem-solving method* in order to answer to a lack of knowledge. Indeed, despite the different conditions and theories from which TEs emerged, they have always been presented as effective means to theoretical aims, ways of consider and test particular answers for specific questions. Indeed, according to Simon, a problem-solving method is *the knowledge of an effective procedure to generate and test the various solutions for a specific problem* (Simon 1997), and TEs have always been used toward this epistemological target. In order to gather some examples to specify the functions of TEs, we can recall TEs which represented controversial explanations for particular phenomena (e.g. *Maxwell's Demon*, *Einstein's Clock in a Box*, *Newton's Rotating Bucket*),[2] theoretical confirmations of a principle or a thesis (e.g. *Lucretius' Spear Through the Universe*, *Stevinus' Chain*, *Thompson's violinist*),[3] or critical (sometimes paradoxical) rejections of theoretical assumptions (*Galileo's Falling Bodies*, Bohr argument against *Einstein's Clock in a Box*, *Gettier's Problem*).[4] Thus, considering TEs as problem-solving methods implies two conceptual specifications of their epistemological status.

[2]For the original version of *Maxwell's Demon* firstly discussed in the letter to Peter Guthrie Tait in 1867 see Harman (1995, pp. 331–332, 2002, pp. 185–186) and the extended version in Maxwell (1872, pp. 308–309); for the references to *Einstein's Clock in a Box* see Bohr (1949); for *Newton's Rotating Bucket*, considered in 1687 *Philosophiae Naturalis Principia Mathematica*, see the recent translation of Bernard Cohen, Anne Whitman and Julia Budenz, (Newton 1999).

[3]For the reference of *Lucretius' Spear Through the Universe*, see *De Rerum Natura* 1.951–987, translated by Bailey (1950, pp. 58–59); for an extended and commented version of *Stevinus' Chain* see Mach (1976); and for *Thomson's Violinist* cf. Thomson (1971).

[4]For the reference of *Galileo's Falling Bodies* see the translation of the 1638 text *Discorsi e dimostrazioni matematiche intorno a due nuove scienze attinenti alla mecanica ed i movimenti locali* by Crew and De Salvio (1914); for Bohr argument against *Einstein's Clock in a Box* see Bohr (1949); and for *Gettier's Problem*, see Gettier (1963).

First of all, we can affirm that TEs provide *targeting solutions* for ignorance problems. In other words, TEs represent the initial proposition and consideration of specific answers to precise questions. Specifically, a TE is the device that an author uses in order to verify if a specific answer has the minimal requisites to be the solution to the given problem. For instance, the famous Galileo's TE of falling bodies was precisely meant to disclaim Aristotelian theory that heavier bodies fall faster than lighter ones, as well as his example of the Sailing ship was aimed at showing *only* the human insensitiveness to Earth's rotation.

The second specification is related to the possibility that TEs can fail. Assuming that the complex know-how to solve a dilemma is not equal to knowing directly the answer, TEs would represent *tentative answers* to questions, just possible solutions to problems. A TE would illustrate the generation and the analysis of an hypothetical solution, which has to be put in a process of testing and verification. From this point of view, a TE is a process "usually best described as heuristic search aimed at finding satisfactory alternatives, or alternatives that represent an improvement over those previously available" (Hogart 1980). In this sense, the generation of possible or alternative answers to replace unsatisfactory but available ones, is not a guarantee of success. We can recall numerous renowned cases of "failed" TEs, confirming that, if TEs are ingenious devices to wrap a convincing answer to a problem, they do represent the ultimate one. Two of the most famous examples of failed (as wrong or inaccurate) TEs are *Lucretius' Spear Through the Universe*, aimed at proving that space is infinite, and *Einstein's Clock in a Box* argument against Heisenberg uncertainty principle, disclaimed by Bohr's improvement of the same TE.[5]

Once assumed that TEs can well represent problem-solving methods employed in scientific and philosophical research, we should deal with a last controversial consideration. Notwithstanding the direct aim and the hypothetical essence of TEs, is quite difficult to examine their inferential structure. As Bishop wrote, "They do not seem to translate easily into standard forms of deduction or induction" (Bishop 1999). Expressed in other words, it is hard to recognize a similar pattern which justifies the relation between the theoretical premises of TEs and the aim they are meant to reach. This is probably the reason why, when taxonomies of TEs were conceived, the problem of the inferential shape of TEs was avoided by considering the specification of the particular goal that various examples of TEs were meant to achieve. For instance, Popper distinguished between *apologetic* TEs (which are constructed in favor of a theory), *critical* TEs (created in order to argument against a theory) and the *heuristic* ones (aimed at illustrating a theory) (Popper 1959). A similar distinction was presented by Brown, who divided between constructive TEs (with the same definition of Popperian apologetic ones), destructive TEs (analogous to the Popperian critical examples), and Platonic ones, which fulfill the

[5]For more reference on Einstein and Bohr's argument on the clock in a box TE, cf. (Bohr 1949).

two-sided function of disclaiming a thesis and generating an alternative one (Brown 1991b). Recently, also Borsboom, Mellenbergh and Van Heerden (Borsboom 2002) and Buzzoni have implemented the set of categories for TEs, respectively adding the "functional cases" and the "empirical" and "theoretical" TEs (Buzzoni 2013).

All these classifications are aimed at speaking about the generation and functionality of TEs (the relation between theoretical premises and aims), but they miss the target. They do not refer to the cognitive processes that lead to the creation of TEs but only to the *effect* the creation of particular TEs provide once they were published and proved within the related theories. In this sense, no one of the definitions provided by Popper or Brown or Borsboom is truly ineffective but none of them covers all the possible types of TE; as proven by the recent increment of the TEs categories, it could always be created a new TE which has an aim that is not included in the previous categories. In this sense, the mentioned descriptions do consider TEs as something theoretically aim-specific but not contextually explicable. They do not speak about TEs as why they are created, and do not speak about which aim they fulfill once they are thought. Therefore, by considering TEs as problem-solving methods we point to overcome this view and analyze the generations of TEs as contextual dependent goal-oriented objects.

By conceiving TEs as effective epistemic and cognitive operations we aim at comprehending their role as part of the inferential processes employed by the scientists or the philosophers who created them. If we consider what Galileo, Newton, Einstein[6] did in physics, or what Thompson, Searle or Nozick[7] accomplished in philosophy, we should be able to see a common pattern that they all performed. They inferred from previous data a hypothesis, they considered its value and presented it as a solution for a specific problem. In short terms, they provided an answer and see if it could solve the problem at hand. The mechanism that they performed obviously escapes a direct translation into forms of deduction or induction. They did not inferred the theoretical hypothesis from a collection of data, nor straightforwardly derived it from an analysis of previously conceived premises. They exposed a case and saw if it could fit as solution for the dilemma. This process is commonly defined by abductive reasoning, and in the next section we will extensively present it as the logico/inferential frame appropriate to describe the generation of TEs.

[6]See, for example, *Galileo's Falling Bodies* in Crew and De Salvio (1914), *Galileo's TE on Inertia* and *Galileo's Floating Boat* in Galilei (1953); *Newton's Rotating Bucket* in Newton (1999), *Newton Cannonball or Orbital Cannon* in Newton (1969); *Einstein's Elevator* in Einstein (2009, p. 510), *Einstein Clock in a box* in Bohr (1949), *Einstein's Twins* in Einstein (2002, p. 198).

[7]See *Thomson's Violinist* example in Thomson (1971), *Searle's Chinese Room* in Searle (1980) and *Nozick's Experience Room* in Nozick (1974).

3 The Generation of Thought Experiment as Performance of Abductive Reasoning

Recently, Igal Galili also suggested that a third logical scheme beyond deduction and induction should be considered in order to explain the inferential power of TEs (Galili 2007). He gave this definition:

> Thought experiment is a set of hypothetico-deductive considerations regarding phenomena in the world of real objects, drawing on a certain theory (principle or view) that is used as reference of validity. (Galili 2007, p. 12)

This description pushes in a good direction, but, from our point of view, also ends in a narrow and old-fashioned road. Galili aimed at combining the different definitions of TEs provided in the last thirty years.[8] The definition which results from his effort effectively overcomes the partial views offered by the previously mentioned authors (Popper, Brown, Borsboom and colleagues and Buzzoni), but it presents two weak points.

First of all, Galili's definition just focuses on the inferential structure of TEs without discussing the problems, mainly raised in the philosophical literature, regarding the creation and the reliability of the scenario. Simply, it does not consider important questions related to the construction and the development of the narration—for instance, if it is an a priori or an experience-sensible model, whether and why it can be considered reliable and reality-related, or whether it could be conducted in an empirical context or using a computer simulation. Secondly, describing the inferential structure of TEs, Galili appeals to a concept—the hypothetico-deductive process—that does not stress the rich cognitive aspects which the analysis of TEs creation should encompass.

Abduction is not only a concept which can define the structure of an explanatory reasoning. Considering the growth of the studies on abduction, its cognitive significance, and the specifications that can now refine its logical form, we can employ it successfully to explain the complexity of TEs, considering both their hypothetical-deductive structures and their tied connection with the scenario construction. Therefore, in order to speak about the creation of TEs as problem-solving methods, which instantiate forms of abductive reasoning, we can start considering the qualities and downsides of abduction as a cognitive tool.

---

[8]Specifically, for its final definition, Galili considers Reiner and Gilbert's claim that a "thought experiment is a design of thought that is intended to test and/or convince others of the validity of a claim" (Reiner and Gilbert 2000), judging it too inclusive. He evaluates Brown's point of view too general, as it defined a TE "a special type of mental window through which the mind can grasp universal understandings" (Brown 1991a). Finally examining the *modelist account* "to perform a scientific thought experiment is to reason about an imaginary scenario with the aim of confirming or disconfirming some hypothesis or theory about the physical world" (Gendler 2004) Brown well exposes the fact that Gendler (as others who shared the modelist view, like Bishop and Nersessian), usually pays more attention to the creation of the imagined scenario in the TEs than to the features of their inferential structure.

3.1 Abduction and the Generation of Hypotheses

Abduction, as an inferential process aiming at finding out explanatory hypotheses starting from a cluster of data, can be described starting from two different points of view. One that focuses on its explanatory power, its efficiency in bringing out hypotheses in order to find the best answer for a determined problem; and another that brings into focus the fallacious form of abduction and its ignorance-preserving trait. Indeed, starting from the latter viewpoint, abduction in its simplest form could be sketched with the so-called "syllogistic" model (Magnani 2009, Chap. 2), in which is presented as the "fallacy of affirming the consequent":

P1: If *A* then *B*
P2: *B*
C: Then *A*

Nevertheless, this straightforward model is relatively meagre in giving off details about how abduction actually works at the cognitive level. An analogous model of abduction, which instead points out both its explanatory capacity and its cognitive significance, is the standard three-steps exemplification provided by Charles Sanders Peirce: (1) *The surprising fact C is observed.* (2) *But if A were true, C would be a matter of course.* (3) *Hence there is reason to suspect that A is true* (Peirce 1992–1998, vol. 2 pp. 226–241). In order to effectively refer to the double nature of abductive reasoning, in 2005 Gabbay and Woods described abduction as "a procedure in which something that lacks epistemic virtue is accepted because it has virtue of another kind", Gabbay and Woods (2005, p. 62). Thus, in the one hand, abduction is the process of inferring certain facts and/or laws and hypotheses that render some sentences plausible, that explain or discover some (eventually new) phenomenon or observation; it is the process of reasoning in which explanatory hypotheses are formed and evaluated and it is a fundamental mechanism to account for the introduction of new explanatory hypotheses in science (Magnani 2009). On the other hand, it derives its excellent explanatory capacity due to the fact that it is a fallacious reasoning which presents an ignorance-preserving or (ignorance-mitigating) character (evidenced by Gabbay and Woods 2005; Aliseda 2005; Magnani 2013, 2015a).[9]

The cognitive significance of abduction in the light of the analysis of TEs has to be sought in its pivotal role in the context of discovery. The agent performing an abduction does not simply select or generate an exploratory hypothesis, but *the*

[9]It is thanks to the GW-schema (cf. Gabbay and Woods 2005), that Gabbay and Woods criticize the so called AKM model of abduction. A primary gift provided by the GW-schema was the opening of the discussion about *ignorance preservation* but also about *non-explanatory* and *instrumental* abduction, considered as not intrinsically consequentialist. Magnani extensively illustrated non-explanatory and instrumental abduction, together with the classical AKM model, in Magnani (2009, Chap. 2), also providing some case studies.

most plausible one, that allows the agent to answer to a particular ignorance problem.[10]

This implies that, as targeting hypothesis finders, TEs enact the structure of abduction as the inference for the best explanation. The creation of an hypothesis or the selection of it in an array of available ones is performed in TEs in the creation and the development of the scenario. The selected hypothesis that must be evaluated is the scenario itself, the answer is embedded in it. A TE must represent a good case or a good situation that enables the listener or the reader to consider plausible or reject a theory. In order to make an example, we can see how the abductive process was performed in the creation of the TE of the Violinist, used by Thompson in order to argument in favor of the permissibility of the abortion (Thomson 1971). She needed a way to explain how an act which was considered by numerous moral philosophers a ending of a person's life, could have been seen as a legit gesture. In the beginning of the TE, she claims that she wanted to start her argumentation from the point of view of the anti-abortion supporters. She affirms "Opponents of abortion commonly spend most of their time establishing that the fetus is a person, and hardly anytime explaining the step from there to the impermissibility of abortion. [...] I propose, then, that we grant that the fetus is a person from the moment of conception. How does the argument go from here?" (Thomson 1971, p. 69). So, she asks what was needed to an anti-abortion supporter to see the sentence "the abortion could be morally permissible" and "the fetus is a person" as two valid assertions. In the frame of the three-step exemplification provided by Peirce we mentioned before (Peirce 1992–1998), an abortion (the killing of a person) considered acceptable by an anti-abortion supporter would correspond to a surprising fact *C*. In order to explain it, she needed to get to the possible reasons that could explain this fact. The hypothetical answer is embedded in the famous moral TE.

In the scene we have a famous violinist who falls into a coma. The society of music lovers determines from medical records that another person (the hypothetical "you") can save the violinist's life by being hooked up to him for nine months. The music lovers kidnap you and hook the unconscious and innocent violinist to you.

[10]Notwithstanding the fact that many standard perspectives on abduction demand two properties, relevance and plausibility, which are presented as possessed by "every" kind of solution for an abductive problem, we should point out the context- and time- dependent character of these requirements, which is claimed by Magnani's Eco-Cognitive model of abduction (Magnani 2015b). This results in the affirmation that irrelevance and implausibility not always are offensive to the performance of a good abductive reasoning. Magnani claims that, in general, we cannot be sure that our guessed hypotheses are plausible (even if we know that looking for plausibility is a human good and wise heuristic), indeed an implausible hypothesis can later on result plausible. Eventually, the plausibility of a guessed hypothesis results a trivial requirement and something similar can be said in the case of relevance. In the case of TEs, the agent performing an abduction select or generates what she thinks is *the most plausible* hypothesis, which depends on her knowledge and her beliefs at her time. Therefore, plausibility and relevance are deemed as strict requirements for the consideration of the guessed hypothesis primarily (and sometimes only) from the author's point of view.

The analogy is clear: according to Thomson, a non-wanted pregnancy is conceivable as the relation between two people, of whom the rights are threaten. The right to live of the fetus can be related to the right to live of the violinist: he is unconscious and he has no deals with the music lovers society; so he is innocent as the fetus, according to the opponents of abortion. Analogously, the rights of the mother to decide what shall happen in and to her body are equal to yours, as the person attached to the violinist. Reframing the context in which the opponents of abortion see it in a circumstance in which the rights of the mother are justified as much as the rights of the fetus, Thomson completed the abductive process. Seeing the person attached to the violinist as fully allowed to also "detach him" (the *A* hypothesis of the Peirce's exemplification) could explain why, even from a point of view which sees the fetus as a person with all her rights, the abortion could be morally permissible. And so, a viewpoint which considers the fetus a person but the abortion permissible, is explained ["hence there is reason to suspect that *A* is true" (Peirce (1992–1998)].

The inferential power of TEs is based on the capacity of abduction to generate and evaluate hypotheses in order to find a new and better solution to the available ones. The hypotheses evaluated are embedded in a scenario which is not a neutral addition to a linear argumentation, but that is related with the aims and the conditions of the logical structure of the TEs. In the next section, we will deal with the modalities through which the scenario is created in connection with the inferential structure of TEs, finally presenting them as instantiation of model-based abductions.

4 Thought Experiments, Abduction, and Model-Based Reasoning

The creation of a TE implies the creation of a narrative structure which canalizes the purposes of the author's argumentation. When Norton claimed that TEs are just "ordinary argumentations disguised in some vivid and picturesque or narrative forms" (Norton 2004) he was not just implying that the imagined scenario was uselessly detailed, but also that the various details were randomly chosen. In other words, he claimed that a determined TE could be disguised with other "picturesque clothing" and it would lead to the same conclusions. This consideration derives from the fact that, now, we can extrapolate the linear argumentation which underlies TEs and replace the details of the scenario originally adopted by the author with some others—as we could replace in our minds *Schödinger's cat* with a dog or *Wittgenstein's beetle* with an orange.[11] But this afterwards possibility does not prove that those details were replaceable in the minds of the authors who thought about them when formulating the related TEs. The details were useful as long as helped the authors to construct the scenario within which they embedded their

[11]See Schrödinger (1983) and Wittgenstein (1958, p. 100).

theories. Considering TEs as problem-solving methods implies also to see the scenario as a functional part of this cognitive operation. Thus, instead of wondering on the "superfluity" of the details of the scenarios we could more simply ask "why did that precise scenario work and no others?" To this question, we can simply answer speaking about the particular qualities of model-based reasoning.

The construction of the scenario has, indeed, a crucial relevance in the development of the logical structure of TEs. As answers to problems, TEs are also a way to *reframe* them. As Brown suggested, summarizing Kuhn's words, thought experiments can teach us something new about the world, even though we have no new empirical data, by helping us *reconceptualize* the world in a new way (Brown 1991a). Through the creation of a narrative scenario, the TE-user can replace the facade of the problem with a more approachable version. To make an example, we can recall the exact words of Turing, who, in order to consider the question "Can machines think?" and wanting to escape the loop of searching the meaning of the two words "machine" and "think", replaced the question "by another, which is closely related to it and is expressed in relatively unambiguous words. The new form of the problem can be described in terms of a game which we call the 'imitation game'" (Turing 1950, p. 433). By doing that, as Turing says later on, he *changed* the question. The question, embedded in the Imitation game, became "Let us fix our attention on one particular digital computer *C*. Is it true that by modifying this computer to have an adequate storage, suitably increasing its speed of action, and providing it with an appropriate programme, *C* can be made to play satisfactorily the part of *A* in the imitation game, the part of *B* being taken by a man?" [p. 441]. It is a less catchy question than the initial one, but it is also more precise and more interesting. In fact, the reason why the Imitation Game can be seen as an effective substitute for the question "Can machines think?" is that after the TE has been presented, the initial question displays a different meaning than before.

TEs, with the replacement of a straightforward argumentation with the production of an imagined scenario, *change* the way the reader can see the meaning of the linear thesis. As Brown advocated, they help to "reconceptualize" it in a new frame. While the initial question "Can machines think?" is too obscure to be investigated before the display of the Imitation Game, the new question asked by Turing has no meaning without the TE. An effective TE as the one proposed by Turing can survive beyond the expression of the linear argumentation because changes the rules of the old reasoning. The sequences of "posing a question", "replacing it with an explanatory scenario" and "reveal the new question and results" that Turing made explicitly is behind the generation of every TEs. The creation of the scenario is a necessary part of the inferential process because it *causes* it. But this is not enough to answer to the question "how there has to be this scenario, instead of any others?" The answer to this question is effectively related to the inferential schema that can be rendered by a model-based reasoning.

If we think about the creation of the scenario as replacing or reconceptualizing the problem at stake, we can see that the tested hypothesis is not inside the TE, but it is the experiment itself. And in that, the thinking activity is just not merely propositional, but effectively model-based. As contended by Magnani, who embraces Peirce's semiotic point of view, "all thinking is in signs, and signs can be icons, indices or symbols. Moreover, every inferences is a form of sign activity, where the word sign includes feelings, images, conceptions and other representation" (Magnani 2009). That is, a considerable part of the thinking activity is model-based. Of course model-based reasoning acquires its peculiar creative relevance when embedded in abductive processes, so that we can individuate a *model-based abduction*. Hence, we must think in terms of model-based abduction (and not in terms of sentential abduction) to explain complex processes like scientific conceptual change. Different varieties of *model-based abductions* (Magnani 1999) are related to the high-level types of scientific conceptual change (see, for instance, Thagard 1992). A TE is nothing more than a hypothesis resulting from an inferential activity which can be traced in the development (as the sequence posing a question, replacing the question, showing the results) of the TE itself. The agent elaborates an inference through the construction of an imagined scenario, which represents the evaluation of an hypothesis and the presentation of it. Indeed this model-based hypothesis is a "mental model" of a theoretical possibility, and the inference that makes the cognitive process working is an abduction, Magnani (2009).

In 1943 the Scottish psychologist and physiologist Kenneth Craik presented a classical but useful definition of mental model, successively borrowed by Gendler, Nersessian, and Bishop in order to develop the mental model account of TEs. In the fifth chapter of his most notorious volume *The Nature of Explanation* he wrote:

> If the organism carries a "small-scale model" of external reality and of its own possible actions with its head, it is able to try out various alternatives, conclude which is the best of them, react to future situations before they arise, utilize the knowledge of past events in dealing with the present and the future, and in every way to react in a much fuller, safer, and more competent manner to the emergencies which face it. (Craik 1943, p. 61)

Essentially, in Craik's and successive elaborations,[12] a mental model illustrates the cognitive process of constructing a simpler and more affordable representation of a complex system or a structure, which maintains the properties of the original phenomenon and helps studying it. The agent employs a mental model in order to carry on the investigation in a flexible structure of the phenomenon itself, which the agent is more familiar with. Indeed, a TE is a device that permits the agent to "manipulate a mental model instead of a physical one" (Bishop 1999). Moreover, through a TE is performed what Kuhn defined as a model of *Gestalt-switch*, a change of perspective, in order to see differently the original problem (Kuhn 1977). Indeed, after the presentation of a TE, the problem is seen through a perspective that encompasses already the hypothesis represented by the TE itself.

---

[12]The difference between Craik's examination of mental models and the "mental model account" of TEs proposed by Bishop, Nersessian and Gendler is extensively displayed in Johnson-Laird (2004).

The creation of the hypothesis, in the case of TEs the creation of the mental model itself (with all its details), is what allows the agent to consider possibilities beyond what has been already conceived. This possibility derives both from the fallacious structure and the creative and heuristic power of abduction. TEs, as model-based abductions, indeed, represent problem-solving methods because they guarantee an highly refined *response* to an ignorance problem. In few words, the production of a mental model is a compass directed to the creation of a more precise answer for the ignorance problem. TEs as model-based abductions are ways to discuss the terms of a theory, and suggest new principles to compose the problems itself.

5 Conclusion

We can now reconsider the original question: how can an imagined scenario bring new information or insights about the actual world? We have presented an answer based on the main assumption that TEs are goal-oriented object, which have been employed as problem-solving methods in the history of science and philosophy. Their epistemological value depends on both their inferential power and the cognitive impact the creation of the scenario represents.

Referring to the inferential structures of TEs as instantiations of abductive reasoning, we justified their explanatory power and displayed their role as methods for discovering a solution to a problem, through the *reconfiguration* of the problem itself. Adopting the definition of mental model, firstly presented by Craik and successively employed by the defenders of the mental model account of TEs, we justified the narrative structure of TEs as necessary design that enables the author to carry out the experiments under more favorable conditions than would be available in the original system. Finally, we presented TEs as instantiation of model-based abduction, describing them as methods trough which it is possible to respond to a problem of ignorance, generating and the evaluating highly sophisticated hypotheses and embed them in adequately elaborated scenarios.

References

Aliseda, A. (2005). The logic of abduction in the light of Peirce's pragmatism. *Semiotica, 1/4* (153):363–374.

Bailey, C. (1950) *Lucretius on the nature of things*. Oxford: Clarendon Press. (translation of De Rerum Naturae).

Bealer, G. (1998). Intuition and the autonomy of philosophy. In M. DePaul & W. Ramsey (Eds.), *Rethinking intuition: The psychology of intuition and its role in philosophical inquiry* (pp. 201–239). Lanham: Rowman & Littlefield.

Bishop, M. (1999). Why thought experiments are not arguments. *Philosophy of Science, 66*(4), 534–541.

Bohr, N. (1949). Discussion with Einstein on epistemological problems in atomic physics. In P. A. Schilpp (Ed.), *Albert Einstein: Philosopher-Scientist, page 201Ð241*. La Salle: Open Court.
Brown, J. R. (1991a). *The laboratory of the mind: Thought experiments in the natural sciences*. London: Routledge.
Brown, J. R. (1991). Thought experiments: A platonic account. In *Thought experiments in science and philosophy* (pp. 119–128). Savage, MD.: Rowman and Littlefield.
Buzzoni, M. (2013). Thought experiments from a Kantian point of view. In M. Frappier et al. (Eds.), *Thought experiments in science, philosophy and arts*. Londo: Routledge.
Craik, K. (1943). *The nature of explanations*. Cambridge: Cambridge University Press.
Crew, H., & De Salvio, A. (1914). *Dialogues concerning two new science*. New York: The Macmillan Company. (Introduction of A. Favaro).
Dennett, D. (1984). *Elbow room*. Cambridge: MIT Press.
Di Paolo, E. A., Nobel, J., & Bullock, S. (2000). Simulation models as opaque thought experiments. In N. H. Packard, M. A. Bedau, J. S. McCaskill & S. Rasmussen (Eds.), *In Artificial Life VII: The Seventh International Conference on the Simulation and Synthesis of Living Systems*, pages 497–506. MIT Press, Cambridge.
Einstein, A. (2002). Anti-relativity company. In M. Janssen, R. Schulmann, J. Illy, C. Lehner, & D. K. Buchwald (Eds.), *The Berlin years: Writings, 1918–1921, volume 7 of the collected papers of Albert Einstein*. Princeton, New Jersey: Princeton University Press.
Einstein, A. (2009). Appendix c. In D. K Buchwald, Z. Rosenkranz, T. Sauer, J. Illy, & V. I. Holmes (Eds.), *The Berlin Years: Correspondence January-December 1921*, volume 12 of *The Collected Papers of Albert Einstein*. Princeton, New Jersey: Princeton University Press.
El Skaf, R., & Imbert, C. (2013). Unfolding in the empirical sciences: Experiments, thought experiments and computer simulations. *Synthese, 190*(16), 3451–3474.
Gabbay, D. M., & Woods, J. (2005). *The reach of abduction: Insight and trial*, volume 1 of *A practical logic of cognitive systems*. Amsterdam: North Holland: Elsevier.
Galilei, G. (1953). *Dialogue concerning the two chief world systems, ptolemaic and copernican* (S. Drake Forewords by A. Einstein, trans.). Los Angeles: California University Press.
Galili, I. (2007). Thought experiments: Determining their meaning. *Science & Education, 18*, 1–23.
Gendler, T. S. (2000). *Thought experiment: On the powers and limits of imaginary cases*. New York: Garland Press.
Gendler, T. S. (2004). Thought experiments rethought—and reperceived. *Philosophy of Science, 71*, 1152–1163.
Gettier, E. L. (1963). Is justified true belief knowledge? *Analysis, 23*, 121–123.
Gooding, D. C. (2002). What is experimental about thought experiments? *PSA, 2*, 280–290.
Häggqvist, S. (1996). *Thought experiments in philosophy*. Stockholm: Almqvist & Wiksel.
Harman, P. M. (1995). *Scientific letters and papers of James Clerk Maxwell* (Vol. II). Cambridge: Cambridge University Press.
Harman, P. M. (2002). *Scientific letters and papers of James Clerk Maxwell* (Vol. II). Cambridge: Cambridge University Press.
Hogart, R. M. (1980). *Judgment and choice: The psychology of decision*. New York: Wiley.
Johnson-Laird, P. N. (2004). The history of mental models. *Psychology of Reasoning: Theoretical and Historical Perspectives, 179*, 417–457.
Kuhn, T. (1977). A function for thought experiments. In T. Kuhn (Ed.), *The essential tension*. Chicago: Univerity of Chicago Press.
Lennox, J. (1991). Darwinian thought experiments: A function for just-so stories. In *Thought experiments in science and philosophy* (pp. 223–245). Savage, MD: Rowman and Littlefield.
Mach, E. (1976). On thought experiments. In *Knowledge and error* (pp. 134–147). Reidel: Dordrecht Holland.
Magnani, L. (1999). Inconsistencies and creative abduction in science. In *AI and scientific creativity. Proceedings of the AISB99 Symposium on Scientific Creativity*, pages 1–8,

Edinburgh, 1999. Society for the Study of Artificial Intelligence and Simulation of Behaviour, University of Edinburgh.
Magnani, L. (2009). *Abductive cognition. The epistemological and eco-cognitive dimensions of hypothetical reasoning*. Berlin/Heidelberg: Springer.
Magnani, L. (2013). Is abduction ignorance-preserving? Conventions, models, and fictions in science. *Logic Journal of IGPL, 21*, 882–914.
Magnani, L. (2015a). The eco-cognitive model of abduction Ἀπαγωγή no: Naturalizing the logic of abduction. *Journal of Applied Logic, 13*, 285–315.
Magnani, L. (2015b). The eco-cognitive model of abduction irrelevance and implausibility exculpated. *Journal of Applied Logic, 13*, 13–36.
Maxwell, J. C. (1872). *The theory of heat*. London: Longmans, Green and Co.
Mellenbergh, G. J., Borsboom, D., & Van Heerden, J. (2002). Functional thought experiments. *Synthese, 130*, 379–387.
Nersessian, N. (1992). In the theoretician's laboratory: Thought experimenting as mental modeling. *PSA, 2*, 291–301.
Nersessian, N. J., Chandrasekharan, S., & Subramanian, V. (2012). Computational modeling: Is this the end of thought experimenting in science? In *Thought experiments in philosophy, science and the arts* (pp. 239–260). London: Routledge.
Newton, I. (1969). *A treatise of the system of the world*. Mineola, New York: Dover Publications Inc.
Newton, I. (1999). *The principia: Mathematical principles of natural philosophy* (B. Cohen, A. Whitman, & J. Budenz, trans.). Berkeley and Los Angeles, California: University of California Press.
Norton, J. D. (2004). On thought experiments: Is there more to the argument? *Philosophy of Science, 71*, 139Ð1151.
Nozick, R. (1974). *Anarchy, State and Utopia*. Chicago: Basic Books.
Peirce, C. S. (1992–1998). *The essential Peirce. Selected philosophical writings*. Bloomington and Indianapolis: Indiana University Press. Vol. 1 (1867–1893), N. Houser & C. Kloesel (Eds.), Vol. 2 (1893–1913) by the Peirce Edition Project.
Popper, K. (1959) On the use and misuse of imaginary experiments, especially in quantum theory. *The Logic of Scientific Discovery* (pp. 442–456).
Reiner, M., & Gilbert, J. (2000). Epistemological resources for thought experimentation in science education. *International Journal of Science Education, 22*(5), 489–506.
Rowbottom, D. P. (2012) Intuitions in science: Thought experiments as argument pumps. In *Intuitions* (pp. 119–134). Oxford: Oxford University Press.
Schrödinger, E. (1983) The present situation in quantum mechanics. In J. A. Wheeler & W. H. Zurek (Eds.), *Quantum theory and measurement*, page part I. New Jersey: Princeton University Press. (Translated by J. D. Trimmer).
Searle, J. (1980). Minds, brains and programs. *Behavioral and Brain Sciences, 3*, 179–212.
Simon, H. A. (1997). *Models of bounded rationality*. Cambridge: MIT Press.
Sorensen, R. A. (1992). *Thought experiments*. Oxford: Oxford University Press.
Thagard, P. (1992). *Conceptual revolutions*. Princeton: Princeton University Press.
Thagard, P. (2014). Thought experiments considered harmful. *Perspectives on Science, 22*(2), 288–305.
Thomson, J. J. (1971). A defense of abortion. *Philosophy & Public Affairs, 1*, 47–66.
Turing, A. M. (1950). Computing machinery and intelligence. *Mind, 59*, 433–460.
Wilkes, K. V. (1999). *Real people personal identity without thought experiments*. Oxford: Clarendon Press.
Wittgenstein, L. (1958) *Philosophical investigations*. Oxford: Basil Blackwell Ltd. (Translated by G. E. M. Anscombe).

Abduction and Its Eco-cognitive Openness

Aristotle's 'Απαγωγή Explained

Lorenzo Magnani

> Unless man has a natural bent in accordance with nature's, he has no chance of understanding nature at all.
>
> Charles Sanders Peirce, *A Neglected Argument for the Reality of God*, 1908

Abstract Aristotle clearly states that in syllogistic theory local/environmental cognitive factors—external to that peculiar inferential process, for example regarding users/reasoners, are given up. Indeed, to define syllogism Aristotle first of all insists that all syllogisms are valid and contends that the *necessity* of this kind of reasoning is related to the circumstance that "no further term from outside (ἔξωθεν) is needed", in sum syllogism is the fruit of a kind of eco-cognitive *immunization*. At the same time Aristotle presents a seminal perspective on abduction, which contrasts with the previous one on syllogismos: the second part of the article considers the famous passage in the chapter B25 of *Prior Analytics* concerning ἀπαγωγή ("leading away"), also studied by Peirce. I contend that some of the current well-known distinctive characters of abductive cognition are already expressed, which are in tune with what I have called Eco-cognitive Model of abduction. By providing an illustration of the role of the method of analysis and of the middle terms in Plato's dialectic argumentation, considered as related to the diorismic/poristic process in ancient geometry I maintain that it is just this intellectual heritage which informs Aristotle' chapter B25 on ἀπαγωγή. Even if, in general, Aristotle seems to sterilize, thanks to the invention of syllogistic theory, every "dialectic" background of reasoning, nevertheless in chapter B25 he is still pointing to the fundamental inferential role in reasoning of those externalities that substantiate the process of "leading away" (ἀπαγωγή). Hence, we can gain a new positive perspective about the "constitutive" eco-cognitive character of abduction, just thanks to Aristotle himself.

L. Magnani (✉)
Department of Humanities, Philosophy Section, and Computational Philosophy Laboratory, University of Pavia, Piazza Botta, 6, 27100 Pavia, Italy
e-mail: lmagnani@unipv.it

L. Magnani and C. Casadio (eds.), *Model-Based Reasoning in Science and Technology*, Studies in Applied Philosophy, Epistemology and Rational Ethics 27, DOI 10.1007/978-3-319-38983-7_25

1 The Eco-cognitive Openness of Aristotle's 'Απαγωγή

To define syllogism[1] Aristotle expressly contends that the *necessity* of this kind of reasoning is related to the circumstance that "no further term from outside (ἔξωθεν) is needed", in sum we can say that syllogism is the fruit of a kind of eco-cognitive *immunization*:

> A deduction (συλλογισμὸς) is a discourse (λὸγος) in which, certain things having been supposed, something different from the things supposed results of necessity because these things are so. By "because these things are so", I mean "resulting through them," and by "resulting through them" I mean "needing no further term from outside (ἔξωθεν) in order for the necessity to come about" (Aristotle 1989, A1 24, 20–25, p. 2).

Contemporary logicians as Gabbay and Woods clearly echo Aristotle's contention, when they say:

> As I have illustrated in a recent article (Magnani 2015a), logic, from its inception, has sought to serve two masters. One is to specify and characterize sets of intuitively logical properties and relations that are definable for propositional structures or for these in relation to abstractively set-theoretic structures. Here the main goal is to get these target notions right, where the question of rightness is intimately bound up with the issue of rightness for. Accordingly, a logic gets consequence right if it is right for sets of sentences taken without reference to *factors of speaker-use and other pragmatic considerations* (Gabbay and Woods 2005, p. 241, emphasis added,).

The canons of strict reasoning are the focus of the new Aristotelian syllogistic logic and of the subsequent developments, that is the extraction and clarification of the inference-rules: truth conditions are guaranteed by appropriate syntactic inference rules. Aristotle favors an intrinsic view of the rules which govern good arguments, to create those constraints that depict classical validity, thought in terms of "necessitation". A good syllogism is a valid argument whose premisses are *non-redundant*, whose conclusion *repeats no premiss*, and whose conclusion is *non-ambiguously multiple* [also, premisses do not have to contain a proposition and its contradictory]. Reasoning by strict (or truth-preserving) consequence "is always a matter of evacuating information already present in premisses. [...] Under Aristotle's constraints every piece of archeological reasoning evacuates its premisses of their total syllogistic information, which is then repackaged in the single proposition that serves as the reasoning's conclusion" (ibid.)

Hence, the theory of syllogism is related to a kind of eco-cognitive immunization. However, it is well-known that Aristotle presents a seminal perspective on abduction: in this case, his ἀπαγωγή exhibits instead a clear eco-cognitive openness, which contrasts the closure indicated in the case of syllogismos. In this article I will describe the role of the method of analysis and of the middle terms in Plato's dialectic argumentation, considered as related to the diorismic/poristic process in

---

[1]Aristotle insists that all syllogisms are valid (by definition) (Woods 2014), there is no such thing as an invalid syllogism. We know the syllogistic tradition began to relax this requirement quite early on.

ancient geometry, showing it as a theoretical heritage which informs Aristotle's chapter B25 of *Prior Analytics*, concerning abduction, I am introducing in this section. Thanks to Aristotle we can gain a new positive perspective about the "constitutive" eco-cognitive character of abduction.[2]

It seems Peirce was not satisfied with the possible Apellicon's correction of Aristotle's text about abduction: "Indeed, I suppose that the three [abduction, induction, deduction] were given by Aristotle in the *Prior Analytics*, although the unfortunate illegibility of a single word in his MS, and its replacement by a wrong word by his first editor, the 'stupid' [Apellicon],[3] has completely altered the sense of the chapter on Abduction. At any rate, even if my conjecture is wrong, and the text must stand as it is, still Aristotle, in that chapter on Abduction, was even in that case evidently groping for that mode of inference which I call by the otherwise quite useless name of Abduction—a word which is only employed in logic to translate the [ἀπαγωγή] of that chapter" (Peirce 1931–1958, 5, 144–145, *Harvard Lectures on Pragmatism*, 1903].

At this point I invite the reader to carefully follow Aristotle's chapter from the *Prior Analytics* quoted by Peirce. In this case the discussion turns arguments that transmit the uncertainty of the minor premiss to the conclusion, rather than the certainty of the major premiss. If we regard uncertainty as an epistemic property, then it is reasonably sound also to say that this transmission can be effected by truth-preserving arguments: by the way, it has to be said that this is not at all shared by the overall Peirce's view on abduction, which is not considered as truth preserving (with the exception of the knowledge-enhancing case—see my article Magnani (2013)—which instead depicts a kind of "casual" truth preserving character).

I want first of all to alert the reader that in the case of the Aristotelian chapter, abduction does not have to be discussed keeping in mind the schema of the fallacy of affirming the consequent, which depicts abduction as a fallacious reasoning, if seen in the light of the classical logic. What is at stake is abduction considered either (1) the classification of a certain "unclear" dynamic argument in a *context-free* sequence of three propositions; or (2) the introduction in a similar "unclear" dynamic three-propositions argument (in this case no longer *context-free*) of few new middle terms. Hence, ἀπαγωγή—leading away (abduction)—is, exactly (in the Aristotelian words we will soon entirely report below)

1. the feature of an argument in which "it is clear (δῆλον) that the first term belongs to the middle and unclear (ἄδηλον) that the middle belongs to the third, though nevertheless equally convincing (πιστόν) as the conclusion, or more so" (Aristotle 1989, B25, 69a, 20–22, p. 100);
2. the introduction of suitable middle terms able to make the argument capable of guiding reasoning to substantiate an already available conclusion in a more

[2]I have described in detail the eco-cognitive model of abduction in Magnani (2015a, 2016).

[3]Apellicon was the ancient editor of Aristotle's works. Amazingly, Peirce considers him, in other passages from his writings, "stupid" but also "blundering" and "scamp" (Kraus 2003, p. 248).

plausible way: Aristotle says in this way we "are closer to scientific understanding": "if the middles between the last term and the middle are few (ὀλίγα) (for in all these ways it happens that we are closer to scientific understanding (πάντως γάρ ἐγγύτερον εἶναι συμβαίνει τῆς ἐπιστήμης))" (Aristotle 1989, B25, 69a, 22–24, p. 100).

It is clear that the first case merely indicates a certain status of the uncertainty of the minor premiss and of the conclusion and of the related argument; the second case, from the perspective of the eco-cognitive openness (eco-cognitive model) of abduction, is much more interesting, because directly refers to the need, so to speak, of "additional/external" interventions in reasoning. It has to be said that Aristotle does not consider the case of the creative reaching of a *new* conclusion (that is of a creative abductive reasoning, instantly knowledge-enhancing or simply presumptive): however, I will illustrate in the following section that this case appears evident if we consider the method of analysis in ancient geometry, as a mathematical argument which mirrors the propositional argument given by Aristotle, provided we consider it in the following way: *we do not know the conclusion/hypothesis, but we aim at finding one thanks to the introduction of further "few" suitable middle terms*.

The following is the celebrated chapter B25 of the *Prior Analytics* concerning abduction. The translator usefully avoids the use of the common English word *reduction* (for ἀπαγωγή): some confusion in the literature, also remarked by Otte (2006, p. 131), derives from the fact reduction is often rigidly referred to the hypothetical deductive reasoning called *reductio ad absurdum*, unrelated to abduction, at least if intended in Peircean sense. Indeed, the translator chooses the bewitching expression "leading away".

> XXV. It is leading away (ἀπαγωγή) when it is clear (δῆλον) that the first term belongs to the middle and unclear (ἄδηλον) that the middle belongs to the third, though nevertheless equally convincing (πιστόν) as the conclusion, or more so; or, next, if the middles between the last term and the middle are few (ὀλίγα) (for in all these ways it happens that we are closer to scientific understanding (πάντως γάρ ἐγγύτερον εἶναι συμβαίνει τῆς ἐπιστήμης)). For example, let A be teachable, B stand for science [otherwise translated as "knowledge"], and C justice [otherwise translated as "virtue"]. That science is teachable, then, is obvious, but it is unclear whether virtue is a science. If, therefore, BC is equally convincing (πιστόν) as AC, or more so, it is a leading away (ἀπαγωγή) (for it is closer to scientific understanding (ἐγγύτερον γάρ τον ἐπίστασθαι) because of taking something in addition, as we previously did not have scientific understanding (ἐπιστήμη) of AC). Or next, it is leading away (ἀπαγωγή) if the middle terms between B and C are few (ὀλίγα) (for in this way also it is closer to scientific understanding (εἰδέναι)). For instance, if D should be "to be squared," E stands for rectilinear figure, F stands for circle. If there should only be one middle term of E and F, to wit, for a rectilinear figure together with lunes to become equal to a circle, then it would be close to knowing (ἐγγύς ἄν εἴη τοῦ εἰδέναι). But when BC is not more convincing (πιστότερον) than AC and the middles are not few (ὀλίγα) either, then I do not call it leading away (ἀπαγωγή). And neither when BC is unmiddled: for this sort of case is scientific understanding (ἐπιστήμη) (Aristotle, B25, 69a, 20–36, pp. 100–101).

This passage is very complicated and difficult. I have indicated words and expressions in ancient Greek because they stress, better than in English, some of the received distinctive characters of abductive cognition:

1. ἄδηλον [unclear] refers to the lack of clarity we are dealing with in this kind of reasoning; furthermore, it is manifest that we face with a situation of ignorance —something is not known—to be solved;
2. πιστόν [convincing, credible] indicates that degrees of uncertainty pervade a great part of the argumentation;
3. the expression "then it would be close to knowing (ἐγγύς άν εἴη τοῦ εἰδέναι)", which indicates the end of the conclusion of the syllogism,[4] clearly relates to the fact we can only reach credible/plausible results and not ἐπιστήμη; Peirce will say, similarly, that abduction reaches plausible results and/or that is "akin to the truth";
4. the adjective ὀλίγα [few] dominates the passage: for example, Aristotle says, by referring to the hypotheses/terms that have to be added—thanks to the process of leading away—to the syllogism: "Or next, it is leading away (ἀπαγωγή) if the middle terms between B and C are few (ὀλίγα) (for in this way also it is closer to scientific understanding (εἰδέναι))". The term ὀλίγα certainly resonates with the insistence on minimality that dominates the so-called classical AKM model of abduction I have illustrated in Magnani (2009).

I favor the following interpretation (Phillips 1992, p. 173): abduction denotes "the method of argument whereby in order to explain an obscure or ungrounded proposition one can lead the argument away from the subject to one more readily acceptable".

In the passage above Aristotle gives the example of the three terms "science" [knowledge], "is teachable", and "justice" [virtue], to exhibit that justice [virtue] is teachable: Aristotle is able to conclude that justice [virtue], is teachable, on the basis of an abductive reasoning, that is ἀπαγωγή. A second example of *leading away* is also presented, which illustrates that in order to make a rectilinear figure equal to a circle only one additional middle term is required; that is the addition of half circles to the rectilinear figure.

I do not think appropriate to consider, following Kraus (2003, p. 247), the adumbrated syllogism (first Aristotelian example in the passage above)

AB Whatever is knowledge, can be taught
BC Virtue (e.g., justice) is knowledge
AC Therefore virtue can be taught

[4]I have already said above that Aristotle insists that all syllogisms are valid; there is no such thing as an invalid syllogism. The syllogistic tradition began to relax this requirement: here, and in the following, I will use the term syllogism in this modern not strictly Aristotelian sense.

just an example of a valid deduction, so insinuating Peirce's interpretation failure. Indeed, it seems vacuous to elaborate on the syntactic structure of the involved syllogism, as Kraus does: the problem of abduction in chapter B25 is embedded in the activity of the inferential mechanism of "leading away" performed thanks to the introduction of new terms, as I explained above. He also says that the second Aristotelian example

Whatever is rectilinear, can be squared
A circle can be transformed into a rectilinear figure by the intermediate of lunes
Therefore, a circle can be squared

still a simple deduction, was questionably supposed by Peirce to be fruit of the correction of Aristotle's original text due to the "stupid" Apellicon, considered responsible of blurring Aristotle's reference to abduction. Indeed, Kraus suggests that, following Peirce, the original text would have to be the following:

Whatever is equal to a constructible rectilinear figure, is equal to a sum of lunes
The circle is equal to a sum of lunes
Therefore, the circle is equal to a constructible rectilinear figure

which indeed fits the Peircean abductive schema. At this point Kraus (2003, p. 248) ungenerously—and, in my opinion, erroneously, as I have already said—concludes "Peirce's argument surely is bad. It begs the question". I disagree with this skeptical conclusion.

We need a deeper and better interpretation of Aristotle's passage. To this aim we need analyze some aspects of Plato's dialectic,[5] ancient geometrical cognition, and the role of middle terms: I am convinced we will gain a new positive perspective about the constitutive eco-cognitive character of abduction, just thanks to Aristotle himself.

2 Geometry and Logic: Models/Constructions and Middle Terms in Abduction

Many researchers (for example Faller 2000; Karasmanis 2011) contend that Aristotle's passage above reworks two examples already given by Plato in the *Meno* dialogue (Plato 1977). The interpretative conundrum is related to the role played by the middle term: first of all Aristotle points out that abduction is such "when it is clear (δῆλον) that the first term belongs to the middle and unclear (ἄδηλον) that the middle belongs to the third, though nevertheless equally convincing (πιστόν) as the conclusion, or more so". This good situation does not always hold. In this last case,

[5]I agree with the following claim by Woods: "Whatever else it is, a dialectical logic is a logic of consequence-drawing" (Woods 2013, p. 31), that is not merely a logic of "consequence-having" (on these concepts cf. my recent article Magnani 2015b on the-so-called "naturalization of logic").

Aristotle says that to have an abduction an act of *introducing* "something in addition" is necessary, and the addition can also be characterized by more middle terms: "That science is teachable, then, is obvious, but it is unclear whether virtue is a science. If, therefore, BC is equally convincing (πιστόν) as AC, or more so, it is a leading away (ἀπαγωγή) (for it is closer to scientific understanding (ἐγγύτερον γάρ του ἐπίστασθαι) because of taking something in addition, as we previously did not have scientific understanding (ἐπιστήμη) of AC). Or next, it is leading away (ἀπαγωγή) if the middle terms between B and C are few (ὀλίγα) (for in this way also it is closer to scientific understanding (εἰδέναι))."

A more careful analysis of the passage requires a reference to some central Plato's ideas about dialectic argumentation. Already in the *Meno* dialogue Socrates "dialectically" reflects upon the various relationships between virtue, knowledge, and teachability and also furnishes the example of a geometrical "analysis" (or "method of hypothesis"), so-called in the literature pertaining ancient mathematics.[6]

2.1 'Απαγωγή and Geometry

The method of analysis in geometry, already employed by Hippocrates of Chios, can involve, to creatively solve the problem at hand, (1) a *diorism*, which resorts to the finding of the definite conditions under which one construction might be inscribed within another, and (2) a *porism*, which refers to direct or intentional discovery through suitable higher constructions related to the finding of indefinite cases, eventually capable of innumerable solutions, so looking for a higher unifying solution. What is important to note is that in the method of analysis new strategic constructions have to be found: translated in syllogistics terms, this means it is necessary "taking something in addition", as Aristotle says in the passage above, that is a new "middle" (or new "middles").[7]

[6]Cf. for example Hintikka and Remes (1974).

[7]Porism is usually translated as lemma or corollary. I am referring here to another meaning that goes deeper into the philosophy of ancient Greek mathematics. In this case porisms are active in solving problems in which it is necessary to adopt new suitable constructions. The most famous collection of porisms of ancient times was the book *The Porisms* of Euclid. This work is lost: the trace survived thanks to the *Collection* of Pappus. Playfair noted that, thanks to porisms, the analysis of all possible particular cases of a proposition would establish that: (1) under some conditions a problem becomes impossible; (2) under some other conditions, indeterminate or related to an infinite number of solutions the problem can be solved. Classical works on porisms are Playfair (1882), Simson (1777). The concept is controversial and still subjected to studies and interpretations provided by researchers in ancient philosophy: a rich reference to the literature available is given in Karasmanis (2011, pp. 39–40).

The activity of finding new geometrical constructions (or new middle terms) is clearly a heuristic process[8]—based on a dynamics of subsequent steps—aiming at discovering new geometrical truths, a process which is a case of ἀπαγωγή, that is of *abduction*, also in the modern sense of the word.[9] In a syllogistic perspective, which regards arguments in general, not necessarily geometrical, the method of analysis still resorts to the activity performed for finding the suitable middle term(s) able to substantiate the reasoning at play.

It is absolutely important to note that in Plato the *logico-dialectical* anticipation of the Aristotelian syllogistic relationship between virtue, knowledge, and teachability is directly derived from the geometrical example, as Socrates expressly says in the *Meno* dialogue (see below, the following subsection). From Hippocrates of Chios to Proclus, ἀπαγωγή is the fundamental pre-Euclidean method for solving problems, as a method of discovery, and at the same time also for proving theorems (Karasmanis 2011), no surprise that it is implicitly central in Plato and still explicitly present in Aristotle's *Prior Analytics*. In this perspective, we will soon see, the English translations "reduction" and "leading away" both stress the fact that the process involves a transition from a problem or theorem to another, which, if known or constructed, will make the original problem or theorem evident and solved (or potentially solved).

2.2 'Απαγωγή, Dialectics, and Logic

Socrates and Meno, by constructing the arguments on whether virtue is teachable are engaged in clarifying the following syllogism (obviously valid, but where both the second premiss and the conclusion are far from being reliable)

AB Whatever is knowledge, can be taught
BC Virtue (e.g., justice) is knowledge
AC Therefore virtue can be taught

Faller (2003a, b) explains that Socrates, exactly thanks to what Aristotle calls a "leading away" (ἀπαγωγή) argument—that is the introduction of new middles—, had established that since virtue is "good" and "there is nothing good that is not embraced by knowledge, our suspicion that virtue is a kind of knowledge would be well founded" (Plato 1977, 87d), consequently, Meno can say: "We must now conclude, I think, that it is; and plainly, Socrates, on our hypothesis that virtue is

[8]I have provided an analysis of heuristics in the light of abductive cognition in Magnani (2014). Heuristics, in so far they can be algorithmically rendered, are still rules-based, even if these rules are weaker from the normative point of view, when compared with the logical rules, and typically closer to what actual human reasoners do.

[9]In Magnani (2009, Chaps. 2 and 3), I illustrate how abductive cognition is characteristically also related to various examples of diagrammatic reasoning (based on porisms, we can say), for example in the case of the discovery of the first non-Euclidean geometries.

knowledge, it must be taught" (89c). In the Aristotelian terms exploited in chapter B25:

AB Whatever is knowledge, can be taught
BC Virtue (e.g., justice) is knowledge
MC Virtue is good[10]
BM Good is knowledge
AC Therefore virtue can be taught

The first premiss is evident, the second uncertain, and the conclusion is even more uncertain. We can arrive—using the Aristotelian words—"closer to scientific understanding (ἐγγύτερον γάρ του ἐπίστασθαι)", with the introduction of a new term "good" and the propositions "virtue is good" and "good is knowledge", which can *possibly* support the second premiss of the original syllogism.

Plato starts from AC, which reflects a situation of ignorance, a hypothesis *to be justified*, instead of its contrary; BC would guarantee the result but it has to be supported. MC, the "leading away" at stake, is the further hypothesis chosen to perform this task: MC results obvious and true.[11] Then Plato establishes that "good is knowledge" (BM) and concludes that "virtue is knowledge". Through this process (dialectical) AC is rendered—again, using the Aristotelian words—"closer to scientific understanding (ἐγγύτερον γάρ του ἐπίστασθαι)".

Meno accepts this conclusion but Socrates is not satisfied: to solve the problem we need—still in Aristotelian words—another "leading away". Indeed Socrates initiates a second argument consisting of a further hypothesis, that "if virtue is teachable, then there would be teachers of it". Unfortunately, because of the empirical fact that there are no teachers of virtue, virtue is not teachable, a conclusion which conflicts with the previous one about teachability. In sum, to prove that virtue is teachable it has been necessary to analyze its nature: "what is virtue"; it has been necessary this method of hypothesis to examine the features of an obscure subject.

First of all we have to note and remember that (1) in the example about virtue Plato adopts exactly the same method used in geometrical "analysis". We also have to stress that (2) we reached two conflicting conclusions (already available, one statement and its negation) and further steps would have to be performed to execute what is usually called *cutdown* process (cf. Magnani 2013) to arrive to be "closer to scientific understanding (πάντως γάρ ἐγγύτερον εἶναι συμβαίνει τῆς ἐπιστήμης)", that is to a unique conclusion (the best result, which echoes abduction as the best explanation). Exactly in the spirit of Peircean original perspective on abduction, we have to select (and so to prefer) one of the two conflicting conclusions.

---

[10]Karasmanis usefully notes that the term "good" is not given in the analogous Aristotelian example I have illustrated in the previous subsection. Aristotle only says that an intermediate term is introduced (Karasmanis 2011, p. 37).

[11]This proposition corresponds to that *arche* (ἀρχή) which was so called, in the case of the geometrical analysis, by Hippocrates of Chios (cf. Magnani 2001, Chap. 4).

Again, let me stress that Plato's argumentation about virtue is the dialectic analogue of a diorismic/poristic geometrical model-based process, which in turn substantiates the Aristotelian "taking something in addition", where various strategies can be activated: various kinds of arguments (for example the reaching of evident higher hypotheses from which the initial one can be deduced), considerations of simplicity, looking for consequences, (for example in terms of empirical ascertainments and testing), which are able not only to create new cognitive perspectives (fill-up aspect) but also to select (cutdown aspect) the multiple or conflicting flow of results.

2.3 *Geometry and Logic Entangled: The Eco-cognitive Openness of 'Απαγωγή*

Geometrical analysis initially transforms a given problem into one that is more abstract and general: even if there are conflicting views in the available literature on the subject,[12] we can say that diorisms and porisms (often consisting in the depicting of locus[13] problems) favor a form of further geometrical cognition devoted—thanks to the study of auxiliary objects—to finding the conditions of possibility of an actual process of subsequent diagrammatic constructions, in turn finalized to solve the problem. Diorisms aim at determining the overall properties of the solutions, and so represent a wide range of mathematical activities, which "lead away" from the problem at hand to other unexplored porismic territories (diagrams and sentential proofs for example, but, for the sake of generality of various cognitive processes, we can also add other model-based or manipulatory activities totally eco-cognitively open). In the diorismic/poristic stage, the geometrician exploits the adopted auxiliary objects to show that a single solution is always possible, or if not, the limitations of the process or how many solutions there may be and how they are arranged (Saito and Sidoli 2010).

The process performs a *reduction* of the problem—caused by the hypothetical question to be solved—to another one (again, it is the Aristotelian syllogistic "leading away"), which we expect will enable us to solve the original problem (I have already said that in the Aristotelian passage above the word ἀπαγωγή is often translated with "reduction", and that we need interpret reduction as the transition to another cognitive sub-process and not as the *reduction ad absurdum*).

In the case of Plato's second problem—the geometrical one—we are to determine whether a certain rectilinear figure could be constructed along the diameter of a circle examining it by means of "a certain helpful hypothesis" (that is by means of an additional term, in Aristotelian terms), expressed in the following passage (see

[12]Cf. above, footnote at p. 7.

[13]It is interesting to note that the term *topoi* (in Latin *loci*) migrates to Aristotle's rhetoric and later rhetoricians' studies, probably parasitic of its origin in geometrical analysis (Faller 2003b).

the emphasis I have added), where a state of ignorance is immediately declared". [I have already stressed in the previous subsection that it is important to note that in Plato it is just the "clarification" of the dialectic relationship between virtue, knowledge, and teachability, which is directly derived from the geometrical example, as Socrates expressly says]:

> So it seems we are to consider what sort of thing it is of which we do not yet know what it is! Well, the least you can do is to relax just a little of your authority, and allow the question – whether virtue comes by teaching or some other way – to be examined by means of hypothesis. I mean by hypothesis what the geometricians often do in dealing with a question put to them; for example, (86e) whether a certain area is capable of being inscribed as a triangular space in a given circle: they reply – "*I cannot yet tell whether it has that capability; but I think, if I may put it so, that I have a certain helpful hypothesis for the problem, and it is as follows: If this area is such that when you apply it to the given line [as a rectangle of equal area] of the circle you find it falls short by a space similar to that which you have just applied, then I take it you have one consequence, and if it is impossible for it to fall so, then some other. Accordingly I wish to put a hypothesis, before I state our conclusion as regards inscribing this figure in the circle by saying whether it is impossible or not*" (Plato, 86e, 87a).

Let me reiterate that Socrates explicitly analogizes his reasoning about virtue to the one used in the geometrical example, and we can reasonably guess that the source of the analogy is exactly the just illustrated geometrical example:[14]

> In the same way with regard to our question about virtue, since we do not know either what it is or what kind of thing it may be, we had best make use of a hypothesis in considering whether it can be taught or not, as thus: what kind of thing must virtue be in the class of mental properties, so as to be teachable or not? In the first place, if it is something dissimilar or similar to knowledge, is it taught or not – or, as we were saying just now, remembered? (cit., 87b).

To determine whether a certain rectilinear figure could be constructed along the diameter of a circle Socrates establishes the hypothesis which I have emphasized in the first passage above from the *Meno* dialogue: the hypothesis needs be worked thanks to a diagrammatic process, a "leading away", which opens up the reasoning to an eco-cognitive dimension, which in our case corresponds to the ἀπαγωγή: an abduction, endowed with its degrees of uncertainty. The echo of this reference to the importance of diagrams in analyzing reasoning is still vivid in Peirce: "I said, Abduction, or the suggestion of an explanatory theory, is inference through an Icon" (Peirce 1997, p. 276).

[14]A strict relationship between geometry and dialectics stills echoes in Proclus: "[…] mathematics reaches some of its results with analysis, others by synthesis, expounds some matters by division, others by definition, and some of its discoveries binds fast by demonstration, adapting these methods to its subjects and employing each of them for gaining insight into mediating ideas. Thus its analyses are under the control of dialectic, and its definitions, divisions, and demonstrations are of the same family and unfold in conformity with the way of mathematical understanding. It is reasonable, then, to say that dialectic is the capstone of the mathematical sciences" (Proclus Diadochus 1873, 43, p. 35).

A brief note on recent cognitive rich research on diagrammatic geometrical reasoning has to be introduced. Fresh studies have shown that false premisses (also due to the presence in models/diagrams of both substantive and auxiliary assumptions, indeed spurious problematic sub-diagrams and new "individuals" can pop-up at any step of geometric constructions (Crippa 2009, p. 105)) are not exploited in the cognitive abductive process, because, in the various heuristics, only the *co-exact* properties are exploited. The notion of co-exact properties, introduced by Manders (2008), is worth to be further studied in fields that go beyond the realm of deductive processes of classical geometry, in which it has been nicely underscored, so usefully touching various discovery cognitive processes.[15] Mumma illustrates that in Euclid's deductive framework diagrams contribute to proofs only through their co-exact properties: I suggest that this is also typical of diorismic/porismic processes and of their creative counterparts, exactly endowed with an objection-refuting role. Indeed

> Euclid never infers an exact property from a diagram unless it follows directly from a co-exact property. Exact relations between magnitudes which are not exhibited as a containment are either assumed from the outset or are proved via a chain of inferences in the text. It is not difficult to hypothesize why Euclid would have restricted himself in such a way. Any proof, diagrammatic or otherwise, ought to be reproducible. Generating the symbols which comprise it ought to be straightforward and unproblematic. Yet there seems to be room for doubt whether one has succeeded in constructing a diagram according to its exact specifications perfectly. The compass may have slipped slightly, or the ruler may have taken a tiny nudge. In constraining himself to the co-exact properties of diagrams, Euclid is constraining himself to those properties stable under such perturbations (Mumma 2010, p. 264).

In the Aristotelian (and Platonic) perspective (see chapter B25 of the *Prior Analytics*) I have delineated in this section we can definitely conclude that the general concept of abduction must be seen as constitutively and widely *eco-cognitive-based*. Indeed, by contrast, we have to remember that Aristotle says, in the passage I have already quoted and that I am reporting again, that a valid syllogism—by necessity—is instead not at all open to something "external": "A deduction (συλλογισμὸς) is a discourse (λόγος) in which, certain things having been supposed, something different from the things supposed results of necessity because these things are so. By 'because these things are so'. I mean 'resulting through them,' and by 'resulting through them' I mean '*needing no further term from outside (ἔξωθεν) in order for the necessity to come about*' " (Aristotle 1989, A1 24, 20–25, p. 2) (emphasis added).

[15]Manders' definition describes the co-exact properties "as those conditions unaffected by some range of every continuous variation of the diagram" and the exact ones as "those which, for at least some continuous variation of the diagram, obtain only in isolated cases" (Manders 2008). "Diagrams of a single triangle, for instance, vary with respect to their exact properties. That is, the lengths of the sides, the size of the angles, the area enclosed, vary. Yet with respect to their co-exact properties the diagrams are all the same. Each consists of three bounded linear regions, which together define an area" (Mumma 2010, p. 264).

Even if in this article I cannot illustrate in detail the diagrammatic constructions, which make possible to afford the geometrical problem illustrated by Plato-Socrates,[16] it is clear that, in syllogistic terms, the geometrical diagrammatic process, as well as the analogue argumentation about virtue, are ways of finding a "middle" ground that solves the problems at hand. Aristotle concludes, in *Posterior Analytics* "Thus it results that in all our searches we seek either if there is a middle term or what the middle term is. For the middle term is the explanation, and in all cases it is the explanation which is being sought" (Aristotle 1993, B, 90a, 5, p. 48).

At this point there is clear evidence that both Socrates' examples are recalled, with slight differences, in Aristotle's celebrated passage about abduction from the chapter B25 of *Prior Analytics*.

Let us come back to the geometrically puzzling example present in the Aristotelian passage, already reported above, involving the effort to square the circle through the lunes, a problem typical, together with the one related to the reduction of the famous Delian problem, of the geometrical research deriving from Hippocrates of Chios

DE Whatever is rectilinear, can be squared
EF A circle can be transformed into a rectilinear figure by the intermediate of lunes
DF Therefore, a circle can be squared
[D = square, E = rectilinear figure, F = circle]

The first premiss is known and true, the second is uncertain, the conclusion even more uncertain: a "leading away", towards the lunes, has to start. In the above syllogism we introduce a new term (N = lune) and two new premisses "EN = the lunes become rectilinear" and "NF = the circle is a sum of lunes": *thanks to and together with* the related diagrammatic constructions, the schema becomes

DE Whatever is rectilinear, can be squared
EF A circle can be transformed into a rectilinear figure by the intermediate of lunes

EN The lunes become rectilinear
NF The circle is a sum of lunes

DF Therefore, a circle can be squared

The new additional premiss, fruit of a "leading away", aims at supporting the second uncertain premiss EF to approximate knowledge and so to solve the problem (Karasmanis 2011, p. 27). Moreover, Aristotle clearly notes, we have to deal with one or few new intermediate terms: the importance of minimality of abductive cognition is prefigured. The concept of abduction is finally established: it is leading away (ἀπαγωγή), Aristotle concludes, if the middle terms between B and C are few (ὀλίγα) (for in this way we are also closer to scientific understanding (εἰδέναι)). The efficacy of the abductive procedure is thus dependent on a minimum

[16]An interesting reconstruction is given in Faller (2003b).

of middle terms, because too many moves will generate excessive distance for the argument to be convincing. In case of multiple kinds of middle additional terms which lead to different conclusion we still have to select/discriminate both the appropriate/productive additional middles and the and best related final result (cut/down problem).

3 Conclusion

Also following Aristotle, we illustrated in this article that abduction is related to local, pragmatic, user-sensitive factors associated to situatedness, that is to factors that are subject to the influence of strong eco-cognitive constraints and chances. On one side Aristotle clearly states that in syllogistic theory local/environmental cognitive factors—external to the inferential process, for example regarding users/reasoners, have to be given up: the related reasoning is "by necessity"; on the other side Aristotle presents a seminal perspective on abduction, which is instead in tune with my EC-Model: Aristotle's ἀπαγωγή presents a clear eco-cognitive openness. I have provided an illustration of the role of the method of analysis and of the middle terms in Plato's dialectic argumentation, considered as related to the diorismic/poristic process in ancient geometry, showing it as a theoretical heritage which informs Aristotle's chapter B25 of *Prior Analytics*, concerning abduction. Thanks to Aristotle we can gain a new positive perspective about the "constitutive" eco-cognitive character of abduction.

Acknowledgements Parts of this article are excerpted from the *Journal of Applied Logic*, 13: 285–315, Lorenzo Magnani, The Eco-cognitive model of abduction. 'Απαγωγή now: Naturalizing the logic of abduction, Copyright (2015), with permission from Elsevier. For the instructive criticisms and precedent discussions and correspondence that helped me to develop my analysis of the naturalization of logic and/or abductive cognition, I am indebted and grateful to John Woods, Atocha Aliseda, Luís Moniz Pereira, Paul Thagard, Woosuk Park, Athanassios Raftopoulos, Michael Hoffmann, Gerhard Schurz, Walter Carnielli, Akinori Abe, Yukio Ohsawa, Cameron Shelley, Oliver Ray, John Josephson, Ferdinand D. Rivera, to the two reviewers, and to my collaborators Tommaso Bertolotti and Selene Arfini.

References

Aristotle. (1989). *Prior analytics* (R. Smith, Trans.). Indianapolis/Cambridge: Hackett Publishing Company.

Aristotle. (1993). *Posterior analytics* (2nd ed., J. Barnes, Trans.) Oxford: Clarendon Press.

Crippa, F. (2009). To prove the evident. On the inferential role of Euclidean diagrams. *Theory of Science, 31*(2), 101–112.

Faller, M. (2000). *Plato's philosophical use of mathematical analysis* (PhD thesis). University of Georgia, Athens, GA, USA.

Faller, M. (2003a). *The origin of Peirce's abduction in Plato's analytic method* (Manuscript, Online).

Faller, M. (2003b). Plato's geometrical logic. In *Proceedings of the Society for Ancient Greek Philosophy* (Online).

Gabbay, D. M., & Woods, J. (2005). *The reach of abduction*. North-Holland, Amsterdam: Elsevier.

Hintikka, J. & Remes, U. (1974). *The method of analysis. Its geometrical origin and its general significance*. Dordrecht: Reidel.

Karasmanis V. (2011). 'Απαγωγή: Hyppocrates of Chios and Plato's hypothetical method of the *Meno*. In A. Longo & D. Del Forno (Eds.), *Arguments from hypotheses in ancient philosophy* (pp. 21–42). Naples: Bibliopolis.

Kraus, M. & Charles, S. (2003). Peirce theory of abduction and the Aristotelian enthymeme from signs. In F. H. van Eemeren, J. A. Blair, C.A. Willard & A. F. Snoeck Henkemans (Eds.), *Anyone who has a view. Theoretical contributions to the study of argumentation* (pp. 237–254). Dordrecht/Boston/London: Kluwer Academic Publishers.

Magnani, L. (2001). *Philosophy and geometry. Theoretical and historical issues*. Dordrecht: Kluwer Academic Publisher.

Magnani, L. (2009). *Abductive cognition. The epistemological and eco-cognitive dimensions of hypothetical reasoning*. Heidelberg/Berlin: Springer.

Magnani, L. (2013). Is abduction ignorance-preserving? Conventions, models, and fictions in science. *Logic Journal of the IGPL, 21*(6), 882–914.

Magnani, L. (2014). Are heuristics knowledge-enhancing? Abduction, models, and fictions in science. In E. Ippoliti (Ed.), *Heuristic reasoning* (pp. 29–56) Springer: Heidelberg/Berlin.

Magnani, L. (2015a). The eco-cognitive model of abduction. 'Απαγωγή now: Naturalizing the logic of abduction. *Journal of Applied Logic, 13*, 285–315.

Magnani, L. (2015b). Naturalizing logic. Errors of reasoning vindicated: Logic reapproaches cognitive science. *Journal of Applied Logic, 13*, 13–36.

Magnani, L. (2016). The eco-cognitive model of abduction II. Irrelevance and implausibility exculpated. *Journal of Applied Logic, 5*, 94–129.

Manders, K. (2008). The Euclidean diagram. In P. Mancosu (Ed.), *Philosophy of mathematical practice* (pp. 112–183). Oxford/New York: Clarendon Press.

Mumma, J. (2010). Proofs, pictures, and Euclid. *Synthese, 175*, 255–287.

Otte, M. (2006). Proof-analysis and continuity. *Foundations of Science, 11*, 121–155.

Peirce, C. S. (1931–1958). *Collected papers of Charles Sanders Peirce*. In Hartshorne, C. & Weiss, P. (Eds.), Vols. 1–6; Burks, A. W. (Ed.), Vols. 7–8. Cambridge, MA: Harvard University Press.

Peirce, C. S. (1992–1998). *The essential Peirce. Selected philosophical writings*. In N. Houser & C. Kloesel (Eds.) (1867–1893) Vol. 1; The Peirce Edition Project (Ed.) (1893–1913) Vol. 2. Bloomington and Indianapolis: Indiana University Press.

Peirce, C. S. (1997). *Pragmatism as a principle and method of right thinking. The 1903 Harvard lectures on pragmatism*. In: P. A. Turrisi. (Ed.), State University of New York Press, Albany, NY (Peirce, C. S., "Lectures on pragmatism", Cambridge, MA, March 26–May 17, 1903, reprinted in [C. S. Peirce, 1992–1998, II, pp. 133–241]).

Phillips, J. (1992). Aristotle's abduction: The institution of frontiers. *The Oxford Literary Review, 14*(1–2):171–196. Special Issue on "Frontiers" G. Bennington & B. Stocker (Ed.).

Plato. (1977). *Plato in twelve volumes* (Vol. II, Laches, Protagoras, Meno, Euthydemus, with an English translation by W. R. M. Lamb). Cambridge, MA: Harvard University Press.

Playfair, J. (1882). *Works of John Playfair*. Edinburgh: A. Constable & Co.

Proclus Diadochus. (1873). *In Primum Euclidis Elementorum librum Commentarii*. B. G. Teubner, Leipzig. ex recognitione G. Friedlein, translated and edited by G. R. Morrow *A Commentary on the First Book of Euclid's Elements*, Princeton: Princeton University Press, 1970.

Saito, K., & Sidoli, N. (2010). The function of diorism in ancient Greek analysis. *Historia Mathematica, 37*, 579–614.

Simson, R. (1777). *A treatise concerning porisms*. Canterbury: Simmons and Kirby.
Woods, J. (2013). Against fictionalism. In L. Magnani (Ed.), *Model-based reasoning in science and technology. Theoretical and cognitive issues* (pp. 9–42). Heidelberg/Berlin: Springer.
Woods, J. (2014). *Aristotle's earlier logic*. London: College Publications. Second revised edition. Originally published by Hermes Science Publications, Oxford, 2001.

Part III
Historical, Epistemological, and Technological Issues

Beyond Telling: Where New Computational Media is Taking Model-Based Reasoning

Sanjay Chandrasekharan

Abstract The emergence of new computational media is radically changing the practices of science, particularly in the way computational models are built and used to understand and engineer complex biological systems. These new practices present a novel variation of model-based reasoning (MBR), based on dynamic and opaque models. A new cognitive account of MBR is needed to understand the nature of this practice and its implications. To develop such an account, I first outline two cases where the building and use of computational models led to discoveries. A theoretical model of the possible cognitive and neural mechanisms underlying such discoveries is then presented, based on the way the body schema is extended during tool use. This account suggests that the process of building the computational model gradually 'incorporates' the external model as a part of the internal imagination system, similar to the way tools are incorporated into the body schema through their active use. A central feature of this incorporation account is the critical role played by tacit and implicit reasoning. Based on this account, I examine how computational modeling would change model-based reasoning in science and science education.

1 Introduction

Modern science deals with entities and patterns that exist at size, time and complexity scales that are not available to human perception and action. Examples include galaxies, gravitational waves, DNA, molecular forces, evolution, plate tectonics, oscillating reactions, biological arms races, complex feedback loops etc. These entities and patterns are described using abstract external representations, such as equations, graphs, models, simulations, theories, etc., and experimentally

S. Chandrasekharan (✉)
The Learning Sciences Research Group, Homi Bhabha Centre for Science Education, Tata Institute of Fundamental Research, V.N. Purav Marg, Mankhurd, Mumbai 400088, India
e-mail: sanjay@hbcse.tifr.res.in

L. Magnani and C. Casadio (eds.), *Model-Based Reasoning in Science and Technology*, Studies in Applied Philosophy, Epistemology and Rational Ethics 27, DOI 10.1007/978-3-319-38983-7_26

investigated using complex and opaque instruments, which themselves embed abstract concepts and mathematical models. Learning modern science (and technology) thus requires learning to:

(1) Imagine detailed mental models
(2) Transform external models and related representations
(3) Integrate the mental models and external models

These three skills form the core of Model-Based Reasoning (MBR), which is now considered the dominant component of scientific reasoning (Hestenes 2011, 2013). Most discussions on MBR focuses on internal models, and not on external models. Until recently, most discussions about MBR-based science discovery (Nersessian 1999, 2010) and learning (Hestenes 2013; Lehrer and Schauble 2006) did not critically examine the media on which the external model is based, particularly the role this factor plays in discovery and learning. This is because most examined external models were based on static media (such as equations, graphs and physical models), and MBR was analysed from the perspective of these static media. Following this static media view, the knowledge encoded in external models was considered persistent and available for examination and analysis.

The current widespread use of computational modeling requires changing these static media assumptions, as computational models are both dynamic and opaque (Chandrasekharan et al. 2012). Following the shift to computational modeling in scientific practice, such models are now used in science education as well (Wilensky and Reisman 2006). Given its unique properties, computational modeling presents a novel variation of model-based reasoning (MBR), particularly MBR based on dynamic and opaque models, and a new cognitive account of MBR is needed to understand the nature of this practice and its implications. To develop such an account, I first outline two cases where the building and use of computational models led to discoveries. A theoretical model of the cognitive and neural mechanisms underlying such discoveries is then presented, based on the way the body schema is extended during tool use. This account suggests that the process of building the computational model gradually 'incorporates' the external model as a part of the internal imagination system. A central feature of this incorporation account is the critical role played by tacit and implicit reasoning.

2 The Nature of Computational Media

One way to understand the impact of science moving to new computational media is to examine other such media transitions in history. A recent and central one is the transition from orality to literacy. This shift, which emerged over 6000 years, changed the nature of cognition. Ong (2013) examines the nature of this shift, and highlights the following points:

1. Oral cultures never "looked up" anything; they only used recall. Writing lowered the need for recall, as well as memory techniques that supported this cognitive process (such as mnemonics, verse and rote learning).
2. Oral thought emphasized redundancy, as the spoken word does not persist. Sparse, linear, analytic thought is thus a product of writing.
3. Oral thought was conservative, as society regarded highly those wise old men and women who specialize in conserving knowledge. This conservation emphasis inhibited intellectual experimentation (a central value of science).
4. Oral thought was close to the human lifeworld, as learning or knowing meant achieving close, empathetic, communal identification with the known. Writing created distance, separating the knower from the known. This set up conditions for "objectivity", in the sense of personal disengagement or distancing.

Taken together, this view suggests that writing is a critical factor that *enabled* the development of science and its supporting values and practices. Hestenes (2011) argues that science and mathematics was made possible by writing. Rotman (2008) takes these points further, examining how the nature of writing is related to western cultural notions of the Self, God, and the Platonic nature of mathematics. Also worth noting is the key power shift associated with the move to writing, where the value of chanting (in Sanskrit/Latin/Arabic) was eroded, paving the way to the 'writing class' replacing the 'chanting class'. More broadly, writing enabled new institutional mechanisms, such as land titles, paper contracts, written law and paper money, which together made possible the economic framework within which science functions. The current pedagogical and institutional mechanisms for education, such as standardised curricula, lecture-driven classrooms, written-exams, and certification, are also shaped by the nature of writing and print media.

Similar to writing and print media enabling and reshaping oral knowledge, learning traditions and associated values, the rise of computing is leading to the emergence of a powerful new media system that is inherently dynamic, interactive, participatory and social—features not readily provided by static print media. These powerful features of new computing media allow re-imagining current discovery and learning practices, particularly model-based reasoning, and institutional mechanisms related to science and science education. Similar to the shift to writing, this move will bring in new value systems. This ongoing shift is widely understood and acknowledged, but what is not clear is the direction of this rapidly unfolding change. An analytic, particularly cognitive, understanding of this systemic shift is critically needed, as this will help society adapt more quickly. This is all the more important because the shift is happening in Internet time ($\sim$50 years), while the shift to writing happened over thousands of years.

As a starting point for the analysis of how new media would change the science and science education landscape, the following list captures some of the features supported by print media (text and graphics) and new computational media. It is worth noting that new computational media include text, which suggests that the transition from print would be different from the shift from orality. Particularly, print will not be replaced, but would be augmented.

Print media	New computational media
Static (i.e. does not move)	Dynamic
Non-manipulable	Manipulable and interactive
Individual focused	Social
Removed from the world	Can be hooked to the world
Linear navigation	Multiple navigation paths and trajectories
Explicit encoding of knowledge	Knowledge emerges from interaction

The following sections outline two cases studies and a theoretical model that could help understand the nature of the shift in science to computational media, and how this is changing model-based reasoning. The first section outlines how the building of a computational model led to a remarkable discovery in an interdisciplinary lab. The second section outlines the way basic science discoveries are made using new crowd sourcing games in biology. The third section examines a theoretical model of the possible cognitive/neural mechanisms involved in these two cases, and how interacting with computational models and games could lead to scientific discoveries. The final section examines the broader implications of this model, particularly one possible trajectory of change for science and science education.

3 Building to Discover

In the fields of biomedical engineering and systems biology, computational models are built to develop insights into the behavior of complex biological systems. Based on this understanding from modeling, new technologies are developed to control biological systems, such as neuronal populations (Chandrasekharan 2009) and metabolic pathways (Chandrasekharan and Nersessian 2015). In such cases, computational models are built to understand highly non-linear systems that are too complex to be modeled using traditional approaches based on equations and graphs. Since the phenomena they model are highly complex and dynamic, the models are highly complex and dynamic as well, which makes an explicit understanding of the multiple interactions between different variables (usually above 10) not feasible. However, fundamental discoveries about the natural phenomena have emerged from such 'opaque' (Di Paolo et al. 2000) models and control systems have been built based on this understanding (Lenhard 2006; Winsberg 2006). What is the nature of model-based reasoning in such cases of discovery and innovation? I briefly outline one such case of discovery below, see Chandrasekharan and Nersessian (2015) for details.

Understanding metabolic pathways (a network of biochemical reactions) is a key problem in systems biology, particularly when seeking to reengineer the pathways to develop new organisms, such as plants that allow cheap production of biofuel. One central problem in the production of biofuel is efficiently breaking down lignin, the key biochemical in the plant cell wall. Developing genetically modified plants with lower amounts of lignin would lead to more efficient biofuel production. Modeling would help in identifying systematic ways to lower lignin levels in plants. In the case we report (Chandrasekharan and Nersessian 2015) G10, an electrical engineer with no background in biochemistry, develops a model of lignin, in two phases, first for poplar, then for alfalfa. Based on these models, he made a series of modifications to the scientific understanding of the lignin pathway. One spectacular finding stood out: The modeling showed G10 that the traditional pathway—used by almost everyone in the field for 20 years—is incomplete, and an element (named X by G10) outside the standard pathway has a significant regulatory effect on the behavior of the lignin pathway.

G10's collaborators found this proposal provocative, and did experiments to test this proposal. The experiments identified a possible candidate metabolite that played the specific roles X played in G10's models. A paper outlining the modeling and experimental results was published in a high-impact modeling journal, and the paper was written jointly with the experimental collaborators. This result illustrates clearly the ideal case of modeling—of the model making a significant experimental prediction, which is then tested and validated by the experimentalists. It shows how modeling can lead to discovery, and the value modeling can provide for experimentalists.

Note that the original goal of the lignin project was tweaking a given pathway so as to make lignin break down more readily for biofuel production, which is an engineering goal. But G10 ended up changing the standardized pathway, the scientific consensus on the mechanism underlying lignin production. This is a basic biological science discovery, generated by an electrical engineer, based on a few months of modeling. The remarkable discovery shows that the built external model is not just a replica of an existing standardized structure (the pathway) for the purpose of tweaking. *The external model, and its building, is a mechanism that affords discovering unknown features of the pathway*. Approaching this discovery event from the point of view of understanding the role of computational models, and more broadly external representations, in science cognition, a key question is: What are the cognitive changes involved in building the external simulation model, and how could these changes lead up to the discovery?

We propose (see Chandrasekharan and Nersessian 2015 for details) that the key cognitive change is that within the course of many iterations of model building and simulation, the external model gradually becomes coupled with the modeler's inner mental system, particularly his imagination (simulative mental model) of the phenomena he is modeling. Based on this coupling, the modeler explores different scenarios. The building process thus slowly creates an "external imagination" that is closely coupled to the modeler's imagination system. This coupling allows "what if" questions in the mind of the modeler to be turned into detailed, and close to actual, explorations of the system.

It is important to note that the model acquires this external imagination role only in a gradual manner, through its incrementally acquired ability to enact the behavior of the system that it is modeling. As it is built over many iterations (such as the first poplar model), using many data sets, the model's output/behavior comes to parallel the pathway's dynamics. Each replication of experimental results by the model adds data, and by proxy, real-world complexity, to the model, and this process continues until the model fits all available experimental data well. At this point, the model can *enact* the behavior of the real system—the pathway that is being examined—and thus support detailed "what if" explorations that are not possible to do in the mind alone (see also Kirsh 2010) or in experiments. Importantly, the model's ability to enact the real system behavior is a very complex judgment made by the modeler, based on a large number of iterations, where a range of factors, such as sensitivity, stability, consistency, computational complexity, nature of pathway, and so on are explored. The gradual confidence in the model is thus a complex intuition about its overall performance, emerging over a long series of interactions and revisions, and does not depend just on data fitting, even though fitting is the most critical process leading to this judgment.

As the enaction ability of the model develops gradually through the building process, the model starts making manifest many behaviors the modeler might have only imagined previously. But, the model goes further, as it also makes visible many details of the system's behavior, which the modeler could not imagine (Kirsh 2010) because of the fine grain and complexity of these details. The gradual process of building creates a close coupling between the model and the modeler's imagination, with each influencing the other. The computational model now works as an external component of the imagination system. This coupling significantly enhances the researcher's natural capacity for simulative model-based reasoning (Chandrasekharan 2009; Chandrasekharan et al. 2012; Nersessian 2010), particularly in the following ways:

1. It allows running many more simulations, with many variables at gradients not perceivable or manipulable by the mind (say 0025 of metabolites a and b). These can then be compared and contrasted, which would be difficult to do in the mind.
2. It allows testing what-if scenarios that are impossible to do in the researcher's mind. Such as, what would happen if I change variable 1 and 2 downwards, switch off 6 and 21, and raise 7 and 11 with a time lag between 16 and 19?
3. It allows stopping the simulation in between and checking its state. It also allows tracking the simulation's states at every time point and, if something desirable is seen, tweaking the variables to get that effect more often and consistently. This "reverse simulation" is impossible to do in the mind or in experiments.
4. It allows taking apart different parts of the system as modules, simulating them, and putting them together in different combinations.
5. It allows changing the time at which some in-between process kicks in (say, making it start earlier or later), and this can be done for many processes, which is very difficult to do in the mind or in experiments.
6. It exposes the modeler to system-level behavior that experimenters would never encounter, as most of the above complex manipulations are not possible in experiments.

The process of building this distributed model-based reasoning system comprising researcher(s) and model leads to the creation of new or enhanced cognitive capacities. We thus propose an "incorporation" account of how computational models leads to discovery (see Chandrasekharan and Nersessian 2015), where the building process leads to two kinds of integration. First, incorporation of real-world data into the model, which allows the model to enact the behavior of the system it parallels. Second, incorporation of the model as part of the imagination system, such that imagined scenarios are tried out in the model, and the results are integrated into the internal model of the system the model parallels. This notion of incorporation is novel, and the cognitive mechanisms involved in this process would be wider than just perception, and would involve cognitive systems relating to the processing and understanding of motor control and tool use (see Chandrasekharan 2014). The possible cognitive/neural basis of incorporation is examined in the theoretical model that follows after the next section.

4 Building with Games

A second example of how new computational representations are radically changing the way scientific knowledge is generated, most notably in the biological sciences and bioengineering, is the case of *Foldit*, a video game (built on top of a computational model) that allows novel protein-folds to be designed by web-based groups of people not formally trained in biochemistry. Using *Foldit*, a 13-year-old player (Aristides Poehlman) designed protein folds that were judged better than the best biochemists' folds in CASP (Critical Assessment of Techniques for Protein Structure Prediction), the top international competition on protein-folding (Bohannon 2009). This remarkable result provides an interesting cognitive insight: the *process of building* new protein folds, using the video game interface, allowed the novice player to implicitly develop an accurate/veridical sense of the mechanics and dynamics of the protein folding problem. In this paper, I provide details of this process more generally, and develop a theoretical account of how discoveries could emerge from building.

The approach of 'crowd sourcing' difficult scientific problems to novices using novel interfaces is now widely accepted, especially after *Nature* published a paper (Cooper et al. 2010) where roughly 200,000 *Foldit* players were included as authors. The paper proposed that harnessing people's implicit spatial reasoning abilities using such model-based games could be a new method to solve challenging scientific problems. This proposal is now confirmed, with *Foldit* players making some remarkable discoveries, including building the structure of a protein causing aids in rhesus monkeys, which was an unresolved problem for 15 years (Khatib et al. 2011). The game is currently being refined to support the development of new drugs by the players. A spin-off game from *Foldit*, *EteRNA*, allows players to build RNA folds, and every week the most promising folds from the gamers are synthesized by a Stanford lab. The synthesis results are then fed back to the gamers, who use these real-world results to improve their designs. This closed loop building process has led to the gamers discovering fundamental design principles underlying RNA structure (Lee et al. 2014; Koerner 2012). Other similar crowd sourcing games include *Phylo* (helps optimize DNA sequences) *Eyewire* (helps map 3D structure of neurons). Eyewire recently helped answer some basic research questions about the way retinal cells detect motion (Kim et al. 2014).

These games mark an important shift in the direction of knowledge flow in science, which has traditionally been from implicit to explicit. For instance, in many areas of biology, the effort is to capture implicit procedural knowledge (such as flight patterns and navigation of birds) in explicit declarative terms (such as aerodynamics and signaling). In physics, procedural knowledge (such as the qualitative understanding of force) is considered to lead to misconceptions, and declarative knowledge (such as Newton's Laws) is used to explain many aspects of phenomenal experience. Given this procedural-to-declarative trajectory of scientific knowledge, the case of *Foldit* and similar games marks a new approach to discovering scientific knowledge, as such cases re-represent declarative knowledge using computational models and a manipulable interface, so that naive participants can use their procedural knowledge to build up novel patterns. At the heart of such games and other similar digital media for discovery is a re-representation—converting explicit conceptual knowledge, developed by science (structure of protein, possible folds, hydrophobic/hydrophilic interactions etc.) to build a *control interface* that can be manipulated using a set of actions. This interface allows building of new representations by novices, using their implicit spatial knowledge. These games thus present a fundamental shift in the practice of science, particularly an acknowledgment of the role played by tacit/implicit sensorimotor processes in scientific cognition (Polanyi 1958, 1966). The success of this approach suggests that there is a close connection between procedural and declarative knowledge.

This is a radical epistemic shift, and it is driven by two irreversible factors. One is the focus on understanding interdisciplinary problems such as climate change, where the phenomena under investigation are spread across many time-scales and spatial levels, and complex feedback loops are standard features of the domain. Existing theory and automated methods are not able to solve the multi-scale combinatorial problems that emerge in such areas. It is also possible that in these domains, as von Neumann (1951) observed, the phenomena are the simplest descriptions possible, and any good model would need to be more complex than the phenomena. A second factor is the emergence of 'Big Data', where petabytes of data are generated routinely in labs, particularly in biological sciences. It is not possible to analyze this avalanche of data without computational models and methods, which themselves fail to work for many problems. A good example is the classification of galaxies using data from the Hubble space telescope, a difficult problem that led to the development of Galaxy Zoo, the first effort to crowd-source science. This web-based citizen-science project has led to at least 30 peer-reviewed papers, and a new astronomical object (Hanny'sVoorwerp) named after the Dutch schoolteacher who identified it.

The crowd sourcing approach to scientific problem-solving is new, but the idea of using the human sensorimotor system to detect patterns, particularly in dynamic data generated by computational models, has been applied right from the beginning of computational modeling. Entire methodologies, disciplines, and phenomena challenging existing models have been built just from visualized patterns on computer screens. These include Complexity Theory (Langton 1984, 1990), Artificial Life (Reynolds 1987; Sims 1994), models of plant growth (Prusinkiewicz et al. 1988; Runions et al. 2005), computational bio-chemistry (Banzhaf 1994; Edwards et al. 1998), computational nanotechnology (reported in Lenhard 2004; Winsberg 2006), and climate change (Schneider 2012). All these novel areas of exploration are based on visualizing data from computational models. Apart from the visual modality, protein structure has been generated as music (Dunn and Clark 1999), and scanning microscope output has been used to generate haptic feedback (Sincell 2000).

This approach to making scientific discoveries, by coupling the sensorimotor systems of a crowd of novice humans to data embedded in novel computational media, raises a number of questions about MBR and cognition. Particularly, what cognitive mechanisms mediate the re-representation (and back) of scientific knowledge as manipulable on-screen structures? What is the relationship between declarative and procedural knowledge, such that this conversion is possible and new discoveries could emerge from this conversion process? At a more applied level, how could the visual and tactile manipulation of model elements on screen, by groups of non-scientists, quickly lead them to build valid structures representing imperceptible molecular entities they have never encountered, especially structures that have eluded practicing senior scientists for many years? What cognitive and biological mechanisms support this manipulation-based discovery process? How can these mechanisms be harnessed better, to develop other collaborative games/interfaces that address more complex and abstract scientific and engineering problems with wider applicability?

Answering these questions is critical for practicing as well as learning this new form of science and engineering. To address these questions, we require a general theoretical account that captures how discoveries could emerge from the building of new computational representations, particularly computational models, and re-representation of data from these models.

In the following section, I propose a novel theoretical account of how building and using such computational models could help in making new discoveries. This account extends the incorporation account sketched in the G10 case above, providing a specific model of the cognitive/neural mechanisms at work in the process of incorporation.

5 Incorporation: The Biological Mechanisms

Since the above cases show how novices can make discoveries in complex scientific domains by building computational structures, the mechanism underlying such discoveries cannot be domain-knowledge based. The computational model is helping the modelers extend their imagination to an external structure in the world, where manipulations can be tried out. The results from these manipulation are coupled seamlessly with the internal imagination system. What cognitive/neural mechanism makes this seamless coupling possible? I suggest that this is made possible by a version of the mechanism that extends the body schema during the use of tools.

A number of studies in monkeys have shown how the body schema is extended to incorporate external objects, particularly tools (for a review, see Maravita and Iriki 2004). One influential study (Irikiet al. 1996) examined the firing of bimodal neurons before and after a monkey learned to use a stick to gather food. Bimodal neurons in the intra-parietal cortex respond to both somato-sensory and visual input on or near the hand. That is, the bimodal neurons coding for the hand area will fire when the hand is touched, as well as when a light is flashed on the hand. Interestingly, this firing happens when the light is flashed not just on the hand itself, but also in the space close to the hand ("peripersonal space"), indicating that the neurons code for the space of possible activity, rather than just the hand. Iriki et al. examined whether this firing pattern changed when the monkey started using a stick as a tool. This investigation was done in three phases (see top panels, Fig. 1, adapted from Maravita and Iriki 2004).

In the first phase, there was no stick and the light was flashed on and near the hand, and the bimodal neuron fired. In the second phase, the monkey passively held

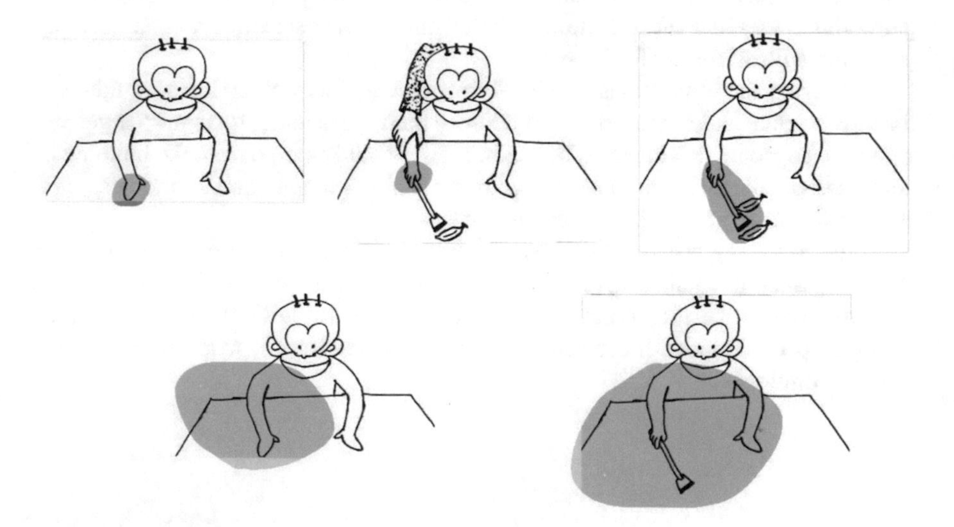

Fig. 1 Monkey with electrodes embedded in the intra-parietal cortex doing the tool task. *Top* panels show the three phases on the task, and how the per-personal space changes. The *bottom* panel shows the way the action-space of the monkey changes

the stick, and the investigators flashed the light near the monkey's hand, as well as at the end of the stick. The bimodal neuron fired only when the light was flashed near the hand. In the third phase, the monkey used the stick to retrieve food from a location that was not reachable by its hand. Immediately after this intentional action, the investigator flashed the light on the hand as well as at the end of the stick. The bimodal neuron now fired for light flashes near the hand as well as at the end of the stick, showing that the peripersonal space (the area of possible activity coded for by the neuron) had been extended to include the area covered by the stick (bottom panels, Fig. 1). The intentional action led to the stick being incorporated into the body, and the monkey's peripersonal space (possible activity space) now extended to the entire area, and objects, reachable by the stick. I will term this "active" incorporation, as the extension occurs only through intentional action. This extension of peripersonal space is important, as it shows that such incorporation is not just about adding an external entity to the body schema. Incorporation expands the range of possible activities the monkey can do—in terms of location of activity, other entities involved, nature of activity, the number of activities, and the permutations and combinations of activities. This expanded range also extends the monkey's understanding/knowledge of the stick, as well as the space around it, which is now understood in relation to the stick. The monkey's cognitive capacities are thereby expanded. Similar incorporation of external entities into the body schema has been shown with humans as well (Farne et al. 2005).

An interesting variation of this incorporation effect (which I term "passive" incorporation) is the rubber hand illusion (Botvinick and Cohen 1998). In this experiment, one hand of the participant is placed on a tabletop, and is visible to the participant. The other hand is placed on the participant's knee, under the table, and is not visible to the participant. The experimenter then places a rubber hand on the tabletop, above and parallel to the unseen hand, and next to the seen hand. The wrist end of this rubber hand is covered with a cloth. The experimenter then touches the unseen hand (under the table) and the seen rubber hand, synchronously, using a brush. After some time, the participant feels the rubber hand as part of his body, and he feels physically threatened if a knife is brought near the rubber hand. This feeling of threat is indicated by a raised galvanic skin response. When the stroking of the unseen hand and the rubber hand is asynchronous, the participant does not report feeling the illusion, and the heightened skin response does not occur. The RHI has recently been extended to induce the feeling of having three arms (Guterstam et al. 2011), and also an "invisible hand effect" when a hand is felt when empty space in front of the participant is stroked in synchrony (Guterstam et al. 2013).

The incorporation of the rubber hand into the body is similar to the incorporation of the tool by the monkey. But it is also different, as the incorporation occurs not through intentional action, but through a dissociation of visual and tactile inputs. One way to understand the relation between passive and active incorporation is to consider the passive as a faint case of the active, where the perceptual effect appears similar to the effect of using a tool, even though no intentional action is executed. In the tool case, the tactile input is seen and felt in a distant manner, but it occurs in synchrony with the visual input of the tool moving. This synchrony could be one of the factors that lead to the tool being incorporated as part of the body schema. In the passive case, a similar synchrony is detected, with no tool present. The brain then "fills-in" the missing tool, by incorporating the locus of the synchrony (the external entity) into the body schema, even though there is no intentional action executed with the entity. Recent results show that such passive incorporation also has cognitive effects. For instance, when asked to bisect a horizontal line midway, most people show a leftward bias (pseudoneglect), which is attributed to the dominance of the right brain hemisphere. This bias is reduced after the rubber hand illusion. This compensatory effect is specific to individuals who report having vividly experienced the illusion (high responders) as opposed to individuals who do not (low responders). Also, pseudoneglect was eliminated only after RHI application to the left hand (Ocklenburg et al. 2012). This suggests that passive incorporation changes the nature of actions that follow, and the cognitive events related to such actions. The extension of the peripersonal space after such incorporation has not been investigated, though the following study seems to suggest that such a change could occur following passive incorporation.

In a further variation of the RHI effect, a remarkable new study has shown that a similar synchronous splitting of the visual and tactile inputs can lead to the feeling of being out of one's body, and owning another body of a different size (van der Hoort et al. 2011). In this experiment, participants lie down, with their head looking toward their feet, while wearing a virtual reality headset that shows the legs of a mannequin lying next to them. An experimenter then simultaneously strokes the participant's legs, as well as the legs of the mannequin, with a rod. This simple manipulation creates a sensory dissociation similar to the RHI: the stroking is felt in one's own leg, but it is seen as happening synchronously in the mannequin's leg. Similar to the RHI, the synchronous dissociation creates the feeling that the feet of the mannequin are the participant's own. Interestingly, the participants then feel like they themselves are the size of the mannequin, and they feel threatened if the mannequin is attacked. This 'out-of-body' experience has remarkable cognitive effects. If the incorporated mannequin is small, the subjects feel short, and when asked to use their hands to judge the size of small boxes shown to them, participants judge the boxes as quite big. Conversely, if the incorporated mannequin is huge, participants feel they themselves are huge, and thus judge really large boxes as small.

Extending this effect further, a similar synchronous dissociation has been shown to create the feeling of being out of one's own body, and being in a point of space outside. This happens when the participant feels the tactile input in her chest, but sees the visual input in a point in space behind her, an illusion achieved using virtual reality goggles. This leads to the incorporation of this (empty) space into the body schema, and the shifting of the visual perspective to that point in space. This effect is quite remarkable, as it shows that the perceptual synchrony can lead to a form of idealized incorporation, where empty space is incorporated into the body (similar to the invisible hand illusion), by shifting the visual perspective to that point in space. This incorporation also has cognitive effects, such as a different judgment of the distance one needs to walk to reach a target (Ehrsson 2007; Lenggenhager et al. 2007). This experiment shows passive incorporation at the level of the whole body, and this type of incorporation seems to alter the nature of cognitive activities performed by the subject, and the space and perspective associated with these cognitive activities. How this global-level incorporation affects possible actions/activities and extension of peripersonal space is not clear, as this has not been explored yet.

These experiments indicate that: (1) Objects are incorporated into the body schema when used as tools, (2) Objects resembling body parts are easily incorporated into the body schema through a synchronous dissociation mechanism, and such incorporation has cognitive effects, (3) Space outside the body can easily be incorporated into the body schema, and this leads to cognitive effects. These results show the possibility of extending your body schema to incorporate external entities and perspectives (and thus knowing them by participation), and how such incorporation can lead to cognitive changes. These are early and indicative results, but taken together with the tool-use case, and the ease with which incorporation occurs, they suggest that such incorporation is possible, and it is very common. The cognitive effects illustrated by these experiments also suggest that such incorporation of external entities and space into the body schema could be a mechanism through which we understand/know external objects—via the new activities, perspectives, or the different ways of doing/examining old activities, which the objects and their features make possible.

The incorporation account provides a new way of understanding how model-based reasoning based on computational models lead to discovery, particularly discovery based on games such as Foldit. Essentially, scientific discovery games work by re-representing conceptual knowledge as a control interface, where global knowledge of the system can be gained through actions on models and feedback from these actions. The above account of how the body schema is extended to incorporate external tools and artifacts suggests the underlying mechanism in the case of Foldit and similar games could be a similar gradual integration of the internal imagination process and the external model, and the implicit understanding of the system's behavior that emerges from this incorporation.

Further, this account could be extended to model-based-learning, where conceptual knowledge is gained through similar actions and feedback, via the manipulation of models and physical artifacts. In mathematics and science education, manipulatives and models are commonly used to improve learning of abstract concepts, such as fraction concepts and area concepts, and unperceivable patterns, such as DNA structure and stereochemistry. More broadly, there are standard approaches to learning based on actions and feedback, such as learning-by-doing and activity-based-learning, and software platforms that promote action-based learning, such as Geogebra, Netlogo (Wilensky and Reisman 2006), and Kill Math, which seeks to promote learning of math and science concepts through manipulations of objects and numbers on screen. The incorporation account of model-based learning allows understanding learning situations involving manipulable models and novel digital media (Landy et al. 2014; Landy and Goldstone 2009; Majumdar et al. 2014; Marghetis and Nunez 2013; Ottmar et al. 2012), and also extend learning frameworks based on modeling (such as Modeling Theory, Hestenes 2006).

6 Beyond Telling

Computational models are complex, opaque and highly dynamic entities that embed experimental data and theoretical concepts. The two cases discussed above suggests that discoveries made by novices using such models significantly exploit *implicit* knowledge, of patterns (case 1) and visuo-spatial structure (case 2). Understanding model-based reasoning using computational models thus require an account where implicit knowledge plays a significant role. The incorporation account (Chandrasekharan 2009; Chandrasekharan and Nersessian, 2015), proposes such a theoretical model, where discoveries based on computational models are based on the gradual development of a coupling between the internal imagination system and the external model. This coupling emerges through the process of building the model and running thousands of simulations and variations. I propose here that the cognitive mechanism underlying incorporation is a reuse/extension of the mechanism involved in the incorporation of tools into the body schema (also see Chandrasekharan 2014).

Since computational models and media are here to stay, what broader implications for science practice and science education are offered by these case studies and the incorporation account? I explore four implications below:

1. From a cognition perspective, a key implication is the wide acceptance of implicit knowledge as a critical component of model-based reasoning and discovery. Computational modelers, in combination with their models, know more than they can tell (Polanyi 1958). Related to this is a focus on the process of building the model, and how building contributes to incorporation, and thereby, discoveries. The building process is poorly understood, and most studies of modeling ignore this critical component, particularly when building is done by communities of modelers, as in the case of *Foldit*. This three-fold combination, of implicit processes, building, and incorporation, could eventually lead to an embodied cognition account of MBR.
2. This shift in scientific practice will be reflected in science education, with the two dominant modes of training in science, apprenticeship and classroom training (which Bruner calls "showing" and "telling" modes), augmented by a modeling-based training. This new "enactive" mode is more social, participatory (as systems such as *Foldit* allow students to work with real problems) and decentralized than the currently dominant "telling" mode practiced in classrooms. While less embodied than the "showing" mode of learning in research laboratories, the enactive mode is more powerful in terms of exploration. The currently dominant telling mode is both enabled by and built around static media such as text and diagrams, and the dynamic nature of new computational media, particularly simulations and visualisations, is already disrupting science education based on this mode.
3. Computational models are constantly revised and expanded, through the embedding of experimental data and theoretical developments. Coupled with their role as generators of counterfactual scenarios and innovations,

computational models develop a complex and constantly changing relationship with the external world. Their correspondence with the real world is achieved, contingent, and constantly evolving. The central role played by computational models in contemporary practice suggests that this nature—achieved, contingent, and constantly evolving—will reshape our understanding of the nature of scientific knowledge, towards science-as-engineered-artifact that becomes part of reality and changes it, rather than (just) the view that science accurately captures pre-existing reality.

4. A central feature of computational models is their extreme ability to generate counterfactual scenarios and mechanisms. This feature makes them ideal for developing new technologies and mechanisms, and this makes computational models one of the key structures supporting the ongoing blending of science and engineering into engineering sciences, particularly in biology. The acceleration of this blending, and the blending of the related distinction between discovery and innovation, is a key practice implication of the shift to computational models and media.

7 Conclusion

The shift from the traditional static media such as text and graphics to computational modeling is set to change the practices of science and science education, particularly model-based reasoning based on external models. I examined two instances of the use of computational modeling to make key discoveries, and proposed an incorporation account of how building these models lead to scientific discovery. This incorporation account was extended further to propose an underlying cognitive mechanism, based on the way the body schema is extended during tool use. I then examined some of the major implications of this account. This account just begins the process of understanding the systemic shift to computational media and its implications for science and science education. A lot more needs to done before we can get a good grasp of the nature of this shift, particularly to design institutional structures around computational media.

References

Banzhaf, W. (1994). Self-organization in a system of binary strings. In R. Brooks & P. Maes (Eds.), *Artificial life IV* (pp. 109–119). Cambridge, MA: MIT Press.

Bohannon, J. (2009). Gamers unravel the secret life of protein. *Wired* Magazine, *17*(05), 17–05.

Botvinick, M., & Cohen, J. (1998). Rubber hands 'feel' touch that eyes see. *Nature, 391*, 756.

Chandrasekharan, S. (2009). Building to discover: A common coding model. *Cognitive Science, 33*(6), 1059–1086.

Chandrasekharan, S. (2014). Becoming knowledge: Cognitive and neural mechanisms that support scientific intuition. In L. M. Osbeck & B. S. Held (Eds.), *Rational intuition: Philosophical roots, scientific investigations* (pp. 307–337). New York: Cambridge University Press.

Chandrasekharan, S., & Nersessian, N. J. (2015). Building cognition: The construction of computational representations for scientific discovery. *Cognitive Science, 39*(8), 1727–1763.

Chandrasekharan, S., Nersessian, N., & Subramanian, V. (2012). Computational modeling: Is this the end of thought experiments in science? *Thought Experiments in Philosophy, Science, and the Arts, 11*, 239.

Cooper, S., Khatib, F., Treuille, A., Barbero, J., Lee, J., Beenen, M., & Popović, Z. (2010). Predicting protein structures with a multiplayer online game. *Nature, 466*(7307), 756–760.

Di Paolo, E. A., Noble, J., & Bullock, S. (2000). Simulation models as opaque thought experiments. In *artificial life VII: The Seventh International Conference on the Simulation and Synthesis of Living Systems* (pp. 497–506).

Dunn, J., & Clark, M. (1999). Life music: The sonification of proteins. *Leonardo, 32*(1), 25–32.

Edwards, L., Peng, Y., & Reggia, J. (1998). Computational models for the formation of protocell structure. *Artificial Life, 4*(1), 61–77.

Ehrsson, H. H. (2007). The experimental induction of out-of-body experiences. *Science, 317*, 1048.

Farne, A., Iriki, A., & Ladavas, E. (2005). Shaping multisensory action-space with tools: Evidence from patients with cross-modal extinction. *Neuropsychologia, 43*, 238–248.

Guterstam, A., Gentile, G., & Ehrsson, H. H. (2013). The invisible hand illusion: Multisensory integration leads to the embodiment of a discrete volume of empty space. *Journal of Cognitive Neuroscience, 25*(7), 1078–1099.

Guterstam, A., Petkova, V. I., & Ehrsson, H. H. (2011). The illusion of owning a third arm. *PLoS One, 6*(2), e1720.

Hestenes, D. (2006). Notes for a modeling theory. In E. van den Berg, T. Ellermeijer & O. Slooten (Eds.), *Proceedings of the 2006 GIREP conference: Modeling in physics and physics education* (Vol. 31, p. 27). Amsterdam: University of Amsterdam.

Hestenes, D. (2011). Notes for a modeling theory. In *Proceedings of the 2006 GIREP Conference: Modeling in Physics and Physics Education*. (Vol. 31).

Hestenes, D. (2013). Remodeling science education. *European Journal of Science and Mathematics Education, 1*(1), 2013.

Iriki, A., Tanaka, M., & Iwamura, Y. (1996). Coding of modified body schema during tool use by macaque postcentral neurons. *NeuroReport, 7*, 2325–2330.

Khatib, F., DiMaio, F., Foldit Contenders Group, Foldit Void Crushers Group, Cooper, S., Kazmierczyk, M., Gilski, M., Krzywda, S., Zabranska, H., Pichova, I., Thompson, J., Popovic, Z., Jaskolski, M., Baker, D. (2011). Crystal structure of a monomeric retroviral protease solved by protein folding game players. *Nature Structural and Molecular Biology, 18*, 1175–1177.

Kim, J. S., Greene, M. J., Zlateski, A., Lee, K., Richardson, M., Turaga, S. C., et al. (2014). Space-time wiring specificity supports direction selectivity in the retina. *Nature*, *509*(7500), 331–336.

Kirsh, D. (2010). Thinking with external representations. *AI and Society, 25*(4), 441–454.

Koerner, B. I. (2012). New videogame lets amateur researchers mess with RNA. *Wired Science*.

Landy, D. H., & Goldstone, R. L. (2009). How much of symbolic manipulation is just symbol pushing? In *Proceedings of the Thirty-First Annual Conference of the Cognitive Science Society*, (pp. 1072–1077). Amsterdam, Netherlands: Cognitive Science Society.

Landy, D., Allen, C., & Zednik, C. (2014). A perceptual account of symbolic reasoning. *Frontiers in Psychology, 5*.

Langton, C. (1990). Computation at the edge of chaos: Phase transitions and emergent computation. *Physica D: Nonlinear Phenomena, 42*, 12–37.

Langton, C. G. (1984). Self-reproduction in cellular automata. *Physica D: Nonlinear Phenomena, 10*, 135–144.

Lee, J., Kladwang, W., Lee, M., Cantu, D., Azizyan, M., Kim, H., et al. (2014). RNA design rules from a massive open laboratory. *Proceedings of the National Academy of Sciences, 111*(6), 2122–2127.

Lehrer, R., Horvath, J., Schauble, L. (1994). Developing model-based reasoning, *Interactive Learning Environments, 4*(3), 218–232.
Lehrer, R., Schauble, L. (2006). Cultivating model-based reasoning in science education. In Sawyer, R. Keith (Eds.), *The Cambridge handbook of the learning sciences*, (pp. 371–387). NY, US: Cambridge University Press, xix, 627 pp.
Lenggenhager, B., Tadi, T., Metzinger, T., & Blanke, O. (2007). Video ergo sum: Manipulating bodily self-consciouness. *Science, 317*, 1096–1099.
Lenhard, J. (2004). Surprised by a nanowire: Simulation, control, and understanding. *Philosophy of Science, 73*, 605–616.
Lenhard, J. (2006). Surprised by a nanowire: Simulation, control, and understanding. *Philosophy of Science, 73*(5), 605–616.
Majumdar, R., Kothiyal, A., Pande, P., Agarwal, H., Ranka, A., Murthy, S., et al. (2014). The enactive equation: Exploring how multiple external representations are integrated, using a fully controllable interface and eye-tracking. In *Proceedings of the Sixth International Conference on Technology for Education* (T4E), IEEE.
Maravita, A., & Iriki, A. (2004). Tools for the body (schema). *Trends in Cognitive Sciences, 8*(2), 79–86.
Marghetis, T., & Núnez, R. (2013). The motion behind the symbols: A vital role for dynamism in the conceptualization of limits and continuity in expert mathematics. *Topics in cognitive science, 5*(2), 299–316.
Nersessian, N. J. (1999). Model-based reasoning in conceptual change. In *Model-based reasoning in scientific discovery* (pp. 5–22). US: Springer.
Nersessian, N. J. (2010). *Creating scientific concepts*. MIT press.
Ocklenburg, S., Peterburs, J., Rüther, N., & Güntürkün, O. (2012). The rubber hand illusion modulates pseudoneglect. *Neuroscience Letters, 523*(2), 158–161.
Ong, W. J. (2013). *Orality and literacy*. Routledge.
Ottmar, E., Landy, D., & Goldstone, R. L. (2012). Teaching the perceptual structure of algebraic expressions: Preliminary findings from the pushing symbols intervention. In *The Proceedings of the Thirty-Fourth Annual Conference of the Cognitive Science Society* (pp. 2156–2161).
Polanyi, M. (1958). *Personal knowledge: Towards a post-critical philosophy*. Chicago: University of Chicago Press.
Polanyi, M. (1966). *The tacit dimension*. London: Routledge.
Prusinkiewicz, P., Lindenmayer, A., & Hanan, J. (1988). Developmental models of herbaceous plants for computer imagery purposes. *Computer Graphics, 22*(4), 141–150.
Reynolds, C. (1987). Flocks, herds, and schools: A distributed behavioral model. *Computer Graphics, 21*(4), 25–34.
Rotman, B. (2008). *Becoming beside ourselves: The alphabet, ghosts, and distributed human being*. Duke University Press.
Runions, A., Fuhrer, M., Lane, B., Federl, P., Rollang-Lagan, A., & Prusinkiewicz, P. (2005). Modeling and visualization of leaf venation patterns. *ACM Transactions on Graphics, 24*(3), 702–711.
Schneider, B. (2012). Climate model simulation visualization from a visual studies perspective. *Wiley Interdisciplinary Reviews: Climate change, 3*(2), 185–193.
Sims, K. (1994). Evolving virtual creatures. *Computer Graphics, 8*, 15–22.
Sincell, M. (2000). NanoManipulator lets chemists go mano a mano with molecules. *Science, 290*, 1530.
van der Hoort, B., Guterstam, A., & Ehrsson, H. (2011). Being barbie: The size of one's own body determines the perceived size of the world. *PLoS One, 6*(5), e20195.
Von Neumann, J. (1951). The general and logical theory of automata. *Cerebral mechanisms in behavior, 1*, 41.
Wilensky, U., & Reisman, K. (2006). Thinking like a wolf, a sheep, or a firefly: Learning biology through constructing and testing computational theories—An embodied modeling approach. *Cognition and instruction, 24*(2), 171–209.
Winsberg, E. (2006). Models of success versus the success of models: Reliability without truth. *Synthese, 152*(1), 1–19.

Is There a Scientific Method? The Analytic Model of Science

Carlo Cellucci

Abstract The nature of the scientific method has been a main concern of philosophy from Plato to Mill. In that period logic has been considered to be a part of the methodology of science. Since Mill, however, the situation has completely changed. Logic has ceased to be a part of the methodology of science, and no *Discourse on method* has been written. Both logic and the methodology of science have stopped dealing with the process of discovery, and generally with the actual process of scientific research. As a result, several first-rate scientists, from Feynman and Weinberg to Dyson and Hawkins, have concluded that philosophy has become useless and totally irrelevant to science. The aim of this paper is to give some indications as to how to develop a logic concerned with the process of discovery and a methodology of science dealing with the actual process of scientific research.

1 Introduction

The nature of the scientific method has been a main concern of philosophy from Plato to Mill. In that period, logic has been considered to be a part of the methodology of science. In particular, Aristotle's *Analytics* is a treatise of both logic and methodology of science. Aristotle refers to it as a single work, *ta analutika* [*The Analytics*], and states that it is an inquiry "about demonstration," so it has a single object, namely, "demonstrative science" (Aristotle, *Analytica Priora*, A 1, 24 a 10–11). The titles *Analytica Priora* and *Analytica Posteriora* were probably added by later editors to designate two parts of a single work.

Since Mill, however, the situation has completely changed. Logic has ceased to be a part of the methodology of science, and no *Discourse on method* has been written. Frege maintains that logic cannot be concerned "with the way in which" new results "are discovered" but only "with the kind of ground on which their"

C. Cellucci (✉)
Department of Philosophy, Sapienza University of Rome, Via Carlo Fea 2,
00161 Rome, Italy
e-mail: carlo.cellucci@uniroma1.it

L. Magnani and C. Casadio (eds.), *Model-Based Reasoning in Science and Technology*, Studies in Applied Philosophy, Epistemology and Rational Ethics 27, DOI 10.1007/978-3-319-38983-7_27

justification "rests" (Frege 1959, 23). The question of discovery is a merely subjective, psychological one, and hence "may have to be answered differently for different persons," only the question of justification "is more definite" (Frege 1967, 5). Similarly, logical positivism maintains that the methodology of science cannot be concerned with the actual process of scientific research, because it "cannot provide general procedural rules for the discovery of a new theory," it can only "formulate precise objective criteria which, for any proposed hypothesis *H* and evidence sentence *E*, determine whether or to what degree *E* confirms *H*" (Hempel 2001, 375). That is, it can only formulate precise objective criteria for the justification of theories already discovered.

As a result, although discovery is a crucial aspect of the process of scientific research, in current textbooks of logic or methodology of science you will hardly find any treatment of the process of discovery, and generally of the actual process of scientific research.

This situation is unsatisfactory. Considering the present condition of logic, Kowalski asks: "Is logic really dead or only just sleeping?" (Kowalski 2001, 2). His answer is: "Even if logic" might be "only half awake today," it "can at worst be only sleeping, to come back with renewed and more lasting vigour in the near future" (ibid., 3). Kowalski's answer, however, seems to be more an expression of hope than a statement of fact, because logic is really sleeping. Having failed to provide a secure foundation for mathematics and a workable basis for artificial intelligence, logic has lost two kingdoms and has not yet found a new role. The question is: How can logic be awakened from its sleep? A similar question can be raised about the methodology of science: How can the methodology of science be awakened from its sleep? The aim of this paper is to suggest an answer to these questions.

2 The Tenet of Classical Analytic Philosophy

At the basis of the assumption that logic and the methodology of science cannot be concerned with the process of discovery of new results but only with the justification of results already found, there is the tenet of classical analytic philosophy, that philosophy does not advance knowledge but only tries to clarify what we already know. Thus Wittgenstein states that philosophy "consists essentially of elucidations" (Wittgenstein 2001, 4.112). We "do not seek to learn anything new by it" but only "to understand something that is already in plain view" (Wittgenstein 2009, § 89). Dummett states that "philosophy does not advance knowledge: it clarifies what we already know" (Dummett 2010, 21). From this it follows that logic and the methodology of science cannot be concerned with the process of discovery of new results, but only with the justification of results already found.

The tenet of classical analytic philosophy, however, has not contributed to the reputation of philosophy. In particular, it has led many first-rate scientists—from Feynman and Weinberg to Dyson and Hawkins—to conclude that philosophy has become useless and totally irrelevant to science. (For references, see Cellucci 2014).

3 Contrast with Aristotle's View of Logic

The view that logic and the methodology of science cannot be concerned with the process of discovery of new results, contrasts with Aristotle's view.

According to a widespread opinion, Aristotle's logic is about deduction. Thus Boger states that Aristotle's "principal concern" is "with deduction" (Boger 2004, 106). Striker states that "Aristotle's logic" is "a general theory of deductive argument" (Striker 2009, xviii). Smith states that "all Aristotle's logic revolves around one notion: the deduction" (Smith 2011).

But it is not so. Aristotle's logic is mainly concerned with the question of how to find premises for solving any problem proposed, and hence is both a logic and a methodology of science. Indeed, Aristotle states that the main task of logic is to tell us "how to reach for premises concerning any problem proposed, in the case of any discipline whatever" (Aristotle, *Analytica Priora*, B 1, 53 a 1–2.). Logic must tell us "by what method we will find the premises about each thing" (ibid., A 27, 43 a 21–22). It must indicate to us "how we must hunt for them" (ibid., A 30, 46 a 11–12). For "surely we ought not only to investigate how syllogisms are constituted, but also have the ability to produce them" (ibid., A 27, 43 a 22–24). In other words, we ought not only to investigate the morphology of syllogisms, but also have the ability to find premises capable of yielding the desired conclusion. For this reason Aristotle states that, while "arguments are made from premises," the "things with which syllogisms are concerned are problems" (Aristotle, *Topica*, A 4, 101 b 15–16). That is, while arguments infer conclusions from the premises, the thing with which syllogisms are concerned is finding solutions to problems. From this it is apparent that Aristotle's logic is primarily intended to be a logic of discovery.

Consistently with this view, Aristotle indicates how to find premises for solving problems. Premises can be obtained from the conclusion "either by syllogism or by induction" (Aristotle, *Topica* Θ 1, 155 b 35–36). 'By syllogism', Aristotle means: by the procedure for "seeking what the middle term is" (Aristotle, *Analytica Posteriora* B 2, 90 a 1). This is the procedure described in *Analytica Priora* A 27–31, that the medievals called *inventio medii* since it is a procedure for seeking a middle term—and hence the premises—for a given conclusion. Such procedure "is the same for all subjects, in philosophy as well as in the technical or mathematical disciplines" (ibid., A 30, 46 a 3–4). For a presentation of this procedure, see Cellucci 2013a, Chap. 7.

4 Factors of the Denial of a Logic of Discovery

At least four factors have concurred in the view that logic and the methodology of science cannot be concerned with the process of discovery of new results, but only with the justification of results already discovered.

The first factor is Romanticism, which exalts intuition and genius. Thus Novalis states that scientific discoveries "are leaps—(intuitions, resolutions)" and products "of the genius—of the leaper *par excellence*" (Novalis 2007, 28). The method of science is the "method of the divinatory genius" (ibid., 100).

The second factor is the development, in the nineteenth century, of theories which employ unobservable entities and processes, and hence cannot be derived from observation. Since such theories cannot be derived from observation, Whewell states that "an art of discovery is not possible. At each step of the progress of science, are needed invention, sagacity, genius; elements which no art can give" (Whewell 1847, I, viii). Discovery "must ever depend upon some happy thought, of which we cannot trace the origin" (ibid., II, 20). This produces a hypothesis which is then "verified, and followed to its consequences" (ibid., II, 41).

The third factor are the foundational problems of the infinitesimal calculus, which suggested that what was urgently needed was a logic of justification rather than a logic of discovery. Thus Frege states that "almost insuperable, difficulties stood in the way of any rigorous treatment" of the infinitesimal calculus, in particular "the concepts of function, of continuity, of limit and of infinity have been shown to stand in need of sharper definition" (Frege 1959, 1). Therefore, today there is need for "rigour of proof, precise delimitation of extent of validity, and as a means to this, sharp definition of concepts" (ibid.). Then logic must not concern itself with the question of discovery, but rather with the question of "how we can provide" a judgment "with the most secure foundation" (Frege 1967, 5).

The fourth factor is the opinion that a logic of discovery should provide an algorithmic method for solving problems. This opinion is widespread throughout classical analytic philosophy. In particular, logical positivism maintains that a logic of discovery is impossible because such a logic should provide an algorithmic method for solving problems. Thus Carnap states that a logic of discovery is impossible because there cannot be a "machine—a computer into which we can put all the relevant observational sentences and get, as an output, a neat system of laws that will explain the observed phenomena," for the purpose of discovery "creative ingenuity is required" (Carnap 1966, 33). Similarly, Hempel states that a logic of discovery is impossible because "there is no generally applicable mechanical routine of 'inductive inference' which leads from a given set of data to a corresponding hypothesis or theory" (Hempel 2001, 31). Scientific hypotheses and theories "are not mechanically inferred from observed 'facts': They are invented by an exercise of creative imagination" (ibid., 32).

5 Logic of Discovery and Logic of Testing

While maintaining that a logic of discovery is impossible because such a logic should provide an algorithmic method for solving problems, logical positivism claims that a logic of testing is viable because there is a machine for testing. Thus Carnap states that, while there is no machine for discovery, there can be a "machine with a much more modest aim. Given certain observation *e* and a hypothesis *h*," it "is in many cases possible to determine, by mechanical procedure," the "degree of confirmation of *h* on the basis of *e*" (Carnap 1966, 34).

But the assumption that a logic of testing is viable because there is a machine for testing, is unjustified. By Church's undecidability theorem, there is even no machine for testing whether or not a sentence is logically valid. As Putnam states, if "there is no logic of discovery" because there is no machine for discovery, then "in that sense, there is no logic of testing, either" (Putnam 1975–1983, I, 268). The "view that correct ideas just come from the sky, while the methods for testing them are highly rigid and predetermined, is one of the worst legacies of the Vienna Circle" (ibid.). In fact, "all the formal algorithms proposed for testing, by Carnap, by Popper, by Chomsky, etc., are, to speak impolitely, ridiculous: if you don't believe this, program a computer to employ one of these algorithms and see how well it does at testing theories!" (ibid.).

Lakatos mockingly states that "primitive men worship algorithms" and feel unsafe if they move "beyond the bounds of ritual," into the wilderness of heuristic methods, therefore they "prefer decision-procedures" (Lakatos 1978, II, 72). But "the Greeks did not find a decision-procedure for their geometry," they "did, however, find a compromise solution: a heuristic procedure" which "does not always yield the desired result, but which is still a heuristic rule, a standard pattern of the logic of discovery" (ibid.).

6 Algorithmic and Heuristic Methods

The opinion that a logic of discovery or testing should be algorithmic is based on the assumption that "the word 'method' is a synonym for an algorism" (Agassi 1980, 187). This assumption is motivated by the wish that a method be always successful, but makes it impossible to develop a logic of discovery or testing. If we want to have such logics, we must consider heuristic methods rather than algorithmic ones.

While algorithmic methods guarantee to achieve a solution to a problem in all cases, heuristic methods do not guarantee that. And yet they greatly reduce the search space, that is, the domain within which the solution to a problems is sought. This makes a solution feasible when an algorithmic method is not available. The formal literature on heuristics tends to suggest that the purpose of heuristics is to formulate mechanical rules that can be programmed on a computer, but this is misleading. The purpose of heuristics is rather to find non-mechanical rules that will guide one to solve problems, even if it takes some skill to apply them.

That methods need not be algorithmic was already recognized in antiquity. Mathematics and medicine were the first areas where the need for methods to solve problems arose, and the earliest methods of which we have notice, namely the method of Hippocrates of Chios for mathematics, and the method of Hippocrates of Cos for medicine, were heuristic methods. Actually, both Hippocrates of Chios and

Hippocrates of Cos used the same heuristic method to solve problems, that is, the analytic method, also known as the method of analysis (see Cellucci 2013a, Chap. 4). This is the oldest method of science that is known to us, and I will argue that it is the only method of science which is still plausible today.

7 The Analytic Method

The analytic method is the method according to which, to solve a problem, we look for some hypothesis that is a sufficient condition for solving the problem, that is, such that a solution can be deduced from the hypothesis. The hypothesis is obtained from the problem, and possibly other data already available, by some non-deductive rule (such as induction, analogy, metaphor, etc.), and must be plausible, that is, the arguments for it must be stronger than those against it on the basis of experience. The solution to the problem is then deduced from the hypothesis. But the hypothesis is in its turn a problem that must be solved, and is solved in the same way. That is, we look for another hypothesis that is a sufficient condition for solving the problem posed by the previous hypothesis, it is obtained from the latter, and possibly other data already available, by some non-deductive rule, and must be plausible. The solution to the problem posed by the previous hypothesis is then deduced from the new hypothesis. And so on, *ad infinitum*. Thus solving a problem is a potentially infinite process.

The reason why, in the analytic method, hypotheses are obtained by non-deductive rules rather than by deductive rules, is that deductive rules are non-ampliative, that is, the conclusion is contained in the premises. For example, in implication elimination (modus ponens),

$$\frac{A \quad A \rightarrow B}{B},$$

the conclusion B is literally a part of the major premise $A \rightarrow B$. Generally, nothing follows from the conclusion of a deductive rule that does not already follow from the premises.

On the contrary, non-deductive rules can be ampliative, that is, the conclusion is not contained in the premises. For example, in induction by enumeration,

$$\frac{A(a_1), \ldots, A(a_n)}{\forall x A(x)},$$

the conclusion, $\forall x A(x)$, is not contained in the premises—if the universe of discourse includes individuals other than $a_1, \ldots, a_n$.

8 The Role of Logic in the Analytic Method

In the analytic method, logic primarily serves to find hypotheses for solving problems, therefore the analytic method involves non-deductive rules. It, however, also involves deductive rules because, while hypotheses are obtained by non-deductive rules, solutions to problems are deduced from hypotheses. Moreover, to see that hypotheses are plausible, one will use the following plausibility test procedure, which involves deduction:

1. Deduce conclusions from the hypotheses.
2. Compare conclusions with each other, to see that the hypotheses do not lead to contradictions.
3. Compare conclusions with other hypotheses already known to be plausible and with results of observations or experiments.

9 Non-deductive Rules

The non-deductive rules by which hypotheses can be obtained are not a closed set, given once for all, but rather an open set which can always be extended as research develops. Each such extension is a development of the analytic method, which grows as new non-deductive rules are added. As Bacon states, "the art of discovery may grow with discoveries" (Bacon 1961–1963, I, 223). (For a list of basic non-deductive rules, see Cellucci 2013a, Chaps. 20 and 21.)

Let us consider a simple example of use of a non-deductive rule in discovery. We say that two things are quasi-equal if, while not identical, they are a very close approximation to each other. Let $a \cong b$: a is quasi-equal to b. Analogy by quasi-equality is an inference by the rule:

$$\frac{a \cong b \quad A(a)}{A(b)}.$$

This rule is non-deductive because two quasi-equal things are not equal, therefore the conclusion is not contained in the premises.

An example of use of analogy by quasi-equality in discovery, is Antiphon's discovery of the hypothesis that the area of a circle is half the circumference times the radius. Simplicius tells us that "Antiphon, having drawn a circle, inscribed in it one of the polygons that can be inscribed" (Simplicius, *In Aristotelis Physicorum Libros Quattuor Priores Commentaria*, 54.20–22). The polygon can be viewed as consisting of n isosceles triangles with the same base b and the same height h, and hence with the same area, $\frac{1}{2}bh$.

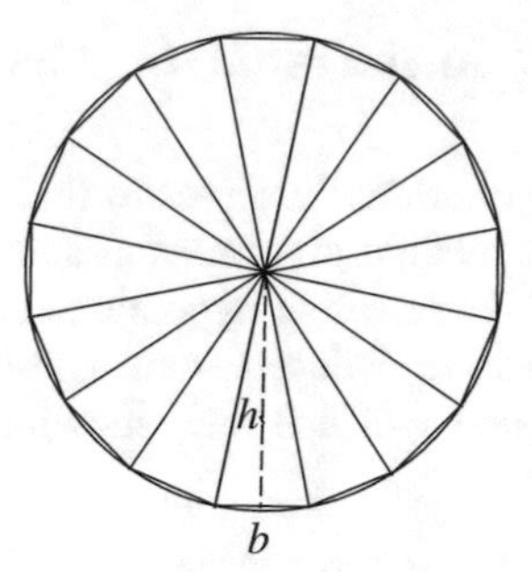

If $p = nb$ is the perimeter of the polygon, the area of the polygon will be $n(\frac{1}{2}bh) = \frac{1}{2}(nb)h = \frac{1}{2}ph$.

Then Antiphon observed that, by increasing the number of sides n, "a polygon would be inscribed in the circle whose sides would by and large coincide with the circumference of the circle" (ibid., 55.7–8). That is, the polygon and the circle would be quasi-equal. From the fact that the area of the polygon is $\frac{1}{2}ph$, by analogy by quasi-equality, Antiphon inferred that the area of the circle is $\frac{1}{2}cr$, where c is the circumference and r the radius.

10 Heuristic Logic Versus Mathematical Logic

Since, in the analytic method, logic primarily serves to find hypotheses for solving problems, the logic underlying the analytic method may be called 'heuristic logic'. There are some basic differences between heuristic logic and mathematical logic. Mathematical logic is based on the following assumptions, originally stated by Frege.

1. The goal of logic is to give a secure foundation for mathematics. Its purpose is "to place the truth of a proposition beyond all doubt" (Frege 1959, 2).
2. Logic pursues this goal through the study of the method of mathematics. This will show "the ultimate ground upon which rests the justification for holding" a proposition "to be true" (ibid., 3).
3. The method of mathematics is the axiomatic method. We start from axioms "expressly declared as such, so that we can see distinctly what the whole structure rests upon," and proceed from them by rules of deduction "specified in advance" (Frege 1964, 2).
4. Logic need only express what is necessary for the axiomatic method. It may "forgo expressing anything that is without significance for the inferential sequence" (Frege 1967, 6). Everything "necessary for a correct inference is expressed in full, but what is not necessary is generally not indicated" (ibid., 12). Then logic will fail to express all aspects of mathematics. But logic "is a device invented for certain scientific purposes," that is, to give a secure foundation for mathematics, "and one must not condemn it because it is not suited to others" (ibid., 6).

5. Logic can actually give a secure foundation for mathematics. Admittedly, logic can give no answer to the question whether the axioms are true, but an answer is provided by intellectual intuition, which is "the logical source of knowledge" (Frege 1979, 267). Intellectual intuition is capable "of grasping a thought," where "what is grasped, taken hold of, is already there and all we do is take possession of it" (ibid., 137).

On the other hand, heuristic logic is based on the following assumptions.

1. The goal of logic is to develop tools for acquiring new knowledge. Only so it can be considered to be fruitful.
2. Logic pursues this goal through the study of the method of mathematics, and science generally. This will show how new knowledge can be acquired.
3. The method of mathematics, and science generally, is the analytic method. One starts from a problem and finds hypotheses to solve it by means of non-deductive rules.
4. Logic must express all that is necessary for the analytic method. It must consider expressing anything that is significant for acquiring new knowledge.
5. Logic can actually develop tools for acquiring new knowledge. For it provides non-deductive rules by means of which one may find hypotheses for solving problems.

These assumptions show the differences between mathematical logic and heuristic logic. From them, it is apparent that mathematical logic is intended to be a logic of justification, while heuristic logic is intended to be a logic of discovery.

11 The Axiomatic Method

As we have seen, mathematical logic assumes that the method of mathematics is the axiomatic method, while heuristic logic assumes that the method of mathematics, and science generally, is the analytic method. Since antiquity, the axiomatic method has been viewed as an alternative to the analytic method. But in the *Republic* Plato criticizes the axiomatic method arguing that, in such method, "the principle," being unjustified, "is not known," and hence "the conclusion and the intermediate steps are constructed out of unknown material" (Plato, *Respublica*, VII 533 c 3–5). Then the principle, the intermediate steps and the conclusion are mere conventions, and "what artifice could ever transform this fabric of convention into a science?" (Plato, *Respublica*, VII 533 c 5–6).

According to a widespread view, in the *Republic* Plato criticizes mathematics, denying it the status of knowledge. For example, Bostock states that "in the *Meno* mathematics had certainly been viewed as an example of knowledge, but now in the *Republic* it is denied that status" (Bostock 2009, 13). This, however, is a misunderstanding, because what Plato criticizes in the *Republic* is not mathematics, but rather the use of the axiomatic method in mathematics, being based on principles, or axioms, that are unjustified. Plato's criticism is similar to Hippocrates of Cos' criticism of the use of the axiomatic method in medicine (see Cellucci 2013a, Chap. 3).

Specifically, what Plato criticizes is a version of the axiomatic method that does not require a preliminary justification of principles, or axioms. Alternatively, as a criterion for the acceptance of axioms, Hilbert requires that axioms be consistent and proved to be consistent by absolutely reliable means. Indeed, he states that "there is a condition" to which the axiomatic method "is subject, and that is the proof of consistency" (Hilbert 1967, 383). Such proof must be carried out without using "any dubious or problematical mode of inference" (Hilbert 1996c, 1139). Hilbert's requirement on axioms can be seen as a response to Plato's criticism of the axiomatic method.

12 The Axiomatic Method and Gödel's Incompleteness Theorems

Without Hilbert's requirement on axioms a science would be a mere convention. But even with Hilbert's requirement the axiomatic method is inadequate. Indeed, it is incompatible with Gödel's incompleteness theorems.

The axiomatic method is incompatible with Gödel's first incompleteness theorem. For according to such method, all true sentences of a theory must be deducible from the axioms of the theory. But, by Gödel's first incompleteness theorem, for any theory in a given field satisfying certain minimal conditions, there is a sentence which is true but not deducible from the axioms of the theory.

The axiomatic method is incompatible with Gödel's second incompleteness theorem. For according to such method, the axioms should be consistent and proved to be consistent by absolutely reliable means. But, by Gödel's second incompleteness theorem, for any theory in a given field satisfying certain minimal conditions, the axioms of the theory cannot be proved to be consistent by absolutely reliable means.

That the axiomatic method is incompatible with Gödel's incompleteness theorems shows that mathematical logic cannot be really a logic of justification. Indeed, in addition to its incompatibility with Gödel's incompleteness theorems, there are also other reasons why the axiomatic method is inadequate; see Cellucci 2013b.

13 The Analytic Method and Gödel's Incompleteness Theorems

Contrary to the axiomatic method, the analytic method is compatible with Gödel's incompleteness theorems.

The analytic method is compatible with Gödel's first incompleteness theorem. For according to such method, the solution to a problem of a given field is obtained from the problem, and possibly other data already available, by means of hypotheses not necessarily belonging to that field. Since Gödel's first incompleteness theorem implies that solving a problem of a given field may require hypotheses from some other fields, Gödel's result even provides evidence for the analytic method.

The analytic method is compatible with Gödel's second incompleteness theorem. For according to such method, the hypotheses for the solution to a problem are not definitive, true and certain, but only provisional, plausible and uncertain, so no solution to a problem can be absolutely certain. Since Gödel's second incompleteness theorem implies that no solution to a problem can be absolutely certain, Gödel's result even provides evidence for the analytic method.

That the analytic method is compatible with Gödel's incompleteness theorems means that the latter are no obstacle to heuristic logic being a logic of discovery. Indeed, in addition to compatibility with Gödel's incompleteness theorems, there are also other reasons why the axiomatic method is adequate; see Cellucci 2013b.

14 Some Misunderstanding About the Axiomatic Method

Several people tend to take for granted that the method of mathematics is the axiomatic method, for at least two reasons.

1. They believe that Euclid's *Elements* are the prototype of all mathematics. But they overlook that Euclid's *Elements* are only a compilation and reorganization of earlier texts for didactical purposes. Indeed, in the *Elements* Euclid "did not bring in everything he could have collected," but only "theorems and problems that are worked out for the instruction of beginners" (Proclus 1992, 69.6–9). He omitted matters that "are unsuitable for a selection of elements because they lead to great and unlimited complexity" (*ibid.*, 74.21–22).

In compiling and reorganizing earlier texts for didactical purposes, Euclid followed Aristotle's indications on teaching. Aristotle states that "all teaching and all intellectual learning come from already existing knowledge" and also "the mathematical sciences are approached in this way" (Aristotle, *Analytica Posteriora*, A 1, 71 a 1–4). They are taught and learned through the axiomatic method, which is not the method of mathematics but the method of the teaching of mathematics. For teaching is carried out through arguments "which proceed from the principles appropriate to each branch of learning," that is, through axiomatic demonstrations, which are the "didactic arguments" (Aristotle, *De Sophisticis Elenchis*, 2, 165 b 1–2). Indeed, didactic arguments are the "demonstrative arguments" which are "treated in the *Analytics*" (ibid., 2, 165 b 9). Demonstrative arguments must not be confused with discovery arguments which, as we have seen, Aristotle bases on *inventio medii* and induction. The distinction between discovery arguments and demonstrative arguments is made clear by Cicero, who states that "all methodical treatment of rational discourse involves two arts, one of discovering" and one of demonstrating or "judging", and while the Stoics "pursued the art of judging" but "completely neglected" the "art of discovering," conversely "Aristotle came first in both" (Cicero, *Topica*, 6).

That Euclid followed Aristotle in assuming that the axiomatic method is not the method of mathematics but the method of the teaching of mathematics, is apparent from the fact that Euclid did not use the axiomatic method in his own research works. In particular, Euclid's *Data*, *Porisms*, and *Surface Loci* were part of "the so-called *Treasure of Analysis*," a "special body of doctrine provided for the use of those who, after finishing the ordinary elements, are desirous of acquiring the power of solving problems," which "is the work of three men, Euclid, the author of the *Elements*, Apollonius of Perga, and Aristaeus the Elder, and proceeds" not by the axiomatic method but "by the method of analysis and synthesis" (Pappus 1876–1878, VI, 634, 3–11). Then it is unjustified to see Euclid's *Elements* as the prototype of all mathematics.

2. They believe that Hilbert identifies the method of mathematics with the axiomatic method, and indeed that Hilbert "has thought mathematicians to think axiomatically" (Dieudonné 1971, I, 311). Thus, however, they overlook that Hilbert views "the axiomatic exploration of a mathematical truth" as "an investigation which does not aim at finding new or more general theorems being connected to this truth, but to determine the position of this theorem within the system of known truths together with their logical connections, in such a way that it can be clearly said which conditions are necessary and sufficient for giving a foundation of this truth" (Hilbert 1902–03, 50).

According to Hilbert, the axiomatic method does not serve to obtain new results, but only to provide a foundation and a justification for already known results. Hilbert aims at "finding a secure foundation for mathematics," and states that the method he follows to this purpose "is none other than the axiomatic" (Hilbert 1996b, 1119). In his view, "for the final presentation and the complete logical grounding of our knowledge the axiomatic method deserves the first rank" (Hilbert 1996a, 1093).

15 The Hypothetico-Deductive Model of Science

The axiomatic method and the analytic method are the basis of two different models of science, the hypothetico-deductive model and the analytic model of science.

According to the hypothetico-deductive model, to formulate a scientific theory about a class of things or facts means to formulate hypotheses, then to deduce consequences from them, and finally to compare such consequences with one another to see whether the hypotheses are consistent, and with the observational and experimental data to see whether the hypotheses stand the test of reality.

Thus Popper states that science "always proceeds on the following lines. From a new idea, put up tentatively, and not yet justified in any way–an anticipation, a hypothesis, a theoretical system, or what you will–conclusions are drawn by means of logical deduction" (Popper 2002, 9). These conclusions are first "compared with one another," in order to test "the internal consistency of the system" (ibid.). Then the theory is tested "by way of empirical applications of the conclusions which can be derived from it" (*ibid.*). Finally the theory is compared "with other theories, chiefly with the aim of determining whether the theory would constitute a scientific advance should it survive our various tests" (ibid.).

However, the hypothetico-deductive model of science is faced with some basic difficulties. Being based on the axiomatic method, the hypothetico-deductive model is incompatible with Gödel's incompleteness theorems. Moreover, the hypothetico-deductive model leaves to one side the crucial issue of how to find hypotheses, it merely states that discovery requires "creative intuition" (Popper 2002, 8). Furthermore, the hypothetico-deductive model is incapable of accounting for the process of theory change, that is, the process in which one theory comes to be replaced by another. For according to it, a theory has no rational connection with the preceding one, except that it agrees with more observational and experimental data than the preceding one. Thus the hypothetico-deductive model leaves to one side not only the crucial issue of the discovery of hypotheses, but also the equally crucial issue of theory change.

In addition to the hypothetico-deductive model, we might consider the semantic model. But the semantic model is faced with as many difficulties as the hypothetico-deductive model; see Cellucci 2016.

16 The Analytic Model of Science

According to the analytic model of science, to formulate a scientific theory about a certain class of problems means to formulate hypotheses that are sufficient conditions for solving them by means of some non-deductive rules, then to formulate hypotheses which are sufficient conditions for solving the problems posed by such hypotheses by means of some non-deductive rules, and so on. All the hypotheses thus formulated must be plausible, that is, the arguments for them must be stronger than those against them.

A theory comes to be replaced by another one when the hypotheses reached at a certain stage are no longer plausible. The hypotheses of the new theory are formulated through an analysis of the reasons why the hypotheses of the old theory were no longer plausible. Therefore, the new theory is rationally connected with the old theory.

Heuristic logic is a part of the analytic model of science, so it is a part of the methodology of science, meant as concerned with the actual process of scientific research.

The analytic model of science is not subject to the problems of the hypothetico-deductive model. Indeed, being based on the analytic method, the analytic model is compatible with Gödel's incompleteness theorems. Also, the analytic model is capable of dealing with the crucial issue of how to find hypotheses. They are obtained by means of some non-deductive rule. Moreover, by what has been said above, the analytic model is capable of accounting for the process of theory change.

Thus the analytic model of science is capable of accounting not only for the crucial issue of the discovery of hypotheses, but also for the equally crucial issue of theory change.

17 Truth and Plausibility

A further problem with the hypothetico-deductive model of science is that it assumes that the goal of science is truth, in the sense of the correspondence theory. Thus Popper states that "science is the search for truth" and "truth is therefore the aim of science" (Popper 1996, 39). Specifically, "science aims at truth in the sense of correspondence to the facts or to reality" (Popper 1972, 59).

But saying that the goal of science is truth, in the sense of the correspondence theory, conflicts with the fact that the history of science offers us many examples of important theories that at one stage were taken to be true but are false according to present scientific theories (see Laudan 1981). Now, a theory that is true at one stage in the sense of the correspondence to reality cannot become false at a later stage. For either there is correspondence to reality or there is no correspondence.

Moreover, saying that science aims at truth, in the sense of correspondence to reality, conflicts with the fact that the correspondence conception of truth does not provide a criterion of truth, namely, a generally non-algorithmic means which allows us to distinguish true statements from false statements. As Kant states, "I can compare the object with my cognition" only "by cognizing it" (Kant 1992, 557). That is, by cognizing the object. But "since the object is outside me, the cognition in me, all I can ever pass judgment on is whether my cognition of the object agrees," not with the object, but only "with my cognition of the object" (ibid., 557–558). Therefore, we cannot know whether a theory about the world is true in the correspondence sense.

The analytic model of science is not subject to these limitations, because it does not assume that the goal of science is truth. Instead, it assumes that the goal of science is plausibility. Then it is not affected by the problem that theories which at one stage were taken to be true are false according to present scientific theories. A theory that is taken to be plausible at one stage may very well become implausible at a later stage, when new data emerge which lead to conclude that the arguments against the theory are stronger than those for it. While truth is an absolute concept, plausibility is a relative concept. Moreover, the analytic model of science is not affected by the problem that the correspondence conception of truth does not provide a criterion of truth. The plausibility tests procedure described above provides a criterion of plausibility.

18 The Goal of Logic

As the hypothetico-deductive model of science assumes that the goal of science is truth, mathematical logic assumes that the goal of logic is truth. Thus Frege states that "all the sciences have truth as their goal, but logic is concerned with the predicate 'true' in a quite special way" (Frege 1979, 128). For "the laws of logic are nothing other than an unfolding of the content of the word 'true'" (ibid., 3). However, by assuming that the aim of logic is truth, mathematical logic is faced with the same problems as the hypothetico-deductive model of science.

Conversely, heuristic logic assumes that the aim of logic is plausibility and hence is not affected by those problems. Replacing truth with plausibility is essential for a logic of discovery. This is made clear already by Aristotle, who states that, in order to show that a hypothesis is plausible, we will "examine the arguments for and the arguments against" (Aristotle, *Topica*, Θ 14, 163 a 38–b 1). Then, "if the difficulties are solved and the accepted opinions are left standing, we shall have proved the case sufficiently" (Aristotle, *Ethica Nicomachea*, H 1, 1145 b 6–7). This is a process of discovery, because "the resolution of a difficulty is a discovery" (ibid., H 2, 1146 b 7–8). Indeed, the process of discovery consists not only in finding hypotheses, but also in showing that the hypotheses thus found are plausible.

19 Conclusion

If logic is to come back with renewed and more lasting vigour and the methodology of science is to be useful, it is necessary to develop a logic concerned with the process of discovery and a methodology of science dealing with the actual process of scientific research.

Heuristic logic and the analytic model of science provide a framework for developing such a logic and methodology of science. In particular, they are not subject to the limitations of mathematical logic and the hypothetico-deductive model of science.

References

Agassi, J. (1980). The rationality of discovery. In T. Nickles (Ed.), *Scientific discovery, logic, and rationality* (pp. 185–199). Dordrecht: Reidel.
Bacon, F. (1961–1963). *Works*. Stuttgart–Bad Cannstatt: Frommann Holzboog.
Boger, G. (2004). Aristotle's underlying logic. In D. M. Gabbay & J. Woods (Eds.), *Handbook of the history of logic* (Vol. 1, pp. 101–246). Amsterdam: Elsevier.
Bostock, D. (2009). *Philosophy of mathematics. An introduction*. Hoboken: Wiley-Blackwell.
Carnap, R. (1966). *Philosophical foundations of physics. An introduction to the philosophy of science*. New York: Basic Books.
Cellucci, C. (2013a). *Rethinking logic. Logic in relation to mathematics, evolution, and method*. Dordrecht: Springer.
Cellucci, C. (2013b). Philosophy of mathematics. Making a fresh start. *Studies in History and Philosophy of Science, 44*, 32–42.
Cellucci, C. (2014). Rethinking philosophy. *Philosophia, 42*, 271–288.
Cellucci, C. (2016). Models of science and models in science. In E. Ippoliti, F. Sterpetti, & T. Nickles (Eds.), *Models and inferences in science* (pp. 95–112). Cham: Springer.
Dieudonné, J. (1971). David Hilbert (1862–1943). In F. Le Lionnais (Ed.), *Great currents of mathematical thought* (pp. 304–311). Mineola: Dover.
Dummett, M. (2010). *The nature and future of philosophy*. New York: Columbia University Press.
Frege, G. (1959). *The foundations of arithmetic. A logico-mathematical enquiry into the concept of number*. Oxford: Blackwell.
Frege, G. (1964). *The basic laws of arithmetic. Exposition of the system*. Berkeley: University of California Press.
Frege, G. (1967). *Begriffsschrift*, a formula language, modeled upon that of arithmetic, for pure thought. In J. van Heijenoort (Ed.), *From Frege to Gödel. A source book in mathematical logic, 1879–1931* (pp. 5–82). Cambridge: Harvard University Press.
Frege, G. (1979). *Posthumous writings*. Oxford: Blackwell.
Hempel, C. G. (2001). *The philosophy of Carl G. Hempel*. Oxford: Oxford University Press.
Hilbert, D. (1902–1903). Über den Satz von der Gleichheit der Basiswinkel im gleichschenkligen Dreieck. *Proceedings of the London Mathematical Society, 35*, 50–68.
Hilbert, D. (1967). On the infinite. In J. van Heijenoort (Ed.), *From Frege to Gödel. A source book in mathematical logic, 1879–1931* (pp. 369–392). Cambridge: Harvard University Press.
Hilbert, D. (1996a). On the concept of number. In W. Ewald (Ed.), *From Kant to Hilbert. A source book in the foundations of mathematics* (pp. 1092–1095). Oxford: Oxford University Press.
Hilbert, D. (1996b). The new grounding of mathematics. First report. In W. Ewald (Ed.), *From Kant to Hilbert. A source book in the foundations of mathematics* (pp. 1117–1134). Oxford: Oxford University Press.
Hilbert, D. (1996c). The logical foundations of mathematics. In W. Ewald (Ed.), *From Kant to Hilbert. A source book in the foundations of mathematics* (pp. 1134–1148). Oxford: Oxford University Press.
Kant, I. (1992). *Lectures on logic*. Cambridge: Cambridge University Press.
Kowalski, R. (2001). Is logic really dead or only just sleeping? In P. Codognet (Ed.), *Logic programming: 17th international conference, ICLP 2001* (pp. 2–3). Berlin: Springer.
Lakatos, I. (1978). *Philosophical papers*. Cambridge: Cambridge University Press.
Laudan, L. (1981). A confutation of convergent realism. *Philosophy of Science, 48*, 19–49.
Novalis (von Hardenberg, G. F. P.). (2007). *Notes for a Romantic encyclopedia. Das Allgemeine Brouillon*. Albany: State University of New York Press.
Pappus of Alexandria. (1876–1878). *Collectio*. Berlin: Weidmann.
Popper, K. R. (1972). *Objective knowledge. An evolutionary approach*. Oxford: Oxford University Press.

Popper, K. R. (1996). *In search of a better world. Lectures and essays from thirty years*. London: Routledge.

Popper, K. R. (2002). *The logic of scientific discovery*. London: Routledge.

Proclus Diadocus. (1992). *In primum Euclidis Elementorum librum commentarii*. Hildesheim: Olms.

Putnam, H. (1975–1983). *Philosophical papers*. Cambridge: Cambridge University Press.

Smith, R. (2011). Aristotle's logic. In *Stanford Encyclopedia of Philosophy*. Stanford: Center for the Study of Language and Information (CSLI).

Striker, G. (2009). Introduction. In Aristotle, *Prior Analytics*. *Book I* (pp. xi–xviii). Oxford: Oxford University Press.

Whewell, W. (1847). *The philosophy of the inductive sciences, founded upon their history*. London: Parker.

Wittgenstein, L. (2001). *Tractatus logico-philosophicus*. London: Routledge.

Wittgenstein, L. (2009). *Philosophical investigations*. New York: Wiley Blackwell.

Ad Hoc Hypothesis Generation as Enthymeme Resolution

Woosuk Park

Abstract To date there seems to be no disciplined way of distinguishing between ad hoc hypotheses and legitimate auxiliary hypotheses. This is embarrassing not just for Popperian falsificationist scientific methodology, for the need for such a distinction seems an important part of scientific practice. Do scientists bother about ad hoc hypotheses at all? Did any towering figure in the history of science care about ad hoc hypotheses? Ironically, the answers to these questions seem to be "Yes" and "No" in both cases. Inspired by Paglieri and Woods' recent proposal for a theory of enthymeme based on the principle of parsimony, I propose to approach the problem of ad hoc hypothesis by interpreting it as a kind of enthymeme resolution. One reason for this interpretative strategy lies in its potential for understanding the pervasiveness and the longevity of the Aristotelian scientific methodology embedded in the scientific practice throughout the ages.

Keywords Ad hoc hypothesis · Enthymeme · Fabio Paglieri · Imre Lakatos · John Woods · Karl Popper

To date there seems to be no disciplined way of distinguishing between ad hoc hypotheses and legitimate auxiliary hypotheses. This is embarrassing not just for Popperian falsificationist scientific methodology, for the need for such a distinction seems an important part of scientific practice, at least insofar as standard introductory textbooks of philosophy of science assume it. But do practicing scientists bother about ad hoc hypotheses at all? Did any towering figure in the history of science care about ad hoc hypotheses? Ironically, the answers to these questions seem to be "Yes" and "No" in both cases. Inspired by Paglieri and Woods' recent proposal for a theory of enthymeme based on the principle of parsimony, I propose to approach the problem of ad hoc hypothesis by interpreting it as a kind of enthymeme resolution. No doubt, this approach is inspired by some recent attempts

W. Park (✉)
Humanities and Social Sciences, KAIST, 291 Daehak-ro,
Yuseong-gu, Daejeon, South Korea
e-mail: woosukpark@kaist.ac.kr

L. Magnani and C. Casadio (eds.), *Model-Based Reasoning in Science and Technology*, Studies in Applied Philosophy, Epistemology and Rational Ethics 27, DOI 10.1007/978-3-319-38983-7_28

to view the problem of ad hoc hypothesis as a case of anomaly resolution. These attempts again are stemming from the efforts to synthesize the recent advances in logic and philosophy of science. To the best of my knowledge, however, no one has hinted at understanding ad hoc hypothesis in terms of enthymeme. I shall show that the different views of Popper, Lakatos, and others on ad hoc hypothesis can be understood as emphasizing different standards for enthymeme resolution: i.e., parsimony and charity. One strong motivation for my interpretative strategy lies in its potential for understanding the pervasiveness and the longevity of the Aristotelian scientific methodology embedded in the scientific practice throughout the ages.

1 The Problem of Ad Hoc Hypothesis

1.1 Popper: Ad Hoc Hypothesis and the Discovery of Neptune

As is well-known, Karl R. Popper was preoccupied with the problem of ad hoc hypothesis throughout his career. In his first book *Logik der Forschung*, he introduces the problem, for example, in the context of examining some objections to his criterion of demarcation of science and pseudo-science, i.e., falsifiability:

> It might be said that even if the asymmetry is admitted, it is still impossible, for various reasons, that any theoretical system should ever be conclusively falsified. For it is always possible to find some way of evading falsification, for example by introducing ad hoc an auxiliary hypothesis, or by changing ad hoc a definition (Popper 1934\1959\1975, pp. 41–42).

However, his most articulate discussion seems to be found in his "Replies to My Critics" published in Living Philosopher series. Here, he explains how he understands the meaning of "ad hoc" as follows:

> I call a conjecture "ad hoc" if it is introduced (like this one) to explain a particular difficulty, but if (in contrast to this one) *it cannot be tested independently* (Popper 1974, p. 986).

He is here referring to the case of the discovery of Neptune, by which he tries to distinguish between ad hoc hypothesis and auxiliary hypothesis:

> In the case of the disturbances in the motion of Uranus the adopted hypothesis was partly revolutionary: what was conjectured was the existence of a new planet, something which did not affect Newton's laws of motion, but which did affect the much older "system of the world". The new conjecture was auxiliary rather than ad hoc: for although there was only this one ad hoc reason for introducing it, it was *independently testable*: the position of the new planet (Neptune) was calculated, the planet was discovered optically, and it was found that it fully explained the anomalies of Uranus. Thus the auxiliary hypothesis stayed within the Newtonian theoretical framework, and the threatened refutation was transformed into a resounding success (ibid.).

Apparently, Popper strenuously tries to distinguish between ad hoc and auxiliary hypotheses. As Bamford perceptively notes, "Popper's attitude to both auxiliary and ad hoc_p hypotheses is negative, however, as both would remove the empirical challenges confronting theories" (Bamford 1999, 375; see also Bamford 1993, 1996). All too probably, Popper would have wanted to view the disturbances in the motion of Uranus as falsifying once and for all the Newtonian theory. However, Uranus' misbehavior turns out to be the momentum for the discovery of Neptune, which is widely hailed as "a triumph without equal for celestial mechanics" (Herrmann 1984, p. 38).[1] Popper's reluctance of distinguishing between ad hoc and auxiliary hypotheses can be indeed detected from his own words:

> For example the observed motion of Uranus might have been regarded as a falsification of Newton's theory. Instead the auxiliary hypothesis of an outer planet was introduced ad hoc, thus immunizing the theory. This turned out to be fortunate; for the auxiliary hypothesis was a testable one, even if difficult to test, and it stood up to tests successfully (Popper 1976, p. 42).

Even though he explicitly points out that the hypothesis of an outer planet was introduced ad hoc, he was enforced to call it not an ad hoc but an auxiliary hypothesis. Commenting on the paragraph just quoted, Bamford caricatures Popper as treating the trans-Uranian Planet hypothesis "as a fellow traveler with ad hoc hypotheses, mystifyingly remarking that it was 'fortunate' the hypothesis was testable" (Bamford 1999, 379).

1.2 Lakatos

The essence of Lakatos's scientific methodology is aptly summarized in his own word as follows:

> All scientific research programmes may be characterized by their '*hard core*'. The negative heuristic of the programme forbids us to direct the *modus tollens* at this 'hard core'. Instead, we must use our ingenuity to articulate or even invent 'auxiliary hypotheses', which form a *protective belt* around this core, and we must redirect the *modus tollens* to these. It is this protective belt of auxiliary hypotheses which has to bear the brunt of tests and get adjusted and re-adjusted, or even completely replaced, to defend the thus-hardened core. A research programme is successful if all this leads to a progressive problem shift; unsuccessful if it leads to a degenerating problem shift (Lakatos 1978a, p. 48).

[1]Herrmann explains rather convincingly the reason why as follows: "Although at the time of this scientific achievement no one doubted the validity of the Newtonian law of Gravity, the understanding of the celestial laws had indeed been demonstrated in a particularly complete and convincing way as a result of this achievement. Astronomical prognoses were at that time nothing new for the researcher, but the discovery of a new large planet by means of the application of a theory understandably made a great impression in the widest circles and was of particular propaganda value, since the confirmation of theory through practice demonstrated the level of understanding of celestial mechanical laws" (Herrmann 1984, p. 38).

Lakatos believes that Netwon's gravitational theory is "possibly the most successful research programme ever". When it was first introduced, according to him, it was submerged in an ocean of 'anomalies'. But Newtonians turned these anomalies (counterexamples or difficulties) into corroborating instances, thereby into a new victory of their programme. In terms of his celebrated metaphor of hard core and protective belt, now Lakatos describes the fortuna of Newton's programme as follows:

> In Newton's programme the negative heuristic bid us to divert the *modus tollens* from Newton's three laws of dynamics and his law of gravitation. This 'core' is 'irrefutable' by the methodological decision of its proponents: anomalies must lead to changes only in the 'protective' belt of auxiliary, 'observational' hypotheses and initial conditions (ibid.).

Against the background of what Popper had to say about ad hoc and auxiliary hypothesis, exactly how are we to understand Lakatos' stance?

1.3 Duhem-Quine Problem

Lakatos is no doubt the most serious critic of Popper's scientific methodology. We should not forget at the same time the fact that Lakatos is still a sort of falsificationist. Indeed, Lakatos himself contrasted his sophisticated methodological falsificationism with Popper's naïve methodological falsificationism (Lakatos 1978a, p. 31f.). In understanding these double aspects of Lakatos against the Popperian heritage, there seems to be no better strategic point than the problem of ad hoc hypothesis.

As Gillies points out, "[i]n the years 1963–4 when he published *Proofs and Refutations,* he was a follower of Popper and a defender of Popperian philosophy":

> The purpose of these essays is to approach some problems of the *methodology of mathematics.* I use the word "methodology" in a sense akin to Polya's and Bernays' "heuristic" and Popper's "logic of discovery" or "situational logic" (Lakatos 1976, p. 3; Gillies 2002, p. 13).

According to Gillies, Lakatos has the same attitude towards Popper in his 1968 article: "Changes in the Problem of Inductive Logic". In this article, Lakatos defends Popper's theory of corroboration against Carnap's theory of confirmation. But, in 1973, only 5 years later, everything changed drastically. Now Lakatos attacks Popper in a ruthless fashion. To quote Lakatos himself:

> Allegedly, Popper's three major contributions to philosophy were: (I) his falsifiability criterion - I think this is a step back from Duhem; (2) his solution to the problem of induction - where I think he is a step back from Hume…; and (3) his literary masterpiece "*The Open Society* by one of its enemies"… what is it called? *The Open Society and its Enemies.* … *The Open Society* is frankly a literary masterpiece: not being a political philosopher I cannot comment on its contents, but I certainly think it is a marvelous book. So, in conclusion, two-thirds of Popper's philosophical fame is based mis-judgement (Lakatos 1999, pp. 89–90).

Now, Gillies must be right in pointing out that Lakatos' criticisms of Popper are based fundamentally on the Duhem thesis:

> In sum, the physicist can never subject an isolated hypothesis to experimental test, but only a whole group of hypotheses; when the experiment is in disagreement with his predictions, what he learns is that at least one of the hypotheses constituting this group is unacceptable and ought to be modified; but the experiment does not designate which one should be changed (Duhem 1962, p. 187; Gillies 2002, p. 16).

Gillies claims at this stage that "Lakatos proposes his methodology of scientific research programmes as a solution to this problem of Duhem's", for, "according to Lakatos, a scientist always works in the context of a research programme, which has a *hard core* or *negative heuristic*." All this can be confirmed by Lakatos' own words:

> The appraisal of large units like research programmes is in one sense much more liberal and in another much more strict than Popper's appraisal of theories. This new appraisal is *more tolerant* in the sense that it allows a research programme to outgrow infantile diseases, such as inconsistent foundations and occasional ad hoc moves. Anomalies, inconsistencies, ad hoc stratagems, even alleged negative 'crucial' experiments, can be consistent with the overall progress of a research programme. The old rationalist dream of a mechanical, semi-mechanical or at least fast-acting method for showing up falsehood, unprovenness, meaningless rubbish or even irrational choice has to be given up. But this new appraisal is also *more strict* in that it demand not only that a research programme should successfully predict novel facts, but also that the protective belt of its auxiliary hypotheses should be largely built according to a preconceived unifying idea, laid down in advance in the positive heuristic of the research programme (Lakatos 1974, 319).

One salient consequence of adopting such a position, especially in connection with Popper's scientific methodology, is striking. Now Lakatos claims that there is no such thing as crucial experiments:

> My modification then presents a very different picture of the game of science from Popper's. The best opening gambit is not a falsifiable (and therefore consistent) hypothesis, but a research programme. Mere 'falsifications' (that is anomalies) are recorded but need not be acted upon. 'Crucial experiments' in the falsificationist sense do not exist: at best they are honorific titles conferred on certain anomalies long after the event when one programme has been defeated by another one (Lakatos 1974, 320).

However, as Hacking convincingly argues, Lakatos tends "to play down the role of experiment too much" (Hacking 1983, p. 254). I will return to this issue in Sect. 4.

1.4 The Different Meanings of Ad Hoc

Lakatos claims that there are "two clearly distinguishable senses" of Popper's usage of the pejorative term 'ad hoc': ad hoc_1 and ad hoc_2. Ad hoc_1 refers to a theory that is "without excess content", while ad hoc_2 refers to a theory that is "without excess corroboration" (Lakatos 1978b, p. 180, n. 1).[2] Further, Lakatos identifies still another unsatisfactory cases that are neither ad hoc_1 nor ad hoc_2: i.e., ad hoc_3 (Lakatos 1978a, p. 40, n. 2; p. 88, n. 2). According to Lakatos, these are the cases of progress "with a patched up, arbitrary series of disconnected theories" (ibid., p. 88).

Nickles counts the third ad hoc-ness rule "most Lakatosian and least Popperian in motivation". Why? Probably because ad hoc_3 is characterized by Zahar, who "explicitly articulates and further develops Lakatos's ideas on ad hoc-ness", as follows:

> [A theory is] ad hoc_3 if it is obtained from its predecessor through a modification of its auxiliary-hypotheses which does not accord with the spirit of the heuristic of the programme (Zahar 1973, 101; see also Nickles 1987, 191).

Contrary to its prima facie Popperian outlook, Nickles claims that the third ad hoc-ness rule "prohibits *any* heuristically unmotivated theory change" (ibid., 192, Nickle's emphasis). One striking consequence of this rule, according to Nickles, is that "blind guesses and Popperian conjectures turn out to be ad hoc_3" (ibid.).

On the other hand, Bamford reports that 'ad hoc' in ordinary English means "for this or the particular purpose", and calls it 'ad hoc_e' (Bamford 1999, 376; Shorter Oxford). He further elaborates that this term is typically used "to describe something which is designed or adopted merely to satisfy a particular requirement, or for the act of doing so". More interesting in Bamford's report is, however, pejorative senses of 'ad hoc' both in ordinary English and in scientific community. In the former, 'ad hoc' "also means 'arbitrary', 'makeshift', 'stop-gap', or the like, while, in the latter, 'artificial', 'cooked up', 'implausible', 'unreasonable', 'unnecessary', 'ugly', or the like".

Bamford aptly characterizes Popper's use of 'ad hoc' as retaining "its original English meaning, including its pejorative overtones". But it seems more significant that it is also based on such a usage of 'ad hoc' in the world of scientists. Bamford draws our attention to Popper (1972, p. 287), where Popper claims that "it is well known that ad hoc hypotheses are disliked by scientists". Popper must be right in his observation, and Bamford could be excused for his failing to give any intelligent comments on it. Rigorously speaking, however, Popper could be begging the question, for he does not provide us with any evidence. Even if he is right, it is not evident which scientists he is referring to. Does he have in mind only the scientists of our time, or Scientists after the 17th century scientific revolution, or all scientists of all ages? Unlike Bamford, then, it might be useful to separate 'ad hoc_s' from 'ad hoc_e'.

Anyway, the sense of 'ad hoc_p' evidently goes beyond 'ad hoc_e' or 'ad hoc_s' by requiring the independent testability. Furthermore, Bamford seems to assume that the reason why scientists dislike ad hoc hypotheses does not have to be investigated insofar as 'ad hoc_e' or 'ad hoc_s' have virtually the same meaning. Even if they have virtually the same meaning, however, due to the special context, i.e., scientific

[2]For ad hoc_1, Lakatos cites Popper (1934), § 19 and Popper (1963), p. 241. For ad hoc_2, he cites Popper (1963), p. 244, claiming that Popper introduces it since 1963 (ibid.).

community, 'ad hoc_s' must add some further elements to 'ad hoc_e'. In other words, 'ad hoc_p' must have been introduced by Popper in order to capture some such 'ad hoc_s'. If I am on the right track here, then there is further need to elaborate why Bamford counts 'ad hoc' as "a tricky term in Popper's lexicon":

> Firstly, the *definiens* for 'ad hoc_p' is a hybrid of the psychological (aiming to remove or avoid a difficulty) and the methodological (non-independent testability). Secondly, a hypothesis can be ad hoc_e without being ad hoc_p, as is any auxiliary hypothesis on his definition above (Bamford 1999, 376).

Based on Popper (1976, p. 42, 1972, p. 244), Bamford shows that Popper sometimes used 'ad hoc' as merely ad hoc_e, another times as ad hoc_p. and still another times as "some stronger unexplicated sense of 'ad hoc' which in fact implies that 'ad hoc_p' is inadequate".

We also noted above Lakatos' distinction between the three different pejorative senses of 'ad hoc': i.e., 'ad hoc_1', 'ad hoc_2', and ad hoc_3. So, we face a plethora of the different senses of 'ad hoc', which is nicely enumerated by Martin Carrier.[3] I have no intention to discuss at length the subtle differences between all these senses of 'ad hoc', though such a distinction seems mandatory for any serious philosopher of science. But we should note the utmost importance of raising the following question: Was 'ad hoc' used pejoratively in the world of scientists, i.e., 'ad hoc_s', prior to Popper used it so? In other words, can we find textual evidence from the scientific literature that 'ad hoc' did have pejorative sense? If so, since when, by whom, and how widely was it used in that sense?

---

[3]Methodological literature provides us with a vast number of conditions for the ad hoc-ness of a hypothesis. Apart from the concepts already discussed (i.e. apart from the predictivistic and the post-Lakatos versions of the heuristic criterion) I am aware of no less than eight different notions of ad hoc-ness (and I do not, of course, pretend to know them all).

(1) No excess empirical content or no independent testable consequences actually *exist* (Lakatos 1971a, 112; Grünbaum 1976, 337).
(2) No such consequences are *known* (Lakatos 1971a, 112; Grünbaum 1976, 336).
(3) No such consequences are *confirmed* (Zahar 1973, 101; Grünbaum 1976, 334).
(4) All such consequences are empirically *refuted* (Lakatos 1971a, 112).
(5) It is assumed that a hypothesis is independently testable, *but* it is further assumed (motivated by the chains of some rival theory) that the hypothesis will fail in subsequent experimental tests (Grünbaum 1973, 717).
(6) The hypothesis is empirically ad hoc in the senses (1) or (2) or (3) *and* has furthermore no theoretical plausibility or sanction (Grünbaum 1976, 333–337).
(7) There are confirmed consequences but the hypothesis is not in accordance with the heuristic of a research programme (Lakatos's 'ad hoc_3') (Lakatos 1971a, 112). In other words, concept (6) holds theoretical plausibility to be a *sufficient* condition for non-ad hoc-ness, concept (7) views it as a *necessary* one.
(8) A necessary condition for an ad hoc hypothesis is its 'non-fundamentality'; it 'fails to go to the heart of the matter' (Leplin 1982, 237; Carrier 1988, 216–217, n. 45).

2 The Problem of Enthymeme Reconstruction

In what follows, I shall sketch an attempt to interpret the controversy between Popper and Lakatos on ad hoc hypotheses in terms of enthymeme resolution. The basic idea is very simple. It seems evident that both the problem of ad hoc hypothesis and the problem of enthymemes are sub-species of the problem of anomaly resolution (see Magnani 1999, 2001, 2009; Humphreys 1968; Brun and Rott 2013; Zenker 2006; Schurz 2013). If so, it is natural and even mandatory to compare problems, concepts, and theories in both areas. Further, we seem to have enough ground to believe that enthymeme has a much longer history than ad hoc hypothesis. If philosophers and scientists before the modern scientific revolution had to face problems comparable to the problem of ad hoc hypothesis, I believe, it would have been one of the most useful scientific methods, devices or procedures to tackle them. To the best of my knowledge, history of enthymeme is yet to be written. Still, there is huge literature on enthymeme, and I have no ambition to do justice to the subtle and sophisticated rival theories. I am simply inspired by Paglieri and Woods' recent proposal of highlighting the role parsimony rather than charity in the interpretation and reconstruction of enthymeme (Paglieri and Woods 2011a, 2011b). The contrast of parsimony and charity in the problem of enthymeme seems to present an entirely new perspective for understanding Popper/Lakatos controversy on ad hoc hypothesis.

2.1 *What Is an Enthymeme?*

Textbooks of introduction to logic usually provide us a short section on enthymeme. For example, Copi deals with it in a chapter entitled "Arguments in ordinary Language", and in a section preceding another section for sorites. According to him, an enthymeme is

> an argument that is stated incompletely, part being "understood" or only "in the mind" (Copi 1982, p. 253).

After pointing out that enthymemes are everywhere in everyday discourse as well as in scientific practice, he turns to the problem of how to test their validity:

> In testing an enthymeme for validity, two steps are involved. The first is to supply the missing parts of the argument; the second is to test the resulting syllogism (ibid., p. 255).

Copi's brief characterization of enthymeme seems to represent what Paglieri and Woods call "the modern conception of enthymeme", which is "something of a hybrid of elements drawn from" Epictetus' and Aristotle's definitions (Paglieri and Woods 2011a, 463, 465–467). One essential feature of the modern conception is reported by Paglieri and Woods as follows:

According to the modern idea, there is some property *Q* that it is desirable for arguments to have, such that an enthymeme lacks it yet its completion has it. Consider validity as an example. Validity is what the enthymeme lacks and its completion has. An enthymeme of this sort is a valid argument-in-waiting (ibid., 463).

Paglieri and Woods count *incompleteness* "as a key feature of enthymeme worth preserving in the modern conception":

We agree with the dominant view that something is missing and yet understood in enthymemes; moreover, that something is essential to their interpretation (ibid., 468).

However, there are many features of the modern conception they found somewhat troublesome. For example, they want to avoid a possible confusion between "interpreting an enthymeme" and "assessing its value". Also, they believe that "something is essential" in interpreting enthymemes does not mean "that, by understanding the missing element, the hearer will be inclined to accept the argument as good" but merely "that the missing element is crucial to make an informed assessment of the enthymeme, be it positive or negative" (ibid.). There are, of course, many other problems and issues in defining an enthymeme, as there are many different varieties of enthymemes. Paglieri and Woods enumerate at least the following different varieties:

(a) Valid and sound elliptic argument: "Socrates is a man, therefore he is mortal".

(b) Purely formal elliptic argument: "All P are Q, so some R are Q".
(c) Unsound elliptic argument: "The mackerel is a fish, so it is colour-blind".
(d) Crazy elliptic argument: "Today I am happy, therefore Mars is not a planet".
(e) Invalid elliptic argument: "Every Catholic priest is male, so John is a Catholic priest".
(g) Defeasible elliptic argument: "Ozzie is an ocelot, therefore Ozzie is four-legged".
(h) Materially valid argument: "The shirt is red, therefore the shirt is coloured".
(i) Complete argument, either valid or invalid: "Socrates is a man and all men are mortal, therefore Socrates is mortal", and "Lassie is mortal and every man is mortal, therefore Lassie is a man".
(j) Isolated statement: "Socrates is mortal" (Paglieri and Woods 2011a, 469–472).

Though it is apparently trivial, one interesting idea is that any invalid argument might become a valid one by adding some appropriate premises. We may consider any arbitrarily selected invalid argument, e.g., P/∴S. As Paglieri and Woods point out, by adding to the premise-set a proposition, i.e., the conditional proposition with P as antecedent and S as consequent, we can produce a valid argument. No one would deny that an enthymeme cannot be "just any invalid argument validated by addition of its corresponding conditional as premiss". But how are we to distinguish between valid and invalid enthymemes? What Paglieri and Woods call the *demarcation problem*, which is a central task for a theory of enthymemes, is nothing but "to preserve the distinction, and to bring it to a decent level of theoretical articulation" (ibid., 467).

Paglieri and Woods suggest the following as the definition of enthymeme:

ENTHYMEME: A is an enthymeme if and only if A contains at least one explicit premiss explicitly linked to an explicit conclusion, and yet A can be assessed according to some standard of argument evaluation if and only if A is first supplemented with some additional premiss P that preserves the relevance of all A's premisses to A + P's conclusion and is selected by applying a general reconstructive principle to A (ibid., 468).

According to them, they included five main conditions in this definition:

(i) A is an argument in a minimal sense, i.e. it contains at least one (explicit stated) premiss that carries the presumption of supporting at least one (explicitly stated) conclusion.
(ii) A is not assessable on some standard of argument evaluation.
(iii) A + P is assessable on some standard of argument evaluation.
(iv) Adding P to A does not make irrelevant any of A's premisses to A + P's conclusion.
(v) The transition from A to A + P is not arbitrary, but rather governed by some general principle (yet to be determined; see Sects. 2 and 3) (ibid.).

We can appreciate how judiciously Paglieri and Woods selected the conditions for an enthymeme to satisfy in this definition. As intended by them, we may focus on the problem of "determining the appropriate criterion that should guide the reconstruction of enthymemes".

2.2 *Charity Versus Parsimony: The Two Rival Theories of Enthymeme Resolution*

According to Paglieri and Woods, we can distinguish between approaches that are committed to the completion-as-amelioration doctrine and those that aren't. Approaches based on unrestrained appeals to charity are clear examples of the former. Among the latter, Paglieri and Woods include their own theory based on parsimony and other analyses of enthymemes that are not committed to the completion-as-amelioration thesis, and yet differ also from their own proposal in terms of parsimony, such as *contextualism* and *anti-reconstructionism*. Paglieri and Woods briefly introduce contextualism as follows:

As the name implies, contextualism appeals to context to help identifying missing premisses in enthymemes. Different versions of this approach focus on different contextual features: among others, intended meaning of the speaker (Ennis 1982), evidence justifying reconstruction of the enthymeme as an invalid argument (Gerritsen 2001), specific commitments endorsed by the proponent of the enthymeme in the dialogical interaction where it occurs (Walton 1998, 2008), and general conversational context against which single arguments are to be assessed and understood by the arguers (Jacobs 1999; Paglieri and Woods 2011a, 496–497).

Also, Paglieri and Woods explain anti-constructionism as follows:

As for anti-reconstructionism, this view does not consider enthymemes as in need of reconstruction, but rather endeavours to define a theory of validity (or lack thereof) applicable to enthymemes without prior supplementation of any "missing" premiss: for an excellent attempt in this direction, see Hitchcock's notion of enthymematic consequence (1998) (ibid., 497).

It is obviously beyond the scope of this paper to do justice to each and every one of the theories identified by Paglieri and Woods. For my present purpose, it would be enough to focus on approaches based on unrestrained appeals to charity as the foremost example of completion-as-amelioration doctrine and Paglieri and Woods' theory based on parsimony. Further, insofar as the former is the mainstream dominant theory of enthymeme (see ibid., 496) and the idea of counting enthymeme as a sort of valid argument-in-waiting is behind the notion of charity (see ibid., 463), what is needed is understanding why and on what ground Paglieri and Woods criticize the theory based on charity and present their own theory based on parsimony as an alternative.

2.2.1 Against Charity

Paglieri and Woods devote an entire section for criticizing the theory based on charity (ibid., Sect. 2 "Against Charity", 473–476). However, the more serious and important criticisms were made even before turning to the section. For they argue that the following widely accepted idea cannot be true:

> GOOD COMPLETIONS AS VALIDATING: A good enthymeme-completion selection will be a premise that validates it (ibid., 472).

They believe that the existence of cases (e) above is enough to conclude that way: (e) Invalid elliptic argument: "Every Catholic priest is male, so John is a Catholic priest". Good completion of this elliptic argument would need to supplementing it with "John is male". But that does not make the argument any more valid. The only possibility left for the defenders of modern conception of enthymemes that Paglieri and Woods can figure out is as follows:

> In contrast, the modern conception of enthymemes would force us to either exclude these cases from the definition of enthymemes (because their completion is invalid, when reconstructed in the most plausible way), or reconstruct them in ways that ensure validity by adding bizarre and false premises (in this example, the conditional "If every Catholic priest is male, then John is a Catholic priest") (ibid., 471).

As a consequence, Paglieri and Woods claim that the doctrine of completion-as-validation conflates two quite different tasks:

> One is the task of spotting a premiss that completes an incomplete argument. The other is the task of finding a premiss that will validate it (ibid., 472).

Encouraged by the possible liberating effect of separating the two tasks, Paglieri and Woods even try to generalize:

> GOOD COMPLETIONS/BAD ARGUMENTS: It cannot be typical of enthymemes that they be properly complete only by premises that make them good arguments, in whatever sense of good fits the particular case. In brief, completion is not amelioration (ibid., 473).

As Paglieri and Woods point out, there is still "the problem of elucidating what guides selection of the appropriate premiss to supplement the enthymeme with". Again, as they do not forget to mention, that is nothing but fleshing out the condition (v) of their definition of enthymemes.

2.2.2 For Parsimony

Paglieri and Woods counts "the pragmatic question of why enthymemes are so frequent in human communication" as another fundamental question that cannot be answered by invoking charity:

> Even assuming charity in interpreting each other's arguments, why do we so often indulge in enthymematic argumentation, and why are we so favourably disposed towards incomplete arguments? If charity were the rule, why would it be so? And why do we systematically fail to speak or write our arguments in a more complete and explicit fashion, relying instead on the charity of the audience? (ibid., 477)

In order to answer these questions, Paglieri and Woods start with Herbert Simon's notion of bounded rationality (see Simon 1955, 1956). They explain that, from the vast literature on bounded rationality, thy draw upon on a single aspect: "how the fact that agents are cognitively resource-bounded affects their dialectical capacities", and "how this same fact should inform a good theory of argumentative rationality" (ibid., 477). As they count resource-boundedness as a simple fact of life, they find the need to be parsimonious in using the finite and scant resources as inherent to the rationality of all our actions (ibid.).

So, Paglieri and Woods' answer to the questions raised above is utterly simple:

> In a nutshell, our hypothesis is that both the frequent use of enthymemes and the principles governing their reconstruction are ultimately motivated by an attempt, by the arguer as well as by the interpreter, to save valuable cognitive resources, without injury to the performance of their respective tasks. Accordingly
>
> PARSIMONY: It is parsimony, rather than charity, that inspires our enthymematic inclinations (ibid., 477–478).

3 Ad Hoc Hypothesis Generation as a Kind of Enthymeme Reconstruction

3.1 Understanding Popper and Lakatos in Terms of Different Enthymeme Resolution Strategies

In order to interpret the problem of ad hoc hypothesis in terms of enthymeme resolution strategies, it must be a prerequisite to identify exactly what enthymeme is at stake. However, it may not be an easy matter to do without inadvertently taking side with a particular theory. Let us use the discovery of Neptune as an example, and focus on only Popper's and Lakatos' views. The enthymeme that would be identified by Popper may have the Newtonian theory of gravitation as the premise, and the unexpected motion of Uranus as the conclusion. On the other hand, Lakatos would have the Newtonian theory of gravitation together with some auxiliary hypotheses as the premise, and the unexpected movement of Uranus as the conclusion.

As we saw above, Popper thinks that "the observed motion of Uranus might have been regarded as a falsification of Newton's theory" (Popper 1976, p. 42; supra, 1.1).

$(N \vdash U) \& \neg U$
$\therefore \neg N$
(Popper's Falsification)

This indicates that the enthymeme being considered by Popper would be something like the following.

N
$\therefore \neg U$
(Popper's Enthymeme)

On the other hand, as with all other people convinced by Duhem-Quine thesis, Lakatos thinks that the negative heuristic of the scientific programme forbids us to direct the *modus tollens* at the 'hard core'. Instead, according to him, "we must use our ingenuity to articulate or even invent 'auxiliary hypotheses', which form a *protective belt* around this core, and we must redirect the *modus tollens* to these" (Lakatos 1978a, p. 48; Supra, 1.2).

$((N\&A) \vdash U) \& \neg U$
$\therefore \neg (N\&A)$
N
$\therefore \neg A$
(Lakatos' Falsification)

So, the enthymeme considered by Lakatos would be rather like the following.

N
A
$\therefore \neg U$
(Lakatos' Enthymeme)

Further, there are apparent problems in applying Paglieri and Woods' distinction between enthymematic parsimony in the arguer and the enthymematic parsimony in the interpreter. For it is not clear who could be the arguer using an enthymeme, for which scientists try to provide an ad hoc hypothesis. Should it be Newton, or a Newtonian astronomer in mid-19th century, or Nature herself? Be that as it may, there would be no serious objection to count any scientist examining the case of the discovery of Neptune as the interpreter of the enthymeme involved. Again, Lakatos' position would be easier to handle. Above all, he is trying to save the core of the Newtonian scientific research programme, i.e., the Newtonian theory of gravitation, from the potential falsifier, i.e., the unexpected movement of Uranus. Also, we may treat Lakatos as implicitly adopting the theory of enthymeme based on charity. The idea of protecting the hard core of the scientific research programme may be understood as an expression of the principle of charity. Also, Lakatos seems to be a fine example of a theorist of charity who equates the good completion of an enthymeme with the validation of the enthymeme.

To be sure, the major hurdle of Lakatos' search for an ingenious auxiliary hypothesis is that the addition of the new auxiliary hypothesis, i.e., the hypothesis of existence of a new planet, to the initial set of auxiliary hypotheses yields an inconsistency. As Sylvestre nicely points out, the initial set of auxiliary hypotheses must have a proposition like "The solar system consists of exactly Mercury, Venus, Earth, Jupiter, Saturn, and Uranus" (Sylvestre 2012, 521). If so, a new auxiliary hypothesis such as "There exists an unknown planet beyond Uranus whose gravitational effect upon it is preventing the calculated positions from fitting the observed data" must cause an inconsistency in the expanded set of auxiliary hypotheses (ibid., 520). What does all this mean? Can we sustain our interpretative hypothesis that Lakatos is an interpreter of the enthymeme championing the principle of charity in this situation? The new set of auxiliary hypotheses, A^*, is unstated in the initial set of auxiliary hypotheses. But shouldn't it be understood and acceptable by the interpreters even then?

Now, let us turn to Popper's case to see whether his position about ad hoc hypothesis can be interpreted in terms of enthymeme resolution. Most philosophers of science would be skeptical about such a possibility, for apparently Popper himself seems to resist to such possibility. Above all, Popper does not suggest any addition to the initial premise-set in order to validate the enthymeme being considered. He has absolutely no interest in validating any argument.

I think, all this is quite understandable. However, that is so, only insofar as we uncritically accept the mainstream dominant view of enthymeme, i.e., the theory of enthymeme based on the principle of charity. Paglieri and Woods' proposal for a theory of enthymeme based on the principle of parsimony may make it possible to reinterpret Popper's position on the problem of ad hoc hypothesis in terms of enthymeme. In fact, this possibility itself is potentially much more interesting than Lakatos' case. Let me explain.

Popper's stance toward an ad hoc hypothesis seems quite similar to "a drastic way of reducing the consumption of cognitive resource … to refuse to process any argument which is not completely and correctly stated by the arguer" Paglieri and Woods discuss. As they point out, this would "imply remaining deaf to all enthymemes". They concede that such a way has definite connection with "a kind of parsimony". However, they deem it as "a very short-sighted one", for "simply disregarding any enthymeme would lead to save cognitive effort at the cost of wasting potentially valuable information" (Paglieri and Woods 2011a, 483).

There is no doubt that Paglieri and Woods' emphasis on the balance between "the cognitive resources that we use to interpret each other's messages" and "the informational resources that we extract from them" is well-motivated. However, I instinctively feel that there might be some further defense for Popper's position. First of all, insofar as Paglieri and Wood's theory of enthymeme based on the principle of parsimony a novel alternative to the mainstream dominant theory of enthymeme based on the principle of charity, it is by no means a small achievement that Popper at least implicitly incorporating the principle of parsimony in dealing with issues definitely related to enthymemes. Furthermore, I think, Popper could defend himself by appealing to the very point of balance, by which Paglieri and Woods criticize him.

Popper may point out that, only by flatly falsifying Newtonian theory of gravitation due to the disturbance in the motion of Uranus, we can ultimately revise our belief in the much older "system of the world", i.e., our belief that "The solar system consists of exactly Mercury, Venus, Earth, Mars, Jupiter, Saturn, and Uranus". In other words, unlike Lakatos, Popper could have denied that our belief in the constitution of the solar system is not one of the auxiliary hypotheses constituting the protective belt. Rather, it belongs to even higher order than Newtonian theory of gravitation. Or, we may say that it is closer to the center of the hard core of Newtonian scientific research programme than the theory of gravitation.

By simply assuming that our belief in the constitution of the solar system is merely one of the auxiliary hypotheses, Lakatos might be begging the question as to the scope of Newtonian theory of gravitation, i.e., as universal rather than local. But it is highly suspicious that Newton himself thought that way about his own celebrated theory. Be that as it may, in validating the enthymeme by revising our belief in the constitution of the solar system, Lakatos is not licensed to introduce a new premise, i.e., that there is another unknown planet. For, insofar as it is unknown, there would be no way for it to be understood and acceptable by the scientific community or the general public.

3.2 *What Were the Discoverers of Neptune Doing?*

If I am on the right track by interpreting the problem of ad hoc hypothesis in terms of enthymeme resolution, it becomes rather obvious that all this should be approached not merely philosophically but more historically. Who were actual arguers introducing an enthymeme at stake? Who were the particular scientists trying to introduce ingenious ad hoc hypotheses? To say the least, we should be well informed enough to understand the episode of the discovery of Neptune historically.

From this point of view Norwood Russell Hanson's study seems to be invaluable. First, he describes the problem situation Leverrier was facing by directly quoting him:

> [In June, 1846], Leverrier wrote: I have demonstrated… a formal incompatibility between the observations of Uranus and the hypothesis that this planet is subject only to the actions of the sun and of other planets acting in accordance with the principle of universal gravitation (Hanson 1962, 361).

According to Hanson, Leverrier was conscious of the fact of Newton's law had been suspect. For example, figures like Huygens, Leibniz, John Bernoulli, Cassini and Miraldi were hostile to the initial formulation the law. Anti-Newtonians seized energetically the theoretical notion of lunar perigee. Then, Moon's secular acceleration was the difficulty for the Newtonians. Nevertheless, he himself never doubts the exactitude of Newton's theory, for, whenever it had been suspect, it had become victorious by distinguished scientists. Lunar difficulty was overcome completely by Clairault and Euler. Laplace resolved the problem of Moon's secular acceleration. Finally, "[a]fter 1826 the Uranian discrepancies the inverse-square law required another term again raised doubts about the inverse-square, and minor modifications were regularly entertained by astronomers" (ibid.).

Hanson seems to be too quick at this stage by jumping to the following report:

> Now the hypothesis of an unseen planet acting on Uranus is publicly entertained. (Note again that this is where Adams began his inquiry). (ibid., 362)

But what is important is, at least, Hanson tries to understand in what situation Leverrier was raising the following question:

> Is it possible that Uranus' inequalities may be due to a planet located in the ecliptic, at a mean distance double that of Uranus? And were this so, where is the planet now? What is its mass?' What are the elements of its orbit? (ibid.)

Hanson emphasizes that Leverrier's and Adams' problem was "the inverse perturbation problem", i.e., "describing the disturbances in Uranus, from which he then infers the mass and orbital elements of the disturbing planet", which should be sharply contrasted with the classical problem of perturbations, i.e., "determining a planet's disturbances in another body with the knowledge of its mass and its orbital elements" (ibid.).

Finally, most pertinent for my purpose, Hanson reconstructs Leverrier's argument as follows:

> (1) Uranus' aberrations are formally incompatible with the New-tonian predictions, (2) But Newtonian mechanics is unquestionably true, (3) And the observations of Uranus' orbit are unquestionably accurate. (4) This tension would be resolved were there some mass having just those dynamical properties required by theory to generate Uranus' observed positions (ibid., 364).

What is interesting is Hanson identifies the pattern of this argument as Peircean retroduction. He writes:

> A further comparison with logic is apposite. Leverrier's question is not "What follows from these premises (Newton's laws and Uranus' positions)?" Nor is it "How can I summarize and generalize these data?" The question is rather, "Given Uranus' aberrant positions as a conclusion, from what further premise (besides Newton's laws) could this conclusion be generated?" The logical structure of retroduction is easily appreciated when viewed against Hempel's analysis of explanation and prediction (ibid.).

Hanson must have very good reason to be impressed by detecting the logical structure of abduction from Leverrier's discovery of Neptune. However, it is rather surprising that he fails to notice the structural similarity between Leverrier's query and that of the charity oriented interpreters of an enthymeme. For they also always pose to themselves the question "From what further premise (besides the explicitly expressed premise) could this conclusion be generate?" It is one thing that the inference by which Leverrier and the interpreters of enthymeme arrive at the conclusion could be abduction. But it is another that the whole context in which they abduce the conclusions is that of enthymeme resolution.

4 Back to Bacon, if not Aristotle

As we saw above, Popper simply assumed that scientists have strong feeling against ad hoc hypotheses. But I am rather suspicious about this assumption. In order to see whether it is true, probably we need to wait for some future empirical study. As for whether it can be grounded firmly on historical documents, again I am somewhat skeptical. Even if there are some supporting cases in history of science, I surmise, they might be rather exceptional. Be that as it may, let us check what antecedent cases have been noticed in the literature.

As Bamford points out, "Popper suggests surprisingly few examples of ad hoc_p hypotheses":

> Popper (1934\1959\1975, p. 83) once claimed the Lorenz-Fitzgerald contraction hypothesis was ad hoc_p, and although he later accepted that it was not, he still suggested that perhaps this hypothesis illustrates 'degree of ad hocness' (Bamford 1999, 377).

There are some other examples like Pauli's neutrino hypothesis Popper discussed in his writings, of course. Still Popper's examples are indeed few. Further, those few cases are from the 20th century science. In other words, Popper seems to have extremely weak evidence for his radical assumption that throughout the history of science scientists have been antagonistic against ad hoc hypotheses. Lakatos seems to agree with Popper that there is strong dislike of merely ad hoc hypotheses among scientists (see, for example, Lakatos 1970, 38–39). Indeed, what Popper has in mind seems to be clearly expressed by the following remarks of Lakatos:

> Why aim at falsification at any price? Why not rather impose certain standards on the theoretical adjustments by which one is allowed to save a theory? Indeed, some such standards have been well-known for centuries, and we find them expressed in age-old wisecracks against ad hoc explanations, empty prevarications, face-saving, linguistic tricks. * [* Molière, for instance, ridiculed the doctors of his Malade Imaginare, who offered the virtus dormitiva of opium as the answer to the question as to why opium produced sleep. One might even argue that Newton's famous dictum hypotheses non fingo was really directed against ad hoc explanations – like his own explanation of gravitational forces by an aether-model in order to meet Cartesian objections. (Lakatos's own footnote)] (Lakatos 1970, p. 32)

No doubt, Lakatos is drawing our attention to some of the most interesting episodes in history of science. Notwithstanding of their rhetorical power, however, these are still quite controversial. No matter how it sounds absurd to Popper and Lakatos, I think, the virtus dormitiva has certain value in the development of science. Any commentator of Newton's thought must have something to say about the celebrated "Hypothesis non fingo". But I do not know anyone else who adopts Lakatos's interpretation of Newton as referring only to ad hoc hypothesis by "hypothesis" in the phrase.

In Hempel's classical *Philosophy of Natural Science* (1966), we find the following as examples of ad hoc hypothesis:

> About the middle of the seventeenth century, a group of physicists, the plenists, held that a vacuum could not exist in nature; and in order to save this idea in the face of Torricelli's experiment, one of them offered the ad hoc hypothesis that the mercury in a barometer was being held in place by the "funiculus", an invisible thread by which it was suspended from the top of the inner surface of the glass tube (Hempel 1966, p. 29).
>
> According to an initially very useful theory, developed early in the eighteenth century, the combustion of metals involves the escape of a substance called phlogiston. This conception was eventually abandoned in response to the experimental work of Lavoisier, who showed that the end product of the combustion process has greater weight than the original metal. But some tenacious adherents of the phlogiston theory tried to reconcile their conception with Lavoisier's finding by proposing the ad hoc hypothesis that phlogiston had negative weight, so that its escape would increase the weight of the residue (ibid., pp. 29–30).

Of course, these are well-known familiar cases persuasive enough for us to ridicule 17th century plenists and the phlogiston theorists. However, none of these cases can be the last word for the historical question whether scientists have always disliked ad hoc hypotheses. Not to mention the fact the plenists and the friends of phlogiston also deserve the honorable title of "scientists", it is simply impossible to draw any conclusion about how those allegedly absurd ad hoc hypotheses were received by the scientific communities and the general public at the time when they were suggested.

Howson and Urbach present us more examples of ad hoc hypothesis by invoking the cases like Velikovsky, Dianetics of L. Ron Hubbard, IQs of different groups of people, and the Neptune hypothesis (Howson and Urbach 2006, pp. 118–121). However, what is truly invaluable in their treatment of ad hoc hypothesis seems to be found in the following:

> These two criteria were anticipated some four hundred years ago, by the great philosopher Francis Bacon, who objected to any hypothesis that is "only fitted to and made to the measure of those particulars from which it is derived". He argued that a hypothesis should be "larger and wider" than the observations that gave rise to it and said that "we must look to see whether it confirms its largeness and wideness by indicating new particulars" (1620, I, 106). Popper (1963, p. 241) advanced the same criteria, laying down that a "new theory should be independently testable…". Bacon called hypotheses that did not meet the criteria "frivolous distinctions", while Popper termed them "ad hoc".* (* The first recorded use of the term 'ad hoc' in this context in English was in 1936, in a review of a psychology book, where the reviewer criticized some explanations proffered by the book's author for certain

aspects of childish behavior: ... (Sprott (1936), p. 249...)) (Howson and Urbach 2006, p. 121; see also Urbach 1982)

Howson and Urbach's historical conjecture of Francis Bacon's anticipation of Popper's negative views about ad hoc hypothesis seems to be supported by recent studies of philosophy of crucial experiments in history of modern science (see Dumitru 2013; Jalobeanu 2011; Anstey 2014). For example, according to Dumitru, we can trace "the development of the concept of the crucial experiment from Bacon's *instantia crucis* to Robert Boyle's *experimentum crucis* and, finally, to Robert Hooke's use of these phrases in *Micrographia*" (Dumitru 2013, 46).[4]

In his influential *Representing and Intervening* (1983), Hacking starts Chap. 15 "Baconian Topics" with the claim that "Francis Bacon (1560–1626) was the first philosopher of experimental science" (Hacking 1983, p. 246). According to Hacking, Bacon was truer and wiser than later philosophers of science in that he was not victimized the false dichotomy between (1) making crucial experiments absolutely decisive, and (2) claiming, as Lakatos does, that 'there have been no crucial experiments in science'. As Hacking emphasizes, "Bacon claimed only that crucial instances are sometimes decisive" (ibid., p. 250). The heart of the matter lies in whether experiments are crucial at the time when they were performed, or only with hindsight. Popperians and Lakatosians may be grasping the first and the second horn of this dilemma respectively. This contrast seems to be quite consonant with the rivalry between the parsimony view and the charity view in enthymeme resolution.

I tend to think this way. It is one thing whether there has been any absolutely decisive crucial experiment in the history of science, or whether such an experiment is logically possible at all, quite another whether some particular scientists or scientific community of their time believe some experiment as a crucial one.

5 Concluding Remarks

Indeed, the problem of ad hoc hypothesis and the problem of crucial experiment seem to be the two sides of the same coin. The different perspectives of Popper and Lakatos on ad hoc hypothesis must parallel their different perspectives on crucial experiments. Popperian falsificationism without crucial experiment seems inconceivable. If so, Popper's reluctance to concede that there can be and sometimes are auxiliary hypotheses in spite of their evidently ad hoc character is quite understandable. As a matter of fact, Popper probably should not have conceded that much. In other words, he should have found some other way of explaining the discovery of Neptune without making such a concession. Once such a concession is

[4]According to Jalobeanu, it is Hooke who coined the term "*experimentum crucis*" (see Jalobeanu 2014, p. 60). However, as Dumitru and Anstey quote, there is textual evidence that Boyle also used the term "experimentum crucis".

made, it is just a matter of time that the possibility of crucial experiment itself is questioned. As Cantor points out, "[p]hilosophers of science now generally downplay the significance of such experiments, denying them any major role" (Cantor 1989, p. 176). Also, as Rowbottom laments, "Popper's work isn't taken all that seriously in contemporary Anglo-American philosophy" (Rowbottom 2010, p. xi). I believe that the observations of Cantor and Rowbottom are essentially correct. And I think it is unfortunate to underestimate the significance of crucial experiment and the lessons from Popperian falsificationist scientific methodology.

To many, the problem of ad hoc hypothesis appears to be a local issue within the falsificationist camp. It is just a problem of how to distinguish between truly bad ad hoc hypotheses on the one hand, and justifiable auxiliary hypotheses regardless of their apparently ad hoc character. In other words, Duhem-Quine thesis caused an irrevocable conflict between Popper and Lakatos. However, our discussion above seems to indicate that the value of the problem of ad hoc hypothesis should be found rather in broader perspectives. In order to understand the problem of ad hoc hypothesis properly, it might be much more meaningful to scrutinize it as a common agenda for the entire scientific community throughout the ages. What is so bad about ad hoc Hypotheses? Was there anyone who anticipated the problem of ad hoc hypothesis? Who should care about ad hoc Hypothesis?

My attempt to interpret the problem of ad hoc hypothesis in terms of enthymeme is merely a tiny venture to answer these fundamental questions in scientific methodology. At the same time, however, it can be understood as an attempt to reinterpret the role and function of enthymemes. Why does enthymeme matter? In what context have we used enthymemes? How are we to fruitfully employ enthymemes in particular context? In other words, I believe that the usual treatment of enthymeme merely as an issue in rhetoric is unfortunate. Such a custom is just a result of historical accident, i.e., that Aristotle discussed in his *Rhetoric*. Even if we understand enthymemes as a rhetorical device, we should expand our perspective, at least to the uses of enthymeme in the scientific context of argumentation. Cantor believes that the reason why contemporary philosophers of science have lost interest in crucial experiment is "partly because they fail to recognize them as rhetorical, dramatic devices" (Cantor 1989, p. 176). Then he points out a revealing fact:

> For those seeking excitement in science, the historical record is packed with these dramatic episodes, the history of optics being no exception. During the optical revolution of the early nineteenth century, when the wave and particle theories of light were seen as menacing rivals, *experimentum crucis* abound (ibid.).

This is revealing at least for two reasons. First, thanks to Cantor, we now know that there was at least one century in which crucial experiments were an essential part of scientific methodology of practicing scientist. Secondly, Cantor's reminder of the importance of the rhetoric in science can support my attempt to interpret the problem of ad hoc hypothesis in terms of enthymeme resolution, for enthymeme has been studied even until today mostly studied in rhetoric.

Fortunately, there are some recent studies that are consonant with this line of thought. Some scholars detect the similarities between enthymemes and thought experiments (Crick 2004). Others pursue the deep connection between enthymeme and abduction (Gabbay and Woods 2005). Still others try to interpret enthymemes in terms of belief revision (Brun and Rott 2013). And, of course, there can be any number of combining these approaches. For example, Schurz's discussion of abductive belief revision seems to quite close to my attempt (Schurz 2013). Furthermore, insofar as the problem of enthymeme resolution is a sub-problem of the general problem of anomaly resolution (Magnani 2001; see also Magnani 2015), we may connect the problem of interpreting ad hoc hypothesis as an enthymeme resolution with some other big issues in scientific epistemology. For example, Thagard's intriguing taxonomy of conceptual revolutions can be reformulated in terms of enthymeme resolution (Thagard 1992). Exactly what differences are there between branch jumping and tree switching? Can we explain the differences in terms of the rules governing enthymeme resolution?

Acknowledgements This paper could not have been written without John Woods and Lorenzo Magnani's moral support.

References

Anstey, P. R. (2014). Philosophy of experiment in early modern England: The case of Bacon, Boyle and Hooke. *Early Science and Medicine, 19*, 103–132.

Bamford, G. (1993). Popper's explications of ad hocness: Circularity, empirical content, and scientific practice. *The British Journal for the Philosophy of Science, 44*(2), 335–355.

Bamford, G. (1996). Popper and his commentators on the discovery of Neptune: A close shave for the law of gravitation? *Studies in History and Philosophy of Science, 27*(2), 207–232.

Bamford, G. (1999). What is the problem of ad hoc hypotheses? *Science and Education, 8*, 375–386.

Brun, G., & Rott, H. (2013). Interpreting enthymematic arguments using belief revision. *Synthese, 190*, 4041–4063.

Cantor, G. (1989). The rhetoric of experiment. In D. Gooding, T. Pinch, & S. Schaffer (Eds.), *The uses of experiment: Studies in the natural sciences* (pp. 159–180). Cambridge: Cambridge University Press.

Carrier, M. (1988). On novel facts. *Zeitschrift für allgemeine Wissenschaftstheorie, 19*(2), 205–231.

Copi, I. (1982). *Introduction to logic* (6th ed.). New York: Macmillan.

Crick, N. (2004). Conquering our imagination: Thought experiments and enthymemes in scientific argument. *Philosophy and Rhetoric, 37*(1), 21–41.

Duhem, P. (1962). *The aim and structure of physical theory* (P. P. Wiener, Trans.). New York: Atheneum.

Dumitru, C. (2013). Crucial instances and crucial experiments in Bacon, Boyle, and Hooke. *Society and Politics, 7*(1), 45–61.

Ennis, R. (1982). Identifying implicit assumptions. *Synthese, 51*, 61–86.

Gabbay, D., & Woods, J. (2005). *The reach of abduction: Insight and trial*. Amsterdam: North-Holland.

Gerritsen, S. (2001). Unexpressed premises. In F. H. van Eemeren (Ed.), *Crucial concepts in argumentation theory* (pp. 51–79). Amsterdam: Sic Sat.

Gilles, D. (2002). Lakatos' criticisms of Popper. In G. Kampis, et al. (Eds.), *Appraising Lakatos: Mathematics, methodology and the man* (pp. 13–22). Dordrecht: Kluwer.

Grünbaum, A. (1973). *Philosophical problems of space and time*. Dordrecht: Reidel.

Grünbaum, A. (1976). Ad hoc auxiliary hypotheses and falsificationism. *British Journal for the Philosophy of Science, 27*, 329–362.

Hacking, I. (1983). *Representing and intervening: Introductory topics in the Philosophy of natural science*. Cambridge: Cambridge University Press.

Hanson, N. R. (1962). Leverrier: The zenith and nadir of Newtonian mechanics. *Isis, 53*(3), 359–378.

Hempel, C. (1966). *Philosophy of natural science*. Englewood Cliffs, NJ: Prentice Hall.

Herrmann, D. B. (1984). *The history of astronomy from Herschel to Hertzsprung* (K. Krisciunas, Trans.). Cambridge: Cambridge University Press.

Hitchcock, D. (1998). Does the traditional treatment of enthymemes rest on a mistake? *Argumentation, 12*, 15–37.

Howson, C., & Urbach, P. (2006). *Scientific reasoning: The Bayesian approach* (3rd ed.). La Salle, IL: Open Court.

Humphreys, W. C. (1968). *Anomalies and scientific theories*. San Francisco: Freeman, Cooper & Company.

Jacobs, S. (1999). Argumentation as normative pragmatics. In R. Grootendorst, F. van Eemeren, J. Blair, & C. Willard (Eds.), *Proceedings of ISSA 1998* (pp. 397–403). Amsterdam: Sic Sat.

Jalobeanu, D. (2011). Core experiments, natural histories and the art of experientia literata: The meaning of Baconian experimentation. *Society and Politics, V-2*, 88–103.

Jalobeanu, D. (2014). Constructing natural historical facts: Baconian natural history in Newton's first paper on light and colors. In Z. Biener & E. Schliesser (Eds.), *Newton and empiricism* (pp. 39–65). Oxford: Oxford University Press.

Lakatos, I. (1970). Falsificationism and the methodology of scientific research programmes. In I. Lakatos & A. Musgrave (Eds.), *Criticism and the growth of knowledge* (pp. 91–195). Cambridge: Cambridge University Press. (Reprinted in Lakatos (1978), pp. 8–101. I quote from the reprinted version.).

Lakatos, I. (1971a). History of science and its rational reconstructions. In R. C. Buck & R. S. Cohen (Eds.), *P.S.A. 1970 Boston studies in the philosophy of science* (Vol. 8, pp. 91–135). Dordrecht: Reidel. (Reprinted in Lakatos (1978a), pp. 102–138. I quote from the reprinted version.).

Lakatos, I. (1974). The role of crucial experiments in science. *Studies in History and Philosophy of Science, 4*(4), 309–325.

Lakatos, I. (1976). *Proofs and refutations. The logic of mathematical discovery*. Cambridge: Cambridge University Press.

Lakatos, I. (1978a). In J. Worrall & G. Currie (Eds.), *The methodology of scientific research programmes* (Vol. 1). Philosophical papers. Cambridge: Cambridge University Press.

Lakatos, I. (1978b). In J. Worrall & G. Currie (Eds.), *Mathematics, science and epistemology* (Vol. 2). Philosophical papers. Cambridge: Cambridge University Press.

Lakatos, I. (1999). In Motterlini (Ed.), *Lectures on scientific method* (pp. 19–112).

Leplin, J. (1982). The assessment of auxiliary hypothesis. *The British Journal for the Philosophy of Science, 33*, 235–249.

Magnani, L. (1999). Inconsistencies and creative abduction in science. In *AI and scientific creativity. Proceedings of the AISB99 symposium on scientific creativity* (pp. 1–8). Edinburgh: Society for the Study of Artificial Intelligence and Simulation of Behaviour, Edinburgh College of Art and Division of Informatics, University of Edinburgh.

Magnani, L. (2001). *Abduction, reason, and science: Processes of discovery and explanation*. New York: Kluwer.

Magnani, L. (2009). *Abductive cognition. The epistemological and eco-cognitive dimensions of hypothetical reasoning*. Berlin: Springer.

Magnani, L. (2015). The eco-cognitive model of abduction 1 απαγωγή now: Naturalizing the logic of abduction. *Journal of Applied Logic, 13*, 285–315.

Nickles, T. (1987). Lakatosian heuristics and epistemic support. *British Journal for the Philosophy of Science, 38*, 181–205.

Paglieri, F., & Woods, J. (2011a). Enthymematic parsimony. *Synthese, 178*, 461–501.

Paglieri, F., & Woods, J. (2011b). Enthymemes: From reconstruction to understanding. *Argumentation, 25*, 127–139.

Popper, K. R. (1934). *Logik der Forschung*. Wien: Verlag von Julius Springer.

Popper, K. R. (1934\1959\1975). *The logic of scientific discovery*. London: Hutchinson.

Popper, K. R. (1963). *Conjectures and refutations*. New York: Basic Books.

Popper, K. R. (1972). *Objective knowledge: An evolutionary approach*. Oxford: Oxford University Press.

Popper, K. R. (1974). Replies to my critics. In P. A. Schilpp (Ed.), *The philosophy of Karl Popper* (Vol. 2, pp. 961–1197). Library of Living Philosophers. La Salle: Open Court.

Popper, K. R. (1976). *Unended quest. An intellectual autobiography*. London: Routledge.

Rowbottom, D. P. (2010). Corroboration and auxiliary hypotheses. Duhem's thesis revisited. *Synthese, 177*, 139–149.

Schurz, G. (2013). Abductive belief revision in science. In E. J. Olsson & S. Enqvist (Eds.), *Belief revision meets philosophy of science* (pp. 77–104). Dordrecht: Springer.

Simon, H. (1955). A behavioral model of rational choice. *Quarterly Journal of Economics, 69*, 99–118.

Simon, H. (1956). Rational choice and the structure of the environment. *Psychological Review, 63*, 129–138.

Sprott, W. J. H. (1936). Review of K. Lewin's. *A Dynamical Theory of Personality Mind, 45*, 246–251.

Sylvestre, R. S. (2012). On the logical formalization of theory change and scientific anomalies. *Logic Journal of IGPL, 20*(2), 517–532.

Thagard, P. (1992). *Conceptual revolutions*. Princeton, NJ: Princeton University Press.

Urbach, P. (1982). Francis Bacon as a precursor to Popper. *The British journal for the philosophy of science, 33*(2), 113–132.

Walton, D. (1998). *The new dialectic: Conversational contexts of argument*. Toronto: University of Toronto Press.

Walton, D. (2008). The three bases for the enthymeme: A dialogical theory. *Journal of Applied Logic, 6*(3), 361–379.

Zahar, E. (1973). Why did Einstein's programme supersede Lorentz's? (I). *The British Journal for the Philosophy of Science, 24*(2), 95–123.

Zenker, F. (2006). Lakatos's challenge? Auxiliary hypotheses and non-monotonous Inference. *Journal for General Philosophy of Science, 37*, 405–415.

The Search of Source Domain Analogues: On How Luigi Galvani Got a Satisfactory Model of the Neuromuscular System

Nora Alejandrina Schwartz

Abstract I introduce the answers given by the current cognitive theories of analogy and by the historical-cognitive method to the problem related to the medium through which scientists can accede to potential analogous models. Then, I argue that it is required a point of view focused on the role of the environment as a supplier of appropriate objects to be employed as analogous models. From this perspective, I analyze a scientific research case: "the animal electricity discovery" by Luigi Galvani. This case shows that the Bolognese physician used the electrophore, the tourmaline and the Leyden jar as models of the seat of opposite electricity within an animal tissue and that he could have recognized those objects as bearers of relevant information from his experimental practice in the laboratory.

Many problems, particularly the scientific ones, can be solved by using analogies. One step that must be taken in order to generate solutions in this way is to represent a similarity between the source domain and the target domain (Gick and Holyoak 1980), which in turn, requires to have an available source analog, i.e., an object or situation that can satisfy the salient constraints of the target. So, a very relevant issue is to determine how this analog can be obtained (Thagard et al. 1990). It is possible to consider some models used within the scientific scope as instances of the source analogues and, therefore, to formulate the question just posed in the following manner: How is it that scientists come to have models that are able to give solutions to target domain problems? This question comprehends others, among which the following ones can be mentioned: Through which media can scientists access to potential analogous models? Can the salient information that these models embody only emerge if they are imagination constructions, or can it be carried by objects available within the environment or within the memory?

N.A. Schwartz (✉)
Facultad de Ciencias Economicas, Universidad de Buenos Aires,
Av. Córdoba 2122, C1120AAQ Buenos Aires, Argentina
e-mail: nora_schwartz@yahoo.com.ar

L. Magnani and C. Casadio (eds.), *Model-Based Reasoning in Science and Technology*, Studies in Applied Philosophy, Epistemology and Rational Ethics 27, DOI 10.1007/978-3-319-38983-7_29

In this paper I will bring in the answers that can be given to these questions from the perspective of the current cognitive theories of analogy and from Nancy Nersessian's historical-cognitive approach. It will be seen that cognitive theories point out that source analogues with an informational structure ready to be mapped can be recognized through memory. On the other hand, historical-cognitive analysis establishes that frequently suitable analog models are not at hand through any medium, and that the information which, in conceptual change episodes, meets the target domain constraints and is new, emerges within imaginary objects laboriously built.

As opposed to the cognitive theories of analogy mentioned before, I will argue that, although humans and, therefore, scientists can retrieve source analogs from memory, they can also recognize them within the environment. As well, I will hold that, as long as the information to be mapped can be embodied in perceivable objects within the environment, scientists can access to it right there, without requiring to build imaginary objects within which to emerge. Thus, I will point out the necessity to take a point of view frankly focused on the environment role in order to understand the whole issue more completely (Sect. 1).

One way to embrace this view is appealing to the study of scientific research historical cases within which plausibly, the environment took the role of supplying potential analogous models. The "animal electricity discovery" offers the opportunity to do it. That is why, after identifying the objects that Galvani used as analogous models of the opposite electricity seat in an animal tissue, I will argue that the Bolognese physician could have recognized those objects as bearers of relevant information from his laboratory experimental practice (Sect. 2).

1 Cognitive and Historical Analysis About the Source Analogues

Traditionally cognitive sciences have assumed that potential source analogues to analogically solve a target problem are pieces of information which are codified or represented in the memory (Vattam 2012, cf. p. 117). Besides, they have held that those source analogs have structures ready to be mapped and generate a solution in the new problem domain (Nersessian 2008, cap. 5). This means that there are informational structures that meet thoroughly the considered relevant features of the target domain which can be get retrieving them from memory. The analogy structure mapping theory by Gentner (1988) and the analogy multi constraint theory by Thagard et al. (1990) are examples of this way of conceiving how one can get source analogues.

Indeed, according to D. Gentner, the base or source situation is represented in a person's long term memory. After accessing to this situation, the selected structure is mapped to the target domain. The multi constraint theory also considers that source analogues potentially useful are retrieved form memory. There the person recognizes those patterns of elements that seem promissory to establish a similarity with the target domain for semantic, structural and pragmatic reasons, i.e., those which satisfy constraints of three kinds: semantic similarity, isomorphism and pragmatic centrality.

On the other side, Nancy Nersessian holds that scientific models used in conceptual change episodes are frequently imaginary constructs obtained along an extended process. Using the "historical-cognitive method", a moderate environmental approach within cognitive studies of science, and differing from the cognitive theories of analogy mentioned before, she argues that often scientific models that display new and satisfying structures are no at hand, and that to have them implies a pretty hard intellectual labour (Nersessian 2005). Nersessian analyzes modelling as an iterative process through which constraints supplied by the respective source domains and by the target domain, are abstracted, interact and are integrated. In this process, each intermediate model that is constructed achieves a higher satisfaction of the constraints of the target domain, and contributes to construct the following model. Intermediate models are hybrid, i.e., they embody not only the constraints of the target domain, but also those of the respective source domains (Nersessian 1999, p. 21). The elaboration of analogous models is carried out by a cognitive system constituted by internal representations, usually coupled with real world resources. They are representational practices that may be interpreted as distributed, i.e., that they can be expanded through internal-external traditional domains. These representations are "(…) created and used in the cooperative practices of persons as they engage with natural objects, manufactured devices, and traditions, as they seek to understand and solve new problems" (Osbeck and Nersessian 2006, p. 8).

Although the way of understanding how humans—and, therefore, scientists—come to get source analogues through memory provided by current cognitive theories of analogy has empirical support; and, on the other hand, although the explanation of the method by which scientists like Faraday have built mental models put in by Nersessian is convincing; I consider that the full comprehension of this issue requires to pay attention to the role played by the environment within which the scientist is, as supplier of appropriate objects to be used as analogous models. One way to do it is studying scientific historical cases that allow to do this exam.[1]

[1]Of course, it is not the same to be able to find a satisfactory source analog within the environment as to build it from the constraints supplied by the environment. I examine here the first alternative, while Nersessian has emphasized the study of the second one.

2 The Environment as Medium of Potential Models: an Analysis Case

The “animal electricity discovery” case allows to analyze the role that laboratories in which Luigi Galvani worked, especially the one that he occupied since 1784, could have played as suppliers of objects that bear an information able to make them representing an anomalous phenomenon.[2] The episode at issue is an instance of conceptual change, mostly motivated by the difficulty to understand the existence of an electrical unbalance within an animal organism—something that did not result physical o physiologically plausible, given the conductive nature of the body tissues.[3] In other words, through his research, Galvani was able to answer successfully to a representational kind of problem.[4]

The path he followed with that objective was to think analogically: he appealed to the domain of electrical tools searching for information that could suggest how to understand the mentioned anomaly within the domain of the animal physiology, i.e., he took and electrophysiological approach. Solving the problem supposed to select an electrical dispositive as adequate model of the neuromuscular system, more specifically, of the places within the animal tissue where there could exist an excess of electricity and a defect of electricity.

In what follows I will try to identify the objects that Galvani used as analog models of the seat of the opposite electricity within an animal tissue. Then I will refer to the issue of the media through which he may have acceded to the information that they embodied, focusing on the function that the environment may have taken in that. It is possible to make this examination taking as primary sources the manuscripts not published by L. Galvani, specifically two memoirs written in October of 1786 and in August of 1787 (Galvani et al. 1967), and the report of his own research published in the *De viribus electricitatis in motu musculari commentarius* (Galvani et al. 1791). In recent years, Marco Piccolino and Marco Bresadola have appealed to these sources in order to reconstruct the path that led

[2]Galvani, as many other Bolognese professors in the XVIII century, developed the experimental research activity in his own house. So, his changes of residence meant changes of laboratories: one was in his father-in-law’s house, the physicist Galeazzi, until 1774; another one, in a house that was property of the religious order of the Teatini until 1774; and, finally, his own residence at the center of Bologna, where he lived since 1784 (Piccolino and Bresadola 2003).

[3]This problem had been posed by the supporters of Haller’s theory of irritability as an objection to the neuro electric theory. Both conceptions presupposed Franklin’s theory of electricity, according to which an unbalanced body is the one which reaches a positive state (*plus*) or rests in a negative state (*minus*), normally by electrification. This process consists in taking some of the only one sort of electrical fluid from a body and bringing it to another. If nerves (like muscles and surrounding tissues) were conductive, there could not exist any electrical imbalance in animal organism, because conductive humors are able to dissipate any electrical imbalance generated within them (Piccolino and Bresadola 2003; Roller and Roller 1964).

[4]Representational problems involve that, in a given situation certain phenomenon escapes understanding and the solver does not know how to obtain the new conceptual resources to understand it (Nersessian 2008, cf. p. xii).

Galvani to establish the hypotheses of the animal electricity. Their description of this itinerary constitutes an invaluable help to begin the research about how Galvani came to have a model of the emplacement of the electrical unbalance within the animal tissue (Piccolino and Bresadola 2003; Piccolino 2008; Bresadola 2011).

2.1 Analog Models

In the memoir of 1786, Galvani mentions the Leyden jar—the first electrical condenser—as a dispositive which takes part in the production of an electrical current, and the electrophore—an electrical generator and condenser. Indeed, he refers to the *circuit* which is formed through the Leyden jar within an electrical discharge experiment as a representation of a flow of an extremely tenuous nervous fluid within the prepared frog, that he had conjectured from the observations and experiments with metallic arches on the animal. Galvani understands that the hypothetical electrical circuit between the frog nerve and the muscle is analogous to the electrical circuit produced by the presence of the two forms of electricity, placed on the inner and the outer metallic plates ("armatures") of a Leyden jar. From this, Galvani infers that the electrical unbalance is placed on the animal parts between which the circuit of nervous fluid is described, so he asserts that "no doubt can subsist that, of the said two forms of electricity, one is situated in muscle and the other in nerve" (Galvani et al. 1967, p. 176).

The other electrical tool mentioned in 1876, the electrophore, is made of two different kinds of components: a metal disk conductor of electricity and other disk, a resinous torte, i.e., metal wrapped in an isolating substance, able to store electricity. Galvani employs it as physical model of the seat of the opposite electricity within the muscle, which has, on one side, conductive matter and, on the other side, non-conductive substance that can hold electricity. With Galvani's words, "(…) there is in muscles a big quantity of substance, which for its nature, may be apt to develop and hold electricity, in spite of the presence inside it of conductive matter (…) This is not unlike what we saw happening in electrophores which are made of analogous substances. If that would appear it would be perhaps justified to call muscle *animal* electrophores" (Galvani et al. 1967, p. 169).

In the memoir of 1787, Galvani introduces a new object with the purpose of representing the neuromuscular system, now at a microscopic level: the tourmaline, and makes a new reference to the Leyden jar. The tourmaline is a semi-precious stone with pyroelectrical properties, which the Bolognese scientist employed as a model of the emplacement of the electrical unbalance within a muscle fiber coupled to a nervous fiber. In this way he reiterates the idea of 1786 about the muscle and the nerve—although now tiny—as the place of the double electricity. On one side, he points out that the stripes or colorless and opaque lines of the tourmaline, visually correspond to the muscle and nervous fibers within the animal tissue. On the other side, he establishes a functional or mechanical analogy between the tourmaline and the fibers, for example, a similarity between, on one hand, the electrical effects both within the whole stone and within its little pieces—when it is heated or cooled—and, on the other hand, the contraction of both the whole muscle and of its little fibers—when the nerve is electrically stimulated in the appropriate way.

> Our electricity has much in common with that of tourmaline stone, for what concerns its localization, distribution, and property of parts. In this stone we observe indeed a double matter, transparent and reddish the first one, opaque and colourless the other; this second one is arranged in stripes. Nobody can ignore that nerves are laid down between the layers of muscular fibres, and when these ones are devoid of blood they are transparent, while nerves are opaque. In tourmaline the poles of the double electricity appear to be situated on the same opaque line, so it is in muscles in the same direction. The double electricity of tourmaline does not belong only to the entire stone, but to every fragment. Similarly, in muscles, the admitted double electricity does not belong only to the entire muscle body, but to every part of it (Galvani et al. 1967, p. 194).

The new reference to the Leyden jar is made in relation to an experimental context. In fact, in this memoir Galvani analyzes the effects on the production of muscular contractions of the frog legs, given the experimental condition that metal foils were applied on which he supposed was the seat of the double electricity within the animal tissue: nerves and muscles. This experiment evoked studies in which a Leyden jar was subjected to a similar condition, i.e., the application of metal foils on the opposite surfaces of the glass container which constituted the body of the Leyden jar (Bresadola 2011, p. 200).

Lastly, in the *De viribus* Galvani models the neuromuscular system as a physical Leyden jar. This instrument has an internal and an external plate; and also, it has a metallic wire, connected to the internal plate and protruding through the Jar orifice, which works only as a conductor. Strictly, obeying to the microscopic perspective that Galvani had adopted yet in relation to the tourmaline in 1787, the final model is a set of minute Leyden bottles.

The central idea is that in the physical Leyden jar the positive electricity and the negative electricity were in the same bottle. In a similar way, Galvani places the electrical unbalance of the prepared animal in the muscle and understands that the nerve only serves as conductor in order to allow opposite charges to put in contact producing in this way the necessary discharge to have the muscle contraction. More precisely, Galvani thinks that the electrical imbalance seat is not the whole muscle, but a muscular fibre penetrated inside it by a nervous fiber:

> It would perhaps be a not inept hypothesis and conjecture, no altogether deviating from the truth, which should compare a muscle fibre to a small Leyden jar, or other similar electric body, charged with two opposite kinds of electricity; but should liken the nerve to the conductor, and therefore compare the whole muscle with an assemblage of Leyden jars (Galvani et al. 1791).

2.2 Memory and Environment

According to the above, Galvani used the electrophore, the tourmaline and the Leyden jar as models, since these electrical tools bore an information that makes them candidates to represent the seat of the electrical imbalance of the neuromuscular system. Lorenzo Magnani proposed calling this kind of information "implicit knowledge" or "tacit knowledge", extending the meaning that Polanyi had given to that expression. Magnani points to "a sort of tacit information "embodied" into the whole relationship between our mind-body system and suitable external representations" (Magnani 2007, 3.2).

Galvani could have appealed to different media in which relevant information of the electrical tools indicated could have been codified. It is plausible that memory had been one of them, since those devices are mentioned in the scientific literature of the time that he consulted (Piccolino and Bresadola 2003, pp. 275–276). Besides, Galvani came from a family of metalsmiths and probably had been able to evoke the knowledge that the jeweler had of the tourmaline. Indeed, they heated that semiprecious stone to know its composition and, therefore, its value. In this way they had realized that when heated, it was electrified and attracted ashes (Bresadola 2011, p. 109).

However, also the environment within which Galvani worked could have supplied the potential models mentioned before. In particular his laboratory was similar to a cabinet of Physics and could have worked as a kind of repository of knowledge embodied in the tools that were there (Piccolino and Bresadola 2003, p. 338).

The role of imagination in getting an adequate model seems to have been limited to modifying the dimensions of the final analogous representation, without introducing changes in the structures and functions mapped to the target analog. So, it was not necessary that Galvani built an imaginary object in which new information could emerge, in order to have an adequate model, he could perceive Leyden jars around him or recall them and recognize in them the relevant implicit knowledge.

Piccolino suggests that Galvani left aside the model of the electrophore and the one of the tourmaline because they did not allow to understand certain experimental results. These consisted in that, when he wrapped the muscle and crural nerve in thin metal foils, and he connected them by a conductive arc, the physiological effects of the animal electricity were stronger than without those foils (Piccolino and Bresadola 2003, p. 343).

However the Leyden jar model allowed to establish correspondences with all the salient aspects of the target analog including these experimental results. So Galvani searched in it the information that could offer a new solution for the representational problem he had faced. He knew that in the Leyden jar, the insulating substance, the glass, is in the interface of the double electricity, i.e., the inner and the outer sides of the glass flask. This is how he explains that in this physical condenser the two forms of electricity (positive and negative) could be without the dispersion of the positive electricity present in the conductive matter (the water inside the flask). Galvani transferred this solution to the target domain in a conjectural manner: may be an insulating substance inside the very muscular fibre, precisely in the interface between the place of the positive electricity and the place of the negative electricity? So Galvani considers:

> It is even more difficult that the existence of a duplex electricity in every muscular fibre itself could be denied if one thinks not difficult, nor far from truth, to admit that the fibre itself has two surfaces, opposite one to the other, and this from consideration of the cavity that not a few admit in it, or because of the diversity of substances, which we said the fibre is composed of, diversity which necessarily implies the presence of various small cavities, and thus of surfaces (Galvani et al. 1791, p. 196).

By this conjecture Galvani was able to face the main objection against the role of electricity in the animal physiology.

3 Conclusion

The problem relative to the manner in which scientists get satisfactory analogous models is complex. One of the questions implied in it refers to the media through which they can bc recognized—if they are already available in some channel of information and, therefore, it is not needed to build them. This aspect of the problem was tackled by different cognitive theories of analogy and by an approach within the cognitive studies of science, the Nersessian's historical-cognitive method.

Nevertheless, I pointed out that these answers alone are not enough and that it is necessary to make a research that emphasizes the role that plays the environment in which scientists are, as a supplier of the suitable objects to be employed as analogous models. On the other hand, I considered the possibility of undertaking such investigation through case studies and began to carry out this task, examining the "animal electricity discovery" by Luigi Galvani.

The case study showed that the Bolognese physician could solve a representational problem selecting electrical devices available in his laboratory as models of the neuromuscular system. These objects could be retrieved from his memory, but also they could have been provided in the environment where he worked. Besides, the case allows to support the idea that visual and functional information embodied in the final model employed by Galvani to represent the electrical imbalance within the muscular fibre—the Leyden jar—was implicit in existent and perceivable objects within the environment or in evocable objects. This means that this information was not an emergent structure within an imaginary object. What is left to be found is an adequate theoretical frame to conceptualize and articulate the results reached here. One way to that is to explore in future works the "ecological" realism tradition proposed by Gibson.

Bibliography

Bresadola, M. (2011). *Luigi Galvani—devozione, scienza e rivoluzione*. Bologna: Editrice Compositori.

Galvani, L. (1791). De viribus electricitatis in motu musculari commentarius. *De Bononiensi Scientiarum et Artium Instituto atque academia comentarii, 7.*

Galvani, L. (1967). *Opere Scelte.* Barbensi G, e. Torino, Utet.

Gentner, D. (1988). Analogical inference and analogical access. *Analogica,* 63–88.

Gick, M., & Holyoak, K. (1980). Analogical problem solving. *Cognitive Psychology, 12*, 306–355.

Magnani, L. (2007). Abduction and chance discovery in science. *KES Journal, 11*(5), 273–279.

Nersessian, N. (1999). Model based reasoning in conceptual change. In L. Magnani, N. J. Nersessian, & P. Thagard (Eds.), *Model-based reasoning in scientific discovery* (pp. 5–22). US: Springer.

Nersessian, N. (2005). Interpreting scientific and engineering practices: Integrating the cognitive, social, and cultural dimensions. In R. Tweney, D. Gooding, & A. Kincannon (Eds.), *Scientific and technological thinking* (pp. 17–56). New Jersey: Erlbaum.

Nersessian, N. (2008). *Creating scientific concepts*. Cambridge London: MIT Press.

Osbeck, L. M., & Nersessian, N. J. (2006). The distribution of representation. *Journal for the Theory of Social Behaviour, 36*, 141–160.

Piccolino, M. (2008). Visual images in Luigi Galvani's path to animal electricity. *Journal of the History of the Neurosciences, 17*, 335–347.

Piccolino, M., & Bresadola, M. (2003). *Rane, torpedini e scintille*. Torino: Bollati Boringieri editore.

Roller, D. Y., & Roller, D. H. D. (1964). The development of the concept of electric charge. In J. B. Conant (Ed.), *Harvard case histories in experimental science* (Vol. 2). Cambridge: Harvard University Press.

Thagard, P., Holyoak, K., Nelson, G., & Gochfeld, D. (1990). Analog retrieval by constraint satisfaction. *Artificial Intelligence*, *46*(3), 259–310.

Vattam, S. (2012). *Interactive analogical retrieval: Practice, theory and technology.* Doctoral Dissertation, Georgia Institute of Technology, Atlanta, GA, USA.

The No Miracle Argument and Strong Predictivism Versus Barnes

Mario Alai

Abstract *Strong predictivism*, the idea that novel predictions per se confirm theories more than accommodations, is based on a *"no miracle" argument from novel predictions to the truth of theories* (NMAT). Eric Barnes rejects both: he reconstructs the NMAT as seeking an explanation for the entailment relation between a theory and its novel consequences, and argues that it involves a fallacious application of Occam's razor. However, he accepts a *no miracle argument for the* truth *of background beliefs* (NMABB): scientists *endorsed* a successful theory because they were guided by largely true background beliefs. This in turn raises the probability that the theory is true; so Barnes embraces a form of *weak predictivism*, according to which predictions are only indirectly relevant to confirmation. To Barnes I reply that we should also explain how the successful theory was constructed, not just endorsed; background beliefs are not enough to explain success, scientific method must also be considered; Barnes can account for some measure of confirmation of our theories, but not for the practical certainty conferred to them by some astonishing predictions; true background beliefs and reliability by themselves cannot explain novel success, the truth of theories is also required. Hence, the NMAT is sound, and strong predictivism is right. In fact, Barnes misinterprets the NMAT, which does not involve Occam's razor, takes as *explanandum* the building of a theory which turned out to predict surprising facts, and successfully concludes that the theory is true. This accounts for the practically certain confirmation of our most successful theories, in accordance with strong predictivism.

M. Alai (✉)
Università di Urbino Carlo Bo, via Timoteo Viti 10, 61029 Urbino, PU, Italy
e-mail: mario.alai@uniurb.it; mario.alai@libero.it

L. Magnani and C. Casadio (eds.), *Model-Based Reasoning in Science and Technology*, Studies in Applied Philosophy, Epistemology and Rational Ethics 27, DOI 10.1007/978-3-319-38983-7_30

1 The No Miracle Argument, Strong Predictivism, and Barnes' Weak Predictivism

Some scientific theories make very detailed predictions about novel surprising phenomena. For instance, Newton's gravitation theory predicted the existence of Neptune with its mass and orbit; Mendeleev predicted new elements with their chemical properties; Fresnel predicted a bright spot at the centre of the shadow of a dark disk; Einstein predicted the bending of light with high precision;[1] etc. When this happens, a large majority of scientists conclude that there must be some substantial truth in the theory, for these successes would be unexplainable if it were completely false. In fact, it would be a "miracle" (i.e., a miraculously lucky coincidence) if a completely false theory successfully predicted phenomena it has not been designed to account for. Now, it is extremely unlikely that miracles or miraculous coincidences happen; so, it is extremely likely that the theory is true (or at least partially and/or approximately so: in the following I will always write 'true' and 'truth' with this implicit qualification). This is the "no miracle" argument for scientific realism proposed by Smart (1968), Putnam (1975, 73), Musgrave (1988), and others, which here I will call more precisely "the no miracle argument for the truth of theories" (NMAT).

The NMAT is also why *predictivists*[2] claim that astonishing novel predictions confirm theories much more than mere accommodation of old phenomena: while the only possible explanation for predictive success is truth, even a false theory may account for known phenomena, if it was built to fit them.

One might object that it is logically possible that false theories have true consequences: among all the conceivable theories, considered as eternal platonic objects, some are false, yet entailing true novel predictions, and the theorist might have found one of them just by luck. But this depends on how *bold* (i.e. precise and informative, hence a priori improbable and surprising) is the prediction. For the more informative a claim is, the fewer are the theories from which it follows (the least informative claims, tautologies, follow from everything, and the most informative ones, contradictions, follow only from contradictions). A weak and unsurprising prediction, like "there is an unknown planet in the universe", would follow from many cosmological theories; but a bolder prediction, like the existence of a new planet with precise location, mass and orbit, follows only from very few. So, among all the possible theories (true and false alike) only a small minority entail predictions as bold as the abovementioned ones; hence the likelihood that a theorist

[1]Michel Ghins reminded me that even Newton, on the basis of his corpuscularist conception of light, believed it was subject to gravitation. But that opinion had been abandoned with the appearence of the electromagnetic theory.

[2]Among whom Whewell (1840), Peirce (1883), Duhem (1906), Popper (1962), Lakatos (1970), Musgrave (1974, 1988, 1999), Kuhn (1962, 1977), Giere (1983), Maher (1988, 1990, 1993), Worrall (1978a, b, 1985, 1989a, b, 2005, 2006), Lipton (1991), Leplin (1997), Zahar (1973a, b), etc.

picks one of them just by luck is negligible. For instance, the value of the magnetic moment of the electron predicted by quantum electrodynamics was $1,159,652,359 \times 10^{-12}$, while the value obtained by experiment was $1,159,652,410 \times 10^{-12}$. So, how many randomly taken theories would made an equally accurate prediction? On some admittedly questionable but reasonable assumptions Wright (2002, 143–144) has figured only 1 over 50^8.

So, if in a certain field F a theory T has been proposed entailing the *bold* new prediction E, the likeliest explanation is not chance, for that would be a "miraculous" coincidence; nor is it purposeful accommodation, since by hypothesis E had not been used in constructing T; rather, it is the following: among *all* possible theories in F only an extremely small minority entail E; but if we consider the subset of theories in F which are *true* and *fecund* (i.e. deep and rich of potential unforeseen empirical consequences), among them a good many (at least a non-negligible minority) entail E. So, the finding of a theory which entailed E can be explained if we assume that the theorist sought for a true and fecund theory, and was enough skilled and reliable in this search to *actually find one*, which therefore had a non-negligible probability of entailing E (Alai 2014a, Sect. 3.2).[3]

Against predictivism, *consequentialists*[4] object that confirmation is a logical relation between the theory and its consequences (pertaining to the context of justification), while novelty is a contingent fact (pertaining to the context of discovery), irrelevant to truth and confirmation. Moreover, they cite historical cases in which novelty appears to have been either not necessary or not sufficient to confirmation (Alai 2014a, 301–302). Therefore some predictivists retreat to *weak* predictivism: they grant that novelty per se has no particular confirming power, but maintain that it is evidence of some other virtue, which in turn increases the likelihood that a theory is true (Lange 2001).[5] For instance, novelty may show that the theory has not been *fudged* (Lipton 1991, 170–184), or that the scientist is reliable (White 2003) or talented (Kahn et al. 1992), or that the data have not been *overfitted* (Hitchcock and Sober 2004). *Strong* predictivism, on the contrary, holds that prediction per se strongly supports the truth of theories, via the "no miracle" argument.

In his thorough and instructive (2008) Eric Barnes calls "the paradox of predictivism" the contrast between the predictivist and consequentialist intuitions on the advantage of prediction over accommodation. He accounts for this contrast by

---

[3]Of course it makes sense to say that the theory found by the scientist had a good probability of entailing E if we refer to it opaquely, as the theory which resulted from his/her efforts. But if we refer to it transparently, as the theory which in fact it is, i.e. T, then it is just a logical fact that E is among its consequences.

[4]Among whom Mill (1843, III, Chap. 14, Sect. 6), Keynes (1921, 305), Rosenkrantz (1977, 169 ff.), Horwich (1982, 108–117), Schlesinger (1987), Howson (1988), Howson and Franklin (1991), Achinstein (1994), Collins (1994), etc.

[5]The labels 'weak' and 'strong' are applied by Lipton to a related but different distinction: according to "weak predictivism" when theories make predictions they are more confirmed because either the theory or the data tend to be different and better; for "strong predictivism", instead, a predicted datum confirms a theory more than *the same* datum would have confirmed *the same* theory if it had been accommodated (1991, 165).

arguing that the *no miracle argument for the truth of theories* (NMAT) I expounded above is invalid, hence predictions cannot offer direct evidence for the truth of theories, and strong predictivism is wrong. In fact, he reconstructs the NMAT as seeking an explanation for the entailment of certain consequences E by a theory T, and argues that it involves a fallacious application of Occam's razor. However, he accepts a *no miracle argument for the truth of background beliefs* (NMABB): first, he distinguishes between *constructing* T and *endorsing* it, i.e., attributing it a high probability of being true;[6] second, he claims that it is not epistemically relevant how T was constructed, but why it was endorsed, and this is what we should explain. Third, he argues that scientists endorsed the successful theory T because they were guided by largely true background beliefs. So, having endorsed a successful theory is evidence that the endorsers held true background beliefs, hence that they were reliable. Besides, their reliability raises the probability that the theory they endorsed is true. Therefore novel predictive success is at least indirectly relevant to confirmation. Thus, Barnes embraces a form of weak predictivism: predictive success per se does not show that T is true, but that T has a virtue (being endorsed by reliable scientists) which makes it likelier that it is true.

I reply to Barnes on a number of points: theory construction is also an important *explanandum,* along with endorsement; background beliefs are not enough to explain the scientists' reliability, scientific method and skills must also be considered; Barnes' weak predictivism can account for the confirmation of theories up to a certain degree, but it cannot explain why some outstanding predictions make theories practically certain; most importantly, the scientists' true background beliefs and reliability by themselves cannot explain novel success, the truth of theories is also required. This shows that strong predictivism is right, and the NMAT is sound.

Eventually, in fact, I show that Barnes' criticisms of the NMAT are based on various misunderstandings: in its most appropriate form the argument does not involve Occam's razor, it takes as *explanandum* the building of a theory which turned out to have novel and bold consequences, and correctly concludes to the truth of the theory, as held by strong predictivism. Thus it accounts for the practically certain confirmation of our most successful theories.

2 From the No NMAT to the NMABB

To begin with, Barnes formulates an anti-superfluity principle (ASP), a version of Occam's razor, stating that *we should not accept more than one explanation of the same explanandum*. ASP is then used in the following reconstruction the NMAT (129):

---

[6]Endorsement is defined as a gradual and contextual notion; but the probability attributed to T must be (i) no lower than an indepedent's evaluator own probability, and (ii) sufficiently high so that any new evidence for T raises the endorser's credibility: pp. 35–36.

If (F) is the fact that a theory T entails a conjunction of true observation statements E, then

(a) (F) stands in need of explanation.
(b) There are just two possible adequate explanations: (1) T was built to fit E, and (2) T is true (for if neither (1) nor (2) obtained (F) would be a "miracle").
(c) Now, if (1) T was built to fit E, then we have an adequate explanation of (F)—and there is no need to endorse (2) (by ASP).
(d) If instead T was *not* built to fit E, then we should endorse (2) the truth of T, since it is the only adequate explanation of (F).
(e) Thus if T was not built to fit E, it is probably true.

To the argument so reconstructed Barnes objects that

(i) while (2) (the truth of T) explains why T entails a true consequence, (1) (that T was built to fit E) does not. For (1) is a contingent fact, hence it cannot explain the entailment relation between T and E, which is a logical fact; nor can it explain the fact that E is true (130–131).

Therefore

(ii) Premise (b) of the NMAT is false, as there are not two possible explanations of (F), but just one.

Therefore

(iii) ASP cannot be properly applied in step (c), for 'built-to-fit' and truth are not explanations of the same *explanandum*.

Therefore

(iv) Conclusion (e) of the NMAT is unwarranted (130–137).

This is a quite puzzling argument, and I shall criticize it in Sect. 4. But two problems can be noticed here. First, claim (i) may engender some confusion: Barnes is right that the fact (1) that T was built for a certain purpose (fitting E) is contingent, and as such it cannot explain the *logical* fact that T entails E. But he should grant that it can explain the contingent fact that

(C) the theorists built a theory which entailed E, as opposed to other possible theories which did not.

In fact, I will soon point out that for his own NMABB Barnes takes as an *explanandum* precisely a contingent fact like (C). Equally, he might have interpreted more charitably the NMAT as explaining a fact of the same sort (for when so reformulated the NMAT succeeds, as we shall see in Sect. 4). Still concerning (i), it might be noticed that the truth of T does not explain why T entails E; but it explains why T's consequence E is also true. I shall come back to these topics in Sect. 4 when offering a better reconstruction of the NMAT.

The second problem with Barnes' criticisms of the NMAT is this: if we concede that (i) the building of T to fit E does not explain (F) (i.e. the entailment of E by T), then he is right that (ii) premise (b) of the NMAT is false, since there is just one possible explanation, viz. the truth of T. But then it is possible to conclude directly to the truth of T, without assuming ASP. So, the NMAT works even better and more simply. Actually, we shall see in Sect. 4 that there is more to say for a correct reconstruction of the NMAT, but Barnes' criticism fails anyway.

At any rate, after rejecting the NMAT Barnes proposes a different no miracle argument, which does not prove the truth of T, but the truth of the *background beliefs* which led to its endorsement (NMABB): first, he argues that the epistemically relevant notion is not theory construction, but theory endorsement: for a theorist Connie might construct T but believe that it is false, while another scientist Endora might accept it as true: if T were subsequently shown to be true, we would consider reliable Endora, who endorsed it, not Connie who constructed it (35). Thus, the relevant novelty notion is *endorsement novelty*: a prediction E is endorsement novel for theory T and scientist S if and only if it is not among the reasons why S endorses T.

One wonders why and how a theorist should construct a theory without endorsing it: perhaps just in order to impress the community or improve one's publication record; but these goals cannot be achieved unless there are good reasons for endorsing the theory. Moreover, theories are not found ready-made, but constructed through subsequent steps or stages, and theorists do not proceed to the next step without at least provisionally endorsing the earlier ones. Of course T's author will not swear on the complete truth of T, and may even suspect that this or that particular assumption of T is false; but no doubt she will endorse the larger part of T. However, in particular cases a theory might be advanced just as an exercise of ingenuity. Moreover, even if authors usually endorse their theories, typically not all endorsers are authors; besides, a scientist might build a theory for accidental reasons (like Kekulé's dream), and then the evidence relevant to assessing the theory will be that used by the endorser, not by the author. In other words, at least in principle, there is an epistemic asymmetry between building and endorsing (i.e., between discovery and justification), and Barnes wishes to preserve it.

Next, he correctly argues that when seeking an explanation for the success of T we are not really asking why T entails E (since this is just a logical fact), nor why E is true (since this is already explained by the meaning of E and the way the world is) (130–131). What we are asking, instead, is how

(EE) scientists came to endorse precisely a theory that has the true consequences E

(as I noticed, therefore, he should have used a similar *explanandum* in his reconstruction of the NMAT). Now, he claims, we would get the required explanation if we found that T was built to fit E, and then endorsed precisely because it fits E (134; but if so, he should acknowledge that the *building* of T is as epistemically relevant as its endorsement). On the other hand, he goes on, if T has not been endorsed because it fits E, then there is just one explanation, besides a *miraculous* coincidence: that the background beliefs which led to the endorsement of T were themselves true, so making the theorists reliable in their endorsement (98, 131, 137–9, 141–144).[7]

This is what I have called Barnes' "no miracle argument for background beliefs" (NMABB). He then goes on arguing that the reliability of the endorsers of T and their true background beliefs make it likelier that T is true; and this, by generalization on all theories with novel success, supports scientific realism. This is why he accepts *weak* predictivism: novelty is not directly evidence for the truth of theories, but only for different virtues (i.e., being endorsed on the basis of true background beliefs and reliability) which in turn are positively correlated with truth.

There is no room here to explore how exactly Barnes understands background beliefs, or why and how in his view they make scientists reliable. But this will not be crucial here: it is prima facie plausible that true background beliefs enhance reliability, and I shall concede it. But I will argue that reliability is not enough to explain scientific success without also assuming the truth of the theory, as the NMAT does.

3 Problems with Barnes' Weak Predictivism

One problem with Barnes' account is that its neglect of theory construction seems to unduly restrict the scope of novelty in theory confirmation: for him T's consequence E is novel just if it has not been used in endorsing T. However, if E has been used in *endorsing* T but not in *building* T, we still need to explain how scientists were able to build a theory predicting E. And if this explanation involves the truth of T (as I show in Sect. 4), E is still *novel* in the crucial sense of confirming T more than if it had been used in building T.

Perhaps Barnes believes that if one built T without endorsing it, that is because she did not have enough good reasons for believing in the truth of T; and if so the prediction of true E would certainly be 'lucky' and thus provide no special evidence for T (beyond the mere fact of E's truth).[8] But, as noticed in Sect. 2, in typical cases one could not build T without having good reasons for believing in its truth, and so at least provisionally endorsing it. Besides, as explained in Sect. 1, if E is really new and bold, striking it by chance is practically impossible.

Another problem is that in framing his NMABB Barnes distinguishes himself from Maher (1988, 1990) and White (2003), who also see novelty as evidence for the reliability of theorists, but trace reliability to their *method*: novel success, for them, is evidence that they followed a reliable method. On the contrary, he claims that 'method' is a very vague term, referring both to specific prescriptions and to very general ones, and it is hard to tell in which sense the most general prescriptions

[7]He also denies that the *empirical adequacy* of background beliefs is a possible alternative explanation (155–162).

[8]I owe this suggestion to an anonymous referee.

could be considered reliable. Besides, creativity is also involved in theory construction, and it is not dictated by any method (115–116). So, theorists are rather made reliable by the truth of their background beliefs.

But to begin with, this claim launches an infinite regress: if we explain the endorsement of theory T by the endorsement of the true background beliefs BB, then we should explain the endorsement of BB by that of further background beliefs, and so on. Barnes stops this regress by claiming that "take-off" theories are not endorsed on the basis of further background beliefs, but only of their simplicity (147–155); but isn't this just reintroducing method (specifically, the methodological prescription of simplicity) as the basic justification for the theorists' reliability?

Perhaps not, since for Barnes background beliefs do not include only factual beliefs, but also methodological and prescriptive beliefs, such as, e.g., "Theoretical inference can be truth conducive in chemistry" (94). So, "take-off" theories could be based on background belief like "Simple theories are probably true". Perhaps in this way Barnes could substitute the vague notion of "method" with more precise methodological beliefs.[9]

But then Barnes' substitution of method with background beliefs would become mainly terminological, or reduce to a different contrast, that between general and specific prescriptions: for the advocates of method-based reliability might simply say that what makes scientists reliable are the specific methodological prescriptions followed by them.

Moreover, if Barnes based the endorsement of take-off theories on methodological *beliefs*, one could ask why *those* beliefs were endorsed, and the regress would start again. On the contrary, it is clear that some factors of reliability do not consist either in true factual beliefs or in sound methodological beliefs, nor can be captured by beliefs, but have to do with skills and behavioural methods which are learned by practice and followed unreflectively, plus individual virtues like ingenuity, sensibility, sensitivity, intuition, etc.

Further, as noticed, it is not enough to explain why scientists are reliable in *endorsing* theories, but also why they are so reliable in *constructing* them to produce theories with true unforeseen consequences. Sometimes Barnes considers theories as eternal abstract objects in a platonic Hyperuranium (135–136), or as generated by random machines (65); but actual theories are never randomly generated; and while they can be abstractly conceived as platonic entities, scientists cannot inspect and choose them ready-made as by a noetic intuition: in actual scientific practice scientists must "construct" them through a step-by-step problem-solving, ideation, appraisal, acceptance or rejection of hypotheses, based on background beliefs but also disciplinary skills, reasoning, creativity, etc.

An even more serious problem concerns Barnes' weak predictivism: at most it can explain why (indirectly) predictions confirm theories somewhat more than accommodations. But there is much more to be explained: *bold* novel predictions do not simply make the truth of theories more probable, but *practically* certain.

[9]I owe also this suggestion to the same referee.

This means that those predictions leave no serious doubts that at least the assumptions essentially used in their derivation are true: for instance, we are *practically* certain that light is a undulatory phenomenon as assumed by Fresnel, that Mendeleev's periodic law is true, that General Relativity is at least partially true, etc.[10] After their bold predictions were confirmed, these theories ceased to be just promising hypotheses and were substantially accepted by the scientific community. Now, simply assuming that T was endorsed on the basis of true background beliefs is far from making practically certain that T is true: *ceteris paribus* it makes the endorsers more reliable, and the truth of T more likely; but reliability is not infallibility, and likelihood is not practical certainty.

Besides, Barnes explains that scientists endorsed T *because they had true background beliefs*. But typically T is endorsed only *after* E is observed (before that, T is just a hypothesis). So, the most plausible explanation is rather that scientists endorsed T *because they observed E*; i.e., because they reasoned by the NMAT: "if T predicted E, and E is novel and true, T must be true".[11]

Other versions of weak predictivism suffer from the same problem: non-fudging and non-overfitting are simply features of sound scientific method, and sound scientific method makes the finding and endorsement of true theories possible, but by no means guaranteed. Even a posteriori accommodation can be achieved on the basis of true background beliefs, without any fudging or overfitting: but then just ingenuity and puzzle-solving ability can explain why the theory accounts for the data, and there is no need to assume that it is (partly) true. The practically conclusive confirmation of some theories can be provided only by the NMAT, and so by strong predictivism: as I suggested in Sect. 1, and will argue in more detail in Sect. 4, when a theory which anticipates surprising phenomena is found, the only possible explanation includes the truth of that theory.

Actually, Barnes' own NMABB could not stand without an implicit appeal to the truth of theories: as we saw, he thinks that scientists can endorse a theory entailing novel predictions because their true background beliefs make them *reliable* (98, 131, 138–9, 141–144). But I suggest that whatever reliability is,[12] it is certainly not like an extra-sensorial perception, enabling to track the bold prediction E *directly*, as a truffle hound's smell for truffles: it can only consist in the ability to find (and endorse) theories which are largely true and fecund, and as such sufficiently likely to have true bold consequences. Thanks to this ability, scientists actually found and endorsed a theory which had enough true content to entail E. So, they found an E-entailing theory by finding a *true* theory (which then they endorsed).

Thus, reliability is necessary but not sufficient to explain a bold novel prediction: it makes it probable, not certain. A reliable scientist might still have come up with a

---

[10]Which of course leaves open both in-principle skeptical doubts, and the possibility of amending these theories on many accounts.

[11]I owe this suggestion to Michel Ghins.

[12]I shall say more on this at Sect. 4. In any case, it is the actual disposition to find and/or endorse successful theories, not the *reputation* of being reliable.

theory which was mainly false, and as such didn't entail E (for, as noticed in Sect. 1, it is extremely improbable that false theories make true bold predictions). Hence, one must in addition assume that the scientist *actually succeeded* in finding a theory with enough true content. So, Barnes' NMABB is parasitic on some form of NMAT, and his weak predictivism on strong predictivism.

All this is shown rather clearly by Barnes' own example of NMABB, concerning Mendeleev's prediction of the new elements Gallium, Germanium and Scandium on the basis of his periodic law (PL). He notices that PL was based on raw laboratory data interpreted by Cannizzaro's atomic weight calculations. So, PL was based on the conjunction of raw data and the background belief

(Km) Cannizzaro's methods of atomic weight calculation are sound.

Hence, the confirmation of novel predictions drawn from PL strongly confirmed the truth of Km:

> Mendeleev' s predictions about the discovery of heretofore undiscovered elements and about the eventual correction of currently accepted values for atomic weights are extrapolations of the same pattern of atomic weight data [provided by Cannizzaro's method]. The probability of the extrapolation being confirmed, if Km is true, is reasonably high, for if Km is true then Mendeleev's atomic weight values are correct, and thus they serve to confirm PL fairly strongly. But the probability of the extrapolation being confirmed if Km is false is quite low - for if Km is false then no conclusion about atomic weights are entailed by Mendeleev's data, and PL is entirely unsupported, leaving us with no reason to expect his predictions to be confirmed. Thus, if we let E = Mendeleev's predicted data, it follows that the ratio of $p(E/Km)/p(E/\sim Km)$ is quite high - and E offers much confirmation of Km (unlike the accommodated atomic weight data) (98).

In synthesis, he claims that the truth of PL would be quite improbable if Km were false; moreover Mendeleev's novel predictions E would be quite improbable if PL were false, hence E would be quite improbable if Km were false; therefore, the observation of E strongly confirmed the truth of background belief Km.

But by the same reasoning, and a fortiori, Mendeleev's predictions should strongly confirm the truth of PL, for the implicit reasoning scheme is

(1) PL presupposes (raw data and Km) [since PL is a bold theory, a priori improbable, and only made probable by (raw data and Km)]
(2) E presuppose PL [since E are a priori improbable predictions, made probable only by PL]
(3) $\sim$Km $\rightarrow$ most probably $\sim$ PL [by 1]
(4) $\sim$PL $\rightarrow$ most probably $\sim$ E [by 2]
(5) E [by empirical observation]
(6) Most probably, PL [by 4, 5, double negation, probabilistic MT]
(7) Most probably Km [by 3, 6, double negation, probabilistic MT].

So, the truth of background beliefs Km is confirmed (at step 7) only by first confirming the truth of the theory PL (at step 6): the NMABB implicitly presupposes the NMAT.

4 Rescuing the NMAT and Strong Predictivism

All this shows that the NMAT and strong predictivism should be preserved, in spite of Barnes' objections. In fact, in Sect. 2 I raised some problems with those objections, and now I will show that his reconstruction of the NAMT is a straw man, and propose a better reconstruction. So, let us consider his reconstruction again:

If (F) is the fact that a theory T entails a conjunction of true observation statements E, then

(a) (F) stands in need of explanation.
(b) There are just two possible adequate explanations: (1) T was built to fit E, and (2) T is true (for if neither (1) nor (2) obtained then (F) would be a 'miracle').
(c) Now if (1) T was built to fit E, then we have an adequate explanation of (F)—and there is no need to endorse (2) (by ASP).
(d) If instead T was *not* built to fit E, then we should endorse (2) the truth of T, since it is the only adequate explanation of (F).
(e) Thus if T was not built to fit E, it is probably true.

Now, first of all, if (F) is understood *literally*, and theories are considered as eternal objects in a platonic realm, it might seem that (F) does not need an explanation, so premise (a) is false. In fact, there are two components of (F): that T entails E, and that E is true; but neither component needs an explanation, since that E is a consequence of T is just a logical fact, and that E is true is just a semantic fact about the meaning of E and the state of things (Barnes 133; White 2003, 660). However, (F) does not reduce to these two components, it rather it consists in the relation between them: it is the fact that T has a consequence *which* is true (i.e., that one of T's consequences[13] is true). Now, for *this* fact we can envisage some possible explanations, although still rather shallow: for instance, that

(ct) T is completely true, hence *all* of its consequences are true;
or, more realistically, that
(pt) a part of T is true, and E follows from that part;

or still other explanations, as I will shortly indicate. Anyway, Barnes is right that (b-1) (i.e., that T was built to fit E) is not one of the possible explanations of (F) *in this literal* reading, so premise (b) (that there are *two* possible explanations of F) is false. But premise (b) and Occam's razor (ASP) are not necessary to the NMAT: on the contrary, without (b-1) and premise (b) the argument works even better and more directly; for then, instead of (b), one can assume that

[13]A consequence referred to opaquely, *qua* consequence of T: not transparently, *qua* E, for then, as just explained, we wouldn't need an explanation of its truth beyond the meaning of E and the way the world is.

(b′) The (partial) truth of T is the *only* possible explanation (except miracles);

so, also the appeal to ASP in step (c) is no longer required. Thus the argument can be better reconstructed, dropping (b) and (c), in the straightforward form of an inference to best (or rather, to the *only plausible*) explanation:

If (F) is the fact that a theory T entails a conjunction of true observation statements E, then

(a′) (F) stands in need of (shallow) explanation.
(b′) (pt) (i.e., that a part of T is true, and E follows from that part) is the *only* possible explanation (except miracles);
(c′) Since the only possible explanation is probably true (by inference to only possible explanation) T is probably at least partially true.

Hence, contra Barnes, there is no fallacious application of Occam's razor (ASP) in this argument. No doubt, however, the reader will feel that something is still wrong with this reconstruction, for (b′) overlooks the possibility of alternative explanations, and the conclusion is reached a bit too easily. This has to do with a further problem of Barnes' original reconstruction: (b) is wrong not only because 'built-to-fit' is not a possible explanation of (F), but also because there can be still another (shallow) explanation: since false assumptions can have true consequences, another explanation may be that

(fce) T has also (or only) false components, E follows from them, and E is true due to its meaning and the way the world is.

So, (b') must be substituted by

(b″) There are two possible explanations of (F): (pt) (i.e., that a part of T is true, and E follows from that part); and (fce).

At this point, however, the argument is blocked, for in Barnes' formulation we find no reason to choose between the shallow explanations (pt) and (fce).

But Barnes' reconstruction does not make justice to the actual NMAT, because the *explanandum* (F) should not be understood literally, but in a richer implicit way, which allows to break this deadlock between (fce) and (pt):

(F′) scientists built (or "found") (and endorsed) a theory (i.e., T), which has some bold true consequences (i.e., E) (White 2003, 662).

In other words, the explananda are not just the logic fact that T entails E and the semantic fact that E is true, for both of which shallow explanations are enough; there is also the epistemic fact that T was built (and endorsed). (F′) is a contingent fact just like the abovementioned facts (C) (which could be explained by the fact that T was built to fit E) and (EE) (which Barnes takes as explanandum in his own NMABB) (Sect. 2). Now, unlike (F), (F′) requires a *deep* explanation, which can be only one of the following:

I. Scientists built T *to fit E*, this is why T entails E, although T is (possibly even completely) false (fce).
II. *By chance* scientists found a theory (i.e., T) which entails E, although it is (possibly even completely) false (fce).
III. *By chance* scientists found a theory (i.e., T) with some true parts, from which E follows (pt).
IV. Scientists *reliably sought* for a *true* and *fecund* theory, and they *did find* one (i.e., T) with enough true content to entail the novel E (pt).

It can be appreciated that the "built-to-fit" explanation, which had no place in Barnes' reconstruction, finds its intuitive role in possible explanation I. Moreover, the deep explanations I and II entail and explain the shallow explanation (fce): they show how scientists built a theory which, though largely or completely false, has true consequences. On the other hand, the deep explanations III and IV entail and explain the shallow explanation (pt): they show how scientists succeeded in finding a theory which has enough true content to entail E.

Explanation I is the most natural one in many cases, but it is not possible if E is *novel*. Explanation II may apply when E is very generic and poor in content (hence, a priori probable, and following from many possible theories), so that it was possible to find a false but E-entailing theory by chance. But if E is novel and *bold* (highly specific and informatively rich), then finding such a false but E-entailing theory by chance would have been miraculous coincidence (Alai 2014a, 307–309). So, barring miracles, also II is ruled out. Therefore, *novel* and *bold* predictions cannot be explained by (fce), and we are left only with explanations III and IV, based on (pt), the (partial) truth of T.

In order to choose between III and IV we must ask how did scientists succeed in building a theory with enough true content to entail E. Since in principle there are infinite (completely) false theories compatible with any body of known data, finding a theory that is even partially true (and besides entails E) *by pure chance* (III) would be a miraculous luck: even more miraculous than finding a false theory compatible with all the known data plus E. So, when E is novel and bold, the only possible explanation is IV, that the theorists succeeded in their search for truth thanks to their *reliability*. The deep explanation of (F') offered by White (2003) is just that the theoretician was reliable, but as we noticed, this is not enough: a scientist might be very reliable, yet fail to find a theory with enough true content to entail E. *Success* in getting truth and fecundity is also necessary: i.e., any deep explanation must entail (pt).

So IV is the only workable explanation; but is it also plausible? I think so, for there is at least a realistically possible story which justifies IV, i.e. which explains, in turn, what makes scientists so reliable to find (and endorse) partially true and fecund theories in a non-miraculous way (Alai 2014b, Sect. 5): it assumes that

(a) scientists have skills, ingenuity, true background beliefs, and employ sound scientific method;

(b) scientific method is truth-conducive, since (i) it is based on the assumption that nature is simple, uniform and intelligible, hence it can be known through analogy, induction, and abduction, and (ii) nature actually *is* simple, uniform and intelligible.

Summing up, a correctly reconstructed NMAT shows that the only plausible explanation of how scientists managed to build theories entailing bold novel predictions involves the assumption that they built (partially) true theories (pt). This fact is explained by the scientists' reliable search for truth and fecundity (IV), which in turn is explained by (a) and (b). So, by inference to the only plausible explanation, in these cases (pt), IV, (a) and (b) can be safely assumed. This accounts for the practically conclusive confirmation of our most successful theories, and for the scientific realist claim that in general we can have cogent reasons to believe in the (partial) truth of theories.[14] So, contra Barnes, the NMAT is correct, and since it supports (pt), also strong predictivism is correct: novel success is *direct* evidence for the truth of theories.

Acknowledgment I thank an anonymous referee and Michel Ghins for very useful comments to earlier versions of this paper.

References

Achinstein, P. (1994). Explanation vs. Prediction: Which carries more weight? In D. L. Hull, M. Forbes & R. M. Burian (Eds.), *PSA* (Vol. 2, pp. 156–164). East Lansing, MI: Philosophy of Science Association.

Alai, M. (2014a). Novel predictions and the no miracle argument. *Erkenntnis, 79*(2), 297–326.

Alai, M. (2014b). Why antirealists can't explain success. In F. Bacchini, S. Caputo & M. Dell'Utri (Eds.), *Metaphysics and ontology without myths* (pp. 48–66). Newcastle upon Tyne: Cambridge Scholars Publishing.

Barnes, E. C. (2008). *The paradox of predictivism*. Cambridge: University Press.

Collins, R. (1994). Against the epistemic value of prediction over accommodation. *Nous, 28*, 210–224.

Duhem, P. (1906). *La théorie physique. Son objet et sa structure*. Paris: Rivière.

Giere, R. N. (1983). Testing hypotheses. In J. Earman (Ed.), *Minnesota studies in the philosophy of science* (Vol. 10, pp. 269–298).

Hitchcok, C., & Sober, E. (2004). Prediction vs. Accommodation and the risk of overfitting. *British Journal for the Philosophy of Science, 55*, 1–34.

Horwich, P. (1982). *Probability and evidence*. Cambridge: University Press.

Howson, C. (1988). Accommodation, prediction, and Bayesian confirmation theory. *PSA, 2*, 381–392.

Howson, C., & Franklin, A. (1991). Maher, Mendeleev and Bayesianism. *Philosophy of Science, 58*, 574–585.

Kahn, J. A., Landsburg, S. E., & Stockman, A. C. (1992). On novel confirmation. *British Journal for the Philosophy of Science, 43*(4), 503–516.

Keynes, J. M. (1921). *A treatise on probability*. London: Macmillan.

[14]Clearly, that we *can* have them does not imply that we *always* have them.

Kuhn, T. S. (1962). *The structure of scientific revolutions*. Chicago: University Press.
Kuhn, T. S. (1977). *The essential tension*. Chicago: University Press.
Lakatos, I. (1970). Falsificationism and the methodology of scientific research programmes. In I. Lakatos & A. Musgrave (Eds.), *Criticism and the growth of knowledge* (pp. 91–195). Cambridge: University Press.
Lange, M. (2001). The apparent superiority of prediction to accommodation as a side effect: A reply to Maher. *British Journal for the Philosophy of Science, 52*, 575–588.
Leplin, J. (1997). *A novel defence of scientific realism*. Oxford: University Press.
Lipton, P. (1991). *Inference to the best explanation*. London: Routledge (2004^2).
Maher, P. (1988). Prediction, accommodation and the logic of discovery. *PSA, 1*, 273–285.
Maher, P. (1990). How prediction enhances confirmation. In J. M. Dunn, & A. Gupta (Eds.), *Truth or consequences: Essays in honor of Nuel Belnap* (pp. 327–343). Dordrecht: Kluwer.
Maher, P. (1993). Howson and Franklin on prediction. *Philosophy of Science, 60*(2), 329–340.
Mill, J. S. (1843). *A system of logic ratiocinative and inductive: Being a connected view of the principles of evidence and the methods of scientific investigation*. London: Parker.
Musgrave, A. (1974). Logical vs. Historical theories of confirmation. *British Journal for the Philosophy of Science, 25*, 1–23.
Musgrave, A. (1988). The ultimate argument for scientific realism. In R. Nola (Ed.), *Relativism and realism in science* (pp. 229–252). Dordrecht: Kluwer.
Musgrave, A. (1999). *Essays on realism and rationalism*. Amsterdam: Rodopi.
Peirce, C. S. (1883). A theory of probable inference. In C. S. Peirce (Ed.), *Studies in logic by members of the Johns Hopkins University* (pp. 126–181). Boston, Mass.: Little, Brown, & C.
Popper, K. R. (1962). *Conjectures and refutations*. London: Routledge & Kegan Paul.
Putnam, H. (1975). *Mathematics, matter and method: Philosophical papers* (Vol. 1). Cambridge: University Press.
Rosenkrantz, R. D. (1977). *Inference, method and decision*. Dordrecht: Reidel.
Schlesinger, G. (1987). Accommodation and prediction. *Australasian Journal of Philosophy, 65*, 33–42.
Smart, J. J. C. (1968). *Between science and philosophy*. New York: Random House.
Whewell, W. (1840). *The philosophy of the inductive sciences*. London: Parker & Sons (London: Routledge, 1996).
White, R. (2003). The epistemic advantage of prediction over accommodation. *Mind, 112*(448), 653–683.
Worrall, J. (1978a). The ways in which the methodology of scientific research programmes improves on popper's methodology. In G. Radnitzky & G. Andersson (Eds.), *Progress and rationality in science. Boston studies in the philosophy of science* (Vol. 58, pp. 45–70). Dordrecht: Reidel.
Worrall, J. (1978b). Research programmes, empirical support and the Duhem problem: Replies to criticism. Ibid. (pp. 321–338).
Worrall, J. (1985). Scientific discovery and theory-confirmation. In J. C. Pitt (Ed.), *Change and progress in modern science*. Dordrecht: Reidel.
Worrall, J. (1989a). Structural realism: The best of both worlds? *Dialectica, 43*, 99–124. Reprinted in D. Papineau (Ed.), *Philosophy of science*. Oxford: University Press.
Worrall, J. (1989). Fresnel, Poisson and the White spot: The role of successful predictions in the acceptation of scientific theories. In D. Gooding, T. Pinch, & S. Schaffer (Eds.), *The use of experiment: Studies in the natural sciences* (pp. 135–157). Cambridge: University Press.
Worrall, J. (2005). Prediction and the 'periodic law': A rejoinder to Barnes. *Studies in the History and Philosophy of Science, 36*, 817–826.
Worrall, J. (2006). Confirmation: Two types, both logical, non historical. In C. Cheyne & J. Worrall (Eds.), *Rationality and reality: Conversations with Alan Musgrave*. Dordrecht: Springer.

Wright, J. (2002). Some surprising phenomena and some unsatisfactory explanations of them. In S. Clarke & T. D. Lyons (Eds.), *Recent themes in the philosophy of science. Scientific realism and commonsense* (pp. 139–153). Dordrecht: Kluwer.

Zahar, E. (1973a). Why did Einstein's programme supersede Lorentz's? (I). *British Journal of Philosophy Science, 24*, 95–123.

Zahar, E. (1973b). Why did Einstein's programme supersede Lorentz's? (II). *British Journal of Philosophy Science, 24*, 223–262.

Traditional East Asian Views and Traditional Western Views on the Heavens: The Discovery of Halley's Comet

Jun-Young Oh

Abstract Traditional astronomy in East Asia developed very differently from that of the Western world, and it was informed by the views of nature inherent to East Asia. This research aims to examine how fundamental forces such as "yin-yang polarity," a core concept of the East Asian view of nature, were reflected in the traditional astronomy of the region. In East Asia, astronomical works that carefully examined the celestial bodies were considered very important, and an astronomical system that was connected to the "human condition" was established. Analogical reasoning or correlative or "associative" thinking were dominant in the region. This connection with the human condition played an important role in the unique astronomy of East Asia, contrasting the mechanical worldview typical in Eastern Europe.

Keywords Yin-yang polarity · Human condition · Associative thinking

1 Introduction

Although traditionally West science is seen as the pursuit of universal knowledge, where observations of astronomical phenomena are used to confirm theories, East Asians interpreted the regular movement of the sun, moon, and stars as representing the state of our world. In particular, they considered solar eclipses as portents of disaster and the appearance of comets as heralding misfortune. Thus, comets and solar eclipses were monitored with great interest (Lee 2012). East Asians sought to observe abnormal astronomical phenomena and understand their meaning in relation to how human society ran. This was grounded in "correlative thinking," where all aspects of the universe were seen as closely interconnected (National Institute of Korean History 2007, p. 34). In the traditional Western world, people believed in

J.-Y. Oh (✉)
Center of Basic Literacy, Hanyang University, Seoul 122-791, Republic of Korea
e-mail: Jyoh3324@hanyang.ac.kr

L. Magnani and C. Casadio (eds.), *Model-Based Reasoning in Science and Technology*, Studies in Applied Philosophy, Epistemology and Rational Ethics 27, DOI 10.1007/978-3-319-38983-7_31

the completeness and invariance of the sky. However, in the East, people did not have such preconceived ideas (Oh 2007, p. 123).

Therefore, the aim of this study is to understand the discovery of Halley's Comet during Korea's Cho-sun Dynasty, based on the traditional Western view of nature.

2 Heaven-Human Correlation Theory (天人感應說) and the Hypothetico-Deductive Method

Abstraction is a crucial feature of rational and scientific thinking, because when comparing and classifying the immense variety of shapes, structures, and phenomena around us we cannot take all of their features into account, but have to select a few significant ones. Thus, we construct an intellectual map of reality in which things are reduced to their general outlines (Capra 2010, p. 27). The difference between abstraction and idealization is the difference between an *epistemic* matter, as when we ignore properties while abstracting, and an *ontological* matter, as when we claim that an object lacks certain properties and idealize it (Nolar 2004, p. 304).

Western philosophy, which is a discipline of thought, is composed of things that cannot be seen, things that cannot be touched, and things that do not exist. If not precisely structured, things cannot be verified. Therefore, the reason why the study of logic is so developed in the Western world is that they developed their thoughts based on "things that do not exist." Because philosophy itself is a structural system of thoughts, the precision of all thought becomes an important issue. However, Eastern thought placed more importance on the validity and communication of experience rather than the elaboration of thoughts. Therefore, logic did not play such a pivotal role (Choi 2015, p. 177).

In the traditional Eastern world, what was ultimately lacking was deductive thinking and a geometrical model. Also, because deductive reasoning at the verification stage of the so-called hypothetico-deductive method was lacking, people had no choice but to stick to the method of discovery. This method significantly weakened the effectiveness of quantitative observation, measurement, and experiments in the study of nature (Yamada 1982). This is in contrast to the case of the Ancient Greeks, who established a geometric model in which the sun, moon, and planets revolve in a circular motion around the Earth, which in turn set the apparent motion of these celestial bodies. This led to success in explaining the observable motion of these celestial bodies. This model was the geocentric theory. Of course, Copernicus' heliocentric theory came to dominate from the 16th century, but the geocentric theory was the beginning of the methodology of modern science, which starts with a hypothesis to explain a phenomenon. In traditional East Asian astronomy, although there was an attempt to directly understand generalities by investigating the apparent phenomena, they did not start with a hypothesis before searching for an explanation for a phenomenon (Yabuwoochi 1970, p. 51).

Who knew that the cold rain would become a monsoon when crops are growing well into autumn? It is because I did not have any virtue that I could not move the heart of the sky (未能感格天心) and brought about such an extreme situation. Therefore, if I am not vigilant nor strive (苦無飭勵之事), how would I impress the sky? (何以格感) It is obvious that I must start with myself (宜自寡始) (translated by Park 2014, p. 24) (Yeong-jo Age, Yeong-jo 4th year [1728], at dawn, July 27).

The preceding kings were used to looking up to the sincerity of revering the heavens (昔年敬天之誠) and the virtue of loving the people from a young age (translated by Park 2014, p. 143) (御製續自省編).

When a comet appears, should we just sit and be a victim to national catastrophe? We shall read the "Gu-dang-seo," and "Cheon-mun-ji." In the Jeong kwan (貞觀) 9th year (635), August 23, Paesung (comets) appeared in Huh-soo (the Emptiness mansion: 虛宿) and Wie-soo (the Roof mansion: 危宿) and passed through Hyun-hyo (Aquarius: 玄枵). It only disappeared after November passed. King Tae-jong (太宗) asked his servants by his side, "What kind of catastrophe is this?" Wu Se-nam (558-638) replied, "From what we hear, if the leader does not cultivate the virtues in governing people, even when the comet appears there is no way to help this leader. If there is no flaw in politics, even if there is an abnormal natural phenomenon, how could it have a negative effect on the national affairs? We ask of you that your highness do not show off his greatness in his achievements that are higher than his predecessors, and do not become careless or haughty in mind just because peace has lasted for a long time. If you refrain from these first, last, and all the time, how could a comet become a thing of worry?" (Oh 2007, pp. 128–129).

Here the leader is being asked to adopt the heaven-human correlation theory (天人感應說) to become a benign ruler. The astronomical phenomena constantly warn the emperor of catastrophe, and interpreting the celestial phenomena accurately and reflecting them in politics is seen as the best way to avoid catastrophes. This is the ultimate goal of "astronomy and calendrical science" (Oh 2007, p. 129). This means that the celestial bodies in the sky and the humans living on the land influence each other and interact (Lee 2012).

The utmost leader of China was called the "son of the heavens" (天子), and this term was based on the unique Chinese political concept that leaders ruled according to the ways of the heavens. Rather than viewing the heavens as a creator god, they thought that the heavens presided over all beings, possessed the utmost morality, and contained the order of nature. Moreover, they claimed that the heavens had their own will and demonstrated this on good and bad politicians through astronomical phenomena as well as natural phenomena on Earth. The son of the heavens incessantly strived to understand the meaning of the heavens through natural phenomena and to rule accordingly. The order of the heavens can be defined as the law of nature, but for Asians the natural laws could not be ultimately explained by human effort. In Greece, which was the origin of modern science in Europe, natural laws were understood through human exploration, and this knowledge was a great source of power in the development of modern science. However, in the Eastern world, including China, the sky, which was an embodiment of natural law and order, was seen as having its own will and could freely create natural phenomena.

This kind of belief was an important reason why Asians never developed the fervor to explore the laws of nature.

3 The Traditional Eastern View of Nature Based on a Traditional Western Perspective: The Philosophy of Plato and Lao Zhang

3.1 Euclidean Geometry, Chinese Algebra

The pioneers of the 17th century, Hobbes and Newton, deduced the laws of nature using mathematical methods, especially Euclidean geometry in Hobbes' case. In Euclidean geometry, all prepositions must follow a logical necessity. Therefore, even in physics they demonstrated that all phenomena ensue from a preceding natural necessity. In other words, a certain cause leads to certain consequences (Bronowski 1977, p. 69).

According to Euclid, each proposition flows from those which precede it by logical necessity; and so in physics, it is argued, each action must flow from those which precede it by natural necessity. Cause must lead unalterably to effect. The laws of nature are thus like the laws of deductive reasoning: by these steps we go from first to last, from cause to last effects, along a path that is unique, certain, and (in principle) predictable in every detail (Bronowski 1980, p. 34).

Westerners thought that if something existed, it existed as only one. It existed in unity and the trait that supported this unity was its "essence." If one had an essence, it was not to be shared with another. Monkeys that are similar to humans should not have any part of the essence of humans. Rationality in humans is the essence that is the foundation of the unity that defines humans, and this is called "substance." This world is perceived to be standing on a certain fundamental foundation. The human ability to understand this substantial world and grasp the rational relationship between humans and such a world is called "rationality" (Cho 2015, p. 178). According to Yugawa, in this kind of scientific methodology, up until the 17th and 19th centuries, science did not really separate its abstract processes from the real world. In the 20th century, only a very small portion of what came out of the mathematical achievements of highly abstract physical theories could be directly verified.

The Ancient Greeks, far more than their contemporaries, speculated about the nature of the world they found themselves in and created models of it. They constructed these models by categorizing objects and events and generating rules about them that were sufficiently precise for systematic description and explanation. This characterized their advances in (and some have said invention of) the fields of physics, astronomy, axiomatic geometry, formal logic, national philosophy, natural history, and ethnography. While many great contemporary civilizations, as well as the earlier Mesopotamian and Egyptian and the later Mayan civilizations, made systematic observations in all scientific domains, only the Greeks attempted to explain their observations in terms of underlying principles. Exploring these principles was a source of pleasure for the Greeks (Nisbett 2003, p. 4). Early in their study of the heavens, the Chinese believed that cosmic events such as comets and eclipses could predict important occurrences on Earth, such as the birth of a conqueror. However, when they discovered regularity in these events, far from building models of them, they lost interest in them.

The Greeks were concerned with understanding the fundamental nature of the world, though in a way their methods changed across eras. The fifth century saw a move toward abstraction and distrust of the senses. Plato claimed that if the senses seemed to contradict conclusions reached from first principles and logic, it was the senses that had to be ignored. Aristotle thought of attributes as having a reality distinct from their concrete embodiments in objects (Nisbett 2003, pp. 8–9).

The abstract characteristic of Ancient Greek philosophy had no counterpart in Chinese philosophy. For the Chinese, the given background for the nature of the world was that it was a mass of substances rather than a collection of discrete objects. All Chinese philosophy shared concerns about harmony, holism, and the mutual influence of everything on almost everything else. However, the Greek philosophers were focused on objects composed of particles (Nisbett 2003, p. 18).

The Renaissance can be characterized firstly by its separation from tradition. It can be defined as the extinction of a particular tradition and the revival of another. This time period broke with the tradition of Aristotelian and Scholastic philosophy that predominated Europe during the Middle Ages for a long time, and instead revived the Platonic view of the world and atomism. In particular, the mathematical tradition stretching from Pythagoras to Plato in Ancient Greece gave a new sense of inspiration to intellectuals in the Renaissance.

3.2 Newtonian Mechanics and Yin-Yang Polarity

An idea born in Greece, solidified by Galileo, and spread through Europe by Descartes and Newton was "mathematics is a language." Supported by continuous discoveries, this led to the mechanistic classical physics of Newton. Mechanical physics explains natural phenomena based on natural cause and effect, mechanistic rules, and the idea that matter is composed of particles.

Euclid theorized an axiomatic system; that is, a system consisting of a small set of axioms and postulates (propositions not proved in the axiom system but assumed to be true within the system) and large theorems derived from the axioms by deduction in adherence to the rules of logic. The most famous theory in physics to be presented as an axiomatic system was Newton's mechanics. The theory originally consisted of three axioms, to which Newton later added a fourth of greater importance. The explanatory and predictive power of Newtonian mechanics was revealed by mathematical deduction of the vast range of regularities from the un-derived four axioms of Newtonian mechanics (Rosenberg 2012, pp. 117–121).

However, East Asians never thought mathematics could be a language and, therefore, did not think that mathematics could be used for explaining natural phenomena; instead, they discussed nature based on yin and yang (Jullien 2005, p. 24).

Because disharmony in the heavens supposedly indicated disharmony in the emperor's rule, astronomy became a matter of national importance at an early point and received official patronage, with an imperial astronomy office being established and operated (McCllan and Dorn 1999, p. 131).

With comets a portent of disaster, Chinese astronomers carefully logged 22 centuries worth of commentary and observations from 613 B.C. to A.D. 1621, including the appearance of Halley's Comet every 76 years from 240 B.C. However, unlike the Greeks, they did not develop explanatory models for planetary motion. They mastered planetary periods without speculating about orbits (McCllan and Dorn 1999, p. 132).

East Asian modes of thought proved inimical to logical, objective scientific reasoning of the sort that developed in the West. Historians have identified a persistent cultural pattern in East Asia variously labeled as "analogical reasoning" or "correlative" or "associative" thinking. This style of thinking, it is said, strove to interpret the world in terms of analogies and metaphorical systems of paired correspondences between diverse things (such as virtues, colors, directions, musical tones, numbers, organs, and planets) based on the fundamental forces of yin and yang and the five "phases" of metal, wood, water, fire, and earth. Yin and yang thus parallel female and male, day and night, wet and dry, the emperor and the heavens. "Wood" is associated with "spring" and the cardinal direction "east," and so on (McCllan and Dorn 1999, p. 138). This way of thinking may have put the Chinese at a disadvantage compared to the logical, objective, and scientific inference developed in the West.

4 Discovery of Halley's Comet in the Western World: Deduction-Induction (Oh et al. 2015; Magnani 2009) Involving Mathematical Abstraction

It was as if the laws of nature became very similar to deductive reasoning. According to this method, we walk along a predestined and solitary path from the cause at the beginning to the result at the end, and by principle, every detailed part of the path can be predicted.

4.1 Deduction

Assuming Newton's law was correct and considering the sun and comet as point mass, Jupiter's perturbation (mathematical abstraction: auxiliary hypotheses), and the times of the last observations of the comet, Halley calculated the time of its next return. He boldly predicted that the comet would be seen in late December, 1758 (expected data).

4.2 Induction

The comet did reappear as predicted, near Christmas of 1758 (observation data), and the only alternative hypothesis was that another comet with the same orbit just happened to appear right around the predicted time 76 years later. That seemed to everyone extremely unlikely, so the data provided very good evidence that the Newtonian model fit.

The concern with abstraction characteristic of ancient Greek philosophy have no counterpart in Chinese philosophy. For the Chinese, the background scheme for the nature of the world was that it was a mass of substances rather than a collection of discrete objects p. 18). All Chinese philosophy hared concerns about harmony, holism, and the mutual influence of everything on almost everything else. However, the Greek philosopher would have been an object composed of particles.

5 Analogical Reasoning in the East

The heaven-human correlation theory of the Chinese Han Dynasty explained the relationship between the sky and humans by describing the way of the heavens through yin and yang theory and arguing for the mutual connectivity between heaven and humans through the so-called principles of analogical reasoning (Lee 2000, p. 32). The observations from Gwansang-gam office of Cho-sun Dynasty, Korea, about Halley's Comet (Nha 2016, p. 155). Comets were an ominous sign to kings, because they are an irregular phenomenon of the sky. Therefore, it was believed that if the tail of a comet pointed to a king's country, that king and his country would have bad luck.

5.1 Analogical Reasoning

Heaven is yin and the emperor is yang.

During the reign of King Young-jo of the Cho-sun Dynasty in Korea, based on observations made in Shandong, China, it was reported that the tail of Halley's Comet was pointing east; that is, at Cho-sun (an astonishing phenomenon). King Young-jo raised the question that, if the tail of a comet is pointed at Cho-sun, would it not be a bad sign to the country, especially the king, and therefore would it not be necessary to observe it more carefully?

5.2 *Question Raised by King Yong-Jo*

The line of reasoning here is as follows: If a comet that appears suddenly in the sky has a long tail, a sudden mishap will lead to serious misfortune for the King; and if it is true that the sky is yin and the king yang, and that the sky and ground influence each other, it must mean that Cho-sun and its king will have a problem. Therefore, would the people of Cho-sun not have to carefully observe the direction of the comet's tail? Figs. 1 and 2.

5.3 *Opinion Based on Observations by Gwansang-Gam*

However, the observations by Gwansang-gam office of Cho-sun Dynasty showed that the tail of Halley's Comet would be pointing at the west in the morning and the East in the evening, without a particular direction. As it was stated:

Therefore, since yin-yang is certainly correct, the observation from Shandong regarding the tail of the comet, which implied misfortune in the Cho-sun Kingdom, was not supported. However, we [should] observe [comets] carefully, because

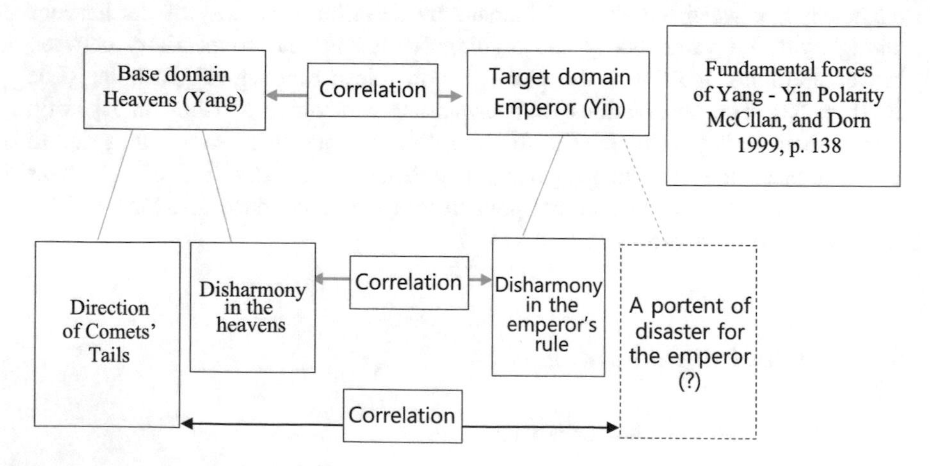

Fig. 1 Analogical reasoning about heaven and earth based on Yin-Yang polarity

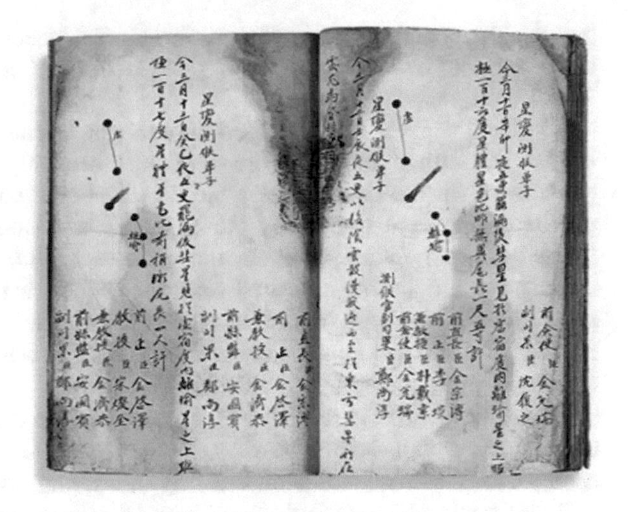

Fig. 2 Sungbyun Dunglok (星變謄錄), observations of Halley's Comet, 07–09 April, 1759 (Ahn 2013, p. 191), 星變測候單子. ©韓國文化財保護 財團

unusual natural phenomena were the warnings against politics, it was not something that could be predicted by humans (translation by Park 2014).

Holism suggests that every event is related to every other event. A key idea is the notion of resonance. For example, if you pluck a string on an instrument, you produce a resonance in another string. Man, heaven, and Earth create resonances in each other. If the emperor does something wrong, it throws the universe out of kilter (Nisbett 2003, p. 17).

6 Conclusion and Discussion

According to Collins (1998), during the 17th-century scientific revolution, math and science, once considered to be significantly different (opposing factions), became very close. In addition, intellectual networks between mathematics and philosophy were established. Mathematics and science involving philosophy were propelled forward in new and surprising directions by the development of technology. Collins states that the East, including India, did not experience this intellectual revolution seen in the West.

This study explored how traditional Western thought spread to the East through the discovery of Halley's Comet. Firstly, in the process of discovering Halley's Comet, the view of nature in the East Asia focused on living, organic nature instead of mechanical nature, which was governed by Newton's laws.

Secondly, the Western world follows the mechanical explanation of Descartes, describing mathematical abstraction that takes into consideration only the important factors according to the intentions of human beings. Descartes' mechanical philosophy explains that the observers on Earth are irrelevant to the phenomena of the sky. However, in the East, because science was specific and experiential, and because the observers on Earth and the celestial phenomena were seen as organically related, the development of thought differed from Western mathematical abstraction.

Thirdly, the West was interested in observing astronomical phenomena because, unlike the Earth, it was possible to identify heavenly bodies' regular, invariable phenomena. In the East, because these phenomena were thought to influence the people on the land (especially the emperor), astronomical observation was performed at the national level. In the Western world, celestial phenomena were usually used to justify a desired scientific model that could be derived through abstraction. In the Eastern world, on the other hand, because science was specific and experiential, and thus mathematical abstraction was difficult to conduct, it was challenging to develop a scientific model.

Conclusively, the Western worldview, typified by simplicity using mathematical abstractions as opposed to East Asia's complexity, are appropriate in the field of science, because simple models are easy to test and revise.

In the traditional East, including China and Korea, the observance of comets was not done for the purpose of discovering a scientific law or theory, but was completely separate from humans, as in the modern West. The surest way for a leader to avoid disasters was through the accurate observation and interpretation of celestial phenomena. Eventually, such phenomena would reflect on one's ruling, based on the heaven-human correlation theory or heaven-human correspondence theory, in which the will of heaven is realized through natural phenomena, with yin and yang as the medium. In the traditional East, the way of thinking was ultimately based more on deductive reasoning than it was in the Western world, and it lacked a geometric basis. Because the deductive reasoning that is the verification stage of the hypothetico-deductive method was lacking, East Asians had no choice but to remain at the observation stage (see Table 1).

Table 1 Characteristics of the science cultures of the traditional west and east

	Science culture of the traditional west	Science culture of the traditional east	
Formation of model	The level of abstraction by simplification is high. They search for the essence of nature. analytical due to abstraction	The level of abstraction by complexification effect is low. Holistic, combined	Nisbett (2003)
Verification of model	Verification by deduction (hypothetico-deductive), Euclid's geometry, Newtonian mechanics, A construct that is built from basic principles and basic concepts like other structures	Understanding of the heaven's intentions rather than verifying.' Fundamental being' does not exist and the explanations are a web of concepts and models interconnected	Yamada (山田慶兒), (1982)
Origin of model	Human's reason, Plato's theory of ideals	Human's instincts and intuition, Yin and yang dualism of Taoism	Cho (2015)
Opinion on scientific knowledge	The knowledge of natural science that excludes the observer is a being that has a fixed structure with mechanical order	The knowledge of natural science including the observers is a changing process that is ecological and organic	Capra (2010)

The mechanistic worldview of classical physics, involving Newton's mechanical rules, is useful for the description of the kind of physical phenomena we encounter in our everyday lives and thus appropriate for dealing with our environment. It has also proved extremely successful as a basis for technology. It is inadequate, however, for the description of physical phenomena in the submicroscopic realm. Opposed to the mechanistic conception of the world is the view of the mystics from the traditional Eastern world, which may be epitomized by the world "organic," as it regards all phenomena in the universe as integral parts of an inseparable, harmonious whole (Capra 2010, pp. 303–304).

Our scientific knowledge can often stay abstract and theoretical. Many of today's theories actively support a society based on the mechanistic, traditional Western worldview, without seeing beyond such a mechanistic world, towards a oneness of the universe that includes not only our natural environment, but also us human beings based in the organic world.

Acknowledgements This work was supported by the National Research Foundation of Korea grant funded by Korean government (NRF-2014S1A3A2044609).

References

Ahn, S. H. (2013). *The history of our comets*. Seoul: ScienceBooks.

Bronowski, J. (1977). *A sense of the future: Essays in natural philosophy* (5th ed.). Cambridge, Massachusetts: The MIT Press.

Capra, F. (2010). *The tao of physics: An exploration of parallels between modern physics and eastern mysticism (35th anniversary edition with a new preface by the author).* Boston: SHAMBIA.

Cho, J.-S. (2015). *Thinking power: No-Ja humanities*. Kyunggi: Wisdom House.

Collins, R. (1998). *The Sociology of philosophies: A global theory of intellectual change*. Cambridge, MA: Harvard University Press.

Jullien, F. (2005/2015). Conférencesurl'efficacité. Paris. Presses Universitaires de France, (K. S. Lee, Trans., 2015, Seoul: KoiSega).

Lee, M.-K. (2000). *Early heavens in Chinese at the world*. Seoul: MiuHak and JiSung Co.

Lee, M.-K. (2012). Seeing heavens based on qi: Role of qi 氣 in east asian astronomy. *Studies in East -western Philosophy, 96*, 391–412.

Maganani, L. (2009). *Abductive cognition: The epistemological and eco-cognitive dimentions of hypothetical reasoning*. Berlin/Hedelberg: Springer.

McCllan, J. 3., & Dorn, H. (1999). Science and technology in world history.

National Institute of Korean History. (2007). *Traditional thoughts about heaven, times, and earth surface*. Seoul: Dusan DongAh.

Nha, I. S. (2016). *Traditional Cosmic World-Views of Korean: Drawing our Sky from our Terrestial*. Seoul: Yonsei University Press

Nisbett, R. E. (2003). *The geography of thought: How asian and westerns thinker differently … and why*. New York: A Division of Simon & Schuster Inc.

Nola, R. (2004). Pendula. *Models, Constructivism and Reality, Science & Education, 13*, 349–377.

Oh, M.-Y. (2007). *The history of east science for students*. Seoul: Duri-media.

Oh, J.-Y., Kim, Y. S., Kim, C. H., Min, B. M., & Son, Y. A. (2015). Scientific inquiries of Galileo's formulation procedures of inertial law. In Lorenzo Magnani, Ping Li, & Woosuk Park

(Eds.), *Philosophy and cognitive science II: western & eastern studies (chapter 11)*. Heidelberg New York Dordrecht London: Springer.

Park, S. W. T. (2014). *Young-jo's statements in the Joseon Dynasty (by Young-jo)*. Seoul: Soulmate.

Rosenberg, A. (2012). *Philosophy of science: A contemporary introduction* (3rd ed.). New York, NY: Routledge.

Yabuwoochi, K. (1970/1997). Scientific literacy of China (S. W. Jeo, Trans., 1970). Seoul: Min-um Co.).

Yamada, G. (1982/1994). The structure of China thought. Cho-IL Newspaper. Co. (S-W., Park, Trans., 1994, Electronic-wave Science Co.).

Search Versus Knowledge in Human Problem Solving: A Case Study in Chess

Ivan Bratko, Dayana Hristova and Matej Guid

Abstract This paper contributes to the understanding of human problem solving involved in mental tasks that require exploration among alternatives. Examples of such tasks are theorem proving and classical games like chess. De Groot's largely used model of chess players' thinking conceptually consists of two stages: (1) detection of general possibilities, or "motifs", that indicate promising ideas the player may try to explore in a given chess position, and (2) calculation of concrete chess variations to establish whether any of the motifs can indeed be exploited to win the game. Strong chess players have to master both of these two components of chess problem solving skill. The first component reflects the player's chess-specific knowledge, whereas the second applies more generally in game playing and other combinatorial problems. In this paper, we studied experimentally the relative importance of the two components of problem solving skill in tactical chess problems. A possibly surprising conclusion of our experiments is that for our type of chess problems, and players over a rather large range of chess strength, it is the calculating ability, rather than chess-specific pattern-based knowledge, that better discriminates among the players regarding their success. We also formulated De Groot's model as a Causal Bayesian Network and set the probabilities in the network according to our experimental results.

1 Introduction

Consider human solving of mental tasks such as theorem proving, symbolic manipulation problems, and classical games such as chess or checkers. A general, widely accepted computational model of solving such problems involves searching among alternatives (Newell and Simon 1972). In games, for example, this amounts

I. Bratko (✉) · M. Guid
Faculty of Computer and Information Science, University of Ljubljana,
Večna pot 113, 1000 Ljubljana, Slovenia
e-mail: bratko@fri.uni-lj.si

D. Hristova
University of Vienna, Vienna, Austria

L. Magnani and C. Casadio (eds.), *Model-Based Reasoning in Science and Technology*, Studies in Applied Philosophy, Epistemology and Rational Ethics 27, DOI 10.1007/978-3-319-38983-7_32

to searching through "my" possible moves, then for each "my" move considering all possible opponent's replies, then again considering my possible moves for each of the opponent's replies, etc. This search may stop when positions are encountered that are estimated without doubt as drawn, or good or bad for one of the sides. The result of search is "my" move (if exists) that guarantees my win against any possible opponent's reply.

In fact, this simple algorithm is *in principle* sufficient to solve problems of the types mentioned above. However, in cases of high combinatorial complexity of problems, such as chess where there are many alternatives at each step, this is practically infeasible because it takes too long for humans, and even for computers. Therefore the search has to be carried out intelligently, relying on search heuristics that are based on the problem solver's knowledge about the domain. Newell coined the term knowledge search (1990) describing the way an agent uses their *directly available* long-term knowledge to bear on the current situation in order to control the search. These heuristics guide the search in promising directions and thus reduce the complexity of search needed to solve problems. The usual, empirically observed relation between the amount of solver's problem-specific knowledge and the amount of search required is: the more knowledge the problem-solver possesses, the less search is needed.

In this paper, we study this trade-off between knowledge and search in human game playing, and investigate their relative importance for success. To this end, we conducted experiments in human problem solving in the game of chess. The problem for the participants was: given a chess position, find the best move for the side to move. The human problem solving model for this case, relevant to our study, was stated in the classical work by De Groot (1946, 1978) and was also used by Tikhomirov and Poznyanskaya (1966).

De Groot's model of chess players' thinking about best moves conceptually consists of two stages: (1) position investigation, and (2) investigation of possibilities, or search. Stage 1, "position investigation" consists of identifying general properties of a position like "Black king is not well protected, so a direct attack on Black king should be considered", and familiar patterns like "White knight is pinned by the Black bishop, so the knight might be in danger". Such patterns make the player suspect that there might be a way to checkmate Black king, and that White knight might be attacked by other enemy pieces and eventually lost because a pinned knight cannot escape from the attack. However, no search among concrete moves is done at Stage 1. This is done at Stage 2, "investigation of possibilities". This consists of the calculation of concrete moves and variations that may lead to the actual exploitation of the spotted motifs. That is, in our examples, to force the checkmate of the king, or force the capture of the knight. Stage 2 is similar to the usual computational problem solving model that involves search described earlier. In this paper, Stage 1 will be referred to as "detection of motifs", and Stage 2 as "calculation of variations". Strong chess players have to master both components of problem-solving skill: detection of motifs, and calculation of variations. The first component is based on the player's chess-specific knowledge, which reflects the player's general understanding of the game. The player has acquired such an understanding through experience and her study of chess literature. The second component, i.e. calculation, seems to be less specific to chess, and is reflected in the player's ability to reliably calculate in his or her mind the possible variations.

The research question explored in this paper is: which of these two components better separates successful players from unsuccessful ones. By successful players we mean those who find best moves more often.

2 Examples of Motifs and Calculation

To further clarify the two stage model of chess problem solving, we now present concrete examples of motifs and calculations.

Figure 1, diagram (a), illustrates "a pin", one of the most common and effective motifs (patterns) in chess. White rook attacks Black knight. The knight cannot escape because this would leave Black king under attack. Black cannot do anything to prevent the loss of the knight, and White wins. Diagram (b) shows how such a motif enables the player to find the right move for White almost without any calculation. The motif itself immediately suggests the winning move rook b1-b7. It then remains to calculate that after all possible Black king moves, White rook can capture the knight. This amounts to searching some 8 positions altogether.

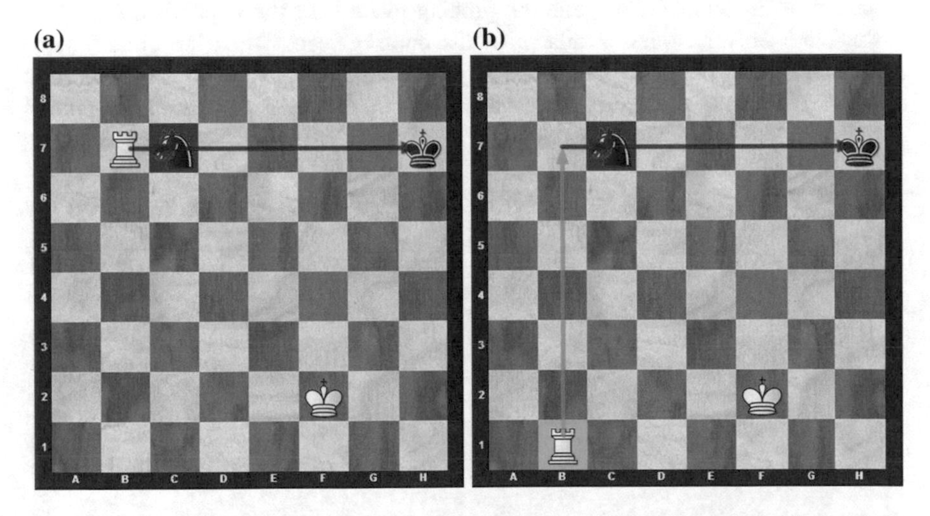

Fig. 1 Diagram **a**. *Black* to move: One of the simplest and most common motifs in chess, called a "pin". *White rook* is attacking the *Black knight*. The knight cannot escape from the rook's attack because if the knight moves away *Black king* will come under the rook's attack (indicated by the *red arrow*) which means a knight's move is illegal. Therefore the Black knight is said to be pinned, and in fact lost in our case. Diagram **b**, *White* to move: The White player in a glance notices an instance of the familiar pin motif (*Black knight* and *Black king* are both on the same line), so the player will immediately perform a little calculation to see whether this motif can actually be exploited. The winning move is: rook moves from square b1 to b7 (*green arrow*) to pin the knight. After any Black's reply, the knight will be captured by the rook

Without the motif, the unguided calculation would have to consider incomparably more positions, something like 30.000. This estimate can be roughly worked out as follows. Let White perform, for example, the breadth-first search. To solve the problem of Fig. 1b, the required depth of search is four half-moves. That is: first level moves by White, then Black's replies, and then the second level moves by White and another level of Black's replies. In each position, White has roughly 22 possible moves (maximally 14 moves by the rook plus 8 moves by the king). Black has roughly 16 possible moves in each position (maximally 8 moves by the knight and 8 by the king). Taking into account gradual reduction of the number of moves with increasing level, this means something roughly in the order of 30.000 positions; compared with 8 when the motif is available.

It should be noted that the pin is a very general concept. It does not have to necessarily involve a rook and a knight and a king, as in our example. A pin occurs between any three pieces where the pinning piece is a long range piece that moves either horizontally, vertically or diagonally (a queen, a rook or a bishop), the pinned piece and the target piece that is indirectly threatened by the pinning piece:

Pinning piece ⟶ Pinned piece - - - - -> Target piece

The arrows show the line of attack by the pinning piece. The pinned and the target piece are of the same colour, and the pinning piece is of the opposite colour. The dashed line only becomes available to the pinning piece when the pinned piece moves away and frees the path to the target piece. Typically, a pin is all the more effective if the target piece is more valuable than the pinned piece. If the pinned

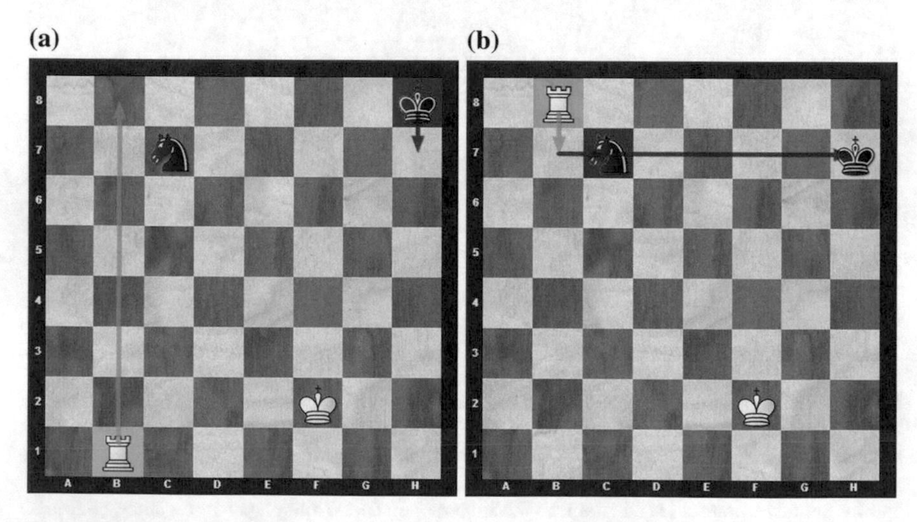

Fig. 2 Diagram **a**: There is no pin yet on the board, but can *White* achieve a pin by forcing the *Black* pieces into the same line? Yes, with *White rook* move from b1 to b8 (*green arrow*). After the move *Black king* is in check and has to move to the 7-th rank, from h8 to h7 or g7. Then the familiar motif appears (diagram b)

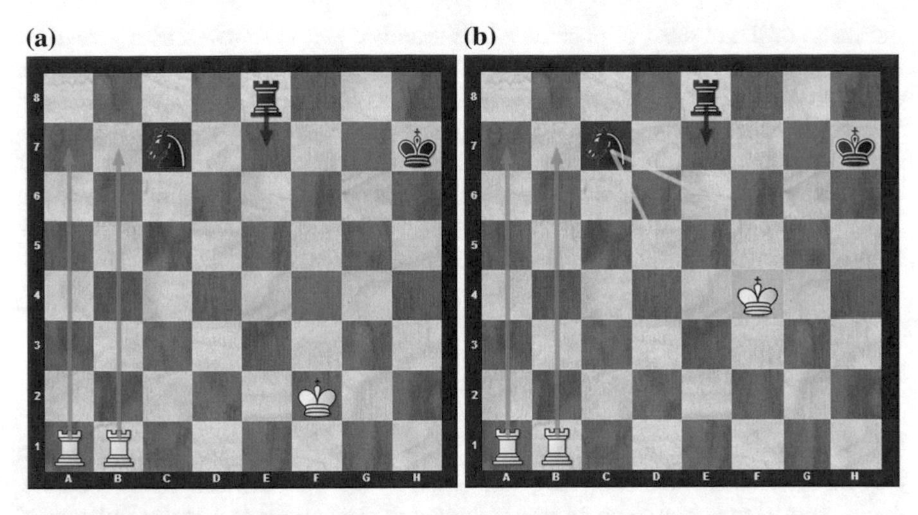

Fig. 3 Diagram **a**: The pin motif in a more complex position with an extra rook for *White* and *Black*. The same idea works for *White* as in Fig. 1, but the calculation here is more complex. White may start with rook from b1 to b7, pinning the knight. Now *Black* has more defensive resources enabled by the presence of *Black rook*. One way is to play Re8-e7, defending the knight with the rook, and interrupting *Black knight* pin against *Black king*. However, the knight is still pinned, this time against *Black rook*. So *White*, seeing this, logically increases the pressure on *Black knight* by playing Ra1-a7, further exploiting the pin. *Black knight* may try to flee to d5, also defending *Black rook*. But the rook is then attacked by two *White rooks*, so White captures twice on e7 and wins. Instead of moving knight to d5, *Black* may try to check *White king* with rook f7 check. Now *White* has to be careful and move the king to e2 which wins eventually, but not to g2 (or g1 or g3) because after that, *Black* can deliver another check by rook f7-g7. After *White king* g2-h2 *Black* has no further checks, but can save himself with Nc7-e6. Suddenly, *Black* pieces have reorganised themselves, *Black rook* is now protected by both the king and knight, and position is drawn. This example nicely shows the role of calculation. The calculation is driven by the motifs, but the final truth is determined by calculation. Diagram **b**: This is the same as diagram (**a**) except that *White king* is now at f4 instead of f2. This small difference offers *Black* additional defensive possibilities and further complicates the calculation drastically. After *White rooks* have doubled on the 7th rank, *Black knight* can move out of attack to d5 or e6 with check. This way *Black* gets out of the pin trouble. But in the process, the knight gets misplaced and cannot return towards his king. Eventually White pieces can trap the knight and win. We do not give concrete variations because they become numerous and very long

piece moves away to escape the attack, the pinning piece may capture the target, possibly winning even more material than just the pinned piece.

Figure 2 shows an example where a pin cannot be immediately created in one move, but can be forced in two moves. The calculation now becomes more demanding, but not significantly so because the motif still guides search very effectively. Any reasonably strong player will immediately look for a way of forcing a pin, and find it in no time.

Figure 3, diagram (a), shows a rather more complex example on the same theme. The pin motif is the same, but two additional rooks now significantly complicate the calculation. But even so, the calculation proceeds in more or less a similar way as in the previous examples according to the familiar mechanisms of exploiting a pin.

Estimates of the number of positions to be searched are: (1) a few tens if guided by the motif, and (2) some 6.000.000 if unguided, which is completely outside human calculation capability. This example illustrates the typical law that the difference between the complexity of guided vs. unguided search grows exponentially with the required search depth. Figure 3b is a small variation of 3a in which White king is at f4 instead of f2. This example shows that, although the main pin motif remains the same as in 3a, the small variation enormously increases the calculation complexity and even brings in several additional motifs in different variations.

3 Experimental Setup

We investigated our research question with an experiment in which 12 chess players of various chess strengths were asked to solve 12 tactical chess problems. A chess position is said to be *tactical* if finding the best move requires calculation of variations, in addition to detecting tactical motifs in the positions.

Our players' chess strength, in terms of official FIDE chess ratings was in the range between 1845 and 2279 rating points. The strength of registered chess players is officially computed by FIDE (World Chess Federation), using the Elo rating system. This rating system was designed by Arpad Elo (Elo 1978). For each player, this rating is calculated and regularly updated according to the players' tournament results. The rating range of our players, between 1845 and 2279, means that there were large differences in chess strength between the players. The lowest end of this range corresponds to club players, and the highest end to chess masters (to obtain the FIDE master title, the player's rating has to reach at least 2300 at some point in their career). There were actually two chess masters among our participants. According to the definition of the Elo rating system, the expected result in a match between our top ranked player against our lowest ranked player would be about 92 % against 8 % (the stronger player winning 92 % of all possible points).

In addition to the differences in chess strength expressed through chess ratings, one could also consider other differences between the players. One such factor might be the chess school where a player was taught, or the particular instructor by whom the player was trained. However, in this paper we did not explore the effects of such additional factors. The 12 chess problems were selected from the Chess Tempo web-site (www.chesstempo.com) where the problems are rated according to their difficulty. Chess problems are rated in a similar way as the players. However, in Chess Tempo a problem's rating is determined by the success of the players when solving the problem. The principle is as follows: If a weak player solved a problem then this counts as strong evidence that the problem is easy. So the problem's rating goes down. If a stronger player solved the problem, the problem's rating still goes down, but not as much as for a weak player. On the contrary, if a strong player failed to solve the problem, this counts as strong evidence that the problem is hard, and the problem's rating increases. In detail, a problem's rating in ChessTempo is determined by using the Glicko rating system (Glickman 1999) which is similar to the Elo system. The Glicko system, at difference with Elo, takes into account the time a player has been inactive. In cases of longer inactivity, the player's rating becomes uncertain. To illustrate the meaning of ratings in ChessTempo, a player with rating 2000 has a 50 % chance of correctly solving a problem with the same rating of 2000. The same player has a 76 % chance to solve a problem with the rating 1800, and a 24 % chance to solve a problem rated 2200.

In our selection we ensured a mix of problems that largely differ in their difficulty. Our selected positions were all tactical chess problems, randomly selected from Chess Tempo according to their difficulty ratings. Based on their Chess Tempo ratings, our problems can be divided into three classes of difficulty: "easy" (2 problems; their average Chess Tempo rating was 1493.9), "medium" (4 problems; average rating 1878.8), and "hard" (6 problems; average rating 2243.5). While the problems within the same difficulty class have very similar difficulty rating, each of the three classes is separated from their adjacent classes by at least 350 Chess Tempo rating points. Some problems have more than one correct solution. To ensure correctness, all the solutions were verified by a chess playing program.

The experimental setup was as follows. Chess problems, that is chess positions in which the participant was asked to find a winning move, were displayed as chess diagrams on a monitor, and the players' solution moves were recorded. Allowed solving time per position was limited to 3 min.

During the player's problem solving, the player's eye movements were tracked by an eye-tracking device EyeLink 1000 and recorded into a database. In the last decades, with wide availability of eye tracking devices, studying eye movements has turned into one of the main methods of research in chess decision making (Reingold and Charness 2005). The processing of recorded eye-movements reveals roughly on which squares of the chessboard the participant was focussing at any time during problem solving.

After the player finished with the 12 problems, a retrospection interview was conducted in which the player described how he or she approached the problem. It was possible to detect from these retrospections what motifs were considered by the player, and roughly how the calculation of variations driven by the motifs was done. Other details of the experiments are described in (Hristova et al. 2014a, b) where the question of automated assessment of the difficulty of chess problems was tackled.

In this paper, we analyse the experimental data with respect to the research question of this paper. The relevant experimental data includes the following. For each player and position, the relevant information consists of: (1) correctness of the solution proposed by the player, (2) the motifs considered by the player in comparison with the motifs needed to solve the problem, and (3) the correctness of the calculation of variations. The motifs considered were found through the players' retrospections, and to some extent verified by the eye movement data, although this verification cannot be done completely reliably.

Figure 4 shows an example of how the eye tracking data can be used. It should also be noted that not all relevant motifs in a position were needed to solve the

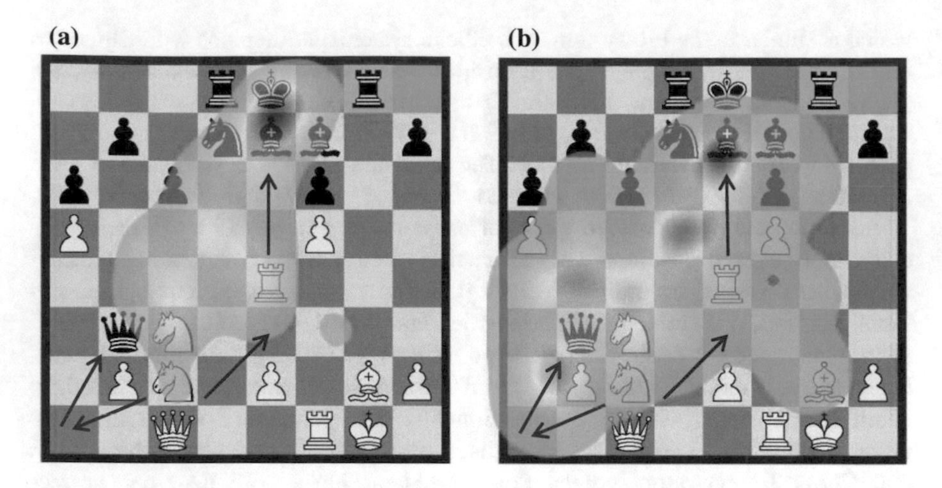

Fig. 4 One of the more difficult test positions. The colouring was extracted from eye tracking data, and shows the intensity of eye focussing on areas of the chess board by a player (*red* is the most intensive, then follow *yellow* and *green*; *white* means no significant focussing). The two diagrams correspond to two players, call them A and B. There are two main motifs in this position: (1) attack *Black king* who is placed very uncomfortably and can be attacked e.g. by the *White rook* at square e4 (*middle* of the board) and *White Queen* moving to square e3. The directions of attack by these moves are indicate by *red arrows*; (2) *Black Queen* is surrounded by *White* pieces and it looks that it might be trapped; some relevant moves by a *White Knight* against *Black Queen* are shown with blue arrows. Calculation of concrete variations reveals that attacking *Black King* does not win for *White*, but trapping *Black Queen* does (by moving the Knight from c2 into the bottom left corner of the board). Player A, according to retrospection, considered both motifs and solved the problem. Player B only considered attacking *Black King* and failed to solve the problem. This difference in motif detection by the two players is indicated in the eye tracking images, although rather subtly

problem. Some of the motifs did not give rise to a winning move. Sometimes there were several alternative winning moves, and there were accordingly several alternative motifs, any of them being sufficient to find at least one of the solutions. Some winning moves were derived from a calculation that required a combination of more than one motif.

These possibilities had to be taken into account when deciding whether the player detected a complete set of motifs needed to carry out correct calculation. To this end, for each position and each possible solution of the position, we defined the "standard" set of motifs necessary and sufficient to find the solution. In defining the standard sets of relevant motifs, we took into account all the motifs mentioned by all the players. In very rare cases when needed, we had to add motifs that fully enabled correct calculation for each possible solution. In doing this, we used our own chess expertise (two of us have chess ratings over 2300 and 2200 respectively). As we verified all the solutions and corresponding chess variations by a chess program, we believe that there would be very little room for defining reasonable alternative standard sets of motifs.

4 Results

4.1 Experimental Data and Basic Statistics

Our experimental data consisted of the performance of 12 players in all 12 positions, that is 144 players' solutions. Out of these, we omitted four cases that were meaningless because the player misunderstood the task (confused the side to move in the position). This gives a total of 140 proposed solutions that we analysed in the sequel (144 − 4 = 140).

Here are some results relevant to our research questions. The players correctly found a winning move in 89 out of 140 problems, i.e. 63.6 %. The players correctly detected all the motifs relevant to a solution in 121 problems (86.43 %). Note that this is the percentage of cases in which the players perfectly detected relevant motifs. The calculations were completely correct in 35.7 % of the problems. It should be noted that a problem was often solved correctly even in the case of imperfect calculation or imperfect detection of motifs.

In the sequel we slightly refine the possible outcomes and introduce the following notation:

- Variable S stands for "success", that is correct solution found.
- Variable M stands for the event "motifs detected". For a problem and a player, M is true if the player correctly detected all the motifs in the problem position that are relevant to a solution of the problem (perfect detection of motifs); otherwise M is false.
- Variable C stands for "calculation correctness". For a problem and a player, C may take one of three possible values:
  - C = CC if the player's calculation in the position is completely correct; that is, it clearly states the critical variations
  - C = CA if the player's calculation is "adequate". That is, the calculation is basically correct, it does suggest a correct move to play (although possibly with a bit of luck), but it is incomplete and/or indicates the player's uncertainty. For example, the calculation is accompanied by the player's comments like "I was not able to calculate everything", "I had to rely on intuition", etc.
  - C = CI if the player's calculation is clearly incorrect, although it may, through sheer luck, even suggest a correct move to play, but for wrong reasons

CC (calculation correct) occurred in 35.7 % of all 140 cases, CA (calculation adequate) occurred in 22.1 % of all cases, and CI in 42.1 %. Note that these do not exactly sum up to 100 % because of rounding errors. Taking these relative frequencies as simple estimates of probabilities we have:

P(CC) = 0.357
P(CA) = 0.221
P(CC v CA) = 0.579
P(CI) = 0.421

Again, the numbers above do not sum up exactly due to rounding errors.

The following results are indicative of the importance of motif detection. Relevant motifs were not perfectly detected in 19 problems. In none of these problems, the calculation was correct; moreover, it was not even adequate in any of them. Estimating probabilities by relative frequencies, we have:

P(CC | ~ M) = 0/19 = 0.0
P(CC v CA | ~ M) = 0.0

From this, one may conclude that it is very unlikely to perform correct or adequate calculation without relevant motifs.

On the other hand, a successful solution may be found with a bit of luck even in the case of incorrect calculation. This happened in 8 out of 59 cases, so relative frequency is:

P(S | CI) = 0.1356

Even more, a successful solution may be found by luck in the absence of detected motifs and under incorrect calculation. This happened in 2 out of 19 cases, giving relative frequency:

P(S | ~ M & CI) = 0.1053

Regarding our question about the relative importance for success between detection of relevant motifs and calculation, correlations between some of the variables in our domain are important. These variables are: success in a player finding a correct move in a given position, detection of motifs, and correctness of calculation. We computed Pearson's sample correlation coefficients, which requires numerical input data. To this end we defined corresponding numerical variables as follows:

- Success = 1 if the problem was successfully solved; otherwise Success = 0
- MotifsDetected = 1 if all the relevant motifs were detected in the position, otherwise MotifsDetected = 0
- CalculationOK = 1 if the calculation was correct or adequate (in our notation CC or CA), otherwise CalculationOK = 0 (i.e. calculation incorrect, CI)

The correlations between pairs of these variables are:

r(Success, Motifs-Detected) = 0.4368
r(Success, CalculationOK) = 0.8870
r(MotifsDetected, CalculationOK) = 0.4643

It should be noted that the correlation between success and correctness of calculation is much higher than the other two correlations above. This confirms that the correctness of calculation is a more important predictor of success than the detection of motifs. This can be explained by the facts that motifs are highly needed for correct calculation, but they are not highly sufficient for correct calculation.

4.2 Causal Bayesian Network Model of Problem Solving in Our Domain

We can formulate a probabilistic model of our chess problem domain as a Causal Bayesian Network. We will use the three binary variables Success, MotifsDetected and CalculationOK. However, in a Bayesian network it will be more convenient to treat them as Boolean variables rather than numerical, so the numerical value 1 will be replaced by true, and 0 by false.

It is most natural to view the causal dependences between these three events according to the problem solving process: first, the player looks for relevant motifs, then she uses these motifs to drive the calculation which results in a successful or unsuccessful solution. This corresponds to the structure of the Bayesian network in Fig. 5.

Note that the link between MotifsDetected and Success cannot be ignored because Success also depends probabilistically on MotifsDetected when CalculationOK is known.

The probabilities for this network are:

P(MotifsDetected) = 0.8643
P(CalculationOK | MotifsDetected) = 0.6694
P(CalculationOK | ~ MotifsDetected) = 0.000
P(Success | CalculationOK ^ MotifsDetected) = 1.000
P(Success | CalculationOK ^ ~ MotifsDetected) = 1.000
P(Success | ~ CalculationOK ^ MotifsDetected) = 0.150
P(Success | ~ CalculationOK ^ ~ MotifsDetected) = 0.1053

It is interesting to check what happens if we omit the link between MotifsDetected and Success, which makes the structure of the model simpler, more intuitive, and better reflect De Groot's basic model of chess thinking (the stage of motif detection is followed by the calculation of variations, then the solver's solution emerges). The structure then becomes:

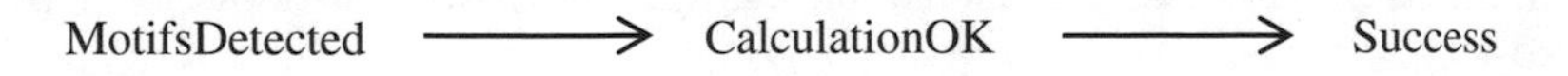

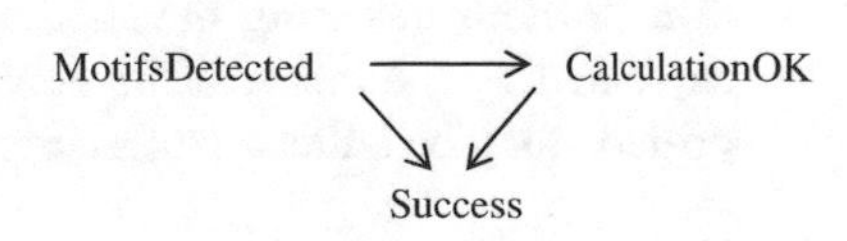

Fig. 5 The structure of a Bayesian network model of chess problem solving

Table 1 M, C, S stand for MotifsDetected, CalculationOK, and Success respectively. P_{full} and P_{simple} are the probabilities of the indicated combined events according to the full Bayes model and the simplified model, respectively

M	C	S	P_{full}	P_{simple}
t	t	t	0.5786	0.5786
t	t	f	0.0000	0.0000
t	f	t	0.0429	0.0387
t	f	f	0.2429	0.2470
f	f	t	0.0000	0.0000
f	t	f	0.0000	0.0000
f	f	t	0.0143	0.0184
f	f	f	0.1214	0.1173

The list of conditional probabilities in the Bayesian network is now also simplified into:

P(MotifsDetected) = 0.8643
P(CalculationOK | MotifsDetected) = 0.6694
P(CalculationOK | ~ MotifsDetected) = 0.0
P(Success | CalculationOK) = 1.0
P(Success | ~ CalculationOK) = 0.1356

The simpler Bayesian model is in fact a good approximation to the full Bayesian model. Table 1 gives a comparison between complete joint probability distribution computed with the full model and with the simplified one. Mean absolute difference between full model's joint probabilities and simplified model's joint probabilities is 0.0021. Mean absolute difference relative to the average probability value in this distribution (i.e. 0.125) is 1.6 %.

5 Discussion

The following observations are indicated by our experimental results:

- The results largely confirm de Groot's problem solving model. The first stage of problem solving is concerned with the detection of motifs, and the second stage is devoted to the calculation of concrete variations to verify whether motifs can actually be exploited. In all the 19 cases when relevant motifs were not detected, the calculation of variations was incorrect. It seems that it is practically impossible to calculate variations correctly without relevant motifs. When the motifs were detected, the calculation was at least adequate in about 67 % of the cases.
- In rare cases, about 10 %, a player did manage to solve the problem successfully even without detecting the motifs and with incorrect calculation. This can be explained by the fact that the number of generally reasonable moves in a chess position may be rather small, so even a random choice may succeed occasionally.

- Relevant motifs were correctly detected in 121 cases, that is 86 % of the total of 140 cases. In these 121 cases, the calculation was completely correct in 43 % of the cases, and calculation was at least adequate in 67 % of the 121 cases. This indicates the conclusion that for our players and type of chess problems, it was easier to detect the motifs than to perform the calculation.

This last point indicates that in our experiment, the dominant discriminating problem solving component between the players' success and failure was the calculation of variations and not detection of motifs (that is domain specific knowledge). Not much difference, in terms of players' problem solving performance, arises from the differences in players' pattern-based knowledge. This may be surprising because our players' chess strength (measured in Elo ratings) varied so much. The results show, possibly against the common intuition, that it is the calculation ability that decisively differentiates between players' chess problem solving success and failure.

To interpret this finding carefully, it should be noted that in our experiment, we used *tactical* chess problems. In contrast to long term strategic problems (also referred to as "positional play"), tactical problems are expected to require more calculation. However, as our results show, calculation alone without pattern knowledge (motifs) is far from sufficient.

Statistical evidence supporting this conclusion regarding pattern knowledge vs. search is also reflected in the correlations between variables. The correlation coefficient between the players' correctness of solutions and the players' success in detecting relevant motifs is 0.4368. On the other hand, the correlation between the correctness of solutions and at least adequacy of calculation is much higher: 0.8870. It should be admitted that the correctness of calculation was, for each player and position, evaluated somewhat subjectively by a chess expert as explained in Sect. 3.1. However, this does not change the overall conclusion that follows from the experimental data. Namely, that the ability to calculate variations is more discriminative among the players regarding their success than is the ability to detect relevant motifs.

The conclusion about calculation ability being more discriminative than pattern knowledge is relevant to a discussion in chess of different views regarding chess teaching and training even at top grandmaster level (Kotov 1971; Wenzhe 2001). The question is which component of chess skill is more important and deserves more attention in teaching and training: chess-specific pattern knowledge, or the ability to calculate variations. There is a common agreement that both of these components are necessary to play really well. However, one side of this discussion, including the Soviet school of chess (Kotov 1971), puts more emphasis on the deep understanding of chess, which includes chess-specific pattern knowledge. The other view is that deep understanding of chess is not sufficiently effective in practical game playing without very reliable support from calculation. Therefore some put relatively more emphasis on the ability to calculate concrete variations. The results of this paper might be interpreted as supportive of the latter approach in respect of this particular dilemma.

The differences between the two approaches are also reflected in chess training of very strong players. This is well illustrated by the following comments by the Russian grandmaster, a former world champion and successful chess coach Alexander Khalifman in the article "Le Quang Liem and the Soviet school of chess" (Chessintranslation.com 2011). Khalifman commented on the young Vietnamise player Le Quang Liem, the surprising winner of a very strong chess tournament in Moscow in 2011. Khalifman explains: "What a school is and what its presence or absence means is something that you can understand very well if you analyse with Asian chess players…. I worked a little bit with Le Quang Liem, and I will say honestly that sometimes my eyes popped out of my head. He is also a very talented boy… and he is trying very hard to grow. But at the moment all he does is calculate and calculate variations. He calculates very well, by the way. But a school is, in my opinion, what you would call a basis of positional principles, playing from general considerations…" This is how Khalifman described his view on the relative deficiency of elements of Soviet chess school in the young Asian player who was, in Khalifman's opinion, overwhelmingly relying on calculation.

6 Conclusions

Our results indicate that the dominant discriminating problem solving component is the calculation of variations and not detection of motifs. This may come as a surprise in the view that traditional chess teaching puts so much emphasis on general chess knowledge which includes detection of chess motifs.

In our experiments, no significant difference in players' problem solving performance arises from the differences in players' pattern-based knowledge (ability to detect motifs), despite large Elo rating differences between the players.

Our results also clearly confirm one aspect of classical De Groot's model of chess thinking. Namely, that satisfactory calculation is not possible without detection of motifs.

It should be noted that these findings are limited to the type of problems used in our experiments, that is tactical chess problems. One question for future work is whether the relative importance of calculation vs. chess pattern-knowledge also extends to non-tactical positions, particularly to long-term positional play. A first intuition on this might be that in such chess positions calculation of variations is relatively less important than in tactical positions. On the other hand, even in sharp positional play it is important to calculate how positional motifs can be realised by concrete sequences of moves.

Another topic of future work is to develop a program for automatically detecting players' motifs directly from the players eye tracking data. This task seems to be rather demanding, but it could help to reduce some uncertainty in interpreting players' retrospections regarding the detection of motifs.

References

De Groot, A. D. (1946). *Het denken van den schaker*. Mij: Noord-Hollandsche Uitg.

De Groot, A. D. (1978). *Thought and choice in chess,* Vol. 4. Berlin: Walter de Gruyter.

Elo, A. E. (1978). *The rating of chess players, past and present*. New York: Arco Pub.

Glickman, M. E. (1999). Parameter estimation in large dynamic paired comparison experiments. *Journal of the Royal Statistical Society: Series C (Applied Statistics)*, *48*(3), 377–394.

Hristova, D., Guid, M., & Bratko, I. (2014a). Toward modeling task difficulty: the case of chess. In *COGNITIVE 2014, the sixth international conference on advanced cognitive technologies and applications* (pp. 211–214). IARIA.

Hristova, D., Guid, M., & Bratko, I. (2014b). Assessing the difficulty of chess tactical problems. *International Journal on Advances in Intelligent Systems, 7*(3 & 4), 728–738.

Kotov, A. (1971). *Think like a grandmaster (english translation)*. London: B.T Batsford.

Newell, A. (1990). *Unified theories of cognition: The William James Lectures 1987*. Cambridge: Harvard University Press.

Newell, A., & Simon, H. A. (1972). *Human problem solving*. Upper Saddle River: Prentice-Hall.

Reingold, E., & Charness, N. (2005). Perception in chess: Evidence from eye movements. In G. Underwood (Ed.), *Cognitive processes in eye guidance*. Oxford: Oxford university press.

Tikhomirov, O. K., & Poznyanskaya, E. D. (1966). An investigation of visual search as a means of analyzing heuristics. *Soviet Psychology*, *5*, 2–15. (Translated from *Voprosy Psikhologii*, *2*, 39–53.).

Wenzhe, L. (2002). *Chinese school of chess: The unique approach, training methods and secrets (english translation)*. London: B.T. Batsford.

Online Resources

Chessintranslation.com, 02.03.2011. http://www.chessintranslation.com/2011/03/le-quang-liem-and-the-soviet-school-of-chess/. Last access January 1, 2016.

Chess Tempo web site. www.chesstempo.com. Last access January 6, 2016.

Explorative Experiments in Autonomous Robotics

Francesco Amigoni and Viola Schiaffonati

Abstract The debate on the experimental method, its role, its limits, and its possible applications has recently gained attention in autonomous robotics. If, from the one hand, classical experimental principles, such as repeatability and reproducibility, play as an inspiration for the development of good experimental practices in this research area, from the other hand, some recent analyses have evidenced that rigorous experimental approaches are not yet full part of the research habits in this community. In this paper, in order to give reason of a part of the current experimental practice in autonomous robotics that cannot be satisfactorily accommodated under the traditional concept of controlled experiment, we will advance the notion of *explorative experiment*. Explorative experiments in this context should be intended as a form of investigation carried out in the absence of a proper theory or theoretical background, where the control of the experimental factors cannot be fully managed from the beginning. We show that this notion arises from (and is supported by) the analysis of the experimental activities reported in a significant sample of papers that have been given awards at two of the largest and most impacting robotics research conferences.

1 Introduction

The discussion on experiments and the effort in developing good experimental methodologies have gained attention in autonomous robotics in the very last years. This field is oriented to develop robot systems[1] that are autonomous in the sense that they have the ability to operate without continuous human intervention, in

[1]Generally speaking, and for the purpose of our presentation, a robot system is an artifact that interacts with the external environment through its sensors and actuators and that is controlled by software programs.

F. Amigoni (✉) · V. Schiaffonati
Artificial Intelligence and Robotics Laboratory, Dipartimento di Elettronica, Informazione e Bioingegneria, Politecnico di Milano Piazza Leonardo da Vinci, 32, 20133 Milan, MI, Italy
e-mail: francesco.amigoni@polimi.it

L. Magnani and C. Casadio (eds.), *Model-Based Reasoning in Science and Technology*, Studies in Applied Philosophy, Epistemology and Rational Ethics 27, DOI 10.1007/978-3-319-38983-7_33

order to work in places hardly accessible by humans or in cooperation with humans in common environments. In autonomous robotics, human operators evolve from being active controllers of the robot systems to being more passive supervisors of the same robot systems.

The increasing attention to experimental issues in this field can be attributed to many factors. For sure, disciplinary and scientific ones play an important role, as autonomous robotics strives for reaching the same methodological standards of other scientific disciplines. Also practical accomplishments are essential in promoting standard ways to measure performance and parameters (e.g., those related to safety). Finally, commercial purposes are emerging with the aim of having standard benchmarks to evaluate products. Accordingly, a number of initiatives have been promoted, ranging from workshop series (Bonsignorio et al. 2015), to special issues of journals (Bonsignorio and del Pobil 2015), to European projects funded under different programs (Rawseeds 2015; RoCKIn 2015), to a generalized interest to experimental issues.

When analyzing the experimental trends emerging in the community, two different tendencies can be observed: on the one hand, the principles of experimental method (such as comparison, reproducibility, repeatability, justification, and generalization) play an inspirational role in the direction of defining a more rigorous approach to experiments; on the other hand, these rigorous approaches are not yet full part of the current research practice in robotics. For instance, from the systematic analysis presented in (Amigoni et al. 2014), it emerges that only few of the experiments conducted in a significant sample of autonomous robotics articles come close to controlled experiments in the sense employed by (Tedre 2015) for computing in general.

To better investigate the nature and the role of experiments in autonomous robotics within this heterogeneous and dynamic context, we believe that the current debate needs to be widened. It has to take into account not only the traditional tools of the philosophy of science, in the form of the philosophy of experimentation, but also other disciplines, both already existing (such as the philosophy of technology) and under development (such as the interdisciplinary field labelled philosophy and engineering). In this paper, we propose to stretch the traditional idea of experiment, with the aim of introducing the (still preliminary) notion of *explorative experiment* to give reason of a part of the current practice in autonomous robotics. To a first approximation, explorative experiments are forms of empirical investigation on the functioning of technical artifacts and on their interaction with the environment, in absence of a proper theory or theoretical background and without the typical constraints of controlled experiments. Our main original contribution is thus a step toward an enlarged framework that can satisfactorily account for all the different forms of experimentation in autonomous robotics.

With the aim of making the discussion more concrete, and without any attempt of being exhaustive, we analyze the papers that in the last years have been given awards at the IEEE International Conference on Robotics and Automation (ICRA) and at the IEEE/RSJ International Conference on Intelligent Robots and Systems (IROS), which are two large and impacting robotics research conferences. Our goal is to show how the notion of explorative experiment emerges from the current practice and how it can give reasons of some current experimental activities in autonomous robotics, in addition and beyond the traditional notion of experimentation.

In the following, we widen the framework of experimentation by considering the many faces of experiments already proposed for computing[2] and the notion of directly action-guiding experiment, and by analyzing the crisis of the traditional experimental paradigm, as it has been conceptualized for experiments with new technologies (Sect. 2). We, then, survey how experiments are conducted in the current practice of autonomous robotics and how they can fit within the already existing categories of experiments in computing (Sect. 3). Finally, we advance a definition of explorative experiments capable to take into account a significant part of the current experimental practice in autonomous robotics (Sect. 4).

2 Widening the Experimental Framework

In this section, we introduce the concepts that enable to enlarge the framework for reflecting on experimentation in autonomous robotics.

The many faces of experimentation. The term ‘experiment’ is used in the field of computing in a variety of ways. As it has been reconstructed in detail by (Tedre 2015), at least five different views of experiments can be recognized in the practice of the field. There are the so called *feasibility experiments* aimed at empirically demonstrating (‘demonstration’ and ‘experiment’ are terms commonly used as synonymous in computing) the proper development and working of a technology. There are *trial experiments*, evaluating some aspects of a system using predetermined variables in a laboratory, and *field experim*ents, aimed at evaluating these aspects of a system outside the laboratory, in the real world. There are also *comparison experiments* devoted to compare different solutions to look for the best one for a specific problem. And, finally, there are *controlled experiments*, those more similar to the traditional notion of experimentation and aimed at achieving generalization and prediction. What is important in this account is not how the notion of experiment should be used, but how it is actually used: “Many would object against calling, for instance, feasibility demonstrations ‘experiments,’ arguing that the term ‘experiment’ has a special meaning in science. They are right. But if one looks at how authors in computing have used the term—not how it should be used—those five uses are easily found” (Tedre 2015, 190). The differences introduced by these categories are surely of great importance in our discussion and in the next sections we will argue for their extension to autonomous robotics at the light of the experimental activities reported in the papers we surveyed.

[2]In this work we use the terms “computing”, “computer science”, and “computer science and engineering” in an interchangeable way to name the academic discipline. While recognizing the relevant difference between the theoretical and practical ends of the computing spectrum, introducing a taxonomy is beyond our scope here.

Directly action-guiding experiments. Besides the traditional notions of experiment, such as the ones just presented, we introduce here that of directly action-guiding experiments, as technological forms of experimentation already present in pre-scientific times. In particular, the difference between epistemic experiments and directly action-guiding experiments, as recently conceptualized in Hansson (2015), can help emphasizing not only that explorative experiments we discuss in this paper are performed on technical artifacts (and not on natural phenomena), but also that they have different purposes than the epistemic ones. An experiment is *epistemic* when it aims at providing information about the workings of the natural world, whereas an experiment is *directly action-guiding* when it satisfies two criteria: (a) the outcome looked for consists in the attainment of some desired goal of human action and (b) the interventions studied are potential candidates for being performed in a non-experimental setting in order to achieve that goal. A clinical trial of an analgesic is one of the examples provided by Hansson to illustrate a directly action-guiding experiment, where the outcome looked for is the efficient pain reduction and the experimental intervention is the treatment that might be administered. A systematic test on an autonomous robot employed to assist an elderly person in her home is also an example of a directly action-guiding experiment: the outcome looked for is the proper interaction of the robot with the person and the experimental intervention consists in the careful tuning of the abilities that the robot must possess to positively achieve this goal.

Exploratory experiments. Directly action-guiding experiments contribute also to introduce an explorative element that characterizes experimentation in autonomous robotics, as we will see in the next sections. The concept of experiment as exploration is not new. For example, in some recent philosophical research, *exploratory experimentation* labels those forms of experimentation in science which are not always guided by theories. One of the first authors to recognize the epistemic importance of exploratory experiments (Steinle 1997) defines exploratory experimentation as driven by the desire to obtain empirical regularities when no well-formed theories or no conceptual frameworks are available. What is important in this characterization (that in this case is based on a detailed reconstruction of the early research in electromagnetism) is that the experimental activity may be highly systematic and driven by the typical experimental guidelines, despite its independence from specific theories. The same term is used with a slightly different meaning in another article appeared in the same year but in the context of some early research in protein synthesis (Burian 1997), where exploratory experimentation is seen as a style of inquiry not guided by theory. These and other similar works are mainly directed to contrast the theory-driven approaches of most of the philosophy of science in the spirit of experimentation as having a life on its own (Hacking 1983). Even if they recognize that exploratory experimentation is typically not completely free of theory, they aim at showing that the epistemic significance of those inquiries are not primarily theory-driven by presenting several detailed case studies. The idea that "the aim of exploratory experiments is to generate significant findings about phenomena without appealing to a theory about these phenomena for the purpose of focusing experimental attention on a limited range of possible findings" (Waters 2007, 5) is probably that serving better as an inspiration for more recent works devoted to provide evidence of the exploratory shift observed in the methodology of some areas of biology (Franklin 2005).

Experimental control. Controlling the experimental factors that are investigated constitutes one of the key issues of the experimental method. To deploy an experimental system, knowledge and control of the interactions between the system and its environment need to be managed. Controlled experiments are usually performed having in mind quite precise expectations of the possible outcomes. The research questions are clearly stated and the hypotheses to be investigated are made explicit. Then, on a general account, experiments are designed and performed varying the different experimental parameters in order to determine which of the different experimental conditions are indispensable and, then, looking for stable empirical rules. For producing stable and repeatable experiments, experimenters vary a number of factors in their experimental systems to examine whether they are relevant or not. The fact that experiments are performed in laboratories responds exactly to this attempt of control.

The crisis of the traditional notion of control. Traditionally the control paradigm for experimentation, as it has been devised in the history of science, relies on two assumptions (Kroes 2015): the experimenter is not part of the system on which the experiment is performed and (s)he is in control of the independent variables and of the experimental set-up. Accordingly, the experimenter is able to intervene both by changing these variables to evaluate their influence on the dependent ones and by varying the experimental set-up. This traditional control paradigm becomes problematic, and a consequent shift in the notions of intervention and control is observed, when considering new technologies as socio-technical systems, namely as hybrid systems composed of natural objects, technical artifacts, human actors, and social entities. The idea of controlling the experimental system from a center of command and control that is outside the system becomes highly problematic (Kroes 2015). Reasons are that the distinction between the experimental system and its environment is critical, but also that the environment is complex, where complexity arises from the co-presence of technical artifacts and natural and social elements.

It is interesting to note that the same crisis in the traditional notion of control can be observed also in a part of the current experimental practice in autonomous robotics. Although the kind of technology we are discussing here does not possess in a full and complete way the features of large-scale socio-technical systems, such as the world civil aviation system (Vermaas et al. 2011), it nevertheless shares some of their characteristics. We could say that the experimental system in the case, for instance, of experiments with autonomous robots is hybrid, in the sense that not just technical components play an essential role for the functioning of the system, and thus have to be evaluated, but also natural objects, human actors, and social entities need to be taken into account (e.g., for their interaction with the robot systems). Moreover, if in the natural sciences it is prescribed that the experimenter should be an outsider of the phenomenon to be investigated, it is not clear how a person developing autonomous robots, namely computation-based artifacts, could be an outsider with respect to a phenomenon (i.e., an artifact) that (s)he has created (Tedre 2011). Except for some significant examples, in autonomous robotics, tests on the artifacts are usually performed by the same people that created them, losing the sort of independence of the experimenter that is prescribed in the classical experimental protocol.

One could ask what is the reason for a robotician to test the artifacts that (s)he has developed and, thus, should know in detail. To answer this question, it is important to recognize at least two sources of unpredictability, arising in the artifact, due to its complex nature and to its interaction with the physical environment (including humans) surrounding it, respectively. This is particularly evident in the case of autonomous robotics, where the goal is that of having robots that do not require continuous human supervision. Autonomous robots are very complex entities composed of interacting modules ranging from sensors, to actuators, to software programs, whose overall behavior is hardly predictable, even by their own designers, especially when considering their interaction with the external physical (and social) world. Not only tests that a given robot is working properly (and possibly better than others) have to be performed without the required independence of the experimenter, but also autonomous robots have to be tested for their proper interaction with environments (including in most of the cases other human beings) that is hardly predictable.

In summary, autonomous robotics, as several other new technologies, can benefit from a wider framework in which its experimental activities can be discussed, as we further argue in the following of this paper.

3 A Survey of Different Experiments in Autonomous Robotics

In this section, we present some considerations emerging from the survey we have conducted on the papers that have been awarded the Cognitive Robotics Best Paper Award and the CoTeSys (Cognition for Technical Systems) Cognitive Robotics Best Paper Award at ICRA and at IROS, respectively, from 2010 to 2015. In total, we consider 11 papers that we deem represent a significant sample of current research on autonomous robots, as they are witnessing the awarded research in two of the main conferences of the field (see Table 1).

As discussed in the previous section, in Tedre (2015) some classes of experiments are identified from the analysis of current practice in computing. According to our sample of representative papers, examples of experiments that fall in these classes are also largely present in autonomous robotics, although some new characterizations of experiments as explorations also emerge.

Table 1 Papers (references and titles) analyzed in our survey

Hoffman and Weinberg (2010)	Gesture-based human-robot jazz improvisation
Grollman and Billard (2011)	Donut as I do: learning from failed demonstrations
Bergstrom et al. (2011)	Generating object hypotheses in natural scenes through human-robot interaction
Thobbi et al. (2011)	Using human motion estimation for human-robot cooperative manipulation
Tenorth et al. (2012)	The RoboEarth language: representing and exchanging knowledge about actions, objects, and environments
Daniel et al. (2012)	Learning concurrent motor skills in versatile solution spaces
Chu et al. (2013)	Using robotic exploratory procedures to learn the meaning of haptic adjectives
Fasola and Mataric (2013)	Using semantic fields to model dynamic spatial relations in a robot architecture for natural language instruction of service robots
Deisenroth et al. (2014)	Multi-task policy search for robotics
Gemici and Saxena (2014)	Learning haptic representation for manipulating deformable food objects
Boularias et al. (2015)	Grounding spatial relations for outdoor robot navigation

Table 2 Some excerpts relative to feasibility experiments, according to the taxonomy proposed by (Tedre 2015)

Tenorth et al. (2012)	"The experiment shows that the system is able to encode the information required for mobile pick-and-place tasks" (p. 1289)
Fasola and Mataric (2013)	"These examples illustrate the ability of the system to parse natural language input, ground noun phrases, infer command semantics, plan, and execute an appropriate solution while obeying natural language directive constraints" (p. 147)

Feasibility experiments. These experiments are basically a form of empirical demonstration, intended as an existence of proof of the ability to build a robot system to perform some task. The outcome of a feasibility experiment is typically binary: positive, if the robot is able to accomplish what it is intended to do; negative, otherwise. Examples of this kind of experiments in the papers we analyzed are reported in Table 2.

Trial and field experiments. These experiments take a step further and evaluate various aspects of robot systems using some predefined variables which are measured in laboratories or in real contexts of use (with some limitations), in the case of trial experiments, or outside the laboratory in complex socio-technical contexts of use, in the case of field experiments. In these experiments, some quantities, like velocity and acceleration of parts of the robots, accuracy and time required for performing a task, or error with respect to a reference (ground truth), are measured to evaluate robot systems. Sometimes, measuring quantities amounts to resort to

Table 3 Some excerpts relative to trial and field experiments, according to the taxonomy proposed by (Tedre 2015)

Grollman and Billard (2011)	"To evaluate our techniques we are concerned not only with whether the task is eventually performed successfully (which it is), but also with the breadth of possibilities that are generated" (p. 3807)
Thobbi et al. (2011)	"Ten trials were performed to test how quickly the algorithm could converge to an optimal policy" (p. 2876) "True velocity and acceleration are derived from the observed position, and are shown in the figure for comparison with the predicted values" (pp. 2876–2877)
Fasola and Mataric (2013)	"To evaluate the ability of our robot system to follow natural language directives, we first analyzed the effectiveness of the semantic interpretation module to infer the correct command specifications given the natural language input" (p. 147)
Boularias et al. (2015)	"Participants were separately asked to point to the goal they would choose for executing each command. The best answer, chosen by a majority vote, is compared to the robot's answer" (p. 1981)

human judgement about the observed behavior of the robots. Some examples extracted from the papers we analyzed are reported in Table 3.

Comparison experiments. These experiments refer to comparing different solutions in some set-ups and are based on some precisely-defined measures and criteria to assess the performance. The compared entities could be different versions of the robot system under testing (for instance, the same robot with or without a specific component) or alternative systems to perform the same task (for instance, systems proposed by other researchers). Table 4 shows some examples of this kind of experiments that are reported in the papers we analyzed.

Table 4 Some excerpts relative to comparison experiments, according to the taxonomy proposed by (Tedre 2015)

Grollman and Billard (2011)	"Because there are more possibilities to explore, in our experiments the donut method took more interactions to succeed than the balanced mean" (p. 3808)
Bergstrom et al. (2011)	"Again, we conclude that point initialization outperforms cluster initialization" (p. 832) "In addition we evaluate how the method in [10] compares to our method" (p. 833)
Thobbi et al. (2011)	"Figure 8 shows the trajectories of both ends of the table for cases where the proposed system was used (case I: with predictions) and the case where only the reactive controller was used (case II: without predictions)" (p. 2877)
Daniel et al. (2012)	"We also compare our approach to the standard unimodal REPS algorithm" (p. 3595)
Gemici and Saxena (2014)	"We compare the performance of our reward based manipulation approach against the baseline algorithms" (p. 644)

Controlled experiments. These experiments are the golden standard of experimentation in the natural sciences and refer to the original idea of experiment as controlled experience, where the activity of rigorously controlling (by adopting experimental principles such as reproducibility or repeatability) the factors that are under investigation is central, while eliminating the confounding factors and allowing for generalization and prediction. In the current experimental practice of autonomous robotics (discussed at the beginning of this paper and in Amigoni et al. 2014), it is hard to find experimental activities that completely fit within this category.

Along with the above categories of experiments, the analyzed papers report evidence of other forms of empirical investigation on the functioning of artifacts and on their interaction with the environment, which are in the direction of explorative experiments and can be roughly organized in the following way according to their purposes.[3] (Note that the categories below are separated for presentation clarity, but their boundaries are rather fuzzy.)

Investigating the role of parameters. Complex software programs controlling robot systems often involve several parameters whose values influence their behavior. For example, the software programs controlling robots could make decisions according to thresholds, or sensor data could be filtered according to factors depending on environment conditions. Often, the designer has only a rough *a priori* idea of the relationship between values of parameters and behavior of robot systems, and experiments are used to elucidate and make more precise this relationship. In a sense, the design of a robot system R requires tests in an experimental setting S in order to be refined with the proper values of parameters P that are good for S. To this end, some experiments reported in the surveyed papers are set up to elucidate the qualitative and quantitative effects of different parameters values on some measurable quantities relative to the behavior and the performance of the robots (Table 5).

Confirmation of expectations or hypotheses. When developing robot systems, the designers consider (and build upon) a set of expectations and hypotheses about the behavior of the artifacts when inserted in their operating environments. Usually, in autonomous robotics, due to the difficulty of building reliable models of the interaction between robots and the portion of the physical world in which they are inserted (Amigoni and Schiaffonati 2010, 2014), these expectations and hypotheses are not based on a solid theoretical ground and can be confirmed only empirically. Schematically, the designer expects a robot system R to show behavior B when it is inserted in experimental setting S, and would like to confirm such expectation. This class of experiments (Table 6 reports a sample taken from surveyed papers) provides a very simple feedback to the design phase in the context of a continuous

[3]Although it is out of the scope of the present paper to investigate the exact positioning of explorative experiments, we believe they represent an orthogonal dimension with respect to the five categories of experiments introduced by (Tedre 2015) and discussed before.

Table 5 Some excerpts relative to experiments devoted to investigate the role of parameters in robot systems

Hoffman and Weinberg (2010)	"We have empirically sampled sound intensity profiles for different solenoid activation lengths, and used those to build a model for each striker" (p. 583)
Thobbi et al. (2011)	"Figure 7 shows the role of the forgetting factor ϕ in determining the confidence" (p. 2877)
Daniel et al. (2012)	"We evaluate our approach with different bounding parameters k for the responsibilities. [...] In a second experiment, we evaluate the influence of importance sampling" (p. 3595)
Gemici and Saxena (2014)	"In this work, we manually tuned the reward functions for our manipulation task for a reasonable level of exploration and exploitation" (p. 644)

Table 6 Some excerpts relative to experiments devoted to confirm expectations or hypotheses

Grollman and Billard (2011)	"Further, as expected, exploration with both techniques increased in the middle portions of both tasks" (p. 3808)
Chu et al. (2013)	"We first analyzed the feature vectors to confirm that they capture meaningful differences in the feel of the objects" (p. 3053) "This relatively low score supports our belief that multiple motions should be combined to increase the recognition of haptic object properties" (p. 3054) "which supports the hypothesis that our methods can produce a meaningful set of adjectives for completely new objects when using all EPs" (p. 3054)

iteration between design and experiments.[4] For example, they are used to provide *a posteriori* justifications on some assumptions, in the sense that the design of the robot system R is based on some hypotheses that are considered valid if R shows the expected behavior B.

Getting insights on the behavior of the robot systems. In the most interesting cases, explorative experiments are used to get intuitions on how robot systems work and on how they perform tasks. Usually, these experiments provide quantitative results that the designers use to inspect, and possibly modify, the design of the internal methods of robot systems (see Table 7 for some examples taken from the surveyed papers). In this case, the qualitative or quantitative influence (or effect, or role) of module M of robot system R on behavior B (observed when R is put in an experimental setting S) are investigated. Note that experiments used with this explorative intention provide a richer knowledge than those used to investigate the role of parameters, ranging from measuring values of internal variables to generating ideas for alternative design solutions.

[4]Note that, in most of the papers we analyzed in our survey, this iteration process is only hinted and only final successful tests are described in detail.

Table 7 Some excerpts relative to experiments devoted to get insights on the behavior of robot systems

Grollman and Billard (2011)	"We believe the decreased agreement at the end of the movement comes from accumulated drift during trajectory generation" (p. 3808) "We believe this behavior (and some of the visual jagginess) arises from our use of gradient ascent in the velocity generation and our initialization" (p. 3808)
Bergstrom et al. (2011)	"The slightly lower performance of the latter indicates that it might be better to let the segments evolve on their own, rather than giving a large bias from the start and having the risk of getting stuck in local minima" (p. 832)
Thobbi et al. (2011)	"Figure 5 shows the variation confidence (C) through the task" (p. 2877) "the trajectory is much smoother when the human is placing the table down as compared to moving upwards [...] It can also be speculated that sophisticated velocity or torque controlled robots would yield smoother motions and offer better improvements in performance using the proposed technique" (p. 2878)
Fasola and Mataric (2013)	"To illustrate the usefulness of the semantic field model towards representing static and dynamic spatial relation primitives for use in path generation and classification, Fig. 6 shows the progression of the *at*, *along*, *away from*, and *in* semantic field values along the execution paths generated for test runs #1–4, respectively" (p. 148)
Gemici and Saxena (2014)	"This means that for most of the objects, one or two information gathering actions was enough to determine the best task oriented action to reach the subgoal" (p. 644)
Boularias et al. (2015)	"We notice that complex commands help finding the right goals because they are less ambiguous than simple commands" (p. 1981)

Assessing the generality of robot systems. The most sophisticate way of employing experiments with an explorative flavor is to gain knowledge about the behavior of robot systems in settings that are different of those considered in their design, in order to evaluate the generality of these systems. In this case, a robot system R that has been designed and developed to perform in settings S is experimentally tested in settings S' (different from S). Table 8 reports some examples taken from the analyzed set of papers. For instance, settings S' could involve noisy data (as opposite to error-free data assumed during development and preliminary experiments) or data that the robot R has never seen before. This last aspect is particularly relevant in the case of learning systems, like those proposed in Deisenroth et al. (2014) and Gemici and Saxena (2014). In other cases, S' could involve special situations (e.g., unexpected behaviors of the humans that are interacting with R) that are excluded from S.

Table 8 Some excerpts relative to experiments devoted to assess generality of robot systems

Thobbi et al. (2011)	"Figure 6 shows a non-typical case where the human chooses to take a pause during the task" (p. 2877)
Fasola and Mataric (2013)	"To demonstrate the generalizability of our approach and its usefulness in practice with real robots in real environments, next we present evaluation results of our robot software architecture using maps of real environments that were generated by physical robots implementing SLAM with onboard laser sensors" (p. 148)
Deisenroth et al. (2014)	"We show that our MTPS approach allows to generalize from demonstrated behavior to behaviors that have not been observed before" (p. 3880)
Gemici and Saxena (2014)	"In order to test the generalization of our algorithm to new object categories, we also included a new category (tofu) not seen during training" (p. 645)

4 Discussion

The examples of the previous section show that if, on the one side, the attempt of autonomous robotics to conform its experimental methodology to that of controlled experiments is not yet fully (and perhaps cannot be) carried out, on the other side the current practice is characterized at various levels by a form of experimentation that seems to deal with exploration. A purely controlled form of experimentation is hardly possible due to the lack of some of the features that in the traditional protocol allow to control the experimental factors. In particular, in the case of autonomous robotics, modeling and predicting the behavior of the robot systems in their interaction with complex environments is not only far beyond the current and near-future technical knowledge, but it is also rather out of experimenter's control due to some intrinsic reasons: the experimenter is part of the system and (s)he is not in full control of the experimental set-ups (Kroes 2015). When we turn, instead, to the idea of exploration, experiments are seen as ways to explore possibilities, to investigate opportunities, and to give back information that is iteratively used to improve the artifacts both in their architecture and in their interaction with complex environments. What is explored is only partially known in advance, and surely not at the level of being expressed in the form of clear hypotheses derived from a strong theory to be tested later in (controlled) experimental campaigns. In a sense, explorative experiments are used to increase the confidence of designers on the behavior of their robot systems in physical environments.

In the context of the widened framework suggested in this paper, we attempt now a still primitive but—in our opinion—promising definition of explorative experiments that is shaped on the analysis of experiments in autonomous robotics, but that could hopefully be extended to other forms of experimentation in computer engineering and, especially, in artificial intelligence. By *explorative experiments* in autonomous robotics we mean experiments that are driven by the desire of investigating the realm of possibilities pertaining to the functioning of a robot system and to its interaction with the environment in absence of a proper theory or theoretical background. More precisely, explorative experiments are a special kind of directly action-guiding experiments which possess the following features:

- They are devoted to testing technical artifacts, meant as artificial entities purportedly built by humans to fulfill a purpose and, therefore, having a technical function.
- Their focus is to iteratively refine the intervention, meant as the union of knowledge and action characterizing experimental practice, and their ultimate purpose is not to test a general theory, but to probe the possibility and limits of the intervention.
- They do not force a sharp distinction between designers and experimenters and, instead, the practitioners often become experimenters.
- The control of the experimental factors cannot be fully managed from the beginning, but is in part carried out after the artifact has been inserted into its environment.

The reason why we use the term ‘explorative’ instead of ‘exploratory’ is to mark our difference from the philosophical work focused on accounting the distinction between exploratory and theory-driven experiments and based on the ways in which experiments depend on theory. In our attempt to characterize explorative experiments we are interested, instead, in the appeal to complexity that has been stressed in the philosophical literature (Burian 1997), where some systems are considered too complicated to be investigated by means of a theory-driven approach. This appeal to complexity certainly applies to biology, but we believe that there are good reasons to extend it to computer engineering as well, in particular when the subjects of the experimentation are not just the artifacts *per se*, but rather the ways in which these artifacts are able to interact with the surrounding physical and social environment. The reference to complexity helps in defining one important aspect we wish to stress in our characterization of explorative experiments: the fact that there is not sufficient information (in most of the cases for the lack of a proper theoretical background and/or previous experience) to provide precise expectations of what investigators will find. Thus, explorative experimentation is a way to find patterns of activities from which scientists could generate novel hypotheses to improve artifacts and gain confidence in their behavior. In this sense, explorative experiments are forms of empirical investigation of novel and interesting ideas or techniques, without the rigorous constraints of typical experimental methodologies. The role of explorative experiments appears thus particularly important in autonomous robotics, because such robot systems are developed to operate in environments that are largely unpredictable and difficult to capture in models, with the consequence that the designers can hardly anticipate the possible outcomes.

5 Conclusions

In this work, we have substantiated the need of reconsidering the traditional notion of experiments within the field of autonomous robotics. The partial, but significant, survey we have presented shows that forms of experimentation as explorations are already performed in the practice of the field. To account for these activities, we have proposed the idea of explorative experiments, as forms of directly action-guiding experiments inspired by the different elements discussed at the beginning of the paper, in order to widen the current experimental framework.

We plan to further refine the definition of explorative experiments, in particular in the direction of considering different forms of control, with respect to those adopted in the classic experimental paradigm, that take place *a posteriori*, after an artifact has been inserted into its environment. Moreover, the feedback that explorative experiments can provide on design of autonomous robots will be investigated in more detail. Finally, the questions relative to the limited repeatability and reproducibility of explorative experiments, which could lead to over-optimistic interpretations of results, will be addressed.

References

Amigoni, F., & Schiaffonati, V. (2010). Good experimental methodologies and simulation in autonomous mobile robotics. In L. Magnani, W. Carnielli & C. Pizzi (Eds.), *Model-based reasoning in science and technology* (pp. 315–322). Berlin: Springer.

Amigoni, F., & Schiaffonati, V. (2014). Autonomous mobile robot as technical artifacts: A discussion of experimental issues. In L. Magnani (Ed.), *Model-based reasoning in science and technology* (pp. 527–542). Berlin: Springer.

Amigoni, F., Schiaffonati, V., & Verdicchio, M. (2014). Good experimental methodologies for autonomous robotics: From theory to practice. In F. Amigoni & V. Schiaffonati (Eds.), *Methods and experimental techniques in computer engineering* (pp. 37–53). Berlin: SpringerBriefs in Applied Sciences and Technology, Springer.

Bergstrom, N., Bjorkman, M., & Kragic, D. (2011). Generating object hypotheses in natural scenes through human-robot interaction. In *Proceedings of IROS* (pp. 827–833).

Bonsignorio, F., & del Pobil, A. (2015). Special issue on replicable and measurable robotics research. *IEEE Robotics and Automation Magazine, 22*(3), 32–154.

Bonsignorio, F., Hallam, J., & del Pobil, A. (2015). Special interest group on good experimental methodologies. http://www.heronrobots.com/EuronGEMSig/gem-sig-events. Last visited November 2015.

Boularias, A., Duvallet, F., Oh, J., & Stentz, A. (2015). Grounding spatial relations for outdoor robot navigation. In *Proceedings of ICRA, 2015* (pp. 1976–1982).

Burian, R. M. (1997). Exploratory experimentation and the role of histochemical techniques in the work of Jean Brachet, 1938–1952. *History and Philosophy of the Life Sciences, 19*, 27–45.

Chu, V., McMahon, I., Riano, L., McDonald, C., He, Q., Martinez Perez-Tejada, J., et al. (2013). Using robotic exploratory procedures to learn the meaning of haptic adjectives. In *Proceedings of ICRA, 2013* (pp. 3048–3055).

Daniel, C., Neumann, G., & Peters, J. (2012). Learning concurrent motor skills in versatile solution spaces. In *Proceedings of IROS, 2012* (pp. 3591–3597).

Deisenroth, M., Englert, P., Peters, J., & Fox, D. (2014). Multi-task policy search for robotics. In *Proceedings of ICRA* (pp. 3876–3881).
Fasola, J., & Mataric, M. (2013). Using semantic fields to model dynamic spatial relations in a robot architecture for natural language instruction of service robots. In *Proceedings of IROS, 2013* (pp. 143–150).
Franklin, L. (2005). Exploratory experiments. *Philosophy of Science, 72*, 888–899.
Gemici, M., & Saxena, A. (2014). Learning haptic representation for manipulating deformable food objects. In *Proceedings of IROS, 2014* (pp. 638–645).
Grollman, D., & Billard, A. (2011). Donut as I do: Learning from failed demonstrations. In *Proceedings of ICRA, 2011* (pp. 3804–3809).
Hacking, I. (1983). *Representing and intervening*. New York: Cambridge University Press.
Hansson, S. O. (2015). Experiments before science?—What science learned from technological experiments. In S. O. Hansson (Ed.), *The role of technology in science: Philosophical perspectives* (pp. 81–110). Dordrecht: Springer.
Hoffman, G., & Weinberg, G. (2010). Gesture-based human-robot jazz improvisation. In *Proceedings of ICRA, 2010* (pp. 582–587).
Kroes, P. (2015). Experiments on socio-technical systems: The problem of control. In *Science and engineering ethics special issue on experiments, ethics, and new technologies, 2015*. doi:10.1007/s11948-015-9634-4.
Rawseeds. (2015). The rawseeds project. http://www.rawseeds.org/home/. Last accessed November 2015.
RoCKIn. (2015). Robot competitions kick innovation. In *Cognitive systems and robotics (RoCKIn)*. http://rockinrobotchallenge.eu. Last accessed November 2015.
Steinle, F. (1997). Entering new fields: Exploratory uses of experimentation. *Philosophy of Science, 64*, S65–S67.
Tedre, M. (2011). Computing as a science: A survey of computing viewpoints. *Minds and Machines, 21*, 361–387.
Tedre, M. (2015). *The science of computing*. Boca Raton: CRC Press, Taylor & Francis Group.
Tenorth, M., Perzylo, A., Lafrenz, R., & Beetz, M. (2012). The RoboEarth language: Representing and exchanging knowledge about actions, objects, and environments. In *Proceedings of ICRA, 2012* (pp. 1284–1289).
Thobbi, A., Gu, Y., & Sheng, W. (2011). Using human motion estimation for human-robot cooperative manipulation. In *Proceedings of IROS, 2011* (pp. 2873–2878).
Vermaas, P., Kroes, P., van de Poel, I., Franssen, M., & Houkes, W. (2011). *A philosophy of technology. From technical artifacts to sociotechnical systems*. San Rafael: Morgan & Claypool.
Waters, C. K. (2007). The nature and context of exploratory experimentation. *History and Philosophy of the Life Sciences, 19*, 275–284.

Fundamental Physics, Partial Models and Time's Arrow

Howard G. Callaway

Abstract This paper explores the scientific viability of the concept of causality—by questioning a central element of the distinction between "fundamental" and non-fundamental physics. It will be argued that the prevalent emphasis on fundamental physics involves formalistic and idealized partial models of physical regularities abstracting from and idealizing the causal evolution of physical systems. The accepted roles of partial models and of the special sciences in the growth of knowledge help demonstrate proper limitations of the concept of fundamental physics. We expect that a cause precedes its effect. But in some tension with this point, fundamental physical law is often held to be symmetrical and all-encompassing. Physical time, however, has not only measurable extension, as with spatial dimensions, it also has a direction—from the past through the present into the future. This preferred direction is time's arrow. In spite of this standard contrast of time with space, if all the fundamental laws of physics are symmetrical, they are indifferent to time's arrow. In consequence, excessive emphasis on the ideal of symmetrical, fundamental laws of physics generates skepticism regarding the common-sense and scientific uses of the concept of causality. The expectation has been that all physical phenomena are capable of explanation and prediction by reference to fundamental physicals laws—so that the laws and phenomena of statistical thermodynamics—and of the special sciences—must be derivative and/or secondary. The most important and oft repeated explanation of time's arrow, however, is provided by the second law of thermodynamics. This paper explores the prospects for time's arrow based on the second law. The concept of causality employed here is empirically based, though acknowledging practical scientific interests, and is linked to time's arrow and to the thesis that there can be no causal change, in any domain of inquiry, without physical interaction.

H.G. Callaway (✉)
Temple University, Philadelphia, PA, USA
e-mail: hg1callaway@gmail.com

L. Magnani and C. Casadio (eds.), *Model-Based Reasoning in Science and Technology*, Studies in Applied Philosophy, Epistemology and Rational Ethics 27, DOI 10.1007/978-3-319-38983-7_34

1 Temporal Symmetry and Fundamental Physics

It is prevalently held that the fundamental or primary laws of physics are symmetric in time and all-encompassing.[1] Part of the plausibility of this understanding of physical law depends on the relationship of physics to the special sciences. There is a plausible sense in which physics is *fundamental* in relation to the special sciences. If exceptions are found to laws or generalizations of the special sciences such as psychology, biology or chemistry, then one reasonably seeks explanation in more basic or underlying scientific laws or generalizations. Psychology may reasonably seek answers to some of its problems in physiology or neurology, biology may look to biochemistry, and chemistry may look to physics, etc.

But in physics the pursuit of underlying explanatory factors or mechanism comes to an end, in the sense that there is no more basic natural science which provides needed explanations or understanding of anomalies or exceptions in physics.[2] I see little ground to question the thesis. This is quite different, however, from the claim that the special sciences are somehow incomplete or defective until and unless they are fully explained by or reduced to details of physics.

Nobel Prize winning physicist, Steven Weinberg, emphasized a related point in his 1992 book, *Dreams of a Final Theory*:

> When we say that one truth explains another, as for instance that the physical principles (the rules of quantum mechanics) governing electrons in electric fields explain the laws of chemistry, we do not necessarily mean that we can actually deduce the truths we claim have been explained. Sometimes we can complete the deduction, as for the chemistry of the very simple hydrogen molecule. But sometimes the problem is just too complicated for us. In speaking in this way of scientific explanations, we have in mind not what scientists actually deduce but instead a necessity built into nature itself.[3]

The sense in which quantum mechanics explains laws of chemistry involves a kind of *projection* from simple cases, where the relationship is clear, to more complex cases, where the actual deduction of chemical laws or properties of complex atoms and molecules is impractical. One reason that chemistry persists in its special modes of understanding and explanation, instead of simply substituting physics, is that the required physics is not fully suited to the complexities; and, in effect, as Weinberg has it, "sometimes the problem is just too complicated for us." Still, where anomalies or exceptions to chemical laws appear, one expects that

[1]See, e.g., Davies (1977, p. 26), who puts it this way: "All known laws of physics are invariant under time reversal,"—though noting the singular exception of processes involving K-mesons and the weak force. Cf. Greene (2004, p. 145): "… not only do known laws fail to tells us why we see events unfold in only one order, they also tell us that, in theory, events can unfold in reverse order".

[2]Cf. e.g., Fodor (1974), "all events that fall under the laws of any science are physical events and hence fall under the laws of physics".

[3]Weinberg (1992, p. 9).

study of the relevant physics might be very helpful. While generally sympathetic to what Weinberg says in this passage, I would emphasize, too, the small contrast I introduced between thinking of this as a matter of reasonable scientific *projection* and Weinberg's talk of a "necessity built into nature itself." Projections from a well understood model can be more or less reasonable, given the state of scientific knowledge, but talk of necessities of nature resists the gradation.

Within physics, there is a prevalent view that various laws, particles or physical constants are more basic, fundamental, primary or foundational, usually in the sense that they play a central role in explaining and ordering a wide range of physical or scientific generalizations. This idea is reflected, for example, in detailed accounts of the fundamental particles of the Standard Model of particle physics.[4] But it is sometimes explicitly held that the fundamental particles of the Standard Model are simply those which are, as yet, not subject to further analysis or explanation. A powerful particle collider, such as the LHC at Geneva, provides energy sufficient to detect quarks within the proton and neutron, but no one has yet detected anything smaller within quarks or electrons. It is important to distinguish, then, between holding that some presently accepted generalizations are "fundamental" *in relation* to what is presently known, and saying, on the other hand, that there must be some system of fundamental physics governing all that will ever come to be known.

Temporal symmetry is a matter of the reversibility of physical events and processes in accordance with physical law. The existence of an arrow of time, it has been argued, "is puzzling," because "all basic theories in physics seem to be time symmetric or time reversal invariant."[5] Or, what is sometimes claimed to be equivalent, is that the fundamental physical laws conserve information.[6] Sean Carroll argues, along related lines, that what is crucial is "our ability to reconstruct the past from the present," and "the key concept that ensures reversibility is conservation of information;" if the information needed to specify any state of the world is always preserved, then "we will always be able to run the clock backward and recover any previous state."[7]

Applying fundamental physical laws, e.g., to atoms, molecules, or other bodies, given their present positions and momentum, one could, in principle, equally calculate their future or past positions and momentum. Given the fundamental laws, information concerning physical units of a system at any given time, carries with it implications regarding the system at any other time. As the British astrophysicist, A.S. Eddington concisely put it, speaking to the issue of reversibility, "when the

[4]See, e.g., the brief account in Penrose (2004, pp. 627–654) and Carroll 2012, *The Particle at the End of the Universe*, "Appendix Two, Standard Model Particles," pp. 293–298.

[5]Savitt (1995, p. 6).

[6]See e.g., Susskind (2008, p. 87): "There is another very subtle law of physics that may be even more fundamental than energy conservation. Its sometimes called reversibility, but let's just call it *information conservation*".

[7]Carroll (2010, p. 121).

[primary] laws are formulated mathematically," then "there is no more distinction between past and future than between right and left."[8] Though similar views are widely held, I want to suggest that this idea has the ring of excessive idealization.

2 Background for Fundamental Physics

In spite of Eddington's appeal to precise mathematical formulation, it seems best for philosophical purposes to follow Einstein and Infield on "fundamental ideas":

> Fundamental ideas play the most essential role in forming a physical theory. Books on physics are full of complicated mathematical formulae. But thought and ideas, not formulae, are the beginning of every physical theory. The ideas must later take the mathematical form of quantitative theory, to make possible the comparison with experiment.[9]

The symmetric, time invariant concept of fundamental physics is rooted, historically, in the interpretation of Newton's laws of motion and his theory of universal gravity. As Einstein put it, writing in 1940, "The first attempt to lay a uniform theoretical foundation was the work of Newton"[10]

> This Newtonian basis proved eminently fruitful and was regarded as final up to the end of the nineteenth century. It not only gave results for the movements of the heavenly bodies, down to the most minute details, but also furnished a theory of the mechanics of discrete and continuous masses, a simple explanation of the principle of the conservation of energy and a complete and brilliant theory of heat. The explanation of the facts of electrodynamics within the Newtonian system was more forced; least convincing of all, from the very beginning, was the theory of light.[11]

Newton's physics eventually gave rise to the "iron determinism" of Laplace's *Philosophical Essay on Probabilities*.[12] According to Laplace, the future is completely determined by the past; and equally, a complete knowledge of the present state of every particle or body completely determines every past configuration. The world is thus a causal "block universe," to use the familiar philosophical term from William James.[13]

Laplace, in an oft quoted passage from his "Essay on Probabilities," provides the following image of the Newtonian world, as viewed by some "immense intelligence":

---

[8]Callaway (2014, p. 76).

[9]Einstein and Infield (1938, p. 277).

[10]Cf. Einstein (1940, p. 488).

[11]Einstein (1940, p. 488).

[12]Laplace and Simon (1820/1902) the Introduction to his *Théorie Analytique des Probabilités*.

[13]See James (1909/2008, p. 47), in the 2008 edition: "Every single event is ultimately related to every other, and determined by the whole to which it belongs." In James' general conception of the block universe, the determination need not be causal.

> We ought to regard the present state of the universe as the effect of its antecedent state and as the cause of the state that is to follow. An intelligence knowing all the forces acting in nature at a given instant, as well as the monetary positions of all things in the universe, would be able to comprehend in one single formula the motions of the largest bodies as well as the lightest atoms in the world, provided that its intellect were sufficiently powerful to subject all data to analysis; to it nothing would be uncertain, the future as the past would be present to its eyes. The perfection that the human mind has been able to give to astronomy affords but a feeble outline of such an intelligence.[14]

For this imagined and vastly powerful intelligence, which came to be called "Laplace's demon," nothing would be uncertain, "the future as the past would be present to its eyes." This is a powerful philosophical image of a world of mechanical, deterministic causation, as nearly comprehensive as the medieval philosophical image of *atemporal* Divine omniscience.[15]

Writing in the late 1920s near the end of the period which saw the intensive development of the twentieth century's two great revolutions in physics, relativity and quantum mechanics, Eddington posed the question and expressed some doubts regarding the comprehensive character of "primary law."

> I have called the laws controlling the behavior of single individuals "primary laws," implying that the second law of thermodynamics, although a recognized law of Nature, is in some sense a secondary law. This distinction can now be placed on a regular footing. Some things never happen in the physical world because they are *impossible*; others because they are *too improbable*. The laws which forbid the first are the primary laws; the laws which forbid the second are the secondary laws. It has been the conviction of nearly all physicists[16] that at the root of everything there is a complete scheme of primary law governing the career of every particle or constituent of the world with an iron determinism. This primary scheme is all-sufficing, for, since it fixes the history of every constituent of the world, it fixes the whole world-history.[17]

Eddington's doubts concerning "iron determinism" and the "conviction of nearly all physicists," are directly connected with the status of the second law of thermodynamics. In any closed system, entropy, understood as a measure of the disorder of physical systems, tends to increase. That is to say that the usable energy, suited or configured to perform work *decreases*. Time's arrow follows a statistical order of development—from less probable, but more orderly configurations toward more probable and less orderly configurations. If the second law, telling us that entropy in a closed system always increases (or at best remains constant) is demoted

[14]Laplace and Simon (1820/1902, p. 4).

[15]Contrast Lloyd (2006, p. 98): "Even if the underlying laws of physics were fully deterministic, however, … to perform the type of simulation Laplace envisaged, the calculating demon would have to have at least as much computational power as the universe as a whole." This is to suggest that the required computation is physically impossible.

[16]Eddington's note: "There are, however, others beside myself who have recently begun to question it." See Eddington (2014, p. 85, Footnote 1).

[17]Eddington (2014, p. 85).

to "secondary" status, and thought of, say, as a practical means of keeping track of developments on larger scales, when the micro-scale details of particles and motions are too complex, then time's arrow, as tracked by the second law, suffers a similar demotion.

On a deeper level, though, the question may be posed of the relationship between the statistical character of the second law and the probabilistic character of quantum mechanics—according to which physical chance is basic. In spite of this possible basis, the idea of a system of "fundamental" or primary and time symmetric physical law has often, even chiefly, survived the challenge posed to it by quantum mechanics and by Heisenberg's uncertainty principle.

To see why this is, we may return to Einstein's related views. Taking into account the varieties of science, focused on more limited domains and closer to human experience, Einstein has it that "from the very beginning there has always been present the attempt to find a unifying theoretical basis for all the single sciences …"[18] Such a "unifying theoretical basis," as he sees it, consists

> … of a minimum of concepts and fundamental relationships, from which all the concepts and relationships of the single disciplines might be derived by logical process. This is what we mean by a search for a foundation of the whole of physics. The confident belief that this ultimate goal may be reached is the chief source of the passionate devotion which has always animated the researcher.[19]

On this approach, fundamental physics aims to encompass all the concepts and laws required for ideal comprehension and unification of scientific results, including all more specialize and practical sub-disciplines; and there is clearly a strong suggestion here of logical reduction of the "whole of physics" to the "fundamental concepts and relationships." To use the literary term, it is very much a matter of searching for thematic unity—of a strictly logical and explanatory sort.

To the end of his life, Einstein worked at a "unified field theory," which would be a comprehensive fundamental theory, unifying the gravitational field of his theory of general relativity, with the theory of the electromagnetic field, starting from Maxwell's equations. It was a desideratum of this program that it would ultimately account for quantum mechanics as correct so far as it went, but non-fundamental.[20] He remained convinced of the "incompleteness" of quantum mechanics, with its intrinsic chance, "quantum jumps," super-positions and non-local quantum entanglements.[21] Einstein's projection of a preferred direction of the development of fundamental physics, however, came into conflict with the actual, historical course of its development. Taking that lesson seriously, one will also understand the tendency to identify "fundamental Physics" with investigation

[18]Einstein (1940, p. 488).

[19]Ibid.

[20]Cf. Sauer (2014, p. 287).

[21]See, e.g., the very influential "EPR paper": Einstein et al. (1935).

of the open problems arising from its best established theories.[22] The best laid plans of mice and men often go astray, and so it is with the best laid scientific projections of fundamental physics.

Einstein's doubts concerning quantum theory are usually viewed as obsolete. Heisenberg's uncertainty principle is taken to supersede Newtonian, Laplacean or Einsteinian determinism, since there is a fundamental uncertainty in the positions and velocities of the basic constituents of the universe. Few are the contemporary physicists who would try to remove the uncertainty relation from physics. Yet, this is not the end of the story, since there is also a quantum conception of determinism, arising from the uniform evolution of the Schrödinger equation and its derivatives. As Brian Greene has put the point, "Knowledge of the wave functions of all of the fundamental ingredients of the universe at some moment in time allows a "vast enough" intelligence to determine the wave functions at any prior or future time."[23] The point shifts determinism from the calculation of outcomes of physical interactions to the calculation of the probabilities for groups of outcomes, while the quantum uncertainty of prediction and measurement of individual results is retained. Whether determinism belongs to fundamental physics comes, in this way, to depend on the emphasis placed on calculation of probabilities vs. the uncertainty of particular measurements.

3 Models in Physics and the Concept of the Graviton

The general wave-particle duality of quantum mechanics and quantum field theory encourages postulation of the graviton, conceived in analogy to the photon as the carrier of the gravitational force. In accordance with hypotheses directly quantizing the gravitational field, the graviton would be a spin2 particle of zero mass and zero electrical charge, traveling at the speed of light. Wave-particle duality is formalized in contemporary physics, telling us that particle momentum p is equal to the Planck constant h divided by wavelength λ.

$$\mathrm{p} = h/\lambda$$

[22]See, for instance, Hewett et al. (2012). The editors of the volume comment that the Standard Model of particles physics "leaves some big questions unanswered;" Some of these questions "are within the Standard Model itself," such as "why there are so many fundamental particles and why they have different masses;" and "In other cases, the Standard Model simply fails to explain some phenomena, such as the observed matter-antimatter asymmetry in the universe, the existence of dark matter and dark energy, and the mechanism that reconciles gravity with quantum mechanics." If what is regarded as "fundamental" is viewed as open to question and inquiry, then the concept is much less problematic.

[23]Green (2000, p. 341).

This formula draws on Einstein's work on the photoelectric effect, the equivalence of mass and energy, and de Broglie's hypothesis of matter waves. Thus, to extend the argument, if there are gravitational waves, of a given wavelength, then there will be corresponding particles, analogous to photons—of zero mass and very small momentum. In Einstein's general theory of relativity, information is never transmitted faster than the speed of light, and on some models of quantum gravity, the graviton is wanted to carry information concerning changing positions of masses and the associated gravitational fields at finite speed and over unlimited range. In effect, the hypothesis formulated in quantum field theory is that masses interact gravitationally by exchange of gravitons.

However, the hypothesis of gravitational waves enjoys much firmer support. Gravitational waves are a prediction of general relativity on most prominent accounts.[24] General relativity has been confirmed in repeated and varied empirical tests for nearly a century. Though gravitational waves have not been detected, as of yet (early 2015), they are firmly expected as a further confirmation of general relativity. Much time, money and effort has been expended on the construction and calibration of sensitive detection devices, such as LIGO (the Laser Interferometer Gravitational Wave Observatory).[25] Following Einstein, gravitational waves are an expected effect of accelerating masses, in analogy with how electromagnetic waves are created by moving charges. Yet gravity is the weakest of the four fundamental forces of nature by many orders of magnitude, and in consequence, the displacements that physicists can expect to measure with instruments such as LIGO, as the signals of passing gravitational waves, are extremely small—on the order of 10^{-18} m, which is 1000 times smaller than the diameter of a single proton.[26] The experimental design depends on reflecting light along two long tracks, set at right angles to each other, and bringing the beams of light back, via mirrors, to a central point where interference can be detected, indicating a slight change in length of one of the two paths.

One expected source of gravitational waves are pairs of small and extremely dense neutron stars in mutual orbit, or again, pairs of black holes spiraling in toward each other and their eventual fusion. Important observational results have shown that identified binary pulsars do in fact lose energy, as they spiral inward, consistent

[24]See, for instance Weinberg (1977/1988, pp. 147–148): "Gravitational radiation interacts far more weakly with matter than electromagnetic radiation, or even neutrinos" and he continues, "For this reason, although we are reasonably confident on theoretical grounds of the existence of gravitational radiation, the most strenuous efforts have so far apparently failed to detect gravitational waves from any source." On the history of the decades long theoretical debate, including Einstein's own occasional doubts, see Kennefick (2007). Approximate solutions of the Einstein equations predicting gravitational waves date to Einstein (1916). See Einstein (1916).

[25]The LIGO project, with major facilities in Louisiana and Washington State, is the largest scientific project ever funded by the U.S. National Science Foundation, to the tune of over $300 million in capital investment and $30 million per year, since the early 1990s.

[26]See "Introduction to LIGO and Gravitational Waves," at the LIGO web pages.

with the prediction from general relativity of their loss of energy by emission of gravitational waves. Astrophysicists Joseph Taylor and Russell Hulse were awarded the Nobel prize for their related work, using the timing of a pulsar as a precise clock, allowing a measurements, over decades, of the predicted energy loss.[27]

The prospect of detecting gravitons, in contrast, is extremely dim.[28] The point has been emphasized in writings of Freeman Dyson, and I quote from the *defenders* of the detection of gravitons. In comparison with the prospects of detecting gravitational waves,

> The possibility of detecting individual gravitons is far more daunting. Indeed Freeman Dyson and colleagues have cogently estimated that it may in fact be infinitely more daunting, namely that it is likely to be impossible to physically realize a detector sensitive to individual gravitons without having the detector collapse into a black hole in the process.[29]

Though both general relativity and quantum field theory are regularly counted to fundamental physics, the graviton does not appear in general relativity or in the Standard Model of particle physics—which does not encompass gravitation. The graviton is instead a postulate which appears in projecting the quantum field theory used in the Standard Model in the hope of unifying its three forces—the electromagnetic force, the strong nuclear force and the weak force—with gravity. A chief point of interest in this, for present purposes, is to understand the contrast of theoretical standing between the use of the word "model" in the description of the Standard Model of particle physics, and the idea of including the graviton in a more comprehensive model.

The standard model of particle physics is admittedly incomplete, since it does not encompass gravity. Beyond that, it is generally recognized that most of the matter in the universe (so-called "dark matter") takes forms outside the standard model.[30] Yet the standard model is also extremely well supported by experimental evidence, and the recent detection of the Higgs boson at CERN counts as further confirmation. To speak of "the Standard *Model*," as a model, is somewhat concessionary, in light of its strong empirical support—with an eye to its incompleteness. In a similar way, general relativity has met every test to which it has been subjected, though the lack of integration of GR with quantum mechanics leaves

[27]See e.g., Taylor's Nobel Lecture, describing his work, 1997. The observed loss of energy is consistent with the generation of gravitational waves, in accordance with solutions of the Einstein field equations.

[28]See Dyson (2012). See also Dyson's review of Brian Greene's *The Fabric of the Cosmos*, in *The New York Review of Books*, May 13, 2004. On the prospect of detection of gravitons in a particle collider, Sean Carroll writes, "Gravitons are only produced by gravitational interaction, which is so weak that essentially no gravitons are made in a collider and we don't have to worry about them." See Carroll (2012, p. 104–105).

[29]Krausss et al. (2014). See also Rothman and Boughn (2006).

[30]See e.g., Carroll (2010, p. 389): "… there must be dark matter, and we have ruled out all known particles as candidates…".

room for new physics. However, the models of new physics which include the graviton, are more theoretical and even speculative; and very little supporting evidence is available. Weinberg writes:

> Because string theories incorporate gravitons and a host of other particles, they provide for the first time the basis for a possible final theory. Indeed, because a graviton seems to be an unavoidable feature of any string theory, one can say that string theory explains why gravitons exist.[31]

But this is a very weak sense of explanation and more a matter of a theoretical, explanatory proposal. String theory predicts the existence of the graviton, as we may understand Weinberg to claim, and *would* explain the graviton, if string theory were sufficiently supported as a "final theory."

While the Standard Model of particle physics and general relativity each belong to more settled "fundamental physics," models including the graviton are more speculative theoretical projections or possible additions to fundamental physics. They are invoked in particular proposals concerning quantum gravity, or the problem of how to integrate general relativity with quantum mechanics; and they treat of quantum field theory as something to be preserved at certain limits, as "effective quantum field theory" and extended to gravitation. Yet models strongly committed to postulating the graviton are only some among a variety of theoretical approaches to the tensions between general relativity and quantum mechanics.[32] As a general matter, Dyson's problem of detecting the graviton reflects the incompatibilities between GR and QFT, centered on the background metric assumed in QFT versus the background independence of GR and the dim prospect of any physical probing of the Planck length. The physical energies required to probe structures at the scale of the Planck length of 10^{-33} cm are so great that they would disrupt the physical geometry of the structures under study.[33] The point casts some doubt on the approach to quantum gravity based on string theory and QFT and opens the door to approaches to quantum gravity based in background independent, non-commutative geometry. From this perspective, even the well-confirmed Standard Model of particle physics, in spite of its considerable strides and impressive empirical support, is one among possible alternative models for new physics beyond.

---

[31]Weinberg (1992, p. 216).

[32]See e.g. Oriti (2009, p. xvi): "I think it is fair to say that we are still far from having constructed a satisfactory theory of quantum gravity, and that any single approach currently being considered is too incomplete or poorly understood, whatever its strength and successes may be, to claim to have achieved its goal, or to have proven to be the only reasonable way to proceed".

[33]Cf. the discussion in Majid (2008, pp. 67–69).

4 Quantum Indeterminacy and Temporal Symmetry

Newtonian mechanics, Maxwell's theory of electromagnetism, and Einstein's revisions of Newtonian theory in special and general relativity count as fully deterministic, in accordance with the claimed temporal symmetry of fundamental physics; but there is also historical and contemporary interest in the challenge represented by the Heisenberg uncertainty principle and the chance or probabilistic element in quantum mechanics. This stands in tension with the strong contrary tendency to think of quantum theory in terms of the uniform evolution of the wave function of the Schrödinger equation and to discount the cogency of indeterministic "quantum state reduction" or the "collapse of the wave function." This is one way in which the thesis of temporal symmetry enters into the complex of issues connected with the interpretation of quantum mechanics.

Regarding the chance element in contemporary physics, we need to ask whether chance will be fully subdued by and assimilated to the uniform evolution of the wave function, or on the contrary, if it might better be regarded as something ramifying through the complexities of the physical world and the wider domains of scientific phenomena. Where isolated, a quantum system evolves in accordance with the Schrödinger equation, which allows for the calculation of probabilities of outcomes of measurements. The other element, however, shows how the state of the system is reduced when an ideal measurement is carried out. While the Schrödinger equation is time reversible, measurement operates only forward in time, and this may be thought of as defining the quantum mechanical arrow of time. But since the Schrödinger equation already determines the probabilities of measured outcomes, via the Born rule, and nothing tells us which outcome will actually be measured on a particular trial, this emphasizes the normal quantum mechanics of the Born rule. To find the probability that the wave function will collapse to a specific state, you take the square of the coefficient of that possible outcome in the Schrödinger equation. The use of the Born rule is empirically adequate, and it does not follow from the Schrödinger equation. In view of these two elements is quantum theory deterministic and temporally symmetric or not?

Amongst the mathematical complexities of the physicists efforts to explain exactly how and why all physical laws are "time-reversal invariant," it seems sometimes to be forgotten that there is a genuine paradox arising from this recurrent motif of fundamental physics. Temporal invariance implies that it is possible, for example, for a tree to shrink down to a shoot and return to the state of a seed, that the old might evolve into younger people and eventually disappear by becoming unborn, or that a dispersion of light could concentrate itself into a narrow beam and return to a flashlight or laser, say. Reverse processes are not frequently observed in nature, though the fundamental laws don't dictate this result. The examples can be multiplied at will, and they represent the physical reality of varieties of unidirectional processes, consistent with our ordinary conception of time pointing from past toward the future. Supposing they are extremely unlikely and not simply impossible, the question persists of why they are extremely unlikely. Why does nature favor processes which increase entropy?

It is held by physicist Lorenzo Maccone, in a fascinating short paper, that

> In fact, the laws of physics are time-reversal invariant. Hence there is no preferred direction of time according to which we may establish a *substantial* difference between the two temporal directions past-to-future and future-to-past.[34]

Maccone's intriguing proposal is that though macroscopic reversals of entropy in isolated systems are statistically and therefore physically possible,[35] they leave no evidence—where, as expected, "thermodynamic entropy is a quantity that measures how the usable energy in a physical process is degraded into heat."[36] Avoiding any "substantial" (or fundamental) conception of temporal asymmetry, Maccone also takes the surprising view that "thermodynamic entropy is a subjective quantity," though "for *all practical situations* this is completely irrelevant."[37]

His argument is that there is a hidden assumption build into the various statements of the second law of thermodynamics, to the effect that whenever an isolated system is obtained by combing two theretofore distinguished systems, "the second law is valid only if the two systems were initially uncorrelated, i.e., if their initial joint entropy is the sum of their individual entropies."[38] But, he holds, it is impossible to know whether a given system is in fact correlated with another in some unknown way, and in consequence, as a practical matter all systems are considered *uncorrelated* without clear evidence to the contrary. Without this assumption, he argues, "it would be impossible to assign an entropy to a system unless the state of the whole universe is known."[39] In spite of the practicality of the second law, then, Maccone's proposals involve a more emphatic version of the distinction between fundamental physics and secondary or derivative law.

The more emphatic character stands out in the general thesis of the article, which states that though "the laws of physics are invariant for time inversion," and "the familiar phenomena of everyday are not," the paradox is solved, according to the argument of the paper, since it argues that "phenomenon where entropy decreases" will fail to "leave any information of their having happened," and this situation "is completely indistinguishable from their not having happened at all."[40]

The position is remarkable, since it preserves temporal invariance by placing any evidence of processes of decreasing entropy beyond possible observation, and in consequence, the second law of thermodynamics, according to Maccone, is reduced

---

[34]Maccone (2009, p. 5).

[35]Cf. the discussion in Greene (2004, pp. 159–163). Greene's point is that the purely statistical reasoning of the second law equally suggests that entropy will be found to increase in the *past* of any system considered, since states of higher entropy are generally more probable.

[36]Maccone (2009, p. 1).

[37]Ibid.

[38]Ibid.

[39]Ibid.

[40]Ibid.

to a "tautology." This is a strong claim, and one may easily suppose to the contrary that reverse processes may be observed—at least under simplified experimental conditions.

Without considering Maccone's central argument, which is based on the physical possibility of reverse processes resulting in decreasing entropy (greater free energy), his position illustrates just how far physicists are willing to go to preserve the claimed symmetry or temporal invariance of fundamental physics. It is a concise and fascinating little paper, and I only aim to suggest doubts about its conclusion. That we do not (or do not frequently) observe macroscopic temporal reversals of physical processes (in closed systems) seems a point too physically significant to want to explain away. The specifics of Maccone's thought experiment are impossible in practice, since it involves control at the quantum level of the results of the forward process, and subsequent erasure of all evidence of it. Part of the interest of the paper is that it rests on an equivalence of quantum processes with information.

Turning to a more prominent view of the relationship between fundamental physics and quantum indeterminacy, I want to briefly consider Brian Greene's discussions of the arrow of time, and in particular, "Time and the Quantum," Chapter. 7, of his 2004 book. This is a fine exposition of the related questions, posed in terms open to the broad, educated public. Greene defends the deterministic, temporal-symmetry orthodoxy. His question in the chapter is "whether there is a temporal arrow in the quantum mechanical description of nature."[41] His conclusion reaffirms temporal invariance of fundamental law, including quantum mechanics, and links the arrow of time to the surprisingly low entropy of the initial condition of the universe, subsequent to the big bang—invoking a cosmological arrow of time. However, while the cosmological arrow of time is widely accepted, it too seems to require explanation.

Greene's provides a concise overview of the theme of quantum decoherence, together with an equally concise overview of the "quantum measurement problem," and his question is posed within this rich theoretical context. In general terms, approaches to the interpretation of quantum mechanics via quantum decoherence can be viewed as a contemporary up-date and revision of a controversial element of the "Copenhagen interpretation" developed, early on by Niels Bohr, Heisenberg, Max Born and associates. The decoherence approach removes the stress placed on "observation" in earlier accounts of quantum-mechanical phenomena.

The "quantum measurement problem" is something of a mare's nest, a formidable complex of old and new arguments and doubts, including on occasion, a continuing, popular fascination with the idea of a special role of observers in quantum mechanics, Einstein's and Schrödinger's early doubts about the Heisenberg uncertainty principle and the Bohr-Einstein debate,[42] Bohm's hidden variable theory, including, again, doubts on the absence of classical determinism in

---

[41]Greene (2004, p. 177).

[42]See, e.g., Born (1954), the Nobel Lecture, p. 256.

quantum mechanics, discussions of the Bell inequalities, doubts about Alain Aspect's famous experimental results supporting Bell,[43] the contemporary proposals for "spontaneous collapse,"[44] "multiple worlds" theories in which there is no collapse, and every possible mixture of these issues and themes.

Greene acknowledge the appeal of the decoherence approach in obviating aspects of the quantum measurement problem. The idea is that there is nothing special about observation or measurement. "Human consciousness, human experimenters, and human observations would no longer play a special role since they (we!) would simply be elements of the environment, like air molecules and photons, which can interact with a given physical system."[45] Measurement is, according to the decoherence approach, simply one more interaction of a quantum system with its environment, in which the wavefunction of the system, and the possibility of interference-effects are reduced or modified. Again, "there would no longer be a stage one—stage two split between the evolution of the object and the experimenter who measures them. Everything—observed and observer—would be on an equal footing."[46] There is no need of an ad hoc, or physically unmotivated distinction between the quantum world and macroscopic objects or measuring instruments; and the lack of quantum weirdness in the macroscopic world falls out as an effect of environmental decoherence. Experimental detection of interference effects of quantum mechanical systems depends on isolating them and considering very small objects such as photons and electrons, as in the classical double-slit experiments, but these idealized and isolated system are not typical of the complex interactions of real-world happenings: "much as adding tagging devices to the double-slit experiment blurs the resulting wavefunction and thereby washes out interference effects, the constant bombardment of objects by constituents of their environment also washes out the possibility of interference phenomena."[47] On the decoherence approach, there is an answer to the puzzle posed by the thought experiment of Schrödinger's cat, since environmental decoherence would have plausibly taken effect long before any observer looks in on the situation.

---

[43]See e.g., the *Journal of Cosmology*, Vols. 3 and 14 on consciousness and the quantum; Bohm (1952), Bub (2010; arXiv:1006.0499v1), Bell (1993), Aspect et al. (1982).

[44]Regarding the "spontaneous collapse" proposal of Ghirardi, Rimini and Weber, Brian Greene remarks that "they introduce a collapse mechanism which does have a temporal arrow—an "uncollapsing" wavefunction, one that goes from a spiked to a spread out shape, would not conform to the modified equations." See Greene (2004, p. 214).

[45]Greene (2004, p. 212).

[46]Ibid.

[47]Greene (2004, p. 210). Cf. Carroll (2010, pp. 253–254) "In the many-worlds interpretation, decoherence plays a crucial role in the apparent process of wavefunction collapse. The point is not that there is something special or unique about 'consciousness' or 'observers' other than the fact that they are complicated macroscopic objects. The point is that any complicated macroscopic object is inevitably going to be interacting (and therefore entangled) with the outside world, and its hopeless to imagine keeping track of the precise form of the entanglement. For a tiny microscopic system such as an individual electron, we can isolate it and put it into a true quantum superposition, but for a messy system such as a human being … that's just not possible".

I have little doubt that discussions of quantum measurement will continue among the physicists for many years to come, simply in virtue of the complexities involved. So long as the notion lingers that the measurement problem is a matter of getting exact predictions on particular experimental runs, however, and the question takes the form, e.g., "Why doesn't the measurement of a particle in superposition result in a superposition of measurements?"—or, "why does something happen in measurement, not precisely predicted by the Schrödinger equation?"—then I suspect that the physicists will be barking up the wrong tree—where they implicitly put the uncertainty principle in question. As Greene puts the point, "Much in the spirit of Bohr, some physicists believe that searching for such an explanation of how a single, definite outcome arises is misguided." Weinberg's recent proposal is of this general character.[48] "During measurement," Weinberg says, "the state vector of the microscopic system collapses in a probabilistic way to one of a number of classical states, in a way that is unexplained, and cannot be described by the time-dependent Schrödinger equation." Weinberg's approach avoids the "many worlds" and "hidden variables" views, and, borrowing from decoherence, avoids the classical "Copenhagen" approach as well.

5 Conclusion: Causality and Indeterminacy

Causality and indeterminacy are fascinating and widely discussed philosophical topics, and they come together in considering the relationships of quantum indeterminacy and fundamental physics to the arrow of time. Since the arrow appears crucial in ordinary and scientific conceptions of causality, where an effect cannot proceed its cause,[49] it would certainly be of interest to the topic of causality and related debates to find a quantum mechanical arrow in support of a thermodynamic arrow. The most promising candidate for a quantum mechanical arrow of time is to locate it in the inherently probabilistic collapse of the wavefunction. Greene puts the point as follows: "if the resolution of the measurement problem that is one day accepted reveals a fundamental asymmetric treatment of the future versus the past within quantum mechanics," then "it could very well provide the most straightforward explanation of time's arrow."[50] This is not the approach Green most favors, however.

The point shows in Greene's expressed doubts about decoherence. "Even though decoherence suppresses quantum interference and thereby coaxes weird quantum

[48]Greene (2004, p. 213). Cf. Weinberg (2012, p. 2): Weinberg proposes a "correction" to quantum mechanics which nonetheless eventuates in "inherently probabilistic collapse" of the state vector, with probabilities given by "the Born rule of ordinary quantum mechanics"; cf Ghirardi et al. (1985, 1986).

[49]The supposition is that this is true, even if, as sometimes argued, causality is an "emergent" phenomenon. See for instance Norton (2003, p. 1), where the thesis is that though causation is not fundamental, it "remains a most helpful way of conceiving the world".

[50]Greene (2004, p. 215).

probabilities to be like their familiar classical counterparts," he writes, "each of the potential outcomes embodied in a wavefunction still vies for realization. And so we are left wondering how one outcome 'wins' and where the many other possibilities 'go' when that actually happens."[51]

But, I submit that Greene's first conjunct here is just what we cannot require of an explanation, if the uncertainty principle is true and quantum chance is fundamental. Regarding the Bohm approach, since, according to Greene, equations are needed that show "how a wavefunction pushes a particle around,"[52] and this would apparently require action superseding the speed of light in case of entanglement at astronomical distances, there is significant physical justification for taking quantum indeterminacy flatfootedly.

Will accepting unidirectional state reduction, quantum indeterminacy and a quantum mechanical arrow help in understanding the observed unidirectional increase in entropy in physical processes, and, on that basis, a thermodynamic arrow of time? One idea is that nature favors state reduction toward conditions of increased entropy and that increases of quantum entanglement, due to diverse interaction increase entropy.[53] But Greene is more intent on the cosmological arrow, based on the initial condition of low gravitational entropy in the theory of inflationary expansion. In spite of that, time symmetric laws cannot explain why the observed world has a comparatively low entropy (contrasting the projected heat-death of the universe) or why it had even lower entropy in the past. Moreover, in related approaches, inflationary expansion evokes the multiverse and emphasis on the "anthropic principle."[54] These are developments which many would like to avoid.

The prospect of finding an explanation of time's arrow within quantum mechanics proper, continues to rest with the Born rule and the idea that in quantum mechanics chance is fundamental. Repeating the same experiment (i.e. the same cause), we get different measurements on different trials. Though resistance to quantum mechanical indeterminacy has sometimes rested on holding onto traditional conceptions of universal causality, fundamental physical support for causality's arrow of time, may ultimately rest on quantum indeterminacy.

The point is somewhat obscured at present, and the obscurity is not unrelated to the role of traditional conceptions of "fundamental physics" in the speculative boom in contemporary physics. This has been stimulated by a number of factors, including the end of the Cold War, the availability of the Large Hadron Collider at CERN, the recent progress of the Standard model, and more basically, the conceptual conflicts between general relativity and quantum mechanics. While there is

[51]Ibid, p. 212.

[52]Ibid, p. 214.

[53]See Lloyd (2006), Chaps. 4 and 5 on thermodynamics, information and quantum mechanics.

[54]Emphasis on the inflationary expansion in the early universe, the multiverse idea and the anthropic principle is even more pronounced in Sean Carroll's recent book, Carroll (2010). But see pp. 339–345, where a range of doubts are discussed.

no contrary evidence to these great twentieth-century paradigms of physics, there has been much theoretical work, of a more speculative character, which aims at models of unification. The possible approaches go off in many diverse directions, though temporal symmetry in fundamental physics is the usual orthodoxy.

References

Aspect, A., Grangier, P., & Roger, G. (1982). Experimental realization of the Einstein-Podolsky-Rosen-Bohm *Gedankenexperiment*: A new violation of Bell's inequalities. *Physical Review Letters, 49*(2), 91–94.

Bell, J. (1993). *Speakable and unspeakable in quantum mechanics*. Cambridge: Cambridge University Press.

Bohm, D. (1952). A suggested interpretation of quantum theory in terms of 'hidden' variables, I and II. *Physical Review, 85*, 166–193.

Born, M. (1954). Nobel Prize lecture, "the statistical interpretation of quantum mechanics". *Nobel prize lectures: Physics, 1942–1962* (pp. 256–267). Amsterdam, London: Elsevier.

Bub, J. (2010). Von Neumann's 'no hidden variable' proof: A re-appraisal. *Foundations of Physics, 4*, 1333–1340.

Carroll, S. (2010). *From eternity to here, the quest for the ultimate theory of time*. New York: Penguin.

Carroll, S. (2012). *The particle at the end of the universe, how the hunt for the Higgs Boson leads us to the edge of a new world*. New York: Dutton.

Davies, P. (1977). *The physics of time asymmetry*. Berkeley and Los Angles: University of California Press.

Dyson, F. (2004). The world on a string, review of Brian Greene, *The fabric of the cosmos*, *The New York review of books*, 51, No. 8 May 13, pp. 16; reprinted in Dyson, F. (2006). *The scientist as Rebel*. New York: The New York Review of Books, pp. 213–228.

Dyson, F. (2012). *Is a graviton detectable?* The Poincaré Prize Lecture. https://publications.ias.edu/sites/default/files/poincare.2012.pdf

Eddington, A. S. (1928/2014). In H. G. Callaway (Ed.), *The nature of the physical world, an annotated edition*. Newcastle upon Tyne: Cambridge Scholars Publishing.

Einstein, A. (1916). Nährungsweise Integration der Feldgleichungen der Gravitation. In *Sitzung der physikalisch-mathematischen Klasse* (Approximative integration of the field equations of gravitation, Einstein 1997, *The Collected Papers of Albert Einstein, The Berlin Years*, Vol. 6, pp. 201–210) (pp. 688–696). Princeton: Princeton University Press.

Einstein, A. (1940). Considerations concerning the fundamentals of theoretical physics. *Science, 91*(2369), 487–492.

Einstein, A., & Infield, L. (1938). *The evolution of physics, from early concepts to relativity and quanta*. New York: Simon and Schuster.

Einstein, A., Podolsky, P., & Rosen, N. (1935). Can quantum-mechanical description of physical reality be considered complete? *Physical Review, 47*, 777–780.

Fodor, J. A. (1974). Special sciences. In *Synthèse* (Vol. 28, pp. 77–115). Reprinted in Fodor (1981). *RePresentations* (pp. 127–45). Cambridge: MIT Press.

Ghirardi, G., Rimini, A., & Weber, T. (1985). A model for a unified quantum description of macroscopic and microscopic systems. In L. Accardi (Ed.), *Quantum probability and applications*. Berlin: Springer.

Ghirardi, G., Rimini, A., & Weber, T. (1986). Unified dynamics of microscopic and macroscopic systems. *Physical Review D, 34*, 470ff.

Greene, B. (2000). *The elegant universe, superstrings, hidden dimension and the quest for the ultimate theory*. New York: Random House.

Greene, B. (2004). *The fabric of the cosmos, space, time and the texture of reality*. New York: Random House.

Hewett, J. L., Weets, H., et al. (2012). *Fundamental physics at the intensity frontier, report of the workshop held December 2011 in Rockville*. MD: SLAC National Accelerator Laboratory, Stanford University.

James, W. (1909/2008). In H. G. Callaway (Ed.), *A pluralistic universe. A new philosophical reading*. Newcastle upon Tyne: Cambridge Scholars Publishing.

Kennefick, D. (2007). *Traveling at the speed of thought, Einstein and the quest for gravitational waves*. Princeton: Princeton University Press.

Krausss, L. M., & Wilczek, F. (2014). Using cosmology to establish the quantization of gravity. *Physical Review D, 89*(4). arXiv:1309.5343v2.

Laplace, M., & Simon. P. (1902) In F. W. Truscott & F. L. Emory (Eds.), *A philosophical essay on probabilities*. London: Wiley.

Lloyd, S. (2006). *Programming the universe, a quantum computer scientist takes on the cosmos*. New York: Knopf, Random House (2007).

Maccone, L. (2009). A quantum solution to the arrow of time dilemma. *Physical Review Letters 103*. arXiv:0802.0438v3.

Majid, S. (2008). Quantum spacetime and physical reality. In S. Majid (Ed.), *On space and time* (pp. 56–140). Cambridge: Cambridge University Press.

Norton, J. D. (2003). Causation as folk science. *Philosophers' Imprint*, *3*(4). http://www.pitt.edu/~jdnorton/papers/003004.pdf

Oriti, D. (2009). *Approaches to quantum gravity, toward a new understanding of space, time and matter*. Cambridge: Cambridge University Press.

Penrose, R. (2004). *The road to reality a complete guide to the laws of the universe*. New York: Vintage.

Rothman, T., & Boughn, S. (2006). Can gravitons be detected? *Foundations of Physics 36*, 1801–1825. arXiv:gr-qc/0601043v3.

Sauer, T. (2014) Einstein's unified field theory program. In M. Jannsen & C. Lehner (Eds.), *The Cambridge companion to Einstein* (pp. 281–305). Cambridge: Cambridge University Press.

Savitt, S. F. (Ed.), (1995). *Time's arrow today, recent physical and philosophical work on the direction of time*. Cambridge: Cambridge University Press.

Susskind, L. (2008). *The black hole war*. New York: Little, Brown.

Taylor, J. H. (1997). Binary pulsars and relativistic gravity. In G. Ekspong (Ed.), (1997) *Nobel lectures, physics 1991–1995*. Singapore: World Scientific Publishing.

Weinberg, S. (1977/1988). *The first three minuets: A modern view of the origin of the universe*. New York: Basic Books.

Weinberg, S. (1992). *Dreams of a final theory*. New York: Pantheon Books.

Weinberg, S. (2012). The collapse of the state vector. *Physical Review A, 85*(6). arXiv:1109.6462v4.

Counterfactual Histories of Science and the Contingency Thesis

Luca Tambolo

Abstract Within the debate on the inevitability versus contingency of science for which Hacking's writings (The social construction of what? Harvard University Press, Cambridge, 1999; Philos Sci 67:S58–S71; 2000) have provided the basic terminology, the devising of counterfactual histories of science is widely assumed by champions of the contingency thesis to be an effective way to challenge the inevitability thesis. However, relatively little attention has been devoted to the problem of how to defend counterfactual history of science against the criticism that it is too speculative an endeavor to be worth bothering with—the same critique traditionally levelled against the use of counterfactuals in general history. In this paper, we review the defense of counterfactuals put forward by their advocates within general history. According to such defense—which emphasizes the essential role of counterfactuals within explanations—good counterfactual scenarios need to exhibit the right kind of plausibility, characterized as continuity between said scenarios and what historians know about the world. As our discussion shows, the same requirement needs to be satisfied by good counterfactual histories of science. However, as we mention in the concluding part of the paper, there is at least one concern raised by counterfactual history of science as used to support the contingency thesis for which the defense based on the plausibility of the counterfactual scenarios does not seem to offer easy solutions.

1 Introduction

Ian Hacking's writings (1999, 2000) concerning the question of whether the results of successful science are inevitable or contingent have provided the basic terminology for the debate between so-called *inevitabilists*, on the one hand, and so-called *contingentists*, on the other hand (see Soler 2008, 2015; Martin 2013;

L. Tambolo (✉)
Department of Humanistic Studies, University of Trieste, c/o via Casona, 7, 40043 Marzabotto, Italy
e-mail: ltambolo@gmail.com

L. Magnani and C. Casadio (eds.), *Model-Based Reasoning in Science and Technology*, Studies in Applied Philosophy, Epistemology and Rational Ethics 27, DOI 10.1007/978-3-319-38983-7_35

Kinzel 2015 for recent overviews). Contingentists maintain that history of science may well have taken a path leading to some alternative S', S'', S''', etc., to our current science S: this is the *contingency thesis*, typically defended by devising—or at least, invoking the possibility to devise—alternative histories of science. Such alternative, counterfactual histories are populated by theories alleged to be as successful as the ones currently embraced by scientists, which according to the *inevitability thesis* are unavoidable stages in the development of science. The more plausible the envisaged counterfactual histories, the contingentists' reasoning seems to go, so much the worse for inevitabilists: if putting forward credible alternatives is actually feasible, then the inevitability thesis will lose quite a bit of its *prima facie* appeal.

Champions of the inevitability thesis find such contrary-to-fact speculations far from compelling. For after all, they ask, how can one ascertain what consequences would have followed, had things gone differently at some juncture in the history of science? Counterfactual history of science then faces the same criticisms levelled at the use of counterfactuals within general history. In what follows, we shall review the defense of counterfactuals put forward by their advocates within general history. According to such defense—which emphasizes the essential role of counterfactuals within explanations—good counterfactual scenarios need to exhibit the right kind of plausibility, characterized as continuity between said scenarios and what historians know about the world. As our discussion will show, the same requirement needs to be satisfied by good counterfactual histories of science.

We shall proceed as follows. In Sect. 2, after some introductory remarks on the current state of the debate on the inevitability versus contingency of science, we shall focus on some essential features of counterfactual history of science as used to support the contingency thesis. In Sect. 3, we shall deal with the defense of counterfactuals within general history devised by advocates of what is variously referred to as "alternative history," "alternate history," "'what if?' history," "allo-history," and "counterfactualism," and emphasize the role of counterfactuals within explanatory models. In Sect. 4, some examples of counterfactual histories of science, highlighting the importance of the plausibility of the counterfactual scenarios and relating such plausibility to the viability of the models put forward by counterfactual historians, will be discussed. In Sect. 5, we shall conclude by indicating one concern raised by the use of counterfactual history of science in support of the contingency thesis for which the above defense of counterfactuals does not offer easy solutions.

2 Inevitability, Contingency, Counterfactual Histories

In order to clarify both the scope of the inevitability versus contingency of science debate and the aim of the present paper, it will be useful to start our discussion with a few preliminary remarks.

First, Hacking's writings have to a large degree shaped the most recent debate: among other things they provide, as mentioned above, the very labels used to designate the opposing camps and their respective claims.[1] Nevertheless, the issue of the inevitability versus contingency of the results of successful science is anything but unknown to twentieth century philosophy of science. For instance, in *The Structure of Scientific Revolutions* (1962/1970) Thomas Kuhn famously characterized paradigms as rival sets of hypotheses competing to win the support of a scientific community. As one of the most well-known practitioners of counterfactual history of science recently put it, according to Kuhn's theory of science, "several of the rivals may have the ability to function effectively and contingent circumstances may influence the outcome of the debate" (Bowler 2013, pp. 26–27) between advocates of competing paradigms. To mention but one more example, the later Paul Feyerabend devoted sustained efforts to a wide-ranging criticism of realism, in which considerations pertaining to the vagaries of historical developments play a central role. In a 1989 paper tellingly entitled "Realism and the historicity of knowledge," later collected in his posthumous, unfinished *Conquest of Abundance* (1999), Feyerabend railed against what he viewed as a key tenet of realism, i.e., the separability assumption, which he formulated as follows:

> what has been found in [a] idiosyncratic and culture-dependent way (and is therefore formulated and explained in idiosyncratic, ad hoc, and culture-dependent terms) *exists* independently of the circumstances of its discovery. In other words, we can cut the way from the results without losing the result (1999, p. 133).[2]

Relatedly, as Sankey (2008) and Kidd (2011, 2016) have pointed out, not only the specific developmental paths followed by individual sciences, but also the very emergence of science as such can be viewed as a phenomenon resulting from various contingencies (an issue to which Heidegger, Husserl, and Wittgenstein, among others, devoted quite some attention). Only recently, however, has the explicit discussion of counterfactual history become one focus of the reflection on the inevitability versus contingency of science.[3]

Secondly, the two conflicting claims around which the inevitability versus contingency debate revolves can be—and sometimes have indeed been—stated in very general and somewhat stark terms. For instance, an often-mentioned champion

---

[1]Although in what follows we shall stick to the "inevitability versus contingency" couple and cognate expressions, some terminological variations in the relevant literature are worth mentioning: for instance, French (2008, p. 572) contrasts "contingentists" with "necessitarians," Henry (2008, p. 552) "contextualists" with "positivists," and Fuller (2008, p. 577) "underdeterminism" with "overdeterminism."

[2]Hacking is of course well aware of such antecedents, discussed at length in his analysis of the debate on social construction (1999). For Feyerabend's take on the contingency of science, see especially Kidd (2016, Sect. 5); for his criticism of the separability assumption, see Tambolo (2014).

[3]See especially the texts mentioned in Footnote 1, which together with Radick (2008) and Bowler (2008) feature in a focus, published in *Isis*, devoted to "Counterfactuals and the Historian of Science."

of the inevitability thesis once commented: "If we ever discover intelligent creatures on some distant planet and translate their scientific works, we will find that we and they have discovered the same [fundamental physical] laws" (Steve Weinberg, quoted in Hacking 2000, p. S66). This phrasing of the inevitability thesis conveys the idea that the results of our science are unavoidable stops along the way to an "imagined end-run science" (ibid., p. S60) of which, according to die-hard inevitabilists, they provide a preview of sorts. An equally strong version of the contingency thesis can be formulated by borrowing Gould's (1989) terminology: nothing is inevitable in the development of a properly conducted physical investigation of the world; therefore, were it possible to "replay the tape" of history of science, the path taken by the scientific enterprise would most likely differ very significantly from the actual historical record.

As the debate proceeds, however, intermediate positions between the above extreme versions of the inevitability and contingency theses are acquiring more and more prominence. This is certainly related to the awareness that science can be viewed as inevitable (contingent) along different dimensions: the social and cultural conditions in which scientific inquiry takes place; the methods that happen to be used by researchers; the evidence available to scientific communities; the standards used to appraise theories; the concepts providing the framework for experiments; etc. In other words, the contingentists' claims concerning the empirical success of putative alternatives to the theories that our scientific communities currently embrace cover only one dimension of the inevitability versus contingency of science issue—although arguably the most important one. Consequently, it has been suggested, in order for the notion of the inevitability (contingency) of science to be a useful analytical tool, the inevitability thesis and the contingency thesis ought to be unpacked into various *inevitability theses* or *claims* and *contingency theses* or *claims*, respectively (see Martin 2013; Kinzel 2015, esp. Section 5; and Soler 2015).

Here we shall not aim at doing justice to the nuances of the debate, for which in the years to come the recently published collection *Science as It Could Have Been* (Soler et al. 2015) will certainly be the mandatory reference. More modestly, we shall focus on one argumentative strategy often deployed—and even more often assumed to be readily deployable, should the need arise—in support of the contingency thesis. The strategy consists in claiming that, although the historical record attests that things went in a certain way, they may well have gone differently. Therefore, today we may well find ourselves championing a science S' which, different from our current science S, would be made of theories enjoying the same amount of empirical success as the ones that we now embrace. In rough outline, then, the strategy used to defend the contingency thesis consists of three steps: (i) some presumably crucial juncture in the past of science, at which things might have gone differently, is identified; (ii) the counterfactual speculation is put forward that, had things indeed gone differently, a different development would have followed; (iii) depending on the author putting forward the counterfactual speculation, such different development is described in more or less detail (sometimes, it is just evoked as a possibility).[4]

Unsurprisingly, champions of the inevitability thesis find such speculations less than compelling. Indeed, alternative histories of science face basically the same objections raised against counterfactual speculations in general history, famously dismissed as "a parlour-game with might-have-beens" (Carr 1961, p. 97) by eminent historians. In the next section, some of the arguments that champions of counterfactual history have put forward in defense of their endeavors will be reviewed.

3 Counterfactuals, Idealized Models and Historical Explanations

It comes as no surprise that counterfactual speculations have attracted quite a lot of criticism from practicing historians: reconstruction of the past as it actually happened, after all, figures prominently in their job description.[5]

One major concern that the devising of alternative pasts typically raises has to do with the very subject matter of counterfactual history. A counterfactual history starts with the identification of a certain juncture in the past at which things—allegedly—might have gone differently, so that the subsequent events might have unfolded in a different way. But given that at such crucial juncture things went exactly as we know they went, the critics ask, is it not the case that the individual who claims that they might have gone differently, thus bringing about different developments, is just projecting her own prejudices on the past? To put it differently: is it not the case that the narrative resulting from a counterfactual approach teaches us more on the historian writing it than on the reconstructed past? The very same criticism can of course be raised against *any* reconstruction of the past; nevertheless, critics of counterfactual history point out, the concern is clearly more pressing in the case of narratives hinging on the choice to assume the past to have been different from how we know it was.

Indeed, as emphasized by Gavriel Rosenfeld among others, "presentist motives" (2002, p. 90) are far from extraneous to the genre of alternate history. In this connection, Rosenfeld argues at length that some typical correlations obtain between the way in which proponents of counterfactual histories view the present and the way in which they recount the past. For instance, in *nightmare scenarios*,

[4]Note that advocates of the contingency thesis do not suggest that one can devise a full-fledged alternative to our science: in view of the long-term collective investments of time, efforts, ingenuity, and resources involved in the emergence of anything as complex as our current body of scientific knowledge, this is an impossible task (see especially Trizio 2008; Kidd 2016).

[5]For a recent critical survey of alternate history, see Evans (2014), which provides the reader with plenty of references to explore the genre. Rosenfeld (2014) and Sunstein (forthcoming), among others, offer critical assessments of Evans' opinionated survey.

the conjured up past compares unfavorably to the real historical record, so that the present is vindicated; while in what Rosenfeld calls *fantasy scenarios*, the alternative past is depicted as superior to the real historical record in order to support the writer's critique of the present.

In spite of the overtly political agendas that often underlie alternative histories, practitioners and defenders of counterfactual history have forcefully argued that its primary goal is not that of criticizing or praising the past but, rather, that of understanding it better. To mention but one example, the editors of the collection *Unmaking the West: 'What-If' Scenarios that Rewrite World History* insist that "not all counterfactual thought experiments are equally subjective and therefore equally speculative" (Tetlock et al. 2006, p. 9). The alternative histories devised in the chapters of their collection, they claim, aim at changing the questions around which the debate on the rise of the West has traditionally revolved. When the question that triggers historical inquiry is counterfactual (e.g., "Why did alternative developments fail to occur?") instead of factual (e.g., "Why did a certain event happen?"), they emphasize, "history looks different" (ibid., p. 5). This is of crucial importance for anyone "interested in the cognitive processes of observing and drawing causal lessons from history as in the historical record itself" (ibid.). To this end, Tetlock, Lebow, and Parker claim, a counterfactual approach is especially useful because it reminds one of the

> many intricately interconnected assumptions scholars need to make to justify claims about the inevitable or improbable rise and fall of civilizations. We see enormous intellectual value—perhaps, indeed, the greatest service counterfactual historians can render—in unearthing the labyrinthine logical complexity of "what-if" assumptions underpinning the often all too confident claims about why the West, and not one of the rest, rose to global hegemony (ibid., p. 9).

Note that, according to Tetlock, Lebow and Parker, shedding light on the historians' implicit assumptions is not the only advantage of a counterfactual approach, and consequently, they go on to provide an in-depth discussion of the various benefits—and challenges—of counterfactual history (on which see also, e.g., Tetlock and Belkin 1996; Lebow 2000; Yerxa 2008). Here, however, we shall not follow the details of their account. For our present purposes, what matters is that the passages quoted above point the reader to a thesis that is emphatically embraced by all defenders of alternative history: the thesis that there is a very close link—one that cannot be severed—between counterfactuals and historical explanations. Indeed, champions of counterfactuals take them to be ubiquitous in historical research, since they are no less than a "by-product of any historical statement that implies causality" (Kaye 2010, p. 38). More specifically, the idea is that whenever a causal claim of the kind "*x* caused *y*" is asserted by the historian who attempts to explain a certain event, the corresponding counterfactual "Had it not been the case that *x*, it would not have been the case that *y*" is simultaneously endorsed, albeit only implicitly. According to their advocates, counterfactuals are thus a fundamental ingredient of the explanatory practice of any historian—even of those who officially dismiss counterfactual history as "history-fiction."

That counterfactuals somehow feature in causal explanations is certainly not an exotic view: as Ronald Giere, for one, put it, "[t]o understand a causal system is to know at least some counterfactuals about that system" (Giere 2015, p. 192). In the case of non-experimental disciplines such as history, however, the point is of special interest. Since the researcher cannot replay the tape of history at will,

> one can hardly discuss the relative importance of causes without engaging in some kind of thought experiment where one removes successively and separately each of the causes in question and evaluates what difference the absence of this cause would have made to the phenomenon in question (Elster 1978, p. 176).[6]

On Elster's account, then, counterfactuals lend crucial support to researchers dealing with the task to estimate, in spite of the scarcity of evidence, the impact of a certain presumed cause on the observed effect that constitutes the object of inquiry. Although Elster himself does not use the notion of model, we may say that the importance of counterfactuals for historians depends on their being a key component of explanatory models. Just like in the natural sciences, the researchers investigating a historical phenomenon can be viewed as proposing an *idealized model* of it, that is, "a deliberate simplification of something complicated with the objective of making it more tractable" (Frigg and Hartmann 2012). Of the countless elements constituting historical reality, only some—assumed to be causally relevant for the historical outcome under inquiry—feature in the model, while others are entirely disregarded.[7] Allowing as they do the manipulation of one or more of the elements featuring in the idealized model, counterfactuals make it possible to study the relationships among the elements within the model. Of course, just like in the natural sciences, model building is subject to constraints, and the idealized model used to investigate a historical phenomenon cannot be manipulated at will, thereby licensing any counterfactual scenario that one could possibly come up with. Quite on the contrary, as our discussion here and in Sect. 4 will clarify, a key constraint on counterfactual scenarios, both in general history and in history of science, is their plausibility.

Unsurprisingly, counterfactuals in history can be defended by means of various arguments. Nolan (2013), for instance, has recently enumerated no less than eight different reasons why practicing historians should care about them, and Tetlock and Belkin have emphasized how counterfactual speculations serve a variety of distinct —although clearly related—theoretical purposes, "from hypothesis generation to hypothesis testing, from historical understanding to theory extension" (1996, p. 16). Nevertheless, it is hard to exaggerate the importance that defenders of

---

[6]The tradition that characterizes counterfactuals in historical explanations as thought experiments dates back to Weber's 1905 essay "Objective Possibility and Adequate Causation in Historical Explanation," published in English in Weber (1949). I wish to thank Marco Buzzoni for pointing this out to me.

[7]As Frigg and Hartmann (2012) put it, idealized models characterized in this way instantiate the so-called "Aristotelian idealization, which amounts to 'stripping away,' in our imagination, all properties from a concrete object that we believe are not relevant to the problem at hand. This allows us to focus on a limited set of properties in isolation."

counterfactuals in history attach to their connection with causal explanations: counterfactuals, causal claims, and historical explanations are, in Johannes Bulhof's words, "three sides of the same strange three-sided coin; you cannot have one without the other two" (1999, p. 147).

In most cases, such enthusiastic embrace of counterfactual speculations goes hand in hand with the acute awareness that not all of them are equal. And although the difference between sound alternative history on the one hand and history-fiction on the other hand is not easily spelled out in the abstract, advocates of counterfactual history have tried to lay down some best practice rules and guidelines. Tetlock and Belkin (1996, pp. 16–31), for instance, proposed a number of criteria for assessing the "legitimacy, plausibility, and insightfulness" of counterfactual speculations, with the aim "to initiate a sustained conversation [...] on what should count as a compelling counterfactual argument" (ibid., p. 17). Martin Bunzl's paper "Counterfactual History: a User's Guide" (2004) is one major contribution to such conversation. As we shall see in the remaining of the present section, Bunzl proposes a solution not only to the problem of telling sound from unsound counterfactuals, but also to that of linking counterfactuals to evidence. The latter problem is no less important than the former. In fact, critics of counterfactual history point out that, since there is no way to ascertain what would have happened, had things gone differently at some juncture in the past, there is simply no point in asking such questions. By definition, the critics insist, contrary-to-fact reasoning does not belong to the domain of proper historical inquiry, which is aimed at establishing what claims concerning the past the available evidence licenses.

Now in a counterfactual conditional—such as, for instance, "Had Adolf Hitler died in the trenches during World War I, no attempt at a final solution of the so-called *Judenfrage* would have been carried out by the German government during the 1940s'"—the antecedent is, of course, false. Those asserting the counterfactual then have to deal with an obvious question: how can one tell what would have happened, had Hitler died in the trenches during World War I? More generally: on what basis can one claim that the consequent of a counterfactual follows from its antecedent? When can a counterfactual conditional be asserted? Bunzl forcefully argues that sound counterfactual reasoning "can be grounded" (ibid., p. 845). In fact, sound counterfactual conditionals are *plausible* ones, which "bear certain evidential markings that we can learn to read" (ibid., p. 849). More specifically, a sound counterfactual conditional can be brought into contact with evidence—if only indirectly—in such a way that its appraisal becomes feasible. This requires one to view the plausibility of the counterfactual as depending not only on the plausibility of the antecedent, but also on the *plausibility of the counterfactual inference*, i.e., the inference that the consequent of the counterfactual follows from the antecedent. On Bunzl's account, in order for such an inference to be plausible, it must be derivable from the conjunction of the antecedent with appropriate background conditions. Among such appropriate conditions, Bunzl lists "established theoretical and statistical generalizations" (ibid., p. 849), possessing the property of projectability, with which the counterfactual conditional has to be consistent.[8] The example in the next paragraph, proposed by Bunzl himself, illustrates the point quite clearly.

Historian Albert Gunns has explained the collapse of the Tacoma Narrows bridge—blown down by winds of 42 miles per hour in 1940, shortly after the construction was completed—as the consequence of poor design. In fact, budgetary constraints led to the design of a two-lane deck, narrower and lighter than the decks of other contemporary suspension bridges that never exhibited the problems that put a premature end to the life of the Tacoma Narrows bridge. Normal design practice, Gunns argues, would have dictated the building of a roadbed more than two lanes wide, which "*would had resulted in a deck that was heavier and more rigid and therefore less susceptible to aerodynamic effects*" (Gunns, quoted by Bunzl 2004, p. 850). Gunns counterfactually claims that, had the prescriptions of normal bridge design practice been followed, the bridge would not have collapsed. The plausibility of the counterfactual conditional asserted by Gunns, Bunzl suggests, depends on the plausibility of the laws of mechanics from which (in conjunction with the antecedent) the consequent can be derived.[9] Since such laws are considered as plausible because of their positive instantiations, the counterfactual conditional is connected to the evidence, if only indirectly, via the laws involved.

The above example admittedly concerns the residual case in which a historian can "borrow" from other disciplines the laws that ground the counterfactual conditional, and therefore make the counterfactual inference plausible. However, Bunzl insists, his defense of counterfactuals in history can be readily extended to cases in which no such laws are available. In these cases, Bunzl hastens to add, "considerations of rationality stand in for them" (ibid., p. 852). Here we shall not follow Bunzl's discussion of such cases in depth. What needs to be emphasized is that on his account, when historians assess causal claims, they deploy the standards of judgment endorsed by their professional community; the very same standards are invoked, although "to a different purpose" (ibid., p. 855), when counterfactual claims are assessed. In short, there are various ways to ground counterfactuals, and one ought not to be misled by the fact that the laws and theories deployed by historians are, in most cases, "what philosophers call 'folk theories'" (ibid.).[10] The crucial point, Bunzl insists, is that sound counterfactuals can be grounded based on generalizations—very often low-level ones, and subject to exceptions—that express

[8]These remarks raise the well-known problems faced by the consequentialist approach to counterfactuals, such as, for instance, the delimitation of the set of background conditions, which Bunzl seems to sidestep. Here, however, we are not concerned with the viability of Bunzl's avowal of counterfactual history; rather, we discuss it at some length because it is one of the most thorough attempts to spell out the solution to the problem of telling sound from unsound counterfactuals in the case of general history.

[9]This specific counterfactual explanation is rendered intuitively even more plausible by the fact that, when the bridge was reconstructed, it had a roadbed with four lanes, and it never suffered from the problems that led to the collapse of the first Tacoma Narrows bridge (Bunzl 2004, p. 851).

[10]This point was famously made by Hempel in his paper "The Function of General Laws in History" (1942).

our beliefs concerning how things go under normal circumstances, or in most cases, and therefore help us to assess how things would have gone, had some circumstances been different.

As hinted above, Bunzl's paper is only one of the contributions to the ongoing discussion on the identification of compelling counterfactual arguments. Unsurprisingly, champions of counterfactuals in history are not a school characterized by unanimity, and there is a continuing disagreement among them concerning how exactly one should spell out the insight concerning the connection between causal claims and counterfactuals. For instance, Nolan (2013, esp. Sect. 2) has challenged Tetlock and Belkin's (1996) and Lebow's (2000) characterization of such connection, and has insisted on the well-known difficulties faced by the consequentialist approach to counterfactuals deployed by Bunzl, as well as by the attempts to analyze causation in terms of counterfactuals.

There is, however, a crucial point of agreement, concerning the fact that the judgments that historians make when trying to ground a counterfactual claim are *continuous* with the ones that they make when trying to justify causal claims. As Nolan (2013, p. 329) put it:

> Provided our ordinary competent reasoning about causation employs counterfactuals, and especially if it does so in a way that cannot be codified easily and is not by explicit algorithms, then a historian relying on that capacity should not also reject reasoning using counterfactuals, on pain of their principles not lining up with their practice. I think this is the best way to argue that the close links between causal and counterfactual judgments mean that historians interested in causation (and most, if not all, should be) should not reject a role for counterfactuals in historical reasoning.

To briefly sum up our discussion in the present section, defenders of counterfactuals in general history take them to be an essential ingredient of historical explanations. Despite their disagreements over the best way to characterize the connection between counterfactuals, causal claims, and historical explanations, and over the best way to tell sound counterfactual speculations from wild flights of fancy, advocates of counterfactuals in history maintain that contrary-to-fact speculations lend themselves to rational scrutiny. The crucial insight here is that, in order for this to happen, counterfactuals need to exhibit the right kind of *plausibility*: the historians' more or less "implicit sense of what is likely to have depended on what" (ibid., p. 328) allows them to discriminate between good and bad candidate counterfactuals. Therefore, although general, formal criteria for spotting sound counterfactual speculations are anything but easy to come by, the bottom line of the defense of counterfactuals in general history can be couched in terms of the *counterfactuals' consistency with what historians know about the world*. It is the historians' admittedly partial and imperfect knowledge of how the world works, and of how historical actors typically behave in this world, that allows them to recognize sound counterfactuals. As the examples discussed in the next section illustrate, the plausibility of the alternative scenarios is a crucial feature also in the case of counterfactual histories of science.

4 On the Uses of Counterfactual History of Science

Just like counterfactual speculations in general history, not all counterfactual histories of science are equal. For instance, they may greatly differ from one another with respect to the kind of divergences from the historical record that they postulate. One may imagine an alternative history in which Charles Darwin dies prematurely and is therefore unable to put forward the theory of evolution by natural selection, but one of his contemporaries—say, Alfred Russel Wallace—readily replaces him as the first proponent of the idea. This would certainly qualify as a counterfactual history of science, since as we know, *On the Origin of Species* was published in 1859 and Darwin died in 1882, aged 73, but in such a scenario, his premature death would be quite inconsequential for the trajectory followed by scientific inquiry. From the point of view of the inevitability versus contingency of science debate, a much more interesting counterfactual history would be one in which Darwin does not feature and his premature death does make a difference to the ensuing development of biology. Historian of science Peter J. Bowler has conjured up one such counterfactual history.

In *Darwin Deleted: Imagining a World Without Darwin* (2013), Bowler devises a counterfactual world characterized by the fact that Darwin prematurely dies in 1832. In Bowler's narrative, however, none of the contemporaries comes up with the exact same idea: competing theories of evolution are put forward (as they were in actual history of science), but natural selection is not "in the air." Natural selection, Bowler insists, "was by no means an inevitable expression of mid-nineteenth-century thought," and in that historical moment only Darwin, by virtue of his unique, "right combination of interests" (ibid., p. 31), could formulate it. In this Darwinless world,

> it would have taken until the early twentieth century for the theory of natural selection to come to the attention of most biologists. Evolution would have emerged; *science would be composed by roughly the same battery of theories we have today, but the complex would have been assembled in a different way*. In our world, evolutionary developmental biology had to challenge the simpleminded gene-centered Darwinism of the 1960s to generate a more sophisticated paradigm. In the non-Darwinan world, the developmental model would have been dominant throughout and would have been modified to accommodate the idea of selection in mid-twentieth century (ibid., p. 9, emphasis added).

It should be noted that Bowler is not particularly interested in the inevitability versus contingency of science debate as such: he openly admits that his investigations in counterfactual history of biology chiefly stem from the desire to defend Darwin's theory by refuting "the claim that the theory of natural selection inspired the various forms of social Darwinism" (ibid., p. 10). According to him, the crucial point that his counterfactual narrative helps to establish is that, had Darwin not published *On the Origin of Species*, "racism and various ideologies of individual and national struggle would have flourished just the same" (ibid., pp. 10–11), their justification being derived from other theories of evolution. In any case, Bowler's counterfactual history has a clear contingentist flavor: the alteration of the historical record at a crucial juncture leads to a science that is (if only slightly) different from the one that we currently embrace.

For our present purposes, what matters is that Bowler's counterfactual history is not a wild, unrestrained flight of fancy. In this regard, he insists that "the viability of the counterfactual world can be substantiated by hard facts," since "we have enough evidence from our own world to show that the alternative could work" (ibid., p. 30). Indeed, the historical record tells us what the rivals to natural selection were, and we know for a fact that such rivals were viable, since they were widely accepted by scientists before the triumph of Darwin's theory.

These remarks suggest that just like in general history, within history of science good counterfactual speculations need to exhibit the right kind of plausibility. In Bowler's case, decades of scholarly engagement with nineteenth century science provide the historian with in-depth knowledge of the institutional setting in which scientific inquiry was conducted, the evidence that was available to the scientific community, the alternatives to natural selection that researchers pursued, the personalities and the interests of the scientists involved, etc. Based on his extensive knowledge of the actual world, and his "sense of what is likely to have depended on what" (Nolan 2013, p. 328), Bowler can not only defend Darwin's theory, but also *explain* Darwin's uniqueness, making a strong case for the claim that had it not been for Darwin, natural selection would have entered the scene much later. In other words, Bowler created a very rich idealized explanatory model, in which a lot of relevant information is included. His Darwinless world is such that the elements featuring within the model are under the control of the researcher, who can therefore manipulate them in a fine-grained way, within the limits imposed by the constraint of plausibility. Needless to say, opinions differ concerning the acceptability of various aspects of Bowler's elaborate account of a world without Darwin.[11] Nevertheless, as Tetlock, Lebow, and Parker remarked, "not all counterfactual thought experiments are equally subjective and therefore equally speculative" (2006, p. 9); Bowler's book, it seems to us, nicely illustrates the point with respect to the case of counterfactual history of science.

In the introduction to the symposium entitled "Counterfactuals and the Historian of Science," published in *Isis*, Gregory Radick commented that "remarkably little systematic attention" (2008, p. 548) has been devoted to counterfactual history of science. More recently, Léna Soler remarked that "counterfactual thinking *about science*" and its history, as opposed to the use of counterfactuals within science, "remains an underdeveloped activity" (2015, p. 9). Our lengthy discussion in Sect. 3, together with the above summary of Bowler's ventures in a Darwinless world, ought to be viewed as an attempt to contribute, if modestly, to the most recent attempts to engage in such an activity.

To the best of our knowledge, only cursory remarks concerning the similarities between the use of counterfactuals in general history and the use of counterfactuals

[11]See, e.g., the review symposium of *Darwin Deleted*, featuring Alan C. Love, Robert J. Richards, and Bowler himself (Love et al. 2015).

in history of science are to be found in the relevant literature. We should nevertheless mention that the first example of alternative history discussed in the present section, in which the explanatory role of counterfactuals is emphasized, is far from exhausting the range of uses to which counterfactual history of science has been put, both within and without the inevitability versus contingency of science debate.[12] Indeed, the examples discussed in the following make it clear that counterfactual history of science has often played *a critical, as opposed to an explanatory, role*, so that occasionally, the motive underlying the deployment of counterfactual histories of science has been an "if only," instead of a "what if" question. Accordingly, alternative histories have been deployed not only in order to challenge the inevitability thesis and the interpretation of the success of scientific theories typically associated with it (that is, scientific realism, or at least some versions of it). As we shall see in a moment, alternative histories are also a tool used to criticize both the epistemic authority with which science is credited and the social order allegedly going hand in hand with it.

The works of some famous feminist scholars provide examples of this latter use. In *The Death of Nature* (1980), historian Carolyn Merchant has claimed that, had the social milieu in which Western science emerged been different, the ensuing trajectory followed by scientific inquiry would have been significantly different. Her work may therefore be viewed as providing a historical backing to the claims put forward by physicist Evelyn Fox Keller, who in a number of papers collected in *Reflections on Gender and Science* forcefully argued that "were more women to engage in science, a different science might emerge" (Keller 1985, p. 76).[13]

Within an analysis of the attitudes towards counterfactuals characterizing historians of science of different philosophical persuasions, John Henry has discussed the case of such "sweeping changes" (2008, p. 557) as a cultural background completely different from that in which Western science arose and developed. In Henry's view, despite the efforts of Fox Keller, Merchant, and other feminist historians and philosophers of science, "it is by no means clear what [the] alternative, and supposedly very different, feminist science would look like" (2008, pp. 558–559). Henry hastens to add that this does not reflect on the central claims of their philosophical agendas, which do not stand or fall with the soundness of counterfactual speculations. Whatever the case, for our present purposes it is important to emphasize that when counterfactual history of science is deployed in a critical role, the counterfactual scenario needs to exhibit the same kind of plausibility that is required when counterfactual history of science is used, as in the *Darwin Deleted* example discussed above, for explanatory purposes. As Soler put it, "what confers plausibility to a counterfactual scientific narrative is its 'close connection' to the actual history of science" (2015, p. 10).

[12] I wish to thank two anonymous reviewers for pressing me to address this issue.

[13] Admittedly, neither Fox Keller nor Merchant explicitly deploy full-blown counterfactual narratives, which they nevertheless plainly evoke.

The notion of "close connection" is of course vague, and different authors disagree on the plausibility of specific instances of counterfactual history of science. In any case, when wide-ranging differences—such as a completely different cultural background—between the actual historical record and the counterfactual narrative are postulated, the resulting scenario is very likely to be impervious to assessments concerning its plausibility (on this, see among others Kinzel 2015, p. 61). For instance, one may ask whether a science emerging from a set of completely different initial conditions would be an investigation of roughly the same phenomena that constitute the subject matter of our current science. In fact, in order for an alternative to our science to qualify as an alternative, it has to deal with the same material with which our science deals: as Trizio vividly put it, "there is little epistemological interest in comparing what our science says about planets with what one might have ended up thinking about viruses" (2015, p. 130). Given a very different developmental history, it is certainly legitimate to postulate a very different outcome, but it is of crucial importance being able to specify, exactly, different in what respects. Wide-ranging differences in the initial conditions make it very difficult to speculate on the trajectory that an alternative science would follow, and such alternative may well end up being simply incommensurable with our current science.

To put the point differently, we may say that wide-ranging differences such as those postulated by Merchant often lead to underdescribed idealized models. In a well-designed idealized model, the researcher can control the elements constituting it, which lend themselves to a fine-grained manipulation allowing one to study their relationships. When wide-ranging differences in the initial conditions are postulated, and the model is not supplemented with relevant information constraining it, at most coarse-grained manipulations of the elements within the model are possible, so that the researcher's control over it is relinquished. Compare such cases with the scenario conjured up in *Darwin Deleted*. In Bowler's model, every other element being equal and under the control of the researcher, Darwin's premature death leads to a counterfactual scenario whose plausibility can be assessed, due to the fact that it is constrained by relevant information, that is, the historian's knowledge of a specific phase of the history of biology. As the reader will recall, the continuity between counterfactual scenarios and what historians know about the world is the bottom line of the defense of counterfactuals within general history. Analogously, the plausibility of a counterfactual history of science can be characterized in terms of its continuity with what historians know about the world. Such continuity, we suggest, is quite difficult to attain in the case of underdescribed idealized models.

Coming now back more specifically to the inevitability versus contingency of science debate, it is far from surprising that counterfactual history of science has been repeatedly used in order to challenge scientific realism. Perhaps the first author to discuss at some length the connections between the various positions that, in principle, can be defended within the debate, Radick has shown that the two couples inevitabilism-realism and contingentism-antirealism "by no means exhaust" (2005, p. 24) the space of possibilities. In a similar vein, Howard Sankey has argued that "scientific realism has no evident implications with regard to the inevitability of science" (2008, p. 259). Nevertheless, it is a fact that realists tend to group around the inevitability thesis, while antirealists tend to favor the contingency thesis. Indeed, one of the most celebrated examples of history of science emphasizing contingency, James Cushing's *Quantum Mechanics. Historical Contingency and the Copenhagen Interpretation* (1994), to which we shall now briefly turn, has a distinctively antirealist flavor.

Famously, in 1952 physicist David Bohm put forward a version of quantum mechanics which, empirically equivalent to the so-called "Copenhagen interpretation"—since the 1920s, the standard version of the theory—is nevertheless incompatible with it from the ontological point of view. According to Cushing, given that the two versions of the theory make the same predictions, the fact that the Copenhagen interpretation became the standard version of quantum mechanics crucially depends on the vagaries of historical development. In fact, based on a detailed examination of the relevant period of the history of physics, Cushing claims that Bohm's version of the theory may well have been devised during the 1920s. However, nobody came up with it at the right point in time, and consequently, the Copenhagen interpretation acquired its status of a standard irrespectively of the merits of Bohmian mechanics, which came to the scene much too late to ever manage to win a comparable support among physicists. In Cushing's reconstruction, then, the order in which ideas are introduced within a scientific debate can decisively influence its outcome: had Bohm's version of the theory come first, history might have unfolded differently, and today the majority of scientists may well be championing Bohmian mechanics as their favorite interpretation of quantum mechanics.

Cushing's narrative, revolving around the role of historical contingency, exhibits an immediate connection to a classic argument against scientific realism, namely, the underdetermination thesis: two competing theories are empirically indistinguishable, and yet posit incompatible underlying ontologies, so that one is left wondering which one of them is true. Besides such connection with the dispute over the merits of realism, what makes Cushing's work of particular interest for the inevitability versus contingency debate is that, as Trizio (2015, p. 129) points out, his case for the possible triumph of Bohm's interpretation of quantum mechanics satisfies a number of demanding constraints, which ensure the plausibility of the counterfactual scenario. First of all, the subject matter dealt with by the theory prevailing in the alternative history of science is the same as that dealt with by the actually prevailing theory. Secondly, Bohm's interpretation of quantum mechanics is as successful as the Copenhagen interpretation. Thirdly, there is a fundamental, irreconcilable disagreement between the two theories. Not unlike Bowler, Cushing created a very rich idealized model, in which a lot of relevant information, providing the necessary constraints on the counterfactual scenario, is included. The crucial element within the model—"the temporal order of events that actually took place in the mind of a handful of researchers" (ibid, p. 134)—is clearly under the control of the historian, and can be manipulated in a fine-grained manner. This counterfactual narrative therefore provides a very effective way to defend the contingency thesis, and to argue that the path followed by scientific inquiry can be influenced by a variety of factors, including the theoretical idiosyncrasies of the researchers involved in the development of a specific field.

Counterfactual history of science covers an extended and uneven territory. Such territory borders on the one side with actual history of science and on the other side with merely logical possibilities, and includes "scenarios that would only slightly differ from our actual history of science, and more creative science fiction (for instance involving twin-earth-like planets or alien beings as the subjects of science)" (Soler 2015, p. 10). As our discussion illustrates, the same kind of plausibility that characterizes good counterfactual speculations within general history will be of great help in navigating the territory of counterfactual history of science.

5 Concluding Remarks

In this paper, we suggested that the plausibility of the devised alternative scenarios is as important in the defense of counterfactual history of science as it is in the defense of counterfactuals within general history. Nevertheless, at least one worry raised by counterfactual history of science as used to support the contingency thesis needs to be briefly mentioned here.

The issue can be introduced by considering Ian Hasketh's (2014) thoughtful review of Bowler's book. In the opening chapter of *Darwin Deleted*, Bowler argues that the contingency of the past means that apparently irrelevant events may lead to unintended, unpredictable, and wide-ranging consequences, and explicitly refers to Gould's claim that, were it possible to replay the tape of life, the outcome would very likely be completely different, due to the intervening contingencies affecting evolutionary history. Nevertheless, Hasketh points out, "there is nothing entirely random or unpredictable or accidental" (ibid., p. 301) in how the events in Bowler's counterfactual history unfold, except for Darwin's premature death, after which no further contingencies play a role within the narrative. Bowler himself suggests that his story "follows a more natural sequence of discovery" (2013, p. 279) than that followed by actual history of science, which was "disturbed" by the appearance of Darwin's theory of natural selection.[14] In brief, Bowler's narrative ends up with a science that is slightly different from ours, but in which natural selection *does* feature. However, as Hasketh remarks, if contingency is taken seriously, then it cannot be used at the beginning of the counterfactual history and ignored at later stages of development: "a truly contingent narrative can only follow a definable course until the next contingency arises" (2014, p. 302; on this, see also Ben-Menahem 1997).

It seems to us that Hasketh's remark generalizes and, as historian Allan Megill nicely put it, "contingency is not a train one can get on and off at will" (quoted in

[14]A short paper by Bowler, anticipating the themes of the book, is tellingly entitled: "What Darwin Disturbed. The Biology that Might Have Been" (2008).

Hasketh 2014, p. 302). In other words, contingency may well end up with undermining counterfactual history of science as used to support the contingency thesis. In fact, after the first alteration in the historical record is introduced, why should one believe that there will be no further alterations, leading the resulting narrative farther and farther away from the historical record? This possibility may be viewed as completely unproblematic, since it is precisely the hardcore of contingentism. Nevertheless, such possibility should worry contingentists wishing to use counterfactual history of science to defend the contingency thesis. In fact, the farther one gets from the historical record, the more difficult it becomes to constrain the counterfactual scenario in such a way as to ensure its plausibility. When contingency is embraced, how can one rule out narratives leading to scenarios that cannot be rationally assessed?

Acknowledgments This chapter is based on material presented at the MBR'015 conference, held in Sestri Levante in June 2015. Thanks are due to the audience, and especially Mario Alai, Marco Buzzoni, and Darrell Rowbottom, for questions and suggestions concerning my presentation. Insightful comments by Gustavo Cevolani, Roberto Festa, and two anonymous reviewers on previous versions of the paper led to important improvements in the final version. Usual caveats apply. Financial support from PRIN grant *Models and Inferences in Science* is gratefully acknowledged.

References

Ben-Menahem, Y. (1997). Historical contingency. *Ratio, 10*, 99–107.

Bowler, P. J. (2008). What Darwin disturbed. The biology that might have been. *Isis, 99*, 560–567.

Bowler, P. J. (2013). *Darwin deleted: Imagining a world without Darwin*. Chicago: The University of Chicago Press.

Bulhof, J. (1999). What if? Modality and history. *History and Theory, 38*, 145–168.

Bunzl, M. (2004). Counterfactual history: A user's guide. *The American Historical Review, 109*, 845–858.

Carr, E. H. (1961). *What is history?* London: Macmillan.

Cushing, J. T. (1994). *Quantum mechanics. Historical contingency and the Copenhagen hegemony*. Chicago: The University of Chicago Press.

Elster, J. (1978). *Logic and society. Contradictions and possible worlds*. New York: Wiley.

Evans, R. (2014). *Altered pasts: Counterfactuals in history*. Waltham: Brandeis University.

Feyerabend, P. K. (1999). *Conquest of abundance: A tale of abstraction versus the richness of Being*. In B. Terpstra (Ed.). Chicago: The University of Chicago Press.

Fox Keller, E. (1985). *Reflections on gender and science*. New Haven: Yale University Press.

French, S. (2008). Genuine possibilities in the scientific past and how to spot them. *Isis, 99*, 568–575.

Frigg, R., & Hartmann, S. (2012). Models in science. In E. N. Zalta (Ed.), *The Stanford encyclopedia of philosophy* (Fall 2012 Ed.). URL http://plato.stanford.edu/archives/fall2012/entries/models-science/

Fuller, S. (2008). The normative turn. Counterfactuals and a philosophical historiography of science. *Isis, 99*, 576–584.

Giere, R. N. (2015). Contingency, conditional realism, and the evolution of the sciences. In L. Soler, E. Trizio, & A. Pickering (Eds.), *Science as it could have been. Discussing the contingency/inevitability problem* (pp. 187–201). Pittsburgh: University of Pittsburgh Press.

Gould, S. J. (1989). *Wonderful life. The Burgess shale and the nature of history*. New York: Norton.
Hacking, I. (1999). *The social construction of what?* Cambridge: Harvard University Press.
Hacking, I. (2000). How inevitable are the results of successful science? *Philosophy of Science, 67*, S58–S71.
Hempel, C. G. (1942). The function of general laws in history. As reprinted in C. G. Hempel, *Aspects of scientific explanation and other essays in the philosophy of science*. New York: The Free Press.
Henry, J. (2008). Ideology, inevitability, and the scientific revolution. *Isis, 99*, 552–559.
Hesketh, I. (2014). Darwinian we are not. Counterfactualism as the natural course of history. *History and Theory, 54*, 295–303.
Kaye, S. T. (2010). Challenging certainty. The utility and history of counterfactualism. *History and Theory, 49*, 38–57.
Kidd, I. J. (2011). The contingency of science and the future of philosophy. *Essays in Philosophy, 12*, 312–328.
Kidd, I. J. (2016). Inevitability, contingency, and epistemic humility. *Studies in History and Philosophy of Science, 55* 12–19.
Kinzel, K. (2015). Are the results of science contingent or inevitable? *Studies in History and Philosophy of Science, 52*, 55–66.
Kuhn, T. S. (1962/1970). *The structure of scientific revolutions*. Chicago: The University of Chicago Press.
Lebow, R. N. (2000). What's so different about a counterfactual? *World Politics, 52*, 550–585.
Love, A. C., Richards, R. J., & Bowler, P. J. (2015). What-If history of science. *Metascience, 24*, 5–24.
Martin, J. D. (2013). Is the contingentist/inevitabilist debate a matter of degree? *Philosophy of Science, 80*, 919–930.
Merchant, C. (1980). *The death of nature. Women, ecology, and the scientific revolution*. San Francisco: Harper & Row.
Nolan, D. (2013). Why historians (and everyone else) should care about counterfactuals. *Philosophical Studies, 163*, 317–335.
Radick, G. (2005). Other histories, other biologics. In A. O'Hear (Ed.), *Philosophy, biology, and life* (pp. 21–47). Cambridge: Cambridge University Press.
Radick, G. (2008). Why what if? *Isis, 99*, 547–551.
Rosenfeld, G. D. (2002). Why do we ask "what if?". Reflections on the function of alternate history, *History and Theory, 41*, 90–103.
Rosenfeld, G. D. (2014). Whither "What If" History? *History and Theory,* 53, 451–457.
Sankey, H. (2008). Scientific realism and the inevitability of science. *Studies in History and Philosophy of Science, 39*, 259–264.
Soler, L. (2008). Revealing the analytical structure and some intrinsic major difficulties of the contingentist/inevitabilist issue. *Studies in History and Philosophy of Science, 39*, 230–241.
Soler, L. (2015). Introduction: The contingentist/inevitabilist debate. In L. Soler, E. Trizio, & A. Pickering (Eds.), *Science as it could have been. Discussing the contingency/inevitability problem* (pp. 1–42). Pittsburgh: University of Pittsburgh Press.
Soler, L., Trizio, E., & Pickering, A. (2015). (Eds.) *Science as it could have been. Discussing the contingency/inevitability problem*. Pittsburgh: University of Pittsburgh Press.
Sunstein, C. R. (forthcoming). Historical explanations always involve counterfactual history. *Journal of the Philosophy of History*.
Tambolo, L. (2014). Pliability and resistance. Feyerabendian insights into sophisticated realism. *European Journal for the Philosophy of Science, 4*, 197–213.
Tetlock, Ph. E, & Belkin, A. (1996). Counterfactual thought experiments in world politics: Logical, methodological, and psychological perspectives. In P. E. Tetlock & A. Belkin (Eds.), *Counterfactual thought experiments in world politics: Logical, methodological, and psychological perspectives* (pp. 1–38). Princeton: Princeton University Press.

Tetlock, Ph. E, Lebow, R. N., & Parker, G. (Eds.). (2006). *Unmaking the west. "What-if" scenarios that rewrite world history*. Ann Arbor: The University of Michigan Press.

Trizio, E. (2008). How many sciences for one world? Contingency and the success of science. *Studies in History and Philosophy of Science, 39*, 253–258.

Trizio, E. (2015). Scientific realism and the contingency of the history of science. In L. Soler, E. Trizio, & A. Pickering (Eds.), *Science as it could have been. Discussing the contingency/inevitability problem* (pp. 129–150). Pittsburgh: University of Pittsburgh Press.

Weber, M. (1949). *The methodology of the social sciences*. In E. A. Shils & H. A. Finch (Eds.). Glencoe: The Free Press.

Yerxa, D. A. (Ed.). (2008). *Recent themes in historical thinking. Historians in conversation*. Columbia: The University of South Carolina Press.

Models, Brains, and Scientific Realism

Fabio Sterpetti

Abstract Prediction Error Minimization theory (PEM) is one of the most promising attempts to model perception in current science of mind, and it has recently been advocated by some prominent philosophers as Andy Clark and Jakob Hohwy. Briefly, PEM maintains that "the brain is an organ that on average and over time continually minimizes the error between the sensory input it predicts on the basis of its model of the world and the actual sensory input" (Hohwy 2014, p. 2). An interesting debate has arisen with regard to which is the more adequate epistemological interpretation of PEM. Indeed, Hohwy maintains that given that PEM supports an inferential view of perception and cognition, PEM has to be considered as conveying an internalist epistemological perspective. Contrary to this view, Clark maintains that it would be incorrect to interpret in such a way the indirectness of the link between the world and our inner model of it, and that PEM may well be combined with an externalist epistemological perspective. The aim of this paper is to assess those two opposite interpretations of PEM. Moreover, it will be suggested that Hohwy's position may be considerably strengthened by adopting Carlo Cellucci's view on knowledge (2013).

Keywords Prediction error minimization · Scientific realism · Analytic method · Perception · Epistemology · Knowledge · Infinitism · Naturalism · Heuristic view

1 Introduction

Prediction Error Minimization theory (PEM) is one of the most promising attempts to model perception in current science of mind, and it has recently been advocated by some prominent philosophers as Clark (2015, 2013a, b) and Hohwy (2015, 2014, 2013).

F. Sterpetti (✉)
Sapienza University of Rome, Department of Philosophy,
via Carlo Fea 2, 00161 Rome, Italy
e-mail: fabio.sterpetti@uniroma1.it

L. Magnani and C. Casadio (eds.), *Model-Based Reasoning in Science and Technology*, Studies in Applied Philosophy, Epistemology and Rational Ethics 27, DOI 10.1007/978-3-319-38983-7_36

Briefly, PEM maintains that "the brain is an organ that on average and over time continually minimizes the error between the sensory input it predicts on the basis of its model of the world and the actual sensory input" (Hohwy 2014, p. 2). Top down predictions and bottom-up sensory signals combine to produce "a kind of internal model of the source of the signals: the world hidden behind the veil of perception" (Clark 2013b, p. 185).

This approach moves along that line of research which looks at the brain as an 'inferential machine', initiated by Helmholtz (1867) and continued, among others, by Gregory (1980), Rock (1983), and Frith (2007).

An interesting debate has recently arisen with regard to which is the most adequate epistemological interpretation of PEM. The debate focused mainly on how the relation between the inner model of the world produced by the brain and the external world should be interpreted.

Indeed, Hohwy (2014) maintains that given that PEM supports an inferential view of perception and cognition, PEM has to be considered as conveying an internalist epistemological perspective.[1] Thus, if we accept that PEM is a reliable description of the mind, we should coherently draw the conclusion that we cannot reach knowledge of the way the world *really* is, i.e. the way it is independently of our mind, because of the indirectness of the relation between our inner model of the world and the modeled world. For example, Hohwy states that "perceptual content is the predictions of the currently best hypothesis about the world" (Hohwy 2013, p. 48). If this is the case, we cannot be sure that our best hypothesis truly corresponds to the world, because the brain cannot "simultaneously access both the internal estimates and the true states of affairs in the world" (Hohwy 2014, p. 4). The brain can only have access to the two homogeneous things that it can compare, namely the predicted and the actual input: "there is no possibility of independent evidence, which would require us to crawl outside of our own brains" (Ibidem, p. 7). Following this line of reasoning, since there is in principle no possibility of comparing our representation of the world to the world itself, it seems fair to conclude that PEM should be considered at odds with Scientific Realism (SR), the mainstream metaphysical view in philosophy of science according to which our best scientific theories are true and we can safely infer their truth from their empirical success (Psillos 1999).[2]

---

[1]On the internalist and externalist conceptions of epistemic justification see Pappas (2014).

[2]It may be objected that the scientific realist view would be better described as follows: supposing that empirical successful theories are true (or approximately true) provides the best explanation for their empirical success. But this 'explanationist' formulation of scientific realism is almost equivalent to that given above. The fact is that scientific realists usually consider Inference to the Best Explanation a valid and truth-conducive inference. For example, Harman describes IBE as follows: "one infers, from the premise that a given hypothesis would provide a 'better' explanation for the evidence than would any other hypothesis, to the conclusion that the given hypothesis is true" (Harman 1965, p. 89). So, if truth is the best explanation of success, and IBE leads to truth, an IBE may be performed to conclude that it is true that a successful theory is true. So, we can infer the truth of a theory from its success. Thus, those two formulations of realism are almost equivalent. I wish to thank an anonymous reviewer for having raised this issue.

Contrary to this view, Clark (2013a) maintains that it would be incorrect to interpret in such a way the indirectness of the link between the world which is modeled and our inner model of the world, because the relation between the world and perception is indeed a direct causal relation. Thus, even if our representations of the world are internal and may be in some sense deemed indirect, the causal connection between the world and our brains which produces such representations guarantees that what is perceived is not just "the brain's best hypothesis. Instead, what we perceive is the world" (Clark 2013a, p. 492). On Clark's view, "biological beings are able to establish a truly tight mind-world linkage [...] by individual learning and evolutionary inheritance," and the inferential functioning of perception does not introduce any sort of "worrisome barrier between mind and world" (Clark 2013b, p. 199). Thus, according to Clark's interpretation, PEM should not be considered at odds with SR.

The aim of this paper is to assess those two opposite interpretations of PEM. In order to do this, PEM is sketched in Sect. 2; then Hohwy's and Clark's interpretations of PEM are presented in Sect. 3; in Sect. 4 those interpretations are evaluated, and it is argued that Hohwy's interpretation is more adequate to account for some of the salient epistemological features of PEM. In Sect. 5, some of the difficulties which still afflict Hohwy's position are underlined. Finally, in Sect. 6, it is suggested that Hohwy's position can be considerably strengthened by relating it to Carlo Cellucci's view on knowledge and science (2013), which will be briefly described.

2 Prediction Error Minimization Theory

PEM is an ambitious theory, which tries to account for the activity of the brain in a unified way. Indeed, according to PEM "prediction error minimization is the only principle for the activity of the brain" (Hohwy 2014, p. 2). PEM basically sees the brain as an organ that continually minimizes the error between the predicted sensory input and the actual sensory input. This view frames the activity of the brain into a wider conception according to which any self-organizing system that is at equilibrium with its environment must minimize its tendency to disorder. In this perspective the characteristic feature of living beings is their attempt to (locally) reduce entropy (Friston 2010). According to this view "biological agents must actively resist a natural tendency to disorder," and "agents are essentially inference machines that model their sensorium to make predictions, which action then fulfils" (Friston 2011, p. 89).

Since "the sum of prediction error over time is also known as free-energy, PEM is also known as the free-energy principle" (Hohwy 2014, p. 2). According to Friston's view, the free-energy principle says that "biological organisms on average and over time act to minimize free energy," and that "brains are hypothesis-testing neural mechanisms, which sample the sensory input from the world to keep themselves within expected states:" as "the heart pumps blood, the brain minimizes free energy" (Hohwy 2015, p. 2). Thus, according to PEM "the brain's main job is to maintain the organism within a limited set of possible states" (Ibidem), and many, if not all, brain functions may be accounted for in terms of free-energy minimization.

In other words, in order to maximize the chance of survival of an organism, the brain has to keep the organism in the range of states which are already known (through evolution, development, and learning) to be compatible with the existence of that organism. In order to do this the brain has to minimize 'surprise', which is a concept from information theory, "defined as the negative log probability of a given state, such that the surprise of a state increases the more improbable it is to find the creature in that certain state" (Ibidem).

The fact is that to accomplish its task, the brain cannot access directly the state of the world in which the organism is embedded. The brain has to create a model of the world, and try to anticipate and predict the incoming states of the world. The brain "must harbor and finesse a model of itself in the environment, against which it can assess the surprise of its current sensory input," because the brain has access only to two quantities, "which it can compare: on the one hand the predicted sensory input, and on the other the actual sensory input. If these match, then the model is a good one" (Ibidem, p. 3). At any stage of this process, the brain deals only with its own 'reconstruction' of what is going on both in the world (*exteroceptive* states) and in the organism (*interoceptive* states). Indeed, even the actual sensory input arriving at the brain cannot be conceived as a direct transferring of a bit of information from the world to the brain. At any stage, there is an 'inferential step' through which the brain models the environment, the organism, and the course of actions. The brain makes hypotheses based on previous knowledge to form a coherent representation of present and future states, but it is also ready to modify or update such hypotheses on the base of the actual input. At any given time t we cannot have the certainty that our hypotheses will not be modified at time $t + 1$.

It has to be stressed that PEM is a new way of accounting for perception, which contrasts the traditional "'passive accumulation' model of the perceptual process", which "depict[s] perception as a cumulative process of 'bottom-up' feature detection" (Clark 2013a, pp. 470–471). According to PEM the role of the *predictions* made by the brain is essential. It is exactly this feature of the brain functioning which accounts for the ability of the brain to relate to the world without having any direct access to it. The brain is seen as a hypotheses producer and verifier, a sort of 'predictive device' which continuously refines (or changes) its predictions.

The idea of the brain as an 'inference machine' was firstly articulated by Helmholtz (1867). PEM follows this line of reasoning and models brain activity in terms of statistical inferences over perceptual hypotheses. According to Hohwy, PEM is "inference to the best explanation, cast in [...] Bayesian terms" (Hohwy 2014, p. 5). The basic idea is that since the brain continuously checks how good its model of the world is by confronting its model with the actual sensory input, its activity may be described in Bayesian terms. Indeed, in a nutshell, Bayes' rule tells us to update the probability of a hypothesis h, given some evidence e, by considering the product of the likelihood, i.e. the probability of the evidence given the hypothesis P(e|h), and the prior probability of the hypothesis P(h). The resulting probability of the hypothesis is the posterior probability of such hypothesis:

$$\mathrm{P}(h|e) = \mathrm{P}(e|h)\mathrm{P}(h)/\mathrm{P}(e) \quad (1)$$

Thus, 'minimize the surprise' for the brain means to maximize the probability of its hypothesis: "if the prediction error is minimized then the likelihood has been maximized, because a better fit between the hypothesis and the evidence has been created. This in turn will increase the posterior probability, P(h|e), of the hypothesis" (Hohwy 2013, p. 46). According to PEM, the probability of a hypothesis h is continuously updated and refined, using the posterior probability of h at time t_n as the prior probability of h in the following inferential step, i.e. at time t_{n+1}: a "neat explanatory circle then seems to transpire: top-down priors guide perceptual inference, and perceptual inference shapes the priors" (Ibidem). The brain tries "to create a closer fit between the predictions […] and the actual sensory input. This corresponds to being less surprised by the evidence given through the senses" (Ibidem).

We can now recapitulate the three main tenets of PEM: (1) in order to account for perception, we should adopt an *inferential* conception of the mind; (2) the division between inner and outer is strict ('inferential *seclusion*' of the mind, see Hohwy 2014); (3) perception, attention, and action have to be conceived as *statistical* inferences.

3 Clark's Versus Hohwy's Interpretation of PEM

PEM is a stimulating and controversial proposal. There are many objections that can be (and have been) raised against this approach.[3] In what follows we will leave aside those objections. We will focus just on Clark's and Hohwy's interpretations of PEM, and on the issue of assessing which one should be preferred in the lights of the epistemological implications of adopting PEM as a theory of the mind.

3.1 The Epistemological Implications of PEM

In order to carry out our inquiry, we will take into considerations (and try to answer) two questions, the first one which can be dubbed 'metaphysical', the second one which can be dubbed 'epistemological'. The first question is: Is PEM compatible with SR? The second question is: Which epistemological position fits better PEM?

[3]Cf. e.g. Rescorla (2015).

For our purposes, we can here define SR as the metaphysical view according to which our best scientific theories are true, in the sense that they tell us precisely what exists in the world. For example, Ellis states that SR can be described as "a two-stage argument from the empirical success of science, to the truth, or approximate truth, of its dominant theories, to the reality of the things and processes that these theories appear to describe" (Ellis 2005, p. 372). Truth is normally intended by scientific realists as *correspondence*.[4] For example, Sankey states that: "correspondence theories which treat truth as a relation between language and reality are the only theories of truth compatible with realism" (Sankey 2008, p. 17). With regard to epistemology, we can here intend it in the broad sense of the philosophical inquiry on what makes some of our beliefs *knowledge*, i.e. justified, or true, or grounded.

3.2 Hohwy's Interpretation of PEM

According to Hohwy, PEM entails an internalist epistemological perspective. Internalism may be intended here in a broad sense as the epistemological view according to which what ultimately justifies any belief is some mental state of the epistemic agent holding that belief (Pappas 2014).[5] Since according to PEM the totality of the brain activity can be accounted for in terms of Bayesian inferences, and since in this line of reasoning knowledge cannot but be conceived as related to the brain activity, the inferential nature of such activity is related to the way in which our knowledge can be considered justified. In other words, since the brain has no direct access to the world, and knowledge is produced by the brain, knowledge cannot but be ultimately justified by the brain's activity itself. According to PEM "the brain is isolated behind the veil of sensory input" (Hohwy 2013, p. 238), and the human mind "appears very indirectly related to the world" (Ibidem, p. 90). Indeed, "mental states do not extend into the environment, and the involvement of the body and of action in cognition can be described in wholly neuronal, internal, inferential terms" (Hohwy 2014, p. 24).

[4]Many positions have been elaborated on the issue of truth, and even if truth as correspondence seems to be the most widespread view among scientific realists, not every scientific realist adopts such view. For simplicity here we will focus on correspondence.

[5]It may be objected that internalism is better described as the idea that justification requires awareness of the process that ultimately justifies a belief. But, in this context, such definition of internalism is equivalent to that given above. Indeed, according to PEM, what we can be really aware of are ultimately nothing but some mental states. So, even if the process that justifies a belief is an 'external' one, we will not be directly aware of such process. We will only be aware of the internal model of such process. So, if internalism is the view according to which justification requires awareness, and according to PEM we can be aware only of some mental states, then in this context internalism may be fairly defined as the view according to which a belief is justified by some mental state of the epistemic agent holding that belief. I wish to thank an anonymous reviewer for having raised this issue.

Briefly, Hohwy's argument runs as follows: we have to adopt an inferential conception of the mind, because otherwise we are not able to account for some very well known phenomena (e.g. binocular rivalry, see Hohwy 2013, Chap. 1). If we adopt an inferential conception of the mind, we cannot avoid to adopt an internalist epistemological perspective, since we cannot eliminate the separateness that characterizes an inferential conception of the mind. Thus, we have to accept that we are in principle not able to avoid some radical skeptical challenge, such as Cartesian skepticism. This is the epistemological price to pay if we want to adopt PEM as a theory of the mind. Indeed, "PEM must necessarily rely on internal representations of hidden causes in the world (including the body itself) in order to predict the sensory input that they give rise to" (Hohwy 2014, p. 17).

As we have already seen, at any given time t we cannot have the certainty that our hypotheses will not be modified at time $t + 1$. Moreover, it has to be stressed that even if the predicted hypothesis and the actual sensory input match, and even if this matching remains stable for a certain amount of time, this does not guarantee us that our hypothesis *truly* corresponds to the state of the world, and so that such hypothesis is true in a strong metaphysical sense. Indeed it could be possible that our sensory system and our internal model of the world both fail to detect and model some features of the world or some modifications of some detected features. Think to a modification that cannot be detected, because its magnitude is below the detectability threshold of our sensory system. In this case, the sensory input and the model would continue to match, while that modification would have nevertheless occurred. The fact is that in order to modify our hypothesis, we need to have some 'clues' that such hypothesis is incorrect. But it is not always easy to have an indication of the inadequacy of some of our hypothesis or recognize to which hypothesis a specific clue refers to. Indeed, perception solves "an *underdetermination problem*. The perceptual system estimates environmental conditions," and it "does so based upon *proximal stimulations* of sensory organs," but the proximal stimulations "underdetermine their environmental causes" (Rescorla 2015, p. 694). Thus, since the environmental causes are underdetermined, also the clues of inadequacy of our hypotheses, which are a subset of the environmental causes, are underdetermined.

The inferential nature of the mind, which makes us constantly prone to error and deception, but which, at the same time, ensures us the only possible way to effectively acting in the world, "should make us resist conceptions" according to which "the mind is in some fundamental way open or porous to the world" (Hohwy 2014, p. 1). If the mind were open to the world, the predictive machinery described by PEM would not be necessary. Indeed, PEM puts "the focus on the evidentiary boundary and the way it forces a clear distinction between internal states, where the prediction error minimization occurs, and hidden causes on the other side of the boundary, which must be inferred" (Ibidem, p. 7).

Moreover, PEM treats the inferential machinery of the mind in Bayesian terms. This means that the inferences that PEM deals with are *statistical* in character. But "any account that ties perceptual content to a statistical model within an evidentiary boundary will wedge apart the statistical model and the hidden causes it models" (Ibidem, p. 9). On this point Hohwy is very clear and states that:

> having access to rain samples and the mean of the rainfall is a very different thing from having access to the actual rainfall, even if the mean carries information about the rain. An explanation of this difference in the case of perceptual inference cannot soften the characterization of the hidden causes so they come to appear somehow unhidden.[6]

Thus, according to Hohwy, the three main tenets of PEM are deeply related: the "seclusion stems from the inferential component such that the upshot of the sub-personal processes is a probabilistically favoured statistical model" (Ibidem), and they jointly entail an internalist epistemological perspective.

This seems to mean that according to Hohwy's interpretation PEM is not compatible with SR. If scientific realists claim that the aim of science is truth, and usually adopt a non-epistemic conception of truth, according to which whether something is true does not depend on our mind, but depends exclusively on the way the world is,[7] this means that PEM cannot satisfy such a realist requirement. Indeed, according to PEM what we perceive or think cannot but ultimately rest on and be justified by our brain activity, and we cannot have any access to the way the world is independent from such kind of activity. As Hohwy states, we cannot "crawl outside of our own brains" (Ibidem, p. 7) in order to compare our model of the world to the world itself.

Thus, if we adopt PEM, we will never be able to assess whether a statement is true because it exactly corresponds to the way the world is independently from us. This is due to the fact that our mind works inferentially and the only things it can compare are homogeneous neuronal inputs. So, even if the realist conception of truth were the right one, if we adopt PEM we will be unable to judge whether we reached the truth, since we will be unable to claim that something is completely independent from our mind. But this contrasts with the fact that the great majority of the scientific realists refutes epistemic skepticism: realists claim that we *do* reach true theories and we *do* know that we have reached the truth. For example, Sankey states that the realist position is "a position of epistemic optimism, which holds against the sceptic that humans are able to acquire knowledge of the world" (Sankey 2008, 3). Since knowledge is usually intended by realists as related to the concept of truth,[8] it becomes clear that in this line of reasoning if we adopt PEM, we are unable to claim to have genuine knowledge. Since, on the contrary, PEM's supporters, and Hohwy among them, claim that we *do* have knowledge exactly

[6]Hohwy (2014, p. 9).

[7]Cf. e.g. Sankey (2008, p. 112): "The realist conception of truth is a non-epistemic conception of truth, which enforces a sharp divide between truth and rational justification."

[8]Cf. e.g. Ibidem, p. 14, fn. 2: "the traditional justified true belief account of knowledge is a minimal condition for a realist conception of knowledge."

through prediction error minimization, we have to conclude that PEM and SR are not compatible, at least because they rest on a different conception of knowledge.[9]

3.3 *Clark's Interpretation of PEM*

According to Clark's interpretation of PEM, it would be incorrect to interpret in the way suggested by Hohwy the indirectness of the link between the world which is modeled and our inner model of the world.

Indeed, Clark admits that following PEM our representations of the world have to be described as internal and may be in some sense deemed 'indirect': PEM "is a challenging vision, as it suggests that our expectations are in some important sense the primary source of all the contents of our perceptions" (Clark 2013b, p. 199). Nonetheless, Clark maintains that "we may still reject the bald claim that 'what we perceive is the brain's best hypothesis'," since "it remains correct to say that what we perceive is not some internal representation or hypothesis but (precisely) the world" (Ibidem).

We can affirm that we perceive 'precisely the world' because of "the brain's ability to latch on to how the world is" (Ibidem). If the brains were not able to adequately 'reflect' how the world really is, we would had not been able to survive. But we survived, so we can affirm that our representations are reliable. Indeed, "it is precisely by such means that biological beings are able to establish a truly tight mind-world linkage. Brains" can be construed as "statistical sponges structured [...] by individual learning and evolutionary inheritance so as to reflect and register relevant aspects of the causal structure of the world itself" (Ibidem). The idea behind such view is that in order to survive the organisms have to produce *true* representations of the world, i.e. representations that 'correspond' to the way the world really is. Thus, it is the causal connection between the world and our brains that produces our internal representations of the world and it is our success in the survival that guarantees that those representations are adequate, i.e. that what is perceived is not just the brain's best hypothesis, but the actual world.

Clark seems to explicitly commit himself to the traditional correspondence view of truth. For example, he agrees with Karl Friston that the "hierarchical structure of the real world literally comes to be 'reflected' by the hierarchical architectures trying to minimize prediction error" (Friston 2002, p. 237, quoted in Clark 2013a, p. 492). PEM "delivers a genuine form [...] of 'openness to the world'," and thus

[9]It may be objected that if someone does not rely on the notion of truth, she is not speaking of knowledge properly, since knowledge requires truth. Thus, it would be nonsense to speak of knowledge without referring to truth. But that knowledge necessarily requires truth is exactly what has been disputed by some of those authors who are unsatisfied with the traditional accounts of knowledge (see below, Sect. 6). Thus, if in their conception of knowledge does not figure any reference to the concept of truth, it seems unfair to conclude that they are not *really* speaking of 'knowledge', for the only reason that we *assume* that knowledge requires truth.

might "be cast as a representationalist version of 'direct perception'" (Clark 2013a, p. 492).

According to Clark, even if PEM adopts an inferential conception of the mind that makes our perception not as direct as the supporters of the direct perception view maintain,[10] nevertheless the close causal relation that obtains between the world and the brain, and the evolutionary and developmental selective processes that shape our priors, consent us to define our perception at least as 'not-indirect'. Clark states that: "If a label is required, it has been suggested" that the metaphysical perspective implied by PEM "may most safely be dubbed 'not-indirect perception'" (Ibidem, p. 493). In this perspective the indirectness of the inferential nature of our mind is tempered by the 'directness' of the externalist justification of our hypotheses somehow measured in terms of success in dealing with the world.

Let's recapitulate the elements of Clark's interpretation of PEM analysed so far which suggest that Clark's view is very close to SR. Clark seems to adopt a correspondence view of truth, which is one of the most widely adopted conception of truth among scientific realists (see above, Sect. 3.1). He also claims that we perceive the world as it really is and not just a hypothesis regarding the world. So he seems to subscribe to a non-epistemic view of truth, which is the view of truth usually adopted by scientific realists (see above, Sect. 3.2). Moreover, Clark seems to justify his claim that we perceive the world as it really is in a way which is analogous to the way in which scientific realists justify their claim that our best theories are true, i.e. developing a 'success argument'. Since our hypotheses about the world are successful in making us surviving, the only plausible explanation for their success is that they are true, i.e. that they are able to reflect how the world really is. This kind of argument is clearly a variant of the 'No Miracle Argument', the argument traditionally used by the realists to support SR, according to which the only plausible explanation for the success of our best scientific theories is that they are true (see e.g. Psillos 1999).

It is worth noticing that in clarifying his view on PEM, Clark explicitly refers to Michael Rescorla's interpretation of the Bayesian approaches to the mind, which include PEM (2015). Rescorla's interpretation of the Bayesian approaches to the mind is straightforward realist. Rescorla explicitly claims to support a scientific realist perspective and advocates that success is a key element in epistemic justification:

> I assume a broadly scientific realist perspective: explanatory success is a *prima facie* guide to truth. From a scientific realist perspective, the explanatory success of Bayesian perceptual psychology provides *prima facie* reason to attribute representational content to perceptual states.[11]

According to Rescorla the Bayesian approaches describe the way in which we estimate the world, and to do this they assume, more or less implicitly, a realist and externalist epistemological point of view, i.e. an epistemological view according to

[10]On the direct view of perception see Soldati (2012). See also Pappas (2014).

[11]Rescorla (2015, p. 705).

which what ultimately justifies any belief is something other than some mental state:

> Accuracy of the percept depends upon accuracy of the individual estimates. By describing perceptual inference in this way, we type-identify perceptual states truth-conditionally. We individuate perceptual states partly through environmental conditions that must obtain for the states to be accurate.[12]

So Clark's interpretation seems to be committed to externalism. Externalism is considered to be the more adequate epistemological position to take in order to support a realist perspective, since it claims that a belief, which is 'internal', is justified by some 'external' element, which is independent from the subject.

For all these reasons, it seems fair to say that according to Clark's interpretation PEM should not be considered at odds with SR. Indeed, according to Clark, PEM gives us knowledge of the way the world really is, i.e. PEM gives us an account of our mind as able to produce *true* representations of world. Since the majority of the scientific realists adopts a conception of knowledge as justified true belief, or some variant of it, and Clark's interpretation of PEM claims that we do have knowledge and that knowledge is related to the truth, in this perspective PEM is compatible with SR.

4 Assessing Clark's and Hohwy's Interpretations of PEM

In order to assess Clark's and Hohwy's interpretations of PEM we will proceed as follows: we will derive some of the most relevant epistemological consequences from the three main tenets of PEM (Sect. 4.1). It is important to stress that both Clark and Hohwy mostly agree on such tenets, so it seems fair to start from them. Then we will try to determine which epistemological position is more compatible with PEM (Sect. 4.2). Finally, we will try to show whether such epistemological position fits better Clark's or Hohwy's interpretation of PEM (Sect. 4.3).

4.1 The Epistemological Implications of the Three Main Tenets of PEM

In what follows it will be argued that if we take into account all the three tenets of PEM described above (Sect. 2), the more adequate epistemological position for PEM's supporters to take is infinitism. Let's see why. Take the first tenet: 'PEM adopts an *inferential* conception of the mind'. It is especially this inferential

[12]Ibidem, p. 702.

characterization of the mind that should lead PEM's supporters to prefer infinitism in epistemology.

This point can be clarified in two ways: by underlining the similarities between the challenges that both an inferential conception of the mind and infinitism have to face (we will take this way in this section); and by showing the incompatibility between what is implied by the three main tenets of PEM and what is implied by the other main epistemological positions (we will take this way in the next section).

With regard to the similarities between an inferential conception of the mind and infinitism, the most striking one is that they have both to face a similar skeptical regress problem. Indeed, as the skeptics deny that is possible to account for knowledge without ending in a regress, so "it seems we cannot explain perceptual inference at all, without ending in circularity or regress" (Hohwy 2013, p. 42). The problem is how to justify the claim that our perception is reliable, i.e. that we perform the correct inferences. As Hohwy states, if perception is an inferential process:

> either the inferential process is constrained or not. If it is not constrained, then there is no robust difference between right and wrong inference, and inference [...] remains unexplained. If it is constrained then the source of the constraints either is already engaged in correct perceptual inference or it isn't. If it is so engaged, then positing the source of the constraints as the explanation of perceptual inference [...] is circular or leads to a regress. If it is not so engaged, then again there is no difference between right and wrong inference.[13]

As in epistemology the main issue is how to connect justification to truth in order to secure knowledge from the skeptical challenge, so for an inferential conception of the mind the main issue is justify the claim that perception is reliable in representing the world even if the mind has no direct access to the world.

The problem is that according to the inferential conception of the mind, the mind can only make inferences in order to represent the world hidden behind 'the veil of perception': the mind cannot go out from what can be called the 'inferential circle'. Thus, according to this view of the mind, our relation to the world is not only possible but actual, but the relation between our representation of the world and the world itself can only be thought in terms of a potentially infinite process, since we can never 'crawl outside of our own brains' and compare our representation and the world in order to definitely state that a relation of correspondence obtains. This makes clear the similarity between such a view of the mind and infinitism. Indeed, infinitism denies the skeptical claim that we cannot have knowledge because we cannot justify our knowledge. According to infinitism knowledge is not only possible but actual, but the justification of our beliefs has to be thought as a potentially infinite process, since we cannot go through the complete chains of reasons that justify our beliefs.

Thus, since PEM adopts an inferential conception of the mind, and such a conception of the mind represents cognition as a potentially unterminated inferential process, if a supporter of PEM tries to determine which epistemological conception fits better her favourite account of the mind, she would probably adopt infinitism.

[13] Hohwy (2013, p. 42).

4.2 *Which Epistemological Position Is More Compatible with PEM?*

To better see the point made in the previous section, let's now turn to the second way of making clear why an inferential conception of the mind leads to infinitism. In order to show the difficulties of making compatible what is implied by the three main tenets of PEM and what is implied by the other main epistemological positions, we have to briefly consider the main alternatives to infinitism that are on the market.

Indeed, it has to be noticed here that both Clark and Hohwy seem to maintain a traditional attitude toward knowledge and justification, in the very minimal sense that they both refute skepticism, and thus have to defend the claims that we do have knowledge, and that knowledge is somehow related to the truth.

In a nutshell, skepticism claims that if we try to justify our beliefs we cannot but end in circularity, *petitio principii*, or infinite regress.[14] According to the skeptics, in any of those three cases we are unable to justify our beliefs. Since in all those three cases our beliefs would be unjustified, and being justified is a minimum (even if insufficient) requisite for a belief to be genuine knowledge, we should conclude that we cannot have knowledge.

On the contrary, those who refute skepticism maintain that knowledge is possible. In order to advocate for this position, epistemologists have negated that one or another of the above reported cases really prevent us to reach genuine knowledge, as skeptics maintain. Thus, traditionally the non-skeptical epistemological options are: *coherentism*, according to which circular patterns of justifications can enable knowledge; *finitism*, according to which finite patterns of justifications can enable knowledge; *infinitism*, according to which infinite patterns of justifications can enable knowledge (Turri and Klein 2014).

Let's briefly analyse them in order to see which one fits better the three main tenets of PEM. Consider coherentism first. The main problem with coherentism is that if we want to defend the claim that we have genuine knowledge and we consider knowledge as related to the truth, coherentism seems to be too permissive. As Klein and Warfield state: "coherence, per se, is not truth conducive" (Klein and Warfield 1994, p. 129). The fact is that not only true sets of propositions may be coherent. For example, Cellucci states that "the propositions of a fable form a systematically coherent whole, though being a fiction" (Cellucci 2014, p. 525). Moreover, if we allow repeating chains of reasons[15] in order to justify a given belief, we do not really *enhance* the justification of that belief, because that very belief would figure in its own justification. Finally, consider coherentism in relation to the first main tenet of PEM, i.e. the inferential nature of the mind. The problem is that the coherence among our inferences would not suffice to assess whether they

[14]See e.g. Floridi (1993).

[15]We refer here for simplicity to 'reasons' even if not every epistemological view requires 'reasons' in order to consider a belief to be justified. See Turri and Klein (2014).

give us genuine knowledge, since those inferences may be internally coherent, but nevertheless be unreliable in representing the external world, which is the main issue at stake in this context.[16]

Thus, it seems fair enough to say that if we adopt PEM we are left with two main epistemological options, i.e. finitism and infinitism. In order to determine which one fits better PEM, take now into consideration the second main tenet of PEM, i.e. the division between inner and outer is strict (seclusion). Consider finitism, i.e. the position according to which finite patterns of reasons can enable knowledge. To see the difficulty of combining finitism and PEM, recall that if we adopt PEM we subscribe to an inferential conception of the mind and that according to PEM the mind cannot have any direct access to the world. Since finitism implies that we should arrive at some *basic knowledge* which does not require justification at its turn, the question arises: How can the inferential circle be 'stopped'? In other words, how could we justify the claim that we have reached the end of an inferential chain where our representation of the world truly corresponds to the world, given the inferential functioning of our mind and that our mind cannot directly access the world?

Consider this issue in Bayesian terms. Since PEM models our mind as a 'Bayesian machine', in this framework the finitist's claim that we arrive at some basic knowledge which does not require to be justified at its turn would amount to know the exact distribution of probability of any possible case without the need of making any new hypothesis in that model. But, as Hohwy clearly states, the brain:

> cannot assess surprise directly from the sensory input because that would require knowing the relevant probability distribution as such. To do this it would need to, impossibly, average over *an infinite* number of copies of itself in all sorts of possible states in order to figure how much of a surprise a given sensory input might be.[17]

Thus, in order to claim to possess some basic knowledge, we should be able to actually go through an infinite inferential performance and assign the exact probability to any possible state of the world. Let's put aside, for the sake of the argument, the philosophical difficulty of making sense of the claim that we can know the exact distribution of probability of all the possible states of the world. The main problem is that finitism has in common with skepticism the idea that it is impossible to actually perform infinite mental operations, given that humans are

---

[16]It may be objected that this is an unfair description of coherentism, since many coherentists usually require in their theories some additional constraint on coherence to account for the truth-conduciveness of coherence. But, as Olsson has clearly underlined, "these theories may be more fruitfully classified as versions of *weak foundationalism* than as pure coherence theories. An advocate of weak foundationalism typically holds that while coherence is incapable of justifying beliefs from scratch, it can provide justification for beliefs that already have some initial [...] degree of warrant" (Olsson 2014, Sect. 1). This means that for our purposes, weak foundationalism, as well as foundationalism, can be fairly considered a kind of *finitism*, since it has to be based on some kind of beliefs that have some basic form of justification, which cannot be accounted for in terms of coherence.

[17]Hohwy (2015, p. 3).

limited beings. It is exactly for this reason, i.e. to avoid infinite regress, that finitism claims that knowledge to be possible must rest on some basic knowledge. But if it is impossible for us to perform infinite operations, then in a Bayesian perspective we are not able to reach the basic knowledge required by finitism to claim to have genuine knowledge. Thus, we should conclude that we do not have knowledge. But PEM's supporters normally claim that we do have knowledge. So there is a clear tension between finitism and PEM.

Let's now consider the last main tenet of PEM: perceptual inferences have to be conceived as statistical inferences. PEM models perceptual inferences using the 'Bayesian decision theory' framework, which models decision-making under *uncertainty*. This choice is due to the fact that according to PEM it is impossible to know the exact distribution of probability of all the possible states of the world. If it were possible to know such exact distribution of probability, there would be no need to continuously update our hypotheses about those states. Rescorla states that:

> The core notion underlying Bayesian decision theory is *subjective probability*. Subjective probabilities reflect psychological facets of the individual or her subsystems, rather than 'objective' features of reality. To formalize probabilities, we introduce a *hypothesis space H* containing various hypotheses *h* [...]. A probability function *p* maps each hypothesis *h* to a real number *p*(*h*), reflecting the agent's subjective probabilities.[18]

This should make clear the divergence between PEM and finitism. When we model perception in Bayesian terms, we construe the hypothesis space's elements as perceptual *estimates*. The goal of PEM "is to describe a statistical inference over *estimates about the perceiver's environment*" (Rescorla 2015, p. 712). It is not easy to see how it could be defended the claim that some 'estimates' are such that they do not need any further justification, and so may be considered as basic knowledge.[19]

4.2.1 A Naturalist Option for Finitism

Let's briefly consider a 'naturalist' proposal which could be made in order to make finitism, PEM and SR compatible. It could be argued that, since priors are given by natural selection, the inferential circle is 'broken'. The world instills in us the correct priors, which are not inferential at their turn and are justified by the world itself, and this fact ends the regress.

The problem is that this *externalist* proposal, which is in line with Clark's approach, just begs the question on what justifies our beliefs. Indeed natural selection deals only with fitness, i.e. survival, and not directly with truth. If we try to connect survival and truth we should adopt a sort of *reliabilism* and maintain an argument that could be roughly described as follows: since some beliefs have

---

[18]Rescorla (2015, p. 696).

[19]Cf. e.g. Hohwy (2014, pp. 2–5): "just as there is a schism between a statistical model and the modeled cause in statistical inference, there is a schism between the prediction-generating models of the brain and the modeled states of affairs in the world."

proved to be successful, they have been selected; selected beliefs are then reliable; reliability is a guide to the truth; natural selection gives us true beliefs. This is a very debated and controversial issue.[20] But reliabilism seems nevertheless to be inadequate to secure a realist *finitist* perspective mainly for two reasons: (1) reliabilism does not completely fill the gap between justification and truth; (2) reliabilism seems to many authors insufficient to account for *human* knowledge.

With regard to (1), it will suffice to recall the words of a realist champion as Psillos:

> In my (1999) [...] I argued that NMA proceeds within a broad naturalistic framework in which the charge of circularity loses its bite because what is sought is not justification of inferential methods [...] but their explanation and defence (in the epistemological externalist sense) [...]. I now think, however, that [...] [what] we should be after are *reasons to believe* that IBE is reliable (and not just an assertion to the effect that *if* indeed IBE is reliable, and we are externalists about justification, we are home and dry).[21]

With regard to (2), here it will suffice to recall the words of Ernest Sosa:

> Admittedly, there is a sense in which even a supermarket door 'knows' when someone approaches [...]. Human knowledge is on a higher plane of sophistication [...]. Pure reliabilism is questionable as an adequate epistemology for such knowledge.[22]

From what we just sketched above, it clearly appears that reliabilism in combination with finitism is at least not an easy option to take for PEM's supporters.

4.2.2 PEM and Infinitism

We discarded coherentism because repeating chains of reasons are objectionably question-begging, and finitism because finite chains of reasons are objectionably arbitrary at their terminus. Thus the only available epistemological option for PEM's supporters is infinitism. Since we have also shown some important convergences between PEM and infinitism, we can conclude that infinitism is the epistemological position that fits better PEM.

4.3 *Comparing Clark's and Hohwy's Interpretations of PEM*

So far we have tried to answer the two questions raised in Sect. 3.1, i.e. whether PEM is compatible with SR, and which epistemological position fits better PEM.

---

[20]See e.g. Vlerick and Broadbent (2015).

[21]Psillos (2011, p. 26). See also Klein (2015, Sect. 1): "reliabilist or externalist responses to philosophical skepticism constitute a change of subject. A belief could be reliably produced [...] but the reasons available for it could fail to satisfy the standards agreed upon by both the skeptics and their opponents."

[22]Sosa (1983, pp. 58–59).

We can sum up our inquiry as follows: Clark's interpretation of PEM involves an externalist perspective and is sympathetic to SR, while Hohwy's interpretation involves an internalist perspective and is less compatible with SR. As we have seen, this also means that Clark's view is more suited for a finitist epistemological perspective, while Hohwy's view seems more suited for an infinitist epistemological perspective. Indeed, it seems not easy to conciliate the realist claims that we do reach the truth and that truth is correspondence with the infinitist perspective on justification. Moreover, we have underlined how infinitists are dissatisfied with an externalist and reliabilist view of epistemic justification, which is exactly the position that characterizes Clark's interpretation of PEM. Then, since we have shown that to account for all the tree main tenets of PEM an infinitist perspective is the more adequate, Hohwy's interpretation of PEM seems to be preferred.

5 On Some Difficulties Still Afflicting Hohwy's Position

In this section we will sketch some of the difficulties still afflicting Hohwy's position: (1) how to model the formation of the hypotheses; 2) the difficulties deriving from infinitism.

With regard to (1), Hohwy says almost nothing on how to model the hypothesis formation process. We intend to refer here to the production of those hypotheses that are not 'innate'. Hohwy just takes for granted that hypotheses are produced and then updated. This is a crucial issue for all the Bayesian approaches to the mind, since Bayesian formalism does not account for knowledge ampliation, it is intended just to model the refinement of the probabilities of given hypotheses.

With regard to (2), first of all there is the question of how knowledge has to be conceived if we adopt PEM. Since Hohwy gives no peculiar account of knowledge, we may presume that he intends knowledge in the traditional sense, i.e. as related to the concept of truth. But we have seen that PEM conveys a view of the mind that may well be considered as taking an 'antirealist stance', given that we have judged it to be at odds with SR.[23] The problem is that there is a tension between the claim that we do have knowledge and that knowledge is related to the concept of truth, and an antirealist stance. Thus, if we adopt PEM the traditional conception of knowledge seems at least to be wanting. Secondly, there are the difficulties deriving from adopting infinitism. Indeed, even if infinitism seems to be the best option when compared to coherentism and finitism, it nevertheless presents several problems. We will sketch just two of the main problems that arise in combining

[23]Hohwy's view can be described as a sort of 'Kantian scientific antirealism', which particularly resembles Bas van Fraassen's scientific antirealism, especially on the issue of 'representation' (see van Fraassen 2008). Indeed, Hohwy's view of the relation between the internal model and the sensory input is similar to van Fraassen's view of the relation between theoretical models and data models. We can at most compare them and make them fit, but this does not guarantee us that they reflect the world itself, since we cannot directly confront our models and the world.

PEM and infinitism. First, if PEM's supporter adopts infinitism, then she has to address the main problem afflicting infinitism, i.e. that of giving some feature which is able to discriminate among chains of reasons without regress. Indeed, infinitism claims that infinite chains of reasons may justify our beliefs. But allowing infinite chains of reasons is insufficient. The problem is how to discriminate those infinite chains of reasons that justify a given belief from those infinite chains of reasons that do not justify that belief:

> The regress condition itself cannot explain the connection between justification and truth, but any additional feature that could explain this connection would undermine the rationale for the regress condition itself [...]. So infinitism must distinguish infinite sequences of propositions that are justification-affording – those upon which actual justified beliefs depend – from those that are not justification-affording, in a way that explains the relevant connection between justification and truth.[24]

Second, there is the problem of *reasoning*. Indeed, PEM deals basically with unconscious Bayesian inference, while infinitists, as we have seen, normally require reasoning, i.e. human conscious reasoning, for something to be qualified as genuine knowledge.

To sum up, in order to strengthen Hohwy's interpretation of PEM it seems urgent to address the following issues: (1) giving an account of the hypotheses formation process; (2) adopting a conception of knowledge more suited to PEM's 'scientific antirealism'; (3) elaborating an anti-skeptical position which is able to avoid the difficulties afflicting infinitism. In the next section, we will suggest that in order to address those issues it could be fruitful to take into consideration Carlo Cellucci's work.

6 The Heuristic View

For reasons of space, it is not possible here to give an exhaustive exposition of the Heuristic View (HV) developed by Cellucci (2013, 2014, 2015). In what follows, we will illustrate just the core tenets of Cellucci's position.

6.1 The Analytic Method as a Model of Hypothesis Production

According to HV, the method of philosophy, mathematics, and the natural sciences is the very same method, and it is the analytic method. The analytic method, which

[24]Cling (2004, p. 110).

goes back to Hippocrates of Chios, Hippocrates of Cos, and Plato, may be described as follows:

> to solve a problem, one looks for some hypothesis that is a sufficient condition for solving it. The hypothesis is obtained from the problem, and possibly other data already available, by some non-deductive rule, and must be plausible […]. But the hypothesis is in its turn a problem that must be solved, and is solved in the same way […]. And so on, ad infinitum.[25]

According to HV, the axiomatic method is inadequate for giving a naturalistic account of how knowledge is pursued. Indeed, the axiomatic method is not able to account for the hypotheses production process, and so it is not able to show the real path that has been followed to reach a given result. On the contrary, the analytic method is the method used in the process of discovery. Indeed, since in order to solve a problem hypotheses are produced by non-deductive inferences, logic is essentially a logic of discovery.[26]

6.2 The Heuristic View and Knowledge

According to HV, the analytic method provides a model of knowledge ampliation. But how knowledge has to be conceived according to this perspective? If the method of philosophy, mathematics and the natural sciences is the analytic method, and the analytic method is essentially characterized by the use of ampliative inferences, i.e. inferences that are not truth-preserving,[27] the problem arises of how to conceive the relation between knowledge produced by means of the analytic method and truth. Indeed, usually scientific realists take the aim of science to be the truth. For example, Sankey states that "the aim of science is to discover the truth about the world" (Sankey 2004, p. 215). Contrary to this view, according to Cellucci the concept of truth has to be replaced with the concept of *plausibility*.[28] Indeed, since the traditional definitions of truth are not able to give us a criterion of truth, i.e. a non-algorithmic means to decide whether a statement is true, they

---

[25]Cellucci (2013, p. 55).

[26]The analytic method has not to be confused with the analytic-synthetic method. According to the analytic-synthetic method as stated by Aristotle, the search for a solution to a problem is a finite process, and once the prime premises have been found, "the only role which remains for analysis is to find deductions of given conclusions from prime premises" (Cellucci 2013, p. 75). On the contrary, in the analytic method there is no given prime premise, the path to find hypotheses is only 'ascending', and it has not to terminate.

[27]Hintikka and Sandu (2007, p. 13).

[28]For a plausibility test procedure, cf. Cellucci (2013, p. 56): "(1) Deduce conclusions from the hypothesis. (2) Compare the conclusions with each other, in order to see that the hypothesis does not lead to contradictions. (3) Compare the conclusions with other hypotheses already known to be plausible, and with results of observations or experiments, in order to see that the arguments for the hypothesis are stronger than those against it on the basis of experience."

cannot avoid the skeptical argument of the criterion (Cellucci 2014).[29] For example, Cellucci states that the concept of truth as correspondence is not adequate as a criterion of truth because, as Kant states:

> according to the correspondence conception, truth "consists in the agreement of cognition with its object," but "I can compare the object with my cognition" only "by cognizing it" (Kant 1992, 557). Then, "since the object is outside me, the cognition in me, all I can ever pass judgment on is whether my cognition of the object agrees" not with the object but only "with my cognition of the object" (557–58). Therefore, we cannot know whether a theory about the world is true in the [...] correspondence sense. This makes truth something that humans cannot reach, and makes the aim of science ultimately unachievable.[30]

Being truth such an *unrealistic* aim, Cellucci takes instead plausibility as the central concept of epistemology:

> the goal of science is plausibility. Scientific theories do not deal with the essence of natural substances, but only with some of their phenomenal properties, and deal with them on the basis of plausible hypotheses. Then a scientific theory is not a set of truths but rather a set of plausible hypotheses. Thus the goal of science is plausibility rather than truth.[31]

Thus, according to HV what we really do, and can do, is producing hypotheses by means of some non-deductive rule, and then assessing the arguments for and the arguments against any hypothesis and provisionally accept or refute such hypothesis.[32]

6.3 The Analytic Method as an Anti-Skeptical Option

Let's now compare HV with infinitism. HV may as well as infinitism be considered an anti-skeptical option, since it claims not only that we do have knowledge, but also that knowledge is necessary to survive. Moreover, HV and infinitism have in common the idea that the fact that knowledge acquisition may be a potential infinite process does not prevent us to consider genuine knowledge that portion of knowledge we reached so far. For example, Cellucci states that:

> Even if, by the finiteness of human capacities, we cannot go through an infinite series, this does not mean that the series of the premises cannot be infinite but only that, at each stage, we can only go through a finite initial segment of the series. And yet we can go through longer and longer finite initial segments.[33]

But HV and infinitism are nevertheless distinct positions. Indeed, infinitism retains the relation between knowledge and truth, and thus has to face the difficulty

[29]On the problem of the criterion of truth cf. e.g. Sextus Empiricus (1976, II.2).

[30]Cellucci (2015, pp. 217–218).

[31]Cellucci (2013, p. 154).

[32]This view is related to Aristotle's definition of *endoxa*, see Cellucci (2013, Sect. 5.7).

[33]Cellucci unpublished, Sect. 3.2.

outlined above (Sect. 6.2). As we have seen, infinitism per se is not really able to face the skeptical challenge and connect justification to truth. On the contrary, HV can safely maintain that knowledge acquisition is a potentially infinite process and that the knowledge produced so far is genuine knowledge, since HV conceives knowledge as plausible and provisional, and does not relate it to truth:

> if the series of the premises is infinite, there will be no immediately justified premises, so no knowledge will be definitive, all knowledge will always be in need of further consideration. But this does not mean that there can be no knowledge. There could be no knowledge only if the premises, or hypotheses, occurring in the infinite series were arbitrary. But they are not arbitrary since […] they must be plausible, that is, such that the arguments for them must be stronger than those against them […].[34, 35]

HV and infinitism diverge also with regard to the 'reasoning requirement' made by the infinitists to consider something as genuine knowledge. Indeed, Cellucci sees the production of knowledge essentially as a problem solving process, which is homogeneous throughout the biological realm. The way in which problems are solved is similar among all organisms, because "knowledge has a biological role, just like other capacities which ensure the survival of organisms […] knowledge is essential for life" (Cellucci 2013, p. 250). Moreover, according to HV even unconscious inferences contribute to knowledge: "in the analytic method, some non-deductive inferences by which hypotheses are obtained may be unconscious" (Ibidem, p. 235). Thus, HV does not require reasoning in order to consider something as genuine knowledge.

7 Conclusion. The Heuristic View and PEM

From the exposition of Cellucci's proposal, it appears that HV could considerably strengthen Hohwy's position. Indeed, HV seems able to provide to those who follow Hohwy's interpretation of PEM: (1) a theoretical account of hypotheses formation, (2) a conception of knowledge decoupled from the concept of truth, and (3) an anti-skeptical position which is able to avoid the difficulties that afflict infinitism, but which at the same time displays those features that made us judge infinitism to be the position more compatible with PEM. Combining Cellucci's view with Hohwy's interpretation of PEM seems then to be a fruitful perspective worth of further investigations.

[34]Ibidem.

[35]It is worth underlining that plausibility has not to be confused with probability (Cellucci 2013, Sect. 4.4). Plausibility involves a comparison between the arguments for and the arguments against, so it is not a mathematical concept. Conversely, probability is a mathematical concept.

References

Cellucci, C. (unpublished). *Rethinking knowledge: The heuristic view*.
Cellucci, C. (2013). *Rethinking logic*. Dordrecht: Springer.
Cellucci, C. (2014). Knowledge, truth and plausibility. *Axiomathes, 24*, 517–532.
Cellucci, C. (2015). Rethinking knowledge. *Metaphilosophy, 46*, 213–234.
Clark, A. (2013a). Expecting the world: Perception, prediction, and the origins of human knowledge. *The Journal of Philosophy, 110*, 469–496.
Clark, A. (2013b). Whatever next? Predictive brains, situated agents, and the future of cognitive science. *Behavioral and Brain Sciences, 36*, 181–204.
Clark, A. (2015). Embodied prediction. In T. Metzinger & J. M. Windt (Eds.), *Open MIND*. doi:10.15502/9783958570115.
Cling, A. D. (2004). The trouble with infinitism. *Synthese, 138*, 101–123.
Ellis, B. (2005). Physical realism. *Ratio, 18*, 371–384.
Floridi, L. (1993). The problem of the justification of a theory of knowledge. Part I: some historical metamorphoses. *Journal for General Philosophy of Science, 24*, 205–233.
Friston, K. (2002). Beyond phrenology: What can neuroimaging tell us about distributed circuitry? *Annual Review of Neuroscience*, *XXV*, 221–250.
Friston, K. (2010). The free-energy principle: A unified brain theory? *Nature Reviews Neuroscience, 11*, 127–138.
Friston, K. (2011). Embodied inference or 'I think therefore I am, if I am what I think'. In W. Tschacher & C. Bergomi (Eds.), *The implications of embodiment* (pp. 89–125). Exeter: Imprint Academic.
Frith, C. (2007). *Making up the mind*. Oxford: Blackwell.
Gregory, R. L. (1980). Perceptions as hypotheses. *Philosophical Transactions of Royal Society Series B, 290*, 181–197.
Harman, G. H. (1965). The inference to the best explanation. *The Philosophical Review, 74*, 88–95.
Helmholtz, H. von (1867). *Handbuch der physiologischen optik*. Leipzig: Leopold Voss.
Hintikka, J., & Sandu, G. (2007). What is logic? In D. Jacquette (Ed.), *Philosophy of logic* (pp. 13–39). Amsterdam: North-Holland.
Hohwy, J. (2013). *The predictive mind*. Oxford: Oxford University Press.
Hohwy, J. (2014). The self-evidencing brain. *Noûs,* doi:10.1111/nous.12062.
Hohwy, J. (2015). The neural organ explains the mind. In T. Metzinger & J. M. Windt (Eds.), *Open MIND*. doi:10.15502/9783958570016.
Kant, I. (1992). *Lectures on logic*. Cambridge: Cambridge University Press.
Klein, P. (2015). Skepticism. In E. N. Zalta (Ed.), *The Stanford encyclopedia of philosophy*. http://plato.stanford.edu/archives/sum2015/entries/skepticism/
Klein, P., & Warfield, T. A. (1994). What price coherence? *Analysis, 54*, 129–132.
Olsson, E. (2014). Coherentist theories of epistemic justification. In E. N. Zalta (Ed.), *The Stanford encyclopedia of philosophy*. http://plato.stanford.edu/archives/spr2014/entries/justep-coherence/
Pappas, G., (2014). Internalist vs. externalist conceptions of epistemic justification. In E. N. Zalta (Ed.), *The Stanford encyclopedia of philosophy*. http://plato.stanford.edu/archives/fall2014/entries/justep-intext/
Psillos, S. (1999). *Scientific realism*. New York: Routledge.
Psillos, S. (2011). The scope and limits of the no miracles argument. In D. Dieks, et al. (Eds.), *Explanation, prediction, and confirmation* (pp. 23–35). Dordrecht: Springer.
Rescorla, M. (2015). Bayesian perceptual psychology. In M. Matthen (Ed.), *The Oxford handbook of the philosophy of perception* (pp. 694–716). Oxford: Oxford University Press.
Rock, I. (1983). *The logic of perception*. Cambridge, MA: MIT Press.
Sankey, H. (2004). Scientific realism and the god's eye point of view. *Epistemologia*, *XXVII*, 211–226.
Sankey, H. (2008). *Scientific realism and the rationality of science*. Burlington: Ashgate.

Sextus Empiricus (1976). *Outlines of pyrrhonism*. Cambridge, MA: Harvard University Press.
Soldati, G. (2012). Direct realism and immediate justification. *Proceedings of the Aristotelian Society, CXII*, 29–44.
Sosa, E. (1983). Nature unmirrored, epistemology naturalized. *Synthese, 55*, 49–72.
Turri, J., & Klein, P. (Eds.). (2014). *Ad infinitum*. Oxford: Oxford University Press.
van Fraassen, B. C. (2008). *Scientific representation*. Oxford: Oxford University Press.
Vlerick, M., & Broadbent, A. (2015). Evolution and epistemic justification. *Dialectica, 69*, 185–203.

Visualization as Heuristics: The Use of Maps and Diagrams in 19th Century Epidemiology

Giulia Miotti

Abstract In this paper, I argue that visualization and the use of figures represent genuine heuristic, knowledge-enhancing tools in scientific inquiry; in fact, visualization shows a distinctive ability in producing genuinely new knowledge by filling theoretical gaps and in solving problems. I show, then, how visualization can be rightfully appraised as a plausible model for the growth of knowledge, gaining a paramount importance when used at the frontier of research. Unrelated here to the notion of intuition, visualization is treated as an ampliative inference and, being obviously related to figures and vision it is also a way of representing knowledge: this double function justifies it as a non-trivial heuristic device, not replaceable by axiomatic-deductive reasoning. A case study is proposed, regarding the London cholera epidemic, spread between August and September 1854. I show how the recourse to dot maps and a "primitive" version of network Voronoi diagrams as instruments of inquiry helped in filling the then existing theoretical gaps consisting in the ignorance of the existence and action of bacteria in disease transmission. Visualization, on the one hand, acted as an effective problem-solving activity, as it permitted the formulation of a successful strategy to stop the spreading of the epidemic; on the other, through the identification of new causes responsible for the spreading of the epidemic, it allowed to surpass the critical theoretical gap at the frontier of epidemiological knowledge.

1 Introduction

In this paper, I argue that visualization, and relatedly the use of figures such as maps and diagrams, shows genuine knowledge-enhancing features with respect to the various fields to which it is applied. According to recent literature, the possible fields of application range from mathematics, where visualization has often replaced

---

G. Miotti (✉)
Department of Philosophy, Sapienza University of Rome,
Via Carlo Fea, 2, 00161 Rome, Italy
e-mail: giulia.miotti@uniroma1.it; giulia.miotti@googlemail.com

L. Magnani and C. Casadio (eds.), *Model-Based Reasoning in Science and Technology*, Studies in Applied Philosophy, Epistemology and Rational Ethics 27, DOI 10.1007/978-3-319-38983-7_37

analytical formalized tools (Grosholz 2007), to cognitive science, where an analysis of the cognitive role played by new computational representations in scientific discovery is strongly called for (Chandrasekharan and Neressian 2014).

Our analysis concerns the role of visualization in 19th century epidemiology.

In the last decades, visualization has been widely studied as a proper heuristic tool, particularly effective when used at the frontier of knowledge: by this expression, I mean a situation of epistemic difficulty in which already acquired knowledge is insufficient to the solution of given problems; a situation, therefore, that needs the introduction of actually new information (Nickles 2009). In this context, the recourse to visualization can provide new strategies of inquiry, acting as a problem-solving activity.

An important *distinguo*: my acceptation of "visualization" is a rather specific one, and it strongly differs from the ones generally proposed by some authors (Giaquinto 2005). Being treated as a problem-solving device, visualization is not related to any kind of intuitive thinking, and its use is patently beyond didactical purposes. Consequently, my analysis highlights the heuristic characteristics of visualization rather than its epistemological ones.

By epistemological characteristics, I refer to the high didactical and explanatory potential usually recognized to visualization. Such definition does not pertain to the current discussion, since it implies that the efficiency of visualization is based on its ability to ease the understanding of already acquired knowledge. According to its epistemological characteristics, visualization *is* in fact a more intuitive access to the understanding of theories and problems than a formal, deductive explanation would be; even though this feature is undeniably useful in didactical contexts, it is deprived of any knowledge-enhancing potential.

By heuristic characteristics, on the other hand, I refer to those abilities and aspects of visualization which justify it as a logical instrument of inquiry. The three most interesting characteristics I consider are:

1. Independence; visualization is a fully independent instrument, not ancillary to formal, deductive means of inquiry. Its use, therefore, is of primary and not of secondary order.
2. Non-triviality; visualization is not an economical way to access knowledge otherwise obtainable; for example, by means of deduction.
3. Ampliativity; being independent and non-trivial, visualization is an instrument powerful enough to produce new information.

The case-study proposed fittingly exemplifies this definition of visualization both as a plausible model for the growth of knowledge and a problem solving strategy.

In fact, the epidemic case taken into account is a "frontier of knowledge" case in accordance with the definition given above: it exhibits a theoretical gap, represented by the ignorance of the notion of "bacterium" and therefore its existence and action in the spreading of diseases. Consequently, it exhibits a problem to be solved, which amounts to the necessity of finding a strategy to stop the contagion.

2 Maps and Diagrams in Epidemiology

In 1854, the London cholera epidemic represented a challenging case for the physicians of the time who tried to stop the spreading of the infection; as a cognitive issue, the epidemic was in fact situated at the frontier of the then contemporary epidemiological knowledge. In accordance with the definition of "frontier of knowledge" as a condition of epistemic crisis in which knowledge available at time *t* proves insufficient to solve the given problem, the London cholera epidemic case shows the inadequacy of the miasmatic theory of disease transmission in preventing the spreading of the infection.

I am going to show how the theories and the related methods employed to face the epidemic obtained no useful results, neither in the identification and explanation of the causes, nor in saving lives by stopping the spreading of cholera. I argue then that the recourse to visualization, in the terms of the map produced by John Snow's inquiry, provided a successful counter-strategy, powerful enough to provide a causal explanation and not a mere description of the phenomenon. Visualization acted, in fact, as an effective tool in leading the inquiry at the frontier of 19th century epidemiological knowledge.

2.1 *The Miasmatic Theory of Contagion*

The miasmatic theory of disease transmission traces its origins back to ancient Greek medicine, initially proposed by Hippocrates and later advocated by Vitruvius and Galen. It remained significantly unquestioned until the bacteriological revolution promoted by Pasteur and Koch around 1886.

According to the miasmatic theory, infective diseases such as cholera, malaria, plague, are caused by infected gaseous particles called *miasmata*. Such particles, in the shape of poisonous vapours exhaled from putrefied, corrupted corpses, propagate through the air, causing contagion when some of these vapours are accidentally inhaled. Air, therefore, is the only recognized means of infection, and since the theory links the very notion of contagion to the action of vapours (or even smells, as some theorists claimed), it does not recognize the action of different pathogenic agents; even in the case of naked-eye visible parasites. In fact, according to this theory, cholera is an air-borne disease, rather than a water-borne one.

Although some physicians were doubtful about the miasmatic theory scientific soundness, in 1854 it was the only theory to be officially accepted by the scientific community. Therefore, in the first period of epidemiological emergency the methods prescribed by the theory were followed, but they led to almost no results. Since the theory described cholera as an air-borne disease and the miasmata (in the form of vapours or smells) as the cause of the infection, it prescribed sanitary measures for the infected areas in order to remove the sources of lethal vapours.

This strategy, although fruitless with respect to its main objective, had the merit of improving the precarious hygienic standards characterizing working-class districts, where cholera was more likely to hit. Nevertheless, as in the 1854 London case, such measures could not stop the spreading of the epidemic, and the reason why districts with similar low hygienic standards were not equally affected by cholera remained unexplained.

2.2 Flaws and Inadequacies

The flaws and inadequacies of miasmatic theory are easier to detect from our viewpoint, a viewpoint with a knowledge advantage: until 1886, in fact, the notion of *bacterium*, its existence and action, was unknown, it is thanks to Louis Pasteur and Robert Heinrich Herman Koch's work that this most critical knowledge gap has been filled. It is now part of common scientific knowledge that the transmission of cholera is caused by a particular kind of bacterium, a vibrio called *vibrio cholerae asiaticae*, and that this particular bacterium, when not "hosted" in human bodies, finds its natural habitat in water.

Therefore, from this advantaged scientific standpoint (if compared to 1854 epidemiological notions), we clearly understand why the miasmatic theory of contagion could not account for the causes of cholera: it is not caused by vapours, but rather by a bacterium. Furthermore, we can also easily understand why the measures adopted during the London cholera epidemic could not prove efficacious: cholera is a water-borne disease, not an air-borne one and the infection takes place through consumption of infected water (or ingestion of food previously contaminated by infected water), not through inhalation of corrupted air.

As already stated, some physicians were doubtful about the plausibility of miasmatic theory and John Snow was among them. He had already experienced a cholera epidemic during 1831 and had started finding the miasmatic explanation unsatisfactory: however, he did not possess the knowledge to convincingly disprove it.

Albeit he could not rely on a theoretical proof for his conclusion not even in the 1854 case, John Snow succeeded in detecting and visually proving that the source of infection was to be found in water, not air. Specifically, in the infected water provided by the pump in Broad Street, one of the most used in the district.

2.3 Visualization: A Counter-Strategy of Inquiry

Such understanding of the cause and nature of the disease was not possible at the time of the London cholera epidemic: as a matter of fact, the above mentioned knowledge gap prevented the formulation of an epidemiological theory that could challenge the miasmatic theory by claiming a better scientific explanation.

This is the main reason why the visual representation offered by John Snow acted as the only possible alternative strategy to this specific sanitary emergency: it did not rely on formal theoretical explanations to show a specific causal connection, but rather on a specific organization and manipulation of data to allow the formulation of plausible alternative hypotheses (Tufte 1997).

The peculiar diffusion of cholera during the 1854 epidemic represented a fortunate occasion for John Snow to analyse thoroughly the modalities of contagion. Deceases and infected cases in general, in fact, were mainly concentrated within a small area South of London, in the district of Soho; this extreme concentration on a small area made the task of mapping easier, and consequently, it facilitated the detection of a common etiological agent. Snow writes:

> Further inquiry, however, showed me that there was no other circumstance or agent common to the circumscribed locality in which this sudden increase of cholera occurred, and not existing beyond it.[1]

The 1854 London cholera epidemic was not the first epidemic case to be visualized in maps, there are, in fact, many interesting examples: one of the first attempts is probably represented by the work of the physician Valentine Seaman during the New York yellow fever epidemic in 1796 (Seaman 1797; Stevenson 1965); another one is the mapping of the 1831 cholera outbreak in the British Isles drawn by the German cartographer August Petermann. Notwithstanding these early attempts, the map drawn by John Snow is the only example of visual explanation in epidemiology (up to 1854) that can be appraised as an actual heuristic model.

The strategy the English physician adopted consists of two steps. The first one calls for a collection of as many data as possible concerning the deadly attacks of cholera; Snow obtained most of them from the General Register Office (relative to 518 deaths) and others through interviews with the deceased's relatives. These data were informative about the presumed date of infection, the date of the eventual decease and the *place* (the street and building) where the infection occurred.

The second step, then, consists in plotting all the collected data on a detailed map of the chosen district; the map reproduces all the buildings of the area (both infected and non-infected), the cemetery (according to the miasmatic theory, a *locus* of infection) and the different water pumps placed within a walking distance in the area considered.

Two main features make the map designed by John Snow of paramount interest to my analysis. The first one: each occurrence of cholera is marked relative to the building the victim inhabited and a thick black line is used to symbolize each decease. The second one: the presence of a dotted line dividing the district in sub-regions and marking distances between each pump plotted in the demarcated area. The first characteristic mentioned, i.e., the symbolization of cholera occurrences as thick black lines, allows the description of Snow's map as a dot-distribution map; while the

[1]Snow (1855) p. 24.

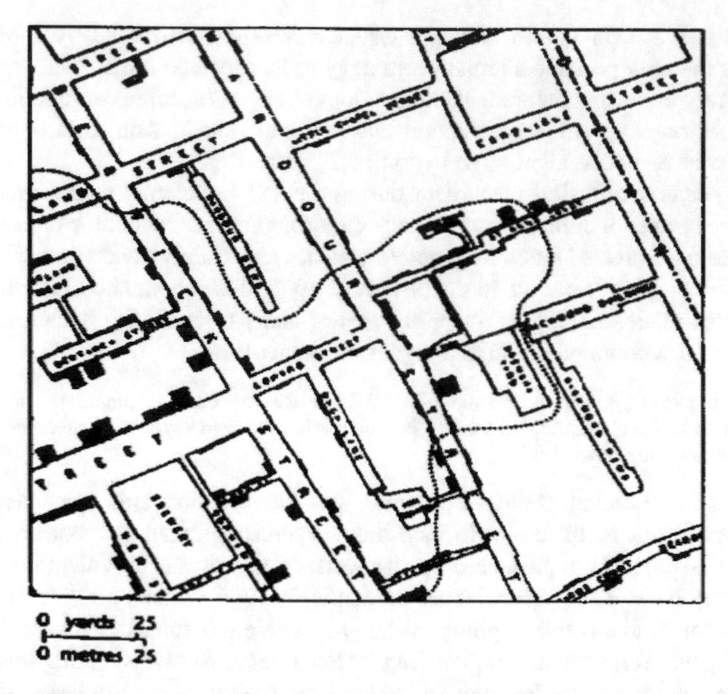

Fig. 1 The cholera map two main features are clearly detectable in this section: (1) the *thin black lines*, pointing out each cholera occurrence and where it occurred and (2) the *dotted line*, dividing the mapped area into sub-regions. *Source* Snow (1855)

second one, the division of the mapped area into sub-regions, allows the description of the map as a primitive version of a network Voronoi area diagram (McLeod 2000).

These two features mark an important difference between the preceding cartographic attempts and this map; while other maps are simple visual descriptions of a phenomenon, this map furnishes an explanation of the phenomenon and the overall information it provides surpasses the information offered by the "raw" data plotted on it (Fig. 1).

2.4 *Snow's Cholera Map as a Dot-Distribution Map*

A dot-distribution map is a device used to describe the geographic distribution of a given phenomenon. In order to do so, it uses a dot-symbol to show the presence of the inquired phenomenon, in one-to-one dot-distribution maps each dot stands for a single occurrence of the mapped phenomenon. Dot-distribution maps are considered one of the first and primary techniques (their use dates back to the 19th century) used to make sense of the global distribution of the mapped phenomenon and, most importantly, to compare the relative densities of different regions on the

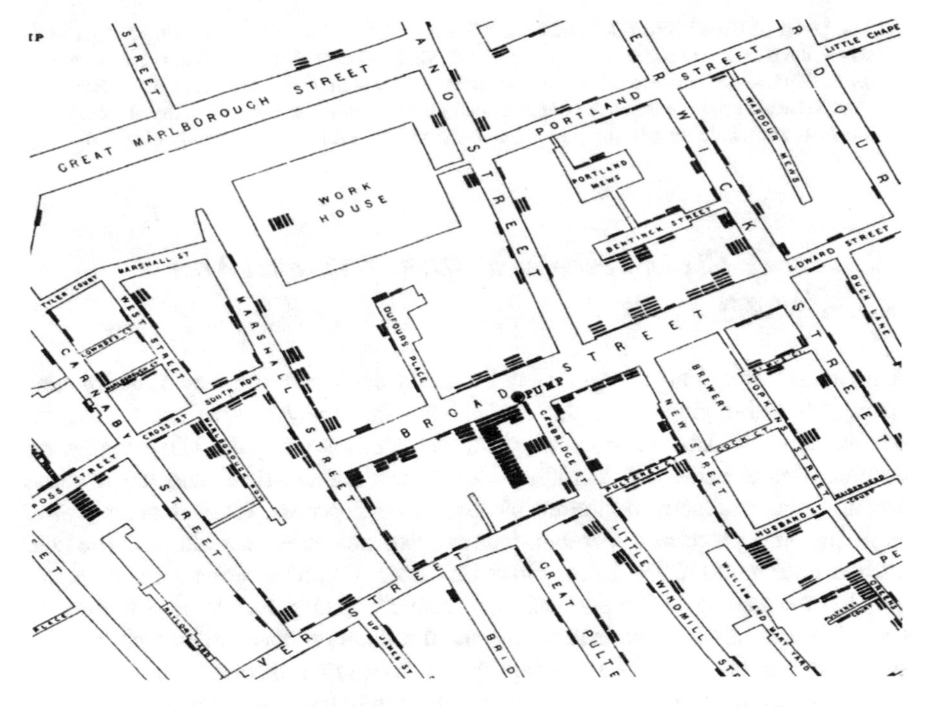

Fig. 2 John snow's map: cholera incidence in South London. *Source* Snow (1855)

map. Therefore, this kind of map is particularly efficacious in displaying data when they relate to phenomena

1. occurring at discrete locations in space
2. changing suddenly at the boundaries.

Since John Snow's map possesses all the characteristics of a dot-distribution map (i.e., the recourse to identical line-symbols to mark each occurrence and the display of data resembling characteristics 1 and 2), it is easily shown why the map he designed can be considered a proper inquiry tool. The use of thin lines for each datum shows the different densities in different map regions with respect to the presence of the phenomenon; in this way, it restricts both the area of inquiry and the possible agents common to the majority of the phenomenon occurrences.

In this way, this process of restriction lets important hidden patterns emerge from the spatial visualization and, furthermore, it permits to narrow down the possible hypotheses concerning the factors of contagion. An example: looking at the map (Fig. 1), the fact that the highest density of dot-symbols cluster around a water pump in Broad Street reveals a hidden pattern that opens up to the hypothesis that the water pump may represent the common feature among the different attacks (Fig. 2). Snow writes:

> The pump in Broad Street is indicated on the map, as well as the surrounding pumps to which the public had access at the time [...] It will be observed that the deaths either very much diminished, or ceased altogether, at every point where it becomes decidedly nearer to send to another pump than to the one in Broad Street. It may also be noticed that the deaths are most numerous near to the pump where the water could be more readily obtained.[2]

2.5 *Snow's Cholera Map as a Network Voronoi Area Diagram*

A network Voronoi area diagram can be briefly described as a way of partitioning space into sub-regions in order to facilitate the analysis and, eventually, the manipulation of data. In the context of this partition of space into sub-regions, different points belonging to different point sets are identified and the relations among them are described in terms of distance or closeness. When data are plotted as points on a plane in a Voronoi diagram, two operations are allowed: the connection of any sub-region to the closest member of another point set and, consequently, the tracing of the shortest path from point to point. As in this case the spatial visualization is obtained by means of a diagram, these operations are also subject to a mathematical description by means of graph theory.

A clear, general definition of Voronoi diagrams, then, is the one proposed by Atsuyuki Okabe: DEF V1 "given a set of two or more but a finite number of distinct points in the Euclidean plane (i.e., a continuous space), we associate all location in that space with the closest member(s) of the point set with respect to the Euclidean distance. The result is the tessellation of the plane into a set of regions associated with members of the point set".[3]

Technically, in spatial tessellation, a network Voronoi area diagram is represented as a network in the shape of a planar geometric graph G(N, L) consisting of a set of *nodes* N = $\{p_1,\ldots, p_n,\ldots, p_{n+1},\ldots\}$ and a set of *links* L = $\{l_1,\ldots, l_k,\ldots\}$, forming a connected component. On G(N, L) the distance from a point "p" on a link in L to a node "p_i" in N by the length of the shortest path from "p" to "p_i" is definable; this distance is called the "network distance" and is denoted by $D_{net}(p, p_i)$. A geometric graph G(N, L) with the network distance is called a "network" and it is denoted by N (N, L).

The difference between a Voronoi diagram and a network Voronoi area diagram resides in the interpretation and measurement of distance between points. While in the first case distance is measured according to Euclidean metrics on a plane, in the second case the distance between points refers to network distance on a graph (Fig. 3).

---

[2] Ibidem, p. 28.

[3] Okabe (2012), p. 43.

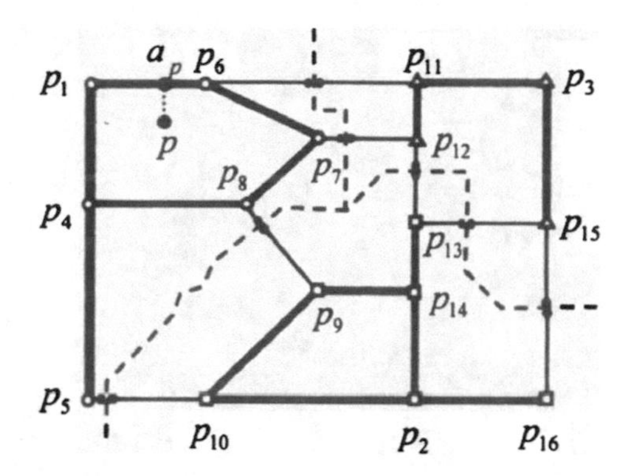

Fig. 3 A formal representation of a network Voronoi area diagram. *Source* Okabe et al. (1992)

John Snow's map can be read as a Voronoi diagram and this interpretation is allowed by the second characteristic highlighted in Fig. 1, the recourse to a dotted line demarcating the area of infection and the sub-division of the data in two different sets whose elements can be reciprocally connected.

A point that should be clarified: "Voronoi diagrams" take their name after Gregorij Voronoi, the Russian mathematician who mathematically defined them towards the end of the 19th century: in the Broad Street map case, then, we do not properly encounter Voronoi devices, many of their applications, in fact, need computer graphic elaborations, implying that even with mathematical well-defined concepts, it was hard to employ Voronoi diagrams. Nevertheless, and far more importantly, the construction of the Broad Street map implies a rather clear use of Voronoi *concepts* (Fig. 4).

Concerning the first feature of the Broad Street network diagram two aspects deserve particular attention. The first one regards the role of the dotted line in the general reading of the map: it does not merely mark the infected areal extension; instead, it works as a boundary marking equal distance between the Broad Street pump and the other pumps in the network. The second interesting feature that allows reading this map as a network diagram, regards the description of the distance within the network: it is measured in terms of distance along the actual street-network of the district, not in terms of Euclidean metric, as would be measured in a "simple" map.

Inside the demarcated area, then, the physician plotted two distinct sets of data: the first set contains all the occurrences of cholera, each one symbolized by a thin black line; the second set contains all water pumps present in the area within walking distance, symbolized by dots. The most interesting fact about this second feature, regards the possibility of associating the members of the two sets, and this is another consequence due to the reading of the map (also) as a network Voronoi area diagram. In this context data are not merely displayed, but rather they can be reciprocally correlated; in fact, contextualizing Okabe's definition for a Voronoi diagram, it may be argued that in the continuous space of the street network (as represented on the map), given the finite set of distinct pumps, we can associate

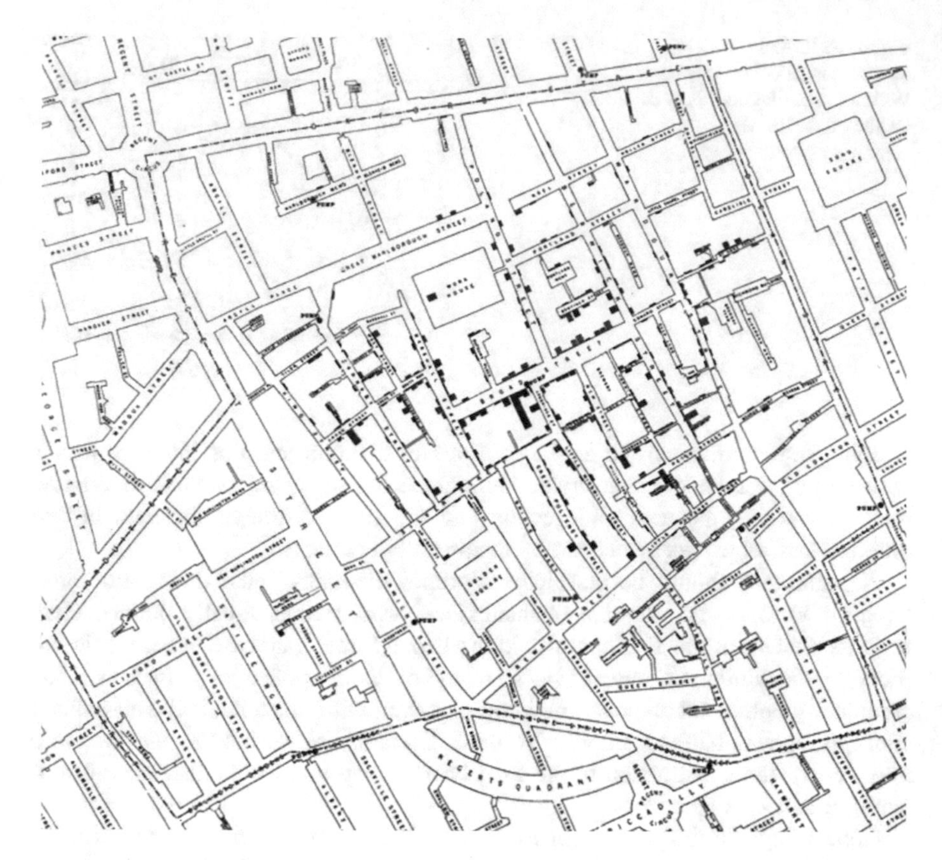

Fig. 4 John Snow's cholera map: a complete section showing all cholera occurrences as *thin black lines*, all water pumps as *black dots* and the *dotted line* running around the district. It runs along Oxford Street on the north side, along Regent Street on the east side, along Dean Street on the west side and along Marylebone Street and Coventry street on the south. *Source* Snow (1855)

each member of the other finite set (containing all cholera occurrences) with the closest pump.

2.6 *The Broad Street Map as a Heuristic Device*

It is now clearer how visualization has worked as a counter strategy with respect to the then existing theory and its theoretical difficulties. The construction of the Broad Street map both as a dot-distribution map and a network Voronoi area diagram allows a "dynamic" analysis of data and phenomena and in a sense this specific strategy of data visualization mirrors a hypotheses-formulation activity.

The first step John Snow took in the inquiry consisted, in fact, in a thorough collection of all cholera related data he could obtain, along with the collected data he plotted different plausible sources of contagion (the cemetery, for example, as expected by the miasmatic theory) and all the other alternative sources (all water pumps within walking distance from the infection focus). Finally, the second step consisted in suggesting a specific reading of the map. As a dot distribution map, in fact, it lets hidden patterns emerge, pointing out clearly a strong clustering of deaths around a specific pump, the one in Broad Street. This achievement notwithstanding, reading the Broad Street map as a dot distribution one would not suffice to draw a plausible conclusion about the actual agent of contagion. The map, then, needs to be read also as a network Voronoi area diagram; the construction of a network marking the relative distances among the different pumps in terms of street-network distance and the connection of all cholera deaths to the closest water pump allow the establishing of causal relations among data.

3 Philosophical Explanation

The abilities shown by the map/diagram John Snow constructed during the epidemic call for a philosophical explanation of why a visual representation of data can be more informative than the collection of data it displays. From a philosophical and logical perspective, our interest is focused on how visualization can be appraised as an heuristic device and how it can produce actually new knowledge that traditional analytic tools fail to provide.

I argue that visualization is a particular kind of ampliative (therefore, non-deductive) inference and that, as such, it is powerful enough to enhance knowledge in heuristic terms, not simply in "epistemic" terms.[4]

Therefore, the only acceptation of "new" I am going to refer to with respect to visualization, is the heuristic one: this definition implies that knowledge obtained by means of visualization as an ampliative inference is "radically" different in content from knowledge obtained by means of deductive inferences.

3.1 Epistemically and Heuristically New Knowledge

In our discussion, the distinction between "epistemically" new and "heuristically" new knowledge is of pivotal importance, since it marks the difference between two acceptations (and two roles) ascribed to visualization. According to the first definition of "new" knowledge, visualization acts as an aid to understanding, a useful device to surpass cognitive difficulties; alternatively, according to the second definition visualization can lead the process of discovery. More radically, these two

[4]This distinction recalls the one between *epistemological* and *heuristic* characteristics of visualization proposed in the Introduction.

definitions of visualization imply two different definitions of knowledge, with different epistemic "status".

In fact, *epistemically* new knowledge and *heuristically* new knowledge can be distinguished and considered as two different kinds of epistemic enterprise since, in a way, they refer to different cognitive objectives.

By "epistemically" new knowledge, we refer to knowledge acquired by means of a deductive inferential process: in this inferential process, conclusions are univocally determined by the content of the premises, therefore, if the inferential process is correctly developed, conclusions will explicitly show what is implicitly contained in the starting premises. The information obtained in deductively derived conclusions is, in fact, new: information "concealed" in the starting premises has been fully developed throughout the inferential process and is clearly shown in the conclusion. Although in this way we certainly obtain an epistemic gain, the quantity of information obtained in the conclusion does not exceed the quantity already contained in the starting premises. Therefore, epistemically new knowledge marks a condition of cognitive equilibrium between premises and conclusions.

By heuristically new knowledge, on the other hand, we refer to knowledge acquired by means of a ampliative (i.e., non-deductive) inferential process: in this process, conclusions are not univocally determined by the premises; on the contrary, they are sensible to the introduction of information external to the inferential process. Therefore, information contained in conclusions ampliatively derived is new since it necessarily exceeds the information contained in starting premises. This surplus of information showed in the conclusions represents an actual knowledge-enhancement; contrary to epistemically new knowledge, heuristically new knowledge represents an increase in the quantity of the information at our disposal, not only an increase in its quality. This feature is the reason why heuristically new knowledge is particularly needed at the frontier of knowledge, a condition requiring new hypotheses to develop novel theories and tackle (yet) unsolved problems.

3.2 Visualization as an Ampliative Inference

Visualization is here treated as a particular kind of ampliative inference with a twofold cognitive function, in the process of knowledge enhancement it plays a role both as a particular kind of representation and as an inference (Ippoliti 2008).

In fact, since it is constitutively related to figures (in the present case-study, maps and diagrams) and more generally to vision, it can be described as a knowledge representation: as such, it allows the visual organization and display of collected data and gives different descriptions of them. The ability of providing different descriptions of a given set of data (and generally, of any represented object) let connections among data emerge: in the case of the Broad Street map, it lets causal relations emerge.

As an inference, starting from the displaying and the different descriptions of data, it allows the generation of hypotheses. By their very definition, hypotheses formulated by means of ampliative inferences permit the introduction of "heuristically" new knowledge. This augmentation in information is possible since ampliative inferences can include information from outside of the inferential process. This kind of inference, in fact, is non-monotonic; i.e., their conclusions are not necessarily derived from the set of initial premises in a straightforward way; on the contrary, they are sensible to the introduction of new information and can also be retracted in the light of further information. In this perspective, the ampliative inferential process can be considered as a never-concluding process since conclusions non-deductively derived, when accepted, act themselves as hypotheses. As a consequence, if, on the one hand, knowledge acquired by means of ampliative inferential processes is to be considered as an actual progress, on the other, it shows an "unstable" character: it is not a final achievement, but only a temporary one, accepted until better hypotheses are found. In fact, since conclusions in non-deductive inferential processes are not univocally determined by starting premises, they are not truth-preserving, and therefore, the new information they contain has a plausible and fallible character (Cellucci 2013).

3.3 Visualization of Data at the Frontier of Research

The "unstable" (plausible and fallible) but knowledge-enhancing character of visualization shows its heuristic potential at the frontier of knowledge. As already claimed, the frontier of knowledge represents a condition of epidemic crisis characterized by a critical theoretical gap, a condition in which available models and theories and classical analytic tools are insufficient to the explanation of phenomena and to the solution of problems. Contrariwise, at the frontier of knowledge visualization, both as a particular kind of representation and as an ampliative inference, provides explanation for yet unclear phenomena and solve problems even in the absence of a well defined theoretical background.

The case-study proposed exemplifies the advantages of visualization in terms of maps and diagrams compared to the recourse to traditional analytic tools on numerical data in terms of a statistical aggregation of them.

Statistical aggregations of data thoroughly display all available information concerning the inquired phenomenon, and they are particularly helpful when a considerable amount of data need analysis. In the London cholera epidemic specific context, though, a statistical analysis of data would have been insufficient for two main reasons:

1. A statistical representation of a phenomenon basically relies on the descriptive abilities of aggregated numerical data and on the abstract representation they provide. The abstract representation of phenomena, though important in the process of their interpretation, cannot grasp the dynamics of a phenomenon strongly characterised in spatial terms.
2. Even when thorough and phenomenon-faithful, a statistical aggregation of data cannot provide any useful insight into the phenomenon unless the statistical

analysis is corroborated by pre-existing theoretical assumptions. With no theoretical background, any statistical description is hardly understandable.

On the other hand, a visual representation overcomes both these flaws:

1. As it relies on maps and diagrams, a visual representation grasps the dynamics of spatially-characterized phenomena, and its very reading and analysis cannot be accomplished without spatial reasoning.
2. Visual representations, as in the case of the Broad Street map, do not suffer from the lack of theoretical background assumptions: they are, in fact, richer than statistical descriptions because, unlike these ones, they are more complex than the phenomenon they describe. This complexity, therefore, not only permits coherent interpretations of data in the absence of theoretical assumptions, but it also works as a "theoretical gap-filler".

3.4 *Knowledge-Enhancing Features*

Visualization, then, owns features that justify it as a logical device, particularly interesting in a heuristic context are: independence, non-triviality, ampliativity.

The first characteristic refers to the independence visualization shows with respect to formal deductive explanations: visualization, in fact, can play an active role in scientific inquiry even when (or, especially when) there is no available deductive explanation for the phenomenon or problem inquired.

Since visualization is independent of formal deductive reasoning, it is also non-trivial. Non-triviality, in fact, refers to the possession of specific proprieties not shared by other formal, deductive instruments of reasoning. An example: as the case-study aptly shows, visualization optimizes spatial reasoning, making it a powerful cognitive resource. Furthermore, visualization is non-trivial since it is not a mere "economical" way to access knowledge which could be otherwise obtainable; the importance of this feature is particularly clear when confronted with frontier-of-knowledge cases.

Finally, being independent of formal reasoning and non-trivial, visualization is also ampliative.

4 Conclusion

As the case-study clearly exemplifies, visualization as a heuristic (i.e., ampliative) device plays a role of paramount importance at the frontier of knowledge. In a context of epistemic and cognitive difficulty, visualization fills theoretical gaps and acts as a problem-solving instrument.

As regards the London cholera epidemic case, the ability of filling theoretical gaps amounts to a cognitive gain: selecting an area with possible factors of contagion, visualization of data permitted the detection of water as the element of transmission of cholera, thus surpassing the theoretical gap represented by the notion of "bacterium". This result implied the slow dismissal of the miasmatic theory of contagion.

Moreover, the construction of a network diagram provided strong visual evidence of the causal connection between water consumption from a specific water pump and transmission of cholera, thus opening to the solution that led to the ending of the epidemic.

The ability of visualization as a representation of displaying and describing (therefore, interpreting) data letting new proprieties emerge, and the ability of visualization as an ampliative inference of formulating hypotheses from data establishing cause-effect relations, describe it as a plausible knowledge-enhancing model at the frontier of research.

References

Cellucci, C. (2002). *Filosofia e matematica*. Laterza: Roma.

Cellucci, C. (2006). The question Hume didn't ask: Why should we accept deductive inferences? In C. Cellucci & P. Pecere (Eds.), *Demonstrative and non-demonstrative reasoning in mathematics and natural sciences*. Cassino: Edizioni dell'Università di Cassino.

Cellucci, C. (2008). *Perché ancora la filosofia*. Laterza: Roma.

Cellucci, C. (2013). *Rethinking logic: Logic in relation to mathematics, evolution and method*. Dordrecht: Springer.

Cellucci, C. (2014). Knowledge, truth and plausibility. *Axiomathes, 24*.

Chandrasekharan, S., & Neressian, N. J. (2014). Building cognition: The construction of computational representations for scientific discovery. *Cognitive Science*. doi:10.1111/cogs.12203

Giaquinto, M. (2008). Visualising in mathematics. In P. Mancosu (Ed.), *The philosophy of mathematical practice*. Oxford: Oxford University Press.

Grosholz, E. (2005). Constructive ambiguity in mathematical reasoning. In C. Cellucci & D. Gillies (Eds.), *Mathematical reasoning and heuristics*. London: King's College Publications.

Grosholz, E. (2007). *Representation and productive ambiguity in mathematics*. Oxford: Oxford University Press.

Grosholz, E. (2011a). The representation of time in Galileo, Newton and Leibniz: Reference and analysis. *Journal of the History of Ideas, 72*(3).

Grosholz, E. (2011b). *Logic, Mathematics, heterogeneity*. In C. Cellucci, E. Grosholz, & E. Ippoliti (Eds.), *Logic and knowledge*. Newcastle Upon Tyne: Cambridge Scholars Publishing.

Ippoliti, E. (2002). Cognitive visualization of some integer-valued Polynomials. *Visual Mathematics, 4*(13).

Ippoliti, E. (2008). *Inferenze ampliative. Visualizzazione, analogia e rappresentazioni multiple*. Morrisville, Lulu Press.

Ippoliti, E. (2011). Between data and hypotheses. In c Cellucci, E. Grosholz, & E. Ippoliti (Eds.), *Logic and knowledge*. Newcastle Upon Tyne: Cambridge Scholars Publishing.

Ippoliti, E. (2014). Generation of hypotheses by ampliation of data. In L. Magnani (Ed.), *Model-based reasoning in science and technology, studies in applied philosophy, epistemology, and rational ethics* (Vol. 8). Berlin/Heidelberg: Springer.

Laudan, L. (1977). *Progress and its problems*. Berkeley: University of California Press.

Magnani, L. (2001). *Abduction, reason and science. Processes of discovery and explanation*. New York: Kluwer Academic.

Magnani, L. (2013). Are heuristics knowledge-enhancing? Abduction, models, and fictions in science. In E. Ippoliti (Ed.), *Heuristic reasoning*. Dordrecht: Springer.

Mancosu, P. (2005). Visualization in logic and mathematics. In K. Jorgensen, P. Mancosu, & S. Pedersen (Eds.), *Visualization, explanation and reasoning styles in mathematics*. Dordrecht: Springer.

McLeod, K. S. (2000). Our sense of Snow: the myth of John Snow in medical geography. *Social Science and Medicine, 50*.

Nickles, T., & Meheus, J. (Eds.). (2009). *Models of discovery and creativity*. New York: Springer.

Okabe, A., Boots, B., & Sugihara, K. (1992). *Spatial tessellations: Concepts and applications of Voronoi diagrams*. Toronto: Wiley.

Okabe, A., & Sugihara, K. (2012). *Spatial analysis along networks. Statistical and computational Methods*. Toronto: Wiley.

Seaman, V. (1797). An inquiry into the cause of the prevalence of yellow fever in New York. *Medical Repository, 1*.

Snow, J. (1855). *On the modes of communication of cholera*. London: John Churchill.

Stevenson, L. G. (1965). Putting disease on the map. The Early Use of Spot Maps in the Study of Yellow Fever. *Journal of the History of Medicine and Allied Sciences, 20* XX, (Vol.8), 226–261. doi:10.1093/jhmas/XX.3.226

Tufte, E. (1997). *Visual explanations. Images and quantities, Evidence and narrative*. Cheshire Connecticut: Graphics Press.

Weisberg, R. W. (2006). *Creativity, understanding innovation in problem solving, science, invention and the arts*. Hoboken: Wiley.

Erratum to: Ideality, Symbolic Mediation and Scientific Cognition: The Tool-Like Function of Scientific Representations

Dimitris Kilakos

Erratum to: Chapter "Ideality, Symbolic Mediation and Scientific Cognition: The Tool-Like Function of Scientific Representations" in: L. Magnani and C. Casadio (eds.), *Model-Based Reasoning in Science and Technology*, Studies in Applied Philosophy, Epistemology and Rational Ethics 27, https://doi.org/10.1007/978-3-319-38983-7_11

In the original version of the book, the new reference "Azeri S. (2013) Conceptual Cognitive Organs: Toward an Historical Materialist Theory of Scientific Knowledge, Philosophia 41(4), 1095–1123" has to be included in Chap. 11 and the citation for this reference has to be included after paragraph 2 in Pg. no. 207 & after paragraphs 2, 3 and 4 in Pg. no. 210. The erratum chapter and the book have been updated with the changes.

The updated online version of this chapter can be found at
https://doi.org/10.1007/978-3-319-38983-7_11

L. Magnani and C. Casadio (eds.), *Model-Based Reasoning in Science and Technology*, Studies in Applied Philosophy, Epistemology and Rational Ethics 27, https://doi.org/10.1007/978-3-319-38983-7_38

Printed in the United States
By Bookmasters